PRODUCTION MANAGEMENT

PRODUCTION MANAGEMENT
INDUSTRIAL ENGINEERING
OPERATIONS RESEARCH
INVENTORY MANAGEMENT

Prof. K.K. AHUJA

B.Sc., B.E. (Hons.) Mechanical
P.G.D.P.M. & I.R. P.G.D.M.M.
M.B.A. M.I.E. D.L.L. Specialised
Course in Management Oxford U.K.

CBS

CBS Publishers & Distributors Pvt. Ltd.
NEW DELHI • BANGALORE • PUNE • COCHIN • CHENNAI (INDIA)

ISBN : 81-239-0185-2

First Edition : 1993
Reprint : 1996, 1999, 2004, 2006, 2007, 2010, 2013

Copyright © Author & Publisher

Published by Satish Kumar Jain and produced by V.K. Jain for
CBS Publishers & Distributors Pvt. Ltd.,
CBS Plaza, 4819/XI Prahlad Street, 24 Ansari Road, Daryaganj,
New Delhi - 110002, India. • Website: www.cbspd.com
e-mail: delhi@cbspd.com, cbspubs@airtelmail.in
Ph.: 23289259, 23266861, 23266867 • Fax: 011-23243014

Branches:

- **Bengaluru:** Seema House, 2975, 17th Cross, K.R. Road,
  Bansankari 2nd Stage, Bengaluru - 560070
  • Ph.: +91-80-26771678/79 • Fax: +91-80-26771680
  • E-mail: cbsbng@gmail.com, bangalore@cbspd.com
- **Pune:** Bhuruk Prestige, Sr. No. 52/12/2+1+3/2,
  Narhe, Haveli (Near Katraj-Dehu Road By-pass), Pune - 411041
  • Ph.: +91-20-64704058/59, 32342277 • E-mail: pune@cbspd.c
- **Kochi:** 36/14, Kalluvilakam, Lissie Hospital Road,
  Kochi - 682018, Kerala • Ph.: +91-484-4059061-65
  • Fax: +91-484-4059065 • E-mail: cochin@cbspd.com
- **Chennai:** 20, West Park Road, Shenoy Nagar, Chennai - 600030
  Ph.: +91-44-26260666, 26208620 • Fax: +91-44-42032115
  • E-mail: chennai@cbspd.com

Printed at :
Nazia Printers, Delhi

PREFACE

Production Management constitutes one of the principal functions of a business regardless of the type of business in which the firm is engaged. The development of Production Management has been extremely rapid. We know that productivity and total productive capacities have expanded tremendously in the recent past. Production Management is not only a set of techniques but on the contrary it is a set of general principles for production, economics, facility design, job design, quality control, inventory management, work measurement, cost and budgetary control, computerisation etc. The present production managemeat calls for a study of mathematical techniques which have been used to develop decision rules for material control at any point of time. The computers have appeared on the scene as a powerful tool in its own right and with it production systems could be, stimulated, modelled after fairly realistic conditions. If a complex system is stimulated on a computer the effect of alternative proposals could be determined quickly without cost and time of actually trying the proposals in practice. On a broader scale, stimulation has been done in business decision games developed by various institutes and now attempts are being made to stimulate through computer based corporate model, the actual operation of the entire firm. The computer has also contributed to the new field of automation. The computer is used to performe to control machine tools through the entire cycle of completed parts without aid of human hand. Now the field of human engineering, human factors or bio-technology also furnishes the basic data of the job design.

India is a developing country. Its vast population is to be provided food, clothing and housing. The country leaders are involved in raising the standard of living of millions of our country-men. Raising of standard of living means raising production to meet the growing demand of goods and services in all spheres. Production can be increased by opening new industries and will help the jobless to become employed. The second way is to buy and install new and automatic machines and process which will increase production and help to reduce price. These methods are capital intensive and may not always be thc best solution of the problem. The third way is to use existing machines, materials and labour of man so that more can be produced from the resources already available. Producing more from the existing resources is called productivity. Productivity is simply the ratio of output to productive operations. Even when the output and input remains steady, if the quality of output is improved, even then the productivity has increased as the intrinsic value of output will be more in this case. Productivity means more goods in the market, more capital for investment, greater employment opportunities and lower prices. In the ultimate analysis it is man who produces material wealth, he delivers the goods and needs to be motivated and directed. We may have a big reservoir of water, but that reservoir by itself has neither strength nor purpose. If however. we can divert this water, channel it and put it through a turbine, we get power. According to author the basic foundation of our world, the most powerful resources "People" is obviously the most unstable ingredient. One of the basic problems in any society is how to motivate people to work ? The book emphasises that human resource is the primary resource and if this resource is polymerised in the organisation it results in cohesive enormous force. Human resource polymerises individual ideas : Desire and Organisational system. It is the human resource which appreciate with time while all other resources add to the cost. Therefore greater emphasis is, therefore, required to be laid on increasing will and skill of the people. The book deals with Operations Research and its importance in decision making. The management process essentially involves decision making relative to various functions of management. Operations Research helps in the decision making task by developing from among available operational alternatives. Operations Research can be applied to operate warning lights when corrective action is necessary. Or helps us in formulating a problem; identifying the parameters that control the system, least the optimality; generating alternatives etc.

In this book the operations Research deals with the areas of Allocation; Inventory Problems; Replacement, Quening Problems; Sequencing and Co-ordination Problems; Competitive Problems and Search Problems etc. The book deals with area of Operations Management which can be defined as management of systems for the provision of goods or services. Operations Management is concerned with the design and operations of systems for the manufacture; transport, supply or service. The role of the operations management is influenced by the objectives which are adopted by the organisation. The need for operations management arise from the fact that operating system must be provided simultaneously with the achievement of effective operations or efficient utilisation of resources. Two operations management is concerned with the achievement of both the satisfactory customers ; service and resources productivity. The book deals with the area of Ergonomics. It is derived from two Greek words : Ergo which means rule or norms. So in simple language ergonomics is tantamount to fitting the job to the worker, Ergonomics aims to provide right kind of relationship between mass and his environment. The contribution of ergonomics is the reduction of the fatigue and monotony of workers : betterment of their health; promotion of physical safty; prevention of accidents in the factory; creation of an overall congenial environment and the ultimate improvement of morale and productivity.

The book emphasises the importance of human resources and Author is of the firm view that investment in human resources is becoming a necessary compulsion for every organisation to ensure growth and success. The Industrial Relations is the personnel relations in the first phase which are essential at the individual level and in the second phase the labour relations.

The book Production Management has been divided into 5 major areas. Part-I deals with Production Management, Production Process, Production Systems, Production Planning and Control, Network analysis, quality analysis, quality control, decision making, maintenance management, cost reduction techniques, product design and product Planning etc. Part-II deals with Inventory Management, Factors Influencing inventory, materials management, purchase management, purchasing procedures, capital equipment purchasing, warehouse management, Stores Systems and stores organisation, Stores control forms and Stores accounting and Transportation and Traffic Management. Part-III deals with Industrial Engineering, Work Study, Application of Predetermined time standards, plant location, plant layout, material handling, human factors in engineering, productivity and system concepts, management information systems, computer fundamentals, use of computers in manufacturing, computer languages. Part-IV deals with Quantitative methods in management, quening theory, lineal programming, simulation and its applications, Replacement Models. The theory of games, maintenance models. Part-V deals with recruitment and selection, motivation, communication, workers participation in management, work and work organisations, organisation, organisation theories, organisation structure, wage and salary administration, wage policy, Wage structure, Absenteeism. The book is a comprehensive text book written for students pursuing post Graduate Diploma in Production Management and operation research, engineering students pursuing production engineering, industrial engineering as one of their paper. The book is useful to students pursuing M.B.A., Post Graduate Diploma in Business Management and also students of Indian Institute of Industrial Engineering. The book is definitely going to help students by providing them an insight into the concepts of production management functions. Functions for better dignosis of the problems to facilitate effective decision making.

COVERAGE OF THE BOOK

The book has been designed to cover all the 5 papers of Diploma in Production Management viz. Production Planning and Control, Inventory Management and Operations Research, Quality Control, Work Study and Productivity Analysis. The book has been designed after going through the curriculum of various engineering institutes, IITs, Various Management Institutes imparting M.B.A. and Post Graduate Diplomas in Production Management, Industrial Engineering and Operation Research.

The coverage of the book is given as under :

Papers covered	Institutes whose syllabus covered
Industrial Organisation and Managerial Economics, Industrial Engineering, Facility Planning and Plant Engineering, Ergonomics and work design. Operations Research, Quality Control and Production Systems, System Design and Value Analysis, Operational Research; Method Study, Work Improvement, Production Planning and Control, Industrial Engineering, Thery of Design and Analysis, Production Management, System Analysis, Production Planning and Control	* Delhi College of Engineering * Indian Institute of Technologies * Indian Institute of Industrial Engineering * B. Tech. Programmes of various Universities * Regional Engineering Colleges **Management Institutes viz.** * Department of Business Management, Aligarh Muslim University * Faculty of Management Studies Delhi University * Department of Management Studies, University of Jammu, Jammu * Indian Institute of Management, Banglore * Department of Business Administration, Lucknow * Faculty of Commerce and Management, Kurukshetra University * Department of Business Management, University of Indore, Indore * Department of Business Administration, Osmania University, Hyderabad * Institute of Management Development and Research, Pune * M.M. Institute of Management, Villiye Parle, Bombay * Birla Institute of Technology, Ranchi University, Ranchi * Moti Lal Nehru Instiute of Research and Business Administration, Allahabad * Jamna Lal Bajaj Institute of Management Studies, Bombay * Department of Business Management, Nagpur University, Nagpur

CONTENTS

1
Production Management

Introduction

The term production quite often means the same as manufacture. To carryout the process of manufacturing we need the following things : Man to do the job, Equipment, Material.

Man to do the job.

Equipment.

Material.

However, if we have a lathe machine, a skilled operator, a metal bar for turning. Will we be equipped for production ? Although we have now all the ingredients to start manufacturing but still we are not ready to manufacture. We do not know what is to make ? When it is to be made ? Do we have customer to buy the product ? What will happen if the lathe machine breakdown ? Are we making the right type and quality of the product ? These questions show us that we need more than just the technical ability to manufacture. We also need : Service activities, Control and functions.

Control and functions.

The service abilities are those which make sure that the manufacturing process can go on and control makes sure that it goes in the right direction. Economics defines production as utility creation. An activity which imparts form utility, time utility, space utility. Economists also define production as the act of making something with the objecl of satisfying human wants thus production is not only confined to goods and commodities but also include services. Consequently the economists definition of production is not restricted to the process of creation of physical items, but the transportation system, banking system, library and hospital can also be considered as a production process. Present views of production are that it is a sequence of operations that transform materials from a given state to a desired state.

Objective of Production. The main object of production is to ensure that it makes a product which can satisfy the needs of the customer by giving the right quality of the product at right place at the right time at right price. To achieve the above objective it is essential that production organisation is well managed.

What is Production Process

Production is a sequence of operations that transforms material from a given to a direct form. The transformation may be done in one or in combination of the following ways :

Transformation by disintegration,
Transformation by integration, and
Transformation by service.

Transformation by Disintegration

It has one ingredient as input and producing several outputs. This transformation is accompanied by changes in the physical shape of the input, for example, making of steel bars from cast ingot, making components from standardised materials on machine tool.

Transformation by Integration

In this we have several components as ihput and one product as output for example producing of household appliances, automobile, T.V. set, computors etc.

Transformation by Service

In this no change in object under consideration is perceptible, only certain operations are carried on the job, to change one of the parameters which define the object for example heat treatment of various jobs to increase tensile strength or other machnical properties of the object, servicing and repairs of automobiles, loading and unloading of trucks etc. Many scrvice operations are not considered as part of industry, but for the planning and control of these industrial operations. The objective of production management is the same as any other area of management that is to realise the overall objectives of the organisatinn.

Production Management Concepts

Production management is the process of designing, manning operating, servicing, and centralising the activities of a manufacturing organisation, responsible for the actual transformation of material inputs into marketable finished goods. Production planning and control which is an integral and important part of the Production Management has to play a vital role. Production management is to decide the objectives of production. This will involve all the functions of planning, organising, directing and controlling the production activities from the short term and long term aspect. The area which help us in setting up our production facilities are :

—Method study,

—Plant lay out,

—Work measurement,

—Production planning and Control,

—Stock Control,

—Quality Control and,

—Maintenance.

Production management is a sub-system of the business organisation. Production management is what a production-manager does. The production manager's role is that of a decision maker. After he has planned allocation of inputs, the creation of value through production process can begin. Once the production processes are started, it must be controlled. The control involves observing results and checking that they confirm to the original plans. The checking is possible by feed-back mechanism. Thus production management is to plan, organise, direct and control the production activities of a business concern. Now we know that production is an art of making something with the object of satisfying human wants which include service in addition to goods and commodities. But the wants of the customers are not satisfied until the right quality of the products are in their hands with the result the marketing process may also be considered as part of the process of production. In certain organisations marketing of products is a minor task in comparison to creation of product while in other organisations marketing process may be of greater importance and may dominate even the creating process. Clearly, the production and marketing are separate but still interdependent functions in a bueiness.

Any business organisation performs functions of production and marketing. Marketiug system discovers the need and transmit

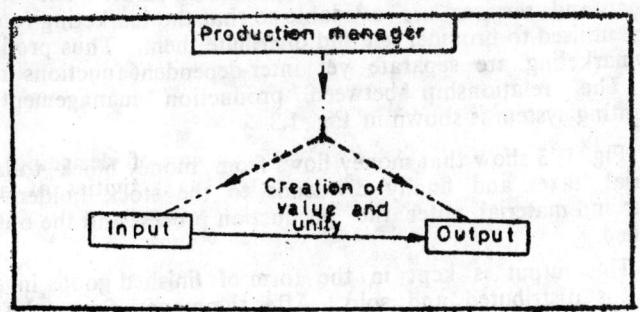

Fig. 1.1   Flow of Activities in Production Management.

need to business while production system supply the needs. Fig. 1·1 gives us an idea about the flow of activities in production.

Production and its Integration with other Areas of the Firm

The interaction of production with various areas of Business. Organisation as are shown in Fig. 1.2. The basic functions of Business Organisation are Production, Marketing and Finance.

Finance provide the necessary funds for the maintehance of Production and Marketing actiuities. Funds not only comes from the sale of goods and services but also acquired through loans— from banks and other financial institutions, sale of stock investment and income. These three functions—Production, Marketing and Finance are to be performed by all the Business Organisations.

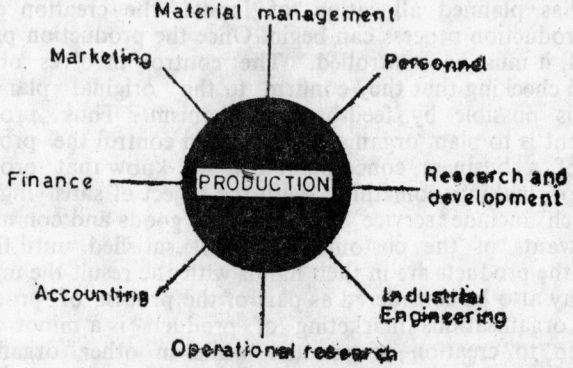

Fig. 1.2   Interaction of Production with other Areas of Busidess Orgenisation

The object of production activity is to provide inputs which include raw materials' men, machine, operating supplies, semi fini-shed products, water, power and place etc. The inputs are assembled and changed to finished goods thereby creating value. The finished products and services are available so that the marketing functions can be utilised to provide, sell and distribute them. Thus production and marketing are separate yet inter-dependent functions in busi-ness. The relationship between production management and accounting system is shown in Fig. 1.3

Fig. 1. 3 show that money flows from money block to labour, material, taxes and finally dividends to the stock holders. The labour and material enter the production process and the output is obtained.

This output is kept in the form of finished goods inventory until it is distributed and sold. After the money from the sale is realised it goes into the money block and the whole cycle is repeated. The production management significantly influence the control of inventories and process flow.

Production and Quantitative Techniques

The more use of analytical and systematic techniques is an indication that production and operation management is on the road to became an applied science. Operation Research (OR) provides the production management how to relate the variable margin ot each product model against productive requirements

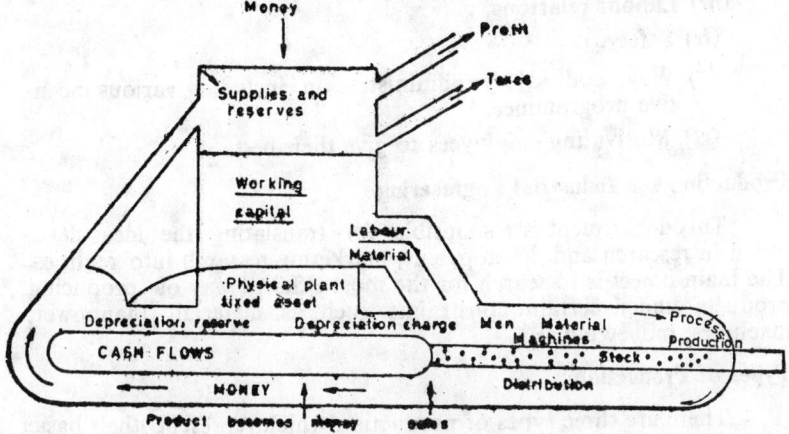

Fig. 1.3 . Relationship between Production Management and Financial and Accounting System.

and to utilise capacity so as to arrive at an optimum marketing goal. The well known (OR) techniques is mathematical programming which has its use in production-sales co-ordination. OR techniques helps us in finding what inventory policies and corresponding quality decisions will minimise inventory cost. The common applications of OR in production management are scheduling, transportation, sequencing, forecasting and network analysis of large projects.

Another important use of OR in production management is to find out the optimum machine-per-operator to minimise total cost. As number of machines-per-operator in any group of machines depends upon following costs :

(*a*) Cost of bad-work produced by machine in the absence of operator attending the machine.

(*b*) Cost of having too many operators in waiting for breakdown.

Personnel and Production

The personnel function in any business organisation is mainly concerned with all matters related to manpower as an input system of business organisation. From the view point of the production manager following are the various areas of mutual interest. Each of these areas have been discussed in detail in the area of Personnel Management in the subsequent chapters :

(*i*) Recruitment and selection.

(*ii*) Training and development of employees.

(*iii*) Labour relations.

(*iv*) Safety.

(*v*) Wage and salary administration including various incentive programmes.

(*vi*) Motivating employees to give their best.

Production and Industrial Engineering

This department is responsible for translating the ideas developed in research and development, marketing research into realities. The main object is to search for the most efficient way of producing products under certain constraints such as material, manpower, machines, money etc.

Types of Production

There are three types of production which represent their basic approaches to production. The three types are :

(*a*) Job production.

(*b*) Batch production.

(*c*) Flow production.

Job Production

In job production the whole product is looked as one job which is to be completed before going on to next. The most common examples are building a ship or a large civil engineering construction job. Job production is hot confined to large projects, it could be the making of a special piece of equipment or a tool.

Batch Production

If qualities of more than one are being made, it is sometimes convenient to split the production into a series of manufacturing stages or operations. Each operation is completed as one of the single items being made, before the next operation is started. In this way a group of identical products, or a batch are made, which move through the production process together.

If more than one types of product is being made, then batches of different products may be moving around the shop floor sometimes requiring operations from the same machine. This leads to problems of how long a machine should be processing a batch of one tope of product before going on to the next process, a different one, or which batch should be worked on first. This type of problem tends to make the planning and control of batch production a difficult task.

Flow Production

When there is a continuous demand for a product, it is sometimes worthwhile setting-up facilities to make that product and no other product. In these circumstances flow production may be the best way of operating. Here the manufacturing is broken down into operations, but each unit moves, or flows, from one operation to the next individually, and not as one of a batch examples are motor manufacturing, fertiliser, pharmaceutical and urea manufacturing. Since only one product 'is being made there are no problems about oriorities, but it is necessary to balance the work load at all stages of manufacture. Examples are motor car manufacturing.

PRODUCTION MANAGER

The person who is given responsibility for the process of production is called Production Manager. Production manager should have both technical and managerial skills. Production manager needs to take two types of decisions :

Problem relatiug to production system like operation and technology.

Problems relating to planning and analysing.

Production manager must make decisions regarding lay out of equipments, techniques of material handling, location and size of the plant, motion and time study, process analysis and quality of the product. Once the production system has been designed and actuated the problems which prodection manager faces is the analysis and co$_e$ntrol of the production system. Routine, leading, scheduling, d spatching expediting are the basic production planning and control activities. The role of the production manager in a production system as shown in Fig. 1.4 . The production manager's role is that of a decision maker, after he has planned allocation of inputs, starts the creation of value through the production process. This is not the end but the beginning of production manager's activities. Once the production processes are started it must be controlled. The control involves observing results and checking that they confirm to the original plans. The checking is possible by feed-back mechanism, So the role of production manager in a production system is that of decision maker to plan and control the production processes.

Role of Production Manager

Production manager is one of the managers in the larger system of the organisation. Now as production management is a

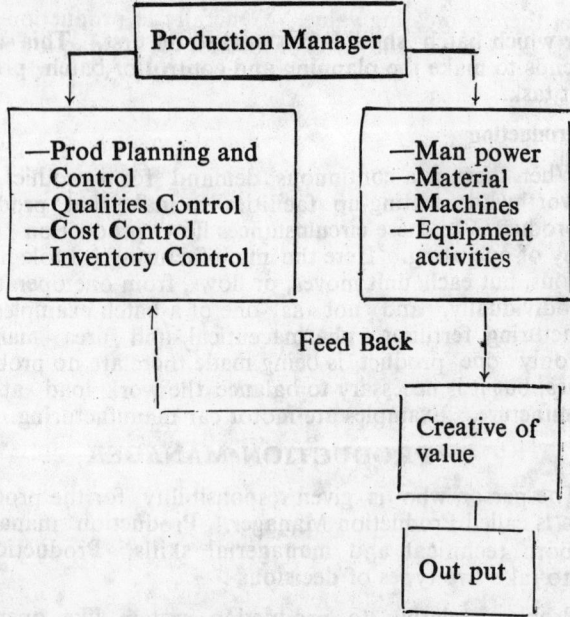

Fig. 1.4 . Role of Production Manager in a System.

sub-system of the business organisation, similarly business organisation is also a sub-system in the environment of the business. The Production manager must perceive his role in all the three systems. To understand his role in a business environment he must examine first the Objective of Business Organisation. The first objective of any organisation is to survive and remain in business and to make reasonable profit. Many organisations do have the objectives of customer's service satisfaction and social objectives. The public-sector enterprises have come to play a significant role in our economy. The objectives of public sector are :

—to check the monopolistic growth.

—to provide the basic services to private sector in the form of infrastructure.

—to create conditions for buyers market.

—to invest in low return and highly capital intensive industries.

Public sector thus has to act as a device and it plays the role of pace setter by function as to quality, design, price and ensure adequate quantity by supplementing the production in various short supply sectors. The first concern of production manager in any production activity is to provide input which include raw-materials, men machines etc. The inputs are then transformed to finished product thereby creating value. Generally to production of products alone is considered the area of interest of production manager but this can also be applied to the production of services also. The production system can be used in any activity which result in products and services such as restaurants, super bazars and automobile repair shops etc.

FUNCTIONS OF PRODUCTION MANAGER

The functions of Production manager, like any other manager involve the following six steps :

—Organising
—Planning
—Directing Operations
—Controlling Results
—Appraising Performance
—Improving Effectiveness

The manager must organise his department to reacn his goals effectively, he must plan his operations so that each activity will fit together in a pattern calculated to produce unified and accurately directed action. He must control his action, and keep it aimeds at planned objectives. At the sametime, he must continuously appraise his results and his goals and further improve his performance by the feedback knowledge, experience, information and properly managed operations.

The management must use an 'Integrated precision management' approach to achieve the proper unification and cooperation of individual efforts and to synchronise the activities of machines, materials and manpower so as to realise company plan and objectives with the greatest possible effectiveness and at the lowest possible overall cost. The production manager can manage the six functions of management in the following manner :

Organising

The first task in developing an effective production planning and control group is to organise all the significant factors affecting the department's activities. The production planning and control function must have a solid framework which has purpose, direction and continuity in order to help maintain the basic structure of the department as operation changes take place over the years. Without this framework, the department would not be properly integrated with company objectives. The following, in brief, enumerates the organising activities :

(a) Obtain a statement of corporate objectives from top management.

(b) Prepare a statement of departmental objectives.

(c) Develop a policy manual for the production planning and control department.

(d) Draw up a department organisation chart.

(e) Evolve departmental job descriptions.

(f) Compile a system and procedures manual.

(g) Introduce a manpower rating and inventory system.

(h) Establish basic criteria for the management of devetopmental performance.

Planning

The role of planning is a multiple one. However, one of the basic purpose that it serves to is relate the organisation to actual operations. Effective direction, control, appraisal and improvement of operations cannot be done, no matter how well the production department is organised, unless the activities are first planned. To try to operate without continuous planning on both—a short -and a long range basis will lead to results which are far less than the optimum that is achievable. The following gives the elements or steps that are part of the overall concept of production, planning and control, hence they should be integrated with each other, as well as with the elements of other managerial skills.

(a) Prepare a statement of long and short term planning requirements.

(b) Develop operating plans (short and long term). This might include some of the following :

(i) Functions and activities

—Sales forecasting and master planning.

—Inventory Control.

—Routing, scheduling and despatching.

—Follow up.

—Preparation of process orders.

—Receiving and warehousing.

—Internal transportation and materials handling.

—Shipping and traffic.

—Marketing trends.

—Expansion and innovation of process.

(ii) Projects and programmes

—Cost control and reduction.

—Scrap salvage and waste control.

—Value analysis and engineering,

(c) Make a departmental manpower forecasts (short term and long term).

(d) Set up a departmental time schedule for the operating plans.

(e) Establish cost and budgetary plans.

Direction

The production planning and control manager must direct the activities of his people within the framework of the total organisation, the manner in which it is done will determine the precision and and effectiveness of the plans themselves, as well as the quality and value of the organisational planning.

Some of the basic elements that are essential to success in directing the activities are :

(a) Effective decision making.

(b) Effective communication and delegation.

(c) Effective motivation and supervision.

(d) Effective coordination and unification.

Controlling

Control serves to ensure that the activities will be carried out in accordance with the Plans. Without effective control of his operations, the manager may find that his planning and organising. are in vain. Basic steps for control are :

(a) Measurement of progress and results.

(b) Comparison of results with plans.

(c) Taking corrective action, if called for.

Appraising Performance

Appraisal is primarily aimed at evaluating the results of operating plans on a continuous basis. In other words, the performance measured in the control phase of management is reviewed, then thoroughly analysed to establish the cauee of any deviation, so that the proper changes can be made in either plans and practices. Appraisal and control, are of course, very closely allied and are often considered to be the part of the same management.

The appraisal of result consists of :

(a) The analysis of variances.

(b) Performance evaluation.

Organisation of Production Department

The Production Department Organisation varies from industry to industry. It depends on the type of product, quantity of product, manufactured, quality of product, number of people working etc. It also depends on the style of management. One typical organisation of Production Department is given below :

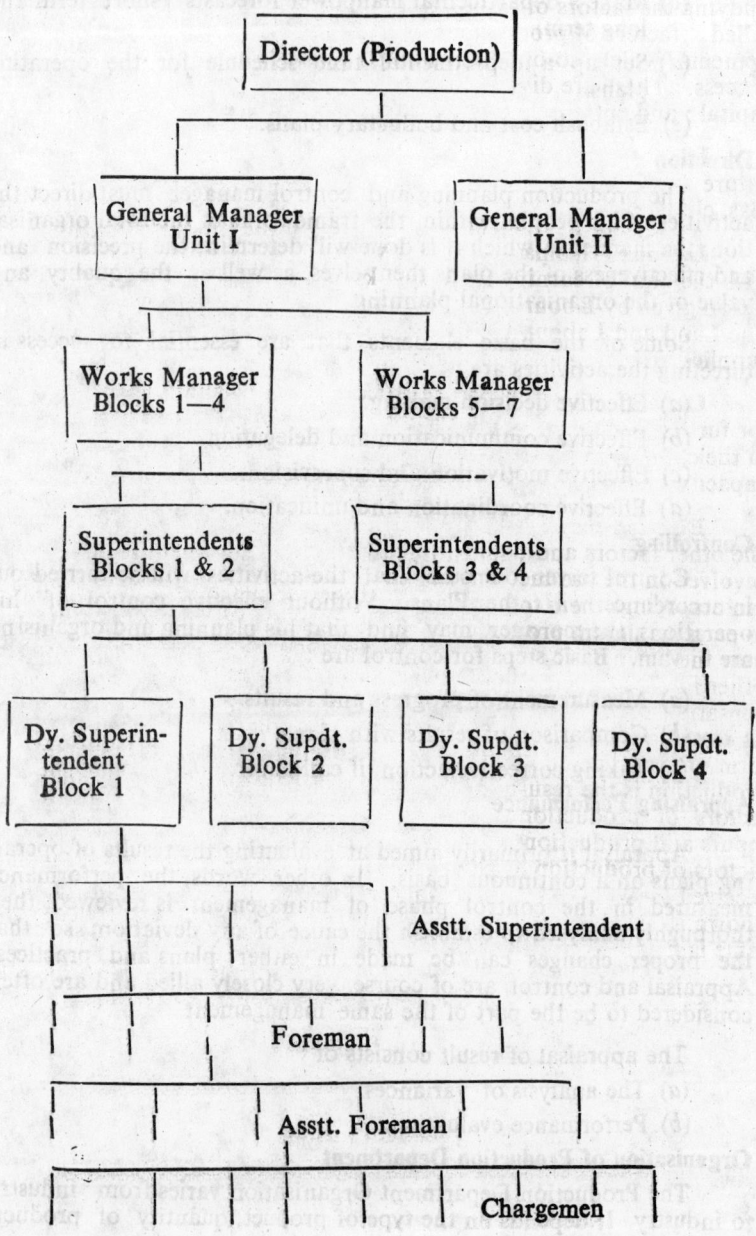

Factors of Production

The economics aspect of production can be analysed by studying the factors of production. The inputs are conventionally called "factors of production". Factors of production are essential elements which co-operate with one another in the production process. These are divided into four categories viz : Land ; Labour ; capital ; and entrepreneurship.

Land : It refers to all natural resources which are free gifts of nature and includes soil, rivers, waters, forests, mountains, mines, seas, climate, air, sun etc.

Labour : Human efforts or work done mentally or physically with the aim of earning income is known as labour. The compensation given by labourers in return for their work is called wages. The Land and Labour are the primary factors of production as their supplies are determined more or less outside the economic system.

Capital : See the inputs or man made goods which are used for further production of wealth are included in capital. An increase in the capital of an economy means an increase in the productive capacity of the economy.

Entrepreneur : An entrepreneur is a person who organises the other factors and undertakes the risks along with uncertainities involved in the production. The entrepreneur hires the other three factors, brings them together, organises and co-ordinates them so as to earn maximum profit. An entrepreneur acts as a boss and decides how the business shall run. The entrepreneur hires and organises other factors for producing goods and services, he pays them compensation in the form of wages to labour ; rent to landlord and interest to the owner of capital. The balance goes to entrepreneur in the form of profit which he gets for rendering entrepreneural services. Production is the result of combined and joint efforts of the four factors of production. The factors of production are known as inputs and production as output. Similarly services rendered by the factors of production are known as factors services and compensation received by them in return as factor income. Factor income is also known as factor payment from enterprise point of view because the same is paid by the enterprise or production unit.

2

Production Process

Concept

Production Process is a process of producing goods and services by combining and utilising the services of factors of production. Production process is a continuous process which has been going on since the dawn of economic history and will continue till this world lasts. If a man has to live and prosper, he must get goods and services to satisfy his multiplying wants. Since goods and services are produced through production process, the latter becomes very essential for human existence. Production process is a continuous process of economic activity which enables the flow of goods and services and their use for satisfaction of human wants. Human wants are unlimited and go on multiplying and are recurring in nature. The urge to satisfy wants keep the wheels of production process going. Similarly the nature of production process has always been changing with the change in the civilisation of human race because of change in the different aspects of production process. The Production Process has different concept in Traditional Economy and Modern Economy. The main features of production process in traditional economy and modern economy are discussed below:

Table I

S. No.	Traditional Economy	Modern Economy
1.	Production of goods and services is primarily for self consumption.	Production of goods and services is for sale in the market.
2.	Use of capital and machinery is nominal as labour intensive methods are used.	Use of capital and machinery is at a large scale and capital intensive methods are used.
3.	There is lack of specialisation and division of labour	Specialisation is at tremendous scale in all lines of production.
4.	There is no uniformity and standardisation of products	Products are uniform and standardised.

5.	Technology is simple and not much change in technology	Technolgy is latest and continuously changing and complex.
6.	The role of Government is limited as Government does not interfere in the economic activities	Government participation in production and economic development is increasing continuously.
7.	Production Process is simple and carried out independently	Production process is complex and interdependent.
8.	Organisation of labour and technology is simple	Organisation of labour and technology is complex.

Modern economy is believed to have come into existence after the industrial revolution in European countries in the middle of 18th century. The technological discoveries in the field of transport and communication; electricity and industrial chemicals has revolutionised the techniques of industrial production. All these led to the emergence of the modern economy in which tremendous changes like production for market, growing specialisation, increasing use of capital technology, large scale production, greater use of money etc. were noticed. The system of production in the modern economy is increasingly capitalistic where capital goods like tools, machinery, raw materials, plants, buildings, railways, trucks, chemicals etc form the basis of large scales production. Thus modern economy where production is for the market and at a large scale, makes extensive use of capital.

What is Manufacturing

Production is a sequence of manufacturing operations that transforms materials into finished products. Manufacturing includes in all the actions and processes which result in the production of finished articles. The production process consists of several stages viz.

— Preparing for production
— Receiving materials
— Making and machining billets and blanks
— Making components
— Assembling units
— Separate articles
— Checking the quality of the parts and assemblies
— Testing the finished articles
— Painting
— Packing
— Shipment etc.

The success of each step depends on type if the organisation and also on an uninterrupted supply of materials, the proper tools, efficient maintenance and accurate stock taking. Due to the great variety of types of production, there is no uniform structure of machine building plants. The most efficient structure for the plant is adopted in each case depending on the specific conditions and preserve only the basic common principles of organisation. The shope, which are the basic production sections of any plant, are subdivided into main and auxiliary shops. In the main shops the parts are manufactured and the units and complete articles assembled. The auxiliary shops (tool repair shops etc.) and services keep the main shops functioning normally.

The division into shops can be made either on the "object" principle, each shop producing a certain variety of parts or units, or on the "technological" principles, each shop being assigned certain operations. The shops are sub-divided into sections organised either on the principle of similar equipment or the production of one type of aricle. The sections in turn are sub-divided into work places each for one work man or for a team. In addition, the plant comprises of a number of departments ; planning, supply, preparation, design engineering, labour and wages, accounts etc.

TYPES OF PRODUCTION

There are there types of production depending on the quantity of articles produced and the nature of their manufacture the types of production can be classified into three categories *viz*. piece meal system lot system and mass production system.

(*i*) **Piece meal production system.** It is characterised by the output of small quantities of articles of diverse nomenclature. On the piece production system each article is made by consecutive operations specified in the flow process charts.

(*ii*) **Lot production system.** It is characterised by the output of recurrent batches of articles. It is sub-divided into small lot and large lot production. The lot production process is based on the principles of parallel—consecutive operations, Sub-divided into separate elements. Lot production is recorded in operating process charts.

(*iii*) **Mass Production System.** It is characterised by manufacturing one type of article in large numbers. The mass production process is based on the principle of "parallel operations". In mass production greater efficiency of labour is achieved due to high specialisation.

Smilarly Types of Production can also be classified as under :

(*a*) **Line Production.** The main characteristics of line production are the use of line layout, and the minimization of the transfer quality. Line production can be used for maching—in all industries —and also for assembly. Although the term line production can be used in a general sense to cover several of variants described below, it is reserved in a particular sense for cases where lines are used for one particular established product. Because the work centres in this case are left permanently set-up, run and set-up frequencies are low. Because the machines are adjacent and in the transfer quantities of one can be used, and transfer frequencies are therefore very high.

(b) **Line batch production.** Line layout can also be used for families of similar items of established design which are all tooled to use the same machines or assembly line in the same sequence. In this case, batches of the different components follow each other down the line, each machine being reset as it completes the last item in a run. The transfer quantity is again minimized but cannot achieve the high transfer frequencies obtained with pure line production. In this instance the capacities of the different station on the line are balanced to the total requirements of different components at different stations on the line, there is generally some ebb and flow of work in progress stock between the machines.

(c) **Group batch production.** Group batch production is characterised mainly by the use of group layout. Because in this case all machining operations for each components in the group family are controlled by one supervisor, it is possible to start some operations before the previous operations are completed, but it is generally impossible to achieve the high transfer frequencies attainable with line production.

(d) **Functional batch production.** Functional batch production is characterized mainly by the use of functional layout. Due to the administrative complication with this type of layout. It is generally essential to fix the same values for order, run, set-up, and transfer quantity for each component processed. This type of production is therefore generally accompanied by very low rates of stock turnover.

(e) **Line jobbing production.** Jobbing production is a type in which the product design is not established, but is only known with the customer's order is received. Where there is specialisation in the type of product for which orders are accepted, line production can still be used. It is much more difficult to achieve a balanced line, and such devices as variable overtime for different parts of the line may have to be used to maintain balance and minimise the stock between stations.

(f) **Group jobbing production.** In this instance group layout is used, each group being equiped to produce a general type of product.

(g) **Functional jobbing production.** Functional jobbing production finally is jobbing production with functional layout. Because production is infinitely variable, it is probably impossible to design a classification system which will cover all possible differences, even in simple production systems. In most companies there is a wide variety of flow types, and the type varies according to the level which is being examined and from one process to the next. In spite of these difficulties the above classification has been found in practice to give a useful indication of the general type of material flow in use.

Theoretical Foundation in Production Management

During the last two to three decades, the focus of attention has moved from traditional techniques such as time study, wage incentive determination, production control, plant layout, work methods design job evaluation, equipment selection, tooling etc. to systems tachniques including or techniques and the use of EDP. Further, it can be noticed that a change of interest from the efficiency of individual, manual work, the use of individual technical resources and the inter-action between the man and machine to methods and techniques related to the process of planning and control of materials flow through a system of man machine components. As a result of this change various disciplines such as production control operations management logiesticks, matetials management and systems analysis concepts have been developed.

BASIC APPROACHES TO PRODUCTION MANAGEMENT

Three main approaches in modern production management are ;

(i) The decision theory approach
(ii) The systems approach
(iii) The contingency approach

(i) **The decision theory approach.** It has adopted elements from decision theory. The approach is based upon the assumption that there is a general decision and problem hierarchv which provides a logical structure to describe the subject.

Problems which fall outside the hierarchical structure tend to be looked as non-production management problems. The approch is oriented towards the decision maker and the ambition is to provide the decision maker with a set of techniques that makes rational decisions possible for example, the *first* decision is the decision as to which process orientation is the best. The proceess can be continuous, intermitted or project. After this major choice is made other process design decisions must follow. This includes selection of equipment, detailed design of the process, make or buy decisions facility location and the way in which the process facilites are laid out or decisions that follow subsequently. Then the focus is shifted to the work station. Once the process has been designd it must be scheduled operate land control. A forecaste is required to draw the schedule. Routine inventory planning and control system must be employed. Finally the quality of product must be periodically sampled to ensure that actual output complies with process objectives (as shown in figure one).

(ii) **The systems approach.** The systems approach is characteri-sed by the use of the general systems theory framework. Systems theory helps us to get in sides into the over all structure of produ-ction management problems. It tends to be a strong instrument for discription and it has some sort of analytical appeal. It emphasis a total few rather than isolated problems.

It has been demonstrated that the most common approach includes the idea that we are dealing with two separate systems : a production system and a decision and information system. The decisions can be categorised into two groups—those related to the design of the system and those related to the control of the system. The design of the system includes problem areas such as the selection and design of products, selection of equipment and process, job design, location of the system and facility layout. The control of the system includes inventory control, production control, maintenance, quality control, labour control and so on.

Table 1. Classification of Production types

Type	Flow System	Batch Frequency				Orders Cycles	Ordering Phase	Remarks
		Order	Run	Transfer	Set-up			
Line production	Line	—	Low	V. high	Low	Single	Single	On line per component
Line batch Production	Line	—	12-100 p.a.	High	Same as run	Single	Single	On line per family
Group batch production	Group	—	12-26 p.a.	High	Same as run	Single	Single	One group per family
Functional batch production	Functional	—	1-12 p.a.	Same as run	Same as run	Multi	Multi	—
Line jobbing production	Line	Same as order	Same as order	High	Same as order	—	—	One line per family
Group jobbing production	Group	Same as customers	Same as order	High	Same as order	—	—	One group per family
Functional jobbing production	Functional	Same as customers	Same as order	Same as order	Same as order	—	✓	—

The above table is highly comprehensive as it integrates the various types of production with Batch frequency, orders cycle and ordering phase.

A more consistant systems approach can be classified as follows :

(*a*) the formation of objectives for the system and its components.

(*b*) the design of a system containing human and fiscal resources to achieve the objectives

(*c*) the allocation of resources to organisational units.

(*d*) the design of planning and control system.

(*e*) the evaluation of components and systems performance.

(*iii*) **The contingency approach.** This approach is characterised by the assumption that there are some important contingencies determining the relevance of different problems and therefore determining the problem structure in a given situation. The contingency approach seems to offer a significantly better instrument for description as well as better framework for explanation and prediction concerning management problems. However, it is not based on imperical dates.

Major Production management Problems Vis-a-vis
Major Production System

(*i*) **Distribution Systems**

(*a*) Determination of the distribution of demand.

(*b*) Forecasting demand.

(*c*) Determination of how much to order at one time.

(*d*) Determination of when to reorder.

(*e*) Determination of service levels and the size of buffer stocks.

(*f*) Design of data-processing systems.

(*ii*) **High-volume Production-Distribution Systems**

(*a*) Forecasting demand and the behaviour of multi-stage inventory systems.

(*b*) Long-range aggregate planning for facilities—plant capability, sizes and location, and warehouse sizes and location.

(*c*) Production facility design.

(*d*) Aggregate planning and scheduling for facilities and manpower.

(*e*) Raw materials procurement.

(*f*) Day-to-day scheduling and adjustment of production levels as demand becomes known.

(*g*) Design of data-processing systems.

(*iii*) **Intermittent Systems**

(*a*) Design and layout of a system to minimize aggregate handling cost.

(*b*) Forecasting demand

(*c*) Aggregate planning for the use of facilities.

(*d*) Scheduling orders to meet promised delivery dates.

(*e*) Scheduling labour and equipment to minimised combined costs of machine set-up, machine down time, labour over time and under time and in-process inventories.

(*f*) Scheduling equipment to utilize most efficient process.

(g) Procuring materials in economical quantities to mesh with the production schedule.

(h) Developing bidding policy and procedure to obtain orders at margins that will achieve a balance between the use of labour and facilities and the desire for profit.

(iv) **Large-Scale Projects**

(a) Planning a network of operations to accomplish the desired result.

(b) Developing schedules of the network of operations such that the critical-patch schedule of the network meets promised delivery dates.

(c) Allocating the use of limited resources of equipment and/or labour in ways that will not interfere with the critical path schedule.

(d) Procuring materials by a schedule that minimizes total inventory costs but meets the needs of the critical-patch schedule.

(e) Developing bidding policy and procedures to obtain contracts at margins that will achieve a balance between the use and maintenance of the stockpile of critica resources (engineers, scientists, skilled labour, key facilitie and so on) and the desire for profit.

Trade off Decisions

The various important trade off decisions indicating Decision Areas ; Decisions and verious alternatives are given in Table I. Similarly Table II indicate the various contraints which should be studied to understand the Economics of the Industry and Technology of the Industry.

Table : Trade off Decisions in Manufacturing

Decision Areas	Decision	Alernatives
Plant and Equipment	Span of process plant size plant location	Make or buy one big plant or several smaller locate near markets or locate near materials.
	Investment decisions	Invest mainly in buildings or equipment or invento-ries or research.
	Choice of equipment	General purpose or special purpose equipment.
	Kind of tooling	Temporary minimum tool ing or production tooling.
Production Planning and Control	Frequency of inventory taking	Few or many breaks in production for buffer stocks.
	Inventory size	High inventory or a lower inventory.

	Degree of inventory	Control in great detailor in lesser detail.
	What to control	Controls designed to minimize machine down time or oabor cost or time in process, or to maximise output of particular products or usage.
	Quality Control	High reliability and quality or low costs.
	Use of Standards	Formal or informal or none at all.
Labor and Staffing	Job specialisation	Highly specialized or not highly, specialized.
	Supervision	Technically trained first-line supervisory or non-technically trained supervisors.
	Wage system	Many jobs grade or few job grades, incentives wage or hourty wages.
	Supeavision	Close supervislon or loose supervision.
	Industrial Engineers	Many or few such men
Product Design/ Engineering	Size of product line	Many customer specials or few specia's or none at all.
	Design stability	Frozen design or many engineering chahge orders
	Technological risk	Use of new processess unapproved by competitors or follow-the-leader policy
	Engineering	Complete packaged design or design as you go approach.
	Use of manufacturing engineering	Few or many manufacturing engineers.
	Kind of organization	Functional or product focus or geographical or other.
	Executive use of time	High involvement in investment or production planning or cost control

| Organisation and Management | Degree of risk assumed | or quality control or other activities. Decision based on much or little information. |
| | Use of staff Executive style | Large or small staff group much or little involvement in detail ; authoritarian or nondirective style ; Much or little contact with organization. |

Table 2
Various Constraints or Limitations in the Important Areas of Production

A. Economics of the Industry

—Labour, burden material depreciation costs
—Flexibility of production to meet change in volume
—Return on investment, prices, margins
—Number and location of plants
—Critical control variables
—Critical functions (*e.g.* maintenance, production control, personnel)
—Typical financial structures
—Typical costs and cost relationships
—Typical operating problems
—Barriers to entry
—Pricing practices
—'Maturity' of industry products, markets, production practices, and so on
—Importance of economics of scale
—Importance of integrated capacities of corporations
—Importance of having a certain balance of different types of equipment
—Ideal balances of equipment capacities
—Nature and type of production control
—Government influences

B. Technology of the Industry

—Rate of technological change
—Scale of processes
—Span of processes
—Degree of Mechanization
—Timber requirements for making change.

Machine Tools

A metal cutting machine tool is a device in which energy is expended in deformation of material for shaping, sizing or processing a product by removing the excess material in the way of chips. A machine tool is a contrivance having a combination of mechanisms whereby a cutting tool is enabled to operate upon a piece of material to produce the desired shape, sizes and degree of finish. In a machine tool the work is being affected by the actual removal or cutting away of material. Often machine tools are described as machines where work pieces are held against the cutting edge of a tool and the material is removed through a deformation process caused by the relative motion between the work piece and tool. Among all the work shop processes, machine tools are the most versatile and almost any product can be produced with them. In fact, the most accurate parts or surfaces are always processed in some forms of machine tools.

Machine Operations Performed in Machine Tools

The purpose of all machining is to generate

(a) forms and

(b) finished surfaces. The form of surface produced in a particular machine tool depends on

(i) the shape of the cutting tool and

(ii) the relative path of motion between the cutter and work piece.

If a work piece is moved past the cutting tool in a linear patch, a straight. Cut plane surface is generated. However, when a work piece is rotated about an axis and tool is traversed in a definite path relative to the axis, a surface of revolution is generated. The different machining processes are named and classified according to

(a) the shape of generated surface

(b) the shape of cutter

(c) the nature of relative movement

(d) the type of finish.

The machining operations are summerised as shown in the table.

Besides these there are other finishing operations which use various relative motions of the job and tools and are termed as, lapping, honing, super finishing, buffing, polishing. Different types of machining operations result in surfaces of various degrees of roughness with a considerable range of variations.

TYPES OF MACHINE TOOLS

Machine tools may broadly be classified into three major divisions. They are as follows :

(a) General purpose or "Basic Machine Tools" such as Engine lathes or centre lathes ; Drilling machines ; Boring machine ; Shaping, slotting and planning machines ; Grinding machines, Milling machines. These machines are able to perform all the metal cutting operations within their range of operations.

(b) Production Machine Tools—such as Multiple— tool lathes ; Multiple head drilling machine ; Capstan and Turret lathes ; Automatic screw machines ; Semi-automatic lathes ; production Milling machines, etc. These machines are specially designed to enhance the rates of production and to reduce the manufacturing cost. These machines are an outgrowth of the 'Basic' machines to do a limited range of work when provided with special accessories or they are designed to machine a given part by employing multiple tools so that a number of tools work simultaneously or so that a number operations are performed automatically and successively on the piece.

Table. Different Machining Operations Performed in Machine tools

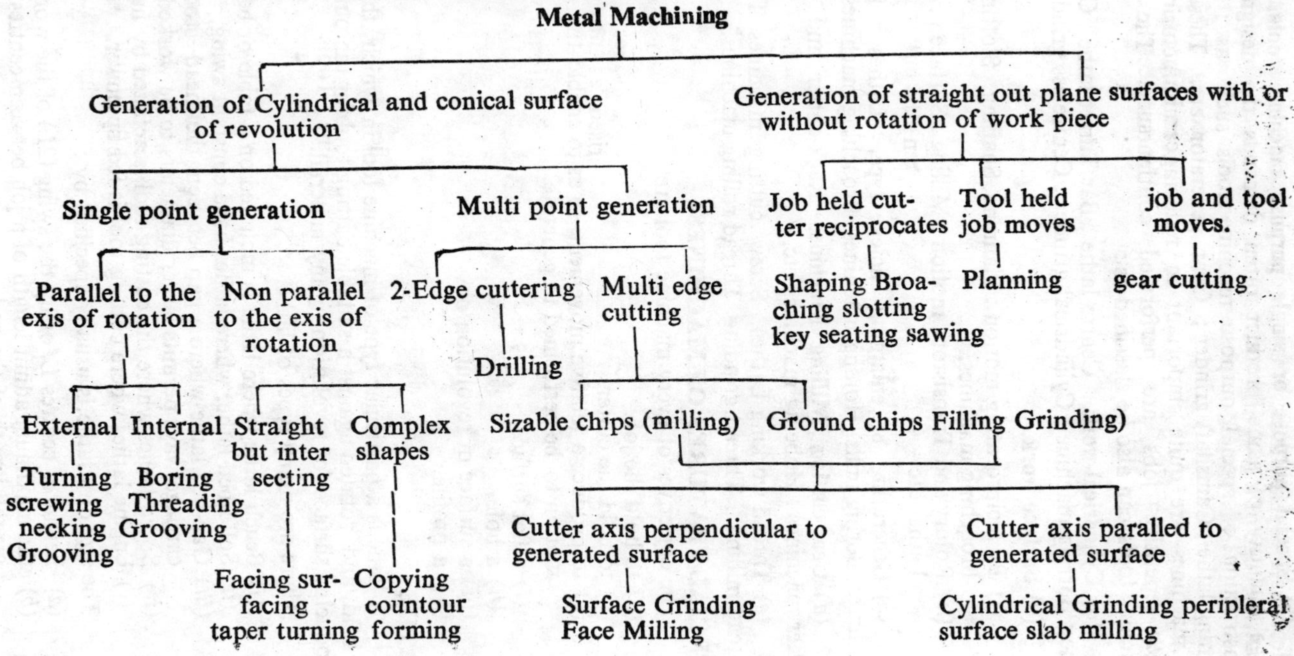

- **Metal Machining**
  - **Generation of Cylindrical and conical surface of revolution**
    - **Single point generation**
      - **Parallel to the exis of rotation**
        - External: Turning screwing necking Grooving
        - Internal: Boring Threading Grooving
        - Straight but inter secting
      - **Non parallel to the exis of rotation**
        - Facing sur-facing taper turning forming
        - Complex shapes: Copying countour turning
    - **Multi point generation**
      - **2-Edge cuttering** — Drilling
      - **Multi edge cutting**
        - Sizable chips (milling)
          - Cutter axis perpendicular to generated surface — Surface Grinding Face Milling
          - Cutter axis paralled to generated surface — Cylindrical Grinding peripleral surface slab milling
        - Ground chips Filling Grinding
  - **Generation of straight out plane surfaces with or without rotation of work piece**
    - Job held cut-ter reciprocates — Shaping Broa-ching slotting key seating sawing
    - Tool held job moves — Planning
    - job and tool moves — gear cutting

(c) **Special purpose or single purpose machine tools.** Very large quantity of same product often demands the design and construction of special purpose machine tools such as : Piston turning lathe ; Camshaft grinder ; Gear generator etc. These types of machines are quite important in reducing manufacturing cost when repetitive jobs are performed continuously. The basic machine tools can also be classified as :

(a) **Cylindrical work.** Centre lathe ; Bar Turret lathe ; Capstan lathe ; Borring machine ; Cylindrical grinder ; Centreless grinder etc.

(b) **Surface work.**

(i) By reciprocating motion : Planning, Shaping, Slotting and broaching machines.

(ii) Rotary and Transistory motion : Milling ; Surface grinding ; Lathe. etc.

(c) **Generating or enlarging holes type.** Lathes, Drilling machines, Boring mills, Honing machines, Broaching machines.

(d) **Gear cutting ; Milling Machine.** Rack-cutter and Pinion-cutter machines ; Hobeing machines etc.

(e) Thread working Lathes, Screw cutting machines, Tapping Thread milling. Thread grinding. Thread rolling machines.

SPECIAL FEATURES OF MACHINES

The objectives of every machine tool are :

(a) to hold the job.

(b) to hold the cutter.

(c) to produce a relative movement, so as to enable the cutting motion to be performed to generate.

(i) a flat surface.

(ii) a hole.

(iii) a surface of revolution or

(iv) a profile.

LATHE

A lathe is a particular type of machine tool in which the work is held and rotated against a suitable cutting tool for the purpose of producing surfaces of revolution in any material.

There are many types of lathe.

(i) Bench lathe where lathe is mounted on a table or bench.

(ii) Solid-bed lathe, where lathes have constant swing.

(iii) Gap-bed lathe where a gap between the head stocks and carriage ways permits larger diameters to be worked upon.

(iv) Speed lathes where the cutting tool is actiated by hand.

(v) Engine lathes where cutting tool is fed by power.

The size of a lathe is usually specified by :

(a) height of centres L. or by the swing (2L) of the work piece.

(b) the maximum admit length of a job between centres or the overall length of the bed.

(c) the power.

(d) range and number of steps of feeds and speeds.

DRILLING MACHINES

Drilling is a term which designates the particular operation of generating holes Drilling machines are classified either according to features of design or application of power.

According to features of construction

(a) Portable or stationary. (b) Pillar or column drill or Radial drill (c) Single spindle or multi spindle.

According to application of power

(a) Hand or power driven spindle.
(b) Sensitive (hand) feed or power feed.
(c) Line shaft driven or individual drive.

Shaping Machines

A shoper, first developed to a machine flat surfaces on small work pieces, is a machine where the cutting tool is hold in a tool holder held on a clapper in the head at the end of a reapproachingram.

There are two types of shapers based on the method of application of feed.

(i) Travelling head shaper.   (ii) Travelling table shaper.

Slotting Machines

The vertical shapers, commonly known as slotting machines, is a machine with a vertical ram which is specially adopted to slotting holds, key ways and grooves. The work is usually supported on a rotating table over a cross. Slide and carriage. Slotting machines are generally specified by the length of stroke.

Planer Machines

A planer, also used to produce flat surfaces on work that piees are too large or otherwise impossible to be handled on a shaper, differs from shaper in that the work table reciprocates while the cutting tool is fed on a cross-rail straddles table. The planning machines are classified according to

(a) Type of housing open side type or double housing type.
(b) Method of obtaining quick return motion.
(c) Method of driving the table.

The sizes of a planning machine is denoted by
(a) Stroke length (or size of table).
(b) Horizontal distance between the housings.
(d) Vertical distance between the rail and table.

Broaching Machines

Broaching is a machining process in which a typical cutter called broach is drawn or pushed past the surface this machining a hole or a grove. Broaching machines are classified as :

(i) Pull or push or continuous type.
(ii) Mechanical or Hydratic drive. (iii) Horizontal or vertical.

Boring Machines

Broaching machines are used to enlarge or bore holes. However. these machines can also perform drilling, milling, facing and similar other operations.

Milling Machines

The term milling as applied to certain kinds of machining operations is based on the action of multiple toothed wheel on the work piece in the same manner as the old "mill stone" acted on the grain in early days Milling machines employ such multiple rotating cutters, called milling cutters. There are various types of milling machines. The most important types classification are :

(a) Column and knee type or fixed bed type.
(b) Plain or universal.
(c) Horizontal spindle (arbour) or vertical spindle.
(d) Mechanical drive or Hydrautic drive.
(e) General purpose of production type.

Grinding Machines

Grinders are the machines which employ an abrasive wheel for the purpose of removing excess stock, at the same time leaving a good finish on the work.

CLASSIFICATION OF MACHINES

Machine Tools are nothing but instruments which have been created for the purpose of manufacturing wide range of products of all categories. Every year more and more machines with better accessoriesness are coming out. The production of all such machines is out come of specialised demand from the various engineering industries. In the manufacture of their products itself, machines are instrumental. There has been an alround development in the machines with respect to the drives control system, materials of construction, power consumption, rigidity, accuracy in machining process, lubrication system, interchangeability in conformance, automation etc.

Classification of Machine Tools

Based on the type of the operation, the machines do, the machines can be classified in eight groups.

Group I Turning. There are various types of machines under this category. Centre Lathes, Capstans, Turrets, Semi-Automatic and Automatic, Lathe, Special purpose machine.

Group II Drilling machines. Horizontal Driller, Vertical Drill, Radial drill, Bench drill, Multi spindle drill, Gang drill, Horizontal or vertical Boring machine, Jig Boring machine, Utrasonic drilling machine.

Group III Grinding and polishing machine. Cylindrical grinder, Universal grinder, surface grinder, Centreless, grinder, Internal grinder Tool grinder, Polishing machine special purpose grinding machine.

Group IV Tooth and Threating machine. Gear cutting, Gear Shaper, Thread miller, Thread rolling, Tooth Slotting, Thread grinding hossing.

Group V Milling machines. Horizontal Machine, Vertical Miller, Universal Miller Machine, Copy Milling Machine, Die Sinking Machine, Cam Milling Machine.

Group VI Planning Machine. Single Sided Planner, Doubled sided planner, Horizontal long planner, Plano Miller, Shaping machine.

Group VII Sawing Machine. Circular Abbrassive wheel, Abrasisive Saw, Band saw, Hack Saw, Circular Saw.

Group VIII Miscellaneous machines. "In Line" Transfer Machine, Rotary line Machine, Special Purpose Machine with unit construction, Special purpose machine used in making ball bearing etc.

Assembly Lines

Assembly lines are concerned with the work of putting pieces together, fastening belts, soldering wires, welding etc. It deals with assembly work, and not with actual fabrication or production. It is the work that men do along lines and not the with equipment or machines. The nature of assembly line work is as follows :

—Requires least skilled men.

—It is a light work relatively and many people like it.

—It is a simple with little learning or thinking to do.

—The jobs are so close that workers on lines can be sociable.

—It does not require any drawings to read to do assembly work.

—The nature of the jobs on the lines is almost same through out in the complexity of the job does not vary from day today or from one job to the other.

—The jobs being so simple, even it is monotonous, the jobs can be easily exchanged from one man to the other and monotony can be avoided.

—Material handling which only adds to the costs, is saved because the items move through conveyor from one job to the other. Further it eliminates inventories from one stage to the other.

—Assembly line are generally capable of handling minor variations in products. As long as the same type of jobs are involved in the assembly of various products, the variety in the sizes, packings, and products can be handled on the same line to a large extent.

—The rate of out put on the lines is high because the rate of output of men and machines are balanced with the desired rate of final output.

—The task broken into minute work elements so that the tasks are combined at each work station to near about one minute job. Every body on the line needs to do one minute job through out otherwise it results into idle time cost which is prohibitive.

—However well balanced, the man cannot work at the uniform rate throughout. With the result the variations in output at each work station will result some imbalance and to avoid such imbalance buffer stock policy is adopted in to some cases.

Process Plants. The scope of production processes covers the entire spectrum from the completely manual task through mant machine systems, automated processes, where labour is either direca or of a vigilance nature. Some production processes have considerable technological base like manufacture of metals, plastics, chemicals, fertilisers, glass, cement, drugs etc.

Processes involves transformation. The basic nature of processing is one of transformation i.e. something is happening or somewhere transforms the thing being worked on. In general, these transforms may effect a chemical change, bulk density basic shape or form, add or substract parts as an assembly change the location of the thing being processed as in transportation operations.

Chemical Processess. Chemical processes are common in industries such a petroleum, plastics, steel making and aluminium etc. Industrially, these processes occur both as batch processes and continuous processes.

Out lines of manufacturing procedures in process plants. The manufacturiug procedures of various processing industries can be basically divided into two units. (1) Unit processes or chemical change (2) Unit operations or physical changes.

Unit Process. The unit process is very useful concept for technical chemical change and has been described as "the commercialisation of a chemical reaction under such conditions as to be economically profitable. This naturally includes the machinery needed and the works involved as well as the physical phases".

The machinery or equipment required to carry out the unit processes are a reactor, a vessel with accessories like a field line, a discharge line, transfer lines, the measuring instruments like temperature, pH etc. a transfer device like a pump or pheumatic conveyor, heating or cooling devices mixing or agitation devices.

Unit operations. The unit operation is a physical change connected with the industrial handling of chemicals or allied materials. It frequently is tied with unit process. These unit processes and unit operation are the common bond between otherwise widely divergent process industries. They are of course are applied differently under the necessarily varying conditions. The main stages involved in the process plants are given below.

(i) **Preparatory work of field materials.** The materials are subjected to various unit operationt where only physical change takes place. They are

(a) Crushing (b) grinding (c) Seiving (d) Mixing of solid with liquids (e) Agitation of liquids to dilute them.

These are all the various operations required in preparing the materials before they are subjected to a chemical change.

(ii) **Chemical conversion of the materials.** The materials thus prepared as per the specifications, are subjected to the unit processes like

(a) Oxidation (b) nut-rilisation (c) chlorination (d) Nitration (e) hydrolysis etc.

At this stage, the materials are converted into different products having different physical and chemical characteristics under varying conditions of temperature, pressure, concentration etc.

Production Process And Value Added

It refers to the addition of value to raw materials and other inputs by a firm by virtue of its production activity. A firm purchases raw materials and services to produce goods and services. During process of production, the firm adds value to the inputs and thus contribute to the flow of goods and services by its productive activities.

In fact it is the value added and not the value of output of the firm which determine the true contribution to the flow of goods and services in an economy. In other words value of output of a firm includes the value of the intermediate products which have already been produced by some other producing units. Thus to find value added, value of intermediate products should be subtructed from value of output. On the other hand if we include value of input in the value added by a firm, it will account to double counting since intermediate products are the output of other firms. Thus value of output differs from value added by the amount of value of input. In other words value added by a firm is broadly gross output minus intermediate consumption. Whenever value is added by a firm by combining and utilising the services of factors of production, income of the same value is said to have been generated. Income generated and value added are two sides of the same coin. Income generated is distributed among four factors of production in the form of rent, wages, interest and profit. It may be borne in mind that income generated consists of factor income only and so it neither includes transfer income nor depreciation. In the production process whatever value is added to the inputs by a firm, income of the same value is said to have been generated by that firm. It is the measure of contribution of the producing unit to the National income. A distinction between the value of output and the value added is very significant in computation of national income. It is not the value of output, but the value added by the various producing units which is taken into account while calculating national income. Thus value added eliminates double counting and avoids inflating the figures of national income. To obtain the net value added at factor cost we subtruct deductable costs viz costs of raw material ; depreciation and indirect taxes from the value of gross output.

3

Production System

Introduction

It would be difficult to over emphasize the crucial role the effective productive systems play in modern society and in our life style. *Webster* defines a system as a regularly interacting or interdependent group of items forming unified whole. Thus a system may have many components and objects, but they are united in the pursuit of some goal. The components of a system contribute to the production of a set of outputs from given inputs which may or may not be optimal or best with respect to some appropriate measure of effectiveness. Systems are often complex although the definition does not specify that they need to be. Invariably, every system is affected by its environment and every system can be thought of as a part of an even larger system. A broad definition of systems runs the risk of leaving out important details involving in the functioning of the system. It is true that some of the most interesting systems for study are complex and that a change in one variable within the system will affect many other variabies of the system. Thus in productive system a change in production system may affect inventory, hours of work, overtime hours, facility layout, and so on.

In a given production system, successful management depends on information system concerning what is actually happening, and how we react to changes in demand, inventory position schedules, quality level and product and equipment innovation. It is a known fact that the high out per man hour economy is thought of as being efficient while its opposite is thought of as being inefficient. But production efficiency is a relative term meaning, essentially, how effectively we employ the appropriate available resources for a given unit of output.

In order to meet the objectives of a company all the subsystems must also be streamlined in the overall production system. In order to meet the production plan, certain realities interfere such as equipment failure, human error, discrepancies in the timing of order flow, quality variation, and so on. Therefore, systems for scheduling maintenance, quality control, and cost control are invented to help retain order where otherwise the system would naturally tend toward chaos.

History of Production System

A formal attention to production economics was paid by the great scottish economist *Adam Smith* in his book "The wealth of the

Nations" in which he emphasised the importance of division of labour. The book was a milestone in the development of production economics because Smith observed that there existed a rational for production. After Adam Smith, an Englishman, *Charles Babbage,* augmented Smith's observations about production organisation and economics. Charles Babbage, crystalised his views in the book, "On the Economy of Machinery and manufactures" concerning the economic advantages resulting from the division of labour. In the years after Adam Smith's and Charles Babbage's observations, the division of labour continued and then accelrated during the first half of the twentieth century.

Frederick W. Taylor was undoubtedly the outstanding historical figure in the development of the production management field. Smith and Babbage were observers and writers, but Taylor was both a thinker and a doer. He was an authoritarian with indomitable will, a fact that caused him to be greatly criticised but, at the same time, may have been a source of his great contributions. Taylor listed four new duties of management which may be summarised as under :

(i) The development of a science for each element of a man's work to replace old rule-of-thumb methods.

(ii) The scientific selection, training and development of workers instead of allowing the workman to choose his own tasks and to train himself as he could.

(iii) To develop the spirit of co-operation between the workman and management to ensure that work could be carried out in accordance with scientifically devised procedure.

(iv) The division of work between the workers and the management in almost equal shares for which they are best suited.

Taylor's work developed in the field of methods engineering, work measurement and human engineering. Taylor's uncomprising attitude in developing and installing his ideas caused much controversy, and he was strongly opposed in many quarters. Taylor's followers were numerous. *Cart Barth; Henry L. Gantt; Harrington Emerson, Frank and Lillian Gilbreth and others worked within Taylor's general framework and philosophy.* Some of the products of Taylor's thinking are viz; Wage Payment plans; Time study methods; and Charts and Mechanical control boards.

System

System is a very popular word with different meaning to different levels of executives. In a business organisation which is composed of formal and informal groups which function both within themselves and as part of the total organisation. Man is a system yet he has sub-systems such as reciprocatory digestive and nervous system. Each system receives stimuli and reacts. The sub-system operating within itself and it receives information from and sends information to other sub-systems. Each sub-system operates in a logical way both internally and as a part of the larger system. The idea of system in a business is the flow of information. Information flows into sub-system and causes certain actions. In a system the information flow is man made. System analysis is the study of the information flow and the development of systems is best serve the whole organisation. The term system can also be referred to as a methodical way of planning, coordinating the work to produce complex project.

The PERT is a system of this type. A system can also means a mathematic model. The system can also be defined as a means of maintaining existing routines work or it is a means of bringing a clear vision of the business to those who operate it. Let us discuss here that part of system which is basically concerned with the transmitting of information to the right place in the right form with right content for taking right action. We spend a lot of money to transmit information, part of the information is really pertinent and part of it is really redundant. Certain types of information do not change for long periods such as engineering design, costing and planning information.

Designing a System

One should know the following steps to design a system.

1. A system is merely for the purpose of mechanising highly routine tasks such as preparation of pay rolls, bills, divident for the shareholders etc.

2. One should work on the system of thought before working on the system modernisation.

3. The systems are known as control systems and are presumably designed for the purpose of providing information to certain key executives so that through them centralised location may exercise broader control of operations.

4. One must know to design an efficient system—the purpose, time span, inherent stability and keeping in view of potential changes.

Challenges to System Approach
Risks

The danger and opportunity to system approach stem from the ability to handle large amount of information quickly. The new tools of data handling and computation which can centralise authority and power in small groups of people. The risk of inflexibility, problem of adequate information, recentralisation, challenges of the near future are the limitations.

Challenges

The growth and the progress of an organisation depends to a considerable extent on simplifying processes to move forward with the complicating and complex processes.

These can be accomplished with executives ability to go forward indefinitely without becoming the casualities of their own complexities. We should devise our systems so that business managers and individual contributors becomes part of the system, rather than the servant of these managers.

System Staff

The integrated system cover the flow of information throughout the company, therefore there is a need to have a staff department to develop and see the operations of the system. The system staff improves business management by applying creative thinking to problems of business communication. The medium of communication is written. Its products are policies and procedures. Systems studies recommend management action, form and flow charts are designed to picture the sequence of staff in a procedure. In a real sense the system staff is responsible for the network of company communication, (written and verbal) between groups and departments, between management and employees linking the operations of the organisation to its future plans. System approach work out the problems from management's point of view but to accomplish this it must be close to companies activities.

Organisation

In system staff there are two types of men :

1. The practitioner—responsible for all efforts of certain departments.

2. The specialist—who goes in certain fields in depth such as production equipment, tabulating systems, special requirements af electronic data processing and so forth.

Definition

A system can be defined as a inter connected complex of functional related components designed to achieve pre-determined goals of the organisation. A system can be considered as a structure of sub-systems each having the following characteristics as shown in Fig. 3.1

Characteristics are :

—Inputs

—Transformation Process

—Output and

—Feedback.

A production system on the other hand is defined as the frame work of activities within which the creation of value can occur. To understand the economics of production the concept of production system should be clearly understood. If we look up a manufacturing plans in normal operation we find that it groups of related activities which frequently cut across the functional or departmental subdivision of the plant. Such group of related tasks are known as

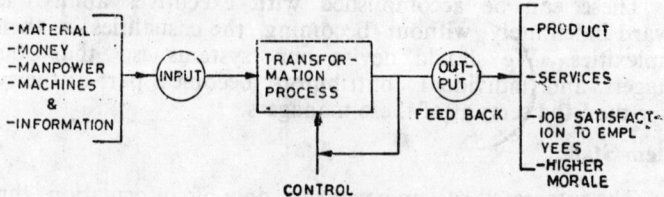

Fig. 3.1   Characteristics of a System.

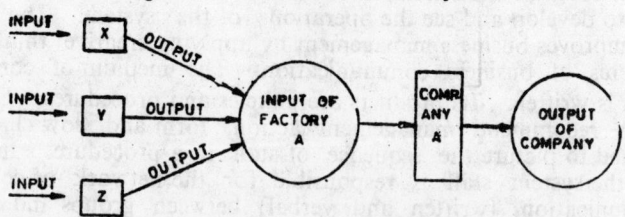

Fig. 3.2   Represents a Production System.

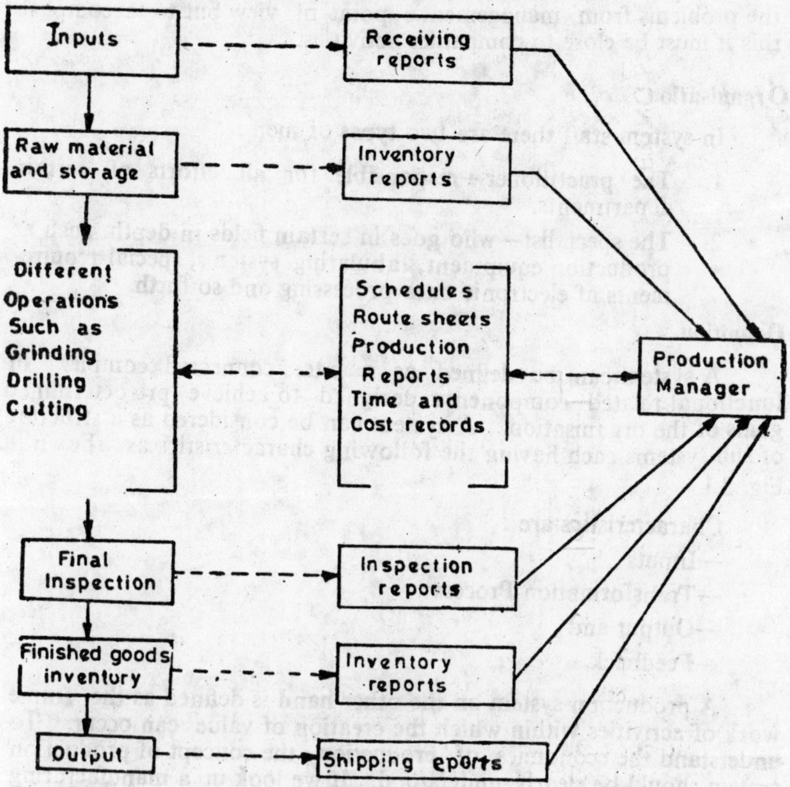

Fig. 3.3   Production System.

MARKETING SYSTEMS

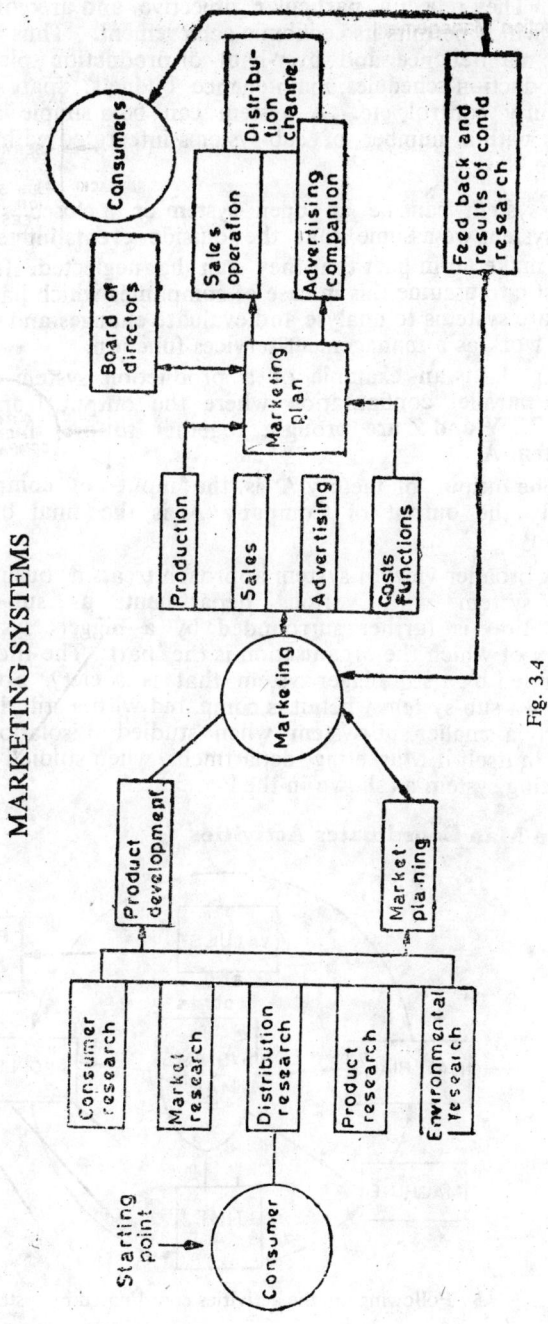

Fig. 3.4

system. They reflect a particular objective and are combined in a manner which permits its economic achievement. Thus a system o periodic maintenance and over-haul of production plant interact; with production schedules, maintenance budgets, spare parts inventory, equity control etc. A system can be a simple and local or complex with a number of sub-systems integrated within an overall structure.

A system can be an open system or a closed system. In a closed system we assume that the outside events influence takes to long to make an impact that they can be neglected. It is although unrealistic to assume this in case of companies which have developed a separate systems to analyse and evaluate changes and trends in the outside work as a management-services function.

Fig. 3.2 is an example of a production system approach. It shows a parallel configuration where the output from three sub-systems X, Y and Z are brought together to form the input to the sub-system A.

The output of factory A is the input of company A and similarly the output of Company A is the final output of the Company.

A broader view of system approach treats a business organisation a system with various departments as sub-system. The organisation is further surrounded by a bigger system, that is Industry of which the organisation is the part. The Industry is also surrounded by a still bigger system that is society. So each system becomes a sub-system when it is compared with a still larger system. Similarly a smaller sub-system when studied in isolation becomes a system in itself if Marketing department when studied it becomes a Marketing system as shown in the Fig. 3.5.

System Man Coordinates Activities

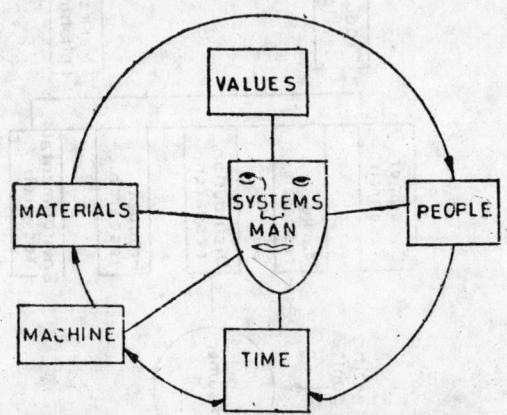

Fig. 3.5. Following are the activities coordinated by system man.

1. Reduction in the operating time cycles.

2. Lowering of inventories.

3. Reduction of errors in predicting cost and delivery dates.

4. Elimination of unnecessary functions and activities.

5. Faster working capital turnover and reduction of working capital requirements.

6. Greater operating flexibility-Faster transfer of top management decisions into action.

7. Elimination of the conflicting systems and operations which work at cross purposes.

8. Increase in the effectiveness of supervision.

9. Strengthening of the organization structure of the enterprise through the disclosure and elimination of practices which violate sound organizational principles.

10. Reducing of the routine cost of performing the planning organizing, controlling and service functions required in the enterprise.

The narrow view of the system approach we treat the system only as a flow of information in the organisation. The information flows into the sub-systems and causes certain action. This generates results and so resultant information flows out. Each sub-system operates in a logical way both internally and as part of the whole. If this is the concept of the system then organisation and system can be compared as follows :

Organisation	Vs	System
1. It refers to individual and group.		It refers what and how things are done.
2. It emphasises designs of organisation structure.		It emphasises communication structure—than of organisation needs.
3. It stresses chain of command authority and responsibility.		It stresses channels of communication — information flow and decisions.
4. It provides compartments of authority and responsibility.		It provides network of questions and answers.

Types of Production Systems

There are two types of Production Systems (1) Intermittent Production (2) Continuous Production.

The intermittent production system examples are machine shop production, building contractor. The continuous production examples are chemical plants, automobile industry etc. Most of the companies cannot be classified straight as intermittent or continuous production, rather in one department of the company continuous production is there while in other departments intermittent production exists. The time required for a continuous production system is always less than the intermittent production systems. The assembly line production of cars or scooters where the product is coming off every few minutes is considered as continuous production. On the other hand in intermittent production systems the products are in a state of partial completion for several weeks or days.

In continuous production systems the most common material handling equipments are belt conveyors, roller conveyors, chutes, rails etc. It is because in continuous production systems one or a few standard products are manufactured with pre-determined sequence of operations with inflexible material handling devices. In intermittent production system portable materials handling equipments are used and various products are produced with greater flexibility in the systems.

Continuous production system require a larger investment than intermittent production system because of fixed path material handling equipments, costly control mechanism and special purpose machines for various operations. Even the marketing techniques also differ for continuous production system and intermittents production system.

Intermittent production system the marketing efforts are directed towards meeting the individual orders for various products while in continuous production the marketing efforts are directed towards developing distribution channels for the large volume of output. The design of a production system starts with the firm and re-occurs intermittently when redesign is necessary. The major decision in the design of production system is the location of plant. Once the location has been decided the next decision relates to layout of facilities. Another problem which concerns the decision of production system is how products are designed and manufactured.

Man-Machine System

The man-machine-system basically consists of three typical systems ; Manual, Semiautomatic ; or Mechanical and Automatic System.

Manual System involve man with only mechanical aids or hand tools. Man supplies the power required and acts as controller of the process; the tools and mechanical aids help multiply his efforts. *Semiautomatic system* involve "man" mainly as controller of the process. He interacts with the machine by sensing information about the process, interpreting it and using a set of controls which may start and stop the machine and possibly make intermediate adjustments. Power is normally supplied by the machine. These are combinations of the manual and semiautomatic systems where the man is also supplying some of the system power. The Man-Machine system is given as in (Fig. 3.6a) and (Fig 13.6b). The figures show the

Diagram of MAN - MACHINE FUNCTIONS

Man (Power Source And Controller)

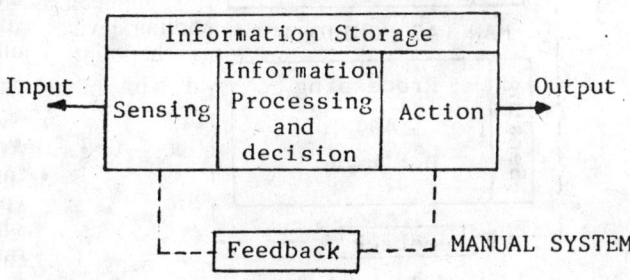

Fig. 3.6 (a) Man As controller

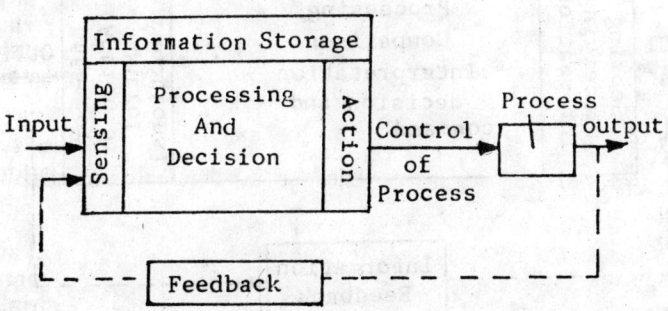

Fig. 3.7 Semiautomatic System

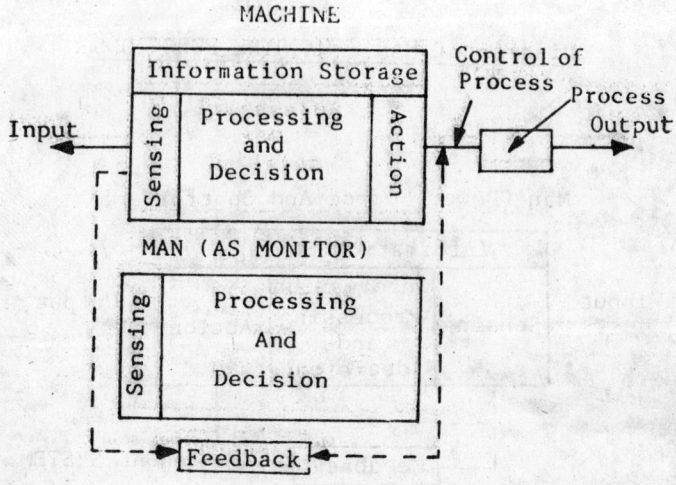

Fig. 3.8 (c) Automatic System

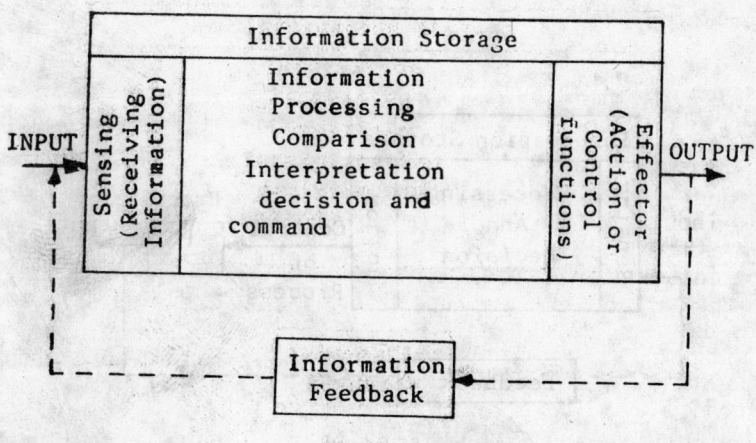

Fig. 3.9

general cycle of activity of "man-machine" system of the semiautomatic type imbedded in the working environment. The 'Man' is supplying some of the system power ; in loading the machine or in some activity in which the man may be involved while machine goes through its cycle in the combinations of manual and semiautomatic systems. *Automatic systems* presumably do not need a man since all of the functions of sensing information processing and decision, and action are performed by the machine Automation at such a level is not economically justified even if the machines could be designed. In this role the man periodically or continuously maintain surveillance over the process through displays which indicate the state of the crucial parameters of the process. Man-Machine system analysis study the information input focussing on visual displays, human control of man-machine systems including human motor activities, human control of systems, man-machiner elationships, analysis of hand motions, and the analysis of man and man-machine cycles; and working environment and its impact on output. The manipulative activity is representative of the dominant features of manual systems; the information input, sensing, and controlling activities are representative of both semiautomatic and antomatic system, and the information input and general monitoring or surveillance activities are representative of automatic systems. The impact of environmen and tne physical relationships of the man to work flow and workplace arrangements are, of course, applicable to all "man-machine" systems.

Functions Performed by Man-Machine Systems

The functions performed by man man-machine systems are represented in Fig. 3.10 The functions are generally comparable to these of the closed loop feed back diagrain. The four basic classes of functions are sensing, information allage, information processing, and action. The information storage interacts with three of the other functions. Information is received by the sensing function. The machine sensing is usually much more special or single purpose in nature than the broadly capable human senses. Information storage for man is in the human memory or by access to records. Machine information storage can be magnetic drum; magnetic tape; punched cards, discs; etc. Input and Out put is related to the raw material or the thing being processed. The output represents some transformation of the input. The process themselves may be of any type, that is chemical, processes to change shape or form, assembly, transport, clerical, and so on. Information feedback concerning the output states is an essential ingredient for it provides the basis for control.

Man Versus Machines

The man has certain physiological ; psychological and sociological characteristics which define both his capabilities and his limitations in the work situations. These characteristics are not static but vary from individual to individual. In performing work, man's functions fall into three general classifications viz ; Receiving information through various sense organs ; making decisions based on information received ; Taking action based on dicisions ; Machines on the other hand perform tasks as faithful servants reacting mainly to physical factors. But man reacts to his psychological and socio-logical environment as well as physical environment. Human beings appear to surpass existing machines in their ability to ; Delect small amounts of light and sound ; store large amounts of imformation for long periods and recall the relevant facts at the appropriate time; Exercise Judgment ; develop concepts and create methods. The exis-ting machines however appear to surpass humans in their ability to ; Respond quickly to control signals ; Apply great force smoothly and precisely ; Perform repetitive and routine tasks ; Perform rapid com-putation ; Perform different functions simultaneously. Where low cost labour is available as in India it is economical to use manual labour even in many tasks in which man is not well suited. Because of relatively high wages in developed country the machines are used much more extensively.

4

Production Planning and Control

Concepts

Production Planning and Control embrace all management decisions, policies, plans and actions which have something to do with optimisation or manufacturing objectives.

Production planning must establish the basic relationship, among (a) Sales forecasts (b) Production capacity (c) Inventory levels (d) Working capital (e) Raw materials requirement (f) Production facilities (g) Product design and its manufacturing sequence (h) Manpower requirement etc. The objection of production planning is to strike the most economic balance between the above factors.

The production planning and control systems are one of the basic activities that determine the effectiveness of a production enterprise. The production planning and control functions and its interaction with other functions is shown below.

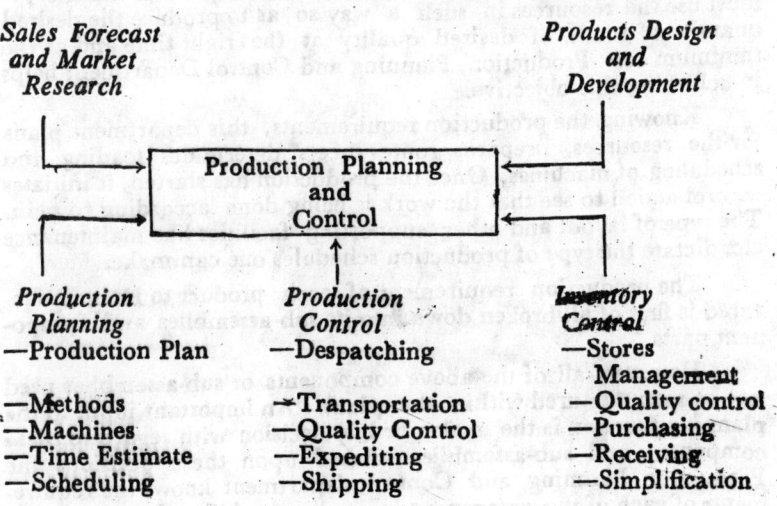

Sales Forecast and Market Research → **Production Planning and Control** ← **Products Design and Development**

Production Planning	Production Control	Inventory Control
—Production Plan	—Despatching	—Stores Management
—Methods	—Transportation	—Quality control
—Machines	—Quality Control	—Purchasing
—Time Estimate	—Expediting	—Receiving
—Scheduling	—Shipping	—Simplification

Importance of Profit

Business and Industrial Organisations have grown in size and complexities with the result the operations has also increasd and the number of decisions to be made have expanded. Not only the number but also the accurate and timely decisions are expected of every business executive to achieve the objectives of an organisation. The prime objective of any business enterprise is to make profit. Profit is the yard stick against which the performance of an organisation can be measured. An organisation which does not make profit becomes "sick" and drag on the economy.

To increase profit one can increase sale revenue by increasing the selling price. However, it may not always be possible to do so because of competition, price control by Government and other factors which are beyond the control of an organisation.

Cost reduction is another way which leads to increased profits. However, cost reduction is also not completely controllable because ore has to pay at the existing wage rates in the market to the employees, buy the raw material at the market price and invest in the plant and equipment.

To reduce cost, the only way is to make the best use of resources which include man, machine and materials. Cost consciousness should be there in all the departments of an organisation which include Production, Marketing, Material Finance and Personnel. The material management has to play a very important role in cutting the costs down because it consumes the maximum resources of a company.

Production Requirements

Production activity refers to the conversion of raw materials into finished product of desired quality and shape. In doing so, it must use the resources in such a way so as to produce the desired quantity of goods of desired quality at the right time and at the minimum cost. Production, Planning and Control Department helps in achieving this objective.

Knowing the production requirements, this department plans for the resources, prepares route sheets determines loading and scheduling of machines. Once the production has started, it initiates control action to see that the work is being done according to plan. The type of layout and other supporting facilities like maintenance etc. dictate the type of production schedules one can make.

The production requirement of each product to be manufactured is first of all broken down into its sub-assemblies and component parts.

However, all of the above components or sub-assemblies need not be manufactured within the plant. An important input to the planning function is the make or buy decision with regard to these components and sub-assemblies. Based upon these decisions the Production Planning aud Control department knows the requirements of each of the components or sub-assemblies that are to be manufactured with the plant.

It is then decided that in what lot size each of these items will be manufactured. The various operations to be performed on these items are determined, which are generally specified by process engineering department. Depending upon the sequence of operation and the layout route sheets are prepared which shows how the product will move within the plant.

The time estimates for all of the operations are obtained from the work study department. This helps in loading and scheduling the machines. While scheduling, due attention must be given to the maintenance policy of the company is following. In case of preventive maintenance the machines should not be loaded when they are to be down for maintenance work. Once the detailed production schedules are made, the manufacturing orders are released to all the concerned people. This is known as despatching.

When the production starts, continuous follow up is done with a view to keep control over production and make any modification in plans if the circumstances demand so. For example, because of a rush order the plans might have to be changed. The Quality Control Department is responsible to see that product of the right quality is manufactured. If we are behind the schedule it may be either due to operations or due to faulty material or machines or wrong planning. In any case Production Planning and Control suggests us to pin point and take corrective action.

Managing of Production

Production is the output of a plant or it is the flow of the product through the plant. The primary function of production planning and control is to manage production and inventories skillfully. This requires participation in the sales forecasts coordinating inventory and production levels with the sales farecasts, planning the product controlling production scheduling which included flow up of work improves so meet scheduled delivery promises, controlling production materials and supplies for purchases and delivery on specific data maintaining physical control of all production materials and supplies. The most effective coordination is achieved when it is under one management namely production planning and control. The managing of production can be broken into three major functional areas :

1. Production Planning
2. Production Control
3. Material Control

The three major functional areas can be further subdivided. Production planning can be divided into long range planning and current planning. Production control can be divided into production scheduling, despatching and follow up. The material control is divided into planned control and physical control. The responsibilities assigned to each of these six functions can be further discussed in their logical sequence of occurrence.

In some companies the work of department called production planning and control is limited to the responsibilities shown in production control and material control with perhaps the additional responsibilities of sales order processing and production releases.

Place of Production Planning and Control as a Department

The place of production planning and control as a department in the management of organisation depends upon the scope of its assigned responsibilities.

Production Planning and control is a joint effort. In production planning and control team every member has some responsibility for developing the progromme and selling it out. Each member should know, understand and accept his responsibility and recognise the way in which it integrates with the responsibilities of others. The planning in many organisation is still left to chance. A successful organisation should think out where it wants to go, why it wants to go, how it can get there and when it should get there and then sets about doing so with minimum of dislocation. In this present era of increased competition and right profits the practices of production. Planning and Control are destined to give a desired progress. No nation can prosper if its industrial companies fail to make adequate profits. The industrial forecasting planning and control enable companies to operate safely and profitably.

Production Planning and Control is the tool that coordinates all manufacturing activities. The management employs this technique to obtain the following targets.

The highest efficiency in production is obtained by manufacturing the required quantity of product of the required quality, at the required time, by the best and cheapest method.

Definition

The understand production, Planning and Control let us define each of these terms :

Production consists of a sequence of operations that transform materials from a given to a desired form. Transformation may be done by disintegration, integration or service.

Planning begins with an analysis of the given data, on the basis of which a scheme of the utilisation of the firm's services can be outlined so that the desirable target may be most efficiently attained.

Control invites and supervises operation with the aid of a control mechanism that feedback information about the progress of work.

Hence, Production Planning and Control may be defined as the direction and coordination of the firm's material and physical facilities towards the attainment of prescribed production goods in the most efficient available way.

PPC—A Management Functions

Production planning involves a system to plan production, cost, quality and time. In a simple language the production planning and control determines when, where, how and by whom the job is to be done.

Production Planning and Control gas the job of sending a continuous stream of directions given to each of the steps and at all parts of the factory. Its directions are again minute and specific. Production Planning and Control is not essentially a management function which comprise the planning, routing, scheduling, despatching, and follow up function in the productive process, so organized that movements of the material, performance of machines and operations of labour, however subdivided and directed and coordinated as to quantity, quality, time and place.

Compariag Production Planning and Control with Human Body

An industrial plant could be compared with the human body, we could say that the building is similar to the skelton because both provide a structural frame work. In the industrial body, the productive departments would be similar to the human muscles which do the work. Service departments such as industrial tool and engineering, maintenance, inspection and transport are similar to glands because they make constant adjustment to the physical body that are both corrective and preventive in their nature, sales, product engineering, purchasing personnel and accounting have functions that are similar to the five necessary organs because they react to external and internal events (sensory organs are touch, hearing, smell and taste).

The events which can and do stimulate our sensory departments are infinite in number, some of these events could be arise or decline in the general economy and/or the company's sales, a new product idea or desire to revise a product design, a rise in product cost or overhead, or an anticipated rise in material prices coupled with an opportunity to buy now at lower price.

The nervous system of the plant is its communication system. This is the complex of routes and channels through which ideas, information, and data are transmitted in speech, writing, signs or symbols by human or mechanical means from one person or point to another. Production planning and Control frequently has been compared with the human brain, which is the controlling centre of the nervous system and more particularly with the cerbellum which is that part of the brain which coordinates the muscular activity and movements of the body. Therefore, we can say that production planning and control department is the controlling centre of the communications systems in a plant which coordinates the physical activities of manufacturing and guides their direction.

PPC—Functions

Production Planning and Control functions are much the same as those of the cerebellum of brain. It receives messages from the sensory departments, interprets and filters out non-essentials and relays these messages to the productive and service departments. In turn, it receives from the productive and service departments, messages of reaction to the original stimulates (feedback) it interprets and filters out the essentials in these responses. In this light, the primary function of production planning and control are receiving, interpretations, decisive and communicative.

To perform these functions effectively the people in the production planning and control must have an objective point of view, an intimate knowledge of products and production processes, and most important an understanding of the persons in the Plant who are responsible for responding to their enquiries. The broader their knowledge of products, processes, and people, the keener their ability to listen, the more effective they will be in communicating the ideas information and facts to other people.

For better production planning and control we must fully understand the major forces that influence them. Most prominent among these forces are :

(a) Management plans and policies

(b) Lead time

(c) Usage during lead time

(d) Plant capacity and flexibility.

PRODUCTION CONTROL

It is an essential element to ensure that the production is carried as per the plan.

Objectives

If is well-known fact that the survival of any business organisation depends on how it can satisfy its customer. Thus, to satisfy the customer by quality and with regard to delivery date is one of the main objectives of any business organisation. This, however, becomes the primary objective of production control.

It is clear from the above discussion that the objective of production control department is very closely linked to the objective of the business enterprise as a whole. Keeping in view the "customer satisfaction", the objectives for production planning and control should be, broadly, as follows :

(a) To deliver the product to the customer when he wants it.

(b) To promise delivery in the shortest time possible if we cannot deliver our product when the customer wants it.

(c) To keep our promises.

(d) To maintain flexibility in the manufactuirng operations so that we can accept and deliver the occasional rush order for the odd product which gives for us a new customer or additional business from an old customer.

(e) To issue coordinated schedules and orders to the production department with complete information to tell them what is expected to meet delivery requirements.

(f) To follow up production schedules to assure that delivery promises will be kept.

(g) To maintain adequate but not excessive inventories of production materials to support the continuous flow of production.

(h) To maintain inventory, production and employment levels that are relatively stable and consistent with the general level of sales.

(i) To plan plant capacities that will provide adequate facilities for future production and sales.

Planning and Control

The planning and control is an integral part of the system. This is explained by a control cycle as shown in Fig. 4.1.

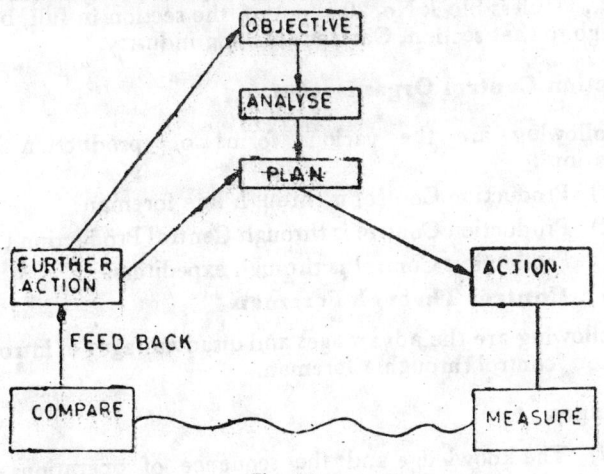

Fig. 4.1

Whatever be the type of the product to be manufactured the basic production control functions have to be done. The difference lies only in what to do, how to do, how often, where is to be done and who should do it. The difference, in the various kind of production control which are classified as :

1. Order Control. This is used for intermittent type of production. Each order is identified separately and all instructions and report form relating to the lot bear this number.

2. Flow Control. This is used for continuous production. Instructions and day-to-day directives are reduced to a minimum. Nearly all the functions requiring directives like· what materials to use, how long it will take, what machnies to use etc. are very carefully performed during the planning of the production time. Plant layout itself involves a lot of production control functions. Production control instructs the manufacturing department as to the planned rate of production and receives their reports showing their actual rate accomplished.

3. Load Control. Similar process industries have their process like continuous industries but their control problem resemble those in intermittent production. The load on the key machines are properly allocated. No separate instruction is issued for every individual lot as all go through the same process.

4. Block Control. Where the products are somewhat varied, but all go through the same operations and where the variations have very little effect on processing time, plant loading and progress control can be simplified by using block control. A block is a collection of orders whose work load adds up to a fixed time duration say one half day. The manufacturing department gets orders only in block. Unless block No. 20 is out of the section in full, block 21 cannot go to that section, Garment making industry.

Production Control Organisation

Following are the various forms of production control organisation :

(1) Production Control is through line foreman.

(2) Production Control is through Central Production Control.

(3) Production Control is through expeditors.

(1) Control Through Foreman

Following are the advantages and disadvantages of introducing production control through a foreman.

Advantages

(i) The knowledge and the sequence of operations is best utilised.

(ii) Complete control over work assignments.

(iii) It builds up the importance of foreman in the eyes of workers.

(iv) No overhead expeditors.

Production Control—Mechanics.

Production control promotes effective shop operation through its direction of activities within the production department. Keeping our production commitment for some period ahead and having assurance that inventory requirements will be cared, we may turn to the details of moving work through the shop itself. There are usually four production control activities viz. *Routing : Loading and Scheduling; Dispatching; and follow up.* **Routing** is necessary where a choice of faliricating method exists. Its function is to list the operations should be performed. The decision on just how to manufacture the product and what methods, tools and equipment can be used most effectively may be made by a section of the production control department, by the engineering department, by an independent production engineering, manufacturing engineering, or methods group, or by industrial engineering personal. Many times routing information is an integral part of the manufacturing order which authorises the shop to do the work. The information carried on the route sheet, operation list or shop order may include any of or all such items as : Identification; Operations Involved; Equipment ; Standards ; Scheduling Information and operating results. Infact the heart of the route sheet is the list of required operations. The route sheet is frequently a convenient vehicle for bringing the information to the attention of the dispatcher, foreman; or production supervisor. The route sheet will sometime provide space for a record of operating results, spoilage, actual times and the like.

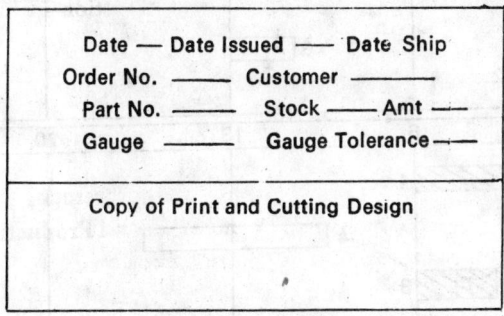

Fig. 4.2 **Route Cards**

Loading and Scheduling, are concerned with the flow of work to the shop and the relationship between shop capacity and outstanding orders. The term "loading and scheduling" use grouped together because it is almost impossible to set a realistic schedule without some knowledge of the load on the shop. A schedule is simply a method of relating facilities orders and time. A simple method for handling schedule, load, and progress information graphically. The Chart A and B are drawn giving form of scheduling. In Chart B the margins of free time come after the component schedules rather than before and so serve as safety factors.

The illustration in fig. (   ) are simple variations of the Gantt Chart, one of the earliest techniques of graphic scheduling. The graphic type of scheduling is most useful in those applications where:

* There are number of variables which need close control.

* It is desirable to highlight the interrelations of many parts or components.

* Quick comparison of progress with plan are required.

* It is important to spot critical items and incipient delays before they have a chance to became very serious.

S. No.	Operations	Department	Time	Order No. Data
1	Operation A	Deptt X		Sept. 91
2	Operation B	Deptt Y		
3	Operation C	Deptt Z		
4	Operation D			

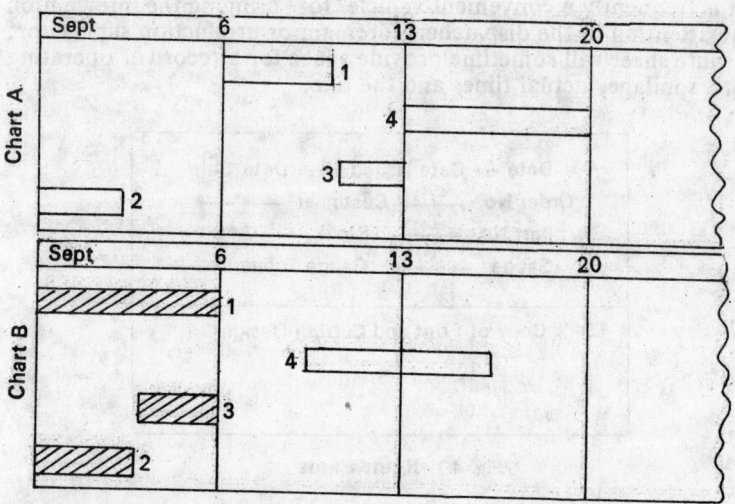

Fig. 4.3 Loading and Scheduling Forms

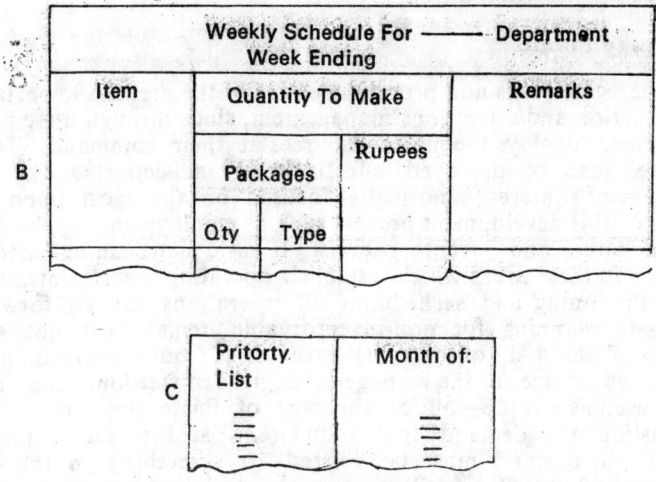

Fig. 4.5 Loading and Scheduling Forms

AGGREGATE PLANS AND PROGRAMS

Most of the production managers want to plan and control operations at the broadest level through some kind of aggregate planning that bypasses the details of individual products and the detailed scheduling of facilities and manpower. What is needed first for aggregate plans is to develop some logical overall unit of measuring sales and output for example number of units produced in some mechanical industries. Second management must be able to forecast for some reasonable planning period, perhaps a years, in these aggregate terms. Aggregate planning increases the range of alternatives for capacity use that must be considered by management. The term aggregate planning includes scheduling in the sense of a program; The aggregate planning concept raise such broad basic questions as : To what extent should inventory be used to absorb the fluctuations in depand that was occur over the next six to twelve months ? Why not maintain a fairly stable work force since and absorb fluctuations by changing production rates through varying work hours ? We know that if inventories are used to absorb reasonable changes in demand, then capital and obsolescence costs, as well as the costs associated with storage, insurance, and handling tend to increase. Similarly changes in the size of the work force affect the total costs of labour turnover. When new workers are inducted, costs arise from selection, training, and lower production effectiveness. The sepration of workers may involve unemployment compensation or other sepration costs, as were as an intangible effect on public relations and public image. By aggregate planning and scheduling decisions regarding various costs are difficult to measure and are not segregated in accounting records viz. Interest cost on inventory mangement ; Opportunity costs ; cost associated with public relations and public image.

Aggregate Planning

Aggregate plans and programmes are of the greatest importance to production and operations management, since through these plans managemen deploys the major resources at their command. These resources can be deployed effectively or ineffectively. The top management's interest is normally focused on the most important aspects of this development process such as employment levels, production rates, and inventory levels. If basic plans can be based on decisions in these areas which establish operating constraints, then detailed planning and secheduling of operations can go forward. Aggregate planning for non-inventoryable items need not take account of the risk of carrying inventories; but it also eliminates inventories as one of the managerial degrees of freedom—they cannot be used as a trade—off for the cost of fluctuating production. Forecasting the demand for unique items at first seems impossible but the demand must be related to something in the economy as a whole. Project Planning and scheduling are, in a sense, aggregate planning and scheduling. The entire project and all of its activities are planned at one time. Such programmes may call for projecting. load one to two years in advance. These loads are not forecasts of demand but actual demands for contracted work. The techniques for something loads in net work schedules to be discussed under deployment of Resources and use in fact attempts to reduce the costs effects of load fluctuations or what we have termed the costs of production fluctuations aggregate planning models. As a matter of fact we can view the custom organisation as holding an inventory of capacity and skills to produce in a field, available on demand, rather than holding an inventory of goods. We must recognise that the part of the high cost of turnover is an opportunity cost of maintaining the going concern value of the organisation. Good managers of such organisations are very senstive to these factors that argue for relative employment stability, even to the extent of altering price policies to maintain the integrity of the organisation in slack times.

Production Line Techniques

The production line is recognised as the way to produce large quantities of standardised items at low costs. Production line is an arrangement of work areas where the subsequent operations are located immediately adjacent to each other, where the material moves continuously and at a uniform rate through a series of balanced operations which permit simultaneous performance throughout, the work moving toward completion along a reasonably direct path.

Principles

Line production takes advantage of the following principles :

(i) **The Principles of minimum Distance Moved** : with work areas immediately adjacent, one operation begins where the preceeding one finishes.

(ii) **The Principles of flow of work.** Flow involves continuous movement at a uniform rate, which the production line provides. Flow is measured by the rate of production rather than a specific quantity.

(iii) **Principles of Simultaneous Operation :** Up and down the line, workers are continuously performing their operations; the first; last, and middle operations are being performed simultaneously without the completion of one operation at a time on all pieces.

(iv) **Principle of Unit operation :** The line is considered as a single producing unit—One series operations or one group of workers solely assigned to one product. The entire line performs as one producing unit.

(v) **Principle of fixed Routing :** The routine is pre-established when the line is set up, and the opportunity for diversion or lost work is minimised.

(vi) **Principal of Minimum Material in Process:** Line production achieves a flowing stream of material—a fixed operation sequence with simultaneous operations.

(vii) **The Principle of Interchangeability :** The interchangeable parts and components are a must-line production takes advantage of this similarity, at the some time being largely dependent on it.

(viii) **Principle of Division Labour :** The most efficient use of labour is to give specific small portion of a job to each of the workers to divide the work and assign one operation or skill to one worker.

How to Plan a Line Operations

Some of the important steps which should be taken to get a line into operation are :

(i) Detail Drawings and specifications.

(ii) Product Analysis and Suggestion for change.

(iii) Rate and volume of production based on sales forecast or contract.

(iv) Design freeze and estimate of tooling ; Layout and operating cost ; and Decision to Make or Buy.

(v) Based on choice of Processes or Production handling of equipment working out the list of Orders and equipment.

(vi) To work out the Routing times and Manpower requirement.

Based on Layout approval to work out the New Tools and Equipment Required.

How to proceed with Installation and Line Balance

Based on Vendor analysis, Purchase Authorisation and Material Delivery schedule to bring out the PILOT LOT and then to make efforts to IMPROVE AND REBALANCE. The various steps to get a production line is given in following diagram.

From the figure we find that fewer the operations, the shorter the line and the less the investment in expensive assembly fixtures. Lining up the operations required and the sequence of operations is the next step in planning the line. Many times it is better to supply a line with equipment that is slower then the most efficient machine might be there is little sense in planning over capacity for one operation. It is generally important that adequate operation analysis and methods study go into the job before it start into production. Production lines call for special equipment but it is usually better that the product design be fixed because with line production just one change may upset an entire sequence of highly synchronised operations. *Line Balancing is the balancing of operations in terms of equal times and in terms of the time required to meet the desired rate of production. The line balance is part and parcel of line production.*

Line Speed and Length : Speed of flow bears a direct relation to rate of production and space per work station.

$$\text{Speed of Line (Meter/hr)} = \text{Rate of production (pieces/hr)} \times \text{Station length or space per piece (Meter/piece)}$$

$$= \frac{\text{Station length or space per piece (Meter/piece)}}{\text{Balancing factor or station time (hr/piece)}}$$

The station length or space is dependent on the size of the part or unit, the size of room, number of equipments and workers and amount of work to be performed.

5

Production Planning and Control Functions

Important Functions

The functions of Production Planning and Control can be described as a systematic line diagram as follows :

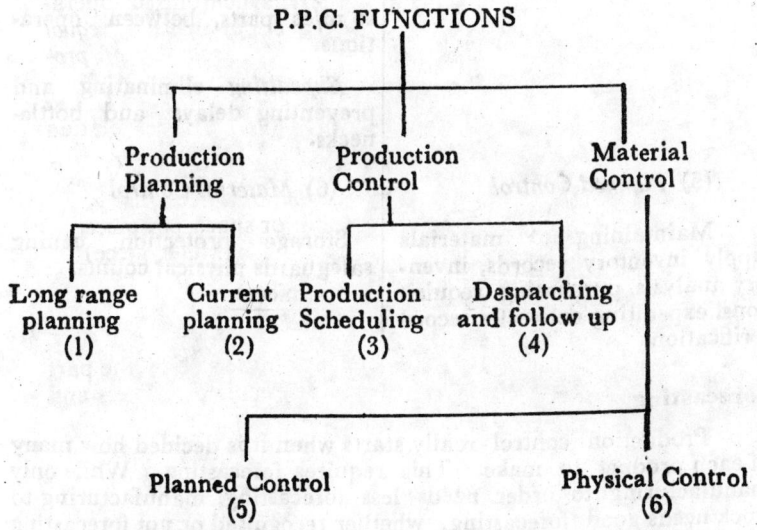

P.P.C. FUNCTIONS

- Production Planning
  - Long range planning (1)
  - Current planning (2)
- Production Control
  - Production Scheduling (3)
  - Despatching and follow up (4)
- Material Control

Planned Control (5)   Physical Control (6)

(1) *Long range Plans*	(2) *Current Plans*
Sales forecast, plant requirement, inventory investment.	Sales forecast items, plant capacity, inventory level.
Short term plans, sales forecast, plant capacities, inventory levels, production levels manpower levels.	Sales order processing and inventory interpretation comparison of current orders, forecast, historical activity control of financial
New product introduction sales estimate, plant requirements inventory investment timing and co-ordination.	goods inventory allocation of production shipping promises, order acknowledgement, order quantity, order timing.

(3) *Routing*

Parts list, Bills of materials, inspection data operation sequences, time allowances set up operation move.

Explosion of production orders. Component requirements, sub-assemblies parts, raw materials, planned control work in process inventories.

Master Schedules, detailed schedules, machine loading.

(4) *Schedule Coordination*

Material requisition, tool requisition job instruction, control papers.

Reporting material used, tools used, progress and delay operations time pieces, utilisation of time operations machines.

Physical control of work in process accountability of pieces etc.

Physical movement materials to m/cs parts, between operations.

Expediting eliminating and preventing delays and bottlenecks.

(5) *Planned Control*

Maintaining materials supply inventory records, inventory analysis, purchasing requisitions, expediting deliveries record verification.

(6) *Material Control*

Storage protection, issuing safeguards physical counts.

Forecasting

Production control really starts when it is decided how many of each product to make. This requires forecasting. While only manufacturing to order needs less forecasting, manufacturing to stock needs good forecasting, whether recognised or not forecasting is done in every company with varying degree of intensiveness.

Demand forecasting is an important input to the managerial functions. It plays a key role in planning for the future decisions. The business executives need to forecast the future with a view to make certain fundamental decisions for product lines, expansion of plant capacity, location of plants and economic activity, employment and labour policy, advertising expenditures as well as in countless other areas. Without an estimate of the future, the decision making is just like stopping in the dark. The decisions are made keeping in view the developments in the National and International economy, the industry and the particular area in which the firm is more interested. Formal forecasting makes the decisions more rational and it helps to focus the attention of the executives on certain areas which may otherwise get omitted.

Demand forecast has specific significance for the production department. Once the demand forecast for various products is known we can find the total demand for each time period and then the detailed production plans can be made. Demand forecast tells the production department what products will be required in what quantity and at what time. Knowing this, the production planning and control department can plan for the resources so as to meet the specified delivery dates. If a forecast is to be of use to production department it must meet the following conditions :

1. The forecast must define expected demand in physical units.

2. The forecast should also include an indication of the probably variation around the expected demand, although admittedly such information is often very difficult to develop.

3. The forecast should be available in time to schedule all tasks required to achieve the necessary output.

4. The forecast must be made repeatedly during future periods to permit the necessary adjustments in materials and production requirement.

5. Finally, the forecast must be reliable because forecasting errors can cost considerable amount of money.

A forecast is essential to plan the production activities. A forecast is necessary to decide when to manufacture each item, how many to manufacture. Similarly a forecast is essential for purchasing new materials, parts or sub-assemblies to find out when to purchase, how much to purchase as item. The forecast is essential even for determining an economic lot size formula.

Sales forecasts are the first approximation in production planning, they provide the base upon which plan may rest and adjustment may be made. For this accurate records sales showing the level and pattern must be known. Sales for each month, quarter and year must be estimated. In case of seasonal sales, analysis may show the need for new products to be sold in other seasons. Annual or monthly figures of production should not be the basis for forecasting sales. The actual sales figure should be employed and adjusted. The sales forecast will not yield accurate results during first few years.

Horizon of the Forecast

Horizon is determined by the lead time necessary to accomplish the purpose involved. In setting re order points for inventory a probablistic forecasts will be needed.

When the economic lot size is used to determine how much of an item should be purchased, a forecast of demand will be needed which will hold as long as the proceeds of the lot will lost.

Forecast Overal Activity

Many production control decisions are based on the forecast of the overall level of activity. This include make or buy decisions, production smoothing etc. When one or a few types of end items are being produced in a given plant, the demand for sub-assemblies, parts and raw materials may be derived from the forecast of the end product. The requirements of various parts and components of a compressor can be worked out from the master schedule of an airconditioning units.

Methods of Forecasting

The forecast may be based on the judgment of the experienced executives. An estimate may be ob'ained from the field salesman whs are in close contact with the customers. The opinion of the traders in the market may also provide an estimate of the future sales. Although these methods are very simple, yet they may not be suitable because these are based on subjective estimates and may be biased. Therefore, it is essential to use some mathematical technique for forecasting.

Moving Averages

This is a very simple method which project the past into the future, if the forces determining the market demand will remain constant over time. A three month moving average is the average demand during the past three months e.g., the demand in the month of April is an arithmatic mean of the demands in the months of January, February and March. Similarly, the demand in May is the average of the demands in February, March and April and so on. This is known as simple moving average.

It may be desirable not to give equal weights to each of the preceding months. The most recent months may be given the highest weight. Then the weighted average may be the correct forecast. There is another way of weighting the demands and it is known as exponential smoothing.

Aggregate Forecast

The aggregate forecast is an important function of the business. It is of an asset to the marketing division as a basis for its planning and to the personnel division for hiring and training programmes. It forms the base for the annual budgets.

The aggregate forecasts is not an activity of the production planning and control.

Individual Forecast

To implement production planning and control for individual components, it is necessary to forecast these items. Unlike the aggregate forecasts, the individual forecasts have the responsibility of the individuals charged with production planning and control.

Derived Forecasts

It is always possible to derive forecasts for material from the overall forecast of demand for the individual end items. A bill of material lists, material required for each of end item produced. The forecast of demand for material on each bill of materials is obtained by multiplying the materials required per end item. The number of the end items to be manufactured.

The bill of materials also lists the specifications and the quantity required for each raw materials needed. The forecasts so obtained should be provided with an appropriate allowance for shrinkage, wastage, during the manufacturing process by applying yield and scrap factors. The final estimate need to be increased by the amount of spare parts required in addition to the number of assembly required. In certain products it is not possible to derive the forecast for individual items form the forecast of overall activities. This problem comes when the forecast of overall activity is not in end product but in monetary units or labour hours. When a series of end items of different specifications are produced it may be different to breakdown an aggregate forecast into end items. The problem of forecasting becomes still more complicated when an item has multiple end item uses. It is true in case of raw materials such as copper or steel bars or angles or sheets of given specifications.

Statistical Forecast

When a manufacturing company makes a wide variety of end items these statistical techniques may be used to develop forecasts only of components but also of end items. Two important techniques of statistical forecasting are :

(1) Opinion forecast

(2) Exponential smoothing.

In both these methods the forecasts are based on the record of past demand its level, trend and variation.

Opinion Forecast

The opinion forecast based on the judgment of the executive and the information obtained from the field salesman who are in direct touch with the customers. The opinion of the traders, distributors are also taken into consideration to estimate the future sales.

These methods are simple yet based on subjective estimate and biases of the various people involved, thus it is essential to envolve mathematical techniques for forecasting. Following are the advantages and disadvantages of opinion forecast :

Advantages

(1) It is less elaborate and requires less technical skill.

(2) It makes use of the knowledge of field personnel, *i.e.* gives first hand information.

(3) The process of forecast passes through the review of number of people with various types of specific knowledge who can assess its accuracy.

Disadvantages

(1) It is a subjective approach.

(2) It takes lot of time with the field and office personnel in making up the forecast.

(3) Generally the final results depend for more on the opinions, of a few influential or persuasive individuals.

Exponential Smoothing

In this case the demand for any period is based on two factors the old demand for this period and the actual demand during the previous period.

Regression or Correlation Analysis

It is likely that the scale of a certain product has some relationship with other products *e.g.* sale of gasoline is related with the sale of automobiles. The sale of a product may also be related with some other indices like material income, income per capita, size of population etc. For example, the time of population and the number of industrial units etc. will affect the demand of electricity in a town. The forecast will be good only if the said indices are coming from reliable sources. The variable which is to be forecasted may be called as the dependent variable and the other one as the independent variable.

In using this technique, first of all the relationship between the variables, is defined in the form of a curve. This is known as fitting the regression line, the line is then used to find the correlation between the variables. If the relationship between the two variables is linear then one can use simple linear correlation the simplest form of a Regression line is given by the equation :

$$Y = a \times bX$$

where, Y and X are the dependent and independent variables respectively we know the actual value Y for each value of X. Whereas the value of the dependent variable as obtained from the regression line is denoted by Y.

The values of coefficients *a* and *b* are calculated so that the following conditions are satisfied :

$$Y-y=0$$

$(Y-y)^2$ = minimum, *i.e.* the sum of the squares of the deviation by the actual value Y from the value y obtained by regression line is maximum. When we have more than two variables we can use multiple correlation. Thus the information about the independent variables can be used for cost the dependent variables.

Combination Methods

Generally the combination of various approaches is used for forecasting. Statistical technique and economic methods may be used together with fluctuations about the trend line with a measure of estimated reliability. The trend line is one forecast of demand and the control limits give an estimate of reliability provided the conditions determining the trend do not change.

Control over Forecasts

Setting up of procedures for making the required forecasts and review or control of forecasts are two important aspects of forecasts.

The review of control aspect can be broken into two parts :

(1) Whether forecasts are made as per the procedures established.

(2) Measurement of the accuracy of forecasts and determination of major causes of errors for improving the quality of effectiveness. Control over the use of procedures is very important in case of short term forecasts.

It is very important to keep the record of the actual information versus, forecasted information for the following reasons :

(1) Knowledge of the reliability of forecasts.

(2) Knowledge of the range, forecast errors and examination of the causes of major errors.

(3) Determination of any systematic bias in the forecast.

Planning means getting ready to make the product. Planning is an intellectual process in which creative thinking and imagination is essential. Planning includes setting objectives, policies, decisions and strategies. Planning is deciding in advance what to do ? Where to do it ? How to do it ? When to do it and who is to do it ? Planning bridges the gap from where we were to where we should go, planning makes things to occur which would not occur otherwise. Following are the important concepts about planning :

1. **Objectivity.** Realistic planning is essential conceptualisation of the objectives, the needs, the priorities, the approach and the activities involved. The objectivity requires a long far sight, correct visualisation and practical anticipation.

2. **Futuristic.** The trend in planning is based on the present and future than on the past. It is very similar to vehicle driving where one drives while looking ahead, anticipating, judging and making decisions and rarely peeping into the rear view mirror.

3. **Planning Time.** At all levels of management one has to devote time for planning. The lack of planning or no planning creates increasing work loads piling up of unattended work which deprives him of his planning time and ultimately resulting in a vicious circle. The time spent on forward planning is more and more as one moves up the ladder of organisation. Let us assume the time spent on advance planning as 100%. then time spent by different level of management is given the following chart :

TOTAL PLANNING PERIOD 100%

Level of Management	Advance Planning							
	Weekly planntng	Monthly	Six Monthly	One Year	II Year	III Year	10 Year	Over 10 Years
1. Top Management (General Manager)	2	5	15	20	25	18	10	5
2. Middle Mangement								
—Functional Managers	7	10	15	30	18	12	6	2
—Departmental Heads	12	15	20	35	10	6	2	—
3. Lower Management								
Foreman	20	20	25	28	6	1	—	—
—Supervisors	35	40	15	8	2	—	—	—

Production Planning and Control
Disadvantages

(i) No expert knowledge is used in the technique of PPC.

(ii) Overburden on foreman.

(iii) Very little opportunity for overall planning and coordination among various departments of the factory.

(iv) No counter check will be possible.

(v) No process inventory will be high.

(vi) No accurate picture of machine loading will be available and there will be difficulty in deciding the delivery periods of future orders.

(vii) Sequence of jobs may not be as required for meeting the deliveries, but they are more likely as per the availability of machines.

(2) Production Planning and Control

Advantages

(i) Accurate picture of present, future load on various machines, could be available.

(ii) Optimum scheduling i.e. the expert knowledge would be available.

(iii) Proper evaluation of the effect of each new order could be possible.

(iv) More accurate delivery commitments can be made.

(v) Inter-departmental coordination would be improved.

(vi) More realistic evaluation of delays and departmental shortcomings.

(vii) Low inventory in process.

(viii) Lead time will be reduced because of optimum scheduling.

Additional Contraints in PPC

(i) Increase in overhead costs.

(ii) Tendency of central office people to loose contract with shop persons.

(iii) Risk of flexibility.

(iv) Danger of encroaching the line authority.

(v) Possibility of giving more excuses for failures.

(3) Control Through Expeditors

Advantages

1. Important jobs would go through in time.
2. Less overhead cost.
3. More flexibility.
4. Opportunity to move some orders through the plant more rapidly.

Disadvantages

(i) Very high indirect cost due to increase in set up cost because of frequent settings.

(*ii*) Limited opportunity for overall planning and coordination.

(*iii*) Overall picture of shop load is not available.

(*iv*) Extreme stress on the system when plant operation approach capacity.

Inter-relationship of Production Planning and Control with other Departments

The organisations structure of a company is an important foctor in determining the nature of production planning and control appropriate to its operations.

Every organisation or a department have a purpose. The purpose must be clearly defined otherwise the functions of a department spread outside the pattern of its framework. So it must have a planned structure suitable to its purpose. The production planning and control department is a service organisation that co-ordinates and secures a smooth, uniform flow of materials into through and out of the manufacturing facility—a flow that provides finished products ready for the customer at the right time and place.

Functions PPC Department

Production planning and control department can be divided into following sections :

(*i*) The material control section.

(*ii*) Cortrol section.

(*iii*) Planning section.

1. The material control section issues purchase requisition at the right time so that material will be on hand when needed. The object remains that neither the inventories becomes tool small to create production delays nor reduce the net working capital.

2. The control section of production planning and control department follows the planning through. This section maintains a close touch with the sales department so that production plan may be adjusted to maintain inventory within the predetermined limits.

3. The planning section of production planning and control department translates sales forecasts into master schedules that make the best possible use of both equipment and personnel. The planning section also prepares the load charts, to find out whether delivery dates can be met and to find out wherther equipment is being over or under loaded.

Production planning and control are co-operative activities shared by several departments instead of confined within one as shown below.

The production planning and control department should cooperate with accounting department also, in order to establish operational budget, financial planning and production schedules.

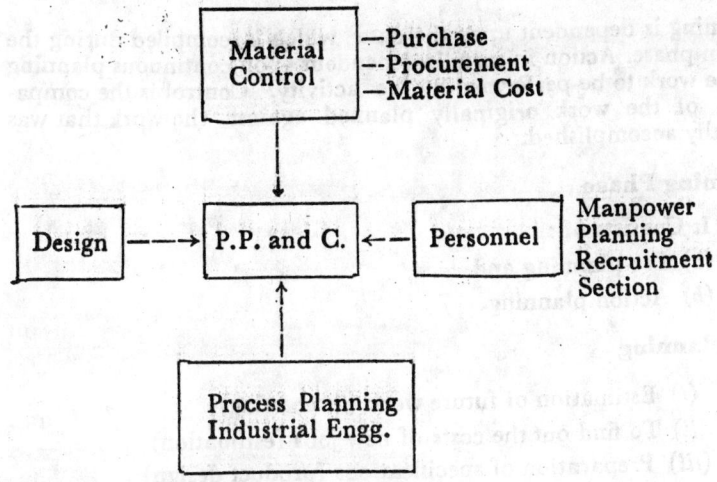

Benefits of Production Control

The advantages of production control may be said to include the following :

(a) Help to stabilise business by getting more and more customers through better service to them ;

(b) Increases the wages to employees as better service to customers results in huge profits ;

(c) Offers job securities to employees as plans are laid to maintain steady flow of production inspite of the seasonal fluctuations in the market demand ;

(d) Increases the value of the goods by way of product development which constantly aims at improving the quality of the product, at the same time tries to bring down the prices ;

(e) Increases the productivity through more effective use of men, materials and machines ;

(f) Ensures better utilisation of capital, as a result of well balanced inventory control at all levels ; and

(g) Assures adequate returns on the capital invested.

Functions of the Production Planning and Control

Functions of Production Planning and Control may be devided as three main phases

—Planning phase
—Action phase
—Control phase

The phases and functions of productions planning and control are mutually supporting and have very close inter-relationship.

Planning is dependent upon the data which is compiled during the action phase. Action in turn, is dependent upon continuous planning of the work to be performed by the activity. Control is the comparison of the work originally planned against the work that was actually accomplished.

Planning Phase

It Consists of :

(a) Pre-planning and

(b) Action planning.

Pre-planning

(i) Estimation of future work (forecasting)

(ii) To find out the costs of new jobs (estimation)

(iii) Preparation of specifications (product design)

Action Planning

(i) Preparation of work details (process planning and routing)

(ii) Determination of requirements and control of materials (material control)

(iii) Determination of requirements of control of tools (tool control)

(iv) Determination of when work is to be done (scheduling)

(v) Determination of requirements and control of men and machines (loading)

Control Phase. It is comprison of :

(a) Action phase

(b) Progress reporting

(c) Corrective action phase.

Action Phase

(i) Starting the work (despatching)

Progress Reporting Phase

(i) Collection of data and interpreting data (Data Processing)

Corrective Action Phase

(i) Making Current Work Corrections (Expediting)

(ii) Making Plan Corrections (Replanning)

6

Production Planning and Control – A System of Management

PPC—A System of Management

The planning has become a requisite for enterprise survival. The attribute of planning is to look into the future. Its tool is the forecasting method employed. The results are always accurate but if the tendencies can be clearly recognised, it is a good attempt. The Planning exercise is continuous. Closer approximation are possible as one travels in time. Complete accuracy is possible only when all the parameters are as nearly constant as this dynamic world will allow.

Since planning is an art, it has the capacity to enthuse people to undertake this work. To peep into the unknown area however tenuous, stimulates thinking of the present in the comparison of the future. Risk taking assumes a purpose. The definition of objective has been one of the most important top management functions, Planning is, therefore, a top management activity par excellence.

One way to narrow the gap between estimates and ultimate actuals is to plan for a long period on the basis of small interval (six months to one year) on a moving basis. Despite its obvious inaccuracies and the dynamism of situation planning occupies a permanent place in management activity. Planes are usually sub-divided as follows :

(i) Perspective—for 10—20 years at a time divided into yearly based estimates.

(ii) Long Term master plans—for 2—5 years at a time divided into monthly broad estimates.

(iii) Short term master plans—for one year divided into monthly close estimates.

(iv) Monthly plans—divided into weekly or daily plans.

Whenever plans are prepared, assumptions underlying the figures adopted are always indicated in good plans.

Production Planning and Control participates as one of the important systems of management in (1) and (2). It is the spear-head and motivating force in (3) and (4) above.

The extent participation of PPC in (1) and (2) depends on the contribution it can make otherwise it is limited to its being the recipient of information on what other systems are contributing. As regards (3) and (4), PPC is one of the basic developers of the plans with the help of all other systems of management. The various stages are already discussed.

(i) Product Engineering

(ii) Process engineering (route sheets, equipment, lists tooling data and personnel·productivity standards)

(iii) Recognition of production requirements.

Master Plan of Production

From these it created the Master Plan of Production. It represents an overall course of action which establishes an objective for the current operations of the company. It is based on a sales forecast and a master production schedule to accomplish it. Purchase programmes, inventory policies, maintenance strategies, marginal investments machine capacities, productivity strategy etc. also enter into the composition of master Production Plan. The assumptions are stated clearly and all the inputs and outputs are standardised for a particular production plan. The most important aspect of Production Master Plan in the compilation of a systematic basis the standards or norms.

Planning in production will be a different procedure depending on the type of production but it has certain basic features common to all types. These are :

(a) Availability of men materials, machines and methods.

(b) Balancing of capacities.

(c) Balancing with market and profit planning.

(d) Optimisation of the Plan.

Balancing

Balancing is one of the most important operations of planning. The strength of the chain is in its weakest link. Optimisation has to take into account the weak features and apply known techniques of analysis to improve them. These improvements in the short term, would be without incurring costs but may mean marginally additional capital costs in the long run.

An essential feature of planning as stated earlier is to recognise that dynamism is a part of life and chanes must occur. Adjustments are, therefore, necessary. A good monitoring system will indicate tends of changes before hand. If they can be recognised, changes will be made in advance otherwise they will overwhelm the planners.

It is important that master plan be kept upto date to spot out any adjustments required to reflect changes in sales prospects, production demands, unforeseen changes in plant and equipment or available manpower.

Repetitive Planning

Getting ready for the first time is *original planning* and doing it over again is *repetitive planning*.

Original planning deals with how products are to made, what materials to be used for various components, what machines and tools are required, etc.. Original planning needs to be done over again for repeat orders only if a product's design or the manufacturing processes change.

After a product is once made, the drawings, processing instructions, the necessary machines and toolings are all available. But still planning is to be done for each repeated order and this is concered with deciding upon the material requirements and man and machine hours requirements for each order.

Prior Planning

Prior planning refers to whatever goes on in the planning phase prior to the first step initiating the activity.

Planning for a product starts when a new idea is first presented. Everything is very tentative at first but different ways of doing things are considered and some methods are discarded while the product is only a rough sketch. Making these choices can be called *prior planning* or *pre-planning* or *early planning*. Thus, to think of how to make products, and how much they will cost are called early planning functions.

The planning is broadly classified as :
(a) Manufacturing planning.
(b) Factory planning.
(c) Production planning.
(d) Financial planning.

Manufacturing Planning

The production planning cannot begin till following two phases of planning are completed :

(i) Manufacturing Planning

(ii) Factory Planning.

When a product comes out of a Pilot plant the manufacturing planning lays the ground for factory and production planning. Manufacturing planning involve following steps :

(i) To revise the specifications of the product to suit the equipments and technique.

(ii) Develop a plan for new process and techniques.

(iii) Determine the operation to be performed in manufacturing.

(iv) Determine the equipments and trained personnel required for each operation.

(v) Determine the time required for each operation.

(vi) Determine the materials required for each part.

(vii) Determine the cast at various levels of activity.

(viii) Finalise the planning based at the level of price and sales value.

Manufacturing planning calculates risks before subsequent plans starts. Manufacturing planning follows research, development and product enineering. It includes rationalising and adjusting the product designate manufacture equipment and methods that are known to be economical and effective. In manufacturing planning the essential manufacturing operations are listed in sequence in which they should be performed. The time required to perform various operations is established by estimates. Costs are estimated for the different methods of manufacturing. The relationship between the selling price and the sales volume will determine whether a given project is practical or not.

Factory Planning

After the manufacturing planning, the considerations will be given to factory facilities for production. In manufacturing planning, the arrangements of equipment the structure to house these equipments are designed. The general activities in factory planning are :

(a) To determine the optimum size of factory.

(b) The geographical study to find out the suitable location.

(c) To prepare the layout.

(d) To determine the type of structure required.

(e) To determine the service facilities.

(f) To determine the construction schedule.

(g) To determine the equipment and installation schedule.

As the size of the organisation grow next thing is to plan the factories for optimum size. After details of manufacturing planning are complete, the factory planning can begin. A schedule for equipment installation is very important. The Gantt charts can be used to schedule both construction and installation.

Production Planning

Production planning translates the sale forecast into master schedule. Production planning utilises the information developed in manufacturing planning and factory planning to perform the following functions :

(a) To prepare production forecast.

(b) To prepare master schedules.

(c) To prepare procurement schedule.

(d) To prepare area schedules.

(e) To prepare personnel schedules.

(f) To establish the finish goods.

(g) To prepare alternative inventory plans for production.

The production planning is used for various operations, to continue the manufacture of goods that are already in production.

Financial Planning

It deals with the preparation of financial budgets. A manufacturing organisation must have table of expenditure and receipts to get a reasonably accurate picture of each position. The working capital is the life blood of an organisation and working capital should be adequate to meet the day-to-day financial obligations such as the pay rolls, pay for raw material, advertising bills and cost of sales promotion. Thus the need for the budget is evident. All financial requirements of a company must be planned in relation to the expected income and the plan is embodied in master budget. Financial budgets are essential in a business organisation for its operation. All budgets, plans or schedules are planned and used continuously for critical comparison of actual with planned performance.

After the engineering and processing of the product to be manufactured have been completed, the production planning and control department is in a position to authorise the works to formulate their programme.

The programme of works has to be undertaken within the limits set by sources as estimated by the PPC department subject to the approval of the top management. In case PPC happens to be a part of the Works management, the Works Manager obtains necessary approval from the management.

The approval is with reference to :

(a) Priorities for customer satisfaction which management considers reasonable and feasible to attend. The normal rule is first come first served unless management considers it otherwise.

(b) Lead times for materials, bought out items and fabricating contracts.

(c) Technological limitations.

(d) Any special instructions which may be given by the management.

The Production Planning and Control Department is in a position to communicate full details of manufacturing instructions. This is done through routing and scheduling document.

Routing

In intermediate flow production the routing instructions are prepared once before the plant is commissioned. They are normally called technological instructions. They indicate in detail the flow of processes type and capacities of equipment used, physical and chemical characteristics of inputs, norms for inputs or raw materials, intermediates, services, individual process times and cycles times of processes where necessary. The main function of routing is to utilise machine at their fullest capacity. *Routing* involves the determination of where each operation in a component part, assemblies or assembly is performed. The aim is to find out route path for the movement of manufacturing lot through the factory. To find this path the emphasis is placed on determining operating data. Operating data includes planning of where any by whom work should be done. It gives the green signal. If not, the exercise is again gone over to see if improvement could by made. The schedule or programme of production must be ready by 15th of the month previous to which it relates. It should then be uptodated continuously and passed on to works well ahead of the period to which it relates. Depending upon the quality of product to be manufactured, production system used routing may be either be generalised or detailed.

In job order planning, routing is confined to each job but scheduling covers work in progress and new orders. Determination of routing is finally the best sequence of operations. Routing in job order planning is based on process engineering for the product. It must either have been worked out by the customer and checked by the manufacturers or the entire work would by done by the latter. In both cases, PPC should get a complete list of manufacturing drawing, bills of materials, process technology and special instruction regarding specific points. They will also get a chart of inspection programme in respect of the product.

Contribution of PPC

PPC makes the following contributions before intimating them to work :

1. The PPC makes comparison between cost factors so that machine assignment can be made in accordance with the best cost choice for the original as well as the alternative machine.

2. The output volume is so calculated that the factors regarding rejection are given proper allowance.

3. PPC makes determination of the most economical production run in case the order as a whole cannot be processed at once.

4. The production papers are fully codified and manufacturing drawings are serialised.

5. Route sheet is so developed that scheduling in facilitated. Information regarding time study data, set up time, tool number and material specifications is written on the route sheet.

The route sheet will also indicate the coverage allowance, *i.e.* the number of pieces to be manufactured will be complemented to compensate for possible rejects for which a standard is fixed before hand by the PPC.

Normally the foreman of the shop can also do this work but in well run shops the foreman is not at all bothered in checking petty details. He should concentrate on supervision of his men.

Role of Routing Supervisor

The routing supervisor has some discretion in indicating set up times and in specifying the types of machines to be used.

Its function is to list the operations required to produce a part or a product and the sequence in which there operations are to be performed. The router must determine the specific machining which should perform specific operation. Router should not only see the machines but should select the machines which perform the job at the lowest cost. The final specification of routing is a production control function. The engineering aspect of the problem are another matter. The decision on what to manufacture, how to manufacture, what methods, tools and equipment used may be made by the production controls, engineering department, manufacturing engineering or methods group or by industrial engineering personnel. The identification section will give the product to which route sheet apply. If the route sheet refers to a particular order it may include such information as order number, number of pieces, delivery date. The heart of the route sheet is the list of required operations.

The route sheet will provide space for the record of operating results spoilage, actual times etc.

If it reaches the planning office before the completion of the job because the office provide information for purposes of control. It helps in compiling the cost data and may serve as a permanent record of the order.

Whenever a plant is engaged in job production new route sheets are required for majority of orders. In case of standardised production it is possible to keep a file of master route cards.

Scheduling information include starting time, finishing time, etc. These time may be in week, days, shifts or in hours or whether they appear at all will depend on the degree of central control involved. The route sheet is a convenient vehicle for bringing these times to the attention of the despatcher, foreman or production clerk.

Loading

Whenever a customer orders for any product his first question is when the product will be delivered to him. The manufacturing organisations which are manufacturing the stock can quote a date for stock on hand or work in process. The job order manufacture need to determine the total order time for each order. The first step in answering the question is to list the available equipment. The machines performing the similar operations are placed in one group. The number of machines in each group multiplied by the number of hours gives the number of machine hours available for loading. Although the availability of machine hours per day is worked out at the time of planning. If we have 20 machines in the group and they are operated on 7 hours shift per day without any already scheduled increase, the machine availability per day is only 140 machine hours or 20 machine shifts. Similarly planning for operators and supervisors must be completed before there can be any additional availability for production planning in terms of machine hours or additional hours.

The Gantt charts can be used to show graphically the relation of availability to machine load. Machine loading can be simplified by using Kardex files equipped with plastic riders to designate how much of the available capacity of each machine groups has been allocated. Certain production executives often make hasty decisions to accept attractive orders without knowing available capacity of their manufacturing facilities. This way production men became frayed to accomplish the goals and even then fails to achieve them. The company earns a reputation for delayed delivery and there loses customers's confidence. The complex problems in equipment loading can be solved by Operations Research and Mathematical Programming. The complex factors such as delivery time orders, shift premiums, overtime premiums, night shift efficiency, overtime efficiency etc. can be arrayed in mathematical programming matrices to obtain optimal solutions.

Sometime the loading problems are so complicated that electronic data processing equipment is required for their solution. The simple act of arrying of loading problem on a mathematical programming matrix provides a valuable aid into the problem.

Dislocations caused by deviations in sales trends and cancellation of orders and requests for certain special orders calls for loading and scheduling to be recomputed.

Machine load charts gives us an idea whether machine or equipment is fully utilised or not. A manufacturer who is involved in jobbing work may find from the machine load charts that his grinders are used only about half of the time while his planners, shapers, drills are overloaded. Now for a rush order involving only the use of grinders the manufacturer will not find it difficult to go about. So the need of machine loading is apparent even in simple situations but it is indispensible in case of complex situations. The machine load charts reveal equipments which are not fully utilised. This enables management to take corrective measures which may help to select orders that will result in balanced and profitable operations.

Scheduling

Scheduling is the pre-requisite to production control phases namely despatching and follow up.

Scheduling is the heart of production control because it integrates the work of other departments to produce a complete product at the promised date as shown :

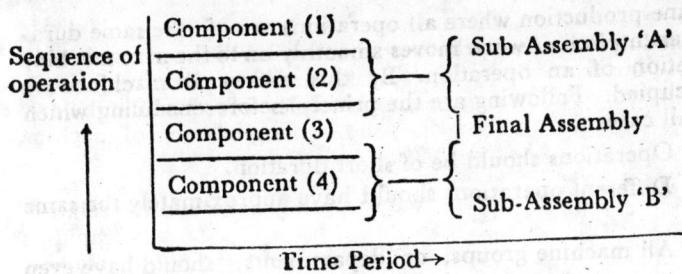

Function of Scheduling establishes the starting and finishing times for production action. The objectives of scheduling are :

(1) Cost should be controlled.

(2) Job should be taken up in proper sequence.

(3) Final product must be delivered in time.

Scheduling problems exist whenever there is choice in which a number or tasks can be performed on one or more facilities by one or more men. The tasks, facilities may represent different things in different situations. Jobs to be processed in a manufacturing plant, patients waiting for a series of tests in hospital, aircrafts waiting for landing clearance, a batch of orders waiting to be processed in a warehouse, bank customers at a row of teller's window and programmes to be run at a computer etc. are some of the examples of scheduling problems.

Scheduling is the last step in Production Planning. At this stage detailed plans are made which specify for each machine, the time schedule at which different products will be processed on these machines. In fact a production schedule is the time table of the Production Department. Before a production schedule is made, the schedular should know the quantity of each product to be manufactured, the delivery dates, type and number of machines and personnel available and an estimate of the processing time. A production schedular aims at choosing the best schedule so as to optimize the use of resources.

To compare different schedules, we must have some effectiveness criterion or the objective function. This criterion may be any one or combination of the following things : minimize the total schedule time, minimize in process inventory meeting, the delivery dates and reducing the total set up cost etc. Production scheduling problems are perhaps as old as the industries themselves. Before the era of scientific management, these problems were solved without getting a "formal solution" and so no one ever felt that there existed a problem at all. It was only in the early 20th century the "formal solution" to production scheduling problems were presented in this form of a graphic representation, more popularly known as Gantt Charts after the name of its originator Henry L. Gantt.

In line production where all operations are of the same duration, a.d each piece of work moves smoothly on to the next machine on completion of an operation. By this means all machines are always occupied. Following are the principles of scheduling which apply to all cases :

(a) Operations should be of short duration.

(b) Different operations should have approximately the same duration.

(c) All machine groups, or departments, should have even load.

(d) As far as possible, work-centres should be employed in the same sequence.

It should be noted that these principles describe accurately the operating principles of a transfer machine, as assembly line, a paper machine and many other forms of continuous production.

Reasons for Scheduling

Following are the reasons for undertaking scheduling :

1. Minimum production costs.
2. Minimum storage costs.
3. Minimum stock investment.
4. Minimum cash outflow.
5. Maximum labour utilisation.
6. Maximum plan utilisation.
7. Maximum customer satisfaction.
8. Maximum operator morale.

In the first situation, the quantity to be made, the rate of production and the resources necessary, are all adjustable, so that the most economic balance can be struck between the various parameters.

In the second, the quantity to be made, the production rate and the delivery dates are all fixed and limited resources have to be used as effectively as possible. This type of production is characteristically "batch" production.

When only one project is being undertaken, it may be possible to use what is called critical path analysis. In some situations, the pre-production activities can be scheduled with critical path analysis, and the subsequent production activities scheduled using one of the other techniques—for example, line of balance. Where a number of projects are being undertaken, critical path analysis is of limited use, although this particular application of critical path analysis is being vigorously explored.

There may be choice, either in the resources used or in the quantities of various products produced. This situation is responsive to the techniques of "Linear Programming".

Types of Schedules

Schedules are of two types : Master schedules and schedules employed to plan the manufacturing and assembly operations required for each product. Two-important points with scheduling functioning to ensure production are :

1. It is simple.
2. It is flexible.

Scheduling problem in a process industry requires the calculations of plant capacities, availabilities of raw materials and forecasts of market demands which determine production lot economic quantities if a number of products are involved. Even where one product is involved and the demand is less than the production capacity, the phasing of production becomes very important. In case demand is more than the engineered plant capacity, various questions of cycle times may have to be investigated with a view to see whether additional production can be secured from the same resources.

Scheduling assumes a different character on an assembly line. In this one job immediately follows the other. The sequence of successive jobs is chosen to balance to total work load at each operation along the line.

Parts and sub-assemblies required for final assembly should be co-ordinated so that they are available when needed. The scheduling problem for these items is to determine the proper start times.

A job shop has a different type of scheduling problem. Various jobs have different sequences of operations. The time to complete a given job depends on the time required for each operation, the time to move from one operation to another but most critically on the time that a job has to wait in line ahead of an operation until other jobs have been completed. For a particular department there are several different plausible way of selecting the next one to be run from among available in terms of criteria such as :

(i) investment in work in progress inventories.

(ii) probability that a given job will be finished in time.

(iii) utilisation of equipment and skilled manpower.

(iv) minimum and maximum time to complete all operations.

(v) flexibility to allow a more urgent job to bypass others.

The scheduling of production is governed by two factors :

(1) External factors.

(2) Internal factors.

The external factors as consumer demand, customer delivery dates and dealer inventory level. The internal factors which are too many and vary with type of the product, the organisation and the culture of management. The important factors are :

(1) Finished goods inventories.
(2) Process time and process interval.
(3) Availability of equipment, personnel, material, facilities.
(4) Economic production runs.

Basic Scheduling Techniques

To co-ordinate and control various activities several basic scheduling techniques have been adopted. Many of these technique are Gatt chart or Bar chart ; the Flow chart, the Line balance, the Milestone chart.

The Flow chart is adopted in work study techniques for work simplification, method improvement and procedure analysis. The line balance is primarily a production technique and is not effective for construction projects. The Milestone chart is almost similar to Bar chart except that it displays the Milestone rather than bars. These scheduling techniques are useful not only in the planning and scheduling of work to be accomplished but also following the progress of work in process.

The Gatt Chart

Henry L. Gantt developed a versatile charting technique. It is based on the principle of relating facts to time. It gives the relationship among different activities in a production process. It depicts on a plan of work in terms of time and serves as an effective communication device.

Gantt or Bar Chart

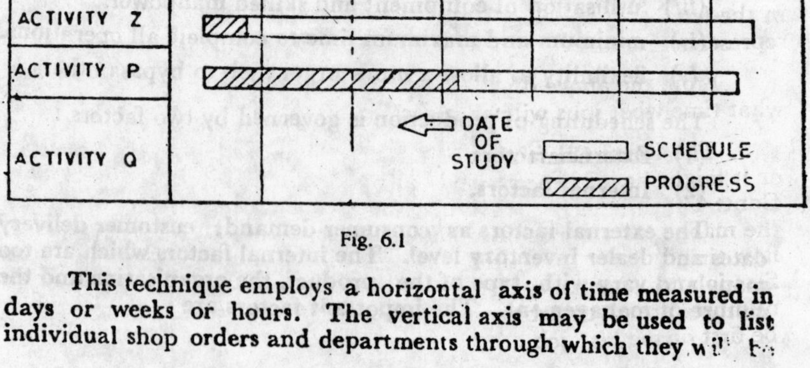

DESCRIPTION OF ACTIVITIES	OCT.	NOV.	DEC.	JAN.	FEB.	MAR.	APR.
ACTIVITY X							
ACTIVITY Y							
ACTIVITY Z							
ACTIVITY P							
ACTIVITY Q							

⟵ DATE OF STUDY

☐ SCHEDULE
▨ PROGRESS

Fig. 6.1

This technique employs a horizontal axis of time measured in days or weeks or hours. The vertical axis may be used to list individual shop orders and departments through which they will be processed. In this form the chart shows both the planned and actual progress of job through several activities or departments as shown in Fig. 6.1 The visual array of information displayed by a Gantt Chart permits the user to absorb the information quickly. The principle of Gantt chart represent that an amount of work actually done in that time to amount scheduled.

Example

Suppose we have three jobs J_1, J_2 and J_3 to be produced on three machines M_1. M_2 and M_3. The Gantt Chart will show the times at which each of these three jobs will be processed on each of the three machines. The basic data needed to construct the Gantt Chart may by provided in the form of the following two matrics namely. The Machine Sequence Matrix (MSM) and the Processing Time Matrix (PTM).

$$\text{MSM} = \begin{array}{c} \text{Job} \\ \begin{array}{c} J_1 \\ J_2 \\ J_3 \end{array} \left[\begin{array}{ccc} M_1 & M_2 & M_3 \\ M_3 & M_2 & M_1 \\ M_1 & M_3 & M_2 \end{array} \right] \end{array}$$

$$\text{PTM} = \begin{array}{c} J_1 \\ J_2 \\ J_3 \end{array} \left[\begin{array}{ccc} 2 & 3 & 1 \\ 3 & 2 & 4 \\ 1 & 2 & 3 \end{array} \right]$$

The machine sequence matrix shows the sequence in which the machines will be required for the three jobs while the processing time matrix gives the processing time required for each operation of each job. Thus job J_2 has to be processed on the machine M_3 M_2 and M_1 in that order required 3, 2 and 4 units of time respectively. The unit of time may be hours, days or weeks etc.

A possible solution for the above problem may be represented in the form of following Gantt Chart, where the horizontal axis represents the time.

From the above diagram we can find for each machine that at what time what jobs will be processed on them.

The Gantt Chart may be an outcome of hit and trial method or it might have been arrived at by a systematic procedure. The Gantt Chart helps in achieving the defined objective to minimize the make span time or also known as scheduled time or maximum flow time. Make span time is the elapsed from the beginning of the first job (not necessary the J_1) till the finish of the last job. The job numbers are given arbitrarily in the problem. Any of the jobs can be first or the last job.

Gantt Chart

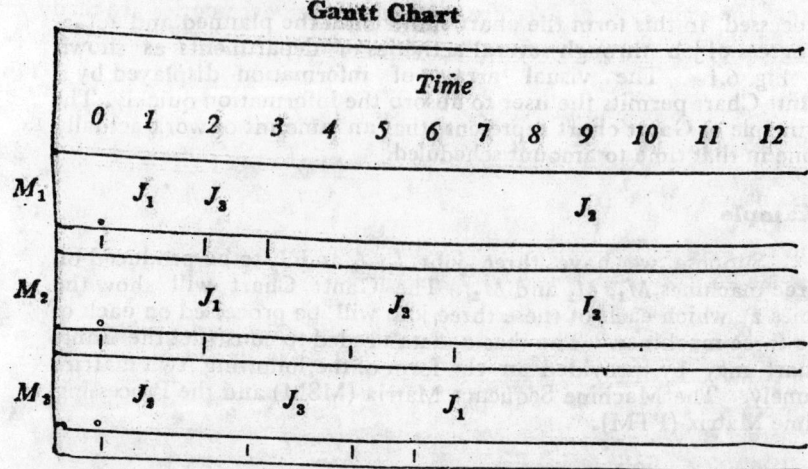

The conventional forms of Gantt Charts are now extensively modified to the visual control boards like production control schedule graphs. The deficiencies of Gantt Chart are :

1. Bar chart can indicate if some work is either behind or ahead of schedule.

2. Bar chart cannot show the inter-dependencies and the inter-relationships between the various activities.

3. It cannot predict the effect of delay in one area over another area.

4. Reporting with this chart require merely guessing or estimating.

5. It can isolate the critical areas of work.

6. Optimising the use of resources is difficult because it is not possible to identify the non-critical areas.

Due to these limitations and inadequacies of traditional planning techniques there is a need for effective planning and scheduling.

Despatching

It translates the paper work into actual production. Despatching co-ordinates and translates planning into actual production. Depatching includes of work orders. Despatching ensures that the right type of jigs, fixtures and other accessories are ready at correct place for a particular operation. Despatching is concerned with the smooth introduction of work to the stop floor.

The despatcher is responsible for a final check on the availability of materials, tools, jigs, fixtures required for the job and may even ensure that these are delivered to the work centre when order is about to run. Most of the techniques used in scheduling are helpful to the despatcher in arranging the work ahead of this department. The individual responsible for despatching an order to production should, varify that before the scheduled starting data all parts, materials for the job are on hand. Many times the despatching activity is controlled from a central point. This happens in case of assembly lines where the product is standardised but not completely uniform. The despatching problems involved are handled by transmitting messages to the stations simultaneously through teletype. The teletype is an electric type writer connected through a telephone circuit. In fact the despatching should be given due importance without which there would be little possibility of completing the production as per the established plan. The work orders issued by despatcher contain the following information :

(A) Name of the product.

(i) Code No.

(ii) Quantity to be produced.

(B) Description of various operations.

(i) Machine to be used.

(ii) Special instructions in operating the machine.

(C) Departments involved.

Thus we see that some kind of despatching is necessary for all types of production; it is essentially needed for the scheduling and control of intermittent production.

Follow Up

After the despatching is completed, production of various operations has been authorised to begin at time as planned by the scheduling department, the follow up is to check on the progress of the order as it is being manufactured from the first operation until the order is covered into finished product. This comparison between the actual and standard progress is made to take corrective action. Follow up functioning checks and measure the effectiveness of previous production control function routing, scheduling and despatching. Expediting is a special form of follow up. Expeditors are used :

to help to eliminate particular problems which are throwing production off scheduling.

=-to speed the production of certain orders.

Purpose of Despatching and Expediting

To summarize, the purpose of active functions of despatching and expediting are :

1. To release the production orders as the appropriate time and facilitate effective flow of information.

2. To record the flow of materials and tools and adjust whenever necessary.

3. To record progress of productions and adjust whenever necessary.

4. To record amount of work in process and verify its effect on the schedule.

5. To record quantities produced and compared with required quantities.

6. To record amount of faulty work and scrap, issue orders for production of replacements.

7. To record machine idle time and check the reason for it.

8. To record stoppage and held up and classify them according to :

(a) lack of drawings or instructions.

(b) lack of materials.

(c) work held up by previous operation.

(d) machine break down.

(e) operator missing are not available.

(f) waiting for inspection to approve work on machine setting.

Effective Follow Up System

The system of follow up is governed by the organisation of the control system and the methods by which routings, scheduling and despatching are accomplished.

The basic aim of follow up activities is that of preventing serious delays. An effective follow up system requires.

1. Recognition of delay.

2. Evaluation of reasons leading to delay.

3. Corrective action to remove delay.

The recognition of delays results from a comparison of planned with actual progress. The information of delay can be obtained immediately from the supervisors and operators concerned.

In case of complex control system the department schedules may be marked in the stop to point out the position of various jobs which have been completed or which are well underway. The route sheet or detachable tickets may be sent back in the same manner when all operations have been completed. These tickets may be used for pay role purposes to provide instructions for the operator.

Follow up is sometimes implemented by special production reports. These reports might be prepared by a foreman, or supervisor, or despatcher at the end of a shift, a day, a week, or a month. Similarly additional information can be obtained from production figures and other observations. In certain industries beside various operations are proceeding as planned, still delays are frequent any inevitable. However, every delay does not necessarily call for immediate action. Any form of follow up is completely ineffective unless it leads to management action. The decision may be involving the attention of staff only production control department or it may require the attention of staff personnel. The first type of decisions normally do not involve much of problem but the success in the second type of decisions depend on the cordial line and staff relationships.

Computer's Role in PPC

Computer based production planning and control systems can aid in the solution of complex problems. These systems are applicable to a broad range of industries but apply most directly to fabrication and assembly operations that make a large number of parts in relatively small lots either for assembly into finished products or for sale to outside customers. Managers are frequently complaining about department overloads, poor delivery performances and to blame these on "poor scheduling" or "poor control". The real cause however may be unbalanced capacity or an unrealistic sales commitment that was made months ago. Let us have an integrated view of the various parts of the production planning and control and the overall view of the whole process.

The dictionary definition of a system includes "a regularly interacting or interdependent group of items forming a unified whole". The items or parts in a production planning and control system fit together in a time dimension and interact in a certain way. Long term strategic plans that commit the company to a configuration of manpower skills, plant and equipment are based on very crude information and analysis. Mereover, these plans constrain the development of more complete, detailed plans closer to actual production dates. In the very short term this control involves putting men in the right places, working on the right jobs and regulating inventory levels. The criteria against which management's production planning and control performance is measured include inventory investments, labour cost, manufacturing cycle time. equipment utilisation and meeting delivery schedules.

Designing PPC System

In designing any production planning and control system which will measure up well against these criteria is to ensure that the plans and guide lines from higher levels should not unduly restrict decision making at lower levels. Flexibility to react to new information and significant deviations from higher level plans must be built into the system at all levels. Feedback [information on actual conditions and performance must flow upward through the system to ensure that the long term plans are based on a realistic assessment of the production organisation.

Production managers usually do not see their job roles in these clear, general terms. When poor performance becomes obvious and at the time of actual production, there is a tendency to find fault in the shortest time dimension and thus focus largely on problems at the lowest level. Many production managers spend their time in a continuous search for information or clerical decision making on the shop floor. So the production managers often find themselves chained to an endless sequence of routine decision making. But recently many production managers have changed the way they manage, utilising new systems for production planning and control. New systems do not eliminate all the crises, but they do point the way toward better control with less management involvement at the detail level. Thus they are making it possible for management to plan, redesign and execute in a more rational manner.

Long Term Capacity Planning

Plant facilities, equipment, skilled labour and working capital to support inventory investment usually cannot be made available to production managers on short notice. Consequently, there should be a long term plans for future capacities. *In a sense, long term planning is the starting point for production planning and control* thus it is the starting point for P.P.C.

The information flow in capacity planning process is shown in Fig. 6.2   Production Planning and Control System can provide a clue as to where the capacity is most needed and the date can help in analysing the effectiveness of alternative course of action for providing additional capacity.

INFORMATION FOR LONG-TERM CAPACITY PLANNING

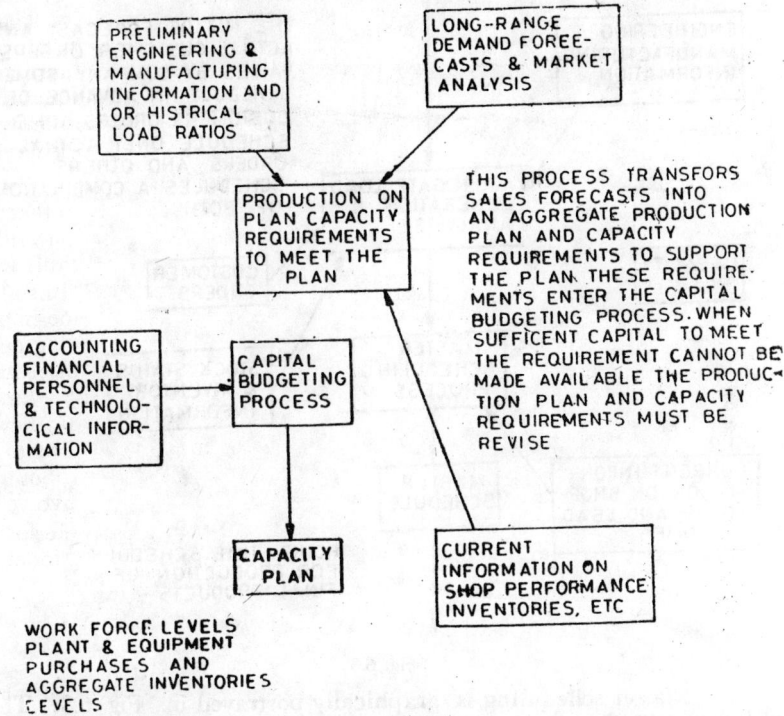

PRELIMINARY ENGINEERING & MANUFACTURING INFORMATION AND OR HISTRICAL LOAD RATIOS

LONG-RANGE DEMAND FORECASTS & MARKET ANALYSIS

PRODUCTION ON PLAN CAPACITY REQUIREMENTS TO MEET THE PLAN

THIS PROCESS TRANSFORS SALES FORECASTS INTO AN AGGREGATE PRODUCTION PLAN AND CAPACITY REQUIREMENTS TO SUPPORT THE PLAN. THESE REQUIRE- MENTS ENTER THE CAPITAL BUDGETING PROCESS. WHEN SUFFICENT CAPITAL TO MEET THE REQUIREMENT CANNOT BE MADE AVAILABLE THE PRODUC- TION PLAN AND CAPACITY REQUIREMENTS MUST BE REVISE

ACCOUNTING FINANCIAL PERSONNEL & TECHNOLO- CICAL INFOR- MATION

CAPITAL BUDGETING PROCESS

CAPACITY PLAN

CURRENT INFORMATION ON SHOP PERFORMANCE INVENTORIES, ETC

WORK FORCE LEVELS PLANT & EQUIPMENT PURCHASES AND AGGREGATE INVENTORIES LEVELS

Fig. 6.2

Master Scheduling

It is the activity that determines the overall production plans for the next several months. Master scheduling is done with the objective to ensure that the actual load in the shop will fall within rather narrow limits, of a few months. Thus the lower limit on a master schedule should be a load (in hours of work) and delivery date requirements which will keep the departments capacity efficiently utilised on current jobs that are neither too ahead nor behind schedule. The upper limit should be the highest load (in hours of work) and the highest load in delivery requirements that can be handled by the depatment and still allow for the inevitable rush job from a highly regarded customer, the last minute engineering change because of scrap losses and the other occurrences which cannot be predicted in advance.

INFORMATION FLOWS FOR MASTER SCHEDULING

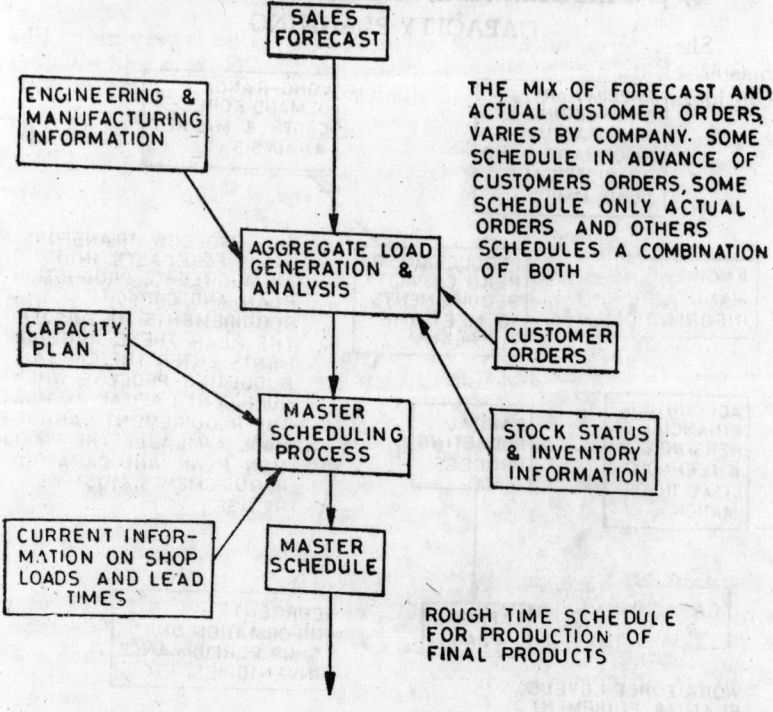

Fig. 6.3

Master scheduling is graphically portrayed in Fig. 6.3. This information has its application in modifying the standard lead times used to convert orders into section load. Lead times vary directly with shop loads. A master scheduling system that overlooks actual section loads and simply uses standard lead times may make commitments which exceed the section capacity to produce.

Short Term Scheduling

Short term scheduling describe the activity which develops the detailed plans necessary to meet the delivery commitments represented by the master schedule. The result of short term scheduling is to set the start and completion time for every component for final products. The short term scheduling is an estimate of the load for the next few weeks for each machine in a department. Machine loading is a term often used for the activity which can be used to assist management in :

(a) assigning men to machines to balance the capacities of various work centres

(b) sporting and reduction bottlenecks in work flow and

(c) planning activities on the shop floor.

Short term scheduling is an activity that is very much like master scheduling but it has a shorter time horizon and involves considerably more detail. The distinction between master and short

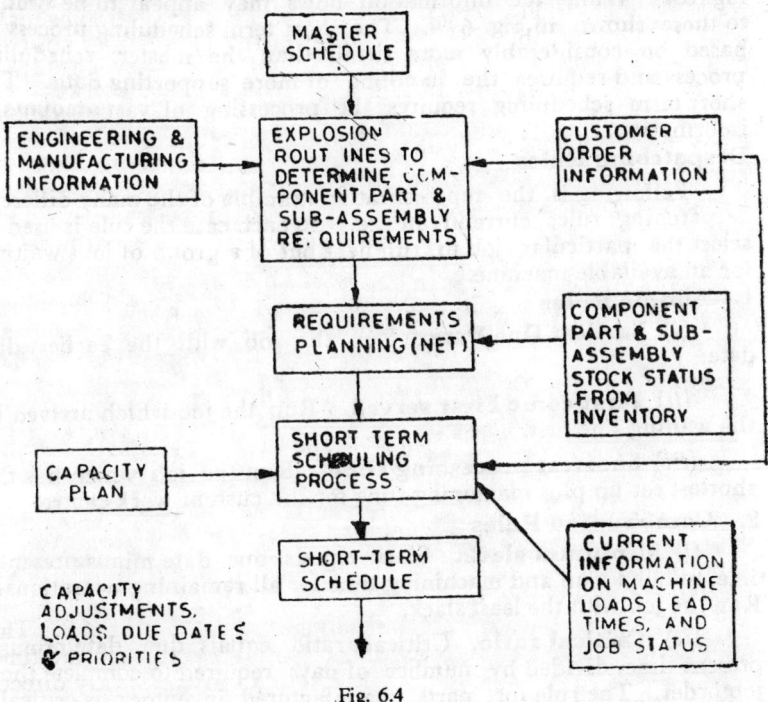

Fig. 6.4

term scheduling may be academic in nature. The computer's ability to make large quantities of detailed information easily available tends to encourage the use of detailed information. The data generally include such things as a bill of materials showing all component parts and assembly sequences, routings or operations sheets for all parts to be manufactured, lists of raw-materials, tools and fixtures required, the information on their availability and their condition. Similarly estimates of existing loads already scheduled against machine in the section or estimates of expected lead times or delays to be encountered as work moves through the shop and estimates of the machining and set up time for each operation on each part.

Capacity Loading

Capacity loading involves starting the number of machines to be manned in each department and the number of shift to be worked. In the process of capacity loading, component requirements are extracted from a bill of materials and the due date for each component is determined by back dating from the master schedules due date. This capacity loading approach is simple to describe, but quite difficult to implement.

The general process of short term scheduling is shown in Fig. 6.4 While the information flows may appear to be similar to those shown in Fig. 6.'4. The short term scheduling process is based on considerably more detail than the master scheduling process and requires the handling of more supporting data. The short term scheduling requires the processing of vast amounts of information.

Despatching Rules

Following is the representative example of the many different despatching rules currently in use. In each case the rule is used to select the particular job to run next out of a group of jobs waiting for an available machine.

1. **Simple Rules**

(*i*) **Earliest Due Date.** Run the job with the earliest due date.

(*ii*) **First come First served.** Run the job which arrived in the waiting line first.

(*iii*) **Shortest processing time.** Run the job which has the shortest set up plus machining time for the current work centre.

2. **Combination Rules**

(*i*) **Minimum slack.** Slack equals due date minus present time, minus set up and machining time for all remaining operations. Run the job with the least slack.

(*ii*) **Critical ratio.** Critical ratio equals due date minus present time, divided by number of days required to complete the job order. The rule for parts manufactured inventory is critical ratio equals available stock over reorder point quantity, divided by standard lead time remaining over total manufacturing lead time. This ratio compares the rate at which stock on hand is being depleted with the rate at which total lead time is being used up.

Despatching and Control

In a large section there are thousands of job orders in process at any given time. Thus there is, not only the problem of limited capacity but also the problem of what individual operation can be done next, due to sequence constraints and material tools and machine availability. Even though every job order in the section may have a scheduled start and finish time, the particular sequence for the individual machine operarions remains to be determined. Often the location of a job is not known, but whether it is ahead or behind schedule and what work remains to be done. Let us see a well designed despatching and shop floor control system can help to organise and rationalize the flow of work through the shop and to ensure that the right jobs are being worked on at all times. Similar the timely reports can greatly assist management in the continuous decision making process that is required to keep a shop going—that is, in controlling the level of in process inventory, tracking down difficult jobs and spotting difficult situations before they develop.

Production Planning and Control and its Place in the Organisation

The functions of production planning and control are to evolve a strategy of maximum production with available resources P.P.C. functions may be called as the planning function to communicate the plan to the manufacturing side and to assess the accomplishments for presentation to the management. Since PPC is concerned with the optimum use of the total resources of the company, it is but natural that it should receive information from all sections of a manufacturing organisation and it should convey relevant information to all concerned. In some cases it merely receives and gives information, in other cases it scrutinises the information before accepting, in some other cases it makes suggestions for improvement of standards based on internal and external evidence. From its all embracing aspect, production planning and control has a place of prestige in the organisation. It is usually regarded as a service function and, therefore, a staff aid to management. Normally it is placed with highest authority in line management as service aid. However, in multi-plant organisations, it might find a berth with the Production Director or Production Manager, Production Planning and Control should be as near the top management as practicable without compromising the planning and monitoring functions inherent in any line management. Like all other organisations, PPC's structure is determined by analysis of its (a) activities (b) decision making pattern and (c) its relationships internally and externally.

In a small organisation, it may also perform the duties of process engineering, purchasing, stocking, inventory planning as well as industrial engineering and costing. In a medium size organisation with distinct job ordering basis, the engineering function and cost may have to be separately organised. In a large organisation, it is useful for it to confine itself with it intrinsic functions, namely of Routing, Scheduling, Despatching, Monitoring, Reporting and Suggestions for work improvement. With its overall sweep on the total performance of the company, an enterprising PPC controller may be tempted to enter into the jurisdiction of other colleagues and thus cause tensions. It is for the management to curb such tendency to be within legitimate limits of the duty of the PPC controller.

PPC cannot function without rational and reasonable top management support. Since it presents the management with plans and their fulfilment reports, its work is always being assessed at the highest possible levels. It has sometime to answer for the mistakes of others.

7
Network Analysis

Network analysis is synthesis of two techniques namely Programme Evaluation and Review Technique and Critical Path Method evolved independently during 1956-57. The two methods have many features in common and are now combined and called Network Analysis. For large non-repetitive operations or projects CPM-PERT and other related network techniques are usefully aids for project managers. A few of the features making the techniques powerful are :

A logical and disciplined basis for planning. CPM and PERT are useful at several stages of project management starting from early planning stages, when various alternative programmes or procedure are being considered, to the scheduling phase, when time and resources schedules are laid out to final stage, *i.e.* in the operational phase, when used as control device to measure actual versus planned progress.

Simplicity. They are straightforward in concept and easily explainable to the layman with no background in network theory. Data calculations, is tedious for large projects, are not difficult. Basic critical path data may be hand calculated with reasonable speed for projects with 500 or 600 activities. Computer programs are readily variable for larger projects.

Improves co-ordination and communication. The network graph displays a simple and direct way the complex manager of various sub-divisions of project may quickly perceive from the graph.

Pinpoints trouble spots and responsibility. Network calculations pinpoint attention to the relatively small subset of activities in a project which are critical to its completion. Managerial action is thus focused on exceptional problems, contributing to more reliable planning and more effective control.

Better management of resources. CPM anables the manager to reasonably estimate total costs for various completion dates. These various trade off possibilities along with other decision criteria, enable him to select an optimum or near optimum schedule.

Versatile and application. CPM and PERT are applicable to many type of projects : from jobs, from new product introduction programme to missile procedures. Moreover, they may be applied at several levels within a given project from a single department working on a sub-system to multi-plant operations within a large corporation.

Scope for unseen changes. As simulation tools, they enable the manager to project into the future effects of planned or unanticipated changes and to take appropriate action when such projections indicate the need for it. Thus for example the manager can quickly study the effects of crash programs and can anticipate in advance potential resource battle. When wisely used by a project manager who understands their strengths and limitations, CPM and PERT can be effective amplifiers of his managerial skills.

Utility

1. It aids the manager in planning, scheduling, controlling the activities.

2. In guiding and directing team efforts more effectively.

3. It permits advance planning, indicates current progress warns of potential future trouble spots when there may still be time to avoid them.

4. It aids in handling uncertainties regarding time schedules, co-ordination of many activities and control of cost involved.

5. It shows its values most strikingly when a special projects of new kind is undertaken.

The Concept of "Planning" and "Control"

Work comprises everything carried out to change the state or position of materials. Simple work may consist of one activity or two and complex work of millions of activities. Program is any work attuned to a target or an objective. Space craft building of which more than 6 million components are required, it is impossible to imagine that such a programme could be executed with speed and efficiency without having a plan of action. The considerations of efficiency make planning necessary for all programmes. *Planning is a comprehensive term which embraces definition of objectives, forecasting and sequencing of resources requirements and marrying them with the resources potential into an action plan.*

Every planning process, howsoever scientific, has to reckon with imponderables. It is necessary to continuously review the plan as work progress and to readjust the resources and objectives as necessary. This is briefly the concept of the term "Control" a function complementary to "Planning". In fact, without effective "Control" 'Planning" in reduced to a simple academic exercise only. "Control" requires feedback information on Progress Status. If managers resist feedback then it often leads to non-acceptance of "planning and control" function. Line managers should instead look to this function as staff assistance and take advantages of the modern management tools and techniques.

What is a Project ?

Project is a term which covers once-through and small-batch programmes. When attempting to determine the completion data for any task/project. Where it is

(1) Construction projects, building of a bridge etc.

(2) Research and development Projects.

(3) Designing of a new pieces of equipment.

(4) Erection and commi sion of Industrial Plants.

(5) Preventive maintenance.

(6) Trial manufacture.

(7) Heavy engineering.

(8) Custom-engineered Products.

(9) Market launching of new Products.

(10) Finalisation of annual accounts.

(11) An inaugural.

(13) A banquet.

(13) A marriage programme.

(14) Preparation for a dinner party.

(15) A picnic party

(16) A picnic outing etc.

Or any other project it is necessary to time-table all the activities which make up the task or the project, that is to say, a plan must be prepared. The need for planning has always been present, but the complexity and competitiveness of modern undertakings now requires that this need should be met rather than just recognized.

Network Techniques of Program Management

The need for management direction on a program or job depends upon two factors—its repetitiveness and complexity. Simple jobs that occur infrequently do not require any involved direction. The techniques for direction of project-type jobs are rather new. However, there has recently been an explosive growth in the family of these technique a number of network techniques have been developed which are named below :

(1) P.E R.T.   :   Programme Evaluation and Review Technique.

(2) C.P.M.   :   Critical Path Method.

(3) R.A.M.P.	:	Resource Allocation and Multiproject Scheduling.
(4) P.E.P.	:	Programme Evaluation Procedure.
(5) C.O.P.A.C.	:	Critical Operating Production Allocation Control.
(6) M.A.P.	:	Manpower Allocation Procedure.
(7) R.P.S.M.	:	Resource Planning and Scheduling Method.
(8) L.C.S.	:	Least Cost Scheduling.
(2) M.O.S.S.	:	Multi-Operation Scheduling System.
(10) P.C.S.	:	Project Control System.
(11) G.E.R.T.	:	Graphical Evaluation Review Technique.
(12) L.O.B.	:	Line of Balance Technology.

The most commonly known and used Network Techniques are PERT and CPM and most of the other techniques are the variations or developments of these techniques only.

Nature of the Techniques

In any project there are usually three types of problems :

(a) *Technical*

(b) *Executive Type*

(c) *Human.*

More complex the project, more difficult the executive type problems—say co-ordination to name one. An example of vastness of this problems in complex projects is given by the polaris weapon system which involve 250 prime contractors and at least 9000 job-contractors. In such projects it is possible that delay in the delivery of as small component might hold up the progress of the entire project. These techniques help to co-ordinate efforts of different agencies :

(i) *To plan resources*

(ii) *To anticipate the occurrence of bottlenecks*

(iii) *To guide corrective action*

(iv) *To forecast the liklihood of meeting target dates.*

The basic functions of management are to make decisions and enforce their implementation. The function still remains, not withstanding the new techniques. The techniques are merely tools in the hand of management—the tools that make the decision process more rational and less painful.

Historical Background to "CPA" "CPM"

(1) *Critical Path Analysis* : CPA

(2) *Critical Path Method* : CPM

(3) *Critical Path Planning* : CPP

History of Net-Work Analysis

Before CPA coming on the scene probably the best-known method of trying to plan was by means of a bar or Gantt chart. There were certain drawbacks and problems in it. *In Great Britain the Operational Research Section of the Central Electricity Generating Board* investigated the Problems concerned with the overhaul of Generating Plant—an area of considerably complexity which was increasing in importance as new higher Performance Plant was being brought into service. By 1957 the O.R. section had devised a technique which consisted essentially of identifying the "*Longest irreducible sequence of events*", and using this technique they carried out in 1958 an experimental overhaul at a Power Station which reduced to overall time to 42% of the previous average time for the same work. Continuing to work upon these times the overhual time was further reduced by 1960 to 32% of the previous average time. The name, "Longest irreducible squence of events" was soon replaced by the name, "*Major Sequence,* and it was pointed out for example, that delays in the "major-sequence" would delay completion times, but the difficulties elsewhere need not necessarily involve extensions in total time. This work of the O.R. group was not made public, although comprehensive reports were circulated internally elsewhere.

Similar development work was being undertaken elsewehere—for example in the U.S. Air force under the code name P.E.P. *Also in 1958, the E.I. du Pont de Nemours Company used a technique called the "Critical Path Method" "CPM"* to schedule and control a very large project, and during the first complele year's use of CPM it was credited with saving the company $ 1 million. Subsequent use underlined the basic simplicity and extra-ordinary usefulness of this method, and by 1959 Dr. Mauchly, who had worked on the Du Pont Project, set up an organisation to solve industrial problems using the Critical Path Method.

Since 1958 considerable work has been carried out, mainly in ihe United States of America, in consolidating and improving these techniques. Much of the effort has been expanded by the Computor Companies, who have devised special names to distinguish their own work.

"CPM" and its advantages

Critical Path Method is a technique employed to schedule and control project. CPM is applicable to both large and small projects. It is widely recognized, most versatile and important management planning.

Some of the important advantages are given below :

(1) It helps in ascertaining time schedule.

(2) With its help control by the Management becomes easy.

(3) It makes better and detailed planning possible.

(4) It encourages discipline..

(5) It provides a standard method for communicating project plans, schedules, time and cost performance.

(6) It identifies most critical elements and thus more attention can be paid on these activities.

Those techniques require a greater planning than required otherwise. Thus these methods increase the Planning cost, but this Cost is easily justified by concentrating attention Critical Paths only and avoiding unnecessary expenses on the strict supervision over whole programme.

Steps in "CPM" Technique

CPM employs the following steps for accomplishing a project planning.

1. Break down the project into various activities systematically.

2. Label all activities.

3. Arrange all the activities in Logical sequence.

4. Construct the arrow diagram.

5. Number all the nodes and activities.

6. Find the time for each activity.

7. Mark the activity times on the arrow diagram.

8. Calculate early and late, start and finishing times.

9. Tabulate various times and mark EST and LFT on the arrow diagram.

10. Calculate the total float for each activity.

11. Identify the critical activities and mark the Critical Path on the arrow diagram.

12. Calculate the total project duration.

13. If it is intended to reduce the total Project duration, crash the critical activities of the network.

14. Optimise the cost.

15. Up-date the network.

16. Smooth the network.

Difference between "PERT" and "CPM"

CPM is a technique used for the Planning and Controlling the most logical and economic sequence of operations for accomplishing a project. The project is analysed into different activities whose relationships, as in PERT, are shown on the network diagram. The network is then utilized for optimising the use of reources, progress an control. PERT is a technique used for scheduling and controlling the projects, whose activities are subject to considerable degree of uncertainty in the performance time.   The method of start-finish critical path and project time of P.E.R.T, and C.P.M. are similar. CPM is an activity oriented technique, while PERT is the event oriented technique.   Event means the beginning or exceeding of one or more activities.   CPM has one time estimate, while PERT has three time estimates.

The fundamental network of PERT and CPM are though identical, yet there are certain differences in details as mentioned below :

PERT	CPM
1. A probabilistic model with uncertainty in activity duration. Expected time is calculated from to, t_m and t_p.	1. A deterministic model with well-known activity (single time based upon past experience. It assumes that, the expected time is actually the time taken.
2. An event-oriented approach.	2. An activity-oriented system.
3. PERT terminology uses words like network diagram, event and slack.	3. CPM terminology employs words like arrow diagram, nodes and float.
4. The use of dummy activities is required for representing proper sequencing.	4. The use of dummy activities is not necessary. The arrow diagram thus becomes slightly simpler.
5. PERT basically does not demarcate between critical and non-critical octivities.	5. CPM marks critical activities
6. PERT find applications in projects where resources (man, materials and specially money) are always made available as and when required.	6. CPM is employed to those projects where overall costs is of primary importance. There is better utilization of resources.
7. Especially suitable in defence Projects and R and D where activity times cannot be reliably predicted.	7. Suitable for problems in Industrial setting, plant maintenance, civil construction projects : expansion schemes etc.
8. PERT has three-times estimates.	8. CPM has one-time estimate.

Thus PERT should be explicit intengible, capable of accepting changes and capable of being monitored.

PERT

Program Evaluation Review Technique

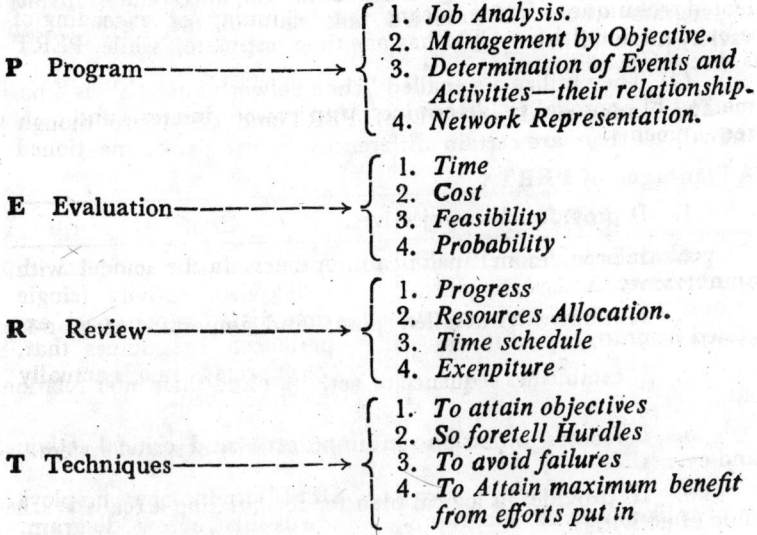

P Program ⟶
1. *Job Analysis.*
2. *Management by Objective.*
3. *Determination of Events and Activities—their relationship.*
4. *Network Representation.*

E Evaluation ⟶
1. *Time*
2. *Cost*
3. *Feasibility*
4. *Probability*

R Review ⟶
1. *Progress*
2. *Resources Allocation.*
3. *Time schedule*
4. *Exenpiture*

T Techniques ⟶
1. *To attain objectives*
2. *So foretell Hurdles*
3. *To avoid failures*
4. *To Attain maximum benefit from efforts put in*

PERT Defined as

1. A Manager's Tool : for defining and co-ordinating what must be done to successfully accomplish the objectives of a project on time.

2. A Technique : that aids the decision maker but does not make decisions for him.

3. A Technique : that presents statistical information regarding the uncertainties faced in completing its many activities associated with a project.

4. A Method : for focussing managerial attention on latests problems that require decisions and/or solutions.

5. A Method : requiring managerial attention procedures and adjustments, regarding time, resources or performance which may improve the capability of meeting target dates.

Distinguishing Features of PERT

These techniques have several distinguishing characteristics :—

(1) They give management the ability to plan the best possible use of resources to achieve a given goal within overall time and cost limitations.

(2) They enable executives to manage "one of a kind" programs, as opposed to respective production situations.

(3) They help management handle the uncertainties involved in programs where no standard time data of the Taylor—Gantt variety are available.

(4) They utilize a so-called "time network analysis" as a basic method of approval to determine Man-power, interests and capital requirements.

Advantages of PERT

1. It provides clear objectives.

2. It provides an analytical approach to the achievement of an objective.

3. It enumerates detailed plans and important programmes events.

4. It establishes sequence of activities and their inter relationship.

5. It focuses attention on important and critical activities and events.

6. It provides a logical plan for formulating a realistic schedule of activities.

7. It provides up-to-date status information of the project and enables quick review of the progress at all stages.

8. It can point out improbabilities and may even predict before their occurrence, and they help remedy the situation.

9. It can foretell the feasibility of a plan and can help in formulating new schedules if the one under consideration be found impracticable or wanting.

10. It avoids slippage of plan and waste of time, energy and money.

11. It is an aid for allocating the available resources for optimum results.

12. It brings about time and cost consciousness at all levels of management.

Introducing PERT into an Organization

As with any other new managerial tool, PERT will require to be introduced into an organisation with care. It is suggested that the following points should be observed.

1. PERT is not a universal tool—there are situations where it cannot be usefully employed. These situations are, in genral those where activity is continuous, for example, flow production. A PERT type situation is characteristically on which has a definable start and a definable finish.

2. PERT is not a life saving drug, it does not cure all ills. Indeed, PERT in itself does not solve any problems, but it does expose situations in a way which will presuit effective examinations both of the problems and of the effect of possible solutions. However, the formulating and implementing of any solution will remain the responsibility of the appropriate manager.

3. PERT must not be made a mystery, known only to a chosen few. All levels require to appreciate the method and its limitations, and an extensive educational programme will be necessary to ensure that knowledge is spread as widely as possible.

4. The person initiating PERT into an organisation must be of sufficient stature and maturity to be able to influence both senior and junior personnel.

5. Whenever possible, the early applications for PERT should be to simple situations. If PERT is first employed on a very difficult task, it may fail, not because of the difficulty of PERT but because of the difficulty of the task itself. However, the failure is likely to be attributed to PERT and the technique will be discredited.

6. PERT will involve committal to, and the acceptance of, responsibilities expressed in qualitative terms. Many supervisors find this difficult to accept, and will often try to escape by creating unreal problems. It is vital to make it quite clear that PERT is not a punitive device ; it is a tool to assist, not a weapon to assult.

7. Difficulties in using PERT are almost always symytous of same Managerial Weakness.

Steps in the Technique of PERT Planning

1. Define clearly the end objectives.

2. Determine the major events that must be completed prior to reaching the end objectives.

3. Analyse and record the work to be done in between events.

4. The project is broken down into different activities systematically.

5. Arrange the events and activities in the order of sequence of their occurrence.

6. The network diagram is drawn. Events and activities are numbered and make sure no activity is left over.

7. Layout and finalize the network in a neat form. May be more than one attempt is needed.

8. Event numbers should be given with the starting event as 1, and progressing successively to the right in the order in which they occur.

9. Any dummy activity should be shown by dotted arrow times on the network.

10. Check your network it should not have loose ends. Network is always closed and all activities lead to the objective.

11. Each activity should have a single location.

12. Using three times estimates, the expected time for each activity is calculated.

13. Standard deviation and variance for each activity are computed.

14. Earliest starting times and latest finishing times are calculated.

15. Expected time earliest starting time and latest finishing times are marked on the network diagram.

16. Slack is calculated.

17. Critical path is identified and marked on the network diagram.

18. Subcritical Path are also identified.

19. Length of critical path or total project duration is found out.

20. Lastly, the probability that project will finish at due date is calculated.

Development of PERT

Project Managers increasingly noted that the techniques of Frederick, Taylor and Henery L. Gantt introduced during the early part of the century for large scale productions, were inapplicable for a large portion of the industrial effort. The work was being undertaken in the U.S.A., and in early 1958 the U.S. Navy Special Project. Office concerned with performance tends on large military development programs set up team of the management consulting firm of Booz-Allen and Hamilton ; to devise a means of dealing with the planning and subsequent control of complex work. This investigation was known as the Programme Evaluation Research Task, which gave rise to (or possible derived from) the code name PERT. By February 1958, Dr. C.E. Clark, mathematician in the PERT team presented the early notions of arrow-diagramming, military drawing from his study of graphics. This early work of Dr. Clark was rapidly published and by July 1958, the first report, PERT Summary Report, phase I, was published. By this time the full title of the work had become "Programme Evaluation and Review Technique", and the value of the technique seemed well established. By October, 1958, it was decided to apply PERT to the Fleet Ballistic Missiles Programme, where it was credited with saving two years in the development of Polaris Missile. Since that time PERT has spread rapidly through out the U.S. Defence and Space Indus'ry. Currently almost every major Government and military agency concerned with space Age Programs is utilising the technique as large industrial contractor in this field. Small business wishing to participate in national defence programs have found it increasingly necessary to develop PERT capability.

Terminologies

Activity. A time consuming element occurring between two events and represented by an activity.

Activity

Start————→finish

1. The arrow is not to be drawn to the scale.

2. It may be curved, bent, horizontal or vertical.

3. Time values can be assigned to activities.

4. An activity may have zero time value.

Event. A meaningful accomplishment or 'mile stone' usually represented by small, circle or box in network.

—Activity start and finish points are events.

—Events have no time values.

Critical path. In network the path with smallest amount of positive slack or largest amount of negative slack. The most rigorous time constraint for this path $T_e = T_l$.

Slak or Float. Difference between last event time and early event time *i.e.* $Tl - Te =$ Float.

Negative slack. The amount of time which is not available to perform the series of activities in a particular slack path and still meet the required completion date.

Positive slack. The amount of additional time which is available to perform the series of activities in a particular slack path and still meet the required completion date.

Network. A flow plan containing the pertinent activities and events.

Dummy activity. An activity with no time value constraining the succeeding activity.

Estimation of activity duration

It is not always possible to estimate the duration of each activity accurately specially when a new project is undertaken. Since to start a network the determination of these durations is necessary, the following method is used—From the people who will be responsible for those activities or from the persons who have undertaken similar works in past, we obtain the following estimates for each activity—

(i) Pessimistic estimated time$=b$

(ii) Most likely estimate time$=m$

(iii) Optimistic estimate time$=a$

The early expected time for completion of activity is

$$Te = \frac{a+4m+b}{6}.$$

If we draw a histogram for frequencies, various estimates of activities against duration it will lead to a bell shape curve which is normal distribution curve.

Variance $\qquad = \left(\frac{b-a}{6}\right)^2$

$$Vt = \sigma_2$$

Deviation $\qquad \sigma = \frac{b-a}{6}$

$$= \sqrt{V}.$$

Te, Vt can be added up for various activities to find out total early expected time TE and total variance VT for a Project, while

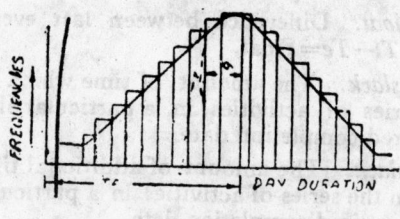

Fig. 7.1

standard deviation σt for activities cannot be added up to give total deviation.

TE = Early expected time for completion.

TL = Latest expected time for completion.

TS = Schedule time for completion.

$$Z = \frac{Ts - Te}{\sigma}$$

When $Ts = Te + 3\sigma$ probability of completion is 99·7%

$Ts = Te + 2\sigma$,,   ,,   ,,   95%

$Ts = Te + \sigma$,,   ,,   ,,   68%

$Ts = Te$,,   ,,   ,,   50%

Network Logic

To organise the activities into network, the following questions are asked concerning each activity :

    (*i*) What activity must produce this one ?
    (*ii*) What octivity can be concurrent with this one ?
    (*iii*) What activity must follow this one ?

Examples

Activity *B* cannot start until activity *A* is completed.

Activity *C* cannot start until both activities *A* and *B* are completed.

Activity *A* must be completed before either *B* or *C*-started.

Both *A* and *B* must be completed before *C* or *D* or both can start.

Both *J* and *K* must be completed before *M* can start, *L* can start only on completion of *J* and is independent of *K*.

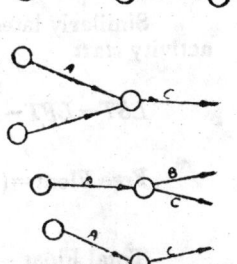

Fig. 7.2.

Dummy Activity

In network the activity *N* is dummy activity and always represented by broken arrow it always has zero time value and used only to describe logical connections.

A loop should never occur in a network as shown in Fig. 7.3

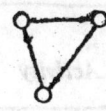

Fig. 7.3

Float or Slack

The term slack is used with an event *i.e.* in case of PERT and float is used with reference to an activity *i.e.* in case of CPM. But still these terms are interchangable. Float or slack means spare time, a margin of extra time over and above its duration which a non-critical activity can consume without delaying the project.

Let us find Earliest Start Time—The time at which an activity can start and worked out by moving from first to last event in the network. The earliest finished time is the earliest possible time at which an activity ends.

$$EFT = EST + \text{Duration of the activity.}$$

Latest Finish Time

It is worked out by moving backward from event 2 to event 1.

Similarly latest start time is 'the possible time by which an activity start

$$LST = LFT - \text{Duration of an activity}$$

Free Float $= (EST)$ Tail Event $- (EST)$ Head event $-$ Activity time

Total Float $= (LFT - EET)$

or
$$(LST - EST)$$

Procedure PERT Analysis

It can be understand in best way by solving illustration given below :

Illustration-I

Activity	Optimistric	Time period in weeks Most likely	Perssimistic
1-2	1	1	7
1-3	1	4	7
1-4	2	2	8
2-5	1	1	1
3-5	2	5	14
4-6	2	5	8
5-6	3	6	15

CHART

Z	Probability
3	0·999
2·8	0·997
2·6	0 995
2·4	0·992
2·2	0·986
2·0	0·997
1·8	0·964
1·6	0·945
1·4	0 919
1·2	0·885
1·0	0·841
0·8	0·788
0·6	0·728
0·4	0·655
0·2·	0·572
0·0	0·500
—0·2	0·421
—0·4	0·345
—1·0	0·159

Find the probability of its completion in 14 weeks

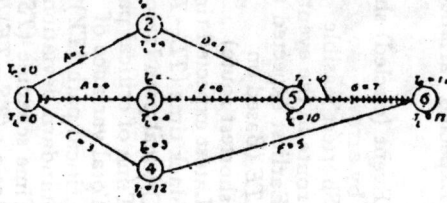

Fig. 7.4 ·

PERT Analysis of Product Development

Time data in weeks

	ACTIVITIES (Arrows)						
	A	B	C	D	E	F	G
1. Activity time (t)	$\frac{1+4\times1+7}{6}=2$	$\frac{1+4\times4+7}{6}=4$	$\frac{2+4\times2+8}{6}=3$	$\frac{1+4\times1+1}{6}=1$	$\frac{2+4\times5+14}{6}=6$	$\frac{2+4\times5+8}{6}=5$	$\frac{3+4\times6+15}{6}=7$ Total
2. Expected variance (U)	$\left(\frac{7-1}{6}\right)^2=1$	$\left(\frac{7-1}{6}\right)^2=1$	$\left(\frac{8-2}{6}\right)^2=1$	$\left(\frac{1-1}{6}\right)^2=0$	$\left(\frac{14-2}{6}\right)^2=4$	$\left(\frac{8-2}{6}\right)^2=1$	$\left(\frac{15-3}{6}\right)^2=4$
3. Event (circled) shown by arrow	2	3	4	5	5	6	6
4. Shortest feasible route to the event	A	B	C	B+E	B+E	B+E+G	B+E+G
5. Earliest expected time TE (Based on shortest route)	2	4	3	10 (4+6)	10 (4+6)	17 (4+6+7)	17 (4+6+7)
6. Latest expected time TL	9	4	12	10	10	17	17
7. Slack time ($TL-TE$)	7	0	9	0	0	0	0
8. Critical path (zeroes)		B			E	G	
9. Time of critical path		4			(4+6=10)	(4+6+7=17)	
10. Total variance of critical path (TV)		1			4		9
11. Standard deviation (TS)							3
12. Time schedule (TS)							14
13. Ratio : $(TS-TE)/\sigma$ $=(14-17)/3$							-1
14. Probability P of fulfilling schedule from chart							16

(1) Enter the estimated time *i.e.* optimistic time, pessimistic time and most likely times, required to complete each of the activities *A, B, C, D, E, F* and *G* as evaluated by the supervisor and managers concerned with each activity.

(2) Enter the expected variance in the estimate. representing the most optimistic and most pessimistic possibilities envisioned. For example activity *B* is expected normally to be completed in 4 weeks, but with unusual 'luck' it might be completed 3 weeks earlier. Also by an unusual amount of bad luck it may not be completed until 3 weeks later.

(3) Show the event indicated by each activity arrow depending on constraints if any, the name of activity generally written above arrow and its value below arrow or above arrow infront of Activity with equal mark. (See figure I).

(4) Find earliest expected time of an activities completion write it, above the event where that activity-completed. At the event where many activities are merging it is the longest path which is takenas *TE*. For example at event (5) activities *A* and *E* are merging giving two paths to event *i.e.* $A+D=2+1=3$ weeks and $B+E=4+6=10$ weeks. As value of $B+E$ is more it is the longest path and is taken as *Te* for event (5).

(5) Find the longest path for final event or adjective of the Project.

(6) The longest route for completion of a project is known as Earliest expected time *TE*. As we cannot delay this the earliest expected time for completion of the project and also the latest expected time (*TL*) in which the objective should be completed.

(7) Do the backward calculation to find out *TL* of various even starting from last event where *TE=TL* by substracting the time of preceding activity or activities. When two or more activities are bursting out from an event the shortest route is taken as *TL*.

(8) Slack time is the difference between latest expected time (*TL*) and the earliest expected time (*TE*). For example for event (4) slack is $12-3=9$ weeks. An activity with slack time is one that has extra resources, need not to meet the schedule of the program from over all view point. Sometimes these extra resources are revised or rescheduled to assist in shortening some of the more critical activities.

(9) Slack time value of zero for various events or activities leader or speaks of critical path which-represent by activities *B, E, G*.

(10) Duration of critical path is 17 weeks represents the amount of time to be expected to complete the program if shortest route is used for latest expected time or longest route for earliest expected time.

(11) As a computational step, add individual variance associated with critical path only. Since this is the path that governs the expected time to complete the program.

$$VT = V_{B_t} + V_{E_t} + V_{G_t} = 1 + 4 + 4 = 9$$

(12) The standard deviation $= \sqrt{V_T} = \sqrt{9} = 3$

(13) Hence $Z = \dfrac{T_S - T_E}{\sigma} = \dfrac{14 - 17}{3} = -1$

Finding the value of probability from chart as 0·159 or 0·16 (Approx.) find the probability of completion of the project in 14 weeks is 16%.

Illustration—II

TABLE I

S. No.	Activity	t_o	t_m	t_t
		Days		
1	1—2	2	5	14
2	1—6	2	5	8
3	2—3	5	11	29
4	3—4	1	4	7
5	3—5	5	11	17
6	4—5	2	5	14
7	6—7	3	9	27
8	5—8	2	2	8
9	7—8	7	13	31

It is given that the scheduled date of completion of project is 38 days.

(1) Find out the probability that project can be completed at scheduled time.

(2) Find out the scheduled time for 94·5% probability of completion of the project.

(3) Find out the probability of completion of path 1-6 7·8

Solution. T_e = Expected time for each activity

$$= \left(\frac{t_o + 4t_m + t_p}{6} \right) \qquad \ldots (1)$$

Standard Deviation $= S_t = \left(\frac{t_p - t_o}{6} \right) \qquad \ldots (2)$

Variance $= V_t = \left(\frac{t_p - t_o}{6} \right)^2 = (S_t)^2 \qquad \ldots (3)$

$$Z = \left(\frac{D - T_e}{S_t} \right) \qquad \ldots (4)$$

Z = Probability that project will meet the schedule or due date.

T_E = Total project duration

D = Scheduled date = 38 (days) given.

Using formula (1), (2) and (3) let us find out the value of t_e, S_t and V_t for various activities.

TABLE II

S. No.	Activity	t_e	S_t	V_t	Critical Path	Total Variance on critical path
1	1—2	6	2	4	—	4
2	1—6	5	1	1		
3	2—3	13	4	16	—	16
4	2—4	4	1	1		
5	3—5	11	2	4	—	4
6	4—5	6	2	4		
7	6—7	11	4	16		
8	5—8	3	1	1	—	1
9	7—8	15	4	16		25

Let us draw network diagram.

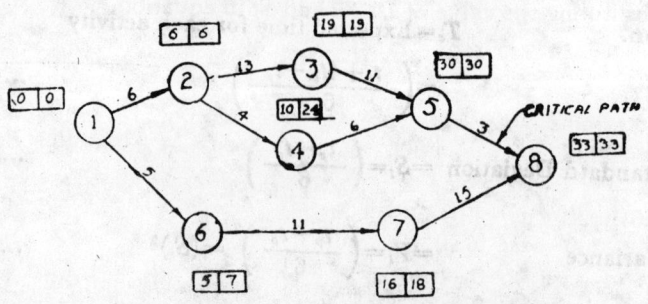

Fig. 7.5. On the Critical Path Total Stock is Zero.

Working out of *EST* and *LST* for the various activities as given in *Table I*.

TABLE III

S.No.	Activity	EST	LST	Total Slack LST—EST	Activity on critical path
1	1—2	0	0	0	—
2	1—6	0	2	2	
3	2—3	6	6	0	—
4	2—4	6	20	14	
5	3—5	19	19	0	—
6	4—5	10	24	14	
7	6—7	5	7	2	—
8	5—8	30	30	0	
9	7—8	16	19	2	—

So critical path is 1-2-3-5-8

Project duration on critical path, we get

$$T_E = 6 + 13 + 11 + 3 = 33 \text{ days}$$

Total variance of critical path = 4 + 16 + 4 + 1 = 25

Total project duration = 33 days

D = Scheduled date = 38 days

$$Z = \frac{D - T_E}{S_t} \qquad \qquad ...(4)$$

$$t = \sqrt{\text{variance of critical path}} = \sqrt{25} = 5$$

Substituting the value of D, T_E and S_t in equation.

$$Z = \frac{38-33}{4} = 1$$

After finding the value of Z, let us read table III we find that probability is 0.841. In other wards there is a probability that project shall be completed in 38 days is 84.1%.

For the 94.5% probability of project being completed. Let us look at the value of Z for 0.945 value of probability.

So $\qquad Z = 1.6$

$$Z = \frac{D-T_E}{S_t} = 1.6 = \frac{D-33}{5}$$

Thus $\qquad D = 41$ days

Assuming critical path as $1-6-7-8$

$$T_E = 31 \text{ days}$$

Variance of the path $= 1+16+16 = 33$

$$S_t = \sqrt{33} = 5.74$$

$$Z = \frac{D-T_E}{S_t} = \frac{38-31}{5.74} = 1.22$$

Now given the value of $Z = 1.22$ and read table I, the probability of meeting the due date is 0.888 or 89.88%.

PROJECT SCHEDULING

The project management calls for a clean pronunciation of the end objectives. It can be analysed into various activities which must be performed, before the end objective is reached. The time essimates for various activities can be obtained and their status and relationship projected in the form of PERT network. The earliest expected time for the end of a project can be forecasted. If the earliest expected time for a project and latest allowable time is known, we can determine critical path on network, and subsequently the square network can be drawn which can be a guide to lead the activity to completion in a scheduled manner. The earliest expected time and latest allowable time for all events follow the normal distribution for any cause causes of variation,

The area under the curve of frequency distribution represents the probability of occurrence of the variable. The area of the curve between any two values X and X_1 represents the probability that any observation chosen at random from distribution will fall between the values Y and Y_1 of the variable as shown in Fig. 7.6

When an event is on critical path, there is no slack. For an event to become critical the condition is,

$$Te = Tl$$

or $\qquad Te - Tl = 0$

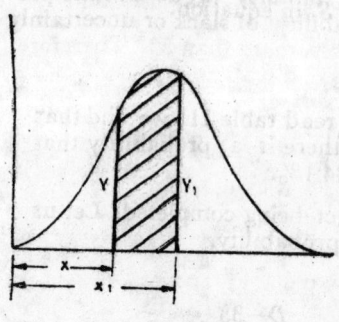

Fig. 7.6

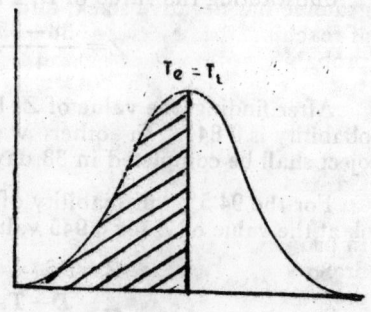

Fig. 7.7

Probability of no Slack

The probability of no slack or probability of success for a critical event is 50—50 if $Tl=Te$. Te when taken as the peak of normal curve divides it exactly into two equal areas and the probability of no slack—represented by shaded area of the curve sho vn in Fig. 7.7.

When T_L is greater than T_E there are good reasons to expect that the project will be completed in schedule time, i.e. there would be more than 50% chances of its completion—Time T_E or the probability of project completing time in this would be more than 5. It will be how much more depends on how for Tl is located from Te on the distribution curve. The unshaded area of the curve to the right to T_L would indicate the probability of 'no slack' and thus will indicate uncertainties. Therefore, uncertainty in measurable term P is

$$P=\text{Probability of no slack}$$
$$=\text{Area } \square$$

It can be physically seed that the programme has got positive slack when T_L is greater than T_E.

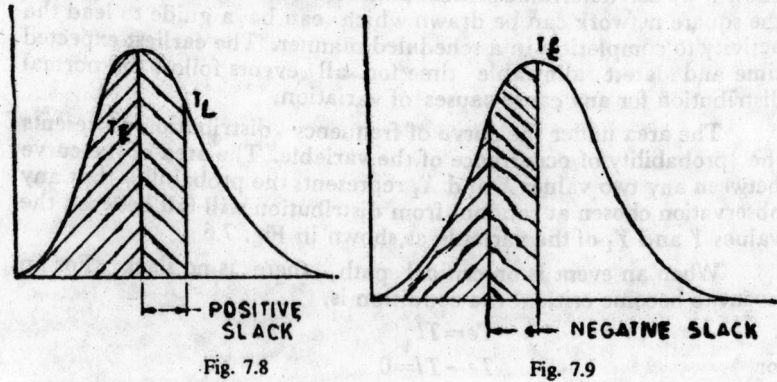

Fig. 7.8

Fig. 7.9

When *TL* is less than *TE*, we similarly argue that the programme has negative slack, and probability of slack or uncertainly of reaching the objective in time will be more than 53. Therefore, probability of no slack or uncertainly (*P*) in his case is

$$P = \text{Area} \sqcap \text{Area}$$

$$= 50 + \frac{(TL - TE)}{TE}$$

Gantt-Chart

In the Gatt chart horizontal bars are drawn for each activity in proportion to the time required for carrying out the activity drawn to the time scale. These bars represent time schedule and progress can be indicated by additional bars along side.

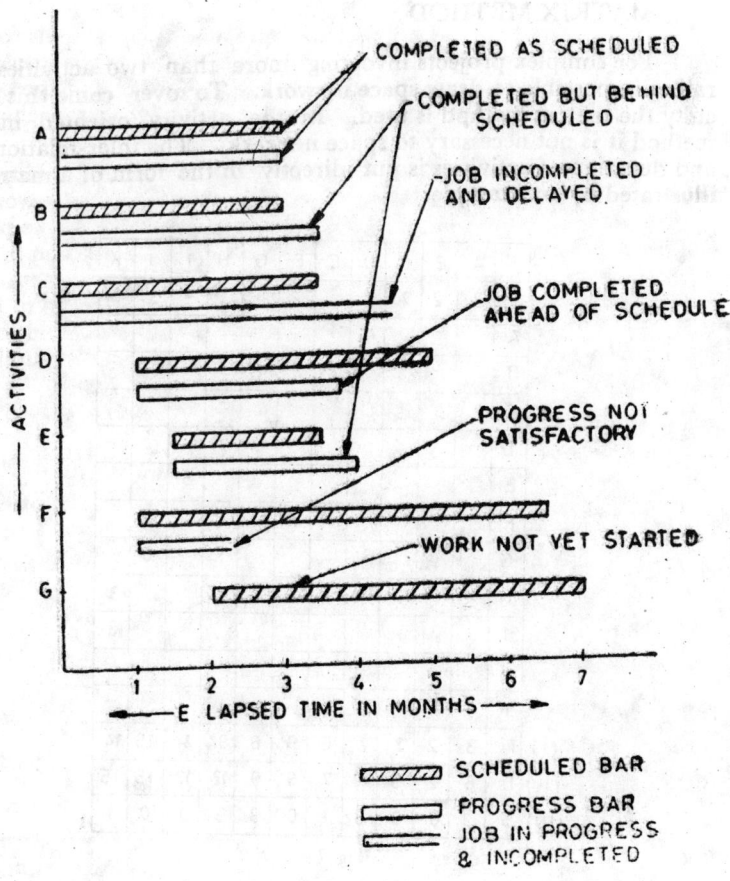

Fig. 7.10

Thus the status of an activity can be observed by progress bar in comparison to schedule bar.

The Gantt chart indicates :

(i) What is to be done ?

(ii) What has been done ?

(iii) What should have been done ?

Gantt chart can be used for any number of activities,

A typical example of Gantt chart for several activities is shown in Fig. 7.10.

Limitation. It does not show the relationship between activities and their impact on project schedule.

MATRIX METHOD

For complex projects involving more than two activities it is rather impossible to draw space network. To over come this difficulty the matrix method is used. In the activity oriented matrix method it is not necessary to space network. The inter-relationship and duration of activities is put directly in the form of a matrix as illustrated by an example.

	A	B	C	D	E	F	G	H	I	J	K
–	3	2	2								
A				4							
B					4	7					
C							4				
D								2			
E									5		
F										6	
G										6	
H											3
I											3
J											
K											
(i) T_E	3	2	2	7	6	9	6	94	11	15	14
(ii) T_L	6	2	5	10	7	9	9	12	12	15	15
(iii) S	3	0	3	3	1	0	3	3	1	0	1

Fig. 7.9

Fig. 11

Illustration III

Interdependency	Activity	Duration
A—D	A	3
B—E	B	2
B—F	C	2
C—G	D	4
G—J	E	4
F—J	F	7
E—I	G	4
I—K	H	2
D—H	I	5
	J	6
M—K	K	3

All the activities are written in columns as well as in rows. in their order as shown in Fig. 7.11   In columns succeeding activities are written while in rows the preceding activities are written.

In column, the succeeding activity duration is written against its constraint indicated in row.

From the inter-relationship table we find that the first constraint is that D is dependent on A. The duration of D is 4. As such 4 is inserted under the column of D against row of A. This way the body of matrix is filled up. After transforming all relationship into matrix it will be observed that no entry is made under column heads A, B and C. This is because A, B and C are independent activities. Thus the values of A, B and C are inserted under column heads A, B and C against a blank row which means A, B and C are preceded by nothing. Similarly there is no insertion against row heads J and K which means that J and K are succeeded by nothing.

Under the principle that an activity [cannot begin until all activities precedent to it in its sequence have been completed. It is easy enough to calculate Te values for each activity as inserted in row (1) at the bottom of matrix. Activities A, B and C are preceded bo nothing (they are initial activities). So their TE values are equal to their respective durations assuming of course they start at zero time. Since D is precedent by A, TE of D is equal to TE of A plus duration of D in row (O). The procedure is followed column headwise from 1st to last activity. One point has to be kept in mind, when a certain activity is precedent by more than one activity the

TE value for all the path have to be calculated and larger value retained. As in this matrix, activity *J* is preceded by *F* and *G*, *TE* for *F* and *G* are 9 and 6 respectively. The larger value 9 is retained and hence *TE* for *J* is 9+6=15.

The largest value obtained in the forward pass computation is 15 which is project duration. Hence project can be planned for completion is not less than 15 units time. If the objective is to allow no slippage in project duration from its early finish time, then the maximum value as obtained in row (*I*) in the matrix is also the latest finish time for final activity.

Whereas the *TE* is calculated and forward path starting from *A* to *K* column headwise. The *TL* is calculated as backward path from bottom to top row headwise. This is because only *TL* known for at the moment of calculation is of activity *J* which is 15. It is necessary to obtain the remaining *TL* of other activities by considering preceding activities. For example activity *I* precedes *K*. *K* being last activity, precedes nothing must have *TL*=15. Therefore, *TL* for *I* must be equal to *TL* for *K* minus its duration, *i.e.* 15—3 =12. Hence *TL*=12 for *I* activity the process is continued until all the values are inserted in row (*II*) below matrix. As in case of forward path in backward pass also we have to be careful when we come across a situation where more than one activities are successors to a single activity *e.g.* activity *B* is succeeded by *E* and *F*. We have to calculate both the values of *TL* for *B* and take the lower value.

Finally in row (*III*) the slack *S* is completed as under :

$$S = TL - TE \text{ for each activity.}$$

Advantages

(*i*) Drawing a network for bigger project by trial and error method is quite laboriours and time consuming. The matrix method obviates all these troubles and gives relevant data of basic scheduling computations mechanically.

(*ii*) It is a better trial for implementation of changes in activities duration and effective project control and review.

However, the value of having a well drawn network of a project in understanding the relationship and dependency of activities should not be under-estimated.

Squared-Network

Gantt chart represents activities in time scale but do not establishes inter-relationship between various activities while space network establishes inter-relationship between activities but do not take time scale into consideration. A combination of these two is developed into the form of a Square Network.

Square Network inpicates the inter-relationship of activities of space network and also mark event to the time scale.

Preparation of Square Network

(*i*) Determine the project duration from space network.

(*ii*) Layout the critical path in straight line against time scale on chart.

(*iii*) Develop secondary paths above and below critical path, showing constraints as vertical broken lines show direction of dummies.

(*iv*) Make all bars equivalent in length to their activity duration and indicate slack by broken line.

(*v*) Schedule all activities for their earliest start time or latest completion time as the need may be.

Advantages

(*i*) It facilitates the review of progress of various activies from time to time.

(*ii*) It is a better tool for reporting progress against schedule to the management.

(*iii*) It affords for suitable adjustments of activity bar with the slack durarion to meet schedule and manpower loading considerations.

(*iv*) It helps in resources levelling.

Illustration II is drawn here on square network as given in Fig. 7.12

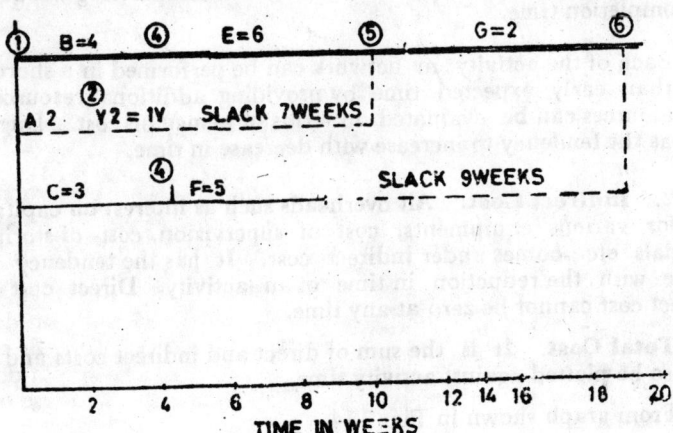

Fig. 7.12 .

Cost Concept of PERT

PERT tells us to achieve the committed objectives in best possible manner and also in the least possible cost and time.

To effect this central, the management must know that :

(*i*) What is required to be done ?

(*ii*) Schedule time required.

(*iii*) The estimated cost.

PERT provides a dynamic total for the execution of a programme and for developing a technique to control time and cost of achievement.

The following functions are required to be carried out :

1. To make a budget of expenditure for given time schedule.

2. To assess the increase or decrease of cost in relation to decrease or increase in time.

3. To arrive at optimum cost.

4. To draw expenditure plan against time.

5. To review the progress in terms of expenditure incurred and time spent.

The cost or expenses to be incurred in the performance of an activity is divided into two sub-heads :

1. **Direct Cost.** The cost incurred in compressing the activity completion time.

Each of the activity in network can be performed in a shorter time than early expected time by providing additional resources. The resources can be evaluated in terms of money or cost. Direct cost has the tendency to increase with decrease in time.

2. **Indirect Cost.** All overheads such as interest on capital, rent for various equipments, cost of supervision cost of storing materials etc. comes under indirect cost. It has the tendency to reduce with the reduction in time of an activity. Direct cost or indirect cost cannot be zero at any time.

Total Cost. It is the sum of direct and indirect costs and it can also be plotted against activity time.

From graph shown in Fig. 7.14

The project duration time X, both direct and indirect costs are equal to Y and, therefore, the total cost of project is $2Y$ which may or may not be minimum cost.

If an activity span is reduced, the direct cost increases. If we plot cost verses duration on a graph we will find a point where the further crushing of activity is not possible irrespective of the amount of resources applied. This cost is known as crash cost and denoted by Cc.

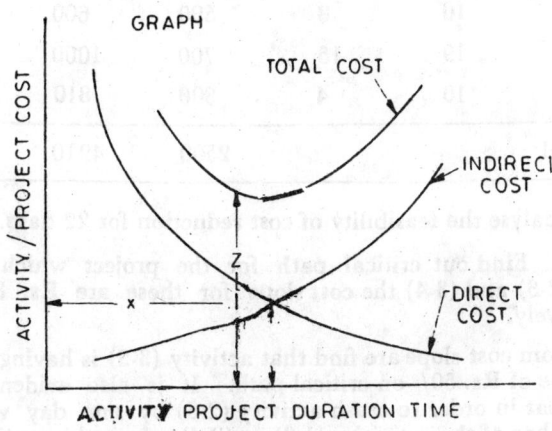

Fig. 7.13

Crash duration

It is the minimum possible duration of an activity which cannot be reduced any more irrespective of resources applied.

Crash Cost

The minimum cost incurred for crash duration.

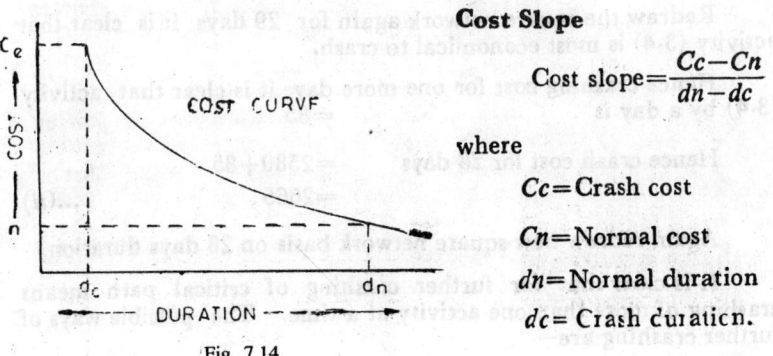

Fig. 7.14

Cost Slope

$$\text{Cost slope} = \frac{Cc - Cn}{dn - dc}$$

where

Cc = Crash cost

Cn = Normal cost

dn = Normal duration

dc = Crash duration.

Illustration IV

Activity	Normal time	Crash time	Normal cost	Crash cost	Cost slope
1-2	10	7	400	640	80
1-3	19	11	600	1160	70
2-3	10	8	500	600	50
2-4	19	15	700	1000	75
3-4	10	4	300	810	85
Total			2500	4210	

Analyse the feasibility of cost reduction for 22 days.

1. Find out critical path for the project which comes as (1-2), (2-3) and (3-4) the cost slope for these are Rs. 80, 50, 85 respectively.

From cost slope are find that activity (3-3) is having minimum cost slope of Rs. 50/- on critical path. It is also evident from the graph that in order to crash activity (2-3) by one day we have to crash either of the activities (1-3) or (2-4) along this activity. The cost of crashing two activities will definitely be more than activity (1-2) alone which is higher only two activity (2-3).

Hence minimum crashing cost for 1 day crashing the activity (1-2)   =Rs. 80·00

Hence crash cost for 29 days   =Normal cost+cost slope
                                =2500+80
                                =2580   ... (i)

Redraw the square network again for 29 days it is clear that activity (3-4) is most economical to crash.

Hence crashing cost for one more day it is clear that activity (3-4) by a day is   =85

Hence crash cost for 28 days   =2580+85
                                =2665   ...(ii)

Again redraw new square network basis on 28 days duration.

It is clear that for further crashing of critical path means crashing of more than one activity at a time. The possible ways of further crashing are—

ALL TO THE TIME SCALE

0 2 4 6 8 10 12 14 16 18 20 22 24 26 28 30

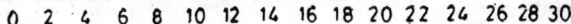

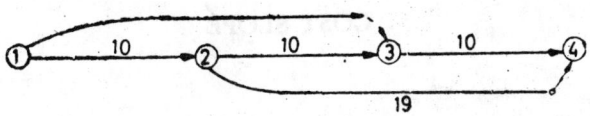

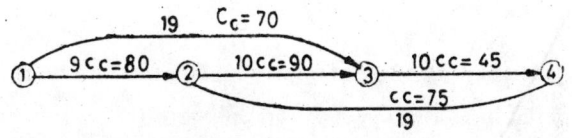

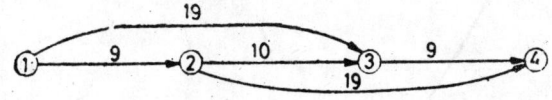

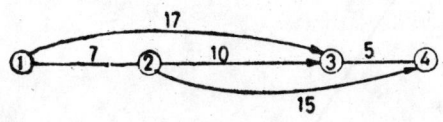

Fig. 7.15

Crashing activity	Total crash cost per day	
(1-2) and (1-3)	80+70=150	...(i)
(1-3) and (2-3) and (2-4)	70+50+75=195	...(ii)
(3-4) and (2-4)	85+75=160	...(iii)

Proposal (i) is most economical. Activity (1-2) can be crashed further only by 2 more days. Hence combination (i) can be crashed maximum for 2 days.

Crashing cost for 2 more days=2×150=300

Crash cost for 28 days   =2665+300=2965   ...(iii)

Further 4 days crashing can be done by adopting proposal (ii) which is 2nd most economical.

Hence crashing cost for 4 more days $= 4 \times 160 = 640$

Crash cost for 22 days $\qquad = 2965 + 640 = 3605 \quad \ldots (iv)$

We also know that for 30 days normal schedule the normal cost is 2500.

COST SLOPE

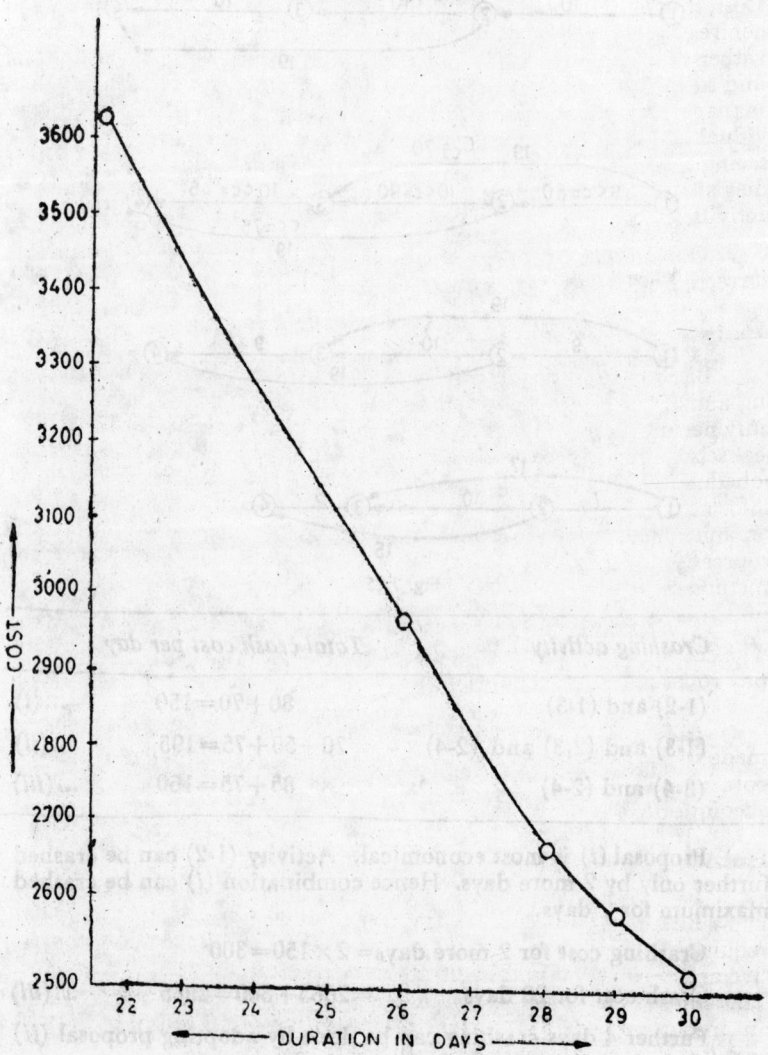

Fig. 7.16

Plotting the crash cost against crash days (from Eq. *i* to *iv*). we get the feasibility curve for cost reduction in most efficient way which is shown in Fig. 7.16

Basic Concept of Network Cost System

The basic concept of the PERT and CPM is simple but importantly different from that of most accounting systems. In PERT/Cost, individual activitities or groups of activities from the micro cost centres for accounting (and therefore managment control) purpose, rather than organisational units (Divisions, departments, sections and so forth). Under PERT or CPM management system, project manager, sub-managers or ordinarily are chosen for supervising individual activities or group activities. Since they are responsible for seeing that activities are completed on schedule, it is argued that they should be responsible for controllable costs associated with activities.

Costs under PERT/Cost system are more easily detected and corrective active are taken more readily.

Analysis and Control of Project Cost

Basic to most cost control systems is a procedure fore comparing actual costs with budgeted costs. The ꓉ERT/Cost syslem not only permit such comparison but because of its integration with project scheduling procedures, also provides a mean of comparing work scheduled with work accomplished. The resulting cost and schedule information allows project managers at both top and intermediate organisational levels to detect problem area and pin point their sources much readily than would be possible with conventional methods of costing.

The costs should be coded according to the net work activities. The cost data should be collected in such a way as to be useful for pay roll and other accounting purpose.

At the same time as costs are reported, an estimate should be made of the percentage of work accomplished. The percentage of cost incurred should match with the percentage of work accomplished.

Human Relations and PERT

It is clear that it is not necessary to be a mathematician to know PERT. The concept of part is to outline the various activities required to be formed, and make an analytical approach to objective achievement. Every activity is done by a group of people. Therefore if every person to whatever group he may belong forms an important pars of the plan. The link relationship of various activities and their effection final objective, once understood by the participants, will ensure their active participants. ꓓarticipation will and enthusiasm in plan ensures success.

PERT and Top Management

PERT is management by objective, and is a top management oriented scheme, with help of PERT/GPM top management combining the projects completion time as early as possible with maximum economy. It can do everything for them, starting from making schedules, and forecasting success to cost control and even can pretell failure and can reason out where it has failed, why it has failed, or how it has failed. It can pin point responsibilities. It can be used for any project.

Even when projects are less complete, and technological break through less frequent it is difficult to look far enough into future to, determine whether the progress to date is sufficient [to meet the schedule and carry all the details of the project always in mind. The management would seek knowledge of some method by which they can assess probabilities, delays and failures. Problems must be recognised advance and either eliminated or planned for. PERT/CPM would need care and attention and patronage to keep it trim and effective. PERT/CPM without top management understanding and support is not worth the attempts.

Management at other levels of control

PERT at best is a mean to an end and not an end itself. It is the person applying it who would get the maximum benefit by way of reducing his effprts to minimum, and he should be enthusiastic in its application. It is not meant to be imposed. Top management, middle management and direct supervisors all have to be assigned clear responsibilities, and can have a separate net work for themselves to follow. This net work, of course, is a part of the entire plan.

PERT planner and PERT Cell

The PERT cell should be created for future planning. Assessment of activities required to be performed should be entrusted to members of PERT cell. The drawing of network, keeping various combinations of circumstances in view, their calculation part should be left to PERT planner. He should high light important activities in the plan. It is his responsibility to keep PERT dynamic and uptodate.

Progress Review

As the work on project progresses, the PERT planner must keep the programme in constant review. Data should be collected by him directly from those who are responsible for executing the activities. Any activity no longer required should be deleted, and any activity that may become necessary should be added to the plan without hesitation. The plan is evolutionary. It is an argumentrative and logical approach, and never originally low. By accepting flexibility of your plan you can lead to success and always adjust to circumstances to resources and to unseeing without giving any chance for failure. PERT should be dynamic in nature.

Replanning

Review and replanning are the corner stones of PERT programming. No labour saving should be thought of when any replanning is to be made. A PERT planner should not hesitate to replan the whole thin if require, and it is then that it becomes an assistance to the executive staff. Many a failure to achieve desired results from PERT are due to PERT planner at times being led away to dictate the plan and avoid labour required for and necessary modification. Any activity requiring re-estimating or rescheduling to care for any programe delays should be readily undertaken.

Data Processing

The effect of various combinations of circumstances should be studied and analysed and their impact on the project to produce optimum results within the stipulated time. The PERT planner can help the management in decision-making, by providing the pros and cons of various combinations. The data regarding progress, resources consumed, on unforeseens to be used in most effective way. They should be reflected in a net work in a simple and facts representative way. If the solutions with increased number of activities and more data to be processed, electronic data processing machines are to be used.

Personal Touch

Though PERT is a management oriented scheme the PERT planner should keep himself in personal touch with first level supervisors and should always be keen to learn their difficulties, if any. This contact should not be for finding the faults and pinpointing the failures and shortcomings, but it should be aimed at finding how these could be avoided, by passed, eliminated, solved or faced.

Project Scheduling without Limited Resources

Until now we assumed that the only constraints for an activity to start and finish are of technological nature. The second equally important assumption made is that of unlimited resources. But in practice most of the managers are faced with the problem of relatively fixed manpower availability, a certain number of machines, or places of equipments and considering money as resource—a limited budget. Jobs which occur on parallel path may complete for the same resources, and even though procedure constraints would not prevent their being scheduled simultaneously, a limited supply of resources might force them to be scheduled sequentially. Heuristic programs for resources scheduling usually take following two forms :

1. **Resource levelling program**. These attempt to reduce peak resource requirements and smooth out period to period assignments, within a constraint on project duration.

2. **Resource allocation program**. These allocate available resources to project activities in an attempt to find out the shortest project schedule consistent with fixed resource limits.

Resource Levelling Program

In some project situations resources can be acquired or released in practically any desired amounts if one is willing to pay the expenses involved in changing resource levels, such as cost of hiring, training, and so on. It is usually prudent, however to maintain relatively stable employment levels and to utilize resources at a more constant rate. In such situation resources levelling program are of most appropriate nature.

In this method the activity slack is used as a means of smoothening peack resource requirements. Start with calculating early start schedule for project and then plotting a resource loading chart for the schedule. In Fig. 7.18 (a) the projects early start schedule graph gives rise to the resource loading chart shown in Fig. 7·18 (b) (The number above each arrow represents its resources requirement in number of man, and since each activity has a different requirement, this number can be used to identify the activity as well).

The peak requirement of 24 men comes on 3rd day when job, 7, 8 and 9 all are activated in early start time. Both 7 and 8 are having slack of 2 and 7 days respectively so these jobs can be delayed upto that slack period without effecting the project schedule. Job 7 is on critical path so it seems reasonable to examine the possibility of moving the other jobs. Since job has the most slack, and could be easily postponed beyond the peak beriod of 3 days and it will be of use to full up the low point on loading cart from 8th day to 11th day. Such a move will result Fig. 7·18 (b).

The man power loading chart has been smoothened considerably and for further improvement days 3 and 4 are peak days now and of the two jobs active on these days only job 8 has slack if we delay job 8 by its full amount of slack, we obtain the result in graph Fig. 7·18 (c).

The peak period now occur days 1 and 2, spannal by job 3, 4 and 6. Job 6 is critical but both 3 and 4 are having slack of 2 days each. The best results are obtained by delaying job 3 as shown in Fig. 7·18 (d).

In the example given above it is ideal giving completely flat manpower loading chart. But in real practice it is difficult to achieve, the approach taken in this example can be used in real projects.

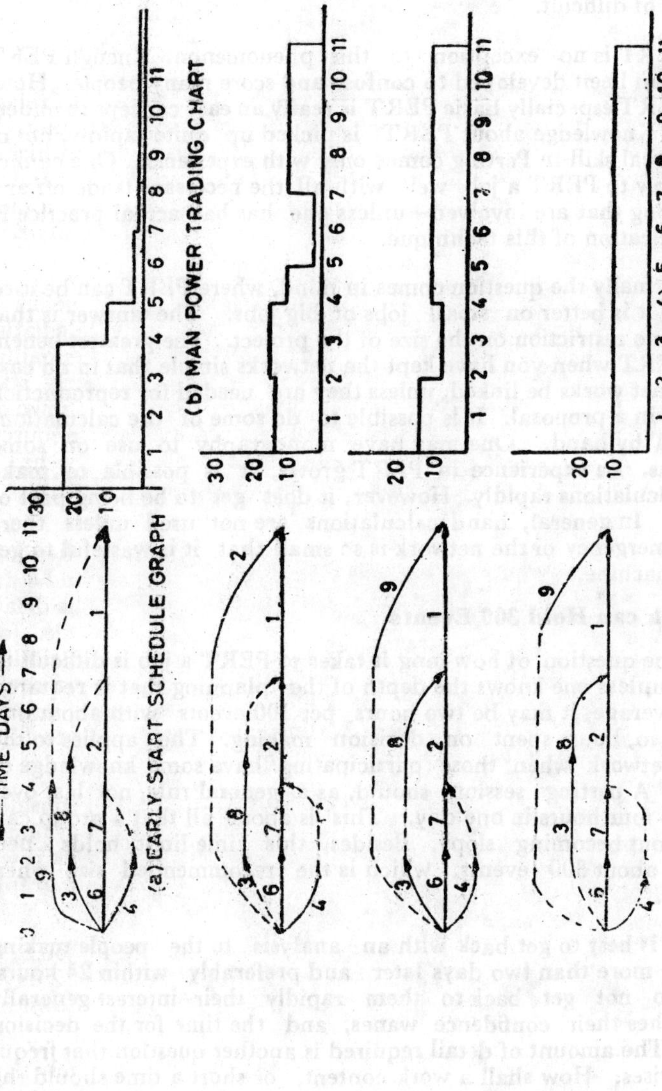
TIME DAYS
0 1 2 3 4 5 6 7 8 9 10 11
(a) EARLY START SCHEDULE GRAPH
(b) MAN POWER TRADING CHART
30
20
10
1 2 3 4 5 6 7 8 9 10 11

Fig. 7.17

GUIDELINES FOR THE USE OF PERT

As every new technique for area of human endeavour makes its appearance, there always arises a certain amount of specialised technology. This generally comes about because it is easier to gene-rate new works than to use old ones and perhaps because it enable those who first came on the scene to make others think something magical of difficult.

PERT is no exception to this phenomenon. Enough PERT jargon can been developed to confuse and score many people. However, PERT especially basic PERT is really an easy concept to under-stand. Knowledge about PERT is picked up quite rapidly, but of course, real skill in Perting comes only with experience. One cannot know how to PERT a job well with all the necessary trade off and replanning that are involved—unless one has had actual practice in the application of this technique.

Normally the question comes in mind, where PERT can be used whether it is better on small jobs or big jobs. The answer is that there is no restriction on the size of the project. The greatest benefit from PERT when you have kept the networks simple that in no case should net works be linked, unless they are needed for reproduction purpose in a proposal. It is possible to do some of the calculations in PERT by hand. One may have monography to use on some occasions. As experience in PERT grows, it is possible to make these calculations rapidly. However, it does get to be being part of the job. In general, hand calculations are not used unless there is an emergency or the network is so small that it is wasteful to get to the machine.

Network can Hold 300 Events

The question of how long it takes to PERT a job is difficult to answer unless one knows the depth of the planning that is required on an average, it may be two hours per 100 events with about one of the two, hours spent on decision making. The applies to the initial network when those participating have some knowledge of PERT. A perting session should, as a general rule, not last over three or four hours in one day. This is about all that a group can do without becoming slopy. Besides, this time-limit holds a net-work to about 300 events, which is the recommended size when possible.

It is best to get back with an analysis to the people making plan no more than two days later and preferably within 24 hours. If we do not get back to them rapidly their interest generally diminishes their confidence wanes, and the time for the decision passes. The amount of detail required is another question that frequ-ently arises. How shall a work content, or short a time should the activities on the networks have ? We can consider that we have planned to the proper depth when we are able to get adequate warning of difficulties that will occur in the future. We should go into whatever detail is necessary to understand the full significance of the activity. If the detail is excessive, however, we will find that we are not getting a true prediction. The only real way to find out it to PERT a few jobs and get a feel for it. There are no hard and fast rules.

One point often overlooked by those learning the PERT technique is that the activity between two events can be broken down into a complete, separate network. If we spot an area that is critical or that looks strange for some reasons or another it is better to separate networks for it.

As far as the minimum time is concerned, it is better to use weeks than days ; but sometimes, when trouble is really at hand using days may be necessary. We must remember that a number of people working a short time get paid as much as few people working for a long time.

Factors should be Considered before Implementing PERT

Engineering department often gets committed to release job on a specific date and must release it regardless of the need for total release by the manufacturing groups. PERT planning has greatly increased understanding between engineering and manufacturing. As long as the manufacturing people understand the overall sequence of release, it does not bother them. They no longer scream for engineering to release unless we are really at fault. Better understanding and better relations with manufacturer save money and keep the inventory low.

1. **Top Management Support.** Following factors should be taken into consideration before impementation of PERT. In the implementation of a PERT system the top management support is in use. However, by support we mean encouragement and not comments. Arbitrary imposition of PERT will hinder its eventual acceptance. This is true even though resistance to PERT is now weaker that it used to be. In fact, PERT is coming into such widespead use that people are becoming afraid not to use it.

2. **Adopt PERT to local requirements.** Requirement of every plant is different, and each has its own peculiar organization. The people are different, the business are different, and the PERT needs are also different.

3. **Replan when necessary.** The basic feature of the PERT technique is that replanning is inherent. We first plan the project ideally and then make the necessary changes to get within the prescribed schedule limitations. There have been instances in which PERT systems have been set up where the plans required formal alteration, notices and formal meetings to get them changed. Such procedures are considered to be too rigid and time consuming and are generally bad practice. If at any time new difficulties arise which

call for a change in the PERT planning, we should make the change and start from the new position.

With PERT we are taking advantage of the law of probability. Even though one event might be missed, we can very well pick up the another one, and if we do not, we can replan later event.

On the other hand, it is something overlooked that replanning in PERT inherenlty increases the risk of not meeting a schedule. The first plan and the first time estimates for a job usually are found to be best. West should not change the estimates once they are made unless the plane are changed. However, in planning alternative, we must be sure to consider the increase in risk that follows.

4. **Leave room for expansion.** When we set up a PERT system, we must be sure to set it up so that it can grow and expand as time processes. We cannot install a PERT system with high calibre people and then leave it for lower-level people to operate. We should always have some knowledgeable individual working with PERT and adapting it to newer requirements. Moreover, the system should not be frozen after it gets working properly. When installing it, we should indicate to the users that it will change from time to time. Like all systems, PERT should be flexible and changed to meet too needs of the users when this is desirable and necessary.

5. **Establish PERT on a permanent basis.** To use PERT just to keep the customer happy is like keeping the front parlor locked up for use only when visitors come.

6. **Do not force formal reporting.** We have for a long time steadily maintained that PERT is basically a planning tool which will report what we want it to. In implementing a PERT operation, some may find that only way the work to be planned and do not want any formal reporting back to management. This as it should (not force formal reporting into) be. We should not force formal reporting to a PERT system unless there is a real need for it. When reports are necessary, they should be geared to the information requirements of each level involved.

Much of the successes of PERT has been atributed to the fact that it just does not take the information, chew it up, and send it back and pretend and as if it were something new. PERT adds new information.

7. **Don't use Inaccurate Estimates.** We can depend to PERT as easily as we can to any other system. But if we will only be wasting time and money. There have been cases where PERT groups should not be fed superfluous data for part of a job. This is obvious when the whole job is analysed. Excessively long estimates will show up on the critical path. There is nothing wrong about

being on the critical path, somebody has to be there. However, if we show up there all the time, there is usually something wrong. The effect of safe estimate is to force management to take increased risks in other areas. Being accurate with PERT estimates pays off.

8. Don't add estimates first. In planning a project with PERT we should first plan the network and then add time estimates. These should we made at random. If we start estimating at one end of the network and workout our way through, we will probably pick up the possible critical paths and shorten the estimates on these paths. The randomizing of the time estimates eliminates this difficulty.

9. Don't demand commitments. One point which cannot be ever-stressed is that we must be sure to separate or estimates from our commitments. An estimate is an estimate and not a commitment. We should use commitments only when absolutely necessary. We must realise that a man is exercising his best judgment, then he gives us estimates, and we should not force him to a commitment. If we, do we will get longer estimate the next time.

10. Don't plan with overtime. When planning with PERT the first time through, estimates should be based on a normal work week, and overtime should be saved for schedule completion when needed. Otherwise the method of shortening the schedule is lost later on when it is needed for handling trouble spots.

11. Don't standardize event titles. One other pitfall to avoid is standardizing event titles. We all agree that standradization is necessary for many facts of our work. In PERT, however, standardization of event titles tends to slow planning and cause confusion. We found that the people doing the planning spent more time deciding on which event title should be used than they did not plan. While doing the work call the events what they wished. This works out very well because then it was part of their work.

12. Resource Allocation Program. The resources scheduling problem is not only one of smoothing requirements but it is also one of allocating resources that are available only in fixed or relatively fixed amounts, the manager may have a limited staff of skilled design Engineers or other difficult to replace, skilled personnel of there may be some fixed number of certain machinery or pieces of equipments or perhaps some construction material can be obtained only at a limited rate. Frequently too, there are budget restrictions. Even if resources could readily be purchased in unlimited quantities the funds to do so may be limited to certain monthly allowances. As a result, activities can be scheduled only as necessary resources become available, even if the effect is to delay the project completion date.

To achieve it, we have to remove the constraint of a fixed due date and push down. If the limits are rather confirming, it is likely that many jobs will be squeezed to the right on ischedule chart, delaying the project due date. In resource allocation we cannot have constraints both on resources limits and on project due date.

Problems In Critical Path Analysis

Ordinarily critical path scheduling techniques do not take explicit account of job resources requirements or of possible limitations on resource availability. Implicitly, it is assumed that only constraints on the start time of a job are technological, *i.e.* a job can be started as soon as its predecessors have been completed. The schedules for early start time and latest start time are established and any job on non-critical path can be delayed upto its total slack period without delaying finish date of project.

Slack in a Limited Resource Schedule

1. The slack concept, though modified, may still be used and retain its utility as a measure of flexibility in a project schedule.

2. Slack is dependent upon both procedue orderings and resource availability.

3. In general resource limitations reduce the amount of slack in a schedule.

4. Since, in general, the early and late start schedules are not unique, so the set of slack valves for a project is not unique. Slack is conditional upon the scheduling rules for creating early start and late start schedules.

5. While a critical path of technologically connected jobs does not always exist in a limited resource schedule, under certain conditions a critical sequence of slackless jobs which span the length of the project can be identified. The jobs are continuous in time, if not in predecessor successor relationship.

8
Quality Control

Quality is never an accident. It is always the result of an intelligent effort. In any organisation the quality function involves all departments and all groups of personnel. Quality control, is a deliberate and planned activity to integrate the quality development, quality maintenance and quality improvement efforts of all the various groups in an organisation. Quality means the very survial in a competitive world. The quality is not the only means for enlarging the share of export and also the sale of goods in the internal market where there is a growing inter firm competition but also a means for providing the right quality of the goods to the customer with satisfaction.

Term 'Quality' conveys that there is something good or desirable in the product. Quality conveys the impression what consumer expect from the product. Quality is the fitness for the pupose at the lowest cost and quality of any product is regarded as the degree to which it fulfils the requirement of customer. Quality can be determined by the characteristics of design, size, material, chemical composition, mechanical functioning, workmanship, finish and so on. Quality signifies the several attributes of product such as small durability, colour, intrinsic properties etc. Quality means the characteristics of materials, parts, assemblies, processes and processing conditions and product including its packaging.

Quality Control is as old as industry itself. From the time man began to manufacture a product there has been an interest in quality of the product. Any industrial situations which will involve the combination of various resources such as materials, money, methods machines, and men. Each of these elements of the combination has intercut or natural variability, plus unnatural variability. The unnatural variability can be solved and therefore can be controlled.

The word quality has a variety of meanings these include :

1. The degree to which a specific product satisfies the wants of specific consumer or it is called consumer's preference.

2. The degree to which a specific product conforms to a design or specification (quality of conformance).

The control of quality is the series of related activities carried out in numerous departments of an organisation to establish the fitness for use. Quality control is an effective system for integrating the quality development, quality maintenance and quality improvement efforts of the various groups in an organisation so as to enable production and service at the most economical levels extending full customer's satisfaction.

Importance of Q.C. in Industry

Quality Control is based in the concept of variability, whether large or small which is inevitable in any manufacturing process. The variability can by classified as one which is inherent to the process, and the one causes for which can be assigned. The knowledge enables us to detect the causes of external variation almost instantly during the course of manufacture, and eliminate them before much damage is done to the process. The approach in quality control is, therefore, to prevent the defects by simply inspecting them after the manufacture.

Benefits of Quality Control

(1) Better understanding of processes and product.

(2) Helps in building information system for improving quality and reducing cost and co-ordinating the activities of various departments.

(3) Increased production under same set-up

(4) Reduction of cost per unit.

(5) Reducion of scrap.

(6) Saving excess use of materials.

(7) Reduction in inspection.

(8) Evaluation of scientific tolerances.

(9) Maintaining operating efficiency.

(10) Less customers complaint.

(11) Increasing quality consciousness.

(12) Building quality into the product rather than inspecting it.

(13) It helps in planned collection and effective use of data in studying cause effect relationships which lead to corrective action on the process.

(14) Reduct in overhead expenses.

(15) Increase in yield and efficiencies.

(16) Increase in goodwill.

(17) Reduction in the idle time loss of machine and equipment as well as operational staff.

The important tool of quality control is the control chart technique. Besides, giving a running commentry on the process with respect to the process level and variability, it indicates the presence of assignable causes and helps in their immediate elimination. Quality control advantage lies in the reduction of waste, rejects and re-work, saving in the consumption of material and quite often reduction in cost of inspection. All these advantages are amply illustrated in the following pages.

Quality control tecnnique are also useful in determining the effectiveness of tolerances. If tolerances are much closer than the permissible variability of the process, rejection and incased cost of production are inevitable. But if the variability of the process is much smaller than the specified tolerance saving in cost of inspection, material or equipment can be affected without any deterioration in quality.

Quality control. aims at achieving quality standards with existing facilities by determining, maintaining and improving. Thus systematic and continuous application of Q.C. would lead to better quality, increased production, reduced cost through reduction in scrap and rework, saving in consumption of material, inspection and above all greater consumer satisfaction. To understand the term quality control let us first understand the term quality and then the term control. The term quality is used in different ways.

Quality Costs

It is an axiom of business that in the long run all costs must be recovered in the price if a firm is to succeed. A firm must provide a product of comparable quality at an equal or lower price than its competitors. Within a narrow band, bounded on the side by the cost of operations and on the other by the price dictated by competition, the firm must operate and, if possible, attain an advantage over its competitors. To achieve this or gain a competitive advantage the firm must provide either a more desirable product at a same price or an equally desirable product at a lower price. This concept resolves itself into a problem of lowering production costs either to allow a lowering of price or to provide the opportunity to increase the quality of the product throgh more expensive materials, more elaboráte design or increased performance. The various costs of a firm can be classified as *Overhead costs ; Design costs ; Production Cost. The overhead costs* include the relatively fixed costs of supervision, clerical, financial, maintenance and facilities. Costs can frequently be reduced in this area, but the probability of reducing costs sufficiently to provide a competitive advantage to an otherwise competitive firm is not too high. *The Design Costs* are critical costs for they establish the base cost of the product. The design defines the materials to be used number of parts and tolerances to be met which establish the theoritical minimum labour and material costs of the product. When the design is fixed, the theoretical costs remain relatively stable and do not normally require continous monitoring. With continous design review and value engineering programmes. *The Prodution Costs* are the variable costs of factory operations, essentially direct labour and material. The differential between the theoretical design costs and the actual production cost is largely caused by in efficiencies in the utilisation of labour, and many of these can be identified as quality costs and losses.

Quality Costs and Losses

Quality losses are simply the costs incurred because of failure to achieve perfection in production. This class of costs includes such items as scrap and rework due to defective material or labour, customer returns, and warranty costs. The significant point is that neither of these costs has meaning without reference to the other. Obviously the cheapest quality control organisation is no quality control organisation at all. It has been demonstrated that these costs tend to receprocate; to minimise, one is to maximise the other; obviously the object is to minimise the total scrap and rework costs.

Quality Motivation

With the coming of the Industrial Revolution, labour became highly specialised. Work that was previously done by individuals was subdivided and spread among many individuals. With mass production the need for standardised parts and, in turn, products engineered in minute detail, further constraining the contribution of people. But with all the industrialisation that has taken place, the role of the individual has never been more important than it is today. Competitive pressures are enormous and labour can contribute greatly to the success of the business. In order to achieve the significant contribution from labour, each person must be made to realise and appreciate the importance of his own role in the production system must be given something tangible in which be can take pride. Recognising there needs, many companies have established formal motivation programmes.

Quality Control Concept

In order to understand the concept of Quality Control, let us understand the term Quality and control seprately and then the two terms together.

Quality

It is the degree to which a specific product satisfies the needs of a consumer; It is the degree to which a class of product prossesses "Potential Satisfaction" for people. The term satisfaction however also includes "Price" "Availability" and "Guarantee". Quality is the degree to which a specific product conforms to specification. The quality characteristics of a product are appeerence, performance, Life dependability, reliability, durability, Maintainability, Taste Odour etc.

Control

The term control is also interpreted in different ways such as :

1. Control is the act of direction, influences and command.

2. Control is the act of varification.

3. Control is the standard of comparison against which to check the results of an experiment.

Quality control should not be limited to inspect the product or material but should be extended to a complete list of functions associated with quality. Virtually all manufacturing industries have a function of measuring the final product and comparing the results with some specifications to decide whether the product should be sold to customers or not. The term "quality control" means the complete bundle of tools and skills used for meeting company's quality objectives. Quality of product is thus a weapon in competition :

Quality has not been quite exploited like other factor such as price or service. The opportunities to exploit quality as a tool includes :

1. Design of a product in such a way as to have high customer appeal from the standpoint of appearance, life, performance etc.

2. Use of "market quality knowledge in striking an optimum between value of quality and cost of quality."

3. Guarantee of quality in a way to minimise any loss of the customer that may arise out of the defectives.

4. Advertising previous performance data through various propaganda channels.

5. Avoidance of any failure that may create a serious blow to quality reputations.

The success of any quality control programme depends to a greater exent on quality objectives and policies designed by and for a management including the following factors :

1. This standard of outgoing quality.

2. The pattern of customer relations.

3. Pattern of vendor relations.

4. The extent of use of impersonal methods of supervision i.e. objectives, planning, reports, goals, charts, controls, audits etc.

Problems of Quality Control

Quality Control is a systematic and scientific system involving the application of all known industrial and statistical techniques to control the quality of the manufactured product. The meaning of Quality Control will be different from one individual to another and also differs from time to time. The various problems of quality control can be grouped into the following three classes,

(a) **Engineering.** The development of a product is basically engineering ; the development of quality evaluation through improved inspection procedures, the knowledge of causes of defects and their rectification is engineering.

(b) **Seatistical.** The concept of the behaviour of a process, which has brought in the idea of 'prevention' and 'control' is statistical ; building and information system to satisfy the concept of prevention and control and improving upon product quality, requires statistical thinking.

(c) **Managerial.** The efficient use of the engineering and statistical concept is managerial. The introduction of quality consciousness in the organisation is also the management. The effective coordination of the quality control function with those of others is managerial.

Benefits

The benefits of quality control can be listed as follows :

(a) Better products design.

(b) Better product quality.

(c) Reducing in scrap, rework and consumer complaints.

(d) Efficient utilization of personnel, machines and materials resulting in higher productivity.

(e) Elimination of bottlenecks in the process of manufacture.

(f) Creating quality awareness in employees.

Principles

The principles of quality control are :

1. Principles of co-ordination.

2. Principles of prevention.

3. Principles of building quality in the product.

1. The concepts of quality prevades all along the length of an organisation and thus it is absolutely necessary to conceive of the quality control function as coordinating the quality activities in

the various departments. Even co-ordination principle is very important in the production cycle of a product. Here the co-ordination is
based on SQC concepts. In any production cycle, there are three
major function—design, manufacture, and inspection. These three
functions were treated as almost independent until recently the
knowledge of capability of process, came to light.

2. The second principle is the principle of prevention. It is
important to consider in any actual manufacture to eliminate the
causes of defects as manufacturing is going on. The routine inspection procedures should sort out the items produced as good. This
approach must be replaced by a new technique of corrective action
as and when manufacturing is continuing.

Following are certain important ideas regarding principles of
building quality.

1. The idea of building quality into the product rather simply
inspecting them.

2. Feed-back information which assists in coordinating the
activities of the various departments.

3. The use of data in studying cause-effect relationships which
lead to taking action on the process.

4. Importance of planned collection and effective use of data.

Function

1. To satisfy the customers by complying with specification
of the product.

2. To be instructive that reoccurrence of mistakes could be
eliminated.

3. To facilitate a procedure where by incentive payments will
not be made for faulty output.

4. To train the operators and make them aware that their
work is being inspected and is of importance to the success of the
organisation.

5. To revise the quality required.

6. To evaluate existing inspection methods and design better
and effective procedures.

7. To avoid the processing of faulty goods.

Objectives

1. Setting various standards.

2. Appraising conformance.

3. Acting when necessary i.e. taking corrective actions.

4. Planning for improvement i.e. developing a continuing
effort to improve the cost, performance and reliability standards.

These objectives may also be divided into two main groups.

A. Objectives for Holding *status quo* :

It includes :

1. Defect levels of vendors.
2. Yield and defect levels of various process.
3. Levels of quality of finished product.
4. Levels of performance for specific quality.
5. Cost of inspection and testing.

B. Objectives for change in quality conformance.

1. To reduce the losses due to defects.
2. To improve the quality of product going to customers.
3. Improving quality mindedness.
4. Reducing the cost of inspection.

TOTAL QUALITY CONTROL

The concept of total quality control which includes in process control also, is therefore followed by many organisations because of their high quality objectives.

TOTAL QUALITY CONTROL

Quality design
1. Setting standards
2. Product research
3. Marketing research
4. Principle of Proto .type model.

Quality process

Inspection in Laboratories
1. Raw Material inspection
2. Final inspection of the product

This is done on sample basis. Low degree of representation.

Inspection of various stage of production (In-process Control)
1. Piloting the quality) of running production.
2. Process conformance.
3. In-process-control data are more important. Representation is high. For further improve

ment and reduction of losses the information gathered during production process is of high value and necessary.

Cost-Value Concepts in Quality Control

If the maketing department has indicated the consumer's requirements with respect to some product, this necessarliy would include at least in some degree an indication of the price which he would be willing to pay to obtain such a product. Fig. 8.1 gives the idea that higher the technical excellence higher shall be cost. As we increase the technical excellence of our product, necessarily the cost of product increases at an increasing rate. The low degree of technical excellence may be attained at a correspondingly low cost of production. It may also carry with it just not a low but even a negative value. That is, a product may be so poor that one would have to pay consumers to take it away.

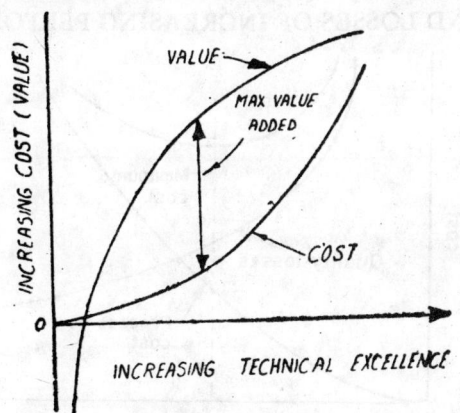

Fig. 8.1

As one moves to the right, that is, as technical excellence is increased, costs increase at an increasing rate while value increases at a decreasing rate. Even though one design may be superior to another in terms of its technical excellence, it may not have increased value in the market-place in terms of the service it renders. There exists some desired degree of technical excellence as shown at the point at which marginal value equals marginal cost, that is, the maximum value added occurs when the slopes of the two curves are equal.

Processes must be designed to produce output according to the desired specifications and these production processes must be monitored to assured conformance. As is shown in figure one may obtain a greater degree of conformance by operating more costly processes, but these more costly processes result in decreasing losses due to lack of conformity at a decreasing rate. It is quite possible to buy processes which are too good, given the most economical specifications as shown in Fig. 8.2 . A process obtainable at lesser cost may result in a higher rate of defective output, that is, low conformance.

Thus the problem of assuring quality consists of two inter-related design phases. First, the design of a product whose reliability is economic in terms of its end use, and second, the design of a set of procedures which will at economic levels assure the attainment of the defined reliability. Establishing proper balance between these two is difficult.

Difference between Inspection and Qualits Control System

In a production system both inspection and quality control will exist as separate systems. In almost all cases they will be separate organizations and, in a large number of situations, bear no organisational relationship to each other except at the highest level in the production organisation.

COST AND LOSSES OF INCREASING PERFORMANCE

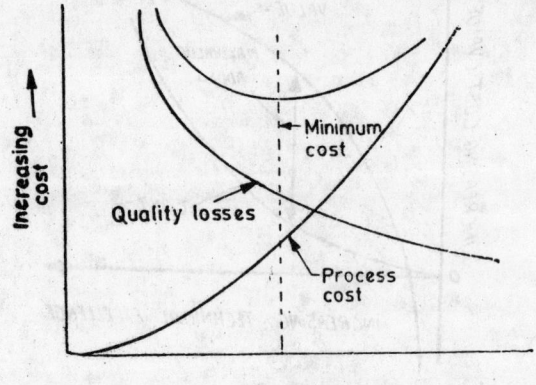

Increasing performance

Fig. 8.2 .

Inspection is concerned solely with the acceptance or rejection of output based on a comparison of the attributes of a unit of output with specifications. Quality control, however, is concerned primarily with determining the capability of processes to meet specifications. Thus quality control is concerned with the prevention of defects while inspection is concerned with their direction.

9

Inspection

Inspection in some degree is an integral part of any production process. The purpose of inspection may be different even at different location with the same process. It is necessary to inspect in order to gain assurance that the process is producing at a level at which restifying inspection is not required. It is certain that quality control is associated with cost. Following are the various costs associated with quality control.

(i) Preventive costs (ii) Appraisal costs (iii) Failure costs.

Prevention Cost. These are such costs as quality control engineering employees' training, and quality maintenance of patterns and tools.

Appraisal Cost. These are the expenses for maintaining company quality levels by means of formal evaluation of product quality. This involves such cost elements as inspection test, quality audits, laboratory acceptance examinations.

Failure Cost. Caused by defective materials and products that do not meet company quality specifications. They includes such loss elements as scrap, spoilage, rework, field complaints. Clifford Paul in one of the international conference on Q.C. gave following esti-mates of the quality cost components for American industries.

1.	Prevention Cost	30%
2.	Appraisal Cost	20%
3.	Failure Cost	50%

In Indian Industries we spend much less on prevention cost, probably nearing to zero. A slight increase in the prevention cost would bring down the appraisal and failure costs substantially.

A we know that the objective of all the Quality Control is to provide quality assurance for the finished product, and to assure optimum quality costs for that product. Engineering encompasses the planning, control, evaluation, and reporting of all quality aspects of the product from its conception through manufacture processing, storage, delivery, installation, maintenance and repair to the end of its service life.

Inspection

Inspection is concerned with estimating the degree to which output conforms with established specification. The process by which degree of conformance is determined may range from the simplest possible comparison of a unit of output with a specification by measurement with a standard instrument all the way to life testing under actual operating conditions of a simple of output followed by rigorous statistical analysis of the results of the experiment. The degree of conformance can be measured by measurement of

(1) Tolerance, and

(2) Allowance.

Tolerance. It refers to any permissible variations in size. For example, a dimension which is specified as 2·0000 inches may be given with many different tolerances. A tolerance of +0·0001 inches would specify that all output within the limits 1·9999 inches 2·0001 inches would be acceptable for use.

Allowance. It refers to a difference in size between parts which fit together. For example, a shaft to be fitted in a hub might be specified with a maximum diameter of 4·001 inches and the hub with a minimum diameter of 4·002 inches. So in this case the minimum allowance would be the difference between these two dimensions, that is, 0·001 inches. So by now we know that the basic quality function is to decide whether a product or material conforms to its specification. This function is generally known as 'acceptance'.

The inspection act constitutes of the following.

(*a*) Interpretation of the specification.

(*b*) Measurement of the product characteristics.

(*c*) Comparison of the above two.

(*d*) Judgement as to conformance.

(*e*) Disposition of materials/products.

(*f*) Recording of the data obtained.

The individual who conducts the above activities is known as inspector and the department which guides this inspector is known as inspection department.

Evolution of Inspection

Following are the several stages of evolution of inspection planning system.

1. Initially, the planning was done by the inspector himself.

2. As the number of inpectors become more and more, inspection "Foreman" appeared on the scene. The planning was then done by these foremen.

3. With the growth of the centralised inspection department, full time planners begin to appear as staff assistants to chief inspectors.

4. Recent trend has been for staff departments to carry out the inspection planning.

Methods of Inspection

Inspection on acceptance basis can be carried out either by :

1. 100% inspection or
2. Sampling inspection.

Sampling inspection is preferred to 100% inspection because of:

(a) Cost   (b) Time

(c) 100% inspection fails to provide 100% assurance against the acceptance of bad quality product due to inspection fatigue. To ensure the correctness of acceptance to the best possible extent, one should ensure the following :

(i) Sampling should be done on random basis representing the total lot.

(ii) Proper sampling sizes.

In order to avoid the shipping of unacceptable products, the buying firm usually wants to defect such defective material as easy as possible.

The purchaser has three alternatives course of action open to him.

(1) Return material to the supplier.

(2) Keep some of more acceptable material and return the rest.

(3) Keep all the material and rework it to the point where it is acceptable.

Economics of Inspection

Materials that are described by performance charactersitics or by detailed specification or perhaps by sample, may well require technical inspection also. Material purchased to unique specifica-tion invariably have many features that are subject to inspection. However same features are typically more critical than others. Consequently technical features are commonly classified as "Critical" "Major" and "Minor", inspecting effort in turn, is allocation in accordance with the criticalness of the feature to be inspected.

When is Inspection Justified

Fundamentally, the decision rests on practical economics of each specific situation.

Two basic Costs involved are :

(1) Cost of Inspection.

(2) Cost resulting from defective parts entering the production or wrong parts reaching the customer.

The object is to minimise the sum of these two costs as the major criterian in arriving at the basic inspection decision.

Single Sampling

Lot-by-Lot Acceptance-Sampling Procedure

In any sampling plan following three values must be specified.

1. The number of units in the incoming lot of material.
2. The number of units to the inspected in the sample.
3. The maximum number of defective items allowable in the sample.

The Procedure of Inspection

1. Determine from the purchase order the applicable sampling plan and lot size.

2. Using this information refers to appropriate sampling take to determine the sample size and the acceptance number to be used.

3. Draw the sample randomly from the lot, several technique can be used in doing this, although the more formal approach is one based on the use of table of random number.

4. Inspect each of the sample items in accordance with the inspection specifications, noted in the purchase order.

5. If critical or fewer defectives are found in the sample the lot is accepted. If more than critical units fail to meet specification the entire lot is rejected.

The Operating Characteristic Curve (OC Curve)

The performance ability of a statistical sampling plan is described by its operating characteristic curve. This helps the manager to know what to expect from a particular sampling plans. Figure shows OC curve for sampling plan with $N=1000$ and $n=135$ $C=3$. The figure was constructed by making this computation individually for a 1% defective lot, 2% defective lot, and 7% defective lot. The OC curve is obtained by connecting the plotted probability points.

OC curve inform us whether to aspect or reject lots of 1000 items. If the incoming lot contains 1 per cent defectives items there is 95% probability that sample will contain 3 or more defectives, thus telling him to accept the entire lot. There is also a 5% probability that the sample will contain more than 3 defects thus telling him to reject the lot.

The OC curve in reality tell the user how will the sampling plan discriminate between good and bad lots. The steeper the slope of the curve the more discriminating is the plan. *Either by increasing the sample size or by reducing the acceptance number.* In practice several points on the *OC* curve have been chosen as bench marks in the design and indexing of sampling plans. The first of these is the *P* value which corresponds with approximately a 95% probability of acceptance, this is commonly called the acceptable quality level (AQL). The second point is the *P* value which corresponds to 10% probability of acceptance this is commonly called the lot tolerance percent defective (L.T.P.D.)

Average Out going Quality (AOQ)

If we assume that all 100% inspected, and that all defectives found are replaced with good products the AOQ for each value of *P* can be given by

$AOQ =$ Number of defective items accepted by the Plan average lot size after 100% inspection.

$$= \frac{P(Pa)\ (N \cdot n)}{N}$$

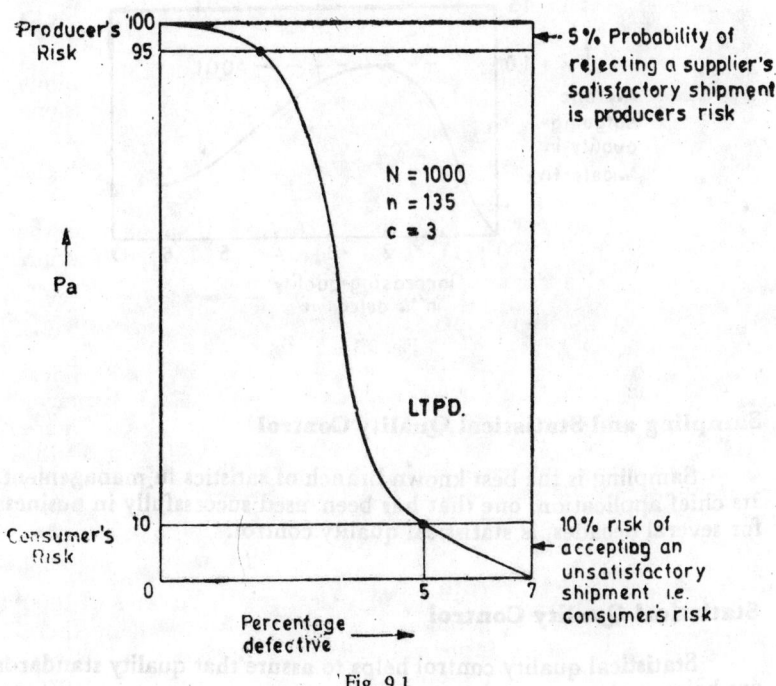

Producer's Risk

5% Probability of rejecting a supplier's satisfactory shipment is producers risk

N = 1000
n = 135
c = 3

LTPD.

Consumer's Risk

10% risk of accepting an unsatisfactory shipment i.e. consumer's risk

Percentage defective

Fig. 9.1

AOQ reaches its bad level then incoming lots contain 2% defectives. This is called the average out going quality limit (AOQL).

Probability that
the incoming lot
will be accepted
by the sample

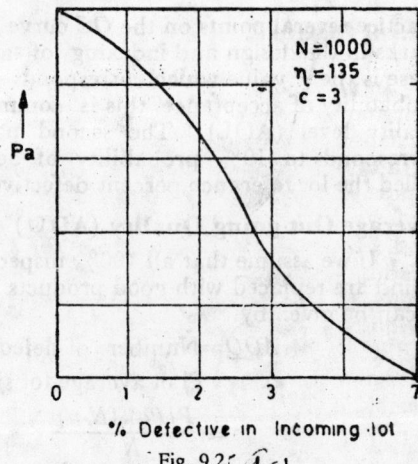

Fig. 9.2

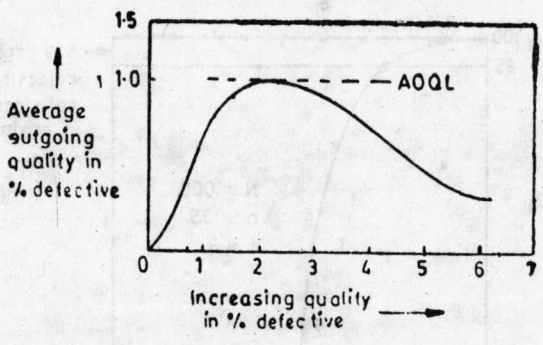

Fig. 9.3

Sampling and Statistical Quality Control

Sampling is the best known branch of satistics in management.
Its chief application, one that has been used successfully in business
for several decades, is statistical quality control.

Statistical Quality Control

Statistical quality control helps to assure that quality standards
are being met.

Statistical Quality Control is concerned primarily with control
charts and acceptance sampling. Control charts are used during
production to determine whether production quality of being
maintained.

Control Charts

In many cases, the most useful control charts deal with the mean and the range. Suppose, again, that we are manufacturing 1·2″ slabs and from past production figures, we calculate the mean length of slabs and call it X or the estimated mean of the universe. Suppose that samples of five slabs are taken at regular intervals and that for each sample, the mean length X and the range are computed. Since no two slabs are exactly the same length, three will be variations in the different X's that are obtained. If these variations are due to chance (random variations), the sample means will be distributed in a statistical pattern. Now let us remember that this is a distribution of sample means and not of individual items. The means of the samples will tend to cluster around the means of the universe.

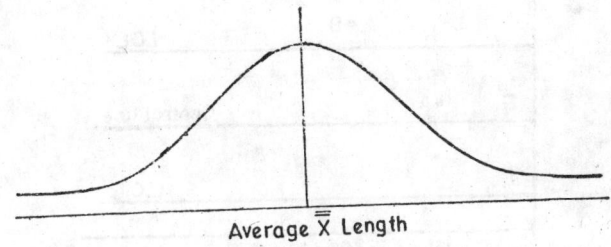

Average $\overline{\overline{X}}$ Length

Fig. 9.4

Now we need an upper control limit (U.C.L.) and a lower control limit (L.C.L.) to tell us when the average length of slab has changed. To illustrate this, assume that $\overline{X}$ is 1·2001 inches. Let

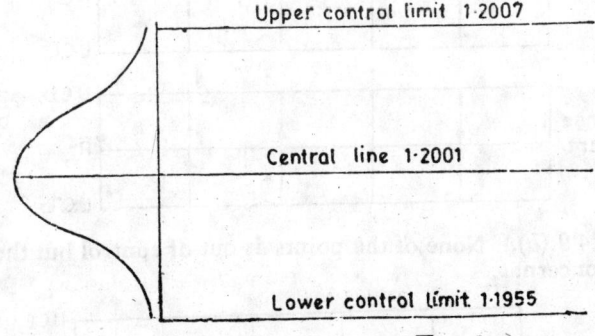

Upper control limit 1·2007

Central line 1·2001

Lower control limit 1·1955

Fig. 9.5 Control Chart for $\overline{X}$.

us take a sample of five slabs from the producing line and the average length of these five slabs happens to be 1·1998. If the point falls above the UCL or below the LCL we assume that this much variation in sample means from their mean is not due to chance ; an investigation is made to find an assignable cause.

   The X Chart alone is not enough because it says nothing about the dispersion of the process.   Changes in dispersion can signal that quality standards are not being met.   So alongwith the $\bar{X}$ chart, we must have a range chart.   An example of an R chart, is shown in figure. The central line is shown as R or the estimated average range of the universe.   This can be computed from past experience.   As in the X chart, there is a UCL and LCL for the R chart.   The R chart gives the limits of the range that can be expected if the dispersion of the universe does not change.   Both the $\bar{X}$ and R charts make use of the exception principle.   The charts are simple enough to be plotted, only when control limits are exceeded is the service of the supervisor or engineer is required.

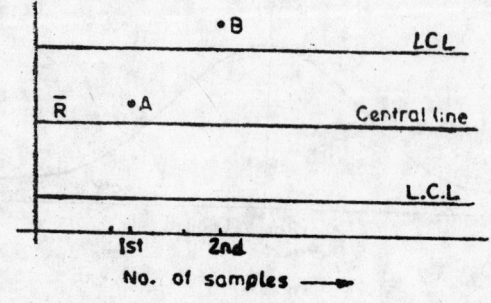

Fig. 9.6

Fig. 9. 6 (*a*).   Samples number 3 and 5 are out of control.

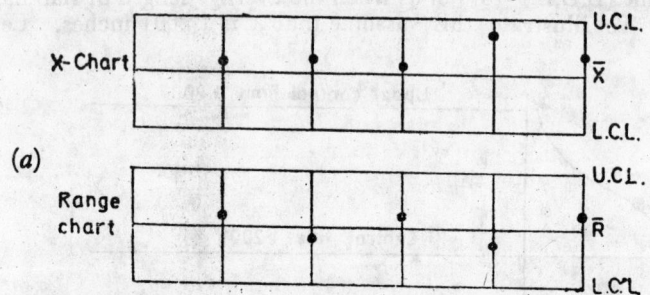

(*a*)

Fig. 24·9 (*b*).   None of the points is out of control but there is cause for concern.

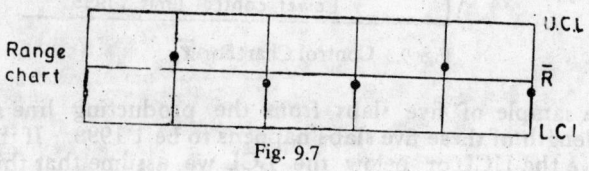

Fig. 9.7

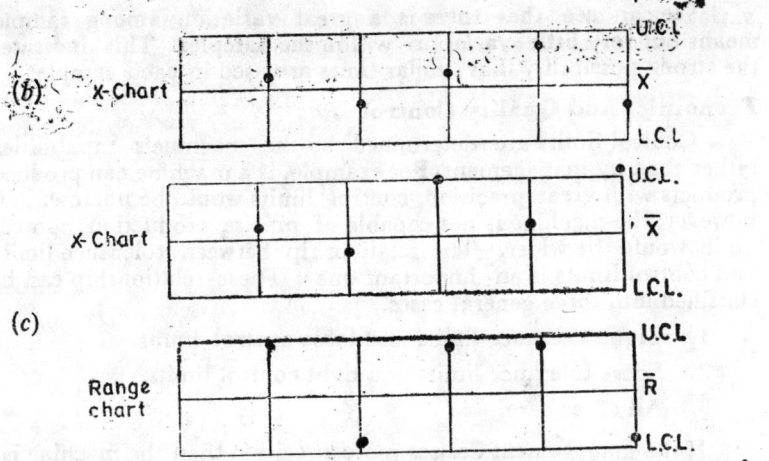

(b) X-Chart

U.C.L.
X̄
L.C.L.

X-Chart

U.C.L.
X̄
L.C.L.

(c)

Range chart

U.C.L.
R̄
L.C.L.

Fig. 9.8 (c). All points in the X̄ charts are out of control. This could mean that the quality cannot be achieved or that improper production methods are being used.

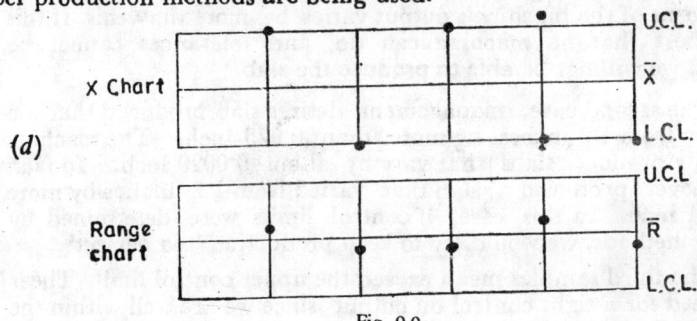

X Chart

U.C.L.
X̄
L.C.L.

(d)

Range chart

U.C.L
R̄
L.C.L.

Fig. 9.9.

Fig. 24·9 (d). All points out of control, but since they all are above the presumed mean of the universe, the actual mean of the universe, has evidently shifted upward.

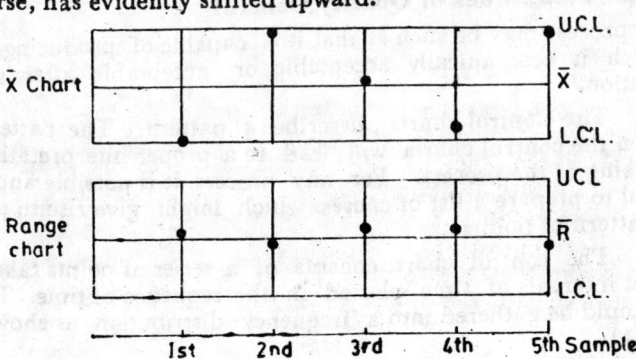

X̄ Chart

U.C.L.
X̄
L.C.L.

Range chart

UCL
R̄
L.C.L.

1st   2nd   3rd   4th   5th Samples

Fig. 9.10.

We can see that there is a great variation among sample means but very little variation within the samples. This indicates the strong possibility that similar times are used in each sample.

Economics and Quality Control

Control limits are determined by the machine's capabilities rather than by management. For example, is a machine can produce products with great precision, control limits would be narrow. If, however, the machine is not capable of precise production, control limits would be wider. The relationship between tolerance limits and control limits is an important one. These relationship can be classified into three general cases.

1. Tight tolerance limits and loose control limits.

2. Loose tolerance limits and tight control limits.

3. All other case.

If the management desires more precision than the machine is capable of producing. For instance, management wants slab produced that will not vary from 1'2 inches by more than '0001 inch. Then after examining the machine's capabilities, it is found that 60 per cent of the machine's output varies by more than this. If this is the best that the machine can do, and tolerances cannot be revised, we will not be able to produce the slab.

The second case, management desires slab produced that do not vary from 1'2 inches by more than 0'025 inch. The machine generally produces slabs that vary by about 0'0020 inch. In fact, it has never produced a slab that varied from 1'2 inches by more than '01 inch. In this case, if control limits were determined by routine methods, we would try to keep production 'too perfect'

The third samples mean exceed the upper control limit. Theer is no need for a tight control on output since we are well within the tolerance limits.

CONTROL CHARTS

The Techniques of Quality Control

A process may be such so that it is capable of producing output which is economically acceptable or acceptable after some rectification.

1. The Control charts , describe a pattern. The pattern of points on the control charts will lead to a proper interpretation of the working of the process. For any process, it is possible and will be useful to prepare a list of causes which might give rise to particular pattern of points.

2. The control chart consists of a series of points taken at different intervals of time plotted in the sequence of time. These points could be gathered into a frequency distribution as shown in Fig. 9-11.

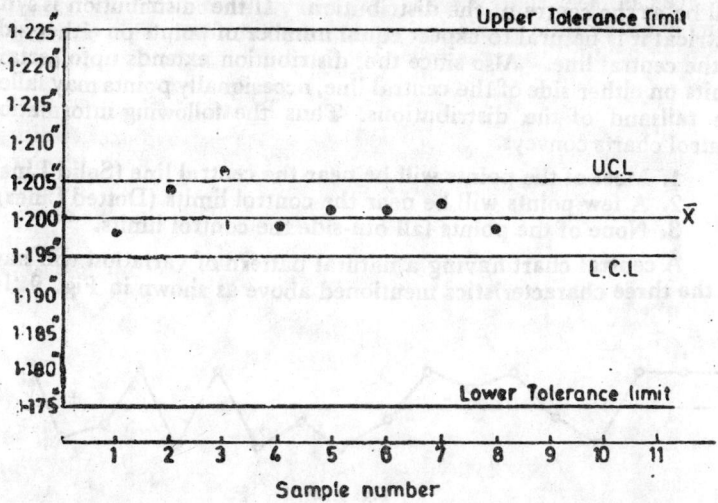

Fig. 9.11

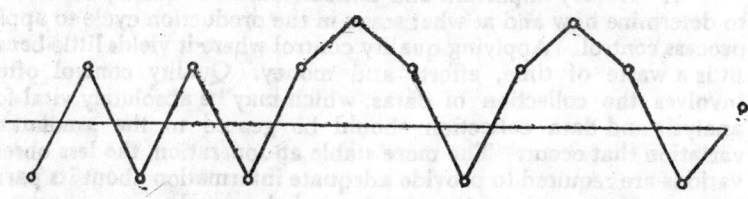

Fig. 9.12.

In order to understand the behaviour of any process, it is essential to understand the pattern it describes on the chart. Ability to interpret the charts depend on the ability to distinguish between the natural and unnatural pattern of the processes as described on the chart.

It has been seen that when the process is in a state of control, the pattern of variation can be represented by a frequency distribution called the Normal Curve. If the same causes are at work, the same pattern of variation will repeat itself day-in and day-out and this situation constitutes a state of control as shown in Fig. 9·12.

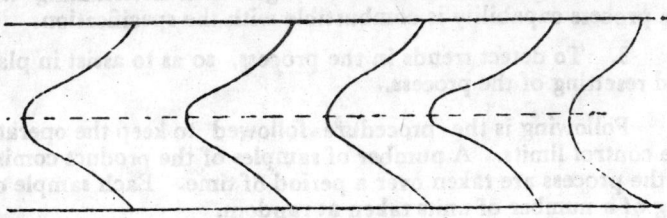

Fig. 9.13

The various type of distribution that most of the points will be at the centre of the distribution. If the distribution is symmetrical it is natural to expect equal number of points on either side of the central line. Also since the distribution extends upto certain limits on either side of the central line, occasionally points may fallen the tailsand of the distributions. Thus the following information control charts conveys.

1. Most of the points will be near the central line (Solid Line).
2. A few points will be near the control limits (Dotted Lines).
3. None of the points fall out-side the control limits.

A central chart having a natural pattern of variation will have all the three characteristics mentioned above as shown in Fig. 9·14.

Fig. 9.14

1. A very important and difficult task in quality control is to determine how and at what stage in the production cycle to apply process control. Applying quality control where it yields little benefit is a waste of time, efforts and money. Quality control often involves the collection of datas, which may be absolutely vital for analysis and data collection should be geared to the amount of variation that occur. The more stable an operation, the less observations are required to provide adequate information about its parameters. How much information is needed depends on :

(i) Degree of quality of the final product.

(ii) The company's policy regarding quality.

(iii) More urgent attention or priority, in other words it depends on an *adhoc* evaluation of the quality situation in the plant.

The Object of Control Charts

1. To establish whether the process is in Statistical control, and the variation are attributed to chance.

2. To guide the production engineer in determining whether the process capability is combustible with the specification.

3. To detect trends in the process, so as to assist in planning and resetting of the process.

Following is the procedure followed to keep the operation in the control limits. A number of samples of the product coming out of the process are taken over a period of time. Each sample consisting of a number of units taken at random.

For each sample the average value $\bar{X}$ of all the measurement and the range R are calculated. The grand average $\bar{\bar{X}}$ and the average range R are then found, and from there we calculate the control limit.

(a) Wide tolerances compared with process capability.

(b) Tolerances comparable to process capability.

(c) Tolerances are comparatively tight resulting in excessive scrap.

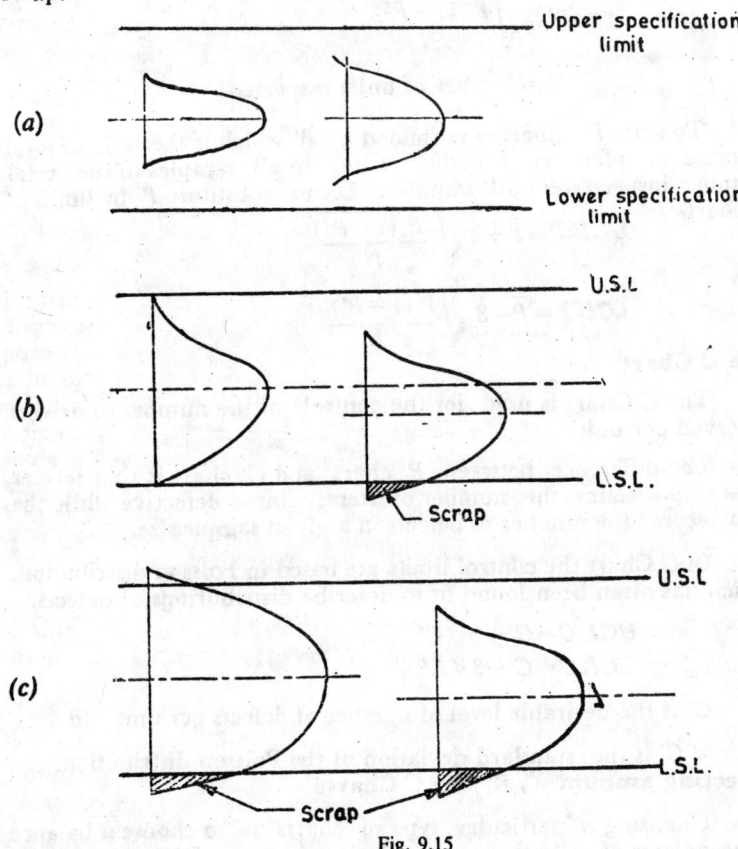

Fig. 9.15

The P-Chart

The fraction defective Chart is used to control the general quality level of the product and to find whether quality level of the product is due to chance alone.

The basic difference between $\bar{X}$ and R chart and the P chart is that former is based on control by attributes. The P chart, on the other hand, keeps a record of the percentage of defective. It can replace X and R chart if say, go-no-go-gauges are used without actual measurement of attributed being recorded.

$$UCL\ P = P' + 3\sigma\ P'$$
$$LCL\ P = P' - 3\sigma\ P'$$

where P' is the ratio of total number of defective to the total number inspected. $\sigma P'$ is the standard deviation of the desirable distribution of fraction defectives and from statisitics theory.

$$\sigma P' = \sqrt{\frac{P'\ (1 - P')}{N}}$$

N = Number of units inspected.

To start P' chart is estimated by P' which is the ratio of total number of defectives actually found in all samples to the total number inspected in all samples. Let us substitute P' by limits of P chart.

$$UCL\ P = \overline{P} + 3\sqrt{\frac{P\ (1 - P')}{N}}$$

$$LCL\ P = \overline{P} - 3\sqrt{\frac{P\ (1 - P')}{N}}$$

The C Chart

The C Chart is used for the control of the number of defects observed per unit.

The difference between P chart and C chart is that former takes into account the number of items found defective while the later records the number of defects in a given sample size.

In C Chart the control limits are based in Poisson distribution, which has often been found fit to describe distributions of defects.

$$UCL\ C = C' + 3\sqrt{C'}$$
$$LCL\ C = C' - 3\sqrt{C'}$$

C' is the desirable level of number of defects per unit and

$\sqrt{C'}$ is the standard deviation of the Poisson distribution.

Selecting amount $\overline{X}$, R, p or C Charts

Choosing a particular type of charts is to choose a balance between cost of collecting and analysing the type of data required to plot the chart against the usefulness.

The $\overline{X}$ and R chart, in which upper control limits are worked-out.

$\overline{X}$ Chart

$$UCL\ \overline{X} = \overline{X} + A_2 R$$
$$LCL\ \overline{X} = \overline{X} - A_2 R$$

R Chart

$$UCL \ R = D_4 \bar{R}$$
$$CLR \ = D_3 R$$

Terms A_2, D_4 and D_3 depend on the Number of items per sample.

CONTROL CHARTS

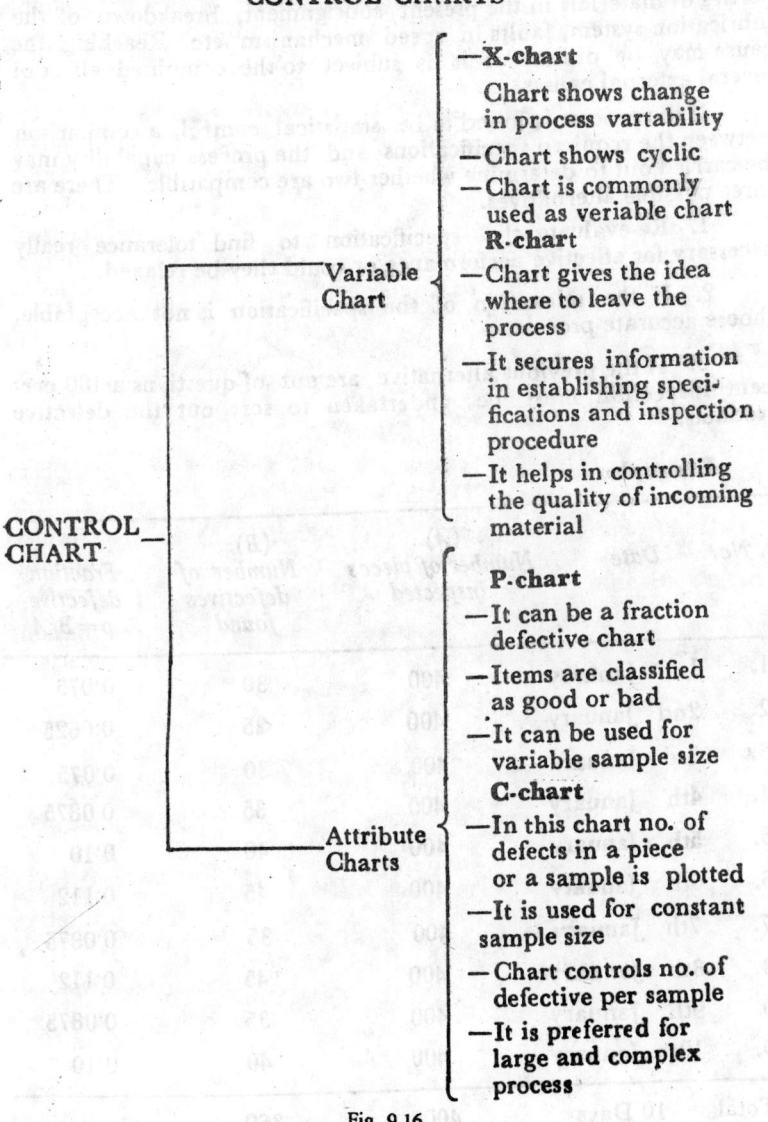

CONTROL CHART

Variable Chart

—X-chart
Chart shows change in process vartability
—Chart shows cyclic
—Chart is commonly used as veriable chart
R-chart
—Chart gives the idea where to leave the process
—It secures information in establishing specifications and inspection procedure
—It helps in controlling the quality of incoming material

Attribute Charts

P-chart
—It can be a fraction defective chart
—Items are classified as good or bad
—It can be used for variable sample size
C-chart
—In this chart no. of defects in a piece or a sample is plotted
—It is used for constant sample size
—Chart controls no. of defective per sample
—It is preferred for large and complex process

Fig. 9.16

As long as the X and R values for each sample are within the control limits the process is considered to be in statistical control.

Soon after drawing the charts, it is necessary to determine whether the process is in statistical control. If it is not, there must be a cause that throws process out of control. Reasons for the process being out of control may be faulty tools, change in properties of materials in the present consignment, breakdown of the lubrication system, faults in speed mechanism etc. Reaching the cause may be difficult if it is subject to the combined effect of several external causes.

If the process is found to be statistical control, a comparison between the required specifications and the process capability may be carried out to determine whether two are compatible. There are three possible alternatives.

1. Re-evaluate the specification to find tolerance really necessary for effective performance or could they be relaxed.

2. If the relaxation of the specification is not acceptable, choose accurate process.

3. If the previous alternative are out of questions a 100 per-cent inspection must be undertaken to sort out the defective products.

Example

S. No.	Date	(A) Number of pieces inspected	(B) Number of defectives found	Fraction defective $p = B/A$
1.	1st January	400	30	0·075
2.	2nd January	400	25	0·0625
3.	3rd January	400	30	0·075
4.	4th January	400	35	0·0875
5.	5th January	400	40	0·10
6.	6th January	400	45	0·112
7.	7th January	400	35	0·0875
8.	8th January	400	45	0·112
9.	9th January	400	35	0·0875
10.	10th January	400	40	0·10
Total	10 Days	4000	360	

Upper control limit (U.C.L.) $= \bar{P} + 3\sqrt{\dfrac{\overline{P(1-\bar{P})}}{n}}$

Lower control limit (L.C.L.) $= \bar{P} - 3\sqrt{\dfrac{\overline{P(1-\bar{P})}}{n}}$

$$\bar{P} = \frac{\text{Total number of defective}}{\text{Total number inspected}}$$

$$\bar{P} = \frac{360}{4000} = 0.09$$

$n =$ Number of pieces inspected on each day $= 400$

$$\text{U.C.L.} = \bar{P} + 3\sqrt{\frac{\overline{P(1-\bar{P})}}{n}}$$

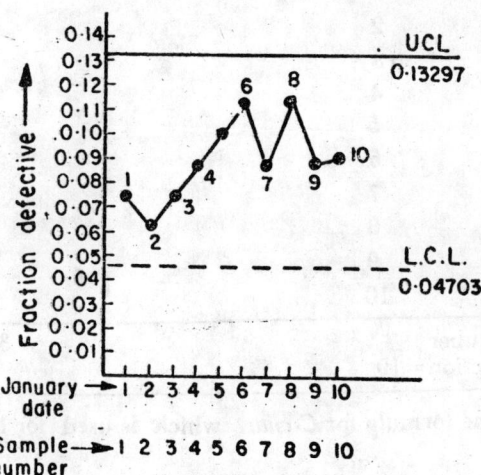

Fig. 9.17

Let us first find out the value of

$$= \sqrt{\frac{\overline{P(1-\bar{P})}}{n}} = \sqrt{\frac{0.09(1-.09)}{400}}$$

$$= \sqrt{\frac{0.09 \times 0.91}{400}} = \frac{1}{20} \times \sqrt{0.0819}$$

$$= \sqrt[3]{\frac{\overline{P}(1-\overline{P})}{n}} = \frac{3}{20} \times \sqrt{0 \ 0819}$$

$$= 0.2865 \times \frac{3}{20} = 0.04297$$

Thus U.C.L. = 0.09 + 0.04297 = 0.13297

L.C.L. = 0.09 − 0.04297 = 0.04703

Let us draw U.C.L. and L.C.L. on the graph paper as per Fig. 24·17.

Problem. In a medium size company carrying out the anodising of various aluminium components inspected the components to locate defects in them. The observations are given below. It is required to plot a *C-chart* and draw the conclusions.

Components inspected	No. of defects found during the inspection
1	2
2	5
3	0
4	5
5	5
6	7
7	2
8	3
9	1
10	7
Total muber of inspection = 10	37 = Total number of defects

The formula for *C-chart* which is used for large and complex parts :

$$U.C.L = \overline{C} + 3\sqrt{\overline{C}} \qquad \ldots(i)$$
$$L.C.L = \overline{C} - 3\sqrt{\overline{C}} \qquad \ldots(ii)$$

$$\overline{C} = \frac{37}{10} = 3.7$$

Putting the values in (*i*) and (*ii*)

$$U.C.L = 3.7 + 3\sqrt{3.7}$$
$$= 9.472$$
$$L.C.L = 3.7 - 3\sqrt{3.7}$$
$$= 2.072 \text{ As the limit is negative.}$$

Let us take it as Zero.

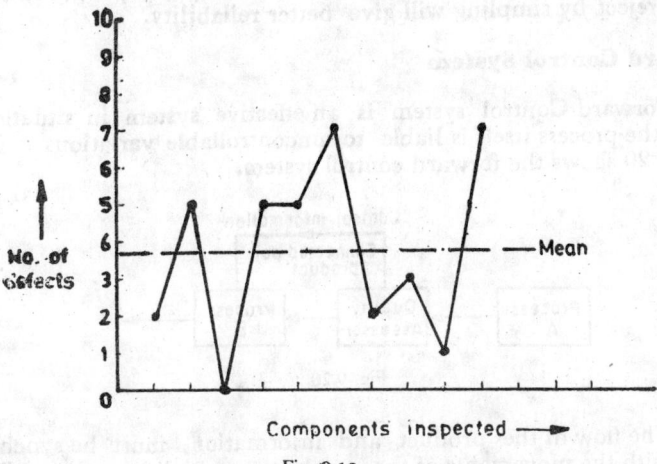

Fig. 9.18

System Concept

Broadly speaking, we have four systems of Controlling the quality of a product :

    (*i*)   Accept—Reject

    (*ii*) Forward—Control

    (*iii*) Backward—Control

    (*iv*) Process—Revision

Accept-Reject System

The accept-reject system of controlling quality involves a quality assessor, who receives the output from Process *A* and sorts out these items that do not fall within his interpretation of the quality specifications. He then passes the rest on to Process *B*. This is shown in Fig. 24·19.

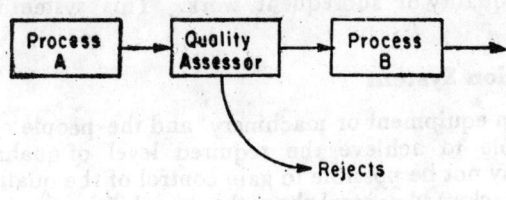

Fig. 9.19

Controlling the quality of product by accept-reject may involve 100% inspecting or it may be achieved by sampling. In fact accept-reject by simpling will give better reliability.

Forward Control System

Forward Control system is an effective system in situations where the process itself is liable to uncontrollable variations. The Fig. 9·20 shows the forward control system.

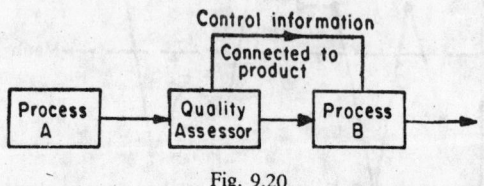

Fig. 9.20

The flow of the product and information, must be synchronised with the movements of the product to which it applies. Process B is adjusted to rectify variations which have been introduced in process A. Food processing is one industry in which forward control can be very effective.

Backward Control System

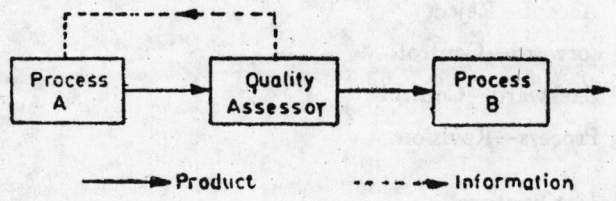

Fig. 9.21

This is the most common type of control. In this the assessment of quality of a particular product is communicated back to the person operating the process. On the basis of this information he is able to decide upon the appropriate action, which will maintain or improve the quality of subsequent work. This system is shown in Fig. 9·21.

Process-Revision System

Production equipment or machinery and the people operating it may be unable to achieve the required level of quality in the product. It may not be possible to gain control of the quality either by forward or backward control then the possibilities remains that process can be changed.

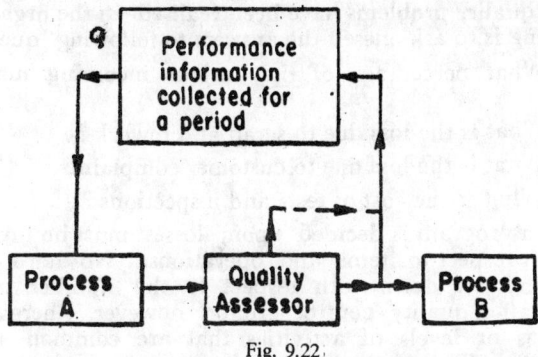

Fig. 9.22

The Present and Future of the Q. C. Movement in India

It was in 1945 that Indian Statistical Institute introduced the First Training Course in S.Q.C. No change or impact was felt immediately. Statisticians had no contact with industrialists who were making large profits were not interested in Quality Control. There was complete lack of large scale participation of this nature, although the need for such involvement was very much because of our backwardness.

In the developed countries the top management regard quality control as one of the vital management functions. Quality control, has in fact, become an integrated part of the organisations. However the Indian enterpreneours are rarely interested in Q.C. because of the seller's market, sheltered by import restrictions.

Quality Reliability

Quality Reliability was the theme of the Asian Productivity Council during a seminar in 1970. In India, as in several other countries of the Asian region, it has been a tremendous task to create an awareness among management of the importance of quality. It is estimated that about 56000 factories are existing in the organised sector. Out of this, about 15,000 are large scale establishments. Out of these perhaps 10% of the factories have been directly or indirectly covered by the SQC training courses. This, of course, has touched only the fringe of the problem. Still lot many efforts are needed to introduce quality consciousness. The productivity councils with its local productivity councils are also heading towards quality consiousness.

Organizing for Quality Control

No programme can be expected to go very far without good leadership. Managerial competence is a definite need in planning a program. The person who is to head a program should have proven ability to take hold of a new project and organize an effective program.

After quality problems have been realised in the organisation, the next thing is to ask oneself the answers to following questions.

(1) What percentage of the various incoming materials is defective ?

(2) What is the loss due to scrap and rework ?

(3) What is the loss due to customer complaints ?

(4) What is the cost of tests and inspections ?

Once a program is decided upon, losses must be further pinpointed with respect to items and operations. No hard and fast rules can be set down with respect to the number of people needed to staff a quality control group ; however, thereare some general types or levels of activities that are common to most organisations. These may be concentrated in a few people or spread over many depending on the size of the company and the nature of the quality problems involved. However, the quality control involves the following activities.

(i) It is the management activity.

(ii) It is the activity of planning.

(iii) It is the activity of gathering data.

(iv) It is the activity of recording and handling the data collected.

Quality Control Engineer

John Lawrence has suggested eight of the most important abilities which the quality control engineer may be expected to possess.

1. Ability to analyze.
2. Ability to synthesize.
3. Ability to sell techniques to the organization.
4. Ability to deal with customers.
5. Ability to make decisions.
6. Ability to apply statistics to quality control problems.
7. Ability to administer.
8. Ability to deal with problems of cost control.

These abilities will be of varying importance in each of the types of activities outlined above.

Quality Control Committee

Any new programme faces resistance in two aspects namely, acceptance and support. Organizing a quality control committee can do much to help clear these hurdles. The committee should consist of the key people from such groups as inspection, production, engineering and design.

This is the group which is to guide the quality control program, to review quality progress, and to recommend the direction of efforts to improve quality. By participation and cooperation the support will tend to come naturally. As results begin to show themselves the group will accept the program and acquire an understanding of it, without necessarily comprehending involved statistical techniques.

Quality-Mindedness

Quality-mindedness is a prerequisite for quality product. Fortunately, most people want to do good quality work and if given the opportunity will do it, but unfortunately, modern mass production has tended to submerge the natural outlet for pride of workmanship experienced by the worker who dealt directly with his customers. The importance of quality was brought home to manufacturer forcibly by his customers ; poor work bring its penalty quickly. In modern production methods, the workman usually has no direct contact with the consumer. Furthermore, great pressure is often brought to bear upon him to increase the quantity of his output without pressure to increase the quality of his work. Often he is paid by the pieces or given a bonus for producing more than a "standard" amount under an incentive system that counts all units of work produced regardless of quality. It is of little wonder under such circumstances that product quality is so commonly a major problem.

Considerations for Quality

The workman has a right to certain considerations if we expect tim to do good work. Following are the important considerations.

1. Satisfactory physical conditions under which he is to work.
2. Good raw materials
3. The right tools which are properly maintained.
4. Adequate instructions and training for the job.
5. Good supervision.
6. The means to check his own work.
7. A reward for good quality work.

Most important of all, the work should be rewarded for good quality work. If there is no reward, there is no incentive. One of the most basic human needs is the need for recognition. If we fail to recognize good workmanship, can we blame the workman for not trying ? Announce of praise for good work is worth more than a pound of blame for poor work.

Introducing Quality Control

A quality control program that overcomes all other hurdles but that does not gain a high level of shop participation is bound to fail. It is therefore essential that the planning be directed toward gaining the support of workers.

Quality control cannot be installed in a shop like a machine. People accept a machine for more readily than they accept new ideas and new ways of thinking. Quality control committee can do much to facilitate acceptance at other management levels. Meetings with formen must be held to win their acceptance. Resistance to new ideas seems to increases as we go down the line to the shop people. This is primarily because we are getting further from the overall objectives as seen by top management and closer to the specific techniques and details of method as seen by the workman.

Quality Control for a New Product

In fact almost all the products can trace back their origin to customer requirements and their success to customer satisfaction. Any product can be called as 'studied in its totality' only when these two basic facts are constantly remembered.

The various stages of a comprehensive quality control scheme for a new product can be enumerated as below :

(a) to analyse and understand thoroughly the market requirements ;

(b) to design the product meeting each one of these requirements ;

(c) manufacture of proto-type and its rigorous testing agains market requirements ;

(d) conducting trial series to solve tooling problems, if any ;

(e) inspection of components during the process of manufacture, at the pre-designed stages ;

(f) testing of the final product against standard specification ;

(g) packing of the product and its shipment to customer ; and

(h) to evolve a system by which customer satisfaction is ensured.

It is not enough to ensure action at all these stages once for all. Even for survival in the competitive market, a regular vigilance over all these stages is absolutely necessary.

To overcome all these difficulties in time, some system should be established to ensure a regular feed-back of information regarding difficulties and prompt attention to each one of them till they are solved. This is the concept of "Total Quality Control" (TQC).

Organisation for Quality

There is a basic conflict between quality and cost, inasmuch as an emphasis on quality and inspection generally results in increased costs, and so higher selling prices. In such cases some balance between quality and costs has to be struck, and an optimum level arrived at beyond which the cost of achieving quality is higher than the benefits derived therefrom.

Total Quality Control

The concept of total quality control is of comparatively recent origin. It means that control of quality should embrace all the stages of manufacturing programme, from the specification of an article to its final shipment.

While in any organisation quality is the concern of everybody, it is usually the responsibility of none. Therefore a separate 'Quality Control Group' should be established within the plant assigned with the responsibility of implementing the programme.

The quality control group should not duplicate such efforts but work with them to fill the gaps, and make the work more effective from the point of view of the organisation. The person in-charge of the department should preferably be an engineer well-versed in the different operations involved in the manufacturing process from the raw materials to the finished product stage. He should also have formal training in modern statistical quality control techniques. Alternatively, a statistician with specialised training in quality control techniques may head the department. But, in such a case, the person should be given sufficient induction within the company to make him familiar with the different engineering processes involved.

Organising Training Programme

It is obvious that quality is affected by the activities of many groups in an enterprise-design, purchase, planning production, inspection, sales and accounts all are concerned with quality and quality control. In fact, implementation of quality control program-mes generally meet with substantial opposition from staff at different levels. Everyone in the organisation should, therefore, be made to feel that 'Quality Control is a helpful policeman and not an unsym-pathetic detective'. Therefore, the first step should be to organise In-Plant training programmes at different levels such as :

(a) Managemet appreciation programmes
(b) Supervisory programmes on quality control techniques
(c) Operative training on methods and reading of charts etc.

The appreciation programmes would help the executives in developing an insight into the techno-statistical aspects of quality, control, enabling them to understand their role in building quality. The intensive programmes for supervisors and operative staff are expected to equip the participants with basic practical skills and thus develop in them the ability to handle quality control problems on a rational basis and analyse them in a scientific manner.

The In-Plant training schemes noted above may be supple-mented by more formal training outside the organisation. There are institutions in the country such as the Indian Statistical Institute, the Indian Society of Quality Control, which are engaged in impart-ing training in the application of statistical quality control techni-ques.

As regards the position of the Quality Control Department in the organisational structure of a company, it would seem that no single set up may be expected to suit all industries. There is a school of thought which considers that the head of the quality control department should be made directly responsible to top management.

It should be appreciated, quality control functions are partly managerial, partly technical and partly statistical. Unless these three aspects are properly blended, quality control techniques will rarely yield the results that are expected of them.

Quality Control and Export

Till recently the export earnings of the country came primarily from export of traditional items such as cotton, jute and leather manufactures, tea, iron-ore, oilcakes, etc. These items still dominate the scene. But in recent years, engineering items have been emerging as potential earners of foreign exchange. The value of engineering exports rose from 3·41 crores in 1956-57 to Rs. 18·90 crores in 1965-66 and to Rs. 84·97 crores in 1968-69. The export target for engineering products in 1969-70 has been set at Rs. 110 crores, and it rose to Rs. 250 crores in 1973-74 which has further increased four folds.

The country is now exporting heavy and sophisticated products involving a high degree of technical skill and workmanship. In the export markets, Indian manufacturers are competing with advanced countries who have years of experience and goodwill, and have already succeeded in securing some valuable contracts against international competition for railway wagons, coaches and track materials, steel pipes and tubes, cranes and hoists, transmission line towers, power-station equipment, cables and conductors etc. The industry is fully equipped not only for exports of industrial plant and machinery, but also for the consultancy and execution of projects for cement, sugar, textile, chemical, pharmaceutical and food processing plants.

To compete successfully in the foreign markets with manufacturers from the highly industrialised countries the Indian manufacture have to build up confidence in the foreign buyers that :

(a) country is in a position to make and deliver products of high degree of engineering skill,

(b) quality of the products is comparable with those from other countries,

(c) prices of the products are competitive.

Quality Control and Value

The challenge to the success of quality control in all its implications is to make "quality" synonymous with "value". It requires an approach that carefully bridges the gap between customer satisfaction and business economics. Many industrial managers have started realising that quality control can be a valuable asset. Quality control, should be capable of causing positive management action and should be dedicated to the theory of prevention, and should be designed not only for profit but also for customer satisfaction.

The essential ingredient that must permeate the entire quality control activity is excellence in management. If quality control is to be capable of constructive involvement in management problems, it must earn respect for its own ability to manage.

We know that industrial worker is not intimately aware of how his skill interacts with others, he can never be sure that he is fulfilling his responsibility. To be knowledgeable is a necessary and important starting point. The second element is the "strength of your own convictions." It means much more : being action-oriented, always taking a clear, timely, and decisive position and giving employees positive support and displaying a willingness to commit oneself to action. The third element of excellence in management is a demand for performance. In quality control, enforcement must be a way of life, both within the quality control organization itself and in those other areas where it is responsible for ensuring control. Knowledge, strength, and a demand for performance therefore are the keys to the initial establishment of excellence in management. The next elements are supplementary to this excellence which are :

(*i*) A clear definition of responsibility.

(*ii*) Discipline.

(*iii*) Development of people (*iv*) Creativity (*v*) Loyalty

(*vi*) Willingness to accept responsibility, to take risks, to accept change, and to learn from experience.

Planning for Quality

Quality planning is the management portion of the total quality control effort, responsible for formulating the policies, procedures, techniques, and detailed plans necessary to achieve all quality control objectives. There are three ways by which total quality costs can be reduced to a minimum by the quality control organization ; these are best envisaged as the three basic objectives of quality planning.

(*i*) to make maximum use of available quality control effort.

(*ii*) to exert maximum influence on the planning of others in an all-out effort to make the control of quality.

(*iii*) to review the results of the total efforts periodically and present them to management in such a way that improvements can be forthcoming.

The evidence of the work of quality planning can be found in inspection plans, in documentation systems, in quality status reports and in procedures for the control of quality. The attitude of other planning groups with respect to the importance of quality control is also very important.

A planned inspection is essentially a trade-off between protecting from errors and risking failure to find errors. Once the parameters to be inspected have been defined, the step in planning is to determine where to perform the inspections. From a manufacturing viewpoint, inspections should be performed at the earliest possible point in the progress.

The last step in planning is to determine how much inspection to perform. Certain critical parameters such as material condition for aircraft structural parts will always require 100 per cent inspection and verification. Planning should be specifically aimed at using today's knowledge to prevent, or at least to reduce, the production of scrap and rework. An equally important ingredient for each inspection plan is the documentation of results.

This device-known as quality log-book. A log-book is the accumulation into one large package of specific pertinent data combined with information on the status of supplementary data physicallay field elsewhere. The use of log-books has simplified the overall quality planning effort and has improved inspector's ability to control total equipment status on a current basis. Without log books as they have been developed, the task would be far more cumbersome and costly, and timeliness would be almost impossible. Inspection planning is not limited to manufacturing and test areas. Quality planning must develop a close working relationship with the broadest possible spectrum of company activity. The magnitude and importance of the total quality control effort make cooperation between quality planning and other company planning groups mandatory. A technical library is necessary, to develop the quality control personnel. Quality planning must be dedicated to the eduction_r of paper-work yet at the same time be trained to take reach di ective seriously. Quality planning, therefore, provides the principal liaison with the customer and has the first opportunity as well as the obligation to develop good customer relations.

The principal objective of quality planning is to create and to develop business and management systems designed to make the quality control function a respected, useful and effective area of management. This objective is accomplished through inspection plans, training, data systems, manual of operations, introducing new techniques, problem-prevention, liasion with associates and customers, and sound administration of the purely business aspects of quality control.

Quality Engineering

The caliber of engineering talent found in a quality control organisation to a large extent determines the quality of the technical contributions that should be expected. When quality control management, attaches a sound business importance to the responsibilities of quality engineering, only than the quality control be able to complete for the best engineering talent available and thus be able to achieve or exceed the required level of technical competence.

The purpose of quality engineering are described as follows :

(1) To bridge the gap between the technology of quality control and the specific technologies of the company's business.

(2) To create a desirable effect. To provide the necessary engineering base that allows the fruits of quality control to achieve the maximum desirable effect on product quality. Quality engineering customarily embodies the groups associated with some of the subject matter already discussed, such as quality planning and corrective action.

Quality Engineering's Responsibility

Quality engineering's responsibility is to support laboratories, such as calibration and materials testing. The minimum role of quality engineering should be understand thoroughly the laboratory function and its potential effect on acceptance. At the same time, quality engineering will again have an opportunity to prevent problems and to improve overall effectiveness through its technical contributions to the laboratory function. The establishment of a failure analysis laboratory thus becomes an important ingredient in the quality process.

Other developments in engineering have led to an entirely new concept known as "systems effectiveness". This represent an attempt to combine reliability, maintainability, human factors, value engineering and systems safety into one composite discipline aimed at improving overall design from these many points of view simultaneously. An overall approach to reliability should be considered most valid when it gives engineering the disciplines and tasks whose purpose it is to improve the inherent reliability of design and gives quality control those which are aimed at maintaining inherent design reliability throughout manufacturing and test cycles.

The most important function of quality engineering is the corrective action. If the control of quality is to be dynamic and it must be dynamic to be effective-corrective action must be the heart. Quality engineering must keep abreast of all technical developments that can aid its cause and allow them to make creative contributions to overall company performance. There will always be a need for new or improved technical controls for which quality engineering has a leading role to play.

Zero Defects (Z.D.)

Zero defects programs have attracted much credit and considerable controversy to the quality control profession in recent years. Zero defects is a motivation program having its concept of working which is beneficial to groups but particularly to the individual. Originating in the aerospace industry during 1961-62, it grew out of the many perplexing problems facing the areospace industry in those days ZD holds that people are conditioned throughout their lives to accept the fact that they are not perfect and will, therefore, make mistakes. By the time they begin working in industry this belief has become firmly rooted and finds its normal expression in the statement : "People are human and thus make mistakes. Nothing can ever be perfect so long as people take part in it." People make mistakes and will keep doing so. Zero defects can challenge each individual to establish a personal goal of superior performance and to strive conscientiously for personal excellence in everything he does. The crux of the challenge is to achieve perfect results by preventing and eliminating defects.

Errors are measures of the importance a man places on things ; he takes better care of certain things than he does of there. In short, man has developed a dual attitude. In some things he is willing to accept errors ; in others he is not. Mistakes have two causes : lack of knowledge and lack of attention. Knowledge can be measured, and shortcomings corrected. through established methods. Lack of attention is an attitude problem ; it must be corrected by the man himself—who must make a conscious effort to do things righ the first time. The most significant feature of zero defects, is the opportunity it affords every employee—from factory floor and office upto top management—to participate in the program and to develop self-improvement. This is the method by which it proposes progressively to achieve overall company improvement. It may be that any success realized to date has been concentrated for the most part at the working shop floor level of the factory. There is little indication of sincere effort on the part of engineering and management to identify and eliminate problems for which they are responsible.

To the majority of quality organizations that have been involved with it. The zero defects probably has come to the organisation both as a boon and a headache. The benefits have flowed from renewed opportunities to discuss with management the configuration of the human error problem and the need for creative new methods of prevention. The headache is associated wholly with implementation. Zero defects has had major success in at least one area. It has captured the promotional talents of many people. Among the methods used for sustaining the ZD program are posters, slogans letters to employees, etc. While all of these have some beneficial value, they are not in themselves sufficient. People need to feel personal satisfaction and receive personal recognition in order to be continually motivated. Consequently, activities involving performance measurement and personal or group awards have been most effective in promoting superior effort.

Objective of Zero Defects

The ZD goals include everything from percentage of discrepancies to employees' lost time. It is obvious that in some programs goals are established merely for the sake of having goals. When performance can be measured in terms of defects, charts based on motivating goals can be most effective. In areas where performance parameters appear to be intangible, zero defects should ask How does one measure performance ? What constitutes improvement in this area. The two basic objectives of zero defects are : It guarantee improvement through the identification and removal of the causes of errors, and it allows every emplopee in the company to participate in the programme freely and effectively. Management can best support the zero defects programme by indicating a willingness to give Zero Defects as much consideration as it gives to the problems that bother its customers.

Statistical design of experiments is a proven technique that continues to show increasing use in the industries. As the Research and Development function comes under increasing pressure to produce fast, accurate results, more scientists and engineers have been using the experimental design.

Benefits of Design of Experiments

The benefits of design of experiments are :

(*i*) It can give more information per experiment than unplanned approaches.

(*ii*) It reduces the lead time of the experimentation.

(*iii*) It improves the efficiency of the experimentation particularly, when many variables are of potential importance.

(*iv*) It is a planned approach for the collection of the data and analysis of information.

(*v*) Quite often the collections from a statistically design experiments are self-explanatory without even statistical analysis.

(*vi*) The reliability of information gathered is increased in the light of experimental and analytical variation.

(*vii*) It has the capability to screen out the interactions among experimental variables. Even the negative interactions can be well seen, in contrast against the belief that experimental variables have an aditive effect.

Stages of Investigations

Process investigations pass through a well define series or stages, whether they are bench studies, pilot plant studies, plant trials. Each study presents its own objectives and difficulties as follows.

(*i*) **Familiarization.** In this stage, the experimenter will be acquainting with the system and trying to duplicate some results or experiments from a prior study etc. Statistical design is not so useful at this stage.

(*ii*) **Variable screening.** Once the objects of the investigation are well set, the next stage is to screen the effective variables from large number of variables which are normally thought of as governing and process. Statistical design is quite useful at this stage, without which the experimenter will end up in large number of variables and experiments and one cannot come to any conclusion.

(*iii*) **Optimisation.** Having screened the effective number of variables that governs the process, the next step is to optimise the number of variables with respect to their levels which can result in the manufacture of best product at the lowest cost, energy, effluent etc. Statistical design of experiments is a powerful tool at this stage.

Principles of Experimental Design

(*i*) The object of any industry is to improve quality and reduce cost per unit. Since, this involves the change of factors that affect the process at the plant level, which is quite costlier, the experimentation becomes a must, either at the *Laboratory* scale or at the pilot plant scale.

(*ii*) The variables that governs the process like temperature, pressure, speed, hardness, etc., are called factors. The specific values of factors are known as levels of the factors. A single factor or a combination of levels of different factors is called a *Treatment*. The numerical results of the experiments made with a particular treatment is known as the *Response* corresponding to the treatment tested.

(*iii*) An estimate of the effect of a single factor obtained independently of the other factors involved in the experiment is called the *Main Effect*. If the effect of one factor is not the same at a different levels of another factor, there exists *Interaction* between the two factors.

(*iv*) It is well known that results of no two experiments will be incomplete agreement despite every effort to maintain the same conditions. This is due to a very large number of factors which are beyond economic control. These differences, *Experimental Errors* introduce a degree of uncertainty into any conclusions that may be drawn from the results of any experiment.

(*v*) If the difference between the effects of treatments are quite marked compared to the experimental error, the results would still show up these differences easily. If however the influence of extraneous factors are sufficiently large, the true effects may be masked.

(*vi*) The experimental errors can be kept in check by following three cardinal principles of experimentation *viz.* Randomisation ; Replication and Local control (*vii*) Randomisation consists of scheduling the experiments with all the different treatments in a random order. It helps to avoid any possible systematic bias so that a particular treatment is not continually favoured. Another, method of avoiding, bias due to a particular treatment being favoured is to have all treatment replicated under different conditions the same number of times. This is known as *Balancing*.

(*viii*) Replication or repetition of observations helps us to estimate the experimental errors which in turn helps to carry out tests of significance, that is, to decide if the observed differences are due to specific treatment effects or only chance. Also such replication increases the *Sensitivity* of the experiment.

(*ix*) The classical method or experimentation is to restrict attention to one factor at a time, the other factors being kept constant. This fails to show interaction effect that may exist between some of the factors consequenc on which optimum combinations are difficult to determined. A single integrated design which permits variation of more than one factor at a time would allow, determination of interaction effects as well as provide more information effects as well as provide more information on main effects.

(*x*) A properly designed experiment will have sufficient replication and local control to minimise experimental errors and randomization to give validity for the results, obtained. Statistical methods are used in the choice of appropriate designs in analysing the data.

(*xi*) To get maximum sensitivity, it is necessary that different treatments are subjected to the same back ground conditions to the extent possible. But inpractice it is not possible due to natural variability of material and the process involved. But it is possible to split a set of experiments into smaller groups with in which such variations are likely to be less than in the set as a whole. This is known as Local Control.

DESIGN OF FACTORIAL EXPERIMENTS

Whenever delebrate changes are made in the experimental conditions to study the variaton of responses, the use of factorial design of experiment is utmost satisfactory. A considerable advantages is gained, if the experiment is so designed that the effect of changing any one variable can be assessed independently of the others. One way of achieving thi object is to decide on a set of levels for each of the factors to be sudied and to carry out one or more trials of the process with each of the possible combinations of the levels of the factors. Such an experiment is called *Factorial Experiment*. This will enable the experimenter to find out the main effects as well as interactions.

Number of Experiments Required

Number of experiments required to be conducted depends upon the number of factors to be studied and the number of levels of each factor to be included in the experimentation. The gerneral formulae for arriving at the number of experiments required is

$$N = (L)^F$$

where
 N = Number of experimental runs required
 F = Number of factors or variables to be studied
 L = Number of levels of each factor to be studied.

Number of Experiments for two Factors with two Levels

$$F = 2,\ L \pm 2$$
$$N = (L)^F = (2)^2 = 4$$

A factor is denoted by a capital letter A, B, C etc.

The lower level of the factor is denoted (I) and higher level of the factors is denoted by the corresponding small letter. For example if there are two factors A and B, their levels are denoted as below.

Table : $(2)^2$ Factorial Design

| S. No. | Treatment | Levels of Factors | |
		A	B
1.	I	—	—
2.	a	+	—
3.	b	—	+
4.	ab	+	+

The notation in the second column represents the treatment combinations. Presence of a small letter means that the factor it represents is at its higher level, while absence of a letter means that the factor stands at its lower level.

Treatment (I) represents that both factors A and B are at lower level.

Treatment (a) refers that factor A is at higher level and factor B is at bwer level.

Treatment (b) represents that factor B is at higher level and factor A is at bwer level.

Treatment (ab) represents that both the factors A and B are at higher level and so on.

Under the column, levels of factors (–) represents the lower level of the factor and (+) represent the higher level of the factor.

In the treatment column (I) and levels column (−) represent the lower level or the absence of the factor. Similarly (a) in the treatment column and (−) in the column of levels represent the higher level of the factor or the presence of the factor.

Example. The experiments are required to be conducted at 2 levels of each factor say pressure and temperature. The lower and higher levels of pressure are 5 kg/cm sq and 10 kg/cm sq respectively where as the lower and higher levels of temperature are 100°C and 150°C as shown below.

Factors	Levels of factors	
	(−) Lower level	(+) Higher level
Pressure P.	5 kg/cm²	10 kg/cm²
Temperature T	100°C	150°C

Since two factors with two levels need four experimental runs under factorial designs, the factorial design for the two factors is given below.

Table : Factorial Design

S. No.	Treatment	Levels of factors		Response R
		Pressure	Temperature	
1.	I	5 kg	100°C	
2.	p	10 kg	100°C	
3.	t	5 kg	150°C	
4.	pt	10 kg	150°C	

Once the experiments are conducted against each treatment the corresponding results or responses are tabulated in the column R and the effect of each factor on the result R can be calculated as shown below.

Table : Calculation of Effect

Sl. No.	Treatment	Levels of factors		Response R
		P	T	
1.	I	5 kg	100°C	60
2.	p	10 kg	100°C	63
3.	t	5 kg	150°C	68
4.	pt	10 kg	150°C	75

Main Effects :

R2−R1=63−60=3 Effect due to variable p when variable T is at lower level

R4−R3=75−68=7 Effect due to variable p when variable T is at higher level

Average effect of P on the response R ; $E_P = \dfrac{3+7}{2} = 5$

R3−R1=68−60=8 change due to factor T when factor P is at higher level.

R4−R2=75−63=12 change due to factor T when factor P is at higher level.

Average effect of T on response R ; $E_T = \dfrac{8+12}{2} = 10$

Interaction :

The average difference in effect due to P when T is at higher level.

R4−R3=75−68=7 effect of P when T is at higher level

R2−R1=63−60=3 effect of P when T is at lower level

Average difference $= \dfrac{7-3}{2} = \dfrac{4}{2} = 2$

Interaction effect=2. Thus after the main effect of P and T and interaction of P and T on the response R are calculated, these effects need to be screened out as to whether the effects of P and T and interaction of PT are significant on R or not. If they are affecting singificantly with what confidence level they are affecting on R. These will be dealt subsequently.

Factorial Designs for three Factors with two Levels

$$F=3 \qquad L=2$$
$$N=(b)^F = (2)^3 = 8$$

Three factors with two levels need eight experimental trials, Let three factors be A, B and C.

Table $(2)^3$ Factorial Design

Sl. No.	Treatment	Levels of Factors		
		A	B	C
1.	I	−	−	−
2.	a	+	−	−
3.	b	−	+	−
4.	ab	+	+	−
5.	c	−	−	+
6.	ac	+	−	+
7.	bc	−	+	+
8.	abc	+	+	+

Treatment (I) refers that all the three factors A, B and C are at lower level (−)

Treatment (a) represents that factor A is at higher level (+) while factors B and C are at lower level (−)

Treatment (ab) represents that both the factors A and B are at higher level (+) while factor C is at lower level (−)

Treatment (abc) represents that all the three ractors are at higher level (+) and so on.

Example. Following are the three factors with two levels, with which eight experiments were conducted by suing factorial design.

Sl. No.	Factors	Level of factors	
		Lower level (−)	Higher level (+)
1.	Concentration : A	5%	10%
2.	Temperature : B	20°C	30°C
3.	PH : C	4	6

The factorial design and the experimental results or responses are given below.

Table : Experimental Results

S. No.	Treat- ment	Levels of factors			Response	$R-60$
		A	B	C	R	
1.	I	5%	20°C	4	63 6	3·6
2.	a	10%	20°C	4	65 0	5·0
3.	b	5%	30°C	4	61·0	1·0
4.	ab	10%	30°C	4	62·0	2·0
5.	c	5%	20°C	6	67·0	7·0
6.	ac	10%	20°C	6	72·2	12·2
7.	bc	5%	30°C	6	61·6	1·6
8.	abc	10%	30°C	6	63·0	3·0

Calculation of Effect of Variables on Response

Main effect : To calculate the main effect of the factor A, the responses corresponding to all the treatments containing a (i.e. at higher level of factors) and all those not containing a (i.e. at lower levels factor A) are averaged and the difference between the two averages is found out. From the above table.

Average response at higher level of A
$$=\tfrac{1}{4}\,(65·0+62·0+72·2+63·0)=65·55$$

Average response at lower level of A
$$=\tfrac{1}{4}(66·3+61·0+67+61·6)=63·30$$

Main effect of $A=65·55-63·30=2·25$

Similarly the main effects of factors B and C are calculated.

Main effect of $B = -5\cdot05$

Main effect of $C = 3\cdot05$

Algebraic Expression for Main Effects

Average response at higher level of $A = \frac{1}{4}(a+ab+ac+abc)$

Average response at lower level of $A = \frac{1}{4}(I+b+C+bc)$

Main effect of $A = \frac{1}{4}[(a+ab+ac+abc)-(I+b+c+bc)]$

By treating (I), a, b, c as algebraic symbols, the above expression can further be simplified

Main effect of $A = \frac{1}{4}[(a+ab+ac+abc-I-b-c-bc)]$

$$4A = -I+a-b+ab-c+ac-bc+abc \qquad \ldots(1)$$

Similarly main effect of B

$$= \frac{1}{4}[(b+ab+bc+abc)-(I+a+ac+c)]$$

$$4B = -I-a+b+ab-c-ac+bc+abc \qquad \ldots(2)$$

Main effect of $C = \frac{1}{4}[(c+ac+abc+bc)-(I+a+ab+b)]$

$$4C = -I-a-b-ab+c+ac+bc+abc \qquad \ldots(3)$$

Interactions

The interaction between the factors A and B is represented by AB. The interaction is one half of the difference between the effect of A when B is at the higher level, and the effect of A when B is at the lower level.

$$AB = \frac{1}{2}\left\{\frac{1}{2}\left[\underbrace{\frac{(abc+ab)-(bc+b)}{\text{effect of } A \text{ with } B \text{ at higher level}}}\right] - \frac{1}{2}\left[\underbrace{\frac{(ac+a)-(c+I)}{\text{effect of } A \text{ with } B \text{ at lower level}}}\right]\right\}$$

$$4AB = I-a-b+ab+c-ac-bc+abc \qquad \ldots(4)$$

$$BC = \frac{1}{2}\{\frac{1}{2}[(abc+bc)-(ac+c)]-\frac{1}{2}[(ba+b)-(a+I)]\}$$

$$4BC = I+a-b-ab-c-ac+bc+abc \qquad \ldots(5)$$

Similarly $\quad 4AC = I-a+b-ab-c+ad-bc+abc \qquad \ldots(6)$

and $\qquad\quad 4ABC = -I+a+b-ab+c-ac-bc+abc \qquad \ldots(7)$

Calculation of Main Effects and Interactions from the Observations (R) of the Example with Algebraic Expression

Since the calculation of effects and interactions is cumbersome with the method of *average* response—an easy method with algebraic expression can be used.

Format for Writing Treatment Combinations and Effects

In all the above calculations of the effects it can be seen that in each calculations four positive and four negative signs are found and only the order in which they appear differ for each effect.

There is an easy method in which the treatment combinations and effects can be written. For one factor with two levels, we can simply write the treatment combinations as (I) and a. For two factors with two levels, add b, ab derived by multiplying the first two (I, a) by the additional letter b. For three factors add c, ac, bc, abc, derived by multiplying the first four (I, ab, ab) the additional factor c and so on.

Table : Calculation of Effects

Sl. No.	1	2	3	4	5	6	7	8	
Treatment	I	a	b	ab	c	ac	bc	abc	
Response R	63·6	65·0	61·0	62·0	67·0	72·2	61·6	63·0	
$R-60$	3·6	5·0	1·0	2·0	7·0	12·2	1·6	3·0	
Effect of 4 A	-1	$+a$	$-b$	$+ab$	$-c$	$+ac$	$-bc$	$+abc$	
	$-3·6$	$+5·0$	$-1·0$	$+2·0$	$-7·0$	$+12·2$	$-1·6$	$+3·0$	$4A=9$ $A=2·25$
Effects of 4 B	$-I$	$-a$	$+b$	$+ab$	$-c$	$-ac$	$+bc$	$+abc$	
	$-3·6$	$-5·0$	$+1·0$	$+2·0$	$-7·0$	$-12·2$	$+1·6$	$+3·0$	$4B=-20·2$ $B=-5·05$
Effect of 4 C	$-I$	$-a$	$-b$	$-ab$	$+c$	$+ac$	$+bc$	$+abc$	
	$-3·6$	$-5·0$	$-1·0$	$-2·0$	$+7·0$	$+12·2$	$+1·6$	$+3·0$	$4C=12·2$ $C=3·05$
Effect of 4 AB	$+I$	$-a$	$-b$	$+ab$	$+c$	$-ac$	$-bc$	$+abc$	
	$+3·6$	$-5·0$	$-1·0$	$+2·0$	$+7·0$	$-12·2$	$-1·6$	$+3·0$	$4AB=-4·2$ $AB=-1·05$

Sl. No.	1	2	3	4	5	6	7	8	
Effect of 4 BC	$+I$	$+a$	$-b$	$-ab$	$-c$	$-ac$	$+bc$	$+abc$	
	$+3·6$	$+5·0$	$-1·0$	$-2·0$	$-7·0$	$-12·2$	$+1·6$	$+3·0$	$4BC=9$ $BC-2·25$
Effect of 4 AC	$+I$	$-a$	$+b$	$-ab$	$-c$	$+ac$	$-bc$	$+abc$	
	$+3·6$	$-5·0$	$+1·0$	$-2·0$	$-7·0$	$+12·2$	$-1·6$	$+3·0$	$4AC=4·2$ $AC=1·05$
Effect of 4 ABC	$-I$	$+a$	$+b$	$-ab$	$+c$	$-ac$	$-bc$	$+abc$	
	$-3·6$	$+5·0$	$+1·0$	$-2·0$	$+7·0$	$-12·2$	$-1·6$	$+3·0$	$4ABC=-3·4$ $ABC=-0.85$

Thus, for three factors the standard order can be written as
(I), a, b, ab, c, ac, bc, abc.

The equation (1) for the effect of A by deleting the multiplier $\frac{1}{4}$, can be arranged in the following standard order

$$-I, +a-b+ab-c+ac-bc+abc$$

By retaining the standard order, the successive signs can be written as

$$A = - + - + - + - +$$

Similarly from the equations (2) to (7) the successive signs can be written and rearranged as shown in the following table.

Table of Signs for Calculating Effects in $(2)^3$ Design

Treat-ment	Total	Effect						
		A	B	AB	C	AC	BC	ABC
I	+	−	−	+	−	+	+	−
a	+	+	−	−	−	−	+	+
b	+	−	+	−	−	+	−	+
ab	+	+	+	+	−	−	−	−
c	+	−	−	+	+	−	−	+
ac	+	+	−	−	+	+	−	−
bc	+	−	+	−	+	−	+	−
abc	+	+	+	+	+	+	+	+

The above table will enable

(i) To know the levels of the factors in each treatment

(ii) To use the signs to the responses in calculating the effects and interactions.

The signs for the treatment combinations for the main effects are obvious. In the expression (1) for the effect of A it can be seen that all treatment combinations containing 'a' have the plus sign (+) and all not containing 'a' the minus sign (−). Same is the case with all main effects.

The signs for any interaction are equal to those obtained by multiplying together the signs for the main effects corresponding to the letters in the interaction. For example, the signs for B and C corresponding to the treatment 'a' are both minus (−) and their product is plus (+) which is in agreement with the sign for interaction BC in the above table. The same holds good for all interactions in generating the signs and hence the most convenient method.

Thus the effects of each factor and their interactions on the response R can be quantified with the above method. Each time an experiment is conducted even under the identical conditions of the experimentation, the variations in the responses are observed due to experimental error. Therefore, it is essential to know the variation in response and hence the effect of factors thus calculated is really significant in comparison with the experimental error. If the effect is insignificant the corresponding factor can be deleted subsequently in the experimentation, on the other hand the significant factors needs to be retained in further experimentation. Hence the significance test based on analysis of variance is required for this purpose.

10

Product Development
and Product Planning

Research

Research unlocks nature's secret. Research has to do with learning nature's laws but industry has to put nature's laws to work and this is where development and design comes in. It is as a result of research, development and design the technologies went on fast changing of late, and a number of products have become absololite. To day we find a number of products as substitutes and even with cheaper in price, better in their functional values. Plastics have replaced may metals excepting with one property that they can not withstand temperatures of high ranges. Particularly research in petroleum has opened up a wide range of products which have revolutionised the way of living itself. The technological innovations from low grade technologies to highly sophisticated technologies have started influencing the socio economic conditions at a rapid rate now. Because of lower pay offs, todays research work needs more careful management than it did in the past. New scientific findings come on so fast that scientists and engineers also find themselves obsolete. For every 5 to 10 years the new sciences Technologies have been developing and now it develops at a faster rate.

Purpose of Research

Generally the industrial research is to try to :

(i) Search for basic chemical or physical relationships, of products and processes.

(ii) Improve their products.

(iii) Find new user for their present products.

(iv) Develop new products.

(v) Reduce the cost of present products.

(vi) Develop tests and {specifications for products and mate- rials.

(vii) Analyze competitors products.

(viii) Find profitable uses of by-products.

Types of Research

These are two types of research (*i*) Basic or pure research (*ii*) Applied research. Pure research deals with nature's basic laws, regardless of law that knowledge can be used. This type of research is usually done at universities and the institutes. Applied research is concerned with solving a problem and usually one that now costs money and will make more money if it is solved. Many industries try to do both type of research but only 10 to 20 percent of resources are spent an basic or pure research. The distinction between pure and applied research will fade away as the time advances, since what is pure research to day becomes an applied one once it finds the application.

Development

Product development changes the products' design, the discoveries of the research may have to be incorporated into the product or into the manufacturing process. New ideas need to be tried out. Ideas stem out from day to day back and forth exchange of information between engineering, sales and production.

Product-Policy Decisions

Product-policy decisions are shaped by a multitude of factors, including the basic nature of the firm's business, its objectives, its resources and its opportunities. Product can be defined as a bundle of physical, service and symbolic particulars expected to yield satisfaction or benefits to the buyer. Today most companies are multi product organisations. Whether large or small, whether in manufacturing, wholesales or retailing, a company generally handles a multitude of products and product varieties. Product policy decisions are made at three different levels of products aggregation (*i*) Product item (*ii*) Product line and (*iii*) Product mix.

Product Item. A specific version of a product that has a separate designation in the seller's list.

Product Line. A group of products that are closely relates either because they satisfy a class of need, are used together, are sold to the same customer groups ; are {marketed through the same types of outlets or fall within given price ranges.

Product Mix. The composite of products offered for sale by a firm or a business unit.

Product mix has three concepts (*i*) Width (*iii*) Depth and (*iii*) Consistency.

The width of the product mix refers to how many different product lines are found within the company. Through increasing in the width of the product mix, the company hopes to capitalise on ood reputation and skills in present markets.

Through increase in the depth of its product mix, the company hopes to entice the patronage of buyers of widely differing tastes and needs. The depth of the product mix refers to the average number of items offered by the company within each product line. The consistency of the product mix refers to how closely related the various product lines are in end use, production requirements, distribution channels or in some other way. Through increase in the consistency of its product mix, the company hopes to acquire an unparalleled reputation in a particular area of endeavor.

Product Life Cycle

As time elapses the product mix of a company will undoubtedly change, new items and lines will be added and old ones dropped. This is because current company products cannot hold their market positions indefinitely. The life time sales of many branded products reveal a typical pattern of development, known as the product life cycle. The five stages can be distinguished as :

(a) Introduction : The product is put on the market. Awareness and acceptance are minimal.

(b) Growth : The product begins to make rapid sales, gains because of the cumulative effects of introductory promotion, distribution and word-of-mouth influence.

(c) Maturity : Sales growth continues but at a declining rate because of the diminishing number of potential customers who remain unaware of the product or who have taken no action.

(d) Saturation : Sales reach and remain on a plateau marked by the level of replacement demand.

(e) Decline : Sales begin to diminish absolutely as the product is gradually edged out by better products or substitutes.

The length of the product life cycle is governed by the rate of technical change, the rate of market acceptance and the case of competitive entry.

Important Phenomena of Product Life Cycle

(i) Products have a limited life. They are boring at some point, may or may not pass through a strong growth phase and eventually degenerate or disappear.

(ii) Product profits tend to follow a predictable course through the life cycle. Profits are absent in the introductory stage, tend to increase substantially in the growth stage, slow down and then stabilize in the maturity and saturation stages and all but disappear in the decline stage.

(iii) Products require a different marketing programme in each stage and management must be prepared to shift the relative levels and emphasis given to price, advertisement. Product improvement and other marketing elements during different stages in the product life cycle.

Product Mix Decisions Stability

Many managements adopt the objective of reasonably stable sales from period to period. High sales variability can be quite costly to the firm. It means that the company has to invest in facilities for peak demand or carry higher inventories to meet peak demand ; hence to pay more interest and greater risk. This increases the number of wrong decisions in forward planning because sales are more difficult to forecast.

Suppose, the company's product mix generates highly variable sales and four different products are under review as possible additions to the product mix. The first product promises highly stable sales. The second product is unrelated to the current products and would produce sales subject to about the same amount of variability as now exists. The third product's sales are expected to show high positive correlation with present total sales. The fourth products sales are likely to show high negative correlation with present total sales. Since the product's sales are highly stable, the addition of this product would not alter the existing level of variability. Since the second product's sales are uncorrelated with current sales the addition of this product would increase slightly the level of sales variability. Since the third product's sales are positively correlated with sales arising from the current product mix, the addition of this product would increase substantially the level of sales variability. Only the fourth product, whose sales are negatively correlated with total current sales, would decrease over all variability. In fact, if a new product were found whose expected sales equaled, in magnitude present total sales and showed perfect negative correlation with them, the addition of this product would produce a new pattern of total sales that would be perfectly stable. The main point is that the objective of sales stability imposes certain constraints on which products might be added or deleted.

Growth

Another objective emphasized by a large number of companies is raising sales through time. The rate of sales growth depends upon where various products in the company's current mix are in their respective life cycles and what plans are made for product additions and deletions. For the company that is committed to a high rate of growth, four basic product market stratigies are available :

(i) *Market Penetration.* The company seeks increased sales for its present products in its present markets through more aggressive promotion and distribution.

(ii) *Market Development.* The company seeks increased sales by taking its present products into new markets.

(iii) *Product Development.* The company seeks increased sales by developing improved products for its present markets.

(iv) *Diversification.* The company seeks increased sales by developing new products for new markets.

The Optimal Product Mix

The company's current product mix is said to be optimal if no adjustment would enhance the company's chances of achieving its objectives. If the company's objective is primarily profit minimisation then the product mix is optimal if profits could not be improved by deleting, modifying or adding products. If the company's objective is primarily sales growth, then the product mix is optional if it yields a rate of sales growth that could not be profitably enhanced by and product mix changes. Typically there are many objectives, and this complicates the problem of defining an optimal product mix.

Any of the following conditions suggests that a current product mix might be less than optional optimal :

(1) Excess productive capacity on a chronic or seasonally recurring basis.

(2) Disproportionality high per cent of total profits from a few products.

(3) Insufficient product width to exploit sales force contacts efficiently.

(4) Steadily declining sales or profits.

We will present a management science perspective on the problem of achieving an optimal product mix. We shall distinguish between the static and dynamic optimal product mix problem.

Static Optimal Product Mix

Given in product possibilities, choose one of them (when $m \angle n$) such that profit is maximised subject to certain constraints being satisfied. Under certain conditions the problem may be solved by mathematical progress ; the most important condition being no demand and cost interactions among the various products being considered. This condition is rarely satisfied in practice.

Dynamic Optimal Product Mix

Management is interested in what will happen to profits, and sales growth as the product mix is changed. The computers contribution would consist of making rapid calculations of profit, stability and growth characteristics of many different possible transformations of the product mix through time.

Some products require periodic modification if they are to remain competitive. Some other products pass on the stage where further modification could not help; they are candidates for elimination. A product modification is any deliberate alteration in the physical attributes of a product or its packing. A number of factors may prompt the manufacturer to alter his product. Since products have such attributes as colour, size, material, functional features, styling and engg. Any one or combination of these attributes would require change. Let us discuss three important and contrasting product modification strategies. Quality improvement, Feature improvement and Styling improvement.

Quality Improvement

A strategy of quality improvement aims at increasing the reliability and durability of the product through better materials of Engg. A premium quality producer often finds it easier to identify potential customers, because income or other customer variables tend to be correlated with an interest in quality.

Feature Improvement

A strategy of feature improvement aims at increasing the number of real or fancied user benefits. It involves redesigning the product so that it offer more convenience, safety, efficiency and versitility. Following are the advantages of feature improvement strategy :

(i) The development of new functional features is one of the most effective means of building a company image of progressiveness and leadership.

(ii) Functional features are an extremely flexible competitive tool because they can be adapted quickly, dropped quickly.

(iii) Functional features allow the company to gain the immense preference of preselected market segments.

(iv) Functional features often bring the innovating company free publicity.

(v) Functional features generate a great amount of sales force and distributors' enthusiasm.

The chief disadvantage of this strategy is that feature improvements are highly imitable ; unless there is a permanent gain from being first, the investment in original may not be justified.

Style Improvement

A strategy of style improvement aims at increasing the aesthetic appeal of the product in contrast to its functional appeal. In the case of packaged food and house hold products, where the opportunities for product styling and featuring are minimal except for colour, texture variations etc. treating the package as an extension of the product may be used.

Optimal Product Modification

In practice, a firm generally pursues some mixture of all three strategies. Just to maintain its competitive position the firm must incorporate the latest developments in quality, styling and functional feature. At the same time, each firm may specialize in one strategy in order to achieve leadership in that area.

Product Elimination

Many sick or marginal products never die, they are allowed to continue in the company's product mix until they "Fade away". In the mean time, they consume considerable resources which may be more fruitfully employed else where. As a result, these marginal products lessen the firms profitability and reduce its ability to take advantage of new opportunities.

The weak products tend to consume a disproportionate amount of managements time. It often requires frequent price and inventory adjustments. If generally involves short production runs in spite of expensive set up times. It requires both advertising and sales force attention that might better be diverted to make the healthy products more profitable. Its very unfitness can cause customer misgivings and cast a shadow on the company's image. There are in fact many reasons for this aversion, logical as well as sentimental.

(i) Sometimes it is expected or hoped that product sales will pick up in the course of time when economic or market factors become more propitious. Here management thinks that poor performance is due to outside factors which will change.

(ii) Sometimes the fault is thought to lie in the marketing programme which the company plans to revitalize. It may be felt that the solution lies in reviving dealer enthusiasm, increasing the advertising budget, changing the advertising theme or modifying someother marketing factor.

(*iii*) Even when the marketing programme is thought to be competent management may feel that the solution lies in product modification. Specially the thinking might be that sales could be stimulated through an upgrading of quality, styling or features.

The foregoing are all logical arguments for retaining weak products in the mix. But there are also situations where the persistence of weak products can be explained only by the presence of vested interests management and consumer sentiments. A lot of people inside and outside an organisation grow to depend on a particular product. Among them are the products manager the employees and certain customers. Eliminating a product from the mix is organisationally disruptive. Personnel may have to be shifted or released.

Process Planning

For the manufacture of a part or product, the exact specifications are required. Based on these specifications the design drawing is prepared. After the release of the design drawing, process planning starts by developing the broad plan of manufacture of the part or product. In fact the selection of materials and process such as casting, forging and die casting etc. are necessiated in the design stage itself.

The relation of process planning to layout and facilities planning. Process planning necessarily blends with the layout of physical facilities. Some process planning takes place during the layout phases of the design of a production system. Subsequently due to limitation in space, or to improve the methods of squence, the original process plan is modified. The process plan needs the route sheets and operation sheets which summerize the operations required the preferred sequence of operations, auxiliary tools, operation required etc. Therefore process plans are inputs to the development of a layout.

The inputs to the process plan are the drawings or specifications which indicate what is to be made is decided by the orders or forecasts. Then an over all planning is made how many and which of the parts are to be bought from outside and how many are to be manufactured in side. For the each parts to be made in side, the plant, tooling a detailed routine is developed. Here the technical knowledge of processes, machines and their capabilities and knowledge of production, economics is required. Since the range of processing alternatives is large, the selection of process is influenced by over all volume, and hence the economics of break even volume with respect the alternative process will be a deciding factor.

The distinction between product design, process planning manufacturing is shown in the following diagram.

Process Planning—A Concept

Process planning must begin during the product design stages where relations of materials and initial forms, such as casting, forgings and die-casting take place. Thus the term process planning is used in its organisational sense and in its functional sense it includes the basic selection of processes necessitated in the design stage. As regard process planning relation to layout and facility planning, process planning necessarily blends with the layout of physical facilities. Some process planning takes place during the layout phases of the design of a production system. To accommodate physical and sequential limitations, to take advantage of available space, or to improve methods or sequence, modifications of the original process plans may be made. The conceptual framework of process planning is shown in Fig. (10.1). The process planning takes as its input the drawings or other specifications which indicate *what is to be made?* How many are to made? and also the forecasts, orders, or contracts. It is a complex assembled product, considerable efforts may go into exploring the product into its components of parts and sub assemblies.

However, for a really complex product such as an air plane or missile, it would be difficult to comprehend the plan for manufacturing without an assembly chart. Assuming that the product is already engineered, we have drawings and specifications of the parts and their diamensions, tolerances, and materials to be used. From the specifications and how many to manufacture. Decisions have to be made as to which parts to purchase which to manufacture in our plant. Finally "Route Sheet" or Operation Sheet (which specifies for each manufactured part, the operations required in the preferred sequence, equipment to be used, special tools, fixtures and gages etc.) are finalised and the imformation can be summarised in the form of an operation process chart. The operation process chart is of great value in the development of a layout.

As time passes, changes creep into manufactring plans because of redesign, the addition or elimination of products and advances in manufacturing technology. Reviews of existing operations are often very fruitful for eliminating duplication and illogical flow.

Product Flow Process Charts

The flow process chart is similar in concept to the operation process chart except that it adds more details and has a slightly different field of application. The flow process chart adds transpotation and storage activity to the information already recorded in the operation process chart. The completed flow process chart can be constructed by actually following the progress of the parts through the machine shop and gathering the required information. It is ofen helpful to supplement the flow process chart by superimposing the flow chart on a floor plan of the work area in order that spatial relationships can be better visualised. The result is called a flow diagram.

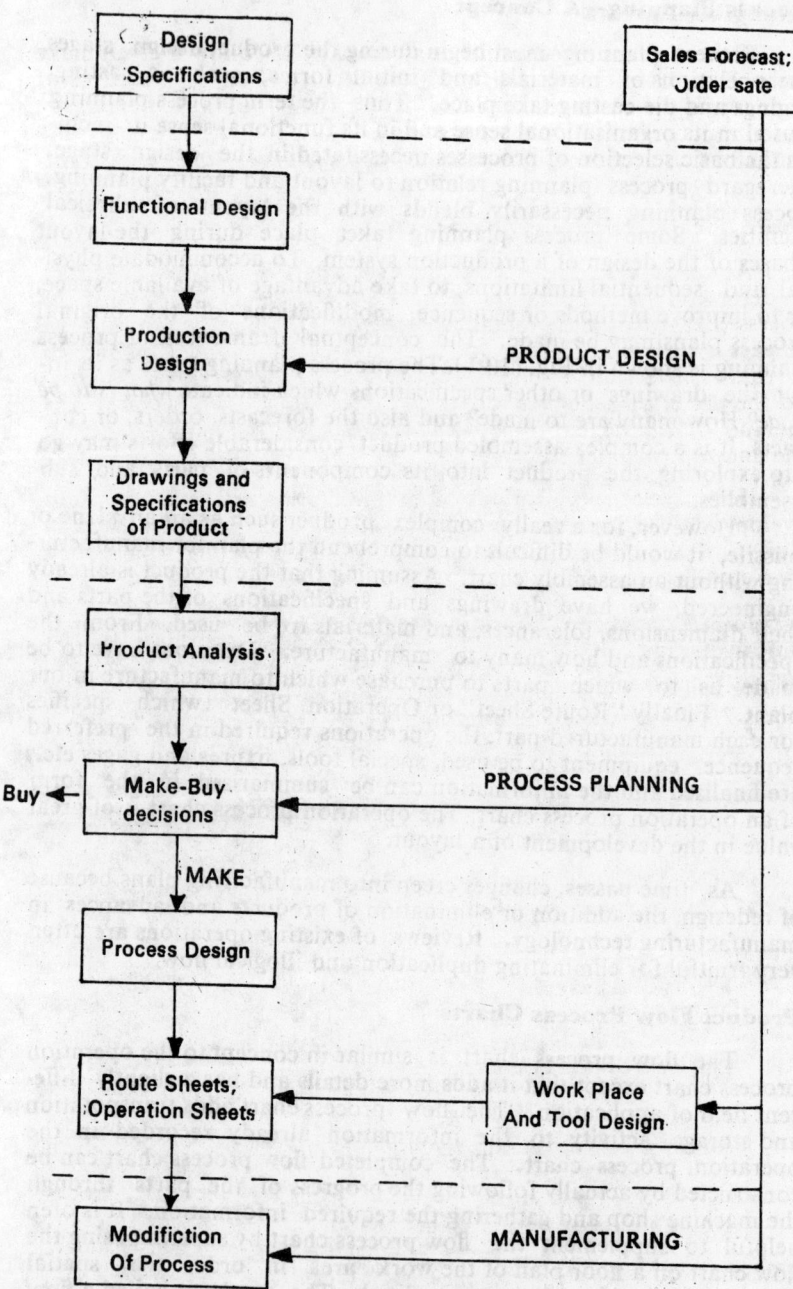

Fig. 10.1 Process Plans

Intangible Factors

Other factors often influence a company to follow a given make buy policy ; Quality, reliability and availability of supply ; control of trade secrets, patents, research and development facilities, flexibility ; and alternative supply sources are some of the factors entering into a make or buy decision in the face of apparent economic disadvantage.

Route Sheets and Operation Sheets

At each stage of its processing, every part is analysed in order to determine the operation required and to select and specifly the process that perform the functions required. Thus, the routing of the part is determined. The information is commonly summarised on route sheets. The route sheet (a) Shows the operations required and the preferred sequence of these operations, (b) specifes the machine or equipment to be used, and (c) gives the estimated set up time and run time per piece.

Here precise specifications of manufacturing methods are often developed in the form of operation sheets, which tell in greater detail how the operation is to be accomplished, in other words, they give a standard method.

Distinction between Product Design ; Process Planning and Manufacturing

The figure (10'2-) shows the distinction between Product Design; Process Planning and Manufacturing.

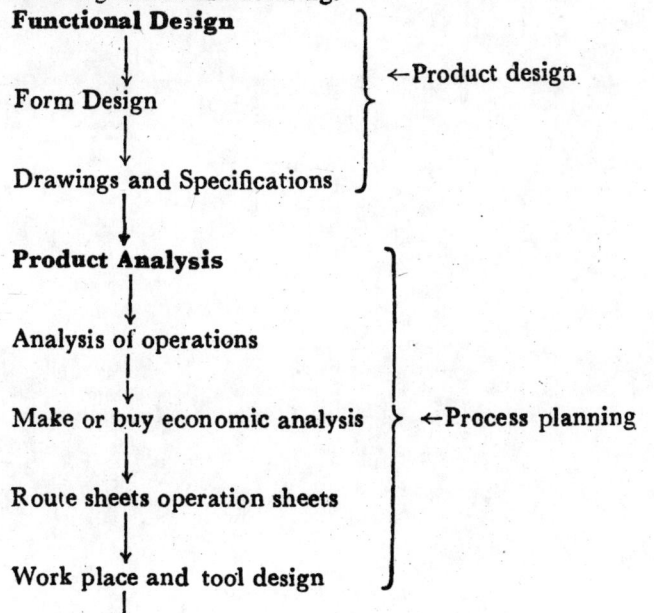

Functional Design

↓

Form Design ⎫ ←Product design

↓

Drawings and Specifications ⎭

↓

Product Analysis ⎫

↓

Analysis of operations

↓

Make or buy economic analysis ⎬ ←Process planning

↓

Route sheets operation sheets

↓

Work place and tool design ⎭

|

Modifications of Process

Plans due to constraints
in the space, quality of
product, availability of
machine etc.   } ←Manufacturing

11

Product Design

Introduction

Product design consists of mainly two features, form and function. The features of form are like shape of the product, appearance of the product etc. The functional features are concerned with the performance characteristics of the products like time required in washing, spinning and drying in a washing machine or number of pieces that can be washed per hour in a washing machine. For the same functional design, different manufacturers make the same product slightly in a different model or shape or appearance so that the advantage is gained by the customer either in the appearance or shape for the price variance he pays for. Some times with little extra additional cost, the functional features are improved to a great extent and it appeals the customers. Here the value analysis and value engineering plays a vital role in product design. Some products are demanded by the customer in different model every year or once in 5 years etc. Some times the manufacturer himself changes the modes periodically with additional features inform and functions to boost up the sales to keep in tune with the taste of the customers. This will lead to a number of innovations.

Factors Responsible for Products Design

—Design of products is quite often imitated from other existing products. To have a completely innovative idea in developing a product is quite costly and time consuming. In the long range product planning the innovative design is very much beneficial. Design by imitation pays quick but cannot last for long since the design of products by innovative resarch can sooner or latter because of its latest technology will throw all the existing designs out of market.

—The decision of a product design is generally based on the team work. The team usually consists the representatives of sales, design engineer, purchase, finance and production. Sales representative will assist in the design features of the product that are required from the customer point of view. Design engineer will look into the extent of possibility of such design required by the customers. Purchase will throw the ideas as to the choice of materials available and the finance will link into the cost aspect of such design. Once the team approves the new product design, as a prospective one, it is taken for the manufacturing and sale.

—After the commercial production, if the costs of the product happened to a more than what the market can absorb, it needs a detailed analysis of costs and remedial measures to arrest the escallation of costs. Finally one needs a value analysis approach towards maximisation of the cost affectiveness so that the product can continue to be in the market.

—When the demand is large compared to the supply, one can think for mass production so that the products can be made at low costs/unit. Every effort is required to study in detail about methods of production and the improvement. Since a minor improvement will result in large savings of resources and enable the manufacturers for better compitetion in the market.

—The design of the product be such that it is easy to repair. Customers prefer it. Alternatively, the concept of modular design is more advantage in the design of the product. Products are assembled from easily detachable sub-assemblies. Whenever a particular part fails, the whole component of which it is a part, can be taken out and replaced with a new component, where the repair or break down costs are too high this modular design is preferred by the customers.

—As the time advances miniaturization of the product is desirable like electronic products. Make it smaller and lighter in weight. It may not be possible in all products. This miniaturisation is done partly by research and partly by design where sufficient knowledge is available. Miniaturization reduces the costs and increased the effectiveness appreciably. Example computer development from Ist stage to 4th stage or 5th stage.

Standardisation

Standardisation of components is essential for interchanging capability of parts which is of vital significance to mass production on assembly line basis. In its simplest form, modern standardisation has been recognised and used since the advent of mass production techniques to which it is a necessary adjunct.

According to the international standard organisation. Geneva "standardisation is the process of formulating and appplying rules for an orderly approach to specific activity for the benefit and with the cooperation of all concerned, and in particular for the promotion of the over all economy, taking due account of functional conditions and safety requirements".

A standard is based on the consolidated results of science, technology, research, development and experience. Normally a standard is established by custom authority or practice as a model to which an action or object can be compared. Hence it should represent the crystallised best thought of the practice of art. Standardisation in its broadest sense permeates almost all fields of human activity.

Standardisation As a Concept

The concepts of standardisation can be traced back to Noah's Ark. We all know that standardisation of components is essential for after-changeability of parts which is of vital significance to mass production on assembly line. In other words importance of standardisation was realised since the advent of mass production techniques. Standardisation is defined by the International Standards Organisation Geneva as the process of formulating and applying rules for an orderly approach to a specific activity for the benefit and with the co-operation of all concerned and in particular for the promotion of the overall economy taking due account of functional conditions and safety requirements.

A standard is based on the consolidated results of science, technology, research development and experience. A standard is established by custom, authority or practice as a model to which an action or object can be compared. Standardisation in its broadest sense permeates almost in all fields of human activity.

Indian Standard Institute

In order to establish standardisation at National level, the Government of India established (I.S.I.) Indian Standard Institute in 1947 with the help of scientific, technical and industrial organisations. I.S.I. is an autonomous body with Union Minister of Industries as *ex-officio* president. In order to provide the assured quality of industrial goods, the I.S.I. Certificate Market Act was passed by the Parliament. This I.S.I. mark is intended to protect the consumer's interest and to achieve optimum utilisation of material resources by producing the quality and product from the right type of raw materials and processes. At organisational level standards deals with product process, methods, materials, specifications, inspection, tests, procedures or other type of requirements.

Advantages of Standardisation

(i) It enables one to have large quantity of few ever items economical buying quantity.

(ii) It reduces negotiation burden and results in a reduction in producurement lead time and better vendor relations.

(ii) It promotes healthy competetion amongst multiple sources avoids shortages, reduces obsolescent items.

(*iv*) It enables one to operate running contracts leading to lower departmental operating costs. It reduces design time and efforts, minimises drafting effort through repetitive use of standard drawings, reduces specification.

(*v*) Writing by repetative use of standard specifications.

(*vi*) It enables to reduce the variety of items and hence results reduction in the inventory of tools and maintanance items.

(*vii*) It results in the improved interchangeability of parts.

(*viii*) It reduces the burden on inspection and quality control departments.

(*ix*) It enables to achieve over all reduction in operating costs and better efficiency of all the departments. For successful application of standardisation in an organisation following aspects should be considered :

(*a*) Dynamic and flexible nature of standardisation.

(*b*) Organisational aspects and responsibility of standardisation.

(*c*) Objectives of company standards.

(*d*) Technique and process of standardisation. In many public and private industries most of the non moving and excess inventory is mostly or account of lack of standardisation. Hence the Govt. gave a serious thought to over come this increasing inventory problem and established Indian Standards Institution (ISI) in 1947, with the active support of scientific, research technical and industrial orgainsations. The I.S.I. has standardised a number of items covering various sectors like steels, Fertilisers, chemicals, automobiles, Electrical goods. The I.S.I. mark is intended to protect the customers interest and to achieve optimum indisation of material resources by producing goods from quality raw materials.

Standardisation is a dynamic process. It can not remain static at one level. It has to move from one level to another higher level is from company standards to industry standards from industry standards to National standards, from National standards to International standards.

Disadvantages of Standardisation

Industry wide standardisation programmes are often only partially successful as standardisation tends to favour large business houses. Small or new companies cannot get a sizable business by m king the same things and selling at the same price as the big companies. They survive by offering something different at the same time close to the same price as standard products. Such type of companies might be said to specialise on specialities and by doing so, to some extent, under cut or standardisation programme. Often customers do not want the standard items, unless it is priced lower. The salesmen also side this view of the customer sand prof make prepay under that "our product is different" If new products are standardised before they have become stable in design standardisation may become an obstacle to progress. In a product development stage, performance standards rather than form or shape standards should be used.

—Standards once set, resist change, thus retarding progress.

—A standard of any given time is more the prevailing style than it is a permanent standard.

Impact of Industrial Robots on Working Life

Robotisation has actually been facilitated by the consensus maintained in undertakings by the trade unions in close and constant collaboration with management. The employees do not object, despite the various job changes brought about by the introduction of robots, since the system of permanent employment ensures that they will always have a post within the undertaking. The Japanese style of management, unlike that in Western countries, has helped to facilitate the increasing tempo of automation in the country. However, this style of management contains the seeds of its own destruction since the increasing automation of factories and undertakings is bringing the system to a standstill, if not to collapse. It is the custom in Japanese firms to make seniority the criterion for promotion and remuneration. In other words merit never takes the precedence over seniority. In a period of rapid economic growth, the seniority system served Japanese firms well. They are all pyramidal in age structure with bases that expand as they grow. There is now a strong trend away from this reliance on seniority and other standards of assessment which are coming into play. In 1977 the importance of merit in the assessment of salary increases was calculated at 42 per cent. By 1982 the figure had become 46 per cent and in undertakings with a staff of over 3,000, it is 53 per cent.

The seniority system is fading in significance in three main divisions of enterprises :

—in research and development departments, creativity and innovation are qualities that are increasingly being expected to accompany technological knowledge and skills.

—in production departments, increasing automation is making human resources increasingly by redundant.

in sales and marketing departments, the widespread system of payment by results takes account not of seniority but of drive and volume of sales.

These changes that are upsetting many established values, including that of permanent employment.

The conventional system of permanent.

The conventional system of permanent employment means that staffs are taken on young and trained in the firm in the light of that firm's needs and practices. The young employee is entirely shaped by his firm, remaining with it and pursuing his career there until he retires. Horizontal movement in the labour market has been practically unheard of at most of big companies in Japan. Any employee leaving one firm to join another would be sure to receive a very low salary. Nowadays, however, radical changes are in course. Although the largest firms can cope with the new situation created by the need to bring together engineers from a variety of fields, small and medium sized firms can only do so by calling on the services of persons with the skills they need but cannot supply themselves. Hence, changes of job may be expected to become an increasingly important feature of the Japanese labour market. A related new factor is that part-time employment is becoming more and more prevalent in Japan. The percentage of workers on part-time employment is a continuing to grow despite the fact that such workers have no real guarantee of a permanent place in the firm employing them. In short, the increasing application of advanced technology has led to a new mobility in the labour market. In order to remain competitive, firms are obliged to call in skills from outside and are no longer in a position to guarantee their old employees a permanent place. Japanese society will thus be compelled to develop a new system of employment to fit in with the technological changes it is experiencing. The seniority system and the system of permanent employment have been the bedrock of long-term survival in Japanese firm since they ensure firms the loyalty and support of their employees, qualities whose benefits need no underscoring. The technological revolution, by making these systems obsolete, has struck a heavy blow at a solid structure that has held firm for decades. The crucial question now is how to make technical innovation and permanent employment compatible and experiments with a transitional system bringing a degree of flexibility to the permanent employment system seem to be under way. The employee who used to be assured of a good retirement allowance when he retired after reaching the age limit—generally between 55 and 60 years of age—may now enjoy almost the same benefits by choosing to retire earlier, between 40 and 50 years of age. Seniority and permanent employment are not the only factors to be affected, however, workers' organisations are also involved.

System of Enterprise and Plant Unions

The Japanese have a strong tendency to behave and express themselves as members of a group and the community always takes precedence over the individual regardless of the context. The union is the guardian of the group spirit within a firm as well as watching over the interests of its members. Long and continuous service in an enterprise which is assured by permanent employment is fundamental to the group spirit or consciousness of belonging to the enterprise. The group spirit or consciousness of the system of enterprise unions, and the obsolescence of the system of permanent employment is eroding the foundations of the enterprise union. The new developments in technology have also forced firms to seek the services of experts outside their staff for specific tasks. Since these experts do not belong to the community represented by the firm they are outside the union's control. At the same time, new techniques and the automation of the factory or office have made the work of many employees redundant. The staff coverage by the unions is declining as the application of microelectronics increases. As a time when the boost that electronics has given to productivity, the use of part-time staff, the spread of the system of staff transfers to avoid unemployment and the need to call in experts for short-time tasks are making the present system inoperatable. Consequently, a new types of union may have to be developed to cope with these new type of employees in a way that is fully adaptable to the new conditions of employment. The unions, however, are not the only structural feature of the Japanese firm threatened by computerisation, automation and robotisation.

The Japanese firms are at present highly competitive and highly productive. However, they are likely to run out of steam in the international market in the near future because of the decline in the traditional factors that the hallmark of that competitivity. There is thus a crucial need for reform and for recognition of the need for a new style of management suited to the new situation created by the microelectronics boom. This is the challenge of the future. The human dimension has been the bedrock of Japanese competitivity. In the new type of management that is taking shape, the human dimension will find its place only in the field of research and development and will be replaced by machines in production and manufacture. The new era of new-style management ushered in by the upsurge in microelectronics calls for a high degree of creativity from research and development departments and, owing to the complexity inherent in the application of advanced technology, will require team work from engineers specialising in different fields. The Japanese group spirit that was the springboard for the tremendous expansion that has taken place up till now is clearly an obstacle to the development of the creative spirit since it crushes any display of individuality in the bud.

Industrial Robots and Computer Aided Manufacturing

Microelectronics technologies such as industrial robots computer-aided manufacturing, flexible manufacturing systems and office automation have brought drastic changes at both shop floor and office level in Japan as well as in other industralised countries. These changes have both good and bad effects on working life. A notable characteristic of the spread of industrial robots is, first, the fact that they are introduced into small and medium-sized companies as well as in big business, one of the main reasons being the relatively low price of industrial robots compared with labour cost. A second characteristic of industrial robot diffusion is the fact that their introduction is now limited in specific industries and jobs. The jobs often taken over by machines are assembly, welding, loading and unloading, cutting metal and painting. The most used types of industrial robots are therefore Numerical Control machines, fixed sequence robots, transfer machines, manual manipulator playback robots and flexible robots. 'Intelligent' robots and other types with the capacity to assemble and to judge are fully expected to spread out to a variety of industries, and also the service sector, in the future. The initial limitation of robots to taking over the 'dirty' work in industry resulted in their acceptance by the workers.

Effects on Employment

The effect of microelectronics technology on working life is its impact on employment, it is not easy to isolate 'pure' effects from the many other factors affecting employment, and there are differences at macro-and micro-economic levels. However, the employment of production workers cannot be so optimistically expected to increase in industries where micro-electronics technologies, and especially industrials robots are widely introduced. One robot is now said to replace 1·5—2·0 workers in the Japanese automobile industry. The introduction of 17 industrial robots in an electric fan manufacturing company in Nagoya, for example, reduced the number of production workers from 64 to 5 and at the same time increased production volume three times. Specific group of workers tend to be strongly influenced by the introduction of industrial robots. Firstly, as already shown in the number of workers in medium sized firms employing 300-999 people decreased on both the shop floor and in the enterprise as a whole, more than in any other size of enterprise. Secondly, middle-aged or older workers are more subject to the influence of industrial robots on employment due to their inability to adapt to the new technology. Thirdly, the number of workers on the shop floor manufacturing and assembling decreased more than in any other department of the enterprise. Fourthly, regular employees are markedly affected by the introduction of industriat robots. Thus, the first impact of industrial robots on employment tends to be concentrated on workers of medium-sized enterprises, at the shop floor level, and on ageing workers and regular employees. However, at this stage of their development their effect on employment does not as yet bring serious social problems in Japan. Among the reasons for a positive attitude by trade unions and blue-collar workers towards the industrial robot is the Japanese way of management, adopting life-time employment informally guaranteeing job security to redundant workers, maintaining a flexible work organisation which makes possible transfer and retaining of redundant employees, consensus-oriented industrial relations, and applying lessons learnt from the experiences with technological innovation during the 1960s when mechanical automation was widely introduced into big compaines. These factors moderating the negative effects of industrial robots may not be expected to survive in the future when exports of Japanese goods are restricted due to international trade conflicts, and when industrial robots and the other microelectronics technologies are more widely introduced, not only in big companies but also in small companies and the service sector. Industrial robots may, sooner or later, cause drastic changes to the labour market structure in Japan. The second impact on the industrial robots on working life relates to the changes of skill and roles of workers operating them in the accompanying work organisation causing serious concern. Management structure may also change to shorter hierarchical levels due to the simplified work organisation on the shop floor. A further effect of industrial robots on working life concerns changes to work content. However, it is not true to declare that microelectronics technology always deprived workers of their machine operating skill. Experience has shown that both conventional skills and new professional knowledge are required to introduce robot-arranging production processes and to teach the procedure of operation to the robots themselves. Therefore, one of the important tasks of personnel managers is said to be to maintain and transmit conventional work skills to workers operating industrial robots.

Responses of Trade Unions to Microelectronics Technology

Attitudes of trade unions toward microlectronics technology can markedly influence the direction and speed of its development and diffusion. In Japan, most trade unions do not actively oppone the introduction of microelectronics technology. According so research by the Japan Institute of Labour in 1983 that, 90·2 petrcent of trade unions studied support its introduction into their workplace, with 53·6 per cent positive and only 36·6 per cent reluctant. The proportion of trade unions supporting increased makedly in comparison with that found in 1981. One of the fundamental policies of Japanese trade unions in dealing with the introduction of microelectronics technology is to secure in advance in joint consultation with the management the Job security. Sohyo, one of the main national centres of Japanese trade unions, reported that they demand it because otherwise they will be later forced to approve working conditions brought about by the new technology unless they checked its effects at the first stage of its introduction. The policy of trade unions to acquire advance information on the effects of mic roelectronics technology on working life through joint consultation systems, and to control its effects through their participation in its introductory processes, may work as well in Japan as in European countries. Another Japanese trade union policy toward microelectronic techonology is security of employment and on improved work environment for their members.

12
Probability

Introduction

General concepts of probability, of waiting line theory, and of incremental cost analysis have provided a rational basis for designing preventive maintenance programs, determining optional crew sizes, and determining capacity requirements so that the reliability of production systems can be maintained.

The maintenance function extends beyond machine and equipment maintenance to include plant maintenance as well. Plant maintenance includes building, grounds plant services such as power plants, heating and ventilation systems, plumbing system and general house keeping. Personnel in the plant maintence function freequently do minor remodeling and relocation of partitions and machines to accomodate changing layout needs. Maintenance department usually include a general machine shop in which needed repair parts can be fabricated. Production management must provide for "break down maintenance" in a way that maintains the reliability of the production system at reasonable levels without "going broke" just keeping machines running. The other important factor in maintenance is to strike a balance between cost factors. When machines break down the various kinds of costs occur : machine down time and possible loss of potential sales, idle direct and indirect labour delays in other processes that may depend for material supply on the machine that is down, increased scrap, customer dissatisfaction from other possible delays in delivering and the actual cost of repairing the machine.

It is not very late when maintenance used to be considered as a necessary evil and attracted least attention of the management. Frequent failures were put-down to old age or to bad machines or sometimes to bad operators. The entire maintenance department use to consist of a skilled foreman with 10 to 12 gangmen and they were mostly left to their derives and operated more or less on "firebrigade" concept, and preventive maintenance was a new word. The situation has, no doubt, undergone a big change since then and management have really started giving much more importance to planned maintenance of the equipment and machinery.

The purpose of maintenance is to make available for production purposes, machinery and equipment to fulfil their technological functions as specified and economically which means that the output quality and quantity from each machine or equipment will conform to specified purposes. For instance, a properly maintained machine tools much produce components to the desired accuracy and quality of surface finish. The purpose of the entire maintenance organisation is to fulfil this function satisfactorily with the machine and equipment utilization factor as high as possible.

Although all machines are thoroughly tested under strict quality control methods at the manufactures works before supply to the user and user also tests these before finally putting them into operation. After some use the machines are subjected to wear and to reduce wear proper care to lubrication, cleaning, timely inspection and systematic and scheduled maintenance is to be given. The machines even rapaired when not in use so that any defect or damage to the part is taken care so that machine starts behaving as new. In this way the maintenance is responsible for the smooth and efficient working of any plant or industry and helps in improving its productivity.

Different System of Maintenance

Maintenance systems can be grouped into five major groups.

(i)   Operator Maintenance.

(ii) Breakdown Maintenance.

(iii) Scheduled Maintenance.

(iv) Planned Maintenance.

(v)   Preventive Maintenance.

(i) **Operator Maintenance.** This type of maintenance system is largely confined to a small engineering industries where supervisors and employees possess the qualifications necessary to control and perform maintenance of their own machines. This system of maintenance is obviously the most elementary one and is perhaps not applicable to anything other than very small factories. Even there the lack of separate control of maintenance records and costs would not reveal the real size of the maintenance problem year after year.

(ii) **Breakdown Maintenance.** This is intended to imply that an independently organised maintenance section acts only as an emergency repair service summoned by the production staff when interruption or restriction of production has actually occurred. It is obvious that in the world of modern industry even this system can be satisfactorily operated in very special conditions and for a limited period. This method will generally be found over the long term to reduce machine availability and give a lower overall mechanical efficiency for production machinery.

(iii) **Scheduled Maintenance.** This is the first stage in the approach to a modern maintenance system. This involves joint discussions in advance by production and maintenance programme so that items considered of use by maintenance staff to need attention can have scheduled periods of availability for repairs. In this way the "where" and "when" of the maintenance programme can be approximately fixed and the most efficient use made of any idle time.

(iv) **Planned Maintenance.** The development is one step ahead of schedule maintenance which decides not only the "when" and the "where" of maintenance but also answers the question" "what" and "by whom". Experience in schedule maintenance can be further used to make more detailed analysis of programmed work so that more precise planning of work and allocation of labour is it are possible. At the stage some method of estimating the work content of the jobs to be done is necessary. This system of maintenance is perhaps the better on which should be adopted by a manufacturing plant of modern complexities.

(v) **Preventive Maintenance.** This system is a further improvement over planned maintenance and answers the "why" for the maintenance work. It involves the policy of upkeep, replacement and modification rather than repair and has invariably to incorporate periodical inspections of the plant in order to diagnose the trouble, if any, and to plan the requirements of maintenance. Many times the work of inspection, maintenance, replacement or modification (by way of accuracy checks, periodic overhand and major overhands) are analysed for a considerably long time, for each important unit of production equipment and is broken down into elements to be further analysed to improve the methods and determine the standard work contents. The techniques of works Study can be applied with advantage under this system.

It is, however, necessary to bear in mind that irrespective of any amount of sophisticated maintenance system being enforced, the breakdown maintenance cannot be avoided altogether. The ultimate goal is therefore to introduce more and more sophistication in the preventive maintenance techniques to minimise the breakdown maintenance in the plant and machinery, since no piece of equipment or machine will have a breakdown without having shown certain signs of malfunctioning earlier.

The preventive maintenance programme cannot be introduced overnight, and has to be based on a detailed study of the earlier maintenance schedule, breakdown hours etc. A detailed study is required to be carried out to cover every important component of the equipment with respect to its condition of working, the past maintenance history, downtime etc. Sometimes inherent design weaknesses can also be identified.

The equipment history card for each equiqment and entity so prepared can be analysed to locate the types of repairs calling for extensive maintenance work i.e. in terms of men hours and spare parts. This would guide with respect to the planning for stocking the spare parts, their quantities, and nature of the work to be executed and can be recast and planned on work study principles. The manufacturers of the equipment generally recommend certain norms, or maintenance frequencies, pre-supposing, a certain working environment. This should be treated as a guide or a cross check, and the maintenance history of an equipment only should show up actual need of equipment in the given environment with respect to the preventive maintenance programme. The higher frequencies of inspection in the beginning can be justified to minimise to breakdown maintenance, but cost over a period would show that such inspections could be reduced to stabilise the preventive maintenance programme, as a maintenance discipline. To object of preventive maintenance is to attain overall lower costs, and hence the need to constantly review its costs *vis-a-vis* the cost of lost production, in order to achieve the optimum maintenance.

Objectives of the Preventive Maintenance

Broadly, the objectives of a systematic maintenance scheme are :

(*a*) To prolong the working life of the equipment and assure optimum availability.

(*b*) By minimising the wear and tear, preserve the value of the plant.

(*c*) To safe guard the investment.

(*d*) To ensure safety of the equipment and the personnel.

(*e*) Keep the productive assets in good working conditions.

Maintenance Functions

The necessity to have a maintenance engineering organisation in a factory is not any more questioned by today's managements but the proper importance of the maintenance functions and the rightful place of importance which the maintenance function should occupy in an organisation is not recognised by all. With the growing complexity of processes and equipment and with the growing magnitude of potential losses suffered in production due to breakdowns, modern managements can no longer look upon the maintenance function as a subsidiary of production but as one of the main tool of plant productivity which must be effectively used to obtain the highest availability of the production equipment commensurate with total maintenance cost. This concept has extended the role of maintenance to include close and active contact during the design stage, full involvement in process selection and continuous coordination with operation of the plant. Maintenance engineer has been responsible for a number of modern techniques being developed to improve the performance of the maintenance function itself.

Top management, should, therefore, realise that even though maintenance engineering is not a primary function, it is a very important support of the primary function of production and on it depends successful production.

The maintenance function would embrace to following activities.

Primary Tasks

(1) Inspection, lubrication, repair and maintenance of all equ'pments.

(2) Maintenance of existing buildings and structures.

(3) Operation and maintenance of utilities, operation and distribution.

(4) Alternation to existing equipments and buildings.

(5) Installations of new equipments and new construction.

In order to fulfil these primary functions there are a number of secondary function which the maintenance department is called upon to perform.

(1) Store keeping.

(2) Plant protection including fire protection.

(3) Waste disposal.

(4) Salvage.

(5) Pollution and noise control.

(6) House keeping in the plant.

Elements of Preventive Maintenance

Following are some of the elements which are to be taken care in Preventive Maintenance.

1. Inspection/Check ups. This is again split up into two categories :

(*i*) External Inspection, and

(*ii*) Internal Inspection.

External Inspection means to check the equipments or machines for external defects such as vibrations and abnormal sound in the machines and checking up of the bearing and bushings for heating up. These defects are felt during the operation of the machine.

Internal Inspection means inspection of internal parts such as gears, cams bearing and other moving parts. It is done when the machine is under shutdown.

Under this heading let us have an idea of planned maintenance which the countries like Russia and Czechoslovakia are following on National basis. In India as well many plants which have come up with the collaboration of above countries following the same procedure with little adjustment here and there according to local plant environments.

The planned Preventive Maintenance consists of :
  (i) Minor Repairs
 (ii) Medium/Intermediate Repairs
(iii) Major/Capital Repairs.

The maintenance system must fit the pattern of maintenance cycle *i.e.* the number of planned overhauls and the order in which they are to be carried out between capital repairs. Both the Russian and Czechoslovakia patterns have adopted repairs or overhauls between 2 major overhauls and they have been adopted at National levels. They are more or less in conformity with each other and pattern is in the following order :

$$I\text{-}I\text{-}II\text{-}I\text{-}I\text{-}II\text{-}I\text{-}I\text{-}III \quad (IV\text{--}V)$$

This is the pattern in the Engineering Industry particularly for machine tools and it is the same for all types of models of metal cutting lathes and all operating conditions. The period between any of these repairs is known as the inter-overhaul time and the period between the major overhauls in known as the maintenance cycle. The manpower and machines utilised for a given type of overhaul have also been standardised in these two countries and approved at the National levels. The machine tools are designated into different categories according to their complexity. Each category is assigned a conventional coefficient and the tool taken for the base standard is a general purpose turning lathe of average complexity coefficient '*R*' has a value of 10. A cylindrical grinding machine has a complexity coefficient of 15 which means 1·5 times more than a lathe.

The average complexity coefficients for certain types of machine tools as given in the Russian Publication is given below.

S. No.	Type of machine tool	Complexity coefficient
1.	Lathes medium size	9—13
2.	Heavy lathes	17—19
3.	Vertical drilling machine	3—8
4.	Radial drilling machine	6—12
5.	Open-side Jig borers	20—35
6.	Horizontal bores (medium)	16—18
7.	Cylindrical grinding machines	10—15
8.	Gear Cutting machines, medium size	10—12
9.	General purpose horizontal milling machines	8—14
10.	Planning machines, medium size	12—15

After establishing the complexity coefficients of these machine tools, the number of man and machine hours required for these preventive maintenance operations are also standardised as follows :

Type of Repair	Man hours	Machine hours
Minor Repairs (I)	4·0	2·0
Intermediate Repairs (II)	10·0	7·0
Major Repairs (III)	26·0	10·0

Thus, the ratio of man and machine hour utilisation for planned overhauls is :

$$I : II : III = 1 : 4 : 6$$

From the knowledge of the complexity coefficient of a machine tool and the man and machine hours utilisation requirement for any type of repair, the time required for any type of repair can easily be found out of shown below.

For a cylindrical grinding machine with a complexity coefficient of 15 time required for major handling would be 15 (26×10) = 540 hours and 360 hours for intermediate repairs.

Such patterns as for engineering industry, can also be worked out for other industries as well.

2. Planning and Scheduling. Every preventive maintenance work should be preplanned in details on the basis of the analysis done on the past records. In order to have effective maintenance planning, certain prerequirements are necessary. Below shows the various maintenance management area and suggested outline of the relationships of basic and defined controls. The failure of defined maintenance systems, such as critical path scheduling or material delivery, can usually be traced to an inadequate base.

Certainly a solid foundation should be established before complicated preventive maintenance or date-processing systems are instituted. There are number of basic requirements.

Areas of Maintenance Management

Work order system

|

Time keeping

|

Materials control

|

Planning and scheduling system

|

Reports
1. Backlog
2. Job costs

3. Performance
4. Budget

Defined controls
1. Preventive Maintenance
2. Cost control
3. Spare parts
4. Stores
5. Materials delivery
6. Maintenance engineering
7. Training
8. Maintenance evaluation
9. Critical path scheduling
10. Data processing
11. Facilities evaluation
12. Others

The production and maintenance department must useful formal work order to specify the work that is to be done and execute the work order before the work is undertaken. Planning and control cannot be applied unless the job itself is controlled through a work order system which provides the basic paper work for preplanning and for individual cost identification of each job.

A typical work order form is shown below :

Work Order

Class of work	Code	Acctt. No.
Repair	—	Labour
Alterations	—	Material
Renewal	—	Issued to Date
New-work	—	Equipment No.
		Work order No.

Description of Work

Requested By	Data wanted	Completed By	Date
Posted to Equipment Record	—	—	—
By	Date	—	—

Reasons

Budget Item No.	Checked By	Estimated By	
Cost Budgeted Estimate	Labour	Material	Total

Project Approved By (Initials)		Date
Maintenance Engg.	—	
Operation. Asstt. Supdt.	—	
Plant Engineer	—	
Chief Engineer	—	
General Manager	—	

A second requirement is for a timekeeping system which identifies the actual labour hours against each work order. The time keeping system should be simple and the basic requirements are only work order number and the actual hours spent against that number. any additional information can be obtained from the work order. The hourly machine should keep his own time and foreman participation should be limited to an approval of the time ticket at the end of the day.

Lastly, materials control is necessary, maintenance stores and spare-parts control must be effective and maintenance planning is also to be effective. A listing of terms carried in stores must be available in a stores and spares catalogue so that materials must be purchased can be ordered before the job is under way. Location of stores items, inventory control, obsolescence problem, and other stores control will seriously effect maintenance performance.

The practical maintenance planning also involves delivery of materials to the job site before the job is started. Since planning includes materials and a good materials-delivery system requires materials availability.

Net work analysis for scheduling and a controlling maintenance work is a powerful tool to economise on the man-hours spent in shut down maintenance and progressive top management should take it obligatory on the part of the managers in charge of Preventive Maintenance or Scheduled Maintenance to involve, net works for each major scheduled maintenance job.

To give an example let us solve one example to understand how PERT/CPM is applied for capital maintenance.

The example is overhauling of huge Reciprocating Pump.

For getting optimum performance out of any individual, activity or group of staff, it is advisable to streamline the actions in such a way that all the activities may be completed in the shortest possible time with maximum results and minimum of wastage. This is possible by making a PERT (Programme Evaluation and Review Technique) chart of the activity.

First the various activities and events were made out as per requirement. The time is calculated as per the general experience and datas available. It was found that by making a PERT chart not only a complete direction of sequence could be given to the staff concerned but the job could be done within the shortest possible time with minimum staff. With a complete picture of the process available it was also possible to make put the needs of spares and machining activities and plan ahead before taking up the job.

The following assumptions were made while drawing the PERT/CPM of capital Maintenance of reciprocating pump in a factory.

1. Three groups of fitters with the following strength are to be employed.

 (i) Group "A" with 4 fitters, 4 helpers.

 (ii) Group "B" with 2 fitters, 2 helpers.

 (iii) Group "C" with 2 fitters, 2 helpers.

2. Three riggers are to be employed to work with the above mentioned three group of fitters.

3. Civil works have not taken into consideration while drawing the PERT and have assumed that the base of Pump is misaligned and that needs rectification by realignment and minor civil work on the foundation bolts.

4. All the materials, spares and manpower would be available at site as and when required to avoid wastage of man hours.

Conclusions

The minimum possible manhours required to complete the capital maintenance of the pump are $= 293$ manhours.

$$\text{Minimum mandays} = \frac{293}{8} = 36 \cdot 6 \text{ or } 37.$$

While working out above mandays it has been assumed that work will be done on 8 hours single shift basis allowance for fatigue etc., may be added locally.

3. **Lubrication.** In lubrication the first and foremost thing is lubrication survey. A complete survey of the plant equipment should be made to ascertain the proper requirement of lubricants. The oil companies render such services which would be well-worth availing of. It is better to rationalise these lubrication requirements by reducing the variety of lubricating oils to the absolute minimum after considering the suitability of alternative brands.

Design of colour scheme or a codification system to indicate all spots which need lubrication, the type of lubricant and the frequency lubrication.

Lubrication should not be left to the lowest paid helpers or untrained illiterate operators. The system of codification, the various types of oils and greases and administering correctly measured quantities of lubricant should be explained at length to mean and should be well understood by them.

Storage and handling of lubricants are of utmost importance and this itself can cover full chapter. But the method of storage and handling of lubricants leaves much room for improvement in many enterprises. The lids of containers are found missing and exposed to dirt, grit and moisture. Workers pick up the grease with their hands which under workshop conditions are not free from dirt and grit.

In addition to the colour coding scheme it is preferable to develop and display on each equipment a lubrication chart. This chart should indicate all the lubrication points, the brand lubricant for each point the method of lubrication the quantity and application frequency. This is because colour marking are apt to become indistinct after a period before they are renewed.

The results of a lubrication system can be studied through the correlation between expenditure on lubrication and the loss of machine hours through breakdowns. Such a study can make the advantages of a planned lubrication system assignable in monetary term.

4. Training of Maintenance Personnel. Proper training of maintenance staff is an aspect which can never be over emphasised. Apart from the initial training of maintenance craftmen, it is necessary to organise refresher courses for craftmen and maintenance supervisors. The advantage of refresher course is that they enable pooling of experience gained by the participants during their years of working. A regular programme for refresher courses in maintenance for the craftmen and supervisors should be drawn up and implemented faithfully by every enterprise. Advantage of specialised agencies can be taken in organising and addressing these refresher course.

Organised instructions are necessary however, to import knowledge about the function of the equipment, its operation and sequence of assembly of the components. Wherever possible working models preferable sectionised, can be of great help to in such instructions and demonstrations. Defects or defective system may or may not always follow the same pattern nor any textbook solution available to problems. An effective maintenance manager is, therefore, who had developed an analytical mind and flaire for improvisation.

5. Motivation Techniques. Job motivation involves the following fragments that impinge on job selection and performance.

(1) Salary and other tangibles.

(2) Titles (includes extension of responsibilities)

(3) Worksite location.

(4) Element of competition.

(5) Opinion of others.

(6) Freedom of decision and movement.

Motivation to work is very much an individual and environmental factor. Some of the factors which go in for this aspect are as follows :

(*a*) Much faith is put in various works entrusted to individuals.

(*b*) Extreme share of consideration is shown by supervisors in case some mistakes or errors in work come to light and remedial training measures are initiated.

(*c*) Proficient planning and scheduling is exercised and most of the work is required to be done at a normal pace in the most suitable atmosphere.

(*d*) Suggestions for alteration and improving methods, working tools, equipment, etc., are taken in the correct spirit by all supervisors and implemented whenever possible. Reward are also given whenever justifiable.

(*e*) Good salary and fringe benefits.

(*f*) Good production incentives scheme.

(*g*) A reasonable good performance rating system.

6. Maintenance of Records. Good record keeping is essential for good preventive maintenance. Management information system for maintenance is to be designed for timely collection of data, its processing to information to appropriate levels of management with required frequency and details so that each manager may plan and control and totality of all efforts is coordinated to have best maintenance at minimum cost.

The following maintenance information system is desirable.

(A) Registers and Codes :

(*i*) Equipment classification (cost centre).

(*ii*) Card (description/identification/Quarterly consumption/ equipment identification).

(*iii*) Machine part Register.

(*iv*) Drawing Register.

(*v*) Equipment classification and numbering of drawing.

(B) Organisation and Personnel :

(*i*) Departments under Engineering Services (Functional distribution).

(*ii*) Department organisation chart.

(*iii*) Service cards (Personal/family details of employees).

(*iv*) Approval and character roll (Officers).

(C) Planning :

(*i*) Monthly shut down planning (Equipment/purpose/effect or product/date and hours of shut down).

(*ii*) Long term planning (capital and major repairs).

(*iii*) Job planning for Maintenance and repair (day/description/type of work/workers allotment details).

(*iv*) Operational Estimates·Consumables stores for financial year (Quantity/Value)

(*v*) Revenue Budget—Spare parts for financial year.

(D) Operational Reports and Control :

(*i*) Daily Breakdown Reports. (Delay hours/nature of break down/causes/steps to avoid reoccurrence.

(*ii*) Daily Delay Report (delay hours/nature and causes of break down/steps/ agency responsible).

(*iii*) *Shift attendance of working Report.*

Strength
Absentee — Reason $\left.\begin{array}{c} \\ \\ \\ \end{array}\right\}$ By categories.

Utilisation analysis

Working Report showing nature of engagements.

(*iv*) Analysis of capital investment, spares stock, spare consumption value, maintenance cost, wages and salaries.

(*v*) Work order for spares/Repairs.

(*vi*) Inspection Reports (for spare parts).

(*vii*) Indents to stores/Purchase.

7. Material Management for Maintenance. Maintenance inventory costs are known to contribute a large proportion of the total inventory costs of plants. Effective control of spares parts inventory is of great importance for good maintenance management and the timely execution of maintenance schedules. Many times a large portion of the downtime towards maintenance is spent is due to the problem of spare parts. The problem could be due to any of the following.

(*a*) Due to uncertainty about the time when required and quantity required.

(*b*) Due to difficulties in physical procuremenr.

(c) Due to difficulties of distribution, storage and control identification.

(d) Due to indifferent quality of supplied spares.

However, the importance of having ready inventory cannot therefore be over-emphasized. Here again, the spare parts for vital and non-vital machines could be treated on different footings as also imported spares. A distinction is also to be made between fast moving spares needed periodically at regular intervals and insurance spares. The spares should be properly codified for identification and standardised to the most possible extent. Attempts should be made for finding better materials for spare parts.

It is advantageous to undertake manufacture of certain simple spare parts like pins, bushes, cams, gears etc., in a centralised maintenance workshop (within the enterprise or in ancillary industry developed by the present enterprise) and maintain small but ready inventories.

Economic Aspects of Maintenance

1. Systematic maintenance procedure offers tremendous possibility for savings in money, materials and manpower. These savings come through

(i) Reduction in down time.

(ii) Reduced loss of materials in process.

(iii) Increased life of the equipment.

(iv) Reduction in overtime.

(v) Timely replacement of spares and machinery.

(vi) Optimum spares inventory.

(vii) Maintenance of product quality.

(viii) Proper running of equipment.

(ix) Optimum operated cost of the machine.

2. It is necessary to decide the degree of maintenance effort needed, since over maintenance increases the maintenance with increases downtime and other losses.

3. The costs involved in a systematic maintenance procedure is more than off set by the advantages gained thereon. However, it is essential to categories the equipment and decide the type of maintenance of each piece of equipment needed based on their criticality.

4. Replacement of machinery is an important aspect of maintenance while availability of equipment and the resources are great limitation, an efficient cost accountings system and the collection of relevant data for decisions regarding repairs and renewals are important. Thus a proper costing system and adequate records are

of extreme importance for the efficient and economic performance of maintenance activity. Allocation of adequate resources for replacement of equipment and for providing maintenance facilities need attention.

5. The capital blocked in inventory of spares is considerable in Indian industries. A proper system storing and control of spares would facilitate release of much wanted capital and also ensure their availability when needed. Preventive maintenance is extremely helpful in streamlining the consumption and storage of spares.

6. In this connection standardisation of spares and equipment, warrants attention of the management and the government, which will help in reducing the variety of spares and standardise the same to the extent possible. In finding better and cheaper substitutes, modifications for better performance and import substitution, value engineering will prove to be a useful technique.

Engineering Aspects

The cost of maintenance accessibility for repairs and maintenance lubrication etc. all need to be thought of. In developing countries most of the equipment is still being imported. Even with in the country climatic and other conditions may differ from place to place. Hence it is necessary to ensure the suitability of the equipment to local conditions before buying the same. At the same time availability of spares and after sales service are important points and for consideration.

Managerial Aspects

Collection and analysis of data and proper planning and control are the causes of efficient maintenance management.

Planning of maintenance work due to the inherent nature of the jobs involved, is far more difficult than production planning. With a system of break-down maintenance, it is almost impossible to plan maintenance activities. However, establishment of a preventive maintenance system greatly facilities planning, by reducing the incidence of breakdowns. Schedules can be prepared for inspection and lubrication and the other activities can be reasonably well planned. After the system is in operation for sometime ; data can be built up and the system can be redefined.

Although accurate measurement of maintenance work is difficult, there are tools and techniques available to get a good estimate of the time required which is necessary for a proper development of labour and control.

Application of network techniques for scheduling of maintenance work has become popular and is being increasingly practised by maintenance managers.

Proper staffing of the maintenance department is another factor which is important. This will have to be based on anticipated workloads, on routine as well as none-routine jobs. While it is necessary to have the right amount of staff, maintenance jobs, being such that a certain amount of wastage is inevitable. It is, however, necessary to get the right type and amount of skills.

Training of maintenance personnel is a subject which should receive some attention of the management. Technical training to improve their skills and also productivity and attitude oriented training are both necessary.

There cannot be any hard and fast rule about the organisation of the maintenance department in an industry. It depends on various factors and varies from company to company depending upon the type, size and other factors. What is important is proper coordination and control among the various functions of maintenance and with other departments.

The maintenance should be independent entity apart from production, with sufficient authority and responsibility with due status, reporting to a fairly level higher level of management.

Maintenance Management will be really effective in an organisation only when there is understanding co-operation and co-ordination between production and maintenance. It is necessary that the production personnel appreciate the importance of maintenance of their equipment. At the same time maintenance should up-to-date its plans in consultation with production.

The performance of maintenance function need to be evaluated from time to time with a view to improve the same. The evaluation may be done through an analysis of costs, downtime hours or losses as compared to the available hours or through frequency of breakdowns.

It is needless to emphasise the priority to be accorded to plant maintenance by the Industrial Management in India. Proper and perspective planning at the National as well as the unit level will pay rich dividends.

Organisation for Maintenance

The success of a maintenance department is largely dependent on its organisation. Each enterprise has to adopt an organisation best suited to its purpose but the following guidelines can be indicated for adopting a particular set-up for specific enterprise.

(i) **Centralised Maintenance Organisation.** It can be adopted with advantage in a small and compact factory when inter-unit communication is speedy. The Chief Maintenance Engineer heads the organisation and should report to the General Manager, who is responsible for overall functioning of the factory.

(*ii*) **Decentralised Maintenance Organisation**. If the factory is large and its various units are situated far apart with difficulties of inter-unit communication, a decentralised maintenance organisation is generally favoured. A separate maintenance group for each unit functions under the administrative control of the Production Superintendent.

(*iii*) **Partially Decentralised Maintenance Organisation**. This set up is an improvement over the second category. Here the running maintenance of the equipment is day-to-day attenation to equipment is entrusted to a group of maintenance workers, who are administratively responsible to the Production Superintendent. Other important maintenance functions *i.e.* planning of scheduled maintenance works, procurement of spares for maintenance, and major over-hauling, operations are kept centralised under the Chief Maintenance Engineer. Such a set up serves the need of Production Superintendent for their day-to-day flexibility of work, and also enables a centralised maintenance policy to be adopted and implemented. Here also the Chief Maintenance Engineer is placed at the same level of authority as that of Production Superintendent, under the General Manager.

13

Decision Making

Management's primary function is to make decisions that determine the future course of action for the organisation over the short and the long term. The decisions may be directed in every conceivable physical and organisational area, they may deal with financial planning mårketing, and personnel, as well as with the operating or production phase. More often than not, decisions cut across these functional lines. Decision theory is directed toward determining how rational decisions ought to be made. It attempts to establish a logical framework for decisions that is firmly rooted in science and mathemetics as well as in the real world for various alternative action paths, risks are assessed so that the decision maker, maker, by his knowledge of the probable results can decide what to do.

STRUCTURE OF DECISION MAKING

The obvious implication of a decision is that alternatives exists, the process of decision making selects from these alternatives the cause of action to be carried out. We have alternatives and purpose to aclieve, and we need criteria for comparision. Each alternative may have both desirable and undesirable aspects ; these conflicting values must be reconciled. Since the results expected are future results, how sure can be of obtaining those results ? Actually, the final desirability of alternatives is the product of the relative benefit obtainable and the probability of attainment. The sophistication of decision making for a given area depends on the level of knowledge with in the area and the complexity of the decisions to be make. The The structure of decision making is given as in Fig. 13.1

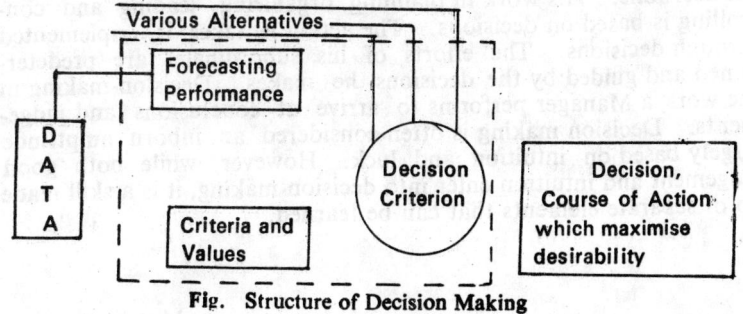

Fig. Structure of Decision Making

Decision making lies deeply embedded in the process of management. Decision making spreads over all the managerial functions and covers all the areas of the enterprise. Management and decision making are bound to go side by side. Whether knowingly or unknowingly the manager makes decisions constantly. The principle of decision making requires that the task of decision making should always be pushed down as far as practicable. On the other hand, the top manager can concentrate on important problems in order to give a effective judgement over the issues leading to decision making.

Most outstanding characteristic of a successful manager is his ability to make sound decisions. Decision making involves choosing a course of action from several alternative courses of action. Decision making—the actual selection from among alternatives, course of action, is at the core of planning. Decision making is a matter of planning organisational objectives and the methods that will be used to achieve them. Managers consider it as their job, because they must constantly choose what is to be done, who is to do it and when is to be done. Decision making must be regarded as a step in planning, even when done quickly and with little thought or when it influences action only for a few minutes. Planning imposes organisational responsibilities on subordinates (managerial and non-managerial personnel). Decision making is also concerned with the problem of motivating subordinates.

Right from the day when the size of business use to be very small to the present day size of the business when ownership is corporate, the importance of decision making has been there. The only difference is that in today's complex business structure, the decision making is getting more and more complex and important. Whatever a manager does, he does through making decisions. Some of the decisions are of routine nature and it might be that the manager does not realise that he is taking decisions. Other decisions are of a special nature that may require years of systematic and scientific analysis. The fact remains that management [is always a decision making process.

What is Decisions Making

A manager's effectiveness is related directly to the quality of his decisions. His work in planning, organising, leading and controlling is based on decisions. The action he takes is implemented through decisions. The efforts of his subordinates are predetermined and guided by the decisions he makes. Decision-making in the work a Manager performs to arrive at conclusions and judgements. Decision making is often considered an inborn amptitude largely based on intuition and luck. However, while both good judgement and intuition enter into decision-making, it is a skill made up of separate elements that can be learned.

PRINCIPLES OF DECISION-MAKING

Three basic principles will help provide a sound basis for management decisions.

(a) **Principle of Definition.** A logical decision can be reached only if the problem is first defined. To solve a problem effectively, you must first know what the p   n is. Time spent in developing a solution to a problem is often wasted because we are treating only the symptoms—we have not come to grips with the problem itself.

(b) **Principle of Adequate Evidence.** A logical decision must be valid in terms of the evidence upon which it is based. It should be possible to support the decision you reach with sufficient verifiable evidence to establish that your decision is both reasonable and logical.

Principle of Indentity

Facts may appear to differ depending upon the point of view and the point in time from which they are observed. The real value of money varies with time. The relative importance of the same facts may differ from year to year.

For these reasons, it is important to identify facts clearly and, in this identification, to make sure that we have taken into consideration differences that may exist because of point of view and point in time. The relative importance of the same facts may differ from year to year.

For these reasons, it is important to identify facts clearly and, in this identification, to [make sure that we have taken into consideration differences that may exist because of point of view and point in time.

Logical Thinking Process

A management decision should be the end product of a series of logically related steps. The logical thought process is made up of five steps.

(a) **What is the Apparent Problem.** When we first see a problem situation, the problem may seem quite apparent. Most often we plunge ahead eager to come to grips with the problem and to get action underway. In solving problems, also, we tend to observe symptoms arising from underlying causes. Often we jump to conclusions based on the superficial view. When this is the case, the probabilities are that our solution will be a treatment for the symptoms, and not for the basic problem. We can avoid this by taking systematic steps to identify the key issues and the central problem. Writing it down at this point will help to clearify our thinking.

(b) **What are the Facts** ? If we are to identify the problem precisely, we must first examine the circumstances surrounding it. This calls for collection and analysis of available facts, determining what the facts means and consideration of assumptions based on these facts.

Data-gathering should be a sharply-focused effort. It is rarely advisable to collect all of the facts bearing on the problem. Some information may be more difficult or costly to secure than the situation warrants. Since the time available for fact gathering is limited, it is wise to determine in advance what information we really need and in what priority we require it. Other-wise, we may find ourselves arriving at conclusions on the basis of large amounts of detail while we have virtually no information on the most significant aspect of the problem. Facts by themselves are not enough. We must determine what the facts mean, what special significance they have in the problem we are considering. Facts are past history. We need to make assumptions of the basis of the facts if we want to develop creative solutions. In fact-gathering, we can focus on the most essential information by investigating systematically each of the decision factors outlined below.

(i) **The Situation Factor**. What happened and precisely how did it happen ? Here we establish a clear-cut picture of the situation as it now exists. So far as possible, we rely upon first hand evidence. The information we gather at this early stage is historical. The should be acceptable to all who are concerned with solving the problem.

(ii) **The People Factor**. Who is involved ? In what way do the people affect the situation ? Consider the point of each person who is important to the problem and determine what special significance this may have.

(iii) **The Place Factor**. Where did the problem occur ? Does the location have any significance ? Are future changes expected which will affect the place factor ?

(iv) **The Time Factor**. When did the situation occurs ? Has the time factor particular significance ? How will the time

(a) **What is the Apparent Problem.** When we first see a problem situation, the problem may seem quite apparent. Most often we plunge ahead eager to come to grips with the problem and to get action underway. In solving problems, also, we tend to observe symptoms arising from underlying causes. Often we jump to conclusions based on the superficial view. When this is the case, the probabilities are that our solution will be a treatment for the symptoms, and not for the basic problem. We can avoid this by taking systematic steps to identify the key issues and the central problem. Writing it down at this point will help to clearify our thinking.

(b) **What are the Facts ?** If we are to identify the problem precisely, we must first examine the circumstances surrounding it. This calls for collection and analysis of available facts, determining what the facts means and consideration of assumptions based on these facts.

Data-gathering should be a sharply-focused effort. It is rarely advisable to collect all of the facts bearing on the problem. Some information may be more difficult or costly to secure than the situation watrants. Since the time available for fact gathering is limited, it is wise to determine in advance what information we really need and in what priority we require it. Other-wise, we may find ourselves arriving at conclusions on the basis of large amounts of detail while we have virtually no information on the most significant aspect of the problem. Facts by themselves are not enough. We must determine what the facts mean, what special significance they have in the problem we are considering. Facts are past history. We need to make assumptions of the basis of the facts if we want to develop creative solutions. In fact-gathering, we can focus on the most essential information by investigating systematically each of the decision factors outlined below.

(i) **The Situation Factor.** What happened and precisely how did it happen ? Here we establish a clear-cut picture of the situation as it now exists. So far as possible, we rely upon first hand evidence. The information we gather at this early stage is historical. The should be acceptable to all who are concerned with solving the problem.

(ii) **The People Factor.** Who is involved ? In what way do the people affect the situation ? Consider the point of each person who is important to the problem and determine what special significance this may have.

(iii) **The Place Factor.** Where did the problem occur ? Does the location have any significance ? Are future changes expected which will affect the place factor ?

(iv) **The Time Factor.** When did the situation occurs ? Has the time factor particular significance ? How will the time factor affect the decision you reach and the action you will carry out ?

(v) **The Causative Factor.** Why did the situation come about ? Why did the problem occur ? Will the factors that caused the present situation remain constant or will they change in the future ?

Information be gathered from many sources. This is one aspect of decision-making that a manager can most easily delegate. The more different sources are considered in securing data, the more varied the viewpoints consulted, the more accurate and objective the information is likely to be.

(c) **The Real Problem.** After we have thoroughly digested the facts and the inferences, we are in a position to review the problem as we first identified it. We may find, in view of the facts, that we had only identified symptoms of a deeper, more fundamental problem.

As much time and effort as necessary should be invested in identifying the basic underlying problem. Our solution will probably not be better than our identification of the problem. Finding the real problem is often difficult because we generally find a number of related problems in the situation. One or more of these may be immediate or emergency problems which require prompt action. Other aspects of the problem may have longer range implications. They require management action, that is, renewed planning, organizing, leading and controlling. In terms of the Principle of Operating Priority, we know that our natural tendency is to ignore the basic management problem and to take operating action without probing deeper. Two factors must be given proper consideration in coming to grips with the basic problem : scope and timing.

Regarding scope much wasted effort may occur if we try to solve problems for other people, or to extent our solutions beyond our own area of accountability. At the same time, we may little our own potential results if we do not take into account in our problem solving the needs and capabilities of other units.

A professional manager considers his problems and solutions in terms of their implications for other units and for the company as a whole. Executive times, unifies and integrates his own actions with those of his superior and with other units. Timing is important. The right solution at the wrong time is better than wrong solution at the right time. We can give full consideration to the time element if we make sure the problem we state encompasses the two aspects : What is the problem now, and what will it be over the long term. Once we have identified the difference, we can begin to look systematically for both short and long term solutions. An important skill of the professional manager is the ability to recognise the difference and to provide for it. The decision as to what really is the problem in matters involving two or more subordinate units, or who own group and other groups work is the area where the manager cannot safely delegate.

(d) **Possible Solutions.** The ready answer will become obvious once the problem is clearly defined. Develop other possible solutions, and carefully analyze those which have promise. Discuss the pros and cons, or advantages and disadvantages of each with subordinates. Consider in each case what will happen if you carry the proposed solution through to its logical conclusion. If need to be analysed after asking questions : Will it give you the results you want ? Has it provided for long range requirements as well as putting out the immediate fire ? This is an excellent stage to secure participation of subordinates perhaps it is not possible to ask your subordinates to help identify the real problem because they did not have sufficient background or an allround viewpoint. But now explain the problem fully to them. Ask for their suggestions. Make sure they back up their proposals with proper study and analysis. They will probably think of some ideas that would never have come to your mind. And remember that the more they know about the problem, the more they recognise its complexities and implications, the more effectively they will carry out the solution you develop with their help.

(e) **Best Solution.** The professional manager decides, not by pulling an idea out of the air, nor on the basis of hung but by rational selection of what promises to be the best of the various possible solutions he has analyzed and considered. Selecting the best solution requires objectivity. Try to get your emotions and feelings out of the picture. At this point let your head guide your heart. Where the decision involves two or more subordinates, or the interests of his unit overall, or other units, the manager himself must make the decision. Because he has been organizationally placed to have best all around perspective for his own unit, the accountability for final decision is his. Once you have made the decision, let people know about it. Tell them why you made it. Some will be disappointed, Perhaps, that you did not follow their ideas more closely. But you can incorporate their suggestions more fully as you take the final step, developing a course of action.

(f) **Course of Action.** Now you decide how to implement the decision. Establish a fully-detailed course of action by planning the necessary activities. Determine :

(i) *Objectives.* What are you trying to accomplish ? Get full participation of those who are to carry out the action.

(ii) *Policies.* What policies apply ? Make sure that you identify these and explain why and how they apply.

(iii) *Programs.* Precisely what action steps are to be taken, and in what priority ?

(*iv*) *Schedules* : Participation is most important in programming and scheduling. So far as it fits within your overall course of action, let each individual determine for himself what he will do and how he will do it, to reach the goals you have agreed upon.

(*v*) *Procedures* : If certain work has to be accomplished in a standardized and uniform manner, specify what procedures are to be followed or make provisions for developing the necessary procedures.

(*vi*) *Budgets* : Are there expense limitations ? Specify what can be spent in terms of the time of people, material and tools and facilities.

Types of Decisions

There are two basic types of decisions, spontaneous and reasoned.

Spontaneous Decisions

These are intuitive decisions. They are conclusions arrived at without conscious thought or analysis. A spontaneous decision is often a hunch-an intuitive reaction to memories or previous conditioning, Spontaneous decisions are, to a large extent, unthinking decisions. Even if they are made after observation of a situation or following review of facts, there is little attempt to determine what the facts mean or to reason logically to a judgement based upon the facts. Often we decide in term of precedent because it requires least effort. Also it is easiest to rationalize.

But situations are rarely identical. All things tend to change with time. So do people. Yesterday's decision may seem best because it is familiar and was successful. But now conditions may call for a revised estimate and fresh conclusions. Spontaneous decisions are useful when dealing with routine. They may be necessary in handling emergencies. But since they do not provide for logical analysis and the development of alternatives. They harness us to the past and offer little opportunity for developing innovations or improving established methods.

Reasoned Decisions

These are based on systematic study and analysis of a problem. The facts are assessed, the alternatives weighed and a decision reached in terms of what is best in the light of all the pertinent factors. Reasoned decision making should be followed wherever possible. It brings into play our full mental resources. It emphasizes the creative aspects of problem-solving and enables us to bring to bear upon our problems a consistent approach to thinking which will minimize the probability of error and give us greater assurance of productive results.

Similarly the types of decisions can be further classified as (*i*) Tactical routine and (*ii*) Strategic.

Tactical routine decisions are those which one makes over and over again following the established rules, policies and procedures. These decisions do not require collecting new data nor conferring with people, nor passing judgement and nor for any originality. Tactical decisions can be taken from the view point of finding a correct answer because here the tactical decisions are one dimensional and at times manager does not realise that he is taking decision. The only action which has to be taken is to accomplish the goals with the minimum of effort and disturbance.

The second type is the strategic decisions. These are the really important ones. They either involve finding out what the situation is, what the resources are and what they should do. A manager is, in fact, concerned with these types of decisions and the higher his level the more of these he is required to make. In this type of decisions, effort should not be towards finding a correct answer only but also towards finding a correct question. Secondly, more finding out a correct answer to a correct question is also not sufficient unless the answer/decision can be implemented. Decisions are not made because they are important or they should be the show pieces of filing cabinets, but the managers are concerned with their performances. Really, this means that the decision must be compatible with each other and also consonant with the goals of the whole business. Therefore, the decisions cannot be taken in isolation of the prevailing situations. For the purpose of making a good decision, limitations must also be kept in view.

Importance of Decision Making

The importance of decision making is very much realised, but the problem is that there is a tendency more towards finding a correct answer rather than a correct question. It is here that the crux of the problem lies. The crux of the problem implies that answer has no importance, but what is of importance is that the manager should find a correct answer to a correct question. Merely finding an answer would be very much misleading and may jeopardise the total objectives since right answer to a wrong question is more harmful than wrong answer to a correct question.

An executive is a man who has to make up his mind quickly and who should always make good decisions. It is also not correct that executives have to make spur-of-the moment decisions all the time. For most of the decisions they get enough time for careful fact finding, consultations, analysis and mulling things over before they have to decide. Deciding is a human process. When one decides he chooses the course which he thinks is the best. It may or may not be the one that is truly the best. Decision making is a blend of thinking, deciding and acting. An important executive decision is only one event in a process which requires a succession of activities and less decisions all alone the way. Decisions also have a time dimensions and a time lag. It takes time to collect facts and to weigh an intelligent decision. More over, after one decides it takes still more time to carry out a decision and often it takes longer before one can judge whether the decision was good or bad. Also, it is difficult to isolate the effects of any single decision.

Process of Decision Making

The process of decision making has shown in Fig. 13·2. It has the following phases :

(i) Defining the problem ;
(ii) Analysing the problem and gathering information ;
(iii) Developing alternative solutions ;
(iv) Deciding upon the best solution : and
(v) Converting the decision into effective action.

Depending upon how the decisions are made, they can either be complete waste of time or best guide to solve the problem. A word of caution here is that sufficient time should be spent on defining the problem, analysing the problem and developing the alternatives. Good time is also necessary to make the solutions effective, but much less time should be spent on finding the right solution because if the preceding steps are taken carefully, finding the right solution will not be much of a problem. Decision making is a cooperative activity performed by executives who occupy positions in a organizational structure. The positions and persons in such a structure are inter-related by a communication system. Therefore, for an effective decision making, importance of a good communication is not ruled out.

(i) **Defining the Problem** Every problem has two aspects. First there is a situation and second there is your dissatisfaction with a situation. When you become dissatisfied with a situation and decide that you ought to do something about it, you recognize that you have a problem. This is the start of decision making.

After recognize that you have a problem, you may try to define and clarify it and to pin-point it. It is not always easy to define the problem, to see the fundamental thing that is causing the trouble and that needs correcting. Yet the manager has to try to find it out and to decide what the core problem is, otherwise he will ask and answer the wrong questions, make wrong decisions and may not solve the real problem. He stays in trouble because he gets answers to the wrong questions. Here again a decision involved and he has to decide what the problem is.

Generally what is experienced is that a problem never presents itself as a case for solution. What at the most can be observed are certain elements or certain symptoms. The first task in decision making is, therefore, to find out what the problem is. To pin point the problem simply by symptomatic diagnosis is not the right answer. In business, we are not dealing with physical sciences that the

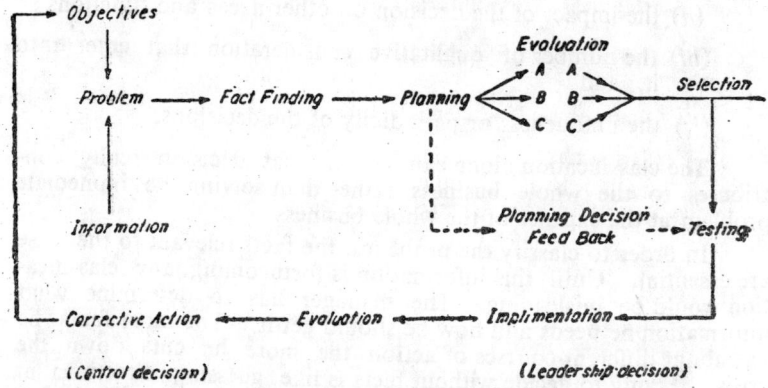

Fig. 13.1 Decision Making Process.

response to a particular situation will always be the same. A manager has to analyse and find out the critical factors of the problem, what are the elements of particular situation which lead to and consitute the problem. Symptomatic treatment of diagnosing the problem is wrong because first the symptoms may not be dependable and even if they are, there is no guarantee that those symptoms will always lead to one particular problem. Therefore, the best way is to analyse the problem rather than diagnose the problem.

The second step in the definition of the problem is to determine the conditions for its solution. In defining any problem, it is of utmost importance that the objectives of the business, both immediate and the long range are kept in view. Side by side, the principle of limiting factor must also be borne in mind. What are the principles, policies and rules of conduct that have to be followed ? To spell out the rules is necessary because in many cases, the right decision will require changing accepted policies or practices. Unless the manager thinks through clearly what he wants to change and why, he is in danger of trying at one and the same time both to alter and to preserve established practice.

(ii) Analysing the Problem. Finding right the question, setting the objectives and determining the rules together constitute the first phase in decision-making, that is to define the problem. The next phase is to analyse the problem ; classifying it and finding facts. It is necessary to classify the problem in order to know who should take the decision, who should be consulted in making it and who are the persons to be informed. Without prior classification, the efficiency of the decision may be jeopardised. By classification we can know how to convert the decision into practice.

There are four principles of classification :—

(i) the futurity of the decision ;

(ii) the impact of the decision on other areas and functions ;

(iii) the number of qualitative consideration that enter into it ; and

(iv) the uniqueness or periodicity of the decisions.

The classification alone can ensure that decision really contributes to the whole business rather than solving the immediate problem at the expense of the whole business.

In order to classify the problems, the facts relevant to the case are essential. Until this information is forthcoming, any classification would be misleading. The manager has to determine what information he needs and how he should get it. The more facts he has about different courses of action the more he cuts down the crisis. Trying to decide without facts is like guessing directions at a cross road instead of reading the highway signs. Gathering facts also takes skill and requires decisions. So a part of any problem is deciding what facts one wants to find out relating to it. Another part of the problem is to decide how to get the desired facts. Finding out facts may be expensive and even difficult. Before one can decide any big problem, he must decide how much time and money he is willing to spend to get the information he needs. It is better to have incomplete information on time than to wait for more complete information that arrives too late to be of use. Since gethering facts costs money, the readily available sources may be used whenever possible. One should use easily tapped files of current information, statistical data, controlled experiments, and the like whenever these are available at low cost Getting facts which bear on a problem does not always require doing a research project. A good many people seem to think that no fact exists unless a survey unearths it and proves that it exists. They seem to forget that facts are there all the time and that a survey does not make them, it just unearths them. Men working on almost any job are actually engaged in informal research all the time. One can use their knowledge, if one can and it may save him from going into an expensive research. When one collects the facts to help solving problems, he wants facts that relate to alternative courses of action. The facts gathering must be solution oriented. One must decide what the several alternatives are and then collect facts which will allow him to compare them. Getting the information is not sufficient, the manager should also know how to use this.

Another aspect of analysing the problem is that it is not always possible to get all the information which is needed for defining the problem. Here the job of a manager is to judge how much the risk the decision involves, as well as degree of precision and rigidity that the proposed course of action can afford. In other words, the decision making process involves precisely judging the unknwon factors.

(*iii*) **Developing Alternative Solutions.** Assuming known goals and clear planning premises, the next step in the decision making process is the development of alternative-solutions for every problem. Without this, one is likely to be caught in a wrong solution. Without resorting to the process of finding alternative solutions, the manager is likely to be guided by his limited imagination. It is rare for alternatives to be lacking for many courses of action. But it times the manager assumes that there is only one way of doing a thing. What the manager has probably not done is force himself to consider other ways, open his eyes and develop alternatives. Unless he does so, he cannot know whether his decision is the best possible. From the above it appears a key planning principle, which might be referred to as the principle of alternatives. In every course of action alternatives exist, and effective planning involves a search for the alternatives representing the best path to a desired goal. Once the manager has decided upon a particular course of action, he thinks that this is the only and the right way of acting. On the other hand, once the manager starts developing alternatives, various assumptions come to his mind which can be a force to compel him to bring those assumptions to the conscious level. A word of caution here may also be added. Alternatives solutions are not a guarantee of the righteousness, but these help the manager to weigh one alternative against another and thus minimise the uncertainties.

Nevertheless, development of alternatives cannot provide a person with an imagination he lacks. But most of us have definitely more imagination than we ever use. The mind's viscous can be trained, desciplined and developed.

It is not always necessary that the alternative solutions will necessitate taking some action. To take no action is also a decision fully as much as to take a specific action. Only few people realise this. It is of imperative nature that in all organizational problems, the alternative of taking no action is considered. For example, if there is an unnecessary post in the organisation, the solution not to fill the post will be the best one. The ability to develop alternatives is often an important as making a right decision among alternatives. The development of alternatives, if thorough, will often unearth so many choices that the manager cannot possibly consider them all. Even with latest mathematical techniques and electronic computers, the analysis of alternatives and their comparison with one another is almost impossible without some preliminary selection.

(*iv*) **Finding the Best Solution.** Once alternative solutions have been thought of, the next step is to analyse each solution and weight one against the other. Decision making implies freedom to choose from among alternative courses of action. If the steps preceding to finding out the best solution have been dealt with adequately, the manager will have several alternatives from which he can make the choice. It is rare situation where there are no alternatives. In fact, if one is faced with such a situation, it will only implies that the preceding steps have not been taken systematically.

There are four criteria for finding the best from among the possible solutions. These are :

(a) *Risk* : The manager has to weigh the risks of each course of action against the expected gains. There will not be any situation where risk is not involved. What matters is the intensity of one risk against another and the expected gains.

(b) *Economy of Effort* : The best manager is one who can mobilise the resources for the achievement of results with the minimum of efforts and the least of disturbance in the organization.

(c) *Timing* : Another important aspect in the process of choosing the best course of action will depend upon the particular situation. If the situation has great urgency, the preferable course of action is one that alarms the organization that something important is happening. If, a long, consistent effort is needed, a 'slow start gathers momentum' may be preferable.

Decisions concerning timing the extremely difficult to systematize. Theyelude analysis and depend on perception. Here, there is one guiding principle whereby managers must change their vision to accomplish something new. It is preferable to present to them the big view, the completed programme and the ultimate aim. Wherever they have to change their habits it may be best to take one step at one time, to start slowly and modesty, to do no more than is absolutely necessary.

(d) *Limitation of Resources* : One of the most helpful principles of planning and decision making may be called principle of the limiting or strategic factor. In choosing from among alternatives, primary attention must be given to those factors that are limiting or strategic to the decision involved. In every area of decision certain factors are strategic in determining whether goals will be attained.

In decision-making the discovery of the limiting factors may not be so easy, since economic factors are often obscure and people and their reactions complex, but the principle is the same. The search for and recognition of limiting factor in planning and decision making nevers ends. Discovery of the limiting factor lies at the basis of selection from alternatives and hence of planning and decision making. It is generally not possible to explore thoroughly every problem and the solution of every limiting factor, so the business manager must exercise judgement in determining where and how research in this area can best be used. In selecting from among alternative three basis for decision are open to the manager—

experience, experimentation and research. The most important resource whose limitations have to be considered, are the human beings who will carry out the decision. The competence, skill, working conditions, abilities and understanding have to be kept in view. A course of action may require the line of competence etc. of the people which is not available in the organisation. Then the obvious preceding step will be to train the people in the organisation or to hire trained people.

Another consideration which is important is that a wrong decision must never be adopted because the resources for implementing the right decisions are lacking. The decision should always the among the alternatives which are genuinely directed towards the solution of the problem. It is no use deciding upon the solution for which the appropriate resources are not present. Such a decision will only be the show piece of the filing cabinet rather than directed towards any better course of action.

(v) **Making the Decision Effective.** The decision will have no meaning unless it can be implemented. The managers are not concerned with the decisions in remoteness, but with their performance. The manager's decision is always concerning what other people should do. For this purpose, it is imperative that people should feel attached towards the performance of the decisions. Any idea of the decisions being sold will be a futile exercise. If this has to be the case, it might also sometimes be that the good decisions have to be subordinated because these connot be sold to the persons concerned. What is right has to be decided by the nature of the problem rather than by wishes, desires and receptivity of the people in the organisation.

In order to make the decision effective, it is necessary that the concerned people understand what the decision involves, what is expected of them and what reasonably they should expect of others. In order to accomplish this, it is desirable that in matters of bringing about change in the behaviour, the principle of slow and steady should be followed. It will be poor decision making which expects the overnight transformation. Another principle which is important is that there should be effective communication. This principle involves that concerned people should be informed only of significant deviation, but this should be in very clear and unambiguous form.

In order to make the people attached to the decision, it will be necessary that there is partic pation by the people in the decision making process. People accept more willingly decisions that they have helped to make. Superiors who discuss problems with their subordinates and who give them a chance to ask questions and make suggestions find more support for their actions than the superiors who decide everything themselves. Of course, subordinates participation is not desirable in every step. For example, they should not participate in the definition of the problem. At this stage, the manager does not know who should participate and who should not because it is not clear whom the decision is going to effect. Their participation is also not desirable at the information gathering stage. The area where the concerned people can participate is in the development of alternatives. If this can be achieved, the people will feel attached to and responsible for taking the definite course of action. Secondly, certain alternatives might come to surface which the manager may not have though of. Participation by subordinates has established itself as one of the most effective methods of decision making. The research in this area shows that participation is helpful in overcoming resistance to the decision of change from one job to another and increases motivation for work. The practice of participation however a complex problem. It is dependent on the nature of decision, the organizational climate and attitude of employees towards participation.

Though the participative decision making in the manner as detailed above is considered an effective approach but opinions also exist in its favour and against. There is some evidence that group decisions are likely to be poorer decision than one man decisions. Group participation does not necessarily improve the quality of the decisions and, in fact, sometimes imairs it. Group decision making is like a train in which every passenger has a brake. People however often seem to feel strongly about the group participation in decision making. They are strongly for it or strongly against the idea. *Harold Smiddy*, former Vice-President of General Electric Company says that there is too widespread drift toward group responsibility 'togetherness', 'other directedness' and a flight from 'apartness'. He thinks that this weakens responsibility and dilutes managerial functions too much. One loses time, encourages managers to evade responsibilities and becomes indecisive and makes it hard to fix responsibility. There is another argument against too much group participation in decision. Not every one wants to participate. Furthermore it is discouraging to participate and suggest and be overruled again and again, whether by equals or superiors. Participative decision making has had good points as well which have been discussed above.

Psychologist *Rober McMurray* says that you can have a considerable degree of authoritarian decision making without upsetting people. He says that employees tend to be contented with what they are used to. Firm decisive and authoritarian leadership does not upset most employees who are used to it. In fact, says *McMurray*, it is preferable unless employees are interested in more participation, or are ready to assume responsibility for it, and have sufficient experience and knowledge, to deal with the problems at hand. He says that most employees are unable or unworthy to participate in management decisions.

Tools of Decision Making

So far we have talked about the importance, types and the steps for taking the decisions. However, much of this process will become cumbersome unless the manager is adequately equipped with certain tools for sorting out many complexities at every stage. Above all, it is essential that the manager understands the process. Firstly a number of mathematical tools have become avaialble which are very useful and powerful. Secondly, with the rapid development in technology, the balance between tactical and strategic decisions is rapidly shifting. Most decisions that have always been tactical, if not routine, are rapidly becoming strategic decisions containing a high degree of futurity.

Management and Non-Management Decisions

Management decisions are concerned with organizational objectives, and managers are the persons who accept the responsibility for making such decisions. The leader of an informal group within the organisation, though he may make decisions that govern the behaviour of others, is not a manager. The same conclusion can be made about a labour leader who participates in the negotiations of a collective bargaining agreement. The decisions made by informal group leaders and labour leaders involve the personal goals of organizational participants.

Top level decisions are series of inter-related and continuing activities. Also the environment at the top level of management hierarchy is a complex relationship. It is at this level where the managers must be a specialist and at the same time a generalist. At this level the job specifications are very broad, overlap of responsibilities is common practice and communication is of vital importance and is frequently carried out on an informal basis. But with in the same managerial hierarchy there are functional identifications that may stand in the way of smooth operation of the decision making process. In situations where an executive identifies strongly with a department, the tendency is for the manager to make decisions that consider the interest of the department first and the total organisation afterwards. At all levels of management the managers must make decisions on the basis of overall organisation objectives as well as departmental goals. It is in such an amorphous environment that the members of top management must make decisions, maintain, control and should carry out their responsibilities to society. There is a tendency among the managers at the top of management pyramid to view objectives, policies, strategies etc. differently than those on the lower levels. This is an indication of organisation weakness. If the lower levels of management do not comprehend the significance of decision made by top management as related to objectives, policies etc., then inefficiency may occur at all levels. This lack of understanding is not only due to mis-communication but also due to not knowing that different points of view exist. In general, managers should view the problems from its many sides to have a greater understanding of the situation.

Analysis of Decision Making

The analysis of any problem for decision making requires a step by step evaluation of the facts and ideas presented before plans or strategies may be developed. It is difficult to set-up a procedure of analysis that would be suitable for every problem.

Several Centres may be critical in the diagnosis of a case. All but one of this diagnositic centres pertain to internal factors which are illustrated in Fig. 13.2. The external environment of the firm is an additional diagnostic center. In a very complex problem all of the critical factors are involved so that it would be necessary to include them when making the diagnosis. In Fig.13.3a the analysis starts with sizing up and continues on to objectives, policy, control,

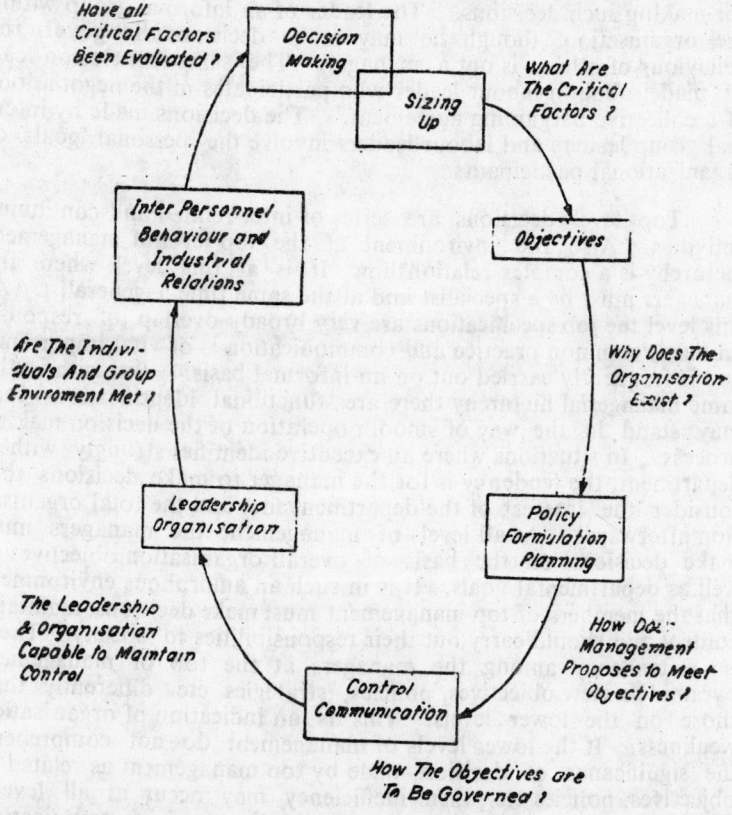

Fig. 13.2 Critical diagnostic centres and paths in decision making.

organisation, industrial relations and finally decision making. Whether a center is critical or not depends upon the scope of the problem and its relationship to other diagnostic centres. Such relationships will be indicated by the connections or paths between them. It should be considered as an analytical frame work, a way of thinking about the relationships between the various factors that bear upon the success of an organisation.

Sizing Up

It is best to size up the situation first by deciding what areas are to be analysed in what order. In analysing, the critical centres should be identified at the beginning. Some of the problems or cases require a complete examination of every factor and going through the complete diagnostic cycle while other problem or cases require only a part of diagnostic scheme. As shown in Fig. 13.3 (a), it requires an analysis of leadership, organisation and inter-personal behaviour ; and policies, planning and strategy are not critical to decisions. Sizing up also means an examination of the evidence, determining the relationship of facts, making some pre-judgements dealing with possible causes and effect.

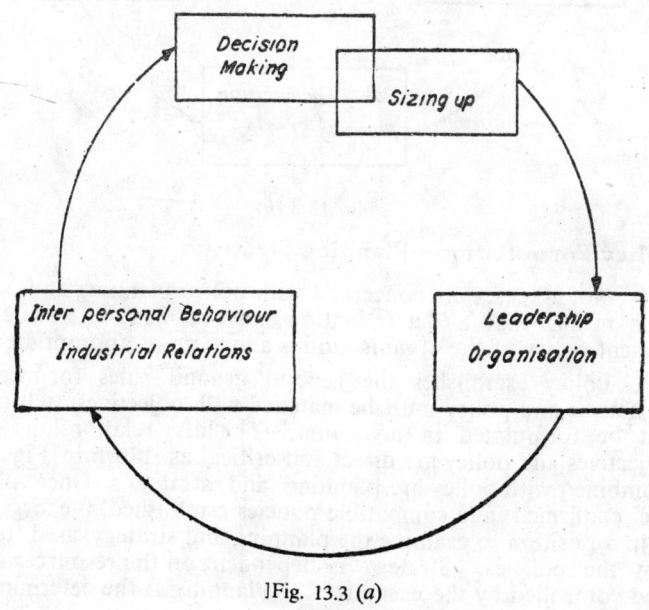

]Fig. 13.3 (a)

Objectives

The objectives of organisation must be defined in workable or useful terms so that they can be related to the executive decisions, Keeping an eye on the profit is of course a very suitable activity for a manager but just as important is the continuous re-evaluation of the product or service objectives. It is of no good for any organization to watch only the profit if the next year it will be out of business because of poor service to the customers. Therefore, a manager must not consider objectives sacred and unchangable but rather is being flexible to meet the dynamics of the business environment.

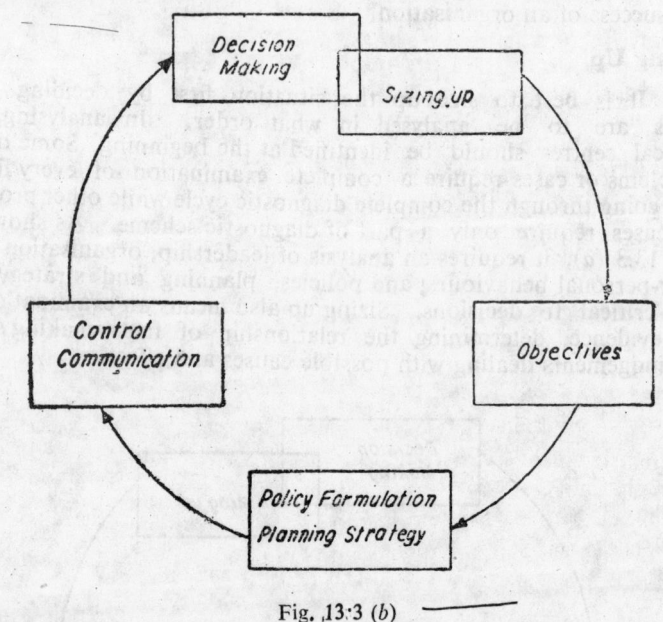

Fig. 13·3 (b)

Policy Formulation—Planning Strategy

Not all cases are concerned with policy planning and strategy. This means that a determination must be made as to whether the present policy of the organisation is adequate or appropriate.

Policy establishes the general ground rules for meeting the objectives and policy must be matched with objectives and it should not be formulated in a vacuum. The inter-relationships between objectives and policy are direct and critical as shown in Fig. 13.3 (b) Combined with policy are planning and strategy. Once objectives are confirmed and compatible policies established, the organisation is in a position to examine the planning and strategy used to carry-out the policies. Strategy is dependent on the resources available and controlled by the executives. Planning is the determination of how best to carry out the strategy.

Control and Communication

As nerves are necessary to the human body, in the same way control and communication net works are necessary to an organisation. It is really important to evaluate the communication aspect because it is an integral part of the control system within an organisation. Reports, memoranda and official notices are part of communication and are also an integral part of the control system with in an organisation. Wordless actions feed-back and lack of communication are also part of communication net work. In diagnosing a problem or a case the manager should examine the controls and communication involved to determine their adequacy and indeed see if the appropriate controls exist. Appropriateness is determined by whether the controls aid in carrying out the plans and strategy and whether they are suitable for the existing organisation and leadership.

Leadership and Organisation

This part of the analysis is related very closely with the control and communication factors. There is such a close integration that it is not always clear as to which center should be diagnosed first, it may be that an investigation of the communication or controls is meaningless without relating them to the organisation and leadership patterns.

Inter-Personal Behaviour and Industrial Relations

These are the other analysis factors where diagnosis of the human element is carried out. The organisation is established by people and is operated by people. It is only in the area of human behaviour where one finds that the legical decisions may not be the correct decisions. It is difficult to relate profits to human activity but as the companies are profit making organisations, human element decisions should be oriented in that direction.

Social and Economic Setting Influencing Decision Making

The external environment is an integral and important part of the problem analysis. The environment may include legal requirements as established by laws, social moves of society and economic conditions. The external environments surround the organisation and since the company is dependent entirely on the society it must continuously refer to the outside world. The social and economic setting constitute the eight critical diagnostic centres as shown in Fig. 13·5. It is no doubt a difficult task for the manager to balance his responsibilities to the individual employees and to the company. If management by itself does not take the initiative in discovering its responsibilities, society will impose the responsibilities upon the company and the managers through legal channels and social pressure.

In carrying out an analysis one should keep in mind that the end result should be a recommendation for action. A decision is to be reached and implemented. The process of analysis should examine the related factors such as :

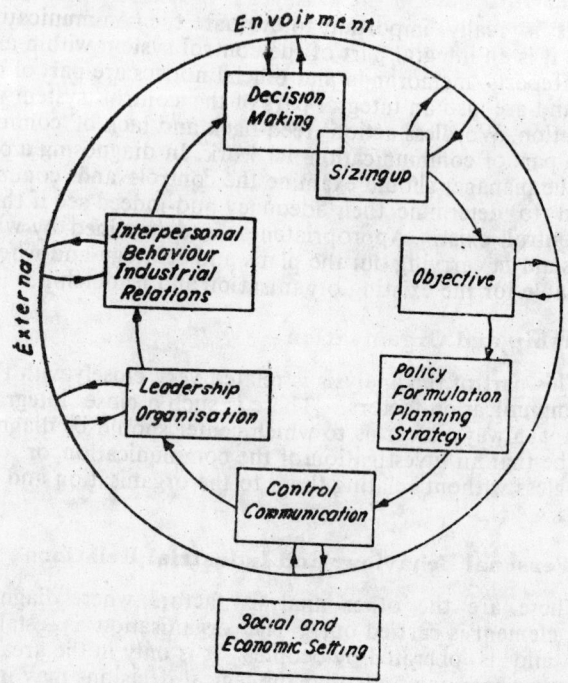

Fig. 13.4 Critical diagnostic centres and external environments.

(*i*) Assessing the risks involved in various courses of action.

(*ii*) Predicting the obstacles of making the various plans of action operational.

(*iii*) Evaluating the financial resources and cost.

(*iv*) Considering the human factor when recommending changes.

The decision making process is described at times in terms of models that contain inputs, variables and outputs. It is simply illustrated in Fig. 13·5.

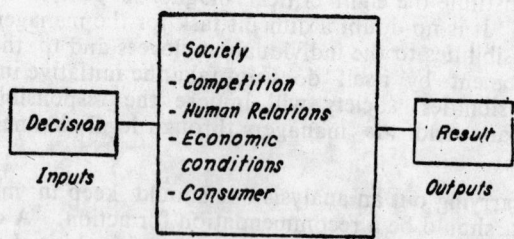

Fig. 13.5 Input-Ouptut Model.

One of the jobs of the manager is to decide what information is to be available for decision making. Sales, cash inventory, accounts payable and receivables are the obvious facts that most companies maintain. Other informations which need to be available are customers preference, employees morale, competitor's position, market trends, changing manpower needs, evaluation of employees performance, changing social norms, etc.

Once the decision has been made, it is important to size up the situation again to be sure that vital information has not been overlooked. It is manager's job to make decisions based on his own evaluation of the facts and upon the judgement concerning the probability of the success of decision in the light of his analysis of critical areas.

Managerial Functions and Decision Making

In analysing the various functions of management ; namely, planning, organising, staffing, coordinating and controlling, we find that each one of them involves decision making. Planning for example, is in itself a decision making process. In planning, we choose between alternatives and in selecting one particular course of action we decide to eliminate other proposals. Similarly, in organising, we 'decide' which activities should be grouped, who should be incharge, and what type of authority should be employed. In the same fashion, when employees are recruited and their efforts directed, the decision making process comes at work. A manager must decide whom to employ or to promote or how best to direct them and lastly, the "controlling function" becomes meaningless without decision. Control implies a "decision" must be made as to what action needs to be taken.

Some Tools to Effective Decision Making

The reason why many people are indecisive is fear of having to assume responsibility for the choices they make. The following rules could help overcome the common fears that enter into decision making.

(*i*) An executive is bound to make a wrong decision occasionally. So what ? As the British stateman Gladstone remarked, "No man ever became great or good except through many and great mistakes. Therefore forget your mistakes, but never 'what they teach you.' They are your best instructors."

(*iii*) Whenever you find yourself hesitating unduly in arriving at a decision ask yourself, "How important to me or other people is this matter ?" Take an objective and impersonal view point as far as possible. If the matter is really an important one, take time to think it over and consult others if you think they can help. But if the problem is minor one, take a decision and then forget it.

(iii) Comfort yourself with the thought that you have added to your stature mentally and emotionally by taking action yourself instead of avoiding the issue or procrastinating or passing the buck.

(iv) Never forget that others will judge you by the long range overall accuracy of your decision, and not by each isolated instance.

The only intelligent attitude to overcome indicision is to develop determination. It may not be possible to overcome it completely because it does not take much strength to do things, but it requires great strength to decide on what to do. It is only by putting forth constant and continued effort and conforming to the process of decision making that any manager can make up his mind as to when and what he should do. Making business refers to a variety of needs and goals. It requires judgement, making it possible by narrowing the range and the available alternatives, giving it clear focus, and providing reliable measurements of effects and validity of actions and decisions.

Need for Creativity in Decision Making

Planning that adds some new and useful element is 'creative'. Every invention of a product, process or machine contains some creative element. Advertisers strain to make their display and slogans new and original ; and sales executives think of new markets, shopping centres, or new packages or merchandise, creativity is need in all socio economic activities viz. medical, science, development of atomic energy and in the discovery of new forms of arts such as literature, music, painting etc.

Because of such a widespread need for creative thinking it is important for a manager to decide whether this quality can be developed by him and encouraged in his subordinates. Some individuals have a greater element of creativeness than their colleagues. It is therefore less imaginative or conservative individual who looks on change as something reasonable. Creative ability are by no means rare. Much of the work in decision making involves creativity. A manager who is full of creative ideas will take better decisions.

From the discussion, we reach to the conclusion that good decisions do not just happen. They must be developed by good analysis, study, understanding and evaluation. In decision making, the first essential is to define the problem. This requires a thorough and careful diagnosis. If we do not know what we are trying to solve, we cannot render a rational solution. Once the problem is defined, the next logical step is the assessment of the situation. We must gather and analyse the related facts. However, one thing we must not overlook is that we should be aware of our personal bias. When a manager fully realises these aspects it will not be difficult for him to measure the situation. This will help him to take objective decisions. Lastly, the decision maker, to find out the effectiveness of his decisions, must answer the question whether or not his alternatives are adequate both in number and in magnitude considering the situation.

Judgement and Decision Making

Judgement is an important element in decision making. When values are clear, information is adequate and risks are reasonably predictable, the decision making may not appear to involve so much of judgement. But this is not always true because in selecting the alternative, comparison of values and relations of things must be mentally formulated and asserted. Let us take a situation where value comparisons are difficult, information spotty and risks are unknown. Here the presence and need for judgement is more evident. Actually, under such conditions, judgement becomes the medium by which conflicting values are resolved, risks assessed and alternatives evaluated.

Decision Making Environment

Our decision making environment, is increasingly becoming more complex and highly interacting and interwoven. Therefore, it is essential to .utilise the modelling approaches which can identify the controllable and uncontrollable situational variables, and can develop appropriate strategic plans which are technically feasible and economically viable. In the initial stages of application of the modelling approaches, we may fail because of 'our inadequacies but no doubt is there that such approaches can lead us to pragmatic decision frontiers where we shall be able to coordinate both technocratic and bureaucratic patterns of behaviour. Its primary role should be of process-change-agent to diffuse power, to divide authority and distribute influence for a successful adoption of the approach or revision of the existing approaches. The analytical findings of the technocratic methods and discretionary judgements of the bureaucratic methods should be amalgamated depending on the environmental requirements. The modelling approach requires that the technocrats, for their part, have an obligation to use more direct, concise terminology, to phrase their findings in terms relevant and understandable to the bureaucrats, and try to understand all elements of the problem facing the bureaucrats. On the other hand, bureaucrats must try to learn more of the languages of technocracies atleast enough to understand the general working of the modelling techniques, if not the details of methodology.

Decision Making: Use of Modelling Approaches

Management science models are potential tools within that design and implementation process. These scientific tools provide only the mathodological insights into the problems to recognise the relevant factors and their systematic interrelationships. But primarily the design process requires the art component of management, *i.e.*, it requires an intuitional amalgam of observation, experience and judgement.

Formalisation of Decision Problem

The decision process is as old as human civilization. It has been recognised progressively that the public policy decision process is a complex one. The complexity assumes different facets ; largeness, multidimentionality and hierarchy, etc. A public policy decision problem, in the formal sense, can be categorised by the following characteristics :

(*a*)   One or a group of decision makers.

(*b*)   a set of possible actions.

(*c*)   a set of outcomes resulting from the choice of a particular possible action or actions.

(*d*)   a set of value dimensions or objectives which enables the value of a particular outcome to be measurable in some sense.

(*e*)   a set of utilities representing the utility of the outcome with respect to its measured value by the decision maker and using the given action.

Behavioural Truths

In the real life situations we observe the various behavioural truths. Behaviourally, people as customers or users of the utilities or outcomes, are the sole causes of any public activity. In this context, they believe in maximum welfare. They judge the welfare by value of growth criteria in terms of use, quality, reliability, service and other benefits.

Behaviourally, people as planners or who take the responsibility of the final users either in the constitutional manner or in the obligatory manner, identify the needs of the users, create environmental conditions for possible actions and outcomes, and make recommendations. In an open democratic society, the behaviour of these three different groups overlap, because the role playing or functional boundaries are not specifically marked. In this paper, our focuss is towards the role of decision makers, and purpose is to elaborate how public decision models are constructed and what are their limitations and potentialities.

Nature of Decision Making

The primary role of decision maker is to make right decisions, which may be defined as decisions made on the basis of perfect information. Perfect information in this context signifies that the decision maker knows exactly the probabilities of outcomes as well as their values to him. Under no circumstances, perfect information can be available for a decision problem solution. The best one can do is to seek for some rational solution under the information constraints. In a democratic pluralistic social framework, the decision maker selects a specific course of action over a series of plausible alternatives put forth by planners. He uses some compromise criteria to judge the best alternative. The compromise criterion, may not be always an optimal criterion, but recognises many side effects of the decision problem for its acceptance by the users.

There are two different types of decision making models, in practice. They may be considered as two different approaches for the solution of decision problems. They are : the technocratic approach and the bureaucratic approach.

In the technocratic approach, one starts from the recognition and analysis of the problem in question. The analysis is carried on in sequential steps and finally feasible alternatives are developed and selection is made by some criterion of optimality.

In the bureacratic approach, the problem is solved in an incremental fasion, *i.e.*, the advances in the solution are made in small steps. In such an approach, there is no absolutely right solution, but a continuous stream of minor decisions on the problem. This is purely a trial and error approach and can be described as a process of muddling through. There are no standard procedures of practices for this type of approach, except being guided by a few traditional and established guidelines.

Strength and Weaknesses of Decision Making Approach

The technocratic approach can either be quantitative or semi-quantitative. It very often disregards irrationalities of socio-political processes and preferences of individual decision makers in view of lack of pragmatism, the solutions although obtained in a systematic fashion are often unacceptable to the policy makers, and if acceptable sometime become very difficult to implement, apart from the problem of convincing the end users about their utilities. On the other hand, bureaucratic approach which is very much linked to our present democratic framework takes into account all those irrationalities and becomes very much appealing to policy makers. In reality, we can observe that in the bureaucratic approach, hardly any attempt is made to incorporate long-term objectives of public policy systems in the analytical framework. Probably, in a country like ours where the political system is more interested in retaining power at least for a period of five years, and where there is no cohesion in political ideologies, a short-term solution of the decision making problem is considered safer, because, a short-term approach can be treated as a satisfactory means in contrast to the process of optimising in the long range.

There are, however, some significant advantages in the incremental method of decision making. In practice, the technocratic approaches are basically target or optimisation approaches where feasible alternatives are generated, evaluated according to some predetermined set of objective values. Such an approach is reasonably accurate and even justifiable if the informations that are necessary as inputs are complete, timely exact. Whereas, in actual practice, such situations never occur, and hence incremental decision approach is sought for to obtain a compromise instead of optimisation, to involve a trade-off between several issues instead of incorporating assumptions, to revise views from time to time instead of waiting for a long time to monitor the solution. In the technocratic approach, ambiguity in solution results are never tolerated and also one cannot incorporate interests and conflicts of pressure groups of public systems managers in the analysis.

Appropriateness of an Approach

In a country like ours, the management policy is bound to be dynamic in nature. A dynamic management policy as either a long term, predetermined equilibrium policy aimed attaining a stable target state subject only to short-term fluctuations ; of a quasi-equalibrium management policy aimed at attaining, after transients, a target state that requires deliberates and continuous corrective interventions to keep the system in quasi-equilibrium with its surroundings and to prevent run-away conditions ; or a non-equilbrium final state while ensuring for a specific period, our ability to utilise the publicly mobilised resources under such conditions.

Decisions making in our country, and also in many developing countries, is at present being done by bureaucratic approach, probably due to historical reasons. However, we feel here that such an approach is not purely irrational. The complexities and inter-connections of decision problems make the attainment of rationality a difficult job. Rationality is bounded by many constraints both internal and external. However, the present mode of decision making can be improved.

We believe that decisions, within the existing organisational framework, can be made on the basis of thorough and systematic analysis. We do not see any reason why a model for every situation cannot be developed to that the current decisions can be linked to the long range objectives. What is necessary is a change of styles of reorientation of strategies or reordering of the problems or restructuring of priorities. What we simply require, is to develop a thinking little more for the future.

Modelling of Situations

In the context of present discussion, let up assume that successful managers are the managers of situations. Situations, we should perceive from the interaction viewpoint of the public policy system. The intensity and diversity of demands on our limited resources have increased considerably during these days. Furthermore, the recognition and acceptance of public policy systems in the fabrics of economic development, increasing social awareness, quality of working life emergence of articulate and diversified special interest groups due to our political systems, and rapidly changing public perceptions of an attitudes to the fundamental principles of management, need a different kind of manager altogether.

Recent developments and contributions by the management scientists in the arena of systems analysis have made it possible to provide the managers with a selected range of alternatives within which decisions may be made. Systems analysis is basically a modelling process, which consists of several tools and techniques to analyse a decision problem. The modelling should be carriedout by technical experts and the results of it should be used as an input to the recommendations by the policy makers. We suggest some of the basic rules of modelling as follows :

(a) A model should be judged from its utility viewpoint, i.e., how one can implement the model in the real world situations. Therefore, a simple model construction is always desirable which can facilitate easy implementation. A simple model is again very much deceptive, because, it cannot incorporate all the relevant parameters and variables. Therefore, it is always necessary to develop some simple models first, and keep the framework flexible to make it sensitive to the changes. A quantitative model may not always satisfy such requirements. Wherever possible, qualitative structuring can be made.

(b) There should be some initial commitment from the decision makers to use a model, and this support should come from the upper echelon of management hierarchy.

(c) Before constructing a model, three important but basic considerations are necessary. They are : technical commitment, which will answer whether a model constrrction is at all necessary ; if necessary, is it possible to construct or not, economical commitment, which will answer about the costs and benefits of modelling work ; and finally the organisational commitment, which will answer, if the other two commitments are justified, and whether the model will be accepted and used.

(d) Modelling and data collection process should be considered as a simultaneous affair. Modelling very often offers a clear insight to the type and quantity of data that should be collected.

(e) Model development should not be judged as one-shot affair, rather attempts should be made to continually up-date the models as and when more information and data become available. Updating should be considered as a regular process, otherwise models will be out-of-date, and they will loose their usefulness and credibility.

(f) Models should be so structured that current decisions can be linked to the long-term plans and programmes. Flexible models are more adaptive than models constructed under rigid boundary conditions without taking into consideration the long-run consequences of the present decisions.

There are some positive and negative factors to consider about the possible use of the modelling process in decision making. Developments in computer technology in recent years and their capability have become so much encouraging that most of the complex problems which require evaluation of a large number of alternatives can be solved without any difficulty. Further more, in our country, we do not find lack of expertise in mathematical modelling. Young engineers, economists and scientists as well as graduates in business management are now entering in large numbers to our labour market, and also are familiar with the new modelling technologies. What our requirements are to develop their abilities by training them to handle real life problems using the modelling approaches. In spite of all these potentialities, we have several other limitations.

Both top management and operating management must recognise their respective roles in the evolution of a modelling approach. Active participation of both the groups in the formulation, administration, and evaluation of the system can bring success to the approach. It is both difficult and foolish to impose an approach on the operating management that has not been a party to its introduction. Anyone who has got experience knows that the best of plans can be so cleverly sabotaged by a group of unwilling personnel that the promulgators will look like fools. One should first clearly realise what sort of managerial judgement will be required before making a move. The data requirements must be ascertained early, and the information collection should avoid long delays. Transitional difficulties must be anticipated and problem boundaries should therefore be made flexible to incorporate the transitional difficulties.

Decision Making is a Management Process

Decision making is so deeply embedded in the management process that it has for all practical purposes become the vehicle for carrying out the management functions. Decision making represents the point at which plans, policies and objectives are translated into concrete actions. On any organisation decisions are to be made to get the wheels of the organisation started and also to keep them running. The management and decision making are bound up inseparably, since they are complementary to each other. It is a mental process by which a manager arrives at his objectives policies, individual decisions or strategies. The various management functions such as planning, organising, directing and controlling can be stated in decision making terms. It is difficult to specify the steps involved in decision making process as sequence of steps can not be rigidly fixed in actual practice. However the steps can be broken down into following sequence.

(*i*) Consciousness of the problem provoking situation.

(*ii*) Recognition of the problem and its definition.

(iii) Search for and analysis of available alternatives.

(iv) Selection of best solution.

(v) Implementing the decision.

A decision is made within the circumstances of a given time and with in the structure of best decisions and actions. The decision making process in an organisation involves a large number of individuals each of whom makes his contribution Consciousness of the situation by the manager will emerge gradually, raising doubts and confusion in his mind and at this stage the manager must keep his head in all the confusions.

Need of Decision Making

Once the manager is aware that something needs attention, he can proceed to a recognition of a problem. Once the manager is aware that something made attention he can proceed to a recognition of a problem. Since it is impossible for a manager to weigh all the facts, he must develop a means of sorting out the relevantness. In the classification of their facts, he can then define the problem in a definate manner. The more completely the problem can be stated the easier the other step will be. Similarly more useful the manager framework of the statement of the problem at hand, the greater the chance that manager can make a decision that is consistent with managers goals. Search for available possibilities is critical to the entire decision making process. In this process the manager must relate the consequences of each alternative which satisfy certain goals, with their effects on other goals. Any decision is a matter if compromise. It is desirable to recognise the relationship between the consequences of the decision on the goals in mind, and the future effect of the decision as a means toward other goals. Seeking alternatives and their consequences would be an extremely large task for even a simple problem if all variations of all alternatives were considered. The means of communicating the decision, the techniques of motivating the group to accept the decision and effects of the decision on the co-ordination of other in the group are phases for implementing the decision.

Characteristic

The managerial decision can be characterized as follows:-

(i) Decision based on rational things.

(ii) Decision is the end process preceded by deliberations and reasoning.

(iii) Decision is selective or choice of the best cause among alternatives.

(iv) Decision is purposive or provides an effective means to the desired goal.

(v) Decisions are based on the concept of commitment.

Classification of Decision

To determine the nature of a decision and the level of authority following are the basic creterias :

(i) The degree of futurity, Decision's impact on other functions; Areas or the business as a whole; Decision is whether periodical recurrent or rase or unique.

Decision may also be proudly classified as major or minor routine or strategic, or policy or operating depending on the issues involved. It can also be classified as programmed or unprogrammed decisions.

Decision Tree

Introduction. Business decisions are made generally in the most complex and compelling situations. The executives make the decisions even at the cost of the business to some extent. Push and pull forces are so many that the executive needs to weigh them and decide upon the best course of action under the circumstances. Generally the business decisions are made under

- Certainty
- Risk

    Uncertainty

    Competitive conditions.

Decision Under Certainty

The decision making uses pay off rows matrix. Throues represent plans, strategies or alternative course of actions etc.

The columns represent the nature or status of the situation etc. like economic recession period, economic boon period or normal period . The figure at the intersection of the row and column represents the out come or net result of the strategy followed under the corresponding status or situation.

Example : The net result of the plans and situations are given in rupees. Since this is under certainty the probability is one.

Plans	Pay-off in Rs. Situations		
	S_1	S_2	S_3
Plan - 1	2000	5000	4000
Plan - 2	5000	7000	5000
Plan - 3	8000	9000	5500

The maximum pay-off is with plan 3 and under the situation S_2 *i.e.* Rs 9000. Thus various alternatives and the outcomes are calculated and arranged as above is the matrix form so that selection of the choice for the maximisation is easy. While calculating financial implications of incentive scheme with various alternative schemes with changes in the parameters of the schemes, this type of approach is very useful. This will enable the management to look into it clearly what are the maximum or minimum financial burden to the company and what are the benefits to the company as well as to the employees under various alternatives. The selection of the alternative becomes easy by this method.

Decision Under Risk

The probability of occurrence of certain situation itself is taken into account for arriving at the decision. Each status or situation is associated with certain probability. The probabilities are based either on the experience or on the past experience of record.

For example it is possible to anticipate the probability of the occurrence of economic recession or economic boon or normal period based on certain indicators or past data. The pay-off matrix for such type of decision under Risk in given below.

	Situation		
	S_1	S_2	S_3
Probability	0.1	0.7	0.2
Plan - 1	Rs. 35000	Rs. 55000	Rs. 5000
Plan - 2	Rs. 45000	Rs. 35000	Rs. 10000
Plan - 3	Rs. 55000	Rs. 15000	Rs. 7000

The out come of each plant with the situations and their probabilities are calculated as

Plan 1 = (0.1) (35000) + (0.7)(55000) + (2.2 × 5000)
Plan 2 = (0.1) (4500) + (0.7) (3500) + (0.2 × 10000)
Plan 3 = (0.1) (55000) + (0.7) (15000) + (0.2) (7000)
Plan 1 = 3500 + 38500 + 1000 = Rs.43000
Plan 2 = 4500 + 24500 + 2000 = Rs.31000
Plan 3 = 5500 + 10500 + 1400 = Rs. 17400

It can be seen that the out come of the Plan 1 in the highest and hence Plan -1 is the best course of action for deciding upon.

Decision Under Uncertainty

The probability of occurrence of the situation or status is not known and hence it is called the uncertainty. There is no past experience for calculating the probability.

There are four creteria in arriving at a decision under uncertainty.

Each type of criterion is named by its profounder.

(i) The Hurwicz Decision Criterion based on optimism.

(ii) The Wald Decision Criterion based on pessimism.

(iii) The sarege Decision Criterion : based on the minimization of regret

(iv) The Laplace Decision Criterion based on the assumption that every state of nature occurs with equal probability.

These four criteria are explained with the following example of pay-off matrix

Plans	States of Nature		
	S_1	S_2	S_3
Plan -1	5000	1000	500
Plan -2	3000	2500	200
Plan -3	1000	1000	1000

(i) As per Hurwicz Criteria : Since nature can be favourable also, one should prefer to select the highest pay-off maximum pay-off This is 5000 in the above. This is a maximum of max.

(ii) As per Wald Decision Criterion : He suggest that the decision maker should select the least pay-off since he will be concious about what could happen worst to come worst.

Then select the maximum of least pay-off

Plan	Minimum pay off
Plan - 1	500
Plan - 2	200
Plan - 3	1000

The maximum of minimum is 1000.

(iii) As per Sevage Decision Criterion, the decision maker should select the minimum of maximum regret pay-off since he will not have to regret, otherwise decision maker regrets after the decision has been made and the situation has changed.

The regret is measured between the highest pay-off value and the given pay-off value in the same status. In the above example 5000 is the highest value with Plan 1 state 1. If Plan 2 and state 1 is selected, the regret= (5000 - 3000) = 2000. Similarly in case of other states.

A regret matrix is arrived below.

Plan	States of Nature		
	S_1	S_2	S_3
Plan -1	0	1500	500
Plan -2	2000	0	800
Plan -3	4000	1500	0

The maximum regret in each plan is

Plan -1   1500

Plan -2   2000

Plan -3   4000

The minimum of the maximum regret in **1500**

(iv) As per Laplace Decision Criterion : All states of nature will have equal cnance of occurring one can not be discriminated against the other. If these are three states of nature, each one has equal probability of occurrence i.e. 1/3

With these probabilities the pay-off in the above example is written for each plan.

Plan -1 = $1/3 \times 5000 + 1/3 \times 500 = 2167$

Plan -2 = $1/3 \times 3000 + 1/3 \times 200 = 1900$

Plan -3 = $1/3 \times 1000 + 1/3 \times 1000 = 1000$

The maximum pay off is 2167 in Plan1, is selected.

Decision under Competitive Conditions

After the plan method or strategies and their states of nature are decided, the next decision is required when the decision makers faces the competetors. Therefore it is called as the decision under competitive conditions. Games theory one of the operations research technique is the extension of the decision theory under competitive conditions. This is dealt seprately in the chapter on Games theory.

Decision Trees

A business organisation would like to go for expanding its market. then it wants to know whether to capture the international market or limit to the national level only. Obviously the expansion of the market depends upon the demand pattern national as well as international level. For convenience purpose the demand can be depicted at three levels say limited demand, average demand and huge demand. The probabilities of these levels of demand and expected sales are given below.

S.No.	Level of demand	National		International	
		Sales level Rs. Lakhs	Probability of demand	Sales level Rs. Lakhs	Probability of demand
(1)	Limited demand	3.0	0.25	1.0	0.20
(2)	Average demand	4.0	0.25	5.0	0.20
(3)	Huge demand	6.0	0.50	8.0	0.60
	Total				

These figures can be depicted in the decision tree as shown in Fig. 13.6

The above calculations reveal that to go for international market is more prospective one as the total anticipated sales in higher than the anticipated national sales.

Anticipated total sales = probabiity × Sales

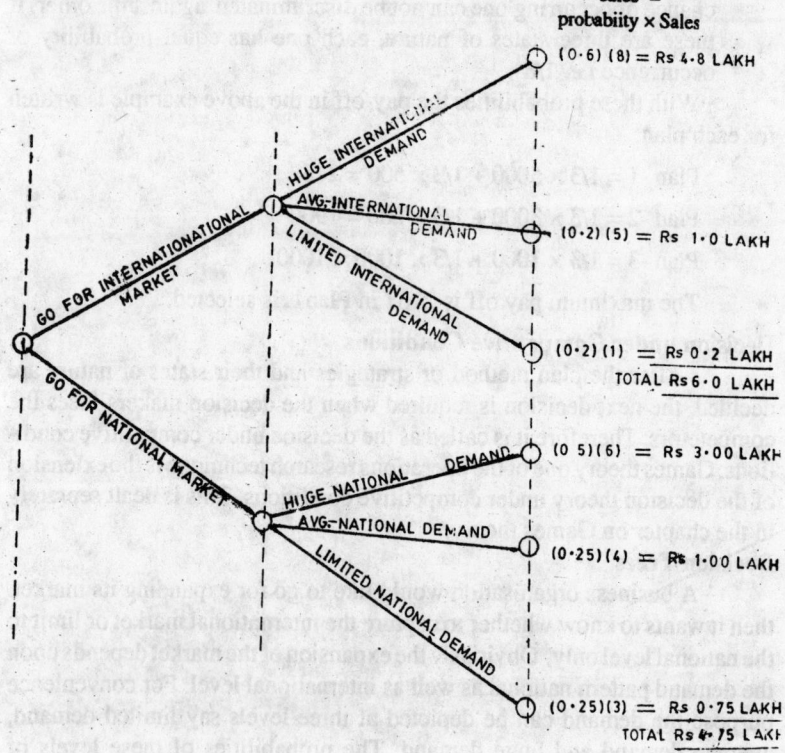

(0·6) (8) = Rs 4·8 LAKH

(0·2) (5) = Rs 1·0 LAKH

(0·2) (1) = Rs 0·2 LAKH
TOTAL Rs 6·0 LAKH

(0 5) (6) = Rs 3·00 LAKH

(0·25) (4) = Rs 1·00 LAKH

(0·25) (3) = Rs 0·75 LAKH
TOTAL Rs 4·75 LAKH

Fig. 13.6

A Decision tree is a visual device or a graphic representation for illustrating all of the choices available at various stages in a multistage decision process and the result of each choice. The starting point in a multistage decision process is convetionally shown by a circular point with line stemming outwards towards the right, each representing a possible choice and each terminated by a circle or point representing the state after the particular choice has been made.

Application of Decision Tree Capital Investment

Example : A manufacturing company produces a special type of soap with the existing equipment at full rated capacity for 8 hours a day. Since the

S.No.	Level of demand	Net revenue with		Probability
		New Equipment	Over time	of demand
1.	20% higher of present demand	Rs. 7.5 Lakh	Rs. 6.8 Lakh	0.667
2.	10% lower than present demand	Rs. 4.00 Lakh	Rs. 5.0 Lakh	0.333

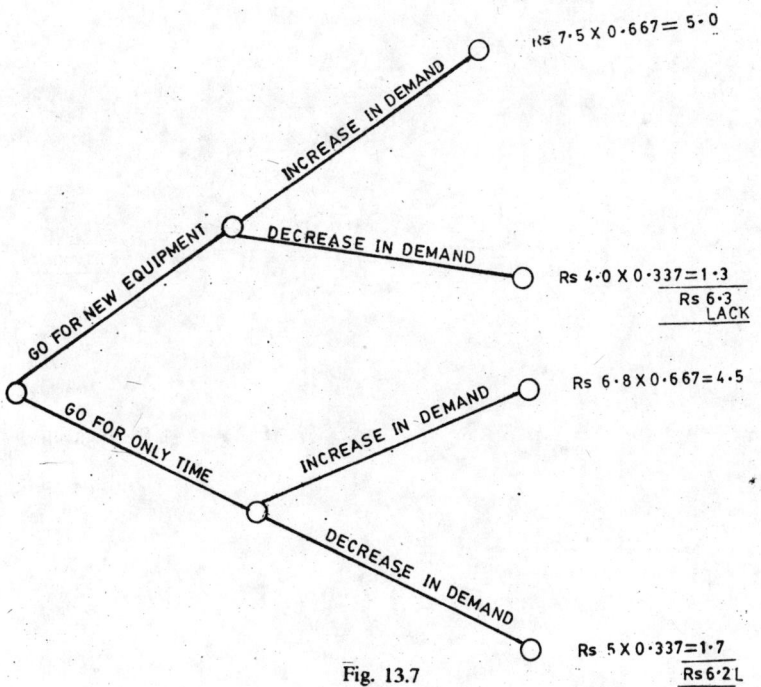

Fig. 13.7

demand has been increasing, the management would like to go for additional one equipment. But the production Manager is of the openion that as and when the demand increases, the additional demand can be met by employing the people on over time. If an equipment is brought and the demand goes down because extra burden where as the overtime can be arrested relatively easily. In such type of situation the decision tree will greatly help. Following are the probabilities of devels of demand and the likely revenues for each course of action for one year period. By using decision tree the solution is obtained.

Though the net revenue with requirement is slightly higher, this increase cannot justify for an additional equipment, since with overtime also the revenue of the same order is obtained. So the production with over time in preferable that too when money is not readily available for the investment.

14

Break Even Analysis

Introduction

A business is said to break even when its income equals its expenditure. The break even point means the level of output or sales at which no profit or loss is made. It represents the position at which marginal profit or contribution is just sufficient to cover fixed overheads when production exceeds the break even the business makes a profit and when production is below the volume of production at break even point, the business makes a loss. The following problems are solved with the help of break even analysis.

—the total profit of business is ascertained at various levels of activity and different patterns of production and sales.

—reporting the top managment the affect on net profits of introducing a new line or discontinuing the existing line.

—where severe competition is being met and it is desired to reduce the selling price, the affect of any reduction on profits can be easily ascertained.

—where the reduction in selling price is intended to increase sales, the increase necessary to allow to earn the previous profit can be calculated.

—the controlability and postponability of expenditure can be worked out from the break even point.

—it helps in planning and managerial control.

—break-even point can be helpful in detecting the effect of gradual changes that may have crept into the operation of budget planning and evaluating new proposals and alternative courses of action.

Thus the utility of Break-even analysis to the management lies in the fact that it represents a cross-sectional view of the profit structure. It highlight the areas of economic strength and weaknesses in the firm.

Assumptions Underlying Break Even Analysis

Following are the assumptions which are taken into consideration before drawing the Break Even Analysis.

1. All costs are perfectly variable over the entire range of volume of production.

2. All costs are absolutely fixed over the entire range of volume of production.

3. All revenue is perfectly variable with the physical volume of production.

4. The volume of sales and volume of production are equal.

5. For multi-products firms the product mix should be stable.

Following are a few terms which are commonly used in Break Even Analysis. These are :

1. Margin of safety.
2. Angle of incidence.
3. Contribution.

Margin of Safety

Margin of safety is the output at full capacity—the output at break even point (it is the difference between output at full capacity—output at break even point) which may be expressed as percentage of output at full capacity. If the margin of safety is small, a small drop in production capacity will reduce the profits considerably.

Angle of Incidence

It is the angle at which income line or sales line cuts the total cost line. If the angle is large it is an indication that profits are being made at a high rate. On the other hand if the angle is small it indicates that profits are being made and are achieved under less favourable conditions.

Contribution

Contribution is the difference between sales and marginal cost. It is also called as marginal profits or gross margins.

The marginal profit provides the contribution towards fixed cost and profit. If contribution is less than the fixed cost the loss is incurred. If the selling price is reduced, contribution will also be reduced assuming variable cost remains unchanged.

Break Even chart is an aid to management in decision making. It is a graphical representation of sale and cost informations. It shows the manager the profitability of an undertaking at various levels of activity.

The break even chart is needed where the business is new or a new product is introduced. In fact the concept of break even assist the management to appreciate the significance of quantity in relation to price. The objective of the chart is to indicate what happens to total costs and total revenues as output and sales change from one level to another. The genuine structure of the break even chart shows on horizontal axis the measure of output such as physical volume of production, percentage of full capacity output or sales volume. The sales lines relates total revenue on the vertical axis to

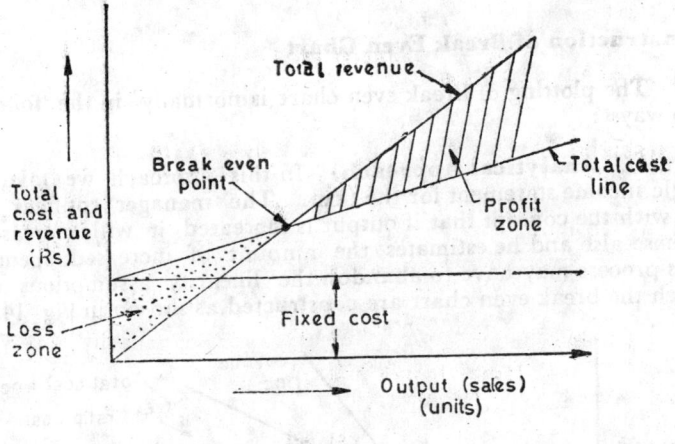

Fig. 14.1

total output on horizontal axis and is always drawn as straight line. There are usually two components variable costs (assumed to vary

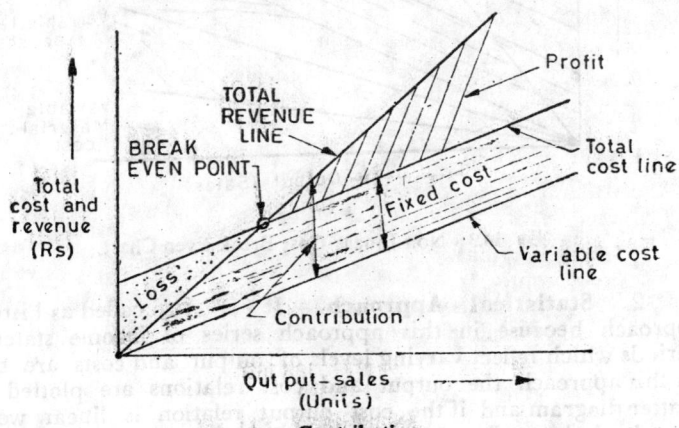

Fig. 14.2 Contribution.

with output) and constant costs (assumed in independent of output) the vertical distance between total sales and cost is total profit at output greater than break even point and at the output less than this represents losses. The break even chart is shown in Fig. 14.1.

Contribution to Overhead and Profit

Fig. 14.2 readily gives us at glance the contribution to overhead and profits. The short term decision making is more concerned with contribution to both overhead and profits than it is to the profit figure alone.

Construction of Break Even Chart

The plotting of Break even chart is normally in the following two ways :

1. **Analytical Approach.** In this approach we take one single income statement for the firm. The manager confines himself with the concept that if output is increased it will increase the expense also and he estimates the amount of increased spending. This process may have to abandon the linearity assumptions upon which the break even chart are constructed as shown in Fig. 14.3.

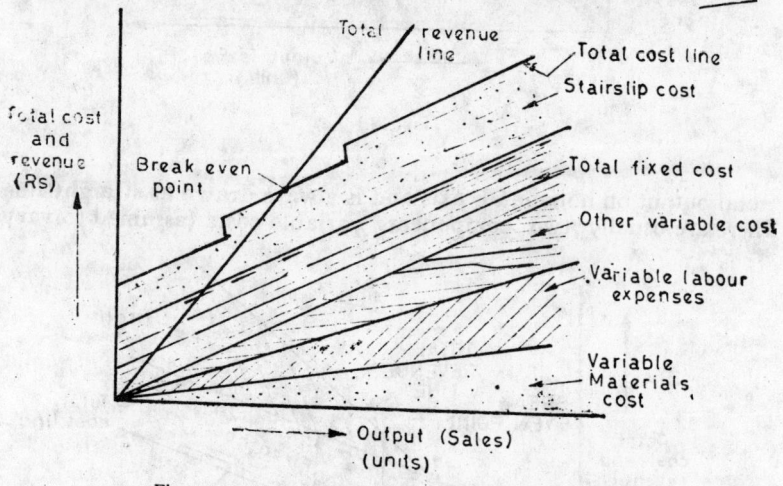

Fig. 14.3. Non-Linear Cost Break Even Chart.

2. **Statistical Approach.** It is also called as historical approach because in this approach series of income statements periods which reflect varying levels of output and costs are taken. In this approach the output and cost relations are plotted on a scatter diagram and if the cost output relation is linear we can draw back the total cost line back to the vertical axis Fig. 14.4

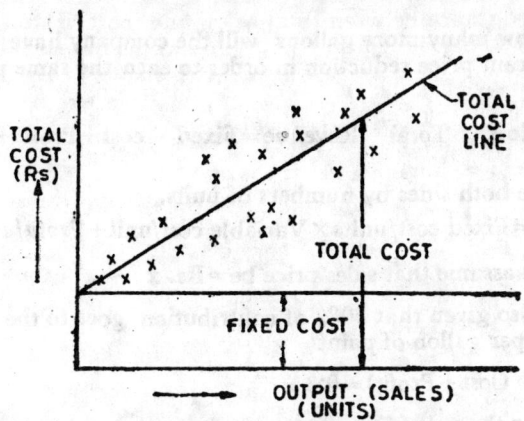

Fig. 14.4. **Finding Total Cost from a Scatter Diagram.**

Calculations of Break Even

Let $S=$Sales price

       $V=$Variable or marginal cost

       $F=$Fixed cost

       $P=$Profit

Let $S-V=P+F$ ...(i)

At break even $P=0$

Equation (i) reduces to

       $S-V=F$ ...(ii)

Let us multiply both sides of Equation (ii) by S

       $S(S-V)=F\times S$

$$S=\frac{F\times S}{S-V}=\frac{F\times S}{\dfrac{S-V}{S}}=\frac{\text{Fixed Cost}}{\text{Contribution per unit}}$$

Sales at B.E.P. $\dfrac{F\times S}{S-V}$
(Rupee-Value)

$$\text{Sales at B.E.P. unit}=\frac{\text{Fixed Cost}}{\text{Contribution per unit}}$$

$$=\frac{\text{Fixed Cost}}{\text{Marginal profit per unit}}$$

Example. A company "A" sold 2,50,000 gallons paints with variable cost of Rs. 4·20 per gallon. Each gallon contribute 30% of its revenue to fixed costs and profits.

This year the company is contemplating a price reduction of 5 per cent.

(i) How many more gallons will the company have to sell at years 5 per cent price reduction in order to earn the same profit this the as lost ?

Solution. Total Revenue=Fixed cost+Profit+Variable cost.

Divide both sides by numbers of units.

Sales price=Fixed cost/units × Variable cost/unit+Profit/unit ...(i)

Let us assume that sales price be=Rs. x

It is also given that 30% of contribution goes to the fixed cost and profits per gallon of paint

$$(\text{Fixed Cost}+\text{Profit})=0\cdot3\ x \qquad\qquad\qquad ...(ii)$$

Putting the value in equation

$$x=0\cdot3x+4\cdot20 \qquad\qquad\qquad ...(i)$$

$$(x-0\cdot3)x=4\cdot20$$

$$0\cdot7x=4\cdot2$$

$$x=\frac{4\cdot2}{0\cdot7}=\frac{42}{7}=6$$

$$x=\text{Selling price}=\text{Rs. }6$$

So putting the value in equation ...(ii)

$(\text{Fixed cost}+\text{Profit})/\text{gallon}=0\cdot3\times6=1\cdot8$

It is given in the problem to cut down the price by 5 per cent So the selling (New) price per unit

$$=\frac{6\times5}{100}=30\text{ Paise}$$

New selling price $=5\cdot70$ Paise

We know that $Q=\dfrac{(F\times P)}{S-V}$ ...(iii)

With change in the selling price

$$Q_n=\frac{(F+P)}{S_n-V}$$

$$Q_n=\frac{1\cdot8}{5\cdot70\times4\cdot20}\times Q=Q\times\frac{1\cdot8}{1\cdot1}=1\cdot2$$

$$=2,50,000\times1\cdot2=3,00,000$$

$$Q_n-Q=3,000,00-2,50,000=50,000$$

In order to reduce the price from Rs. 6 to Rs. 5·70, we are to produce 50,000 additional units.

Principles of Break Even Analysis

The break even analysis involves following principles.

(1) To find the safety margin with proposed volume.

(2) To find the quality needed to have desired profit.

(3) To find the effect of change in price.

(4) To find the effect of change in costs.

(5) To find whether to accept an order or not.

(6) To find whether to add or drop a product line.

(7) To find whether to make or buy.

1.   Finding safety margin. The safety margin refers to the extent to which the company can afford to decline in sales before it starts incurring losses. The safety margin can be represented by the percentage ratio of sales over break even volume to volume of sales.

$$\text{Safety Margin} = \frac{(\text{Sales} - \text{B.E.P.})}{\text{Sales}} \times 100$$

2.   Quality needed to have desired profit. This is one of the most common application of break even analysis.

The formula used to find the desired quantity is :

$$\text{Target quantity} = \frac{\text{Fixed cost} + \text{Target profit}}{\text{Contribution per unit}}$$

3.   Effect of change in price. This problem with which the management is most often faced whether to reduce prices or not. To take this decision the management will have to consider the following points.

Reduction in price leads to reduction in contribution margin and thereby increasing the sales volume. But at times reduction in price may not always lead to proportional increase in the volume of sales which depends on the elasticity of demand of the product. The information about elasticity of demand may not be easily available. The change in the break even point due to change of price is drawn below.

A different price levels, total revenue lines can be drawn as shown in Fig. 14.5. From the chart one can then estimate the volume that will be sold and produced at each price and thus profit at each price value.

$$\underset{\text{(B.E.P. in units)}}{Q} = \frac{F+P}{S-V} \qquad \qquad \dots (iii)$$

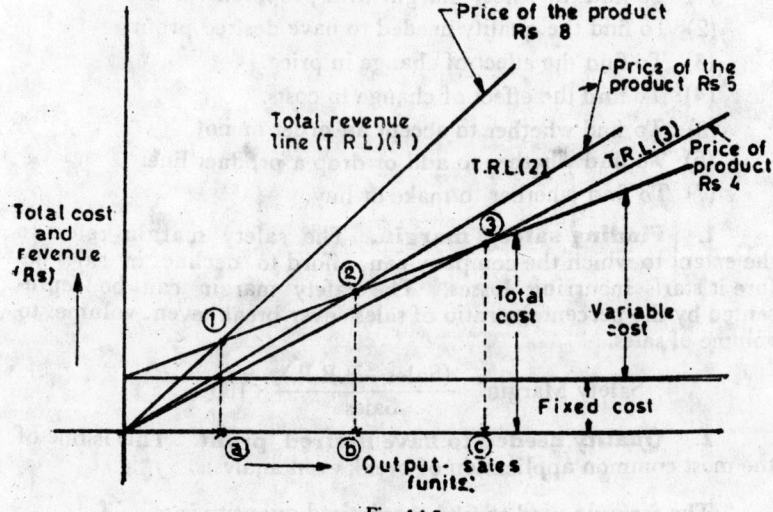

S = Sales price
V = Variable cost
F = Fixed cost
P = Profit

Fig. 14.5

As price changes equation (*iii*) changes

$$\underset{\text{(New B.E.P. in units)}}{Q_1} = \frac{F+X}{S_n - V}$$

S_n = New sales price
R_1 = New B.E.P. in units

4. Effect of change in costs. The change in costs can be grouped as change in the variable costs and change in the fixed costs.

Variable Cost Change

An increase in variable cost leads to reduction in the contribution margin. Now immédiate question in mind is what should be the sales volume to maintain our present profit without change in the price or what should be the new price to maintain the present profit without any change in the sales volume.

$$Q = \frac{F+T}{S-V}$$

$$Q_1 = \frac{F+T}{S-V_a}$$

Q_1=New break even points
V_n=New variable cost

The new selling price will be $S_n = S + (V_n - V)$
These relations can be graphically shown as follows Fig. 14.6

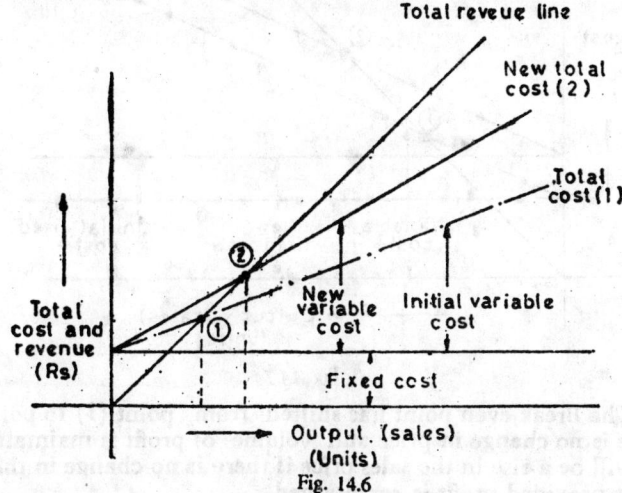

Fig. 14.6

The break even point have shifted from point (1) to point (2). Thus higher quantity is to be sold to maintain the present profit assuming no change in the price of the product.

Fixed Cost Change

An increase in the fixed costs a company may be due to change of the external environments in which the company is operating. The fixed cost change may be due to increase in executive salaries, property taxes etc. An increase in fixed cost will bring a change in Break even point assuming no change in the price of the product.

$$Q - \frac{F + P}{S - V} \qquad \qquad \dots (iv)$$

$$Q_n - \frac{F_n - P}{S - V} \qquad \qquad \dots (v)$$

Solving (iv) and (v) $Q_n - Q = \frac{F_n - P}{S - V}$ \qquad \dots (vi)

$$S_n = S + \frac{F_n - F}{Q} \qquad \qquad \dots (vii)$$

Q_n=New Break Even Point.
F_n—New Fixed Cost.
S_n—New Selling Price.

This can be represented graphically as shown below Fig. 14.7

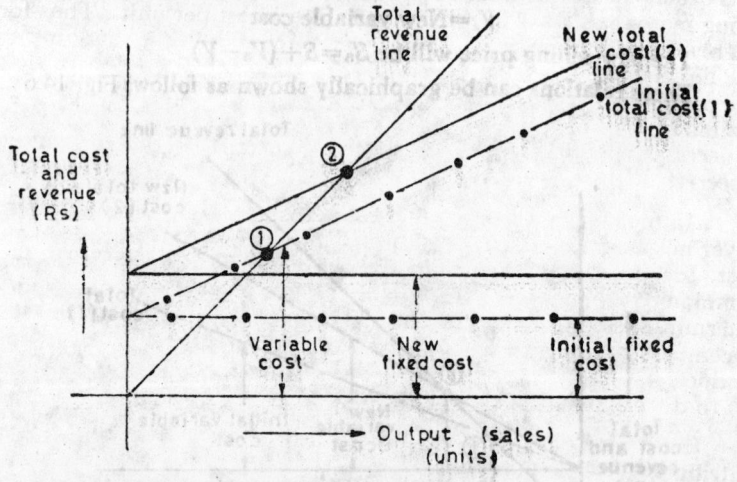

Fig. 14.7

The Break even point has shifted from point (1) to point (2) if there is no change in price and volume of profit is maintained or there will be a rise in the sales price if there is no change in the sales volume provided profit is maintained.

5. To accept an order or not. Quite often the management is faced with an alternative to accept an order at a price with less than the selling price if extra capacity to produce a product is available. Company can accept an order which meets not only the direct cost but also contribute towards common cost.

A company manufacturing a certain component was operating at 50 per cent of the capacity. A foreign importer offerred to purchase 5,000 units of the component at Rs. 5 per unit (The average cost of the product is Rs. 6·00). The question was whether the order should be accepted. The cost data were as follows :

Total capacity	20,000	components
Actual output	10,000	,,
Direct material	Rs. 10,000	,,
Direct labour	Rs. 10,000	,,
Fixed expenses	Rs. 40,000	,,
Total cost	Rs. 60,000	,,
Average cost	Rs. 6 per unit.	

If we look into the problem deeply, we find that the additional cost of producing 5,000 units would be only Rs. 10,000 (Rs. 5,000 for materials and Rs. 5,000 for labour). The incremental revenue would be Rs. 25,000 and the firm would gain Rs. 15,000 by accepting this order. In fact, it would be profitable for the firm to accept anything more than Rs. 2·00 which is the direct cost per unit. Thus for highly competitive situation, what is more important is that at least the direct costs must be covered. But if the customers learn that the component is being sold at a lower price, they may also demand a similar price cut or else threaten to cancel their orders. So what is necessary is that the acceptance of this order should be kept a secret.

In the long-run, the aggregate revenues from all products must cover not only the direct costs but also contribute towards common cost. Ideally, each product should make a significant contribution to common costs, but it is not possible to state any general rule for determining satisfactory or unsatisfactory contributions. If factors of competition prevent a firm from setting a price of one of its products that will cover direct costs, there may be no alternative but to discontinue that product.

If a competitive price does cover direct costs and yields some contribution to common costs, the question arises 'how high must that contribution to be justify long-run continuance of that product in a company's line'? The question can be answered in the light of availability. But if the product is discontinued, there are others which must be substituted for it so that there is a higher contribution to common costs. The elimination of one product from the line may cause the loss of sale of other complementary products. Product-princing decisions should be made with a view to maximise company profits in the long-run.

6. **Whether to add or drop a product line.** Any executive in a business organisation will be confronted with following problems when he decides among the two alternatives of add or drop a product line :

1. Should a new product be added in view of the estimated revenue and cost ?

2. If the product is dropped from the product line, what would be the consequent effects on revenues and costs. The break even analysis is quite helpful in deciding the question of product planning.

In a medium size industry the following datas are given :

Product	Price Rs.	Variable Cost Rs.	%Sales Units
A	6	4	30
B	10	6	20
C	20	12	50

Fixed cost per year = Rs. 7,500

Total sales of the three products = 2,50,00 units

It was decided by the management to drop the Product B and introduce the product D. To find out whether it is worth while to drop product B and introduce product D.

Product	Price Rs.	Variable Cost per unit Rs.	% Sales Units
A	6	4	50
D	16	6	10
C	20	12	40

Total fixed cost per year = Rs. 7,500

Sales for the current year = Rs. 26,000

Solution. Product A contribution is Rs. $(6-4)$ = Rs. 2. We have the sales price of Rs. 6 or in other words for the selling price of Rs. 6 we get a contribution of Rs. 2. The contribution for 33% sales of product A shall be

$$\frac{6-4}{6} \times 30\% = 1/3 \times 30\% = 0.10$$

Similarly for Product B,

$$\frac{10-6}{10} \times 20\% = 0.08$$

For Product C,

$$\frac{20-12}{20} \times 50\% = 0.20$$

Thus the contribution ratio for product A, B and C is 0.38

$$= (0.10 - 0.08 - 0.20) = 0.38.$$

Total sales in Rs. 25,000

$$= 25,000 \times 0.38 = 9,500$$

Profit $\qquad = 9,500 - 7,500 = 2,000$ profit

Similarly the profit or contribution on new product line of products A, D and C,

$$= \frac{6-4}{6} \times 50\% = 0.17 = \frac{16-6}{16} \times 10\% = 0.06$$

$$= \frac{20-12}{20} - 40\% = 0.16$$

Thus contribution ratio is 0·59

Total sales   =26,000=0·39=10,140

Profit   =10,140—7,500=Rs. 2,640

Hence the proposed change is worth taking.

7. Make or Buy. Break even analysis helps us to a greater extent in resolving the decision on Make or Buy. Make or Buy decisions are vertical integration decisions. Make or buy is specific. It is used to decide whether to utilise the production facilities or not.

Example. A Car manufacturer buys a component at Rs. 10 each. If he decides to make it himself his fixed and variable cost will be Rs. 10,000 and Rs. 5 per component respectively. To decide whether to make or buy the component.

$$\underset{\text{(Quantity)}}{\text{B.E.P.}} = \frac{\text{Fixed Cost}}{\text{Purchase Price—Variable cost}}$$

$$= \frac{10,000}{10-5} = 2000 \text{ components.}$$

Now interpretation is that if the car manufacturer wants more than 2,000 components it should be manufactured in the plant and for more than 2,000 components should buy from the same supplier.

Scope

Break even analysis can be most helpful in detecting the effect of gradual changes that may have crept into the operating relationship, checking the proposals, soundness, budget planning and evaluating new alternatives.

Budget Planning

The detection of creep can be accelarated by applying break even analysis in budget planning, and the overall soundness of a new budget can be tested as well. The approach under these circumstances would be through a break even chart, comparing the break even point determined from flexible budget figures with the break even point of past operations, and comparing budgeted profit with the profit that should be earned at the budgeted volume according to the break even chart.

Evaluating New Proposals and Alternatives

Proposals to add or drop a product line, changes in selling prices, changes in a manufacturing process, makes a part instead of purchasing it, or take similar courses of action should first be evaluated in terms of the effect on profits at expected operating levels. A secondary consideration however, should be effected on the "cushion" between expected operating levels and the break even point. Knowledge that a change will drastically reduce the cushion should receive careful consideration in determining whether such reduction is too great a price to pay for anticipated benefits.

The Break even analysis is useful in the area of managerial planning and review, including detection of unplanned change, overall review of budget planning, and evaluation of new proposals and alternative actions. The technical problems of break-even analysis, we know that no consideration has been given to such matters as changes in production mix, price-level changes, allocation of fixed costs between plants or products, or non-linear cost factors. These factors are, however, of major importance and must be given due consideration in break even point determinations.

Limitations

The static break even has serious limitations which we normally ignore. The limitations are :

(1)   Errors of estimating the true static cost-function.

(2)   Over simplification of static revenue function. The dynamic forces shift and modify the static functioning and managerial adoption to altered environment.

(3)   Profits are residuals, the profit function gets the full impact of these inaccuracies.

(4)   The correlation between output and selling does not mean a stable relation. Because the salesmen's commissions may be function of sales. Similarly advertising appropriation is determined by certain percentage of expected sales. Thus correlation between output and selling is not a true functional relationship. The output may depart from forecasts which are causing changes in advertising plan.

(5)   Plotting Company's past cost and revenue, is plotting its migration path, this has little validity for current profit forecasting.

(6)   The other drawback of break even chart is the wage and materials price increase causing trouble because they are non-reversible.

(7)   The lines drawn between cost and income are straight lines but these lines should be wary line. We known that if production goes beyond a certain point we have to give overtime, may have to put second shift and add more supervisors and resulting much higher cost at once. The straight line assumption may also be fallacious on the extreme right side of our charts because of our capacity limitations.

(8)   The lines on break even chart should be bands or zones extending across the chart and not these lines only.

(9)   Output indices used in most break even analysis are not very satisfactory.

(10) The break even chart is based on the past data and information which are not adjusted for changes in wages, changes in price of raw materials. These adjustments are quite cumbersome. We should use analytical techniques to avoid these adjustments.

(11) The chart assume that the price of the output are given. This assumption of horizontal demand curve holds true only for a perfect competition only.

(12) Break even charts are based on arbitrary assumptions as to the relative importance of different products in multiproduct firms.

(13) The simple break even chart makes no provision for corporate income tax. If company cause profit and suffers loss every alternative year then the adjustment for taxes is quite difficult.

(14) The break even chart concentrate mainly on short run cost functions and thus neglecting the other cost elements of the enterprise cost structure which in turn effect the accuracy of short run cost function.

(15) The break even chart is an over simplification of expected profits at various levels of output. The assumption that profit is a single valued function of output is wrong. The profit indeed varies with output, but it also changes with change in production plants, the intensity of selling efforts.

According to Elwood Buffa, the break even point is really a bolt on the chart and not a point.

Break even analysis may be useless for certain companies because of the wide fluctuation in price and predominent cost. When product mix varies greatly, profit margins differ among products, advertising or sales promotion are highly shifting, product design and technology change cautiously over short periods then it is very difficult to use break even analysis.

15

Occupational Health and Safety

Introduction

Health and safety of the workers is the responsibility of any government having a strong socialistic structure. A number of provisions have been made in the Factories Act to prevent health hazards and accidents in Industry. Textile Industry is one of the oldest Industry of our country. The workers engaged in the Textile Industry suffered from diseases such as Asthma and T.B. on account of the low subsistence level in the earlier period and the minor ailments went unnoticed and uncared for. With the mechanization of the Industry, the health and safety hazards have increased manifold and with the advancement of medical science health consciousness has also assumed importance. Today, there are diseases of the respiratory system arising out of flying dust which consists of fragments of cotton fiber, leaf scales, cotton husk; diseases occurring on account of high temperature, humid atmosphere and constant wet process; diseases on account of excessive noise; diseases on account of poor lighting conditions and diseases resulting from careless handling of corrosives, dyes, colors and chemicals. It has been generally observed during the inspections of factories that adequate attention is not given by the managements to provide corrective and preventive measures against these health hazards. For example, the ventilated area is rarely according to the norms fixed; exhaust fans are either not provided or they are too insufficient; use of masks is more of less unknown to the managements and the workers; temperature control is hardly maintained; noise is something which is considered as natural to the working conditions; natural light is considered as more than sufficient and no precaution is used in handling of dyes, colours and chemicals. Machinery and equipment are generally old and not much attention is given towards their preventive maintenance so long as they go on giving service. Proper guarding and fencing of machinery is sometimes rare, let alone its timely oiling and lubrication.

It has also been noticed that the workers do not use the precautionary aids wherever they have been provided to them. The workers are generally averse to the use of masks, earmuffs, tight–fitting clothes, gloves, gumboots, goggles etc. etc. The constant influx of the workers of the agricultural sector to the industrial environment necessarily means lack of knowledge and lack of adjustability on the part of workers and there is thus a dire need for a constant training and motivation. An occasional visit of a group of workers to some Industrial unit where all these precautionary measures have been provided and actually used by workers can be of great help in the promotion of occupational health and safety. While there is need for education and motivating the workers to use these preventive aids in the interest of their health and safety, the management cannot be absolved of their responsibility and they should not take up the plea that it is no use spending on these items when the workers do not make use of them.

An increasing rate of accidents in any industry creates a sense of insecurity and fear amongst the workers and impairs their efficiency on the shopfloor with resultant deterioration in there overall performance. It is also not always easy to find a substitute if a good worker is disabled or loses his life. It is, therefore, utmost interest of the management also to ensure all safety measures; as an accident not only disrupts the working but also result in loss of goodwill of its employees notwithstanding all the money spent on amenities and facilities provided to them.

It is a joint venture where both the management and the workers should participate and show adequate interest in safety. Without worker's whole-hearted co-operation at all levels the management would be left with no alternative and the provisions of Factories Act would be meaningless when translated into action. As discussed earlier, Industrial workers being mostly from the agricultural background and ignorant about the modern Industrial technology, it is very necessary that a new worker should be properly trained in his job and made conscious of the hazard, he is likely to encounter in his vocation. This should be a continuoes process and it would be desirable to organize short-term courses for the workers from time to time.

Health Hazards. Health means an adjustment of an individual to physical, psychological and sociological environments. The various factors that effect an individual at workplace are :

(i) Physical factors: Noise, Light, Heat.

(ii) Chemical and Biological factors : Toxic Gases, Fames, Dust, Bacterias etc.

(iii) Social and psychological factors : Job stress, organisation style, culture and leadership, behaviours and attitudes of co-worker.

The various physical, social and psychological factors have been discussed in the chapter of Human Engineering. Let us discuss some of the important chemical factors :

(a) *Dust Hazards:* Grinding/Polishing, drilling, Blasting, Cement industries, Handling of insecticides and pesticides, Handling of asbestos, coal dust etc.

(b) *Gas, vapour hazards:* Sulphurdioxide, Chlorine and Hydrochloric acid gas, Carbon monoxide and Hydrogen Cyanide gas, Benzene, Methylchloroform; Ethylene Tetrachloride vapours.

(c) *Smoke hazards:* Gases resulting from incomplete combustion and carbonaceous materials.

(d) *Fumes, mist hazards:* Solid particles size varying from 0.2 to 1 micron. Mercury and zinc fumes, paint fumes, Mist of sulphuric acid.

Chemical contaminants. The chemical contaminants enter human system mainly through three avenues *viz.,*

(a) Inhalation, (b) Absorption, and (c) Ingestion.

Inhalation process is through the intake of air-borne contaminants in the industrial environment. These contaminants enter into the lungs by process of breathing producing serious injuries and diseases, when they are deposited and accumulated in the sufficient amount. Similarly Absorption through the skin takes place when many organic sub stances such as Phenol, creosote, nitrobenzene, tetraethyl, hydrogen cyanide are absorbed to a limited extent through intact skin by way of the air spaces to the gland and gland cell. The Ingestion of toxic materials may result from many sources such as contaminated food beverage or from putting fingers or other contaminated objects into the mouth.

ILO's assistance to prevent factory hazards. Following the Bhopal Gas disaster on the 2nd and 3rd December, 1984 night causing a major human tragedy in the history of mankind, the International Labour Organisation has offered assistance to India in the field of control and management of industrial hazards and promotion of occupational safety. Seeing this disaster, the ILO had sent expert teams to India to study the situation and identify areas of assistance both immediate and long-term. According to the ILO area office in New Delhi, a number of projects and recommendations have been formulated. Twenty-seven Factory Inspectors from the States and four Industrial Hygienists from the Central and Regional Labour Institutes in India are being sent to Perth (Australia) on a 10-week intensive training in Industrial Hygience and Safety. The ILO sponsored course will include monitoring and control of major hazards and environmental pollution. The training is being funded from a UNIDO/ILO project on strengthening of factory inspection and advice services and will cover a wide range of safety and preventive methods in major hazard control.

Safety Management in Developed Countries. There is always an element of risk in the working life particularly so in an industry. Health and Safety Commission Report in Britain submitted during 1984-85 revealed that the total number of deaths in all industries due to accident and· occupational diseases were 1934. The break up of deaths due to accidents and occupational disease are given in the following table :

Table -I

All Industries

Total deaths due to accidents	645
Deaths due to Occputaional disease	889
	1534 Deaths

This data has been obtained from the Compensation paid to the employees on the deaths. Table II is dictating the total number of major injuries year-wise in all industries excluding fishing. The total major injuries during 1981 to 1984 were 18,892.

Table-II

All Industries

In Percentage

Years	Fatal & Major Injuries	Manufacturing
1981	60.3	70.8
1982	62.1	74.6
1983	64.4	79.3
1984	60.9	87.7

Toxic Material

During the past 4 years there has been a tremendous increase in Toxic Materials use in industry. In order to control the use of toxic material there should be a Threshold Limit Values (TLV) beyond which exposure should not increase or Threshold Occupation Exposure Limit (OEL) should be maintained. A study was carried out by Health and Safety Commission, a tripartite body in Britain for Vinyal Chloride Manomer (V.C.M.) which causes blood cancer and it was found that the following is the threshold limit value during various years :

1930's	—Acute Effects Noted	
1962	—T.L.V. of 500 P.P.M	
1971	— „ „ 200 P.P.M.	(Parts Per Million)
1975	— „ „ 25 P.P.M.	
1978	— „ „ 5 P.P.M.	

In order to reduce risk more emphasis should be laid on design of equipments and procedures of production. The design should include intrinsic safety measures containing design guards and easy access for maintenance. Procedure should involve Planned maintenance of Safe system of the work and permit procedures etc. The safety department should maintain a file and should plan the education and training of work-force, trade union leaders and supervisors. The involvement of trade union leaders and supervisors is very important for introducing health and safety measures. The studies have revealed that Asbestos fibre are very dangerous to health. Similarly dust is also dangerous but if it contains silicon it is very dangerous. In every organisation the management should carry out the systematic assessment of Toxic substances, carrying out their chemical audit, maintaining register and safety factors and develop a policy and priorities as given in Figure 15.1. Similarly the following diagram indicates how the Asbestos fibre hazards should be handled by management. Asbestos fibre causes lung disease and its impact can be understood by the example of a man who died at the age of 35 years and who never worked in Asbestos Factory. On analysis it was found that as a 14 year old boy he was exposed for 1 week to Asbestos Fibre. To solve such problems it is the question of deciding the priority. Only solution is the management and labour to sit and decide the solution. Regarding health and safety the developed countries of Europe and also Japan are following double standards. Certain multinational companies shift the hazard units from the developed country to other developing countries or Third World countries.

The Health and Safety Policies should be formulated and should religiously be followed by the companies. The Health and Safety policy should include the following elements :

(i) Policy should be unique to the organisation–it should be specific to company circumstances.

(*ii*) Signed by the managing Director—it should be the part of the company policy and it should specify which director is really responsible for implementation of policy.

(*iii*) It should be clear to the management that Health and Safety is a managerial responsibility.

(*iv*) Policy should be clear about the safe design, construction and operation of plant and substances.

The generalized model of Health and Safety management should be arrived at after understanding the nature of problems of chemical substances, keeping in view laws, codes and standards. Model procedure for chemical substances is given in figure (15.1). The model for Asbestos substance is given in figure (15.2) and Health and Safety management is shown in figure (15.3).

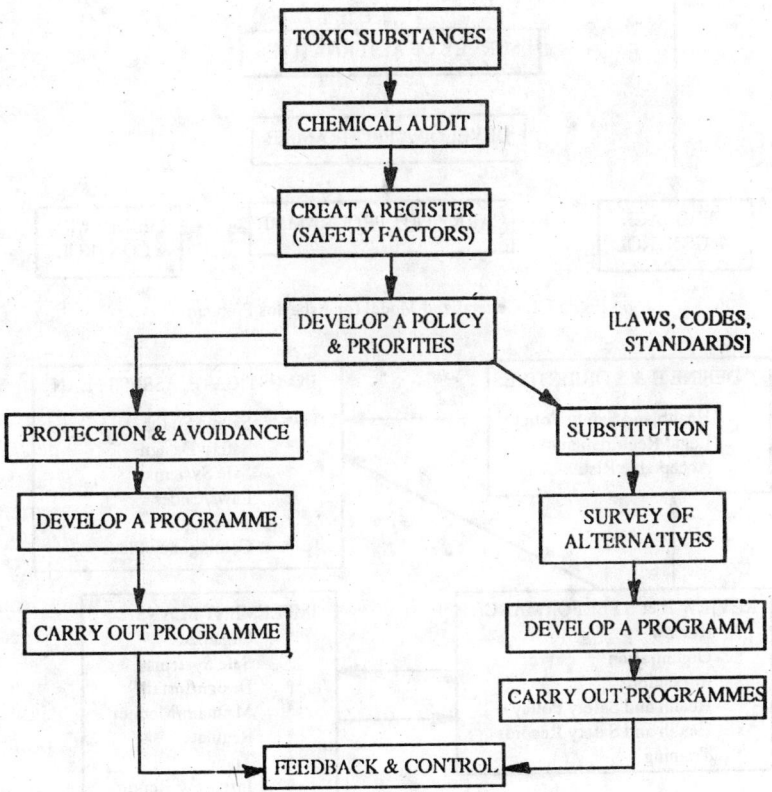

Fig. 15.1 Generalized Model for Chemical Substances.

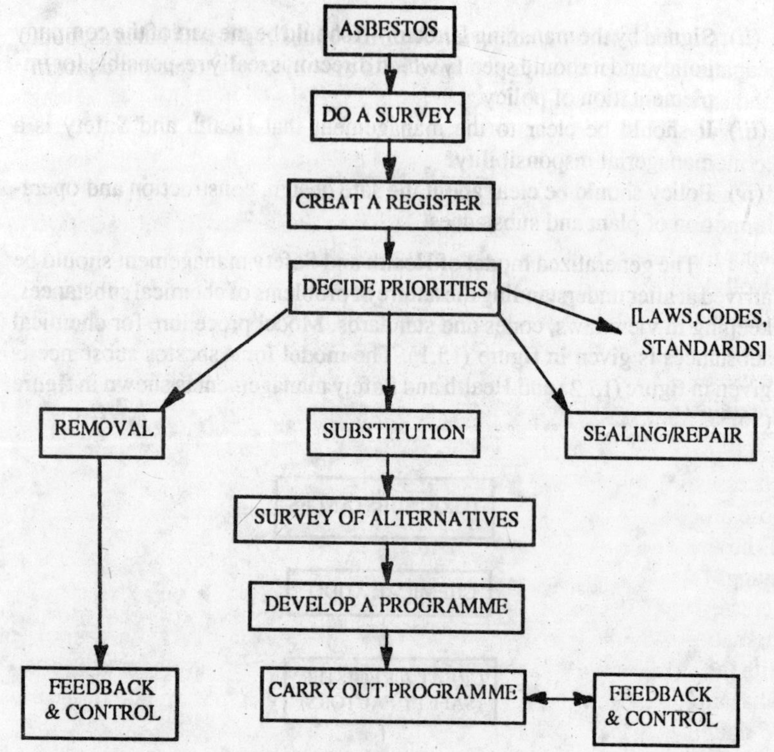

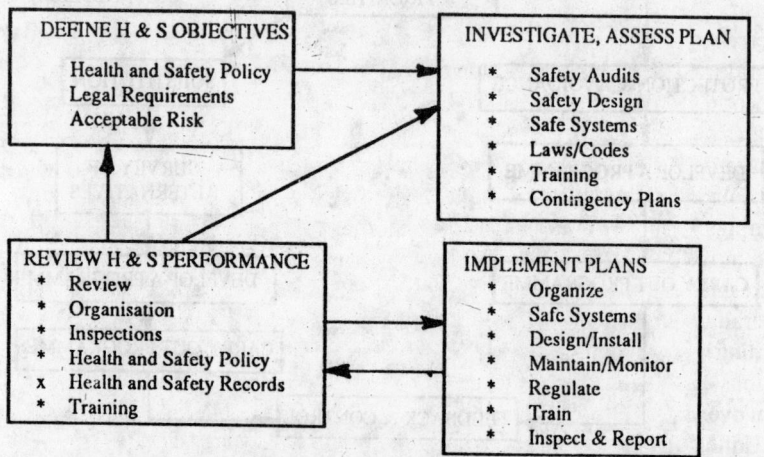

ASBESTOS

DO A SURVEY

CREAT A REGISTER

DECIDE PRIORITIES → [LAWS,CODES, STANDARDS]

REMOVAL   SUBSTITUTION   SEALING/REPAIR

SURVEY OF ALTERNATIVES

DEVELOP A PROGRAMME

FEEDBACK & CONTROL   CARRY OUT PROGRAMME ↔ FEEDBACK & CONTROL

Fig. 15.2. Generalized Model for Asbestos Problem.

DEFINE H & S OBJECTIVES

Health and Safety Policy
Legal Requirements
Acceptable Risk

INVESTIGATE, ASSESS PLAN

* Safety Audits
* Safety Design
* Safe Systems
* Laws/Codes
* Training
* Contingency Plans

REVIEW H & S PERFORMANCE
* Review
* Organisation
* Inspections
* Health and Safety Policy
x Health and Safety Records
* Training

IMPLEMENT PLANS
* Organize
* Safe Systems
* Design/Install
* Maintain/Monitor
* Regulate
* Train
* Inspect & Report

Fig. 15·3. Health and Safety Policy.

Double Stands in Industrial Hazards. Industrial health hazards, both occupational and environmental have been the object of increasing concern in industrialised Nations. Control requirements and compensation liabilities have favoured the emergence of alternatives to asbestos, carcinogenic dye intermediates, mercury and other very hazardous materials. In 1965, Imperial Chemical Industries (ICI), in England, closed a plant making the dye intermediate, alpha-naphthylamine. ICI in effect conceded that baldder cancer from naphthylamine dye intermediates could not be avoided even in a well-designed plant. When imports of the dye intermediate went up following the closure of the British factory, the immorality of the implicit double standard was sharply questioned by Dr. Robert Case. Britain banned the use of dye intermediates including Benzidine in 1967, but still imports Benzidine-derived dyes today.

Asbestos Hazards. In "hazard export" the hazardous industrial processes are set up in one country to serve markets in another country where those processes would not be allowed to function. In such cases, the export of the uncontrolled or poorly controlled hazardous process nets enough savings (in fixed costs, operating costs, and liabilities) to more than offset increased shipping costs. An integral part of the hazard export cycle is an international double standard in industrial hygiene and pollution control. With the industrialization of the Third World, However, clear-cut examples of hazard export may be far less numerous in the future than simple cases of the double standard. With the toxic substance controls being imposed in these countries the stage is now set for moving certain production facilities to developing countries to serve markets in developing countries. A likely industry for such a shift in production is pesticides. When multinational enterprises discover and take steps to control health hazards (as from vinyl chloride and acryllonitrile) in only some of their far-flung operations around the world, the double standard arises. The double standard may exist even in the area of warning and educating those at risk. Fundamental is the labelling of hazardous substances with appropriate warnings and educating those at risk. Fundamental is the labelling of hazardous substances with appropriate warnings. Typically, when the large Australian asbestos firm, James Hardie, Ltd., moved to set up the South Pacific Asbestos Association, a company spokesman admitted to using warning labels only on products sold in Australia. Hardie products sold in Indonesia and Malaysia bore no such warnings, and mounting liabilities, the use of asbestos in industrialised Nations has declined significantly. Japan's annual asbestos use dropped from over 300,000 metric tons in 1976-77 to 235,000 metric tons in 1978. In the United States there was a 36 per cent drop in consumption from 1979 to 1980, amounting to 359,000 matric tons.

TABLE
DOUBLE STANDARD CASES

Industry	Location	Type of Hazard Reported	Maltinational Affiliation
1	2	3	4
Asbestos milling	South Africa	Children with severe asbestosis	Cape Asbestos (U.K.)
Alpha-nophtylamine manufacture	Outside the U.K.	Bladder cancer	Imperial Chemical industries (U.K.)
Benzitine dye manufacture	Outside the U.K.	Bladder cancer	—
Asbestos textile manufacture	Agua Prieta and Juarez, Maxico	Not informing workers, not providing clothes change, neighbourhood pollution	Amatex (U.S.)
Asbestos friction product and textile manufacture	Bombay, India	Numerous workplace hazards uncontrolled, failure to inform workers and tell them of medical exam. Findings.	Turner and Newall, Ltd. (U.K.)
Asbestos	Countries without labelling requirements	Failure to affix warning labels	Entire asbestos industry (Asbestos International Association)
Asbestos cement manufacture	Ahmedabad, India	Water pollution, solid waste dumping no warnings on products.	John-Manville (U.K.)

Contd...

1	2	3	4
Asbestos textiles	Republic of South Africa	Lack of workplace dust controls	Deutsche Kap Asbestwerke
Asbestos textiles	Quebec, Canada	Eczema, Cancer	Asbestos Corpn. Ltd.
Epoxy spraying	Shipyards outside of Denmark		Not known
Chromate and dich-romate manufacture	Lecheria, Mexico	Waste dumping workplace exposure producing nasalseptum perforation.	Bayer (W.Germany)
Dye Manufacture	Bombay, India	Water pollution	Montedison (Italy)
Mercury-Cell chlorine plant	Managua, Nicaragua	Mercury poisoning, water pullution	Pennwalt Corpn. (U.S.)
Steel-making	Malaysia	Air pollution work-place hazards	Nippon Steel (Japan)
Polyvinyl chloride manufacture	Malaysia	High workers exposure to (carcinogen) venlychloride	Japanese Companies

The substitution of asbestos in major markets in the United States is a commercial reality today. A conference on asbestos substitutes was held by the U.S. Environsmental Protection Agency in July 1980. Dr. Anthony Robbins, former Director of the National Institute for occupational Safety and Health (NIOSH), has said that he knows of not one use in which asbestos is essential today. National Institute for Occupational Safety and Health (NIOSH) and its counterpart regulatory agency, the Occupational Safety and Health Administration (OSHA), have re-evaluated the medical literature on asbestos and concluded that it should be eliminated if possible ; and if not, the limit for occupational exposure should be lowered to 100,000 fibers per cubic meter of air. The biggest new market for asbestos is in construction in developing countries, where Chanadian asbestos mining firms are seeking new customers. In India, the affiliates of major British and American asbestos companies operate facilities that are 50 years behind the standard of practice these firms observe at home today. At Hindustan Ferodo, owned 74 per cent by Turner and Newall, Ltd., (U.K.), the dust is very poorly controlled, and workers are not told of the mortal hazards of asbestos. In Dortmund, West Germany, the United Asbestos Workers brought suit against the firm Techno-Einkauf for importing brake pads from South Korea. It was apparently not disputed that the products were manufactured under unsafe and substandard working conditions compared to the safety measures required by law in West Germany. The double standard arise when one country bans the use of a hazard substance and other countries do not. When Denmark banned opoxy spraying in its shipyards, the International Metal Workers Federation immediately broadcast the warning that this cancer-suspect, enzema-producing work was being diverted to shipyards in countries where there were no serious restrictions on epoxy spraying. Cotton dust, noise, chemical dye, and safety hazards were overwhelming in the factory of India-Malaya textile owned by Indian leading textile group. Workers there who complained were told thing were even worse in India, Newman reported. Developing countries that do not want to be dumping and testing grounds for hazardous technologies would be well advised to forster the development of expertise in government and trade unions in order to confront this threat.

16

Industrial Safety and Good House Keeping

Introduction

Industrial accidents are mainly due to human failure—somewhere in the claim of circumstances which lead to injury, we find the human factor. Specialist sections of management—engineers, electricians, designers, work study offices—are often lacking in an appreciation of the need to apply the principles of accident prevention to their work. "Unsafe" method accounts for a very high proportion of accidents. Therefore safty must be an integral part of any industrial undertaking, it must be built in at the design stage, in production planning and in operator training. The challenge of true of mind by efficient and regular safty training for all levels of management and worker. The aim must be to make every individual in industry have more regard not only for his own personal safty, but also for the safty of others.

Features of a Safty Organisation

Many persons influence safty directly. The chief executive, production managers, maintenance personnel or industrial relations employees, the plant physician the supervisor and production employee can affect safty qualitatively. Factors that govern plant safty effectiveness include choice of equipment, procedures and materials; budgets, the selection of personnel, the kind of leadership provided, and the work habits of the individual. When a safety specialist is employed to direct the plant safety programme his function usually is to assist line managers to control the safety of their operations. Ordinarily he does not assure any management prerogative but, develops information and materials which enable managers to optimise their authority is the pursuit of safty. Some of the features of safty organisation is the chemical Industry are:

A Management participation in safty programme.(Declaration of company safty policy).

B Supervisory responsibility for employee safety. (Each member of the supervisory staff is held responsible for safety).

C Assignment of staff function to safety personnel.

   (a) Safety department, safety safety engineer or other designated personnel given recognition and responsibility.

   (b) Duties include correlation of accident-prevention activities, promoting the safety programme.

D   Provision for safe working conditions, building and equipment.
   (a) Establishment of engineering standards.
   (b) Review of plans for new processes or equipment from safety and fire-protection engineering point of view.
   (c) Industrial hygiene surveys to evaluate potential chemical exposure hazards.
E   Safety education and safety training.
   (a) Education of supervisory personnel to assist them in carrying out their assigned safety responsibilities.
   (b) Employee education (New employee safety induction programmes).
   (c) Development of safety consciousness in all employees (use of bulletins, posters, films, and other visual aids).

Safety department should have a reasonable "Central Location" near the dispensary if possible, and with sufficient space available to provide for group meetings. Facilities must be available for strong and displaying samples of safty equipment, posters, safe-practice references, equipment catalogs and for the necessary records, files ad charts of the department.

Safety committees. A safety committee or group of plant committees is common to many plant safety organisations. They are subject to the usual problems of committee operation. When safety committees are established they require executive initiative. The committee size should be kept small for effective work but large enough to include all involved in plaint's operation. Co-ordination of the committee activities, encouragment of their effectiveness, and provision of assistance in their studies are leading functions of chief executive of safty.

Accident Prevention Programme
Basic steps in safety Programme

In industries, many plants, some of them far reaching, have been formulated for the prevention of accidents. Some of the useful plans are based on fundamental principles which start with simple steps undertaken usually in the following order:

(i) **Obtain co-operation of plant manager:** The manager's desire for safety achievement must be clearly visible through his action to achieve it.
(ii) **Obtain co-operation of other Heads of Departments:** The other heads of departments must make safety an integral part of the operating organisation.
(iii) **Analyse Accident records:** Accident reports for the past year or two should be analysed to earn, if possible, the how, who, where, when, and why of each accident.

(iv) **Hold meetings of Operating Executives:** All supervisors, sectional heads and operating heads should then be summoned to a general meeting presided by chief of safety.

(v) **Organise Educational Work:** Formulate a programme to maintain interest and supply information on safety to management supervisors and workers.

Role of Safety Director. On man must be responsible for development and codification of safety information, counselling managerson safety implementation methods, and developing progress measurement data in every plant regardless of its size or the type of safety organisation. He may be called safety engineer, safety specialist, safety inspector, safety manager or safety director. The chief operating executive in effect directs the plant safety effort. A safety director's actual position in the plant organisation varies with the general organisation of the individual plant. The following are general functions performed by the safety director whether he is a line executive or head of a seprate department:

(i) Responsible for formulating and administering the safety programme.
(ii) Advice management on all safety matters.
(iii) Maintain accident records.
(iv) Supervise or advise on safety training.
(v) Personal plant inspection.
(vi) Co-ordinate with medical department.
(vii) Consult with other agencies on safety problems.

Good House-Keeping

It is essential wherever chemicals are stored or handled. Chemical burns caused by workers tripping or slipping while handling containers can after be avoided if floors are kept in good repair and free from rigid system of orderliness in the plant and in the operations of the workers is essential to reduce the frequency of injuries. Labels, warning tags and clear instructions can prevent improper handling and storage of dangerous materials.

Controlling Atmosphere Hazards

Following are some of the factors reducing the dangers associated with atmospheric cartaininants.
(i) Substitute less hazardous compounds.
(ii) Revise process or operation.
(iii) Segregate hazardous process.

(iv)   Provide local exhaust systems.
 (v)   Design, after, and maintain building and equipment to achieve better control.
(vi)   Require personal protective equipment.
(vii) Educate employees to use established safe working methods.

For good housekeeping accumulations of dirt and refuse must be removed daily from floors and storage equipments. The cleaning policies turn around two be used. How to clean might start or contract cleaners. The advantages of using contract cleaners appear to be; They have cleaning expertise; They can purchase cleaning materials more cheaply etc.

Housekeeping Check list

It may be useful to have a check list which covers all factors associated with health, safety and house keeping.

Cleaning

★   Do we accumulate dirt and waste for excessive periods before it is removed?
★   What cleaning policy?
★   Is it effective?

Ventilation

★   Is there appropriate ventilation? Does the ventilation provide sufficient are changes?

Lighting

★   Is the lighting sufficient for the tasks being performed?
★   Is the quality of lights appropriate?

Toilets and Washing facilities

★   Are sufficient provided per the factories Act?
★   Safety clothing and equipment
★   Do we provide adequate safety equipment for:
★   Welding
★   Use of abrasive wheels
★   Power presses
★   Use of toxic fluids.

Fire and safe access

★   Is the fire certificate up to date and in line with current factory layouts?
★   Do we have regular fire drill with the fire alarm sounded at least every three months?
★   Is fire fighting equipment supplied and well maintained?
★   Are employees instructed in how to fight a fire?

Welfare

- ★ Do we provide adequate:
- ★ Toilets and washing facilities
- ★ Dressing and drawing rooms
- ★ Water drinking facilities?

17

Cost Reduction Techniques

Cost Reduction and Materials Management

On a manufacturing industry material takes a major share (50 to 70%) of the total cost. Therefore comapany must aim at a material cost reduction. It has been observed that material cost gets enhanced by defective design of the product; wrong selection of raw material ; poor manufacturing methods and this can be controllled by shop supervision. The responsibility for controlling direct material cost can be divided amongst; designer; purchaser; and supervisior. However, the greatest scope for cost reduction throughout industry lies at the stage of product design. The cost of manufacturing, warehousing and distributing will depend greatly on the decisions taken at this stage. The effect of design upon product cost makes it imperative that systematic consideration should be given at this stage at all aspects contributing to its cost. Some of the factors are: standardisation and variety reduction; compactness of the finished product leading to economy of packing; use of materials considering their cost; influence of design upon manufacturing cost interms of tolerances, tools, jigs and fixtures and processing time.

Standardistaion

Value analysis programmes are usually designed to reduce the cost of an item. Standardisation programmes may eleminate the item entirely. Standardisation is essential to a modern mass production economy. Almost every major industry has standards to classify its products. A standard is definied as that which has been established as a model to which an object or an action may be compared. The purpose of a standard is to provide a criterion for judjement. Companies are concerned with standards both for the products they design and for the materials designed by their suppliers. Standardisation is therefore required not only for ensuring procurement of the right quality of incoming material, but also for cost reduction.

Aims of Standardisation

The aims of standardisation should be to have uniform standards for similar items, and the standards evolved should take cognizance of the indigenous availability of materials to the maximum extent possible. Standardisation enables the materials manager to achieve overall economy and ensures inter-changeablitiy of parts. Since more than one manufacturer can supply standard items, it will imply better availability, better price and better delivery. Standardisation also implies routinising purchase efforts, less stock and hence less obsolete items. Standardisation also means less inspection efforts. To understand the concept of standardisation, let us understand the term standard. Standard is defined as a model or general afreement of a rule established by autority consensus or custom created and used by various levels of interest. A firm may similarly prepare, by consulting different departments, a standard for guiding the activities. Similarly, at the industry level, the related industry may also prepare industrial standards. At the national level, by consulting manufacturers, scientist, users and government departments, national standards are evolved. Such national standards are discussed to form international standards.

Standardisation and Materials Management

We all know, the fewere the items that need to be controlled the simpler and more efficient the materials management process. It is in the interest of materials management department to promote standardisation and simplification of specifications. The engineering groups are primarily responsible for standards and specifications but materials personnel can make a substantial contribution streamlinign the methods and standardisation of productivity and thus reduce cost.

Simplification

The process of standardisation logically leads to simplification or variety reduction. This implies unnecessary varieties and standardising to the most economical sizes, grades, shapes, colour, types of parts and so on. The items can be analysed by their frequency of usuage or movement analysis over the last few years. On the basis of analysis, a firm could set standards to these items. The setting up of standard depends on the effect the diamension variation has on the performance of the product. Simplifying work methods requires caseful study of each operation in order to find shorter or more advantageous ways to reach the same goal.

Benefits of Standardisation

Quite a large number of Public Sector organisations have started using standardisation as it leads to variety reduction. Standardisation has been successfully used in forms, purchase orders and contracts. Standardised sampling and testing has also resulted in reduced inspection costs in a large number of organisations. Some of the important benefits are:

(i)   Standardiation reduce inventory cost.
(ii) It helps in better communication.
(iii) It helps in inventory analysis.
(iv) It makes quality control more firm.
(v)   It helps in the promotion of export and creating confidence in the international market.

Linear Programming

Sometimes costs can be reduced without making any changes whatever in design, price or number of items carried in inventory. Operations research techniques can help the companies to have a competitive edge. Linear Progamming and Simulation techniques of operation research are very helpful to materials management. In materials management, linear programming can be applied to make or buy; Inventory management; Scheduling and physical distribution problems. The most important application of operations research has been the transportation and physical distribution.

Linear Programming is an extremely useful technique used for the purpose of maximising some objective such as profit or minimising, some objective such as costs in order to achieve the objectives of the organisation. Linear programming is now used increasingly to solve management problems. Linear programming is used in situations where numerous activities are competing for limited resources. The person responsible for developing linear programming was George B. Dantzig, who in 1947 developed the simplex method of Linear Programming. Since World War-II; the linear programming models have been used increasingly to solve management problems. Teh term linear reveals that the technique is mathematical and involves linear functions and the term programming refers to certain mathematical procedures to get the best solution to a problem involving limited resources. Linear Programming is a technique of numbers and the application of the numbers to arrive at a better decision. It is a tool in the hands of managers to take better decision among the number of alternatives.

Definition

Linear Programming is a technique of selecting the best possible (optimal) strategy among a number of alternatives:

(i)   Linear Programming is an important tool used by the management, involved in choosing most appropriate (best possible) alternative among a number of alternative actions.

(ii) Linear Programming helps the management discover whether the maximum profit or the least production programmes open to it.

(iii) It is applied to problems which involve interaction between alternatives.

Uses and Applications of Linear Programming

(i)   Linear Programming helps in determining the optimum combinations of several variables with given constraints.

(ii) It provides additional information for proper planning and control over various operations in the organisation.

(iii) It also helps the managers to have better understanding about the phenomenon and various activities of the organisation for the organisation construction of suitable mathematical model visualizing the relationship between variables if any and making improvement over them.

(iv) It also contributes the development of executives through the technique of model building and corresponding interpretations.

(v)   It provides a clear specification of the management problems by defining:
(a)   the objectives to be persued by the organisation.
(b)   the various restrictions imposed by the system.
(c)   the various alternatives and relationship between various variables.

(d) the contribution made by each alternative to the objectives of the organisation.

(vi) Linear Programming has found useful avenues in the fields of agricultural, industrial, military, space and research etc.

(vii) Specific areas where it can be fruitfully applied are:

    (a) Production scheduling and inventory control satisfying the demand and minimising the cost of production and inventory cost.

    (b) Blending operations and problems—where basic components are combined to produce a product that has definite set of specifications and one may be interested in calculating best possible combination of these components which maximises the profit or minimises the cost.

(viii) Other important applications of linear programming techniques can be in the area of:

* Purchasing and materials management.
* Routing and scheduling.
* Assignment problems.
* Marketing.
* Space exploration.

The problem of linear programming can be formulated with the help of the following terms.

The Transportation Problem

Many firms buy materials from several suppliers that is shipped to several plants. In considering how business is to be divided, they must take in to account not only the price charged by each supplier but also the cost of shipping the material to each plant. Transportation model helps to solve the problem to decide on the qualities to be purchased from each vendor and where they are to be shipped so that plant's requirement will be met at overall minimum cost.

Transportation Method: With this method problems of Plant location; Machine assignment material transportation and Products Mix can be solved.

Example

Plant	Warehouses				Output (Weekly production in plants A & B)
	1	2	3	4	
A	3	4	9	2	23
B	6	5	8	8	27
					0
Demand (Customer demand at each warehouses	12	13	15	10	50

3. 4. 9, 2 and 6, 5, 8,8 are the costs in Rupees per tonne of distributing product from plant A and B to respective warehouses 1,2,3,4.

Solution

The first step in the transportation method is to load the basic matrix without bothering whether the solution formed is optimal or not. There are various mehtods to do it but we shall be considering only tow common mathods:

(1) Northwest Corner Rule
(2) Vogel's Approximate Mehtod.

Northwest Corner Rule

In this method the initial assignment starts by loading the cell located in the upper left hand corner of the matrix and filling the non-requirements to the right and downward until all the requirements of the problem have been satisfied.

Ist Solution

Plant	Warehouse 1	Warehouse 2	3	4	Output
A	3 (12)	4 (11)	9 0	2	23
B	6	5 (2)	8 (15)	8 (10)	27
Demand	12	13	15	10	50 / 50

The transportantion cost involved in the first solution can be evaluated as
$$= 3 \times 12 + 11 \times 4 \ 2 \times 5 + 15 \times 8 + 10 \times 8$$
$$= 36 + 44 + 10 + 120 + 80 = 290 \text{ (rupees)}$$

In order to ascertain whether the solution I is optimal or not the empty cells in the solution must be evaluated. Following is the procedure to evaluate the empty cell:

1. Shifiting 1 unit from loaded cell into the empty cell.
2. Comparing the cost of orginal schedule with the cost of new schedule involving the empty cell.
3. The loaded cell which is selected must have following two charactcristic:

(i) Loaded cell must be situated in the same row or colemun of the empty cell.

(ii) The route used for evaluation would not require more than one emply cell during same evaluation.

The effect of shifiting 1 unit from plant B to warehouse 1 would be
$+6-3+4+5 =$ Rs. 12

The plus sign indicates that the total costs cannot be reduced by adding one unit to B_1 combination.

Evaluation of empty cell B-1

Plant	1	2	3	4	Output
A	⌊3 ⑪ –	⌊4 ⑫ +	⌊9	⌊2	23
B	⌊6 ① +	⌊5 ① –	⌊8	⌊8	27
Demand	12	13	15	10	50

Further Evaluation
Evaluation of Cell A-4

Plant	1	2	3	4	Output
A		⌊4 ⑩	⌊9	⌊2 ①	23
B	⌊6	⌊5 ③	⌊8 ⑮	⌊8 ⑨	27
Demand	12	13	15	10	50

Evaluation of cell A_4 follow the route
$-4+2-8+5 =$ Rs. –5.

(Assigning positive signs to the cost of those cells which gained one unit and negative signs to those cells which loses one unit.)

Evaluation of A_3 Cell

Plant	1	2	3	4	Output
A	⌐3 (12)	⌐4 (10)	⌐9 (1)	⌐2	23
B	⌐6	⌐5 (3)	⌐8 (14)	⌐8 (10)	27
Demand	12	13	15	10	50 / 50

Moving one unit to A_3, we remove one unit from A_2. In order to restore one unit to warhouse 2, we shift one unit from B_3 to B_2.

The cost associated with these moves are :

$-4 + 9 - 8 + 5 = +2$

Thus adding one unit to cell A3 would not reduce the total costs. Thus evaluation of empty cells give the following inference:

   (i) If the evaluation is positive, then the cost of new solution would be greater than the cost of present schedule.

   (ii) If the evaluation is negative than the cost of new solution would be less than the cost of present schedule of the evaluation result into zero value than the cost will not change by using these empty cells.

The second feasible solution and optimum solution afte analysing empty cells is given below. Now the evaluation for all empty cells involve costs greater than zero. So any change in the solution would result into increase of total costs.

Plant	1	2	3	4	Output
A	⌐3 (12)	⌐4 (1)	⌐9	⌐2 (10)	
B	⌐6	⌐5 (12)	⌐8 (15)	⌐8	
Demand	12	13	15	10	50 / 50

If the evaluation revealed negative costs then another feasible solution could have been constructed. The evaluation process is complete when all evalutions are positive. Thus an optimal schedule has been obtained.

(2) Vogel's Approximate Method

This method was developed first by W.R. Vogel. This method gives a good approximation and provides an optimum solution to most of the simple problems. Following are the various steps used for Vogel's approximate method:

1. Construct the basic matrix
2. Find out the difference in each column and row between the minimum and the next higher to minimum.
3. Select a row or column containing the largest difference.
4. Select maximum unit in that square or column which carries the lowest cost.
5. The steps (2), (3) and (4) are repeated till all units are distributed satisfying the capacity and requirement constraint.

Vogel's Method

Let us find with the help of Vogel's Method.

Ist Step   **ware-houses**

Plant	1	2	3	4	Output (Weekly production of the products A & B difference)	
A	$\lfloor 3$	$\lfloor 4$	$\lfloor 9$	$\lfloor 2$ ⑩	23 23 – 10 = 13	4 – 3 = 1
B	6	5	8	8	27	6 – 5 = 1
Demand	12	13	15	10	50 / 50	

Difference 6 – 3 = 3   5 – 4 = 1   9 – 8 = 1   8 – 2 = 6
                                                                    ↑

IInd step

Plant	1	2	3	Output	Difference
A	⌐3 ②	⌐4	⌐9	13 13 – 12 = 1	4 – 3 = 1
B	⌐6	⌐5	⌐8	27	6 – 5 = 1
Demand	12	13	15	40 40	

Difference 6 – 3 = 3   5 – 4 = 1   9 – 8 = 1
                 ↑

IIIrd Step

Plant	2	3	Output	Difference
A	⌐4 1	⌐9	1 1 – 1 = 0	9 – 4 = 5
B	⌐5	⌐8	27	8 – 5 = 3
Demand	13	15	28 28	
Difference	13 – 1 = 12 5 – 4 = 1	9 – 8 = 1		

Difference   13 – 1 = 12

                        5 – 4 = 1   9 – 8 = 1

IVth Step

Plant	2	3	Output	Difference
B	$\lfloor 5$ (12)	$\lfloor 8$ (15)	27 / 27	$8 - 5 = 3$
Demand	12	15	27	

Let us combine step I, II, III and IV to get the initial solution.

Plant	Warehouse				Output
	1	2	3	4	
A	$\lfloor 3$ (12)	$\lfloor 4$ (1)	$\lfloor 9$	$\lfloor 2$ (10)	23
B	$\lfloor 6$	$\lfloor 5$ (12)	$\lfloor 8$ (15)	$\lfloor 8$	27
Demand	12	13	15	10	50 / 50

Let us evaluate the empty cells, B_1, B_3 and B_4. Evaluating square B_1.

Plant	Warehouse				Output
	1	2	3	4	
A	$\lfloor 3$ (12) $-$	$\lfloor 4$ (1) $+$	$\lfloor 9$	$\lfloor 2$ (10)	23
B	$\lfloor 6$ $+$	$\lfloor 5$ (12) $-$	$\lfloor 8$ 15	$\lfloor 8$	27
Demand	12	13	15	10	50 / 50

If we transfer 1 unit from square B_2 to B_1.

The change in the cost of transportation will be

$+6 - 3 + 4 - 5 = +2$

Thus transferring one unit to cell B_1 is not economical.

Evaluating empty sq. A-3

Plant	Warehouse				Output
	1	2	3	4	
A	⌐3 (12)	⌐4 (1) – — +	⌐9 +	⌐2 (10)	23
B	⌐6	⌐5 (12) +	⌐8 (15) –	⌐8	27
Demand	12	13	15	10	

Transferring one unit from square A_2 to square A3. the new cost of transportation will be

$9 - 4 - 8 + 5 = +2$

Thus the new set of assignment is not economical.

Evaluating empty sq. B-4

Plant	Warehouse				Output
	1	2	3	4	
A	⌐3 (12)	⌐4 (1) +	⌐9 —	⌐2 (10)	23
B	⌐6	⌐5 (12) –	⌐8 (15) —	⌐8 +	27
Demand	12	13	15	10	

We are not considering the occupied cell B_3.

Transferring one unit from square A4 to sqaure B4, the new cost of transportation will be

$8 - 5 + 4 - 2 = +5.$

Thus transferring one unit from square A4 to square B4 is not economical. The evaluation for all empty cells involve cost greater than zero. So any change in the solution would result into increase of total costs. We can also conclude that Vogel's approximate method gives a good approximation and provides an optimum solution to most of the simple problems.

Simulation

One of the most important functions of management is to make decisions. While some of these decisions may be taken on a day to day basis and be only significant in the short term, other decisions may be of critical importance to the long term future of the company. The simulation with decision model provides two important advantages to the decision maker:

(i) It offers an inexpensive method for analysing the behaviour of the system under different conditions.

(ii) It attempts to predict what would happen in real life.

Simulation duplicates the esential features of a system without actually obtaining reality. This duplication may take one of the following model forms.

(i) **Physical models:** Examples of theses model forms include the ways in which a praticular design or operation might work. The purpose of building such models is two fold: firstly to have a better understanding of the design relationships and secondly, to identify any area of weakness in the model so that the design may be corrrected before the actual real-life product is built.

(ii) **Analogue Models:** Analogue models represent the essential characteristics of the problem under analysis by physical representation. Physical analogue models have also beeb used in economics.

(iii) **Symbolic Models.** Symbolic models make use of mathematical and statistical notation to represent problems. The symbolic models can be further expanded into:

(a) *Analytical Models:* with there models the solution can be derived directly from the model, as is the case when simulation is carried out by hand.

(b) *Numerical models.* The solution is given in the form of numerical output as is the case when simulation is carried out by computer.

As with any model, simulation focuses on the important characteristics of the system and hence by implication largely ignores those parts of the system which is assumed to be unimportant. One of the main areas of difficulty, therefore is constructing a simulation model to know what to leave out. Simulation can be applied even by smaller materials management organisation. Simulation is a device that permits the materials manager to predict what will happen to a system when demand changes. At any given time the materials management is concerned with efficient use of available resources subject to fluctuating and often unpredictable demands. Simulation is the device that permits a company to measure what will happen when a particular group of resources is subject to input that varies at random. Materials manager simulates operation of the system by subjectign it to random variations of input using the Monte Carlo technique.

Organising Cost Reduction

There is no way to accomplish cost reduction or no one method that is going to lead to the ultimate in a cost reduction programme. The success of the cost reduction programme is assured provided there is organisation of people behind the cost reduction derive. Effective organisation for cost control is based on the psychology of the individual. Reducing cost and improving the profit picture are invariably the results of many small measures properly applied. Organisation for cost reduction may be defined as a systematic arrangement of a dynamic element essential in operation of any successful business. Broadly speaking, there are seven basic areas in a company which are subject to cost reduction analysis viz. Finance and Control, Sales and Marketing; Manufacturing products and service; purchasing and personnel. The areas mentioned concern every facet of the business organisation. When considering cost reduction there is no business area which can be eliminated. Therefore, cost reduction programme should involve every member of the business organisation and should reach at all levels in the organisation hierarchy. In otherwords company should take the approach to cost reduction through consideration of all its people or cost reduction has to go to every employee in the organisation.

For the success of an organisation of cost reduction following point must be considered:

(a) An/executive has to be responsible.

(b) The top management must have interest, cooperation, consideration and firm belief that cost reduction is worthwhile and necessary for the success of an organisation.

(c) The cost reduction awareness must be inculcatéd in the employee and must encompass all employees—each person that is part of the organisation.

(d) Cost reduction is a continuosu operation. It cannot be started and stopped at will but must be worked on constantly. There must be constant effort to improve, and to think of new methods.

18

Inventory Management

Introduction

The pressure for operating capital has made business an increasingly aware of inventory as a form of earning investment. Why are we always out of stock? A complaint from a large numbers of businessmen faced with the dilemmas and frustrations of attempting simultaneously to maintain stable operations, provide customers with adequate service, and keep investment in stocks and equipment at reasonable levels. In order to get at the answer to this question as a basis for taking action, it is necessary to step back and ask some rather different kinds of questions: Why do we have inventories? What affects the inventory balances we maintain? How do these effects take place? From these questions a picture of the inventory problems can be built up which shows the influence on inventories and cost of various alternative decisions which the management may ultimately want to consider.. The basic problem of inventory policy is to strike a balance between operating savings and costs of capital requirements associated with larger stocks. Striking this balance is easier to say than to do.

What Is Inventory

Inventory constitute one of the most important elements of any system dealing with the supply, manufacture and distribution of goods and services. The concept of inventory control is very old but it came in light when Harris F.W. published his work on classical order size model. This work was extended by Raymond F.E. (1931) and Wilson R.M. (1934). But only after the second world war with the development of operational research and computer technology that the theoretical concepts got a practical application. The basic purpose of inventory holding stocks in a material flow system is to decouple successive stages of system. Following are important purpose of inventory:

(i) Existence of time lags between manufacturing, and transport operations.
(ii) The need to schedule various stages of the system independently.
(iii) The need to meet the fluctuation in demand and production rates.
(iv) The need to maintain control over the quality of the finished product.
(v) The need to exercise influence over changes of material prices particularly basic raw materials.

All the above factors involve costs. The inventory control mainly concerned with making optimum decision with respect to the variables which are subject to control. Inventory control is a multi-item, and multi-stage in nature. *Inventory is an idle resource which is useable and have value. The idle resource may be men, money, materials, plant acquisition.* The manpower, capital and plant acquisition problems are essentially inventory control problems.

The scientific inventory management helps the purchase department to determine when to buy and how much to buy. The problem before the management is to balance the following opposing costs.

S. No.	Cost to have Inventory	Cost not to have Inventory
1.	Return on Investment	Stock out costs
2.	Storage cost	(Demand × Profitability × Future)
3.	Handling cost (Labour and Industry)	Loss of customers.
4.	Handling Equipment	Down time cost.
5.	Obsolescence	Idle Labour and idle Production.
6.	Spoilage and Shelf life	Capacity and other cost.

The scientific control can also be defined as a system of internal check to control. We know that inventory tie up a sizable part of working capital and also cost a good deal to carry them. Money resources are scarce and efforts should be directed to conserve it by all means. Inventory is money but is it not at all like money in the bank but it is the money in which the we pay interest instead of earning interest. Arriving at a level between too much and too little stock is a real challenge to the management. The uncontrolled inventories are dangerous and at times it is called as grave year of business.

The inventory control exist in some form in every plant. the system in a particular company may be formal or informal, cheap or expensive. In one case control may depend upon the efforts of a large control staff while in another the responsibility may rest with the plant manager and the foreman. Even systems used in similar plants, differ in outward appearance at least. Production planning coordinates the production department with other department of the business. Production control promotes effective shop operation through its control of activities within the production department itself.

Inventory control determines the levels of composition of inventories of parts, materials and products which will protect most effectively the production, sales, and financial requirements of the business. A system of production planning and control is not and end in itself. It can only be justified because it:

(1) Provides the means for effective coordination of the Production department with other departments of the business; and

(2) Promotes effective shop operation through its control of activities within the production department itself.

In any given company a good system of planning will strike the best balance possible between the required amount of control and the overhead cost entailed in achieving it. If tight control cost no more than loose control management would of course try to attain the maximum control possible in a particular situation. This is not the case. The tighter the control a system produces, the more overheads is required to create it.

Every businessman must have faced the problem of either being out of stock or a large amount of money tied up in the form of inventories. Inventory is an evil which cannot be eliminated. On the other hand large level of inventories hamper the growth of the company. Inventory control is to determine the optimum level o inventory which is essential to carry out the various operations of the organisation efficiently and effectively.

Inventory control is essential for effective plant operations. The inventory carrying charges from 12% to 22% of inventory value with the average near 18%. A company with restricted working capital is always in a dangerous position and there is no easier way to dissipate working capital than to build up excessive inventory.

Definition of inventory

(1) Inventory is wider sense is defined as any idle resource of an enterprise. It is commonly used to indicate materials—raw, in process, finished, packaging, spares and others—stocked in order to meet an expected demand or distribution in the future. Even though inventory of materials is an idle resource in the sense it is not meant for immediate use, it is almost a necessity to maintain some inventories for the smooth functioning of an organisation.

(2). Inventory is made of all those items ready for sale or of items which keep the process running.

Why inventories are essential

In some circles inventories are thought of as a sign of wealth, even excess inventories in relation to the magnitude of production and distribution function are considered advantageous. On the other hand, the wise businessmen place more emphasis on having working capital in the form of cash and securities. In recent years a greater emphasis has been placed on having the means of purchasing materials than having the materials themselves. Excessive inventories have been the death of many a business and very high inventory levels have tipped the scale in our economy.

The finished product after packaging are first stored and sent to the market as the need arises or orders are received. Also the raw materials needed for the finished product cannot be directly fed to the production department from the market. These have to be stored first after procurement. These things put together give an idea of storing these items. This process of storing is called inventory.

Necessity and Advantages of Inventory Control

There are various reasons why organisations should maintain inventories of goods. The fundamental reason for doing so is that it is either physically impossible or e economically unsound to have goods arrive in a given system precisely when demands for them arises. Without inventories customers would have to wait until their orders were filled from a source or were manufactured. In general, however, customers will not or cannot be allowed to wait for long periods of time. For this reason alone the carrying of inventories is necessary for almost all organisation that supply physical goods to customers.

There are, nevertheless, other reasons for holding inventories. For example, the price of some raw materials used by manufacturers may exhibit considerable seasonal fluctuations. When the price is low, it is profitable for him to procure a sufficient quantity of it to last through the high price seasons and to keep it in inventory to be used as need in production. Another reason for maintaining inventories, a reason particularly important to retail establishments, is that sales and profits can be increased it one has an inventory of goods to display to customer. Furthermore to even out the work loads on the shops in the event of fluctuation loads and a protective buffer against transportation delays and production rate changes.

Advantages

(i)   Introduction of a proper inventory control system helps in keeping the investment in the inventories as low as feasible.

(ii)   Ensures availability of material by providing adequate protection against uncertainties of supplies and consumption of materials.

(iii)   Allows full advantage of economies of bulk purchases and transportation.

(iv)   Reduces changes of going out of stock.

(v)   Leads to reduction in inventory levels.

(vi)   Releases more of capital for other operations.

(vii)   Increases profitability of an organisation.

(viii) Adequate customer service.

(ix)   Advantage of price discounts by bulk purchasing.

(x)   Providing flexibility to allow changes in production lines due to changes in demands or any other reason; and

(xi)   Even out the work loads on the shops in the face fluctuation demands.

Evils of Excess Inventory

★   Essential though they are, inventories also mean lock up capital of an inventories which could be invested in more profitable operations.

★   Excess inventory adds to the cost of carrying the inventory more store space, equipment and personnel, insurances, taxes, pilferage etc.

★ Excess inventory invites risk of deterioration and obsolescence.

★ Changes in the prices of inventory materials sometime go unfavorable.

☆ Maintenance of cost money.

Control Aspects of Inventory

A company which finds that different portions of its inventory present different problems will have three general methods of control to choose from. These are:

★ Elimination of certain inventories

★ Semi automatic routines.

★ Periodic review.

Elimination of Inventories

At first glance, elimination inventory may seem a strange form possible. It is more feasible to eliminate the inventories entirely if little advance information is available on exact specifications and quantities. Steel plate is one item required in the manufacture of every unit, if the firm did carry inventory this would be one material it would have to stock, however where one customer specifies I" plate, another calls for 1.5". In the view of these facts the company does not order steel plates until the actual specifications for each order are at hand. This solution is possible because the manufacturing interval is very long and vendors can provide the required plate in a reasonable period.

Semi-automatic routines

The semi-automatic types of control are usually variations of the maximum minimum system. They are most applicable for standard items with fairly stable prices and steady usage. For materials which do not possess these characteristics, maximum-minimum control offers few advantages. This system is usually established by considering the minimum balance desirable, the re-order point, the standard order quantity, and the maximum allowable the minimum balance is the point below which the minimum represents is determined by the rate at which an item is used, its size and cost its size and cost its similar factors, importance in the process, whether substitutes are available and by Ideally, a new shipment should arrive just as stock reaches this minimum. Slight variation in usage or delivery intervals will prevent such exact operation in practice. The order point is located somewhere above the minimum. Its exact location will depend upon the delivery or manufacturing intervals required after an order is placed and upon the usage rate of a particular item. When the available balances reaches the order point, a clerk will note this fact and place an order. The standard order quantity is that amount of material which will be requisitioned each time available balances drop to the order point. This quantity is determined by considering such things as vendors' discounts, economic lot sizes, storage and handling costs, order processing costs, the danger of obsolescence and desired rates of turnover. The maximum minimum type of systems offer semi-automatic should be emphasized because it is the key to successful operation of the system. The precise amounts established as minimum order points and adhering to them, give the system an appearance of exactness and automaticity which it does not always possess.

Periodic Review

The third type of inventory control might be called 'periodic review' for want of a better term. All the factors considered under the maximum-minimum system are applicable here. But definite points are established in advance. The interval will be shortest where:-

★ Price of usage fluctuates widely.
★ cost in high.
★ shortages are apt to be critical.
★ there is considerable risk of obsolescence.
★ Materials are bulky.
★ delivery or manufacturing cycles are long.

These three methods of control are not mutually exclusive. One company may use any or all the them to control different categories of inventory. Some commodities, such as special motors,may not be ordered until their use and specifications can be established exactly.

The inventory control system is a composite of many routine operations.

Types of Inventory

1. Transaction inventory
2. Speculative inventory
3. Precautionary inventory

(1) **Transaction inventory.** It is formed basically of those items which are basically needed for transaction for example transaction of finished saleable products or transaction of raw materials.

(a) *Delinking Production with purchasing Department* is one of the function of this for example receipt of raw material is stored in the form of inventory and whenever production department needs raw material it put indent to inventory organisation rather than to purchase department.

(b) *Delinking production with sales organisation.* Similarly it delinks sales and production departments.

(2) **Speculative Inventory.** It is basically kept as a measure of speculation of increase in the price of raw material or increase the sale price of finished goods. It is generally resorted to before the budget proposals.

(3) **Precautionary Inventory.** The machines are to be kept running. There may be breakdown of machines at any time due to damage of any part due to wear and tear. At that time it cannot be purchased immediately. Therefore,such items which are essential to keep the process running are to be stored like machine parts, tools, etc. It is also called maintenance repair and operation inventory.

Classification of Inventory

There are another way of classifying various types of industrial inventories such as:

(1) **Production Inventories.** Raw materials parts, and components which go into the manufacturing process of the company products.

(2) **M.R.O. Inventories.** Maintenance, Repair and Operating supplies which are consumed in the production process but which do not form a part of the product (often known as consumables).

(3) **In-process Inventories.** Semi-finished products at various stages in the production operation.

(4). **Finished goods Inventories.** Completed products ready for shipment.

Inventory Control Systems

Control over inventories means good long range and intermediate planning of production operations good production scheduling and good methods of control. A comprehensive inventory-control system, including production planning, scheduling and control must be closely coordinated with other planning and control activities such as cash planning, capital budgeting, and sales forecasting, since it impinges on a wide range of production, sales, and the financial policy and operating decisions. The essentials of an inventory-control system can be grouped into three broad classes:

(i) **Long-range planning:** In order to budget capital for facilities and inventory investment, to arrive at a balanced capital budget in view of long-range business forecasts and possible errors in these forecasts.

(ii) **Intermediate policy making and planning as a basis for short-term scheduling:** Decisions should be made on what money is currently worth, what current service requirment are. Therefore short term scheduling consistent with inventory policy lays the ground rules.

(iii) **Short-term scheduling:** In short-term scheduling of work assignments to keep facilities and men employed and stock balanced in view of demand for output as it actually materializes. This must be done within a consistent framework of policies governing the levels of production and employment to be maintained.

The inventories serve as cushions in each of there stages of planning to absorb the shocks of demand forecast errors to permit more effective use of facilities and staff in the face of demand fluctuations and to isolate one part of the system from the next to permit each to work more effectively. Fig () analyses the three planning functions and the flow of information. The figure illustrates, the long range plan making use or demand forecasts and preliminary policy decisions. On capital allocation and value and on the amount of risk to be assumed to show the implications of policy choices. The purpose of production planing is to resolve the general planning questions within the limitations of operating resources, with the objective of maximising profits or return on investment rather than to define precisely how much of what item to be made on what machine.

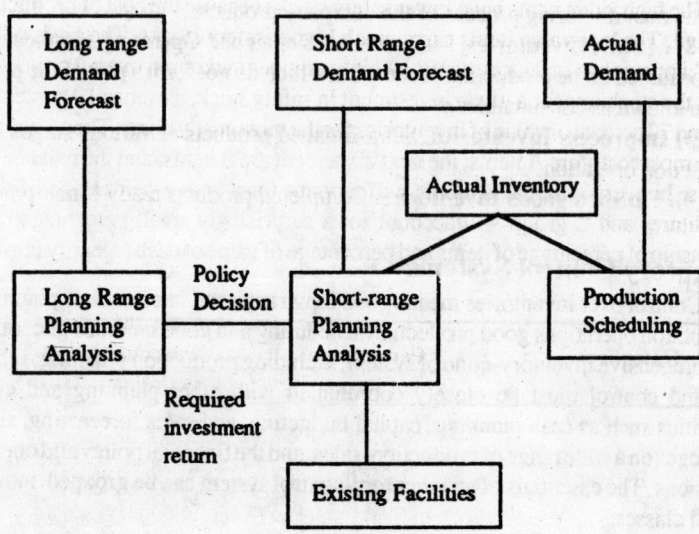

Fig. 18.1 Inventory planning.

Role of Forecasting

Fig. 18.1 shown in detail are essential in inventory management. The forecasts may be hidden, crude, intuitive, but they are there. The important forecasting period is that just ahead, the time needed to make any current production or purchasing decision effective. In case of seasonal items of forecast at least through the next cycle is needed if stock are to be build in anticipation. Forecast errors cost money; the bigger the errors, the bigger must inventories to be guarded against them. But forecast errors or limits of error, is not enough . Sometimes specifying a maximum sales forecast, the maximum which the production and distribution organisation will be required to service, is a satisfactory way of accounting for forecast error.

The ABC Analysis

Introduction

The basic reason underlying ABC analysis is to exercise selective control on inventories thereby economising the efforts and cost involved in inventory management without materially reducing the extent of control exercised. Materials managers rarely, worry about stock outs of some stationary items. This is not in case with many other items, particularly production parts and materials. In many cases a stockout of even a minor production part can be costly. Safety stocks almost always are better value for low cost items then they are for expansive ones. Modern inventory control system take this into account by classifying items by value of usage The high value items have lower safety stocks because the cost of production is so high. The low-value items carry much higher safety stocks. The basis of the "ABC" approach to inventory control which provides maximum overall protection against the stockouts for a given investment in safety stock, assumes 10 percent of items and 70 percent of value of inventories for the most costly items. The 10 percent that are most costly are A items; the next 20 percent are B items; and the balance are C items. In every case the A items will amount for a heavy percentage of total expenditures and C group will account for a surprisingly small percentage. The relationship of percentage of items and percentage of value of each category is given as under:

Category	Percentage of items	Percent of value
A	10	70
B	20	20
C	70	10

Although here are one seventh as many A items as C items, seven items as much is spent A items as on C items. Thus the average expenditure of an A item is 49 times greater than the average expenditure for a C item. Investment is safety stock for A item must be 49 times grater than for a C item in order to afford the same protection against stockouts. The ABC system permits selective inventory control. Safety stocks are kept low for high-value items, which should be subject to extremely close control by materials personnel; stock-outs are prevented by maintaining much higher safety stocks. With ABC control, it is possible to risk fewer stockout and reduce investment in inventories. This is proved by comparing ABC control with overall control inventories are divided into three classes: The expensive A items account for 10 percent of the total and roughly 70 percent of the value; the B items for 20 percent of the total and about 20 percent of the value; and the C items for 70 percent of the total and about 10 percent of the value. In other words on costly items, tight control is substituted for the protection of inventory. This is economic since it permits reductions is inventory investment with the low cost items, it is cheaper to carry inventory than to pay salaries of the personnel needed for close control. It would take seven times as much effort to maintain tight control.Over the 700 C items as it would for the 100 A items, and it would be possible to reduce investment for C items by only one seventh as much as the reduction in the investment for A items.

With the B items, a middle of the road policy would probably be followed. There would be some control, but the company would also rely to a greater extent on inventories to protect against stockouts than it would with the A items.

Objective of ABC Analysis

The object of carrying out ABC analysis is to develop policy guidelines for selective control ABC analysis enables the materials manager to exercise selective control when he is confronted with a large number of items. The tighter and accurate procedures are essential for 'A' value items. The degree of control should be regourous for 'A' items and should be minimum for 'C' items. ABC analysis is helpful to rationale the number of orders and thus reduce the overall inventory cost. It is also common to further subdivide 'A' items as A_1 and A_2 or A^+ and A^- and similarly categories for 'B' and 'C' items for exercising finer control. For 'A' items there should be weekly control statements; Rigorous value analysis; Accurate forecasts in materials planning; Central purchasing and storage; Multiple sources and should be handled by senior officer while for 'C items there should be a follow up and expediting is exceptional cases; Decentralized purchasing; Minimum value Analysis; Rough estimates for planning and can be fully delegated.

Methodology of Handing ABC analysis

Following mechanics can be used for classifying 'A'; 'B'; and 'C' categories:

(i) Calculate the Rupee value for each item is inventory and the annual consumption.

(ii) Arrange all the items is descending order; value wise.

(iii) Prepare a list of items according to annual rupee value.

(iv) Compute a total running total item-by-item.

(v) Computer print for each item the cumulative percentage for the item could and cumulative annual issue value. A graph can be drawn based on the above data and we can visually see the ABC categories as ABC analysis is based on the principles: Analysis based on annual consumption value; Dose not depend on the importance of the item and limits for ABC categorisation are not uniform but will depend upon the size of the undertakings.

Concept of ABC Analysis

Any average medium-size organisation consumes a few thousand tons of stages. An organisation choosing high degree of control of each item would, therefore, neither be practical, considering the work involved, nor worthwhile since not all items are of equal importance. Hence it is desirable to classify or group the items depending upon their importance and subject each class or group the items depending upon their importance and subject each class or group of items to control, commensurate with importance. This is the principle of selective control as applied to inventories and the technique of grouping is termed as A.B.C. analysis or classification which it said to be "Always Better Control". The basis of analyzing the annual consumption cost goes after the principal "Vital Few"—"Trivial Many" and criterion used here is the money spent and not the quantity consumed. The Following Fig 18.2 brings out clearly the concept of the ABC Analysis.

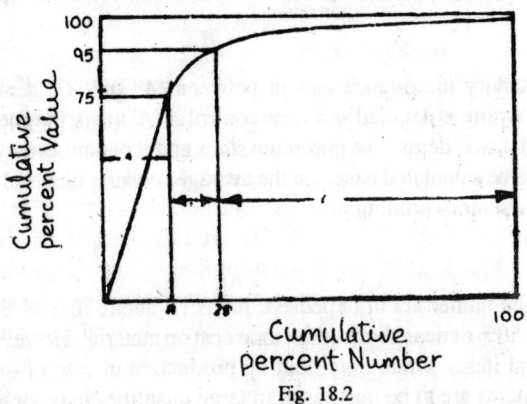

Fig. 18.2

'A' Items

From the graph and analysis it may be found that about 8% of items cost more than 75% of the cost on inventory. This is classified as 'A' items. These are the most important items form the control point of view.These should be ordered and scheduled for receipt on a monthly, weekly or even daily basis in order to keep the investment on them as low as possible. Their movement through the shop should have high priority and the movement time kept at a minimum. Safety stock or reserve stock on such items should be kept to an absolute minimum. A close watch on he ordering, follow up stocking and issue of these items are a must for economy on Inventory.

Control policy for A Items

The measure to taken on 'A' items can be briefly put down as follows:

1. Annual or half yearly contracts for supplies with as frequent staggered deliveries as is economical.

2. Develop and revise more often ordering quantities, reorder points and safety stocks for items not covered by long terms contracts as indicated in items I.

3. Frequently reviewing of stock position and consumption patterns and maintenance of records of issues and stock to enable to get the up-to-date position of stocks at any time.

4. Precise quality specifications or material standard.

5. As far as possible, two or more suppliers should be sought for each item so that the dependency on one supplier is not here. Due to any contingencies the other supplier could be approached.

6. The control of 'A' items should be directly under the charge of Purchasing Manger himself.

7. As these items are to be stocked to a minimum level, the purchase department should pay maximum attention to expedite the delivery of the items.

8. Waste control measures to reduce the scrap, rejection, rework and sub-standards.

9. Continuous development work or research carried out whenever possible.

'B' Items

These are secondary importance and in between 'A' and 'C' items. Even though, they do not require as detailed and close control as 'A' items, they need more attention and control than C items. The minimum stock and economic order quantity of these items should be calculated based on the average consumption lead time for prominent, cost of inventory holding etc.

'C' Items

'C' items are the numerous in expensive items i.e about 70% of the items contributing to only 50% or near about of the total cost on material. However, these items may be critical items which may hold up production in case of their non-availability. These items are to be purchased in large quantities in order to secure purchase discounts and minimise time and effort required by order clerks, store keepers and accountants, the company should carry say six months to one year stock of these items taking into consideration storage facilities available and storage costs.

Advantages of ABC Analysis

1. Preference for keeping inventory can be placed properly after ABC analysis.

2. Stores personnel are placed better with this analysis i.e. their time can be utilized better.

3. Storing, handing and delivery of material to production department become better.

4. We known I'C' items take more time in review as they are more in number. Review can be decided accordingly.

VED-Analysis

The analysis specially pertains to the classification of maintenance spare denoting the essentiality of stocking spares.

V—Stands for vital-items when go out of stock or when not readily available, completely bring the production to a halt.

E—Stands for Essential-items without which temporary losses of production or dislocation of production work occurs.

D—Stands or Desirable-all other items which are necessary but do not cause any immediate effect on production.

SDE-Analysis

For developing countries and especially where certain items are in scarce supply, this analysis is very useful.

S—Stands for Scarce items, especially imported items and also those which are very much in short supply.

D—Stands for Difficult items which are available in the market but not easily available. For example, items which have to come far off cities.

E—Items which are easily available; mostly local items.

HML-Analysis

The cost per item (per piece) is considered for this analysis. High cost items *(H)* Medium cost items *(M)* and low cost items*(L)* help in bringing controls over consumption at the departmental level.

FSN-Analysis

Here the quantity and rate of consumption is analysed to be classified as Fast-Moving *(F)*, Slow Moving *(s)* and Non-Moving *(N)* items.

19
Factors Influencing Inventory

Introduction

Materials account for over 50% of the production costs. In addition, frequently as much as one third of company's total investment is in the form of work-in-process, finished goods and stores inventory. Out of the three inventory forms, the stores inventory usually represents the largest share and this is the area in which the purchasing department, either directly or indirectly, can contribute significantly to company profit by efficient management of quantity and timing of purchases. There is one right quantity to buy for any given transaction but since there are many different kinds of transaction, the determination of correct quantity is a complicated matter. Similarly there is only one right time to purchase the right quantity. The issue is important because, if too small a quantity is purchased the unit cost will usually be higher, shortages are likely to increase, expediting work will necessarily be greater, and the relationships between vender and purchaser will suffer. On the other hand if too large a quantity is purchased, the excess inventory will raise costs, obsolescence will become a more serious problem and the need for additional storage facility will create investment problems.

Inventory Control

It may be defined as planning, ordering and scheduling of materials used in the manufacturing process. These are four major reasons for maintenance of sound inventory control procedures:

(i) Unit Cost : quantity purchases permit lower unit costs.

(ii) Operating Costs : quantity purchases permit efficient use of manpower; machines and facilities.

(iii) Customer Service : quantity purchases assure optimum customer service and provide for efficient scheduling of internal options.

(iv) Efficient use of : The balancing of unit cost; operating costs; Invested Capital customer service with cost of capital.

Right Quantity

Buying the right quantity is one of the most important as well as the most complicated tasks of purchasing. A company's ability determine the right quantity and is influenced heavily by basic managerial planning, organisation, co-ordination; and control. An effective inventory control system must be integrated with other company wide planning and control activities such as cash flow planning; capital budgeting; sales forecasting and production planning, scheduling; and control.

Types of Inventory Control Systems

There are two basic systems controlling inventories: (i) Cyclical ordering system and (ii) Fixed order quantity system.

Cyclical Ordering System

It is a time based system which involves scheduled periodic reviews if the stock level of all inventory items. When the stock level of a given item is not sufficient to sustain the production operation until the next scheduled review, an order it placed replenishing the supply. The frequency of reviews varies from the company to company. Stock levels can be monitored by physical inspection; by a visual review of perpetual inventory cards, or by automatic computer surveillance. In operations where a small quantity of material is involved, the simplest method is a periodic physical count of stock. In practice, the cyclical ordering system is well suited for materials whose purchases must be planned months in advance because of established and infrequent production schedules maintained by suppliers. The cyclical ordering system tends to peak the purchasing work loads around the review dates. This disadvantage can be avoided to some extent by regulating frequency of reviews. In this system the unexpected changes in secular trends are so gradual that they rarely have any effect on short term materials forecasting and planning. Errors in cyclical forecasting are far more likely to affect materials management. An unexpected shift in the business cycle works like a two edged sword on materials management: it affects both the lead time and the lead time usage.

Fixed Order Quantity System

It is based on the order quantity factor rather than on the time factor. The design of this system recognizes that each item possesses its own unique optimum order quantity. The major advantage claimed for the system are that (i) each material can be procured in most economical quantity (ii) Purchasing of items/materials is when required (iii) Positive control can easily be exerted to maintain total inventory investment at the desired level simply by manipulating the planned maximum and minimum values. The system functions correctly only if each of the materials exhibit reasonable usage and lead time. In fact irrespective of the system inventory policy for different groups of items 'A'; 'B'; and C there are two basic problems: what quantity of an item should be ordered each time; and when should an order be placed. In order to find a solution to these problems we should try to analyses the four cost factors: Cost of Material; cost associated with keeping the materials in stock (Inventory Carrying Costs); Cost associated with placements of purchase order (ordering costs); Carrying Inventory: The cost of material is the most important and for items with frequent unexpected price variation. While cost of materials and inventory carry costs are totally dictated by the price of the items. The inventory carrying cost arise out of the following factors.

:   Interest charges (4 to 5%)
:   Insurance Cost (1 to 3%); Protection against theft and fire etc.
:   Properties Taxes (1 to 3%)
:   Obsolescence and deterioration cost (4 to 10%)
:   Stationery and consumables used in stores
:   Depreciation and repairs cost.

These cost come nearly 20 to 25%. These costs are arbitrary and vary from industry to industry and from item to item even in the same industry.

Ordering Cost or Acquisition Cost of Items

Placement of a purchase order for a material is associated with certain obvious cost due to advertising, consumption of stationary and postage, telephone charges, telegrams etc. In fact all the annual expenditure of the purchasing deptt. of a company can be considered to be on the purchase orders it places during the year. The cost associated with ordering would, therefore consist of

(a) Salaries of the staff n administration and purchase deptt.
(b) Rent for the space used by the purchasing deptt.
(c) The postage, telegram and telephone bills.
(d) The stationary and other consumable required by the purchasing deppt.
(e) Entertainment charges on vendors.
(f) Travelling expenses.
(g) Lawyers and court fees due to any legal matters arising out of purchase.

These all usually come to 20-25% as ordering cost. Obviously more the number of orders placed in a period, the more would be the stationary and postage consumed, more staff and officers will be required for handling the work, the more will be the space required for accommodating them and so on. Thus the total expenditure on purchasing or ordering would depend on the numbers of orders placed. It is assumed that the expenditure on ordering of material is directly proportional to the number of orders placed. The ordering cost is expressed as the cost/order and is calculated by dividing the total ordering cost during a year (or a period). These are about 20-25%. Following are the components or elements of Inventory carrying cost and Procurement cost which are the most important costs in inventory management.

Components of Inventory Carrying Cost

The components of inventory carrying cost are:

(a) The salaries and wages of storing, receiving and issue of the material personnel.
(b) The loss by way of interests in the capital invested in the stocks of materials.

(c) The rent of the stores premises.

(d) Taxes on inventories.

(e) The depreciation and repairs cost for the stores facilities and handling equipments.

(f) Loss of materials through pilferage and deterioration, obsolescence.

(g) Stores insurance cost.

(h) Stationery and other consumables used by the stores.

The cost generally varies from 10 to 20%.

Procurement cost

This includes:

(a) The salaries and wages of the purchase staff purchasing the materials.

(b) The expenditure on travelling etc.

(c) The expenditure on the use of Post and Telegraph services such as postage, telegrams and telephones.

(d) The expenditure on stationery and other consumables.

(e) Entertainment expenditure on receiving of suppliers.

(f) Rent for the premises used by the purchase department.

Certainly it can be sad that more the number of orders are placed the more shall be the above expenditures.

The ordering cost is generally dependent on the purchasing procedures of the individual organisation. The procurement or ordering cost generally have no relation with the value of an order.

Carrying Inventory

Cost of carrying is not the same as carrying inventory. Carrying inventory results where there is stock left on hand after the demand for the item has terminated. This cost is called the overstock cost. The inventory problem can be static or dynamic. Static inventory problem is with retailer who wishes to sell greeting cards. His sale season is very limited and few cards left over at the end of season will have a very little or no salvage value. Thus if he has too much stock he will have a cost due to the loss he will suffer for each greeting card he overstocks. The dynamic inventory problem could be for household items such as towels etc. There is always a demand for towels so any towel left over after one time period can be sold in the subsequent periods thus there is no overstock cost to a greater extent. The similar examples, holds true for high fashion items, woman clothes. If only one order can be placed we have the static inventory problem if several orders can be placed during a season then we have dynamic problem. There will be no overstock cost involved until last order period of the season.

Procurement and Consumption Cycle

A simplified procurement and consumption cycle for an item having a steady consumption all through the year and which is available immediately on placing an order is shown below:

At time A the stock is zero and hence an order would be placed and the delivery being immediate, the stocks would be brought upto a level say Q. These would be steady issued upto time B and another order would be placed at B to bring up the stock to Q and so on. The inventory at the beginning of the period A-B would be Q and at the end of the period zero or the average inventory would be $Q/2$ and orders would have to be placed of A, B and C etc.

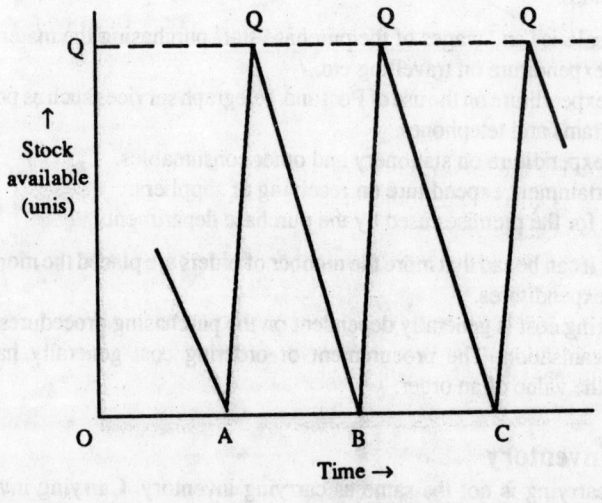

Fig. 19.1

Economic Order Quantity (E.O.Q.)

It is discussed here as to what should be the quantity Q ordered each time to keep down both the inventory carrying cost and ordering cost in this consumption and procurement cycle.

Consider the procurement of the entire annual requirement of an item under the procurement and consumption cycle. For keeping the inventories and inventory carrying costs low it would be better to procure the item in as small consignment as possible. But this would mean larger number of orders and so more ordering cost. This has been graphically shown below. The requirements are thus conflicting and there is a particular quantity at which the sum of both the ordering and inventory carrying costs is minimum and this very quantity is called EOQ (Economic Ordering Quantity).

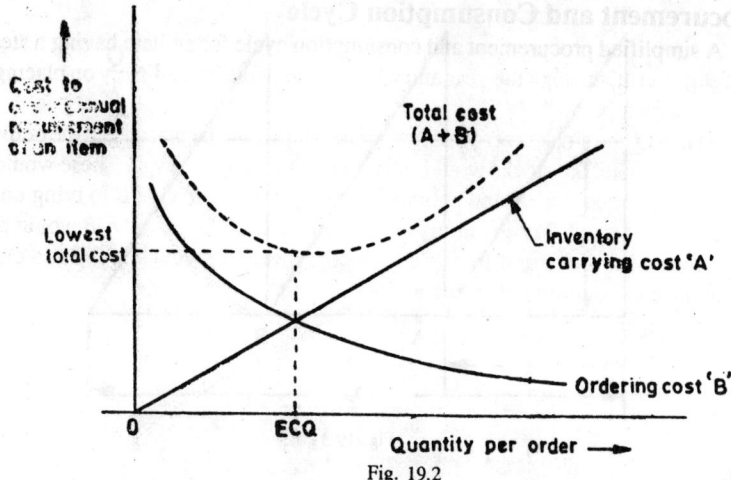

Fig. 19.2

Let us now derive the formula:

Let A = Ordering cost/order

Q = Quantity/order

λ = Total quantity ordered in the period under consideration.

IC = Inventory carrying cost per unit.

TIC = Total procurement and inventory cost or total inventory cost or No. of orders in the period under consideration would be = λ/Q.

TIC = Ordering cost + inventory carrying cost

= Cost per order × No. of orders + (average inventory × inventory carrying cost per unit)

$$TIC = A \times \frac{\lambda}{Q} + \frac{1}{2} Q.I.C. \qquad \ldots (1)$$

Differentiating (1) with respect to Q, for the total cost to be minimum and then equating to zero.

$$\frac{d(TIC)}{dQ} = -\frac{\lambda A}{Q^2} + \frac{1}{2}IC = 0$$

or

$$\frac{\lambda A}{Q^2} = \frac{1}{2}IC$$

or

$$Q^2 = \frac{2\lambda A}{IC}$$

$$Q = \sqrt{\frac{2\lambda A}{IC}} = EOQ$$

$$EOQ = Q^* \times \sqrt{\frac{2 \times \text{Annual consumption} \times \text{ordering cost}}{\text{Inventory carrying cost as a fraction}}}$$

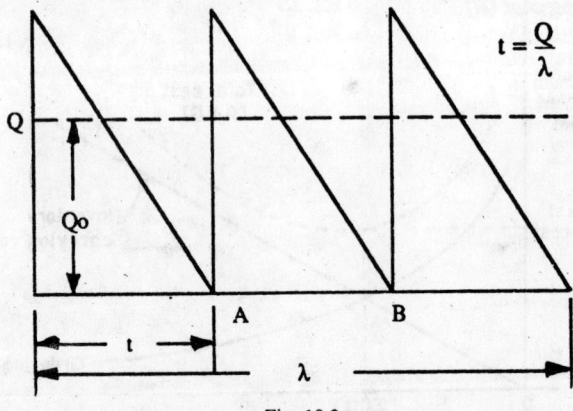

Fig. 19.3.

It can be proved that at *EOQ* the ordering cost and inventory carrying cost are equal.

$$Q = \sqrt{\frac{2A\lambda}{IC}}$$

or

$$Q^2 = \frac{2A\lambda}{IC}$$

or

$$\boxed{\frac{QIC}{2} = \frac{\lambda A}{Q}} \text{ at } EOQ \qquad \qquad \dots (2)$$

Equation 2 indicates that at *EOQ*, inventory carrying cost and ordering cost are equals.

It may be noted here that shape of the total cost curve is fairly flat near $Q = EOQ$ in other words if little more or little less is ordered than *EOQ*, the total cost only increase slightly. In practice 25% less or more than *EOQ* makes very little difference in the total costs. Also the formula is quite general and 'A' need not necessarily be annual consumption. 'A' could very well be for 6 months consumption cost. When simultaneous consumption is there and the rate of which is X then the formula for E.O.Q. becomes,

$$Q = \sqrt{2A\lambda/IC(1-\lambda)}$$

Problem 1

Given

Annual consumption (λ)   = 36,000 units
Inventory carrying cost (*I*) = 20% per annum

Ordering cost (A)   = Rs. 25
Cost/unit (C)   = Re. 1
Find Q at EOQ and total cost of Inventory.

Solution

$$Q = \sqrt{\frac{2A\lambda}{IC}}$$

Here

$\lambda = 36,000$

$A = 25$

$C = 1$

$I = 0.2$

Putting the values of λ, A, I, C,
We get,

$$Q = \sqrt{\frac{2 \times 36,000 \times 25}{0.2 \times 1}}$$

$$= \sqrt{36,000 \times 25} = 600 \times 5 = 3000$$

This means that for the order quantity 3000 units company will incur the minimum total cost on ordering and carrying inventory. This is explained below:

Qty. per Order (Q)	No. of Orders	Inventory levels (Rs.) (Q/2)	Ordering cost (Rupees) (A)	Inventory Carrying cost (Rs.) (B)	Total cost (A × B)
3000	12	1500	12 × 25 = 300	1500 × 0.2 = 300	600
2400	15	1200	15 × 25 = 375	1200 × 0.2 = 240	615
3600	10	1800	10 × 25 = 250	1800 × 0.2 = 360	610

When 12 orders are placed and inventory is maintained at an average of Rupees 1500, the total cost is the minimum is

$$(300 + 300) = \text{Rs. } 600$$

Problem 2

In the above example, it is forecast that the next year consumption will increase to 144,000 units *i.e.* 4 times. On this basis calculate *EOQ* and *TIC*.

Solution

Given

$$\lambda = 144{,}000$$
$$A = 25$$
$$Q = 6000$$
$$I = 0.2$$
$$C = 1$$

$$EOQ \text{ (for new consumption)} = \frac{2A\lambda}{IC}$$

$$= \sqrt{\frac{2 \times 25 \times 144{,}000}{0.2 \times 1}}$$

$$= 5 \times 1200 = 6000 \text{ units.}$$

Note. Sometimes deviations from economic order quantity have to be made for variety of reasons. In such cases some believe that deviations towards a quantity less than the economic order quantity is more desirable than deviations towards a quantity greater than EOQ. This however is not correct. In our example if the deviation is great than the EOQ by 20% total cost of ordering and carrying inventory increases by Rs. 10. If the deviation is less than EOQ by 20%, the increase in total cost is Rs. 15. Thus we have noted that deviation towards a greater quantity than EOQ increases total cost of ordering and carrying inventory by lesser amount than what the same % deviation of 20% towards the lower side would increase the total cost of ordering and carrying inventory.

$$TIC = \frac{A\lambda}{Q} + \frac{1}{2}QIC$$

$$= \frac{144{,}000 \times 25}{6000} + \frac{1}{2}6000 \times 0.2 \times 1$$

$$= 24 \times 25 + 600$$
$$= 600 + 600$$
$$= 1200$$

Problem 3

A sewing m/c manufacture buys from a nearby foundry the body casting in which the m/c is assembled. The foundry provides usually good service and always makes delivery on 8th day of the receipt of the order. The normal consumption of casting is at an average rate of 100/day. The consumption varies from 80—120 casting/day. The management decides to set an order point and safety stock so that under the present operating conditions they do not run out of stock.

Solution

Average number of casting needed = 8 × 100 = 800
Maximum number of casting needed = 8 × 120 = 960
Buffer stock or safety stock   = 960—800 = 160

The management has concluded here that they should not run out of stock by establishing this order point at the level of 8=960 units. The safety level stock is 160 unit.

Lead Time—6—10 days
Average no. of casting used during the lead time = 8 × 100 = 800
Minimum ,,   ,,   ,,   = 6 × 80 = 480
Maximum ,,   ,,   ,,   = 10 × 120 = 1200
Safety Stock   = 1200 – 800 = 400
Order pt—minimum use   = 1200 – 480 = 720
Order pt—maximum use   = 1200 – 1200= 0

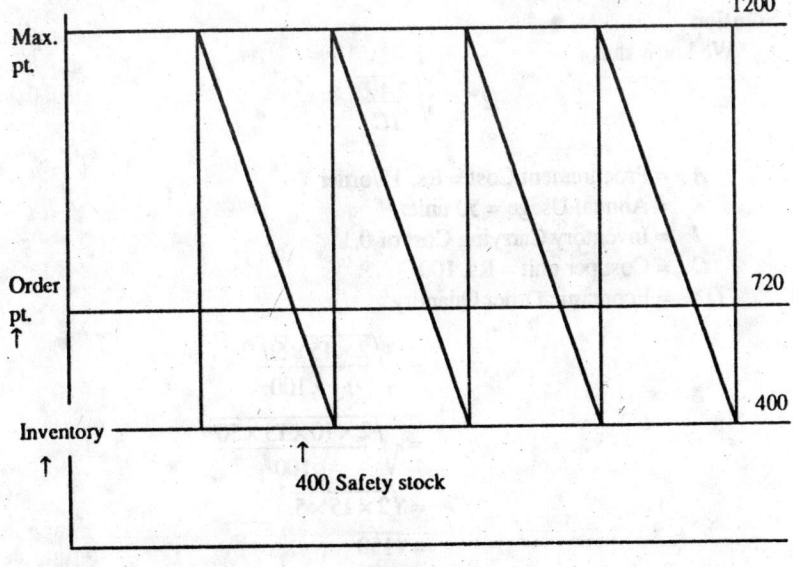

→ Time

Fig. 19.4

Problem 4

An oil engine manufacturer purchases lubricators at Rs. 42/piece from a vendor. The requirement of these lubricators is 1800 per year. What should be the order quantity per order (A = Rs. 16 and I = 0.2)

Annual requirement is = Rs. $1800 \times 42 = 75600$

$$A = \text{Rs. } 16$$
$$I = 0.2$$

$$EOQ = \sqrt{\frac{2A\lambda}{IC}}$$

$$Q^* = \sqrt{\frac{2 \times 1800 \times 16}{0 \cdot 2 \times 42}}$$

$$= 82.8 \text{ or } 83 \text{ lubricators}$$

Problem 5

Calculate the value of Economic Order Quantity. Given the following:

Annual usages	= 50 units
Procurement Cost	= Rs. 15 per order
Cost per price	= Rs. 100 per item.
Inventory Carrying Cost	= 10%

Solution

We know that

$$Q^* = \sqrt{\frac{2A\lambda}{IC}}$$

A = Procurement Cost = Rs. 15/order
λ = Annual Usage = 50 units
I = Inventory Carrying Cost of 0.1
C = Cost per unit = Rs. 100
Q^* = Economic Order Quantity

$$= \sqrt{\frac{2 \times 15 \times 50}{0 \cdot 1 \times 100}}$$

$$= \sqrt{\frac{2 \times 10 \times 15 \times 50}{100}}$$

$$= \sqrt{2 \times 15 \times 5}$$

$$= \sqrt{150}$$

$$Q^* = 12.21$$

Taking the whole number. The most Economic Order quantity where the inventory carrying cost and inventory procurement cost are minimum are 12 Nos. per order.

Problem 6

The annual consumption of an item is 12000 per year. The procurement cost is Rs. 15 per order and inventory carrying cost per unit per year is Re. 1.

Find out the most economic order quantity and the total cost of inventory also.

Solution

We known that

$$Q^* = \sqrt{\frac{2A\lambda}{IC}} \qquad \text{... (A)}$$

A = ordering cost per order.

Q^* = Economic order quantity.

λ = Annual consumption.

I = % of inventory carrying cost.

C = cost of item.

If we substitute the value in (A) we find the economic order quantity is
$$Q^* = 600.$$

So everything order is placed the number of units to be ordered are 600. Since the total requirement is 12000 per year, 20 orders at an interval of 18 days are to be placed assuming 360 days in one year. The total cost of the inventory consists of two parts:

(1) Total procurement cost.

(2) Total inventory carrying cost.

Total ordering cost = Rs. 20 × 15

= 300.

Similarly total inventory carrying cost

= 300

Total inventory cost is Rs. 600

Any other policy will increase the total cost. If each item cost Rs. 10 then total usage value per year is Rs. 120,000 inventory carrying cost I is 0.10 and Q^* can be calculated with the formula given before.

In calculating the economic order quantity only the cost of ordering and stock holding have been taken into account. In certain cases the ordering quantity might have to be adjusted taking into consideration:

(a) Bulk.

(b) Available space.

(c) Package size.

(d) Shipment size.

The economic order quantity is suited in a static equilibrium or when there is no uncertainties.

To deal with economic order Quantity in the real situation following terms should be discussed.

(1) Lead time

(2) Re-order point or level

(3) Safety stock.

Lead Time

The procurement and consumption cycle has got two elements in it—(1) Lead Time (2) Safety Stock.

In the earlier simplified cycle the supplies of material were assumed to be immediate on order but in practice this is not so. *From the time the requisition for an item is raised, it may take several weeks or months before the supplies are received, inspected and taken into stock. This time is called as lead time* and involves the time for the completion of all or some of the following activities.

(a) Raising of a purchase requisition.
(b) Inquiries, quotations, scrutiny, and approval, (Import license procedure for imported item).
(c) Placement of an order on supplier/suppliers.
(d) Suppliers time to make the goods ready (may have to be manufactured or supplied ex-stock).
(e) Transportation or clearing.
(f) Receipt of goods at the company.
(g) Receiving inspection.
(h) Taking into stock.

Obviously, in order to receive supplies before the stock reaches zero level, it is necessary to order the materials much in advance i.e. when the stock available is sufficient to last during the lead time. This is shown graphically in Fig. No. given on next page.

Suppose an item has a lead time of 15 days and the monthly consumption of the item is 600 units, then a recorder must be placed when the stock available is sufficient to last 15 days is 300 units.

In graph above Q = Order Quantity.

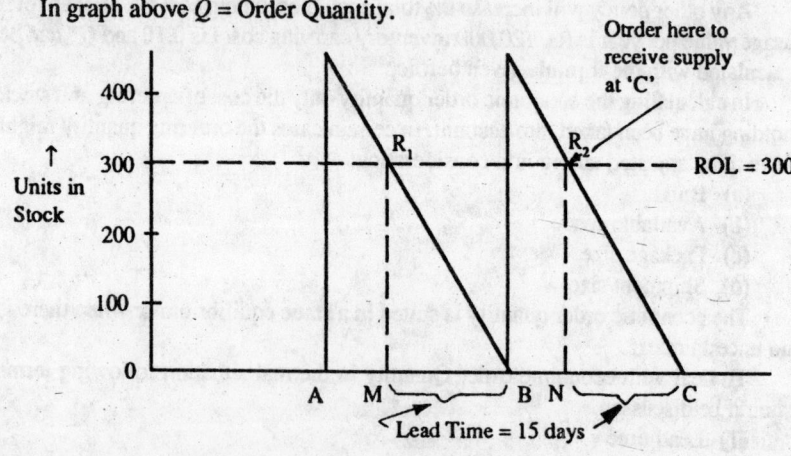

Fig. 19.5.

NC   = MB = 15 days is lead time Re-order level
(ROL)   = stock sufficient to last during the lead time is 300 units.

Safety Stock (Buffer or reserve)

It is well known that neither the consumption rate of a material is constant throughout the year nor is the lead time. Hence in the earlier example though recorder is placed at a stock level of 300 units, the consumption rate may rise subsequently and the stocks may well be exhausted in 7 days instead of 15 days or it may be that the supplier fails to supply after 15 days as expected. In either case a stock out would be experienced resulting into hampering of production to guard mainly against these uncertainties in consumption rate and lead time, an extra stock is maintained all along the this is called as buffer stock or safety stock or reserve. This stock also comes in use when:

(1) Any excess in process rejections.

(2) Rejection at the time of receipt or goods due to damages or substandard quantity.

Since safety stock is a part of inventory, it should be maintained just sufficient to guard against the uncertainties and not too excessive (especially for A items).

Before deciding how much the safety stock should be analysis of the following aspect is essential :—

(1) 'Is the variation in consumption more predominant in the lead time.

(2) If the variation in consumption is more predominant way it is so and can it be forecasted in advance.

(3) If the variation in lead time is more predominant is it restricted to a particular period (say monsoon) or spread all over the year?

In most cases it is found that the variations in consumptions can be predicted fairly in advance accurately by good production and maintenance planning and does not present much of a problem. But in the present Indian conditions, it is the lead time variation that is more erratic and unpredictable.

Uncertainty

The safety stock is needed because of uncertainty. The uncertainty can consist of:

(a) Uncertainty of demand i.e., how much the actual demand differs from what you expected.

(b) Uncertainty of delivery, i.e., how long the lead time is going to be.

(c) Uncertain quantity, i.e. how many scrap or imperfect items the ordered quantity is going to contain. Uncertain quantity is usually small and will be neglected in the discussion.

Determining Safety Stock

One simple method of determining safety stock in such cases is to approximately estimate the maximum lead time and the normal lead time for an item in consultation with the purchasing personnel and from the past records. The safety stock should then be sufficient to last the periodic difference between maximum and normal lead time period.

Suppose for an item monthly consumption is 100 units, the normal lead time is 15 days and maximum lead time is estimated as one month. The safety stock would be,

(Maximum lead time in a month—Normal lead time in a month)—(monthly consumption)

$$= (1 - \frac{1}{2}) \, 100$$

= 50 or say 50 to be on safe side.

Or for an imported item with a normal lead time of 1 year and maximum estimated lead time 15 months, the safety stock would be

(15—12) = 3 months or say 4 months for additional safety.

It should be well understand that safety stock is meant only to provide for above normal lead time and above normal consumption rate. The normal lead time and consumption rate is already taken care of in setting Re-Order Level (ROL). If normal lead time consumption is 300 and safety stock is 100, the reorder level is set at 400 units (i.e. safety stock = lead time = ROL).

Optimum Safety Stock

If the safety stock maintained is inadequately low, the inventory carrying cost would be low on the safety stock but stock-outs will be very frequently experienced and the stock out cost would be very high. As against this, if safety stock maintained is rather large stock-outs would be rather rare but the inventory cost would be high. An optimum stock is better. Stock outs may give rise to following cost:

(1) Customer loss or customer dissatisfaction.

(2) Loss of production.

(3) Idling of machines or man.

(4) Emergency purchases at high prices.

(5) Extra transportation changes for speedier modes of transportation.

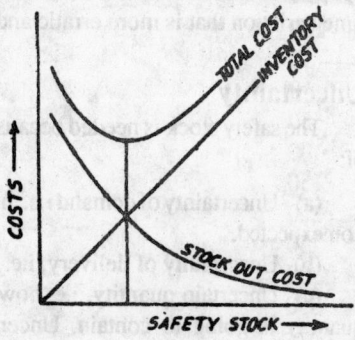

Fig. 19.6

In fact the cost of holding high levels of stock must be weighed against the cost of having stock out. The problem is of calculating the two costs. The cost of running out of stock is often difficult to assess. If the management decides it could not run out of stock on more than 2% of the orders. In other words, management policy is that 98% of orders must be met from stock, i.e., the defined service level is 98%. We must know what service level is defined before one can workout what safety stock should be hold.

Forecast

The expected demand is a forecast of demand. In this difference between the actual demand and the expected (forecast) demand is known as the forecast error. To compensate for this forecast errors we hold safety stock and so the level of safety stocks is, to a large extent, based on the extent of the forecast errors.

Service level

We hold safety stock to make sure that we can satisfy the demand but because of the uncertainty we won't be able to satisfy all the demand all the time. So we must decide what proportion of the demand we want to satisfy, i.e., how often we will allow stock-outs to occur. In other words we must decide on the service level we desire.

Forecast Errors

Since the safety stock depends on forecast efforts and the desired service level, we would like to find the forecast error. For this we study the past data on forecast efforts and try to see what pattern they fall into. In statistical terms, we want to find out the distribution of the forecast efforts. Once we have found how the forecast efforts are distributed, we can go on to draw conclusions about the probability of getting errors of any size.

Suppose we have a record of the last 30 forecasts of the demand made for an item. We can group the errors and draw up a table given below.

Forecast Error			Frequency	% age frequency	Cumulative
— 100	to	— 50	12	40	40
— 50	to	— 0	6	20	60
0	to	50	5	16	76
50	to	100	2	7	83
100	to	150	1	3	86
150	to	200	0	0	86
200	to	250	1	3	89
250	to	300	0	0	89
300	to	350	1	3	92
350	to	400	0	0	92
Over	to	400	2	7	99
			30	100	(Approximate)

From the table we can see that 18 times the actual demand was less than the forecast. It means that more than half the time the actual demand was less than the forecast demand. We are not much worried about the occasion when demand turns out to be less than what had been the forecast, since the safety stock is not affected.

The table also shows that 89% of the errors were less than 250. Hence if we had safety stock of 250 units we would have been able to satisfy the demand 89% of the time *i.e.* the service level would be 89%.

If the distribution of forecast errors is one of several well known statistical distributions or can be approximated, then we can calculated the safety stock for any required service level quite easily from statistical tables.

Techniques of inventory control (Ordering System)

Having discussed all the basic elements of a procurement and consumption cycle we can now depend on various ordering systems.

1. **The fixed internal System/cyclical review System/Replenishment System.** The basic type of state control is known as the fixed internal system. In this system the stock level is reviewed on a regular cycle say once a month. The interval between review is known as the review period. From the cycle shown in the figure below can be explained how the system is determined.

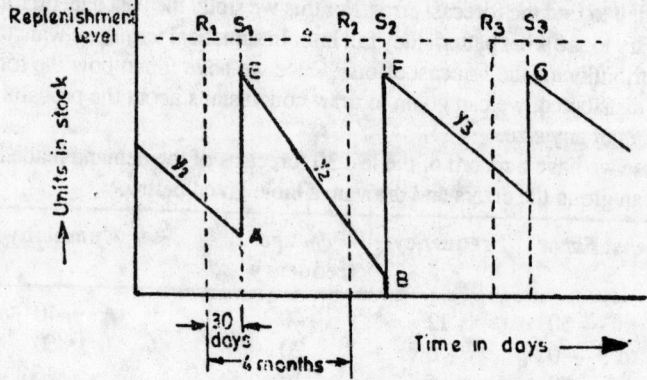

Fig. 19.7.

Suppose 4 months is fixed as the review period for an item and lead time for the item is 30 days. Then the order would be placed every 2 months *i.e.*, at ordinates R_1, R_2, R_3 etc. and the respective supplies would b received at ordinates S_1, S_2, S_3 etc. (30 days after R_1, R_2, R_3 etc.). At R_1 the stock available to be Y_1 then this stock together with the quantity ordered at R_1 (supplies received at S_1) should be sufficient to last till the next supplies are received at S_2 *i.e.* to last for a period of 5 months (1 Review + 1 lead time). In addition some buffer/safety stock would also be necessary to take care of any increased consumption or increase in lead time.

To find out how much to order, we calculated a figure called stock required. The stock which we have in hand and an order immediately after a review date must be good enough to last until the order which we lace at the next review date is available. This order would not arrive till one lead time after the next review date. So stock required = (Lead time + review period) × usage + safety stock on each review date the stock in hand plus that an order is compared with "Stock Required" and an order is placed to bring to stock upto the "Stock Required" level.

In a fixed internal system the combined cost of stock holding and placing orders will always be higher than a fixed order quantity system because the stocks have to last only over the lead time but also over the review period.

2. **Fluctuation In Stock Level.** The following diagram, in a simplified form should the stock level of an item fluctuates.

The amount used over a period is the usage in that period. As the stock is used up over the period, the stock level falls until it reaches a level at which an order is placed. If × is the re-order level. When the stock falls to re-order level, an order is placed for a certain amount. This is known as the order quantity and is shown as *OQ* in the diagram.

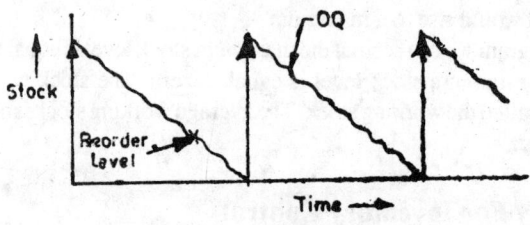

Fig. 19.8.

To take a couple of extreme examples. Suppose an item whose consumption is 10 number per month and its lead time is one month and perhaps it might be possible that 150 units are left over from a construction job. After 6 months the store keeper orders what ever the *EOQ* tells him which might be 20. But in 6 months he has used only 60, so he is reordering at 90. With usage of 10 per month and 1 month lead time store keeper will still have 80 on hand when the ones he ordered arrives. That is 8 months supply.

The other extreme : if the store keeper waited until he got down to 10 before reordering he would run out of stock just about the time the newly ordered ones arrive. No protection against variation in lead or in usage. If these were never any variation in usage or lead time he would reorder when stock was depleted to 11 units. Which is a best set up; minimum inventory and never a stock out. But there are bound to be fluctuation in lead time and in usage so there is a need to set the Reorder point high enough to provide protection against those fluctuations.

If the reorder point were 15, for instance, still 5 are on hand in the normal course of events (10 per month usage 1 month lead time). The inventory would vary between 25 and 5 averaging 15 of which 10 is "working stock". The 5 is called safety stock? A normal 2 weeks' supply in the example which might be adequate if there were very little variation in lead time or usage, but might not be nearly enough wide variations.

Reorder Point should be whatever one determines one should have as a safety stock for each item, plus the number of that item which normally be used during the lead time. ($RP = SS$ + Usage during lead time) obviously, the more uniform these factors are the fewer pieces needed as Safety stock protection and wider the fluctuation the more is needed for comparable protection. In the example given above suppose once every 5 years, the monthly usage would exceed 15 units. By setting the Reorder Point at 15 protection have been provided equivalent to one stock out in 5 years.

It an order quantity does not arrive immediately after is has been ordered. The time which elapses between the stock reaching the reorder level and the arrival or order quantity is known as the lead time. Stocks go on falling during the lead time until the order quantity arrives and is in the stores. In an ideal situation, the stock should reach zero just as the new stocks arrived. Then when an order quantity come in the stock level would rise to a maximum.

From the diagram we can set that the maximum stock level is equal to the order quantity and the minimum stock level is equal to zero. The stock we have been talking about, is called the working stock. The average working stock is equal to half the order quantity.

Responsibility For Inventory Control

The basic responsibility of Inventory control lies with top management. The effects of poor inventory management will not be directly visible on the operating statement as a composite cost of inventory management. Management can only hope to control inventories on an overall basis. On of the best ways to do this is through standards used in conjunction with an overall materials plan. For example, suppose the plan assumes that production and sales will be Rs. 10 lakhs per month for the first three months of the year and then will increase steadily to a peak of Rs. 15 lakhs in June. Obviously some inventory built up is necessary to accommodate the sales increase. However, it is not necessary to increase inventories in proportion to increase in output. Plants can operate with higher stock turnover ratios as production increases to near-capacity levels. What happens in fact is that materials move through the plant more rapidly when production increases. Thus the need for inprocess inventory may not increase as rapidly as out put. Raw-material inventories may not change at all. The extra demand causes stocks to drop more rapidly to their re-order points. Outstanding purchase commitments would grow, but, if everything goes accordingly to plan, actual stocks of purchased increase. It has been seen that routinization of the daily operation often camouflages the importance of sound management in this area.

Nevertheless in most companies these indirect costs, dispersed and hidden throughout the operating statement can have a significant influence on profit. Top management should, for this reason, carefully formulate and periodically review the basic political, mans and forecast which constitute the frame work with in which the daily inventory control operation functions.

The Inventory control functions are usually assigned to materials management department or production control department. A production inventory is the reservoir which operationally connects these two departments. The materials management department, feeds the inventory reservoir and production control utilises the inventory by utilising production facilities.

To summarise the need is to have high degree of co-ordination between materials and production control department relative to inventory control policy and operations. This is essential for optimum utilisation of results and efforts of materials and production groups.

20
Materials Management

Materials are an important determinant of the total cost of production as it occupies something like 20% to 35% of the total cost which in certain cases may be 60%. The materials mean raw materials components, sub-assemblies and finished products. Materials are classified into two heads viz; Direct; and Indirect materials. Direct materials goes directly into product and forms a part of the end product while indirect materials are used in processing or packing the direct materials. Materials management involves number of issues like the determination of quantity and quality ; purchasing ; store issuing and despatching. Materials management today is a distinct area of Industrial Management and plays a very vital role in production and productivity. This is a concept, the aim of which is cost reduction as a result of integrated approach towards the management of materials at all stages viz Planning ; Purchasing; Receiving ; Stocking and Disposal.

Materials Management is concerned with the planning and programming of materials and equipments, market research for purchase ; pre-design value analysis, procurement of all materials including capital goods, raw materials, components and assemblies, finished materials, packaging and packing materials stores control and inventory control ; transportation of raw materials and materials handling ; value analysis, disposal of scrap, surplus and salvage. Materials Management is basically a most importanant functional area of any organisation or undertaking to achieve the best results so far as profitability is concerned. Materials Management covers a much wider field and deals with all aspects of materials supply and utilization as well as costs. It is concerned with the entire range of functions which affect the flow, conservation, utilization, quality and cost of materials. It covers that aspect of Industrial Management which is concerned with the activities involved in the acquisition, storage and flow of all materials directly and indirectly employed in the production and marketing of finished goods.

The term "supply and disposal" used by some governments and "logistic management" by the Armed Forces aim at representing the same concepts as those underlying materials management. But the term *"Materials Management"* has found wider acceptaece covering all the inter-related functions mentioned earlier. Materials Management should therefore, be considered as a function of prime importance for our industrial economy.

Materials management provides tools and techniques, most of which are very simple, to reduce material costs substantially such as inventory control. In addition to reducing material costs, efficient materials management can also bring about :

(1) Reduction of foreign exchange requirements by getting the maximum value out of the available foreign exchange or by reducing the value of imports, thus contributing substantially towards narrowing the foreign exchange gap.

(2) Reduction of cost finished goods of high quality and maintaining the quality at reduced cost.

Industrial managers have been slow in appreciating the importance of materials management as they have been preoccupied with other matters, such as finding new markets, labour problems and so on. Material management bring about increased profits, results in increased productivity of capital by preventing large amounts capital being locked up for long periods in inventories.

Increased productivity is the only answer for the peculiar problems of today's Indian Economy of shortages, stagnation and the threatening recession. Productivity can be increased in a variety of ways such as increased production with fixed quantities of raw materials or same quality of finished goods and services with less use of capital and resources. Modern materials management tries to accomplish increased productivity through optimum use of capital employed.

Functions of Materials Management

One of the basic functions of the management is to employ capital resources as efficiently as to yield the maximum results. This can be done in either of two ways or by both :

(1) By maximising the margin of profit.

(2) By maximising production with a given amount of capital.

This means that the management should try to make its capital work as hard as possible. Much time is devoted to make only labour work harder and in this process the capital turnover and hence the productivity of capitally is often neglected.

The various functions of Materials Management are :

(i) Materials Planning and Programming.

(ii) Purchasing.

(iii) Storekeeping.

(iv) Inventory control.

(v) Receiving and warehousing.

(vi) Value Analysis and Standardization.

(vii) Pre-Design Value Analysis.

(*viii*)   Production control.

(*ix*)   Transportation.

(*x*)   Material handling.

(*xi*)   Disposal of scrap, surplus and salvage.

Definition

"Materials Management is a body of knowledge which helps the manager to improve the productivity of capital by reducing material costs, preventing large amounts of capital being locked up for long periods and improving the capital turnover ratio".

Materials management is a comparatively new concept in our country. Materials Manager must derive its objectives through understanding from the very depth of local environments, at the same time, taking into consideration the human factors, their attitude, sense of responsibilities and their mental make up and approach to new ideas. These factors togethr with the prevalent methods with their good and bad points must be studied in depth before a new procedure is evolved and forced down the throat of the staff concerned.

It may be easy to convince the higher managements about the efficacy of a sophisticated method but down the line with others who are connected with the work, it is not that easy. If the implementations of Materials management is taken up with an open mind it would be easier to achieve the result sooner or later. While the field of Materials management covers all aspects of material costs, supply and utilization, opinion is divided as to what functions should be included under the unified direction of the Materials manager. Even among undertakings which have accepted Materials management as a distinct management function there exist wide differences regarding its scope.

Objectives of Materials Management

The fundamental objectives of an integrated management approach can be viewed as :

(*a*)   Procuring Better Value.

(*b*)   Obtaining Standard or better yield.

(*c*)   Reducing investments in stock through Inventory control and materials flow time broadly discussed as follows :

(*i*)   Survival and growth.

(*ii*)   Maximum service to customers.

(*iii*)   Good working conditions for employees.

(*iv*)   Advancement of the employees.

(*v*)   Technological lead over competitors.

The objectives of Materials Management as such should be supported in every way by :

(*i*) Maintaining continuity of productive operations by ensuring a uniform flow of materials.

(*ii*) Reducing materials costs by systematic use of scientific techniques.

(*iii*) Releasing working capital for productive purposes by efficient control of inventories.

(*iv*) Increasing the competitiveness of end products by ensuring right quality at the right price, especially in foreign markets.

(*v*) Saving foreign exchange through economic use of foreign purchases and import substitution.

(*vi*) Establishing good buyer-seller-relations.

(*vii*) Ensuring low departmental costs and high efficiency.

(*viii*) Setting high ethical standards. In this way it is clear that Materials management covers all aspects of materials, including flow, cost, quality supply, conservation and utilization.

The prime objective is to supply the user department with the required quantity at a constant rate as well as uniform quality so that production or service rendered is not held up. At the same time materials manager has to ensure the optimum usage of facilities like capital storage space and other aspects of materials management.

Materials management offers a wide scope for reducing costs, saving the foreign exchange, conserving scarces materials, improving productivity—(specially of capital) and increasing profits.

It will be readily understood that it is much easier to reduce materials costs than to reduce labour costs or overheads, also as mentioned earlier, as material costs predominate in the total cost of a product, the saving is substantial.

By careful financial analysis, it can be shown that a 5% reduction in materials costs will result in increased profits equivalent to a 36% increase in sales. The Break-even chart shown in Fig. 20.1 that a substantial increase in profits can be obtained with a small decrease in material costs which reduce the variable costs appreciably as material costs from the bulk (cover three-fourth of the variable costs).

The chart also shows that profits starts coming of much earlier as the Break-Even point is shifted break-wards with the same total production. Increased production and sales are necessary to bring about the same increase in profits which result by reducing material costs or increasing production and sales to the extent of 30% to 40% which is not easy. Increased production may need increase in fixed assets and increasing sales to that extent, may not always be possible without extra sales efforts like *Additional Advertising* which increase expenses. Further it depends more on external-environments over which the management has little control.

The primary task of the modern materials management, with an integrated view is purchasing—

- —Right Quality and
- —Right Quality of
- —**Right Material**
- —**At Right Price**
- —Form Right source
- —With the Right contractual obligations
- —At the Right Time
- —Using Right Mode of Transport.

Storing the materials in the different warehouses with appropriate methods also forms an important aspect of materials management. Since "uncontrolled Inventory is Industries Cancer," the Materials manager is also responsible for the right quantity of inventory norms, i.e., for fixation of safety stock, re-order point, reorder quantity, etc.

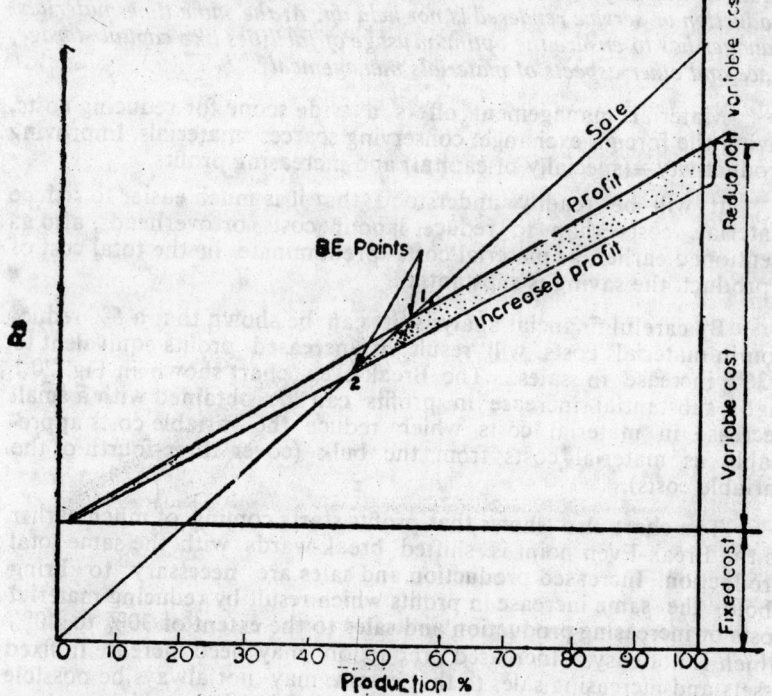

Fig. 20.1.

Functions of purchases, stores and inventory control primarily fall in the realm of the materials management, there are a large number of techniques and functions which can be applied only with the cooperation of executives in the computer, finance, user, maintenance, traffic and other departments. These techniques briefly mentioned below, are normally tackled on project-team basis with cooperation of others.

Materials Planning

"Materials Planning or planning the future requirements is based on the user's needs or forecasts and materials consumption control tends to formulate the broad norms of consumption for individual items of regular use and spare parts.

Important substitution assumes great significance in today's foreign exchange, crisis situation and hence source development becomes very important and prominent. Source Development obviously depends on the ability of an organisation to provide the technical know-how and finance as well as to guarantee a minimum off-take etc. The above lead to vendor being 'rated' according to his performance as well as to more meaningful buyer-seller relationships involving ethical principles reflecting in corporate-image of an organization. In this context, one has to decide, whether to "*make*" the item in the workshop or "*Buy*" from outside, after duly considering factors such as economical viability, atmosphere of industrial relation, additional investments and relevant costs, which implies that the other executives are also involved in the "*Decision making process*".

The technique of *Codification*, '*Standardisation*, *Variety Reduction*, *Simplification* and *Incoming Materials Inspection* cannot be fruitfully employed without the whole hearted cooperation of the user '*Value Analysis*', which aims at reducing the cost of the product without attaching the functional utility by challenging the specification, can be implemented only through the project team-approach. It is desirable to consult colleagues of other departments for savings in costs due to *Materials Handling*, *Packaging*, *Transportation and Shipping*.

The analysis of "*Lead Time*" assumes great importance particularly in government departments as the fluctuation in Lead-Time affect the inventory. In Government department and Public Sector undertakings, the lead time for available spare parts is over 12 months, whereas in Private sector to produce the same item, it is in the range of six months. The reduction in Lead Time with resultant decrease in the capital looked in Inventory, can be achieved only through proper systems and produces for which the cooperation of the executives in the *Administration*, *Transportation*, *Production*, *Inspection*, *Maintenance*, *Finance* and *Audit* are necessary.

Cost Effectiveness

The planning team designs equipment, purchasing team buys and engineers install before it is built into a viable plant. But then production people looks for supply of materials. Many production planners are very fond of aiming at good production, plans, but forget about the time and cost. The time dimension (and the total cost) should in no way be overlooked in effecting good operational design.

The good Materials management is invariably linked with the maintenance and control of materials flow, it will probably require more concentration and effort on those few areas where cost effectiveness will have significant impact on better performance and results.

Just as important a responsibility is for the Materials manager to keep himself aware on the issue of waste and thus can bring in savings at all stages of materials purchaing e.g., the contribution which appropriate packaging can make to avoid waste by damage in materials handling or have we stopped to think of the waste caused by corrosion to metals in stores ?

Trends in Materials Management. Decades ago, when the importance of Materials management was not realized, the storekeeper in India used to report to the production manager, whereas the purchase man was usually a small person in the management hierarchy working in the finance department. Gradually, due to increased size of organization and competition as well as complexity of the business, there was an awakening and in the recent past, the jobs of purchasing and stores were combined and placed under one executive. With the realization that the improved productivity needs streamlining the purchase, stores and related functions as well as application of Modern Material Management techniques, the concept of Integrated Materials Management came into being very recently.

The "Integrated Materials Management" concept is fairly new in the lexicon of Business Management even though 'Materials' constitutes the most important of the M's like Money, Man, Machine, Materials, etc., and Man has been using the materials" since the pre-historic days. The concept in hardly two decades old, even in the developed countries and in India it is being practised in some progressive undertakings only since last ten years, the result of the integrated approach to materials management, in these organisations are obviously visible and remarkable indeed.

The Materials manager should have sufficient authority to economise purchase and effect cost reduction in all aspects of materials function including optimising inventory holdings. Hence, organisationally, the materials manager, to derive full benefits from the Materials function, should be on par with the other departmental heads, i.e., Finance, Marketing, Personnel and Production and should report to the General Manager or Chairman.

The modern tools and techniques which have been evolved and come to stay must be brought home to the staff concerned in simple language. These being the part and parcel of an approach to an integrated and objective control, essentially the system must fix certain standards or norms. Thus for the success of this new area, the materials manager has to educate the top management of the crucial and vital role that the department plays in effectively contributing to the corporate growth of any organisation. He has to initiate suitable reporting systems of the results achieved in order to get appreciation of the top management and his colleagues in any organisation. The degree of sucess of materials management in any firm obviously depends on the initiative of the materials manager, culture of the organisation as well as inter-relations between different departments in the form of cooperation, collaboration and communication as well as machinery for Conflict Resolution.

Non-moving Items

It is defined as particular items which are not requisitioned by the using department over a specific period of items. These items are subject to obsolescence, deterioration, pilferage, inventory carrying cost. Thus material management should resort to study of such items at regular intervals. It has been seen that determining a norm for non-moving item is a difficult task. In some organisations the norms for slow moving and non-moving items are not rightly understood and hence not followed in practice. The non-moving items can always be examined under different categories such as raw materials, bought out components, spare parts and tools and items manufactured for exports. Most of the large projects in India are outcome of collaborations with foreign companies. Regarding technical know-how we have had to depend upon them but for decisions regarding quality and quantity of material to be procure depend upon us. With the result we faced a situation in which on one hand the plant starved of essential spares and on other a large number of parts remained untouched. In certain public sector certain parts or components did not move for more than 15—20 years. The collaborators recommend a certain inventory level depending upon the conditions prevailed in their country.

Vendor Rating

The objective assessment of vendors demands a systematic analysis of their past performance. But re-assessment at regular intervals is a time consuming process and thus did not receive adequate attention. Recently adequate attention has been given to find out the objective standards and procedure to compare and thus evaluate the suppliers. The systematic assessment of suppliers ability to meet the quality, delivery shedules, product price and service giving an appropriate weightage to each factor can help us in designing the vendor rating system. The overall rating could be prepared to evaluate the vendors. Following are the three methods used for rating vendors.

    (1) ' The categorical method

    (2) The weighted-point method

(3) The cost-ratio method

1. The categorical method. This is a functional approach method, giving emphasis on the concept of value analysis. In this method the functional reality is given more importance in vendor selection. We can device a system with three levels namely-High value, Middle value and Law value. In each of these value system a list of factors is made for the purpose of evaluation. For example the delivery, price and quality are given high value. The geographical location, suppliers reputation are of middle value. The reciprocating and minor services are of low value. This method relies heavily on judgment of the individual carrying out the rating.

(2) The weighted point method. The General Electric Company of U.S.A. is responsible for bringing out the weighted point method. In this method the number of factors such as the objectives of the organisation, its products and the economic conditions of the organisation are included. The relative worth of these factors as compared to each other will give a composite performance index. The relative worths of these factors vary from product to product, organisation to organisation The following are the maximum but average points for the best performance.

Factors	Average Points
(1) Quality	35 Points
(2) Price	30 ,,
(3) Deliver	20 ,,
(4) Service	15 ,,

These points are allocated based on the nature of product. For example for toys quality is less important but the attraction and price are of main considerations while for aircraft components quality aspect is very important in relation to the price.

Cost Ratio Method

This method attempts to reduce the subjective elements common to other methods. In this method a list of suppliers services is established in relation to the price. In this method the object is to evaluate the suppliers on the basis of proceeding considerations which may not be practical consideration. The small differences in the point score among various suppliers may not be of a great significance to justify their relative merits, for example suppliers X, Y, Z, have scored 60, 62 and 63 points respectively. In such cases place the orders with one of the supplier as long as his performance is satisfactory.

If the first supplier fails to meet the schedule the second supplier is selected as so on. The table I will indicate the relation of cost factor with the amount spent.

TABLE I

| Name of the vendor... |
| Product to be supplied... |
| Quantity to be supplied..............................say γ |
| Period of Supply... |

S. No.	Cost Factor	Amount
1.	Visit to Suppliers plant	—
2.	Sample approval	—
3.	Incoming inspection	—
4.	Value of the projected part	—
5.	Reworking cost. (if any)	—
6.	Miscellaneous cost.	—
7.	Follow up cost	
	Total Cost	X

Total cost of purchases $= \gamma X$

Hence, quality cost ratio $- \dfrac{X}{\gamma X} \times 100$

$- \dfrac{1}{100\gamma} \%$

It has been observed that organisational objectives some time limits the free play of the supplier evaluation. The human element is normally a responsible factor in preventing objectivity in suppliers selection.

MATERIALS MANAGEMENT RESEARCH

The increasing complexities in the procurement of industrial raw materials and components has given birth to new technique and method for increasing the system effectiveness. It depends to a greater extent on how effectively these techniques are utilised depending upon the availability of their own resources. All over the world large size organisations are spending a considerable amount on research in materials management. The research on availability should be attached more attention and importance. The availability research provides information concerning the availabilty of raw materials and components in National and International market at a right price.

The purchase management really cannot function. Availability research should be the direct responsibility of the officer who is reporting to material manager as shown in Fig. 20.2.

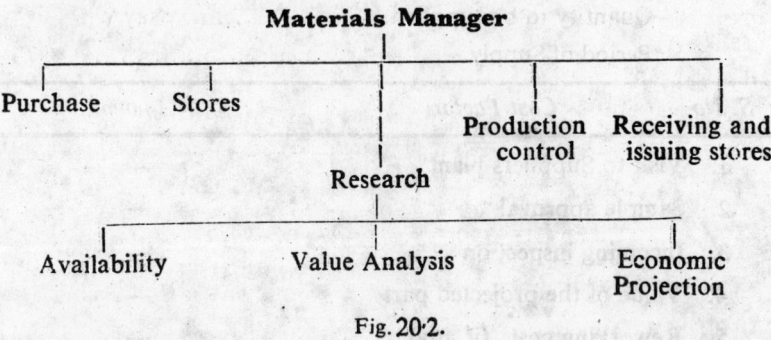

Fig. 20·2.

The executive responsible to availability research should have the following attributes.

(1) He should be aware of his organisation's objectives and its products.

(2) He must have a logical and analytical thinking.

(3) He should be aware of market trends.

(4) He must be a creative man.

(5) He should not mind his criticism.

(6) He should encourage two-way communication.

In addition to this if these activities are to be run in a systematic way then certain amount of autonomy has to be given to the person carrying the availability research.

Development of Information

Analysis of Products	Utility	Reliability
Product Information	Value	System
Source	Relations	Forecasts
Availability	Movement Economics	Confidence
Movement Pattern	Variables	Constraints

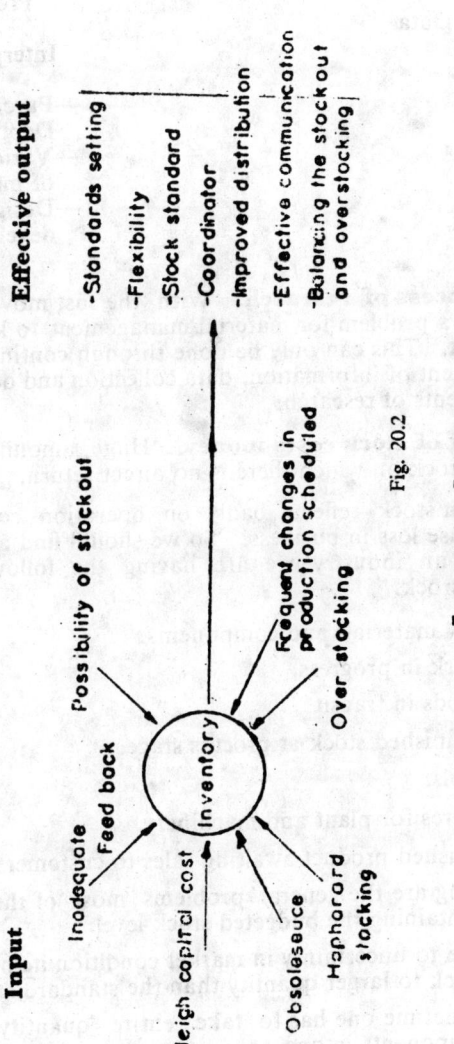

Fig. 20.2

Inventory Management

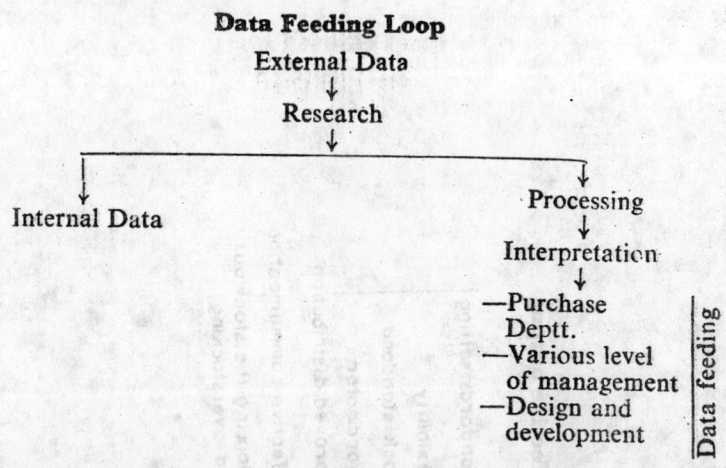

Data Feeding Loop

The process of research. With the fast moving technology it has become a problem for material management to keep pace with this movement. This can only be done through continuous research. The development of information, data collection and data feeding are essential elements of research.

Holding of stock costs money. Huge amount of money is tied up with stock on which there is no direct return.

Too high stock reflects badly on operation cost and too low srock may cause loss in business. So we should find a healthy mean for this. In an industry we are having the following principal categories of srock.

(i) Raw materials and components.

(ii) Work in progress.

(iii) Goods in transit.

(iv) Unfinished stock at process stage.

(v) Tools.

(vi) Spares for plant and machinery.

(vii) Finished product awaiting sales to customers.

Following are the general problems most of the industries are facing in maintaining the budgeted stock level.

(a) Due to uncertainty in market condition items are sometime taken into stock to larger quantity than the standard stock policy.

(b) Sometime one has to take entire quantity specially the imported components in one go.

(c) Uncertainty in getting import licences which has got direct bearing on production and inventory level.

(*d*) Sometime the requirement beig very small and it become difficult to procure exactly as per the requirement.

(*e*) Sudden and unwarranted increase in selling by the large scale/medium scale manufacture.

(*f*) Excess of the wrong materials in the wrong place.

Thus to maintain the budgeted level of inventory one needs to review the inventory, level and to compare it with budgeted and action should be taken to bring possible improvements. This can be done by inventory board consisting of the following persons :

(1) Materials manager,

(2) Finance manager,

(3) Production manager, and

(4) Industrial Engineer.

Because of increasing complexities larger organisations are taking the help of computive for the problems related to stock control and material flow. Fig. 20·3 will illustrate how inventory management is associated with certain elements.

Standardisation

The concepts of standardisation can be traced back to Noah's Ark. We all know that standardisation of components is essential for inter-changeability of parts which is of vital significance to mass production on assembly line. In other words importance of standardisation was realised since the advent of mass production techniques. Standardisation is defined by the International Standards Organisation Geneva as the process of formulating and applying rules for an orderly approach to a specific activity for the benefit and with the co-operation of all concerned and in particular for the promotión of the overall economy taking due account of functional conditions and safety requirements.

A standard is based on the consolidated results of science, technology, research, development and experience. A standard is established by custom, authority or practice as a model to which an action or object can be compared. Standardisation in its broadest sense permeates almost in all fields of human activity.

In order to establish standardisation at National level, the Government of India established (I.S.I.) Indian Standard Institute in 1947 with the help of scientific, technical and industrial organisations. I.S.I. is an autonomous body with Union Minister of Industries as *ex-officio* president. In order to provide the assured quality of industrial goods, the I.S.I. Certificate Márk Act was passed by the Parliament. This I.S.I. mark is intended to protect the consumer's interest and to achieve optimum utilisation of material resources by producing the quality and product from the right type of raw materials and processes. At organisational level standards deals with product, process, methods, materials, specifications, inspection, tests, procedures or other type of requirements. Since standardisation involves inter-departmental activities so it should be outside the influence of any single department. For the successful application of standardisation in an organisation the following aspects should be considered :

    (*i*) Dynamic and flexible nature of standardisation.

    (*ii*) Organisational aspects and responsibility of standardisation.

    (*iii*) Objectives of company standards.

    (*iv*) Technique and process of standardisation :

What is Standardisation

Standardisation is a managerial concept which seeks effectiveness in operations through the acceptance and systematic employment of established and agreed upon procedures and practices. Management in its direction and control of enterprises uses the standardisation process to draw upon past experience and proven engineering, economic, and administrative throught to organise, evaluate, and formulate, the current practice that will most contribute to the success if its activities. The standardisation is pursued not only in engineering and manufacturing but also in other functions of a business.

Definition

Standardisation is the process of establishing standards. It is the organised procedure we go through to decide what standard will be and to have it accepted and used.

Specification

The term specification as applied to purchasing, manufacturing and marketing has become acceptable. It is a statement or description as complete and definite as possible of particular requirements, Characteristics, or quality that establish with certainty the terms of a contract, the details of construction, the composition of materials, the controls of processes, the sequence in procedures, and the like. In industry, the term is applied to materials, jobs, products, and so forth, as a matter of custom and should in these instances be considered synonymous with standard.

Simplification

Simplification is the reduction of variety by eliminating the unneeded. It is one of the procedure used in standardisation and consists of removing from a list of those items that are deemed excessive, unnecessary or redundant.

Effectiveness of Standardisation

To evaluate the effectiveness of standardisation the following factors must be considered :

* Creation of standards
* Application of standards
* Maintenance of standards
* Computerising the standards
* Co-ordination with external agencies

21

Purchase Management

Introduction

The scarcity of raw materials has practically put the people in purchasing department in a very tight position. The Purchasing department can be in a better position if the designers take a little pain to consult the purchasing personnel about the technological capabilities of the vendors. It has been observed that drawings are sent to purchase department for components with high degree of close tolerance. But we should ensure that high degree of tolerance is really needed because close tolerances not only make procurement difficult but also contribute to high price. The main outset of company's cash goes in materials purchase thus purchasing people should take greater responsibilities and should analyse the existing procurement policy and should tune with the overall organisational objectives and policies. The various activities associated with Purcahse are shown in Fig. 21.1. Materials supplies and equipment are the lifeblood of any industrial concern whether Govenrment or Commercial. These are used for converting raw material into forms that have higher utility and value to the consumer. Thus a value is added by the manufacturer which result into profit. Teh efficiency of any business activity depends upon having materials, supplies and equipment available in proper quantity with proper utility at the proper place and time and the proper price.

In India approximately 60 paise of every rupee of manufacturing is spent on

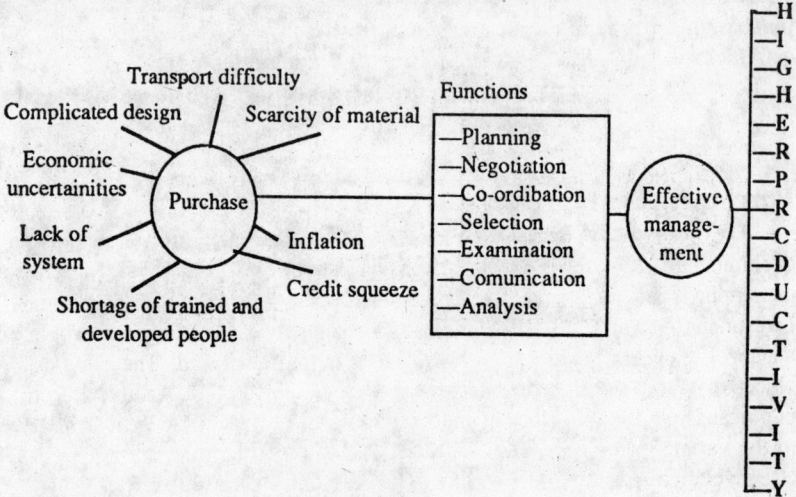

Fig. 21.1

material and about 90% of net working capital of industry is tied up in inventory. The improvement in purchase management by the help of standardisation, value analysis, material substitution, transport saving, cost reduction by packing modificatin and scrap reclamation all can contribute to the profitability of an organisation. Purchasing in its narrow sense refers merely to the act of buying an item at a price. A broader meaning of purchasing makes it a managerial activity that goes beyon the simple act of buying and includes the planning and policy activities, research and development, service selection etc.

Importance of Purchasing.

Purchasing Management, best reflects the role of purchasing in corporate operation and organisation. There is a trend toward introducing materials management in industry today, but to say that it has superseded purchasing would be to anticipate the future more than we feel is warranted. Purchasing is still a clearly identifiable function in most business organisations. The term "management" suggests that purchasing decisions involve the weighing of alternatives possibilities, and many of these alternatives involve the influence of other functions on the purchasing decisions. No organisation can operate without materials, supples and equipment. The efficiency of any business activity is contigent upon having materials, supplies and equipment available in proper quantity with proper quality at proper place and time, and at the proper price. Purchasing is a managerial activity that goes beyond the simple act of buying and includes the planning and policy activities covering a wide range of related and complementory activities such as research and development; proper selection of materials and sources from which those materials may be bought. The follow up to ensure proper delivery for inspection of incoming shipments to ensure both quantity and quality compliance with the order; the development of proper procedures; methods and forms to enable purchasing department to carry out established policies; the coordination of the activities of purchasing department with other internal dividiosn of the concern as traffic, receiving, storekeeping and accounting so as to facilitate smooth operations; and the deyelopment of effective communication system in the company so that a true picture of the performance of the purchasing function is presented.

Purchasing Objectiives

The objectives of purchasing is to conduct purchase function so as to minimise or eliminate disruption in production resulting from lack of any materials equipment or supplies. Further more this objective must be achieved with a minimum investment in reserve inventories. Another objective of purhasing is the maintenance of adequate standards of quality for items purchased. The purchasing objective is lowest ultimate cost rather than the lowest initial cost. Another objective of purcahsing is the avoidence of duplication, waste and obsolenscene with respect to the various items purchased. The mission of purchasing organisation is the effective commitment of the company's funds. Its objective is the economic success of the business organisation. Sound procurement is itself the proper management of the materials rupee investment. In other words the objective of purchasing is not so much to procure the raw materials at the lowest price but to reduce the cost of the final product. To ensure these objectives a large number of parameters such as right price, right quality, right contractual terms, right time, right source, right material, right place, right mode of transportation, right quantity and right attitude have to be considered jointly. The objectives of purcahsing is to procure goods that are best suited than those that rank highest in quality. Avoid duplication, waste and obsolensce with respect to various items purchased. Maintain competitive position of the company, develop good relations with suppliers and develop internal relationship that lead to understanding and harmony among the various organisational units within the company.

According to A.G. Pearson the prime objective of Purchasing Organisation is "Effective commitment of the company's funds" and having goods of the Right Quality, form Right Source, at Right Price and at Right Quantity and Time.

(i) **Right Quality.** Cost and Quality are critical dimensions. The inter-action between two is very complex. A right quality is not necessarily a best quality. To a large degree manufacturers determine the quality of goods by the desired quality of product to make. The considerations are basic materials, workmanship, grades, sizes, designs, colours, patterns,durability. For the purpose of purchase quality of raw materials etc., need to be established.

The quality must be described precisely so that vendor should understand what is exactly needed. The exact specifications of items have to be given, preferably in terms of market grades, brand or trade names, commercial standard, on blue prints, chemical or physical characteristics, material and method of manufacture. Use of national standards shall be made to the extent possible. Alternately, materials standards shall be fixed by the committee set up for the purpose in the organisations. Simplification (variety reduction) is a great boon for purchase department and is repaid in terms of reduced inventory investment, greater quality discount, better competitive prices, reduced clerical and handling costs, besides less recording, controlling, receiving and inspecting. It is prudent to obtain samples along with question or a sample of such material be available to supplier before quoting the price. This will help in building buyer supplier relations besides, scrutiny and decisions of choosing the supplier.

(ii) **Right Quantity.** Quantity to be purchased varies with the production strategy and planning. For running operations, P.P.C. determines the volume of production in advance, and in such cases it is desirable to enter into annual or bi-annual contracts specifying periodical supplier at regular intervals. This leaves the burden of ensuring the uninterrupted supply with the suppler and lot of energy of purchase department is saved. In case of varied strategy of manufacture, the techniques of E.O.Q. is followed to save the producer from the danger of stock outs as well as carrying cost of surplus inventory. Other strategical considerations in determining quantities, are combination of items to reduce transportation costs, anticipation of market conditions both of the raw materials/spares as well as the minimum quantitity of finished goods.

(iii) **Right Source.** The source of supplier is determined normally by calling quotations and the lowest bidder is selected provided he has quoted as per the requirement of producer in terms of quality, quantitiy and period of delivery. But such ideal situations do not appear in all cases and selection of supplier involves a strategic considerations of various factors. It is always prudent to select a manufacturer in case of patents standard products even if it means a little extra transportation. Secondly, the past records, financeial capacity, technical ability, and other resources play important role in the selection of supplier. Testing facilities of supplier eases, inspection of goods and all other problems arising out of inadequate supply at the producer's end.

The purchase department has to maintain a vocabulary of suppliers of different goods and develop it further from the knowledge/data collection through journals, bulletins, and newspapers, trade directory, and exchange of information through buying associations.

(iv) **Right Price.** The right price is the worth in terms of quality, time and adequacy of supply of an item obtained. It is no doubt easy and safe to go for standard products at a higher price but then one has to keep in mind the utility of item in the ultimate worth of product.

There is a tremendous growth in the number of varieties specially spares, fixtures and such other goods after the Industrial Revolution and innovations have put forth cheaper brands in the market. This has developed a situation where the new industries have to use their skill in selection of new brands who are attempting continuously towards recognition and standard brands who have proved their utility and due to increased demand are constantly pushing up their prices.

(v) **Right Time.** The ideal time of purchase period would be the minimum time for which the goods remain unconsumed. This could be achieved if the stockist or manufactureer of raw materials supplies the day-to-day requirements in regular instalments. This would save the storage. But the geographical and market conditions do not permit and it is where the ordering system comes into existance. The timing policies will depend upon fluctuating prices as well as problems arising out of monopolistic trade and sellers' maturity.

For large industries, located in interiors, "hand to mouth" buying is not practicable and as such long lead times have to be given to allow cheaper mode of transportation and economical bulks. It is the practice to remain in touch with the production department to know the expeditions/delays in schedule. The modern purchasing needs follow up actions right from the beginning of demand to ultimate supply stage including despatch and chasing of transporters through special messangers. These costs have to be weighed in terms of actual delays in production and purchase department cannot just afford to rest with telegraphic reminder or trunk calls.

Types of Purchasing

Purchasing is primarily a buying activity and therefore types of purchasing vary with the purpose of buying. There are four types of buyers:

1. Industrial buyers,
2. Buyers for wholesalers,
3. Buyers of merchandise for retail stores,
4. Ultimate consumers buying from the retail stores.

As such purchasing fundamentally is aimed at resale and consumption or conversion. Above four types will fall in one or both of these categories *viz.* Industrial buyers, and consumers (buying from retail stores) will purchase for consumption or conversion whereas Buyers for wholesalers and merchandise for retail stores aim at distribution or resale. As for the purpose it is evident that buyers of type (2) and (3) above are opportunist and mainly concerned with the market conditions. Ultimate consumers, a large percentage have to go by a necessity and the quantity purchased is small. The main activity of purchasing in terms of technique or scientific management rests with industrial buyers.

'Induastrial Purchasing' is defined by Dr. Watte as *"the procurement by purchase of the proper materials, machinery, equipment and supplies or stores used in the manufacture of a product adopted to the marketing in the proper quantity and qality at the proper time and a the lowest price consistant with quality desired."*

In most organisations normally the consumption of materials anges between 60% to 70% of the cost of products (Sales Value). These percentage vary from industry to industry, *e.g.* in the mining industry the consumption of materials would be proportionately lesser than another production unit but all the same the supply function remains vital and important. Besides Industrial Undertakings Central and State governments and other local bodies have to make large purchases. The important aspecthere is public accountability in the sense that those who spend public money must be responsible to the public representatives. This leads to added precaution ensuring that all purchases are made without any favour and form the most competive biddders. The very nature of purchase activity places the officials concerned in contact with the suppliers and thus call for judgement and decisions which are very important not only from purchasing angle but also from the fact that those concerned have to display the highest sense of recognition.

Purchasing Principles

Folowing are some of the important principles:

(a) The acceptance of order is to be specific and not implicit. Preferably this should be acknoledged by either side before it is binding.

(b) Ambiguity and mis-interpretation should be avoided specially in respect of quality, delivery, service, discount or any other such changes.

(c) The Indian Sale of Goods Act lays down that unless a specific provision is made regarding the time of payment or any other stipulation as to time of payment or any other stiputaltion as to time being essence of contract, the same is not implied. It has to be stipulated.

(d) The supplier takes warranty of goods as per specifications laid down in tender and not whether these are fit for intended use. If it is so required the detailed information has to be stipulated well in advance or through negotiations.

(e) The other features are buyer's right to reject unwanted goods, liquidation for damages, passing of ownership etc., are covered in legal aspects.

In case of major contracts, involving large sum of money or, extending over long periods of time it is advisable to avail of professional and legal opinion. Familiarity with law provisions will enable personnel engaged in purchasing to decide in which situation it will be desirable to take legal opinion. Summarising with *ethics* of purchasing, the following code be kept in mind by purchase manager:

(i) To be aware of company's interest.

(ii) To be recepient to advice from his colleagues and be guided by such counsel without imparing the dignity and responsibility of his office.

(iii) To buy without prejudice-seeking to oobtain maximum value for each rupee.

(iv) Eager to seek knowledge of materials and process at the same time establish practical methods for conduct of his office.

(v) Accord prompt and couteous reception.

(vi) To respect his obligations.

Role of Purchasing Manager

Unlike production, manager, purchase of capital equipment is decided by a committee constituted by top management normally called as purchse committee. The committee usually consists of members representing production, Design and Engineering, Finance and Purchase. The recommendations of this committee are then finally approved by the General Manager or Chief Executive of the company before the capital equipment is purchased. The next important role of a purchase manager is to apply the qualitative information regarding the image, the technical capabilities, the financial soundness and the reliability of the supplier. He should also find out whether the supplier is capable of providing technical know-how mainly in regard to the installation, commissing and after sales service. A purchase manager has to get complete details regarding offers of potential suppliers, tabulate them and supply them to the finance department so that they can make a thorough analysis. He shold also in collarberation with te finance department, come up with a suitable payment terms with the supplier. On the whole, the role of purchase manager in the procurement of capital equipment is that of a liasion man.

A New Approach to a Better Purchasing Management.

We all know that the responisbility of minimising cost revolves on purchasing management. It has been observed that for a steel plant the cost of converting raw materials into saleable products is only 8% while purchased materials, supplies and services account for 37%. It is therefore evident that much lies in the economy of purchase. In case of purchase, various factors that come into operation included purchase procedure, freight, quality of goods, vendor selection etc. In order to study the effects of each one of these factors a common measure is needed. In many organisations particularly Government, one is accustomed with the method of purchasing from the lowest tenderer. It remains to be seen how fallacious is the criterion of purchasing from the lowest-tenderer. The lowest tenderer need not necessarily be supplier of the right type of material and that too at the right time and in right quantity. This may at times lead the purchase manager to face situations which results in stock outs and defective products.

Visible and Invisible Costs. There are visible and invisible costs associated with purchasing. Visible costs include net price, freight, delivery date, etc. where as invisible costs are related to quality and delivery date which are hidden in the vendor's past and future expected performance. In order to study the influence of invisible costs, one can start with the terms of payment. At times the vendors offer discount on large quantity purchases and facilities for deferred payments. In such situations, large scale purchase leads to locking up of capital in inventory. One has to balance between discount for bulk purchase with that of cost of borrowing money for the purchase. In case of freight cost, one faces alternatives of carrying the inventory by different modes of transportation. In this case also one has to predecide as to whether the materials shall be shipped by premium or lowest cost transportation. The cost of transportation may also vary form time to time. In case the purchaser has to balance between the advantages of different modes of transportation with that of cost of borrowing money. The next factor which is of importance is quality of raw materials. Bad quality raw materials leads ultimately to the cost of prefering claims. Here again the cost of litigation for claims have to be balanced against the cost of borrowing money. The early shipment cost as well as late delivery cost also causes trouble for the purchaser. A new technique called Vendor Cost Imporvement Contribution is also used by enlightend purchaser. This involves helping the supplier in reducing the cost of raw materials through alternatives by suggesting for Make or Buy, standardisation, inventory, transportation, respecification etc. In effective purchase management a technique known as "**Hidden Cost Ratio**" has also become quite popular. This is a measure which enables the purchaser to compare the cost of purchasing from various suppliers. The hidden Cost Ratio shows the cost of freight, Claims, early shipment minus cost of improvement contribution over the rupee value of total purchase. If the hidden cost ratio of a supplier is Rs. 0.05 it means that the cost of doing business with this supplier is Rs. 1.05 for Rs. 1.0 quoted price. Such an analysis helps the purchaser to select a supplier. The hidden cost ratio can be calculated when the data pertaining to the behaviour of the suppliers over a long period of time is available. Therefore, for successful purchasing management the various techniques of determining the influence of factor like Vendor Selection; Transportation, Improvement Contribution, Quality etc are determined through the common measure of cost. The success for implementation of scientific purchase method will largely depend on the performace of the vendors which can be ascertained inter alia on maintenance of computerised historical performance data of vendors.

22

Purchasing Procedures

Introduction

Purchasing procedures refers to the way in which a purchase transaction is carried through from its inception to its conclusion. Purchasing policies outline the broad objectives to be accomplished and guidelines within which the procedures must accomplish the desired results. Procedures outline in detail the functions to be performed by the people involved in the purchasing operation. Forms and records used to implement procedures and policies. Since there are wide variations among industries, firms, products and personnel with the result there are different set of procedures that apply to all cases. Some of the common steps to complete purchasing transaction are Recognition of need; Description of requirement; Selection of possible sources of supply; Determination of price and availability and placing of order; Follow up and expediting of order; checking the invoice; Processing discrepancies and rejections; closing completed orders; maintenance of records and files. Let us examine these basic steps one by one to understand purchasing procedures.

Recognition of the Need. Any purchase transaction starts with the recognition of the need for an item by someone in the company. The need may often be satisfied by a transfer of materials from another department or store room, however the store room must also replenish its supplies of the item. For many items a well run purchasing department, will anticipate the needs of the using department, anticipating advance needs is one of the consideration that determines the size of the order for an item. In such cases purchase officer buys a quantity sufficient to reduce the likelihood of such orders in case of sudden increases in the rate of consumption of item. A purchase officer has to develop alternate source of supply and potential alternative materials.

Description of the need. Once the need has been recognized it must be so accurately described that all parties will know exactly what is wanted. An improperly or poorly described need can be costly. Securing an adequate and complete description of needs calls for the exercise of tact or skill on the part of purchasing officer. He must have enough knowledge of the item being purchased to be able to recognise inadequate descriptions.

Selection of the Source. The next is the selection of the source for requisitioned items. For branded or patented items there may be but a single source but in most cases there will be number of alternative suppliers. For regular purchase consideration like goodwill, price, company's policies etc. are considered for selection of source.

For items that has not been purchased with regularity, there are several things a purchase officer may look for viz:

(a) Prepare list of supplier, by reference from salesman, user department, vendors catalogue etc.

(b) Narrow the list by discussing with the salesmen, interview, investigations.

(c) Negotiate the price with one or more based on company's policies.

The process consists of selecting the desired number of suppliers, in accordance with established guidelines from whom quotations will be requested. For non routine purchases, the procedure involves a careful survey of potential sources of supply.

Ascertaining the Price. Purchase department must ascertain price information on the item being purchased. Price will be one of the important factor on which the final choice is based. For many items it s possible to keep current price information, in the form of manufacturers catalogue, price lists and discount schedules. Such source of information must be kept uptodate. A second method of securing price information is to negotiate with the seller until agreement is reached or price and terms. This method is specially suited to goods made to the specification of the buyers. The third method is competitive bid. This method is widely used by governmental purchasing departments because of statutory requirements but also applied by industrial Purchasing department.

Placing of order. All orders should be in writing and should be on the buyers purchase order to avoid the possibility of level difficulties. When order is placed by telephone it is the practice to confirm the order by sending the supplier a regular order.

Follow up of the order. It should not be supposed that once an order has been placed, the purchasing department has no further responsibilities. Every purchasing department has the responsibilities for following up orders that it places e.g. seller may not receive order even though it has been mailed. Thus it is necessary to institute a follow up procedure for all orders placed with supplier. Responsibility for follow up is generally assigned to the buyer who placed the order. Follow up consist essentially of holding a supplier to his promise of delivery. Follow up procedures should be employed whenever the costs or risks resulting from delayed deliveries or non-deliveries are greater than the cost of follow up procedures.

Checking invoices. Invoices checking is a part of purchase function since receipt of invoice constitutes the usual notification that the suppliers has made shipment. Further it is the responsibility of buyer to see that his orders are accurately filled and billed though it is an accounting procedure yet in order to avoid delay it should be done by purchase department.

Maintenance of Records and Files. purchase orders are legal contracts and as such should be preserved for as long as they have legal significance. Requisitions and similar documents should be preserved by the purchasing department, since they constitute the authority on which purchasing department had taken its actions to a given item.

. **Maintenance of Vendor relations.** Good relations are based on mutual trust and confidence, these grow out of the dealings between buyers and sellers over a period of time. The purchasing department effectiveness is measured by the amount of goodwill with suppliers.

Automatic Purchasing Procedures

The large volume of orders processed through the purchasing department, many of which are repeated several times during a year has led purchasing executives to seek ways of minimising the paper work. The automation of paper work has made great strides in the field of purchasing, maintenance of inventory stock record cords. Several machines and techniques have been developed to simplify and speed up the entering of inventory receipts and disbursements and to secure reports and disbursements and to secure reports on inventory levels. The use of such equipment saves considerable time and ministerial efforts and brings to the attention of the purchasing authorities current information inventories of materials. The other development in the area of automatic purchasing procedures is **travelling requisition.** This is a form that contains identifying information with respect to an item that is repeatedly purchased and includes the names and addresses of approved suppliers as well as prices paid on previous purchases. Executives have a common opinion that in the field of purchasing are in agreement for major purchasing innovations are being developed in automatic purchasing procedures.

Specialised Purchasing Procedures

Most specialised purchasing procedures relate to maintenance, repair and operating supplies. Specialised purchasing procedures deal almost exclusively with reducing the cost of handling small purchase orders. The small-order problem is not one that can be solved by eliminating all small orders. Better planning and scheduling can reduce such small orders by combining several into a single-purchase transaction. Whatever procedures are adopted by a firm to meet its small-order problems, such procedures must be designed to meet the specific-needs of the firm. No standard answer can be applied to all firms and all situation. It has been observed that many small orders occur because of carelessness, errors and ignorance on the part of purchasing personnel or related departments. Improvements in internal efficiency and management of purchasing function would automatically improve this problem at the same time that other purchasing practices are being corrected. The proper approach to solving the problems created by small orders involves at least three steps.

(i) Analyse and categories the small orders that have been processed in the department for a specific time to determine whether the orders had to be small orders.

(ii)   Determine if any procedure could be adopted that would reduce the costs of handling such orders.

(iii) Exercise ingenuity in proposing and testing alternative procedures for handling small orders.

The majority of purchase orders involve small sums of money, since a substantial part of the operating cost of a purchasing department which involves the paper work required to place orders. Some of the procedures to be described completely eliminate the carrying of inventory by purchasing firm and thus eliminate this cost for the purchaser.

Blanket Orders

The blanket order is the most popular alternative to the single item, fixed-price order. A blanket order may be an agreement to provide a designated quantity of specified items for a period of time at an agreed price. A second type of blanket order is an agreement to furnish all of the buyers needs for particular items for designated period of time. Under this type of blanket order the quantity is not fixed until the time period has elapsed. The blanket order is best for items with low unit value, but high annual usage, whose rate of usage can not be accurately planned. The blanket order may be written to cover categories of goods that are broadly described, for example fasteners. Price is handled in number of ways in a blanket order. Firm prices may be negotiated for each item covered. The blanket order may specify market price and include a method of determining such price. The advantages of Blanket Order Procedure are:

(i)   It releases the buyer from routine work.

(ii)   It requires fewer purchase order which reduce the clerical work in purchasing, accounting and receiving.

(iii) It often achieves lower prices through quantity discounts.

Stockless Purchasing

Under this system the buying company has no financial responsibility for inventory of the good being purchased. The inventory is owned by the supplier. The goods may be located either at the supplier's or buyer's location. Under stockless purchasing system it is probable that prices will be slightly higher, since the suppliers assumes most of the warehousing and inventory costs. However the cost at the point of use may will be lower, because the seller may be able to perform these functions more economically than the buyer since he is a specialist in the products and handles his own inventory in any care. Stockless buying is advantages to both the parties and therefore likely to continue to spread and become a standard rather than specialised purchasing procedures:

(i)   Buyer gains significantly as he no longer is required to tie up captain inventory.

(ii)   Purchasing routine is significantly reduced.

(iii)   Vendor's salesman do not have to solicit orders, they can devote more time to aiding buyers through service.

(iv)   Lead time is considerably reduced and may even be eliminated under consignment buying.

(v)   Buyer is lively to get a lower price because the seller is willing to make price concessions when he knows that he will have all customer's business for contract period.

(vi)   The supplier can also eliminated more rapidly from his inventories spare parts that may became obsolete.

Re-ordering
Systems Contracting

A system contract is a total corporate technique designed to assist the buyer and seller to improve re-ordering of repetitive-use materials or service with absolute minimum of administrative expense and with the maintenance of adequate business controls. There is some confusion as to whether a systems contract is different from blanket order buying, automatic purchasing arrangements, and similar ideas. The essential difference is that in practice the system contact has been long-term in its operations, where as the other arrangements are if shorter duration, with changes in suppliers being relatively frequent. Following are some of the points which differentiate between systems contract and other methods of purchasing:

Choice of Vendor

Under the systems concept not only will the agreement be of longer duration but much more formal method of selecting the vendor will be employed to eliminate personal considerations. The chosen vendor should be a specialist in the materials covered by the contract so as to be able to offer maximum quantity to a large buyer of such items and discounts to the buyer.

Materials Covered by System Contract

In this system the prior rate of usage of particular products must be determined as well as the frequency of reordering over some pest period. The records show not only transactions with that supplier but also previous purchases from competitive sources. One could argue that transactions have no bearing on the choice of system contractor which has already been made.

Standardization

Under the system contract the buyer receives only the brands produced or sold by the contractor. This means that standardisation programme must be adopted.

Requisitions

In the system contract, the vendor guarantee delivery and, to remain a contractor, there is no need to requisition more than the buyer's immediate needs: There is no need for multiple copies of a requisition as is the care with most purchasing procedures.

Order Filling by Vendor

In system contracts the vendor assigns a number to requisition when it is received. The vendor prices the requisition, since it comes to him unpriced. He selects the method of shipment, since on-time delivery is a crucial element in the systems contracts arrangement.

Advantages of Systems Contracting

The following are some of the advantages of systems contracting:

(i) Systems contract substantially reduces store records, requests for quotations, bids, purchase orders; follow up and expediting procedures shipping notices, invoices and so on.

(ii) There is reduction in store room requirements, it nets a saving in personnel and space cost.

(iii) Opportunity for errors in ordering is reduced.

(iv) Obsolescence due to changes in requirement, is reduced. In addition without risk of materials in inventory, changes that would have obsoleted such materials are made more rapidly.

(v) A systems contract has significant advantages to the vendor as well as the buyer.

(vi) There is savings in paper work for the vendor under a systems contract.

(vii) In systems contract the supplier can concentrate attention on new ideas and materials not covered under the contract.

Small Order Procedures

All organisations have a significant volume of purchase transactions that are non recurring nature and involve insignificant sum of money for an individual transaction. It is difficult to anticipate the need for small, non recurring purchases because the situations that generate such needs are usually unrelated to one another. Frequently the item involved can be rough only from a single supplier, thus precluding the possibility if combining orders to increase total value.

Petty Cash Systems

It is system that virtually eliminates paper work. Petty cash is a monthly set aside to meet minor expenses of a business. Where this technique is employed, the general petty-cash fund or a special find for small purchases must be under the control of the purchasing department to preserve the advantages of centralized purchasing.

Telephone Orders

The use of telephone ordering of small value, non recurring purchases is thoroughly logical. Frequently when buyer receives a requisition from a using department, he must in any event, telephone a supplier to determine the availability and price of the item.

23

Capital Equipment Purchasing

Capital Equipment

Capital equipment refers to those items of machinery and equipment whose long life and high value require that they be carried in balance sheet and depreciated over a period of time. Capital equipment are boilers, pumps, electric generators, machine tools, mechanical conveyors, rail road engines and cars, steel rolling mills and generators.

Characteristics of Capital Equipment Purchasing

The purchase of capital equipment differs from the purchase of raw materials, supplies and other items in that each purchase is the result of careful negotiations, requiring the consideration of many more vendors than is usual for other purchases. The other difference between capital equipment and other purchases concerning the lead time, is the interval between the placing of an order and the delivery date. As a general rule, the lead time requirement for capital equipment is much greater than for other materials and supplies. Similarly the specifications for standard capital equipment frequently are not so flexible as the specifications for raw materials and supplies. If a buyer insist on rigid specifications, he would reduce competition and probably pay a higher price as a consequence.

We know that materials manager is both a user and buyer of equipment and relies almost entirely on the Judgement of users of the equipment for engineering evaluation. The financial evaluation of equipment is primarily a responsibility of the organisation's controller. However, materials management should also play a major role in the economic evaluation of proposed equipment purchases. **The materials manager should also serve as catalyst to acquint others in the organisation that might help reduce costs.** The materials department normally buys capital equipment to specifications developed by company's manufacturing engineering department. The engineering department plays a more active role in buying equipment than they do in the purchase of parts and materials. The materials department has a special interest in capital goods purchases, however, because of the opportunities they offer for cost reduction.

The buying decisions for capital equipment are usually based on an analysis prepared by the finance department which in turn, is based on data submitted by manufacturing engineering and purchasing. The unit value of capital equipment is high because equipment contains a higher proportion of labour to raw material than almost any other kind of purchase. The high unit value is most significant for financing. The top management must usually budget and authorise such expenditure before they can be made.

Procedure for purchasing capital Equipment

The first step in the purchase of capital equipment is the evaluation of need.

Evaluation of Need

The recognition of the need originates with the using department and the evaluation of this need requires the study of alternative methods, a cost analysis of alternative methods, a search for equipment that will do the job, and a second cost study to determine the savings made possible by the use of the proposed equipment.

Specification

Once the need has been established and the type of equipment determined, the next step is to draw up specifications. The manufacturer's representatives are also consulted in designing the specifications.

Negotiations

The materials management department, which has so far arranged the contacts with Vendors, determine which vendors to solicit for quotations. The specifications are provided to selected vendors and they are invited either to quote or to send a representative to survey the job before quoting.

Ordering

After deciding the supplier, it is necessary for the purchasing department to work out with that vendor all details of the purchase order. An experienced buyer knows that, although he may be reimbursed money for any damages incurred he can save himself much trouble, cost and work by taking precautions before placing the order.

Follow-Up

Because the purchase of capital equipment usually covers an extended period of time, follow up of the order is an important responsibility of the purchasing department. The follow up of an order should be performed by the executive who placed the order.

Factors Favouring the Purchase of Capital Equipment

The various considerations favouring the purchase of New Equipment are:

Long Life Expectancy

One of the important reasons for buying new equipment rather than used is that the life expectancy of new equipment is longer.

Uniform Life Expectancy

The nature of a particular project may justify the purchase of new equipment. It would be unwise to install an item of used equipment with an uncertain life which could cause a breakdown operations.

Technological Innovations

The new equipment is usually of more modern design. The manufacturers are constantly striving to improve their products, not only because of competition for the product, but also in an effort to improve their own manufacturing efficiency.

Reduced Maintenance

The new equipment quite naturally requires less maintenance than used equipment. This is strong argument for the purchase of new equipment. Related to lower maintenance cost for new equipment is the fact that lessor stock of spare parts is needed for new equipment. Similarly it is much easier to evaluate competitive bids for new equipment than to compare bids on used equipment or bids that involve some of each. Similarly for new equipment, specification play a very important part in the purchase of major equipment. Once careful specifications have been worked out, it is a relatively simple matter to relate them to new equipment being considered.

Factors Favouring Purchase of Used Capital Equipment Cost

Low Cost is perhaps the major consideration favouring the purchase of used equipment.

Delivery. Sometimes a compelling reason for the purchase of used equipment is its immediate availability. When production is required on short notice, and the only possible way to achieve it is by the purchase of used equipment. Similarly when the equipment long life is not required it is satisfactory to buy used equipment. Another factor for the choice of used equipment is that the buyer can inspect the equipment before purchase.

Economic Analysis

Approaches to Capital Equipment

The company considering the purchase of new equipment makes a careful cost analysis to compare the operation of the proposed equipment with his present equipment. Because of the large size of capital equipment expenditures, most companies use detailed economic analysis procedures to calculate the expected return from the purchase against the expected cost.

Payback Period

Payback period is calculated by relating the cost of the equipment to the profit it earns and then estimating the number of years it takes for the equipment to pay for itself. The basic objective is to purchase the equipment with fastest payback period. Payback period is simply the time it takes to cover the cost of the equipment and is expressed in the formula $P = \dfrac{C}{R}$ where P = The payback period; C = cost and R = the cash return from the investment, or the present value of the future return. This simple approach assumes the return each year to be constant, does not allow for depreciation or taxes, and ignores returns after the original investment is recovered.

Discounted cash Flow

According to this approach the management assumes that a minimum return must be made on their capital, and an investment is not made if this standard is not met. For example, a lot of companies assume that they must earn a 20 percent return after tax on new investment in plant and equipment. They then discount the value of future cash flows to the present at this rate. The discounting factor assumes that the return is spaced out evenly during the year. The discounted cash flow refers to the rate of discount which when applied to the future cash flows will equate their sum to the supply price of the assets. It is given in the formula.

$$C = \frac{Rs}{1 + r} = \frac{R_1}{1 + r} + \frac{R_2}{(1 + r)^2} + \ldots + \frac{R_n}{(1 + r)^n} + \frac{S}{(1 + r)^n}$$

where r = discounted rate of return

$R_1; R_2;$ = cash flow after taxes in years 1, 2, ... 4.

   n = life of association

   s = salvage value

After the discounted rate of return on the investment has been calculated it is compared with the cost of the capital to be invested. If the return is greater than the cost, the purchase is economically sound, because the discounted return is greater than the return that could be obtained from the invested capital in alternative use.

Present Value

Any investment calculations involve two things: the expected pay off; or savings, which extends over many years, and the cost of the capital invested. Because capital equipment will usually produce savings over a long period of time, it is necessary to determine the worth of the future savings today. The present value of anticipated savings can be obtained by several means, the simplest of which is:

$$V = \frac{R_1}{1 + i} + \frac{R_2}{(1 + i)^2} + \ldots + \frac{R_n}{(1 + i)^n} + \frac{S}{(1 + i)^n}$$

where V = Present Value

      i = the interest rate on capital

$R_1; R_2; R_3; \ldots\ldots\ldots R_n$

                = cash flow after taxes in years 1, 2, 3, ..., n

      n = life of asset

      s = salvage value in year n.

If the present value of the investment, V, exceeds the cost of the equipment, the purchase is economically sound. The determination of the present value of estimated future savings is an integral part of any capital equipment economic analysis.

Return on Assets

The approach relates the cash savings anticipated to result from the purchase to the amount of money invested.

$$\text{Return of Assets} = \frac{\text{Present Value of Savings}}{\text{Rupee investment}} \times 100$$

This approach is subject to all conceptual difficulties already mentioned viz. Estimated if of equipment; estimated future returns, etc.

Therefore variety of economic analysis techniques is great, and they often involve quite detailed and complicated procedures. It a new machine has a greater out put or a faster rate of operation, one might assume this to be a greater value. But this is nob necessarily true if there is no need for the faster or greater output.

Leasing

At times the capital equipment or items of major equipment may be leased instead of being purchased. The principal difference between the two types of transactions is that title to the equipment remains with the original owner.

Factors Favouring Leasing

One of the important reasons for leasing equipment is the temporary nature of certain needs. The example of items normally taken on leasing are; Metal scaffolding; Wheel barrows. Pneumatic tools and many other such equipments whose life is longer than the project. Similarly leasing for temporary use is the rental of power equipment, such as transformer, for emergency service until other equipment can be delivered or repaired. Companies with very high opportunity costs of capital often prefer to lease many of their assets. For example, a company that earns 20 percent on its own capital might prefer to lease a fleet of cars for its salesmen rather than buy them. The leasing company might be willing to settle for 10 percent return capital. Leasing has grown enormously in popularity in recent years. The tentative percentage of leasing by different industry as showed by a survey is; Transportation Equipment (23%); Materials-Handling Equipment (18%) Machine tools (17%) and other machines and equipment (11%).

Advantages

Companies that lease do so both to avoid the responsibility of ownership and to make capital available for other purpose. On the other hand if a company owns equipment it must capitalise its costs. Depreciation charges deductible as operating expense must be spread over the life of the equipment. Normally lessor pays not only an adequate amount to cover probable maintenance, depreciation; taxes; and other expenses but also a healthy profit for the lessor. Decisions to make or buy; or lease affect both manufacturing and materials activities and should not be made without the approval of both the manufacturing and the materials manager.

Sometime the nature of the equipment may make outright purchase impracticable. Similarly the fact that certain equipment has a high rate of depreciation is another reason for leasing. Companies operating fleets of trucks and passengerscars often resort to lease for this reason. It has been found that, because of high depreciation of such automatic equipment, it is often more economical to rent the fleet and leave the problem of replacement to the lessor. It should be remembered that, by leasing equipment, the buyer is spared the large cash outlay required for out right purchase.

There are some unfavourable aspect of leasing that should be taken into account before decision to leasing equipment is made. Over an extended period of time it probably will cost a company more to lease than to buy equipment. This is so because the lessor expects to make a profit proportionate to his risk, and the risk supplier is greater from leasing than form selling. A lessee should also recognise that he will have to grant the lessor access to leased equipment for repair and maintenance. Such free access can prove embarrassing to a company that has methods and processes which it desires to keep private.

Mechanics of Leasing

Equipment available under lease may be available directly from the manufacturer, an outside sales agency, and firms specialising in financing and leasing equipment. Leasing arrangements are evidenced by a contract. The contract describes the equipment so that it can be positively identified, defines the duration of the lease, sets the rental rate and terms of payment, and defines the responsibility for maintenance of the equipment. Frequently the lease agreement includes an option of outright purchase with the provision that rental payments may be applied against the purchase price.

Warehouse Management

Receiving And Storage
Introduction

Receiving and storage activities constitute the final link in the materials supply chain. The receiving activity is responsible for expeditious receipt and general inspection of all in coming materials. It also completes the communication activity by promptly notifying all interested parties of the receipt and condition of incoming materials. The stores organisation is responsible for the physical identification and safe storage of materials. The way in which receiving and stores activities are conducted directly influence inventory carrying costs and, to some extent, the direct labour costs of production. A sound stores operation does a thorough job of current and future planning for the storage methods, layout and equipment it employs. The planning process can produce benefits such as: Ready accessibility of major materials; A high degree of flexibility of arrangement; a reduced need for materials handling equipment; Minimum of material pilferage, easy physical counting and efficient utilisation so storming space. Since the materials management department is responsible for materials up to the point where they are ready for production use materials personnel should have interest in the receiving and stores operations. If the stores department is independent of the purchasing department, the relationship between the two should be closer and more continuous than those between any other two departments. On all stock items the stores department initiates the purchase requests on which the buyer acts. The buyer's decision concerning a purchase is based on such factors as rate of use, number of defective parts, and trends in the rate of use. The buyer must keep the stores department informed on minimum stocks and reorder points so that the stores department can deep its inventories at proper levels.

The system of store keeping and stores inventory control is very important. If inaccurate stores keeping exist in an organisation, it will result in greater opportunity for their; waste; over or under ordering wastage of space and labour and excessive inventory investment. The poor store keeping results in a larger investment in inventory, the firm's working capital will be less liquid and effectively less adequate in amount than would otherwise be the case. Given poor control and bad store keeping, there may also be a problem of product quality deterioration.

Definition of Stores Management

Stores management has acquired greater importance in today's business organisations. In this age of cut throat competition and multiple products no industry can maintain favourable position if the cost of production is not maintained at a competitive level. The present concept of storekeeping has been given stature because of changing tastes, technical complications, attitude of buyers, increasing specialisation in industries. One can define storekeeping as air and water of modern industry. Store keeping refers to the safe custody of all materials stocked in stores for which the store keeper acts as a trustee. It simply means that the materials are to be stored in stores in such a manner that there is least possibility of theft, fire, damage and they may be easily located and issued whenever required for use.

Storage of material, spares, goods and equipments is an integral parts of the process by which the materials management department of an organisation maintains a continuous flow of these items form the point of their origin outside the organisation to their point of use inside the organisation. It will thus be clear that stores cannot be treated as a dumping ground but an active organ of the Business concern. Store keeping embraces all the activities right from the receipt of supplies of raw materials, spare parts, equipments, their proper storage and issue to used department. This also includes the storage of finished product before despatch to dealers. These activities involve maintenance of proper records of all the transactions.

Aims of Stores Management

Minimisation of cost of production: The main aim of storekeeping function is to minimise the various costs involved, while providing efficient and effective storekeeping service. Over and above the costs of materials as required for product are the costs of storage itself. These include the cost of capital investment in materials and supplies, the cost of preservation and handling of stores, the cost of record keeping, cost of use of storage equipment, and store building, and expenss for arranging light, heat, ventilation and paying taxes, insurance, losses due to spoilage and improper location of stores. Since these costs directly and indirectly influence the production cost the objective of store-keeping is to make efforts in the direction of reducing them to their minimum so as to minimise the cost of production.

The primary aims of storekeeping are to provide service to the organisation by:

(a) Making available a balanced flow of raw materials, equipment, tools and other components to the product line within the least possible time, and that too of right quality and quantity.

(b) Providing spare parts, general stores, and other materials required by the production and proper functioning.

(c) Receiving from production unit and issuing to the sales unit the finished products.

(d) Accepting and storing the scarp and other discarded materials and making arrangement for their disposal, in consultation with the appropriate departments and authorities.

Coordination with other departments: The stores department for its very existence has to establish coordination with the materials control department. But it does not mean that it can ignore other department of the organisation. The following departments are closely linked to the stores department. The smooth and efficient functioning of the stores department cannot be thought of without the active co-operation of each of the following departments:

★ Production Department,
★ Inspection Department,
★ Design and Engineering Department,
★ Finance and Accounts Department,
★ Purchase Department
★ Sales Department, and
★ Transport Department

The stores department is dependent on each of them for its day-to-day operations. It may not be out of place to mention here that without proper functioning of any one of thee departments, the smooth functioning of either the stores department or any other department will come to a standstill, hence the need for a positive correlation between various departments has lately been recognised by stores managers as one of the objectives of the stores department.

Functions of store-keeping

The following are the important functions of store keeping:

(i) Receipt of materials into store
(ii) Storage and preservation
(iii) Record-keeping, and
(iv) Issue of materials.

(i) **Receipt of materials into store.** The raw materials, tools, equipment and supplies which come from outside are delivered to the receiving section where they are inspected and then moved into appropriate storage areas.

(ii) **Storage and preservation.** Storage means custody of materials in a scientific way so as to help in the maintenance of regular flow of materials in and out of the store-house for the smooth and efficient running of the entire organisation with which that storehouse is connected. For proper storage, a storehouse should be well laid out and equipped with bins, racks, boxes and other in order to facilitate access, identification and storage for each item of stock. On the other hand, preservation refers to the safe custody of the materials so as to help the maintenance of its value and quality. Reservation helps in making available the materials in perfect serviceable condition. Proper preservation of stores over a period of time able condition. Proper preservation f stores over a period of time may entail elaborate measures, depending upon the nature of the material, the length of time in storage and the rate of deterioration.

(iii) **Record keeping.** Maintenance of records of goods carried in and out of the store is one of the most important functions which storekeeping has to perform. Proper maintenance of records helps the management in minimising the cost both of storage and production. This in turn assists in planning the future course of production and purchasing. Records also extend a helping had to the Costing Department.

(iv) **Issue of Materials.** Ultimately the materials received, stored and preserved are to be issued to various departments on proper authorisation and requisition as and when demanded. This is one of the most important functions of storekeeping because the materials are supplied to the stores department by the suppliers on the purchase orders of the purchase department for their issue.

Factors Contributing to Successful Functioning of Stores

The following factors lay down the foundation of the successful functioning of a stores department:

Location of the Store: On the location of the store depends the successful performance of storekeeping functions. If the storeroom is centrally situated, it may be easily accessible to other departments resulting in better coordination, greater facilities for the receipt and issue of materials, and making the storage more economical.

Layout of the Store: The layout of a store offers an opportunity to the storekeeper for an efficient store keeping. To ensure free movement of materials, timely inspection and day-to-day verification of the receipts and issues, it is desirable that store should be free from congestion. Proper grouping, classification and coding of stores, proper arrangement of storeroom equipments, proper indexing for easy accessibility, maximum possible utilization of available space, and proper placement of receipt, issue, record and inspection counters are the essential of a good layout of a storeroom. If the layout is good, it will go a long way in making the storekeeping functions a success.

Records. Proper, up-to-date and complete records results in reduction or elimination of wastage and misappropriation in the stores department. It also acts as a check on dishonest workers in the store and removes the possibility of theft and pilferage. Efficiency and economy, protection from all kinds of damages, settlement of disputes with creditors, debtors, insurance companies and government, checks against overstocking and under-stocking, are some of the advantages of store records which make storekeeping more successful and efficient than it would be in the absence of records.

Preservation. The underlying idea of storekeeping function is to protect the stores from all kinds of damage so that the materials carried in the store may not lose their quality and value for a long time to come. This requires a bright, airy, fire-proof and well-ventilated storeroom in which materials can be kept in safe custody. Proper arrangement for keeping the materials safe is also required. If the materials are preserved properly, naturally it will make the storekeeping function successful and efficient.

Verification. There should be a constant check on the movement of materials and on the balances of materials in bins. The physical verification of stores makes storekeeping successful since timely detection of fraud, mismanagement, misappropriation, and laziness on the part of storekeeper helps in prompt action. The verification, that too perpetual, may prove to be a boon for successful storekeeping.

Management. There cannot be two opinions if it is said that the success or failure of an action is largely dependent on its management. The storekeeping functions cannot, therefore, be an exception. The internal management, the policy level management the distribution of different functions of storekeeping among main store, sub store and different classes of officials can definitely make a success of failure of it.

From the foregoing discussion it is clear that factors underlying successful store-keeping are to be always kept in mind if the stores department is to secure more and more advantages from its day-to-day operations.

Benefits of Good Storekeeping

The success of storekeeping will be judged by smooth running of production department. Following are the advantages that accrue from a successful and efficient storekeeping:

(a) Location of materials in the storeroom becomes easier, thereby avoiding unnecessary delay and confusion.

(b) Smooth running of the production department is ensured because of regular and timely supply of materials to the department.

(c) Blocking of capital by avoiding overstocking becomes out of question.

(d) Hold ups and delays in factories are avoided by ensuring against understocking and its resultant disadvantages.

(e) Physical verification becomes the order of the day and renders all its subsequent advantages to the stores department.

(f) Accounting of all the materials is facilitated.

(g) Minimisation of wastage of time, labour and money due to leakage, theft, pilferage, damage, etc., is achieved.

(h) Availability of statistical information and proper recording makes the data more reliable.

(i) Ascertaining of cost of storage and cost of production comes within reach which in turn equips the sales department for better performance. This also helps in correct costing and market study.

Types of Storehouse

The store depots according to their functions and utility may be divided into the following broad categories:

(a) Main or Central Stores Depot.
(b) Sectional or Branch Stores Depot.
(c) Tool Stores Depot.

Central Stores Depot

The functions of the main or central stores depot are:

(1) to receive all the materials, tools, equipments, etc.
(2) to issue the materials to sectional or branch stores depots, tools to the tool stores depots and equipments to the various departments (generally production department) as and when required by them.
(3) to take steps for the replenishment of stock. The main stores depot should be under the charge of Stores Officer. Of course, the actual handling of stores would be done by the storekeeper and his assistants. Thus the central store is a wholesale supplier to other departments, and sub-store depots, who operate on a retail basis and are engaged in issuing goods directly to the users of the materials. The following are the advantages of centralised storage system.

Advantages

★ Minimum possible stock of spares, equipments, tools and fixtures are to be kept, thus avoiding unnecessary blockade of money.
★ Range of goods becomes wider so as to benefit the users more than can be possible under smaller stores depots.
★ Scientific and better method of stock control becomes possible.
★ Storage cost can be reduced, since bulk stores always take less space.
★ Efficient and successful organisation of inspection and verification can become possible.
★ Standardisation becomes easier.
★ Deterioration in stores can be minimised through more regular turnover of stocks.
★ Coordination between the central store and the different departments can be improved.

Disadvantages

However, central store system also has its own disadvantage which are enumerated below:

★ Because of duplication of work, handling and transport costs are increased.
★ More staff may be required to handle huge stocks.
★ Recurrent shortages at the point of usage may occur, if the organisation is not efficient.
★ A tendency towards rise in the stock investment prevails, if the sub-stores are not rigidly controlled.

★ Increased documentation work may involve extra cost.

★ Risk of deterioration, fire, pilferage and left in increased mainfold.

Branch Stores Depot

The sectional or branch stores depot is located in each section, unit, department or branch from where the material are drawn by the letter group. This depot, in turn, draws its requirements from the Main or Central Stores Depot to which it places its requirements form time to time. The main stores depots issues the materials to the sectional stores depots for a certain period say, a fortnight or a month. These sub-stores order to facilitate the different production centers with issues of materials different in specification and qualities, and in time. The following are the main advantages of a sectional stores depot.

(a) Prompt and efficient execution of demand requisitioned by the department becomes a possibility,

(b) If regular supplies are received, unnecessary hold ups and delays in work are avoided.

(c) In fact the advantages received from a sectional store are the disadvantage faced by the central stores depot.

Tool Stores Depot

Tool stores depot, even if there is a centralisation of storage facility, is generally maintained as a separate depot in large industrial enterprises. This store is responsible for issuing tools, spare part and other accessories to different sections (department) in general and production department, in particular. The tools are drawn from the main store once and then they are issued by the stores depots as and when required by the departments. The main advantage of such store is to relieve the main stores depot from the head of executing the retail requisitions of the departments. In this way it helps in smooth and efficient running of all the departments without any stoppage of, or delay in work. Another advantage is that proper and uptodate records of tools, which are issued on loan basis temporarily, can be easily maintained than would be done in the case of central stores.

Warehouses

Warehouses are the godowns which take the responsibility for keeping and storing goods and provide ancillary service in order to help the small and medium sized traders and manufacturers who, due to technical and economic reasons, may not like to have their own storehouse. These warehouses, undertake to preserve the goods in a scientific and systematic manner so as to maintain their original value, quality and usefulness, in exchange of which they charge certain prescribed rent at a fixed rate in advance. Thus, we may say that a warehouse is a store depot not in the strict sense of the term but in a broader sense. Its functions are:

(a) To keep the goods in safe custody till these are not claimed back by their owners.

(b) To preserve the goods so that they do not lose their value, quality and usefulness.

(c) To undertake these duties against agreed charges, known as rent, paid by the owners of the goods.

(d) To work as an agent of the owner (trader and/or manufacturer) by—

(i) undertaking the job of loading and unloading the merchandise on the form ships, railways, etc.

(ii) arranging carriage for transporting the merchandise.

(iii) classifying packaging, packing, dealing with purchasers under instructions of the owner, and

(iv) providing other facilities in export and import trade as and when required by the owner.

(e) To function is bankers of the owner of goods stored in the warehouse by:

(i) Advancing loan on the security of the goods carried in the warehouse,

(ii) getting clearance from customs authorities on behalf of the owner,

(iii) keeping the sale proceeds (if authorised) till it is not claimed by the owner, and

(iv) honouring hundis, bills, etc., on the security of goods warehoused.

Importance of Warehouse

Warehouses enjoys importance of their own. The traders and manufactures have started realising the role which warehouse can play in the advancement of their business. The services of warehouses are being utilized advantageously bu import and export houses, and they have not to bother for the complicated procedures for clearing off the materials. The trader has only to instruct the latter. Thus he saves his time, energy and money, without caring for the arrival of goods at the dock, and can devote his full attention to his business which no doubt brings him increased profits, and does a good service to the community by supplying the materials without any break. Likewise in export business the trader is relieved of the botheration of loading, taking care of customs authorities, arranging and waiting for ships, etc. leaving t entirely to the care of the warehouses.

Those traders who are interested in taking advantage of the price fluctuations, if the trend is bullish, and those who are interested in avoiding the current market trend because of its bearish tendency, find warehouses as the great refuge under such circumstances. Manufactures who have produced in anticipation of demand have no alternative but to wait for the favourable market for them, the warehouses are the source of great solace. The merchants or manufactures (generally monopolists), who want to influence the price curve in the market by their action, take the help of warehousing facilities. Warehouses also offer facilities of packaging, grading, packing, etc., which naturally are the added advantages of warehousing system, that too economic ones, in the sense that these facilities make the goods more acceptable to general public. The warehouses can broadly be divided into two categories.

1. Ordinary warehouse, and
2. Bonded warehouse.

Bonded Warehouse

Ordinary warehouse takes the responsibility of keeping, storing and preserving the goods against certain fixed charges. The bonded warehouse is licensed and authorised by the customs authorities for storing the goods till the import duty due on it is paid. The bonded warehouses may be either Government owned or privately owned. The following are that advantages and privileges which a bonded warehouse as compared to the ordinary one, enjoy and offers to its customers.

(a) The goods which are kept in bonded warehouse can be inspected, handled, sampled, packed, packaged and blended by the owners.

(b) The goods in bond can be shown to the buyers and sold out to them from the warehouses itself.

(c) The customs authorities may supervise and inspect the good in bond, i.e., on which the duty is still not paid.

(d) The owner is offered the facility of removing the goods in bond in instalments and duty is payable on the removed part of the goods.

(e) The customs authorities can specially permit delaying in the goods on which duty is still unpaid.

Layout of Stores

The stores is not a dumping ground, but part of the economy of the business and one of the resources of production. The stores should be designed for quick and simple operation and for economic use of space. It should be situated when deliveries can conveniently be made bot by suppliers to the stores and by the stores to its customers. It equipment and layout should be such that incoming goods can be off-loaded, checked and put to stores and out going goods collected and issued without any delays. Before detailing the dimensions of the storage area, the methods of storage will have to be worked out. The materials may be piled up, stacked, placed in bins, racks or shelves, depending upon the size and nature of the individual item.

For achieving that aims of a good layout first of all the storeroom should be divided into rows and columns or bays with proper numbers allotted to them so as to make easy the location of materials in the storeroom. Further, storeroom should be divided into sections according to the stature of classification of materials like chemicals, tools, paints, iron and steel, stationery, machinery, etc., Articles of like nature should be grouped together. Small items, such as nails, screw, bolts, nuts, etc., should be stored in small but closed bins, whereas for the storage of lengthy and bulky goods, open or wide space should be utilized. The best of storing these things are the containers in which they have been received from the suppliers. The use of unit load containers is recommended.

There should be proper arrangements for receipts and issue counters, and a place should be provided for record-keepers and personnel of the inspection department. It is advisable to provide separate counters, rooms, endorsers for receipt, issue, inspection and records. Adequate space in between two rows for the movement of men and materials is a must for ensuring prompt execution of orders of the production department. Adequate provision for expansion should also be made. Proper placement of fire extinguishers sand bags, buckets, etc., is also one of essential requirements of a good layout of a storeroom. Entry of unauthorised persons into the storeroom should be strictly watched.

Environment Conditions Effecting Stores Functioning

Stores play a vital role in the operations of a company. Stores is often equated directly with money as money is locked up in stores. It is important that stores are constructed with a futuristic orientation, so that sufficient flexibility for expansion needs is inbuilt. The activities of receiving the goods, stocking in appropriate locations, material handling and issues must be done swiftly and economically. Comfortable working conditions must be provided to the stores personnel to get maximum efficiency and morale. Some of the environment conditions such as Noise; Improper lighting; and Extreme Heat; Coal and Dust have direct impact on the working of stores.

The improvement in these conditions contribute the fatigue alleviation which results into higher morale, low accident r ate and less absenteeism. The conditions which affect fatigue and are responsible for effective working atmosphere are:

(i) Noise Industrial Music.
(ii) Illumination.
(iii) Temperature
(iv) Humidity
(v) Ventilation
(vi) Safety

(i) **Noise.** Noise is unwanted sound and it increases fatigue. If the noise is too much and continues for a along time it might cause deafness, for example, boiler maker deafness, sheet iron worker and blacksmith deafness. The effect of noise depends on the kind of noise and kind of work. Mental work is affected more than manual work by noise. The effect of noise is greater when it is irregular. Sounds of either very high or low tone qualities are more irritating than those in middle zone. In a study of mental arithmetic metabolism was mush higher during the first noisy period but reduces with adjustment.

The noise of moderate intensity serves as a psychological inducement to work. One study of the reduction of noise concludes that it is an investment which under ordinary conditions will increase productivity by 10 to 15 per cent. But if the individuals who are exposed to very high pitch noise such as boiler makers and aeroplane pilots they suffer permanent deafness.

(ii) **Industrial Music.** Music in factories and stores has become fairly common and therefore, requires an evaluation. some of the studies made in England where researchers found the initial interest for dealing with monotony. The effects of musical programmes on simple assembly operations showed production increase up to 6 per cent, when production on days with music was compared to that on days without music. In a radio-tube assembly factory in which music was customarily played, the effectiveness of fast, slow and mixed programmes was studied. The scrap rate was found to be less when either fast or slow music was played than when there was non music or when fast and slow musical programmes were alternated. Musical programmes also had a beneficial effect on employee morale.

The increase in output varies considerably with the job and the shift. Simple repetitive jobs that can be performed well while talking seem to benefit most by music. The research findings indicate that music seems to benefit most by music. The research findings indicate that music seems to encourage young, inexperienced workers to increase production on routine jobs. No significant effects were found for the older skilled workers. The general finding is that employees like music. At least 75 per cent strongly favour it and only 1 to 2 per cent oppose it. Slow music seems more desirable than first music, variety is essential to keep young and older employees satisfied with the programmes. Vocal music may be distracting for some employees, and music with storing beats may disturb or help depending on how it fits into the movement rhythms of a job. Spaced programmes seem to be favoured. At present, it appears that the music should be on about 12 per cent of the time for the day shift and as much as 50 per cent of the time for the night shift. Most favourable effect of music its influence on boredom. It takes the mind from the work as well as frees the brain of the obligation of initiating the activity.

(iii) Illumination. Many industrial undertakings take this advantage of natural light during the day time. But it should be remembered that the illumination of the work place by means of artificial modern lights is far better than taking the natural day light in carrying out any work efficiently. Any natural light which the industry uses should be carefully utilized in the working conditions. North light is commonly used in industries because the variation of light is not much during the day time in a factory. Engineers are utilizing modern techniques in the plant layout to avoid such changes of illumination during the day in the workshop either by putting the transparent shades on the top of the factory or cutting down lights from all directions except from the north. Some jobs need special light which needs more illumination. Because of the uniform light in the work pace, the workers do not find any visual difficulty while coming out from a specific work situation which is illuminated brightly.

The entire work area should be uniformly illuminated. Areas outside the immediate focus of observation, wall surfaces, for example, if unevenly illuminated cause continual adjustment changes and consequently higher visual fatigue. Since the pupil of the eye must adjust itself when focusing on a bright spot and then on a dark spot, expansion and contraction of the pupil sets up an increased strain and fatigue in the worker.

Indirect lighting in the best method of producing uniforming. Eye muscles tend to pull towards areas of high illumination. Where the high point of light is not in the work area, a tug-of-war exists between the muscles that are concentrated on the work area and those that veer to the higher illumination. The human eye adjusts so readily to changes in illumination that the effects of poor lighting are usually not immediately observable to the untrained eye. Errors in judgement are often found when people's reports on brightness are checked against readings on light meters. Poor illumination causes fatigue and irritability and is a source of errors and industrial accidents. Money spent to correct improper lighting will be more than repaid in higher output per worker, lower production costs, and a generally happier, healthier and probably safest work force.

The colour of surrounding objects and surfaces is related to lighting. The use of white or light pastes improves illumination by reflecting high percentage of light, while deeper tones have a considerably lower reflective value. The occurrence of glare should however, be avoided. Several studies made on the use of coloured light show that people prefer the light of daylight colour. The use of this so-called while light is said to assist materially in visual efficiency.

   (i)   Various jobs requires different intensity of light above 50 foot candles.
   (ii) The severe and prolonged visual tasks, such as fine engraving, and discrimination or inspection of time details of two contrast requires 50 to 60 foot candles.
   (iii) The prolonged critical visual task such as drawing, fine assembling, fine machine work, and proof reading, require 50 to 60 foot candles.
   (iv) The visual tasks such as skilled bench work, require 19 t 20 foot candles.

(iii) **Temperature.** Indian are accustomed to working in a temperature which will be considered as high. At several places like Calcutta, Madras, Bombay etc. where the employees work in a moderate climate throughout the year, there the temperature does not rise or drop to the extremes. The efficiently goes down during the extreme temperature conditions.

(iv) **Humidity.** Variation in humidity play very important role in industry. Rise of humidity in low temperature to a moderate extent gives relief to workers from severe dry cold, whereas, rise in humidity adds further discomfort to the rise of temperature. It is more discomfortable to withstand wet heat and dry cold.

Low temperatures can also affect productivity. Ability to make precise movements with the hands and fingers decreases. discomfort has been within a room, that is, cool air at floor level and warm air at head level.Furthermore, where sizable temperature differences existed within a building and the nature of the work required worker's movement from the area to area, discomfort was noted.

(vi) **Ventilation.** Temperature and humidity can be controlled to a great extent if adequate ventilation by the circulation of fresh air is arranged in the factories. The circulation of fresh air is generally done by fans, ventilations, the blowers. Artificial arrangements are also made in chemical industries to avoid pollution of air, through gases and fumes. In cement factory where the stones are crushed the operation crates lot of dust. Similarly in the spray shop the spray particles pollute the air. In such cases, exhaust fans are not adequate enough to supply fresh air and so gas marks are provided to workers to avoid fumes and dust during work. Dust and fumes are constant health hazards in the Indian industries.

1. During winter the temperature for light work should be 65°F, for active yet light work 60 to 65°F. For work involving more muscular exertion, 55°F to 60°F.
2. During summer the temperature should be kept below 68°F.
3. The introduction of fresh air should not be less than 1,000 cubic feet per person per hour.
4. In summer adequate air circulation should be arranged. The efficiently increases if the air movement is moderately high.
5. During summer the relative humidity should be 70 per cent at least.

(vii) **Safety.** This factor is perhaps the most important aspect. In stores a large number of goods are handled every day. Accidents considerably reduces the morale and effectiveness of the system. Following are some of the measures necessary if accident are to be checked:

(a) Safety appliances such as goggles etc. should be provided and their use must be encouraged.
(b) Good house keeping is essential. Stocking must lie in appropriate locations so that handling is minimum.
(c) All stores equipment must be kept in good order.
(d) Safe actions should be stimulating by installing "safety awards".
(e) Provisions of fire fighting facilities is necessary especially where inflammable materials are stored and handled.

25
Stores Systems and Stores Organisation

Introduction

There are two basic systems for controlling stores materials : (a) Closed stores system and (b) An open stores system. A company normally employs both systems. The application of each system depends upon the nature of the production operation and the way in which materials are used in production.

Close System

Closed stores system is one in which all materials are physically stored in a closed or controlled area. Whenever possible, the general practice is to maintain physical control by locking the storage area. This system is designed to afford maximum physical security and to ensure tight accounting control of inventory material.

Open System

Open system represents the second major type of stores system. It finds its widest use in highly repetitive types of operations. In plants, using the open stores system, no storeroom as such exists. Each material is stored as close to the point of use as is physically possible. Materials are stored in bins, on shelves, in racks, on pallets and so on. However the storage configuration at each work station is arranged to fit the available space. The open system is designed to expedite production activities. It places little emphasis on the physical security of materials. The open system also places less emphasis on the accounting control of materials. No perpetual inventory records are kept in an open system and the accounting charges are determined indirectly rather than directly. As a result the system provides control over material usage only if it is used in conjunction with an accounting system employing standard cost technique.

The open system is must applicable in situations where a repetitive production operation produces standardised products. Materials handled in an open system should not be subject to pilferage, nor should they be easily damaged. If production requires delicate or pilferable item, they should be controlled in a closed store room.

Stores Classification; Identification

In any factory storeroom the problem is of properly locating the various materials with in the store room. A proper storing arrangement requires: A method for classifying and identifying the items stored; A workable system for identifying and working storage spaces; and a system for assigning the actual location for materials to the stored in the store rooms. Once all locations in the store room have identifying symbols and all materials and other stored items also have identifying symbols or numbers, all that is needed is a set of index cards showing which materials are put where in store room.

Identification

In normal way of identifying an article in the storehouse is by its simple description, but this is neither satisfactory for stores purposes nor this will help in doing the operation speedily and effectively. The description is sometimes vague and may be confusing. Accurate identifications of an article, sometimes, demands lengthy descriptions which may generally be complicated and thus adding to the confusion in understanding, A "chair" if said as such is not the identification. An arm chair, plastic caned back, revolving, steel frame, foam cushioned and so on may be the accurate description of the chair in question.

In order to identify correctly and on a logical basis and to avoid multiplication of items several times and to save time and labour, to facilitate easy location and proper functioning of the storehouse, a proper codification is needed. The importance of classification can be explained by taking library as an example, considering a well organised library where books are classified, numbered, and properly stocked. One can find any book he wants easily and also can find out whether the book is in the library or not. A sense of orderliness is introduced with the result the functioning and management of library is greatly simplified and effective control could be established.

Need of Classification and Codification

The need for classification and codification is even greater in an industry. Classification is the operation of bringing alike things together and coding is the allocation of a symbol (alphabetical and/or numerical) which **uniquely defines an article or a subject and places it in its correct classification. The cost accountant needs a logical code reference to each part and of each process and each piece of plant and equipment. The material controller must be given the basic information upon which all his controls are built in a code form which is complete and systematic. The major areas where classification and coding could be applied are:**

(i) Materials in stores.
(ii) Parts, subassemblies and assemblies.
(iii) Equipment, dies and tools.
(iv) Drawings and other design reports.
(v) Process and operations.

(vi) Identification of standards and other documents.

Principles of Classification

(i) **Comprehensive Principle:**
The system of classification must cover the entire range of items for which it is devised and at the same time allow reasonable scope for extension.

(ii) **Mutual exclusiveness:**
There must be only one place for anyone thing. As the user progresses in his search from main class to sub-class, from sub-class to group, and so on through consecutive classification, he is able to discard vast information and is able to restrict his search to a very narrowing range.

(iii) The appropriate order of complexity must be maintained in the arrangement of classification criteria i.e. from cheap to expensive, from soft to hard, single to complex.

Codification of Materials

One of the most important factors in dealing with the materials is the human element, which cannot be ignored or by passed in devising any method of identifying or codifying the materials. Hence any stores management has to keep into the problem considering 'human being' as one of the important elements. Though the normal way of identifying an article in the storehouse is by its simple description but this is neither satisfactory for stores purposes nor this will help in doing the operation speedily and effectively. The description is sometimes vague and may be confusing. The 'dust bin' may be known as 'refuse container' 'rubbish box' etc. This may result in confusion and out of which duplicate ordering or overstocking may become the order of the day. In England, an electrical firm had as many as 12 names of an items depending upon its function in various using departments. Will this not results in utter confusion, inefficiency to the storehouse as a whole? Accurate identification of an article, sometimes, demands lengthy description which may generally be complicated and thus adding to the confusion in understanding. A 'Chair' if said as such is not the identification. An arm chair, plastic caned back revolving steel-frame, foam cushioned and so on may be the accurate description of the chair in question. The whole story of the article is to be told if accurate identification is needed.

Merits of Codification

In order to identify correctly and on a logical basis to avoid multiplication of items several tines, to save time and labour, to facilitate easy location and proper functioning of the storehouse, a proper codification is to be evolved so as to obtain the following benefits to: avoid the long and unwieldy description; have accurate and logical identification; prevent duplication; standardise the items,; reduce the varieties; have efficient purchasing department; have efficient recording and accounting; simplify and facilitate the mechanical recording; simplify and facilitate the pricing and costing; have proper system of location and indexing; assure correct and efficient inspection and assure production as planned and as required.

Long and Unwieldy Description. There is no denying the fact that if there is no logical system codification then for identification as store one has naturally to depend on the description of the material which has to be in greater detail. If one is interested in properly and correctly identifying the materials carried into the store. It is, but obvious that these descriptions will be lengthy enough, which will naturally become unwieldy. Unfortunately, they will not bear fruit. Instead that may further complicate the situation. A scientific codification system, thus, will avoid all these and make the storehouse more efficiently and effective operating organisation.

Accurate sand Logical Identification. A separate code is allotted to each of the items available in the storehouse indicating different sizes, quality, price, usability special characteristics, specification, etc., of the item concerned. This distinguishes, one item from the other, even if nomenclature is the same. This obviously help in accurate identification, which is logical in the sense that cannot override each other. If it does, it will result in the defeat of the very purpose of codification.

Prevention of Duplication. All items are separately codified and are arranged in certain logical order. Similar materials are grouped together such as stationery item, and is given a code, e.g., 07. Once a code is allotted to a particular group in no case the code number is changed. This prevents duplication. Neither duplicate ordering is possible nor because of having different nomenclature materials pile up in the storehouse due to different places, bins and nomenclature. As in any codification system it is not the name given to an item is taken into account but the code is considered and action taken accordingly. Thus, duplication at all stages is prevented with the help of proper and scientific system of codification.

Standardisation and Reduction in Varieties. Standardisation is the order of the day and codification's main and important aim is to achieve standardisation. For codification grouping of similar identical item is a must. This enables the store authorities to examine the entire range of stores. This facilitates in eliminating those varieties in place of which other varieties of the like quality and use can profitably be used by adhering to lesser varieties, thus, reducing the number of varieties to the minimum. By this process standardisation can be achieved and investment in items can also be reduced to a greater extent thereby administering economy in purchasing. Inventory carrying cost may also be prepared to examine the range from details and length points of view. If it is found on examination that it is again detailed enough and there is a further scope of reduction, then an examination may be carried out for further reducing the range making it more standardised and less cumbersome.

Efficient Purchasing Department. The filling up of purchase requisition particularly the specification column is a task in itself in the absence of a code which may easily indicate the materials required. Buying instructions to the suppliers become easy and quick. Items are grouped for facilitating the allotment of codes and in turn section wise organisation of purchase department based on the code groups becomes a possibility, which in centralised purchasing serves well in dealing with the purchase orders and in taking the advantages of bulk purchasing.

Efficient Recording and Accounting. The codes improve chances of effective stock control efficiency recording and result yielding accounting. Since due to codes everything is clear and numerical codes, if used make the store-room operation efficient, accounting become up-to-date, scientific and perfect. Chance of mistake in minimised. Presence of codes minimises the fraud, takes less time and energy in recording detect the defects with quick and ease, pricing and valuation become more accurate and reliable.

Facility in Mechanical Recording. Mechanical Recording has come to stay in today's accounting process and since it is neat and clean, free of mistakes, less time consuming and less fraudulent, people have started more in believing mechanical records then in pen kept records. It is also less costlier. But the mechanical recording would have been a distinct possibility in the absence of codes. Keyboard, purchase cards, computers, etc., would stop working if no codes are there. The use of machines is limited so far as the long description is concerned, because that will be more time consuming and hence costlier thereby defeating the very purpose of mechanical recording. But codes solve the problem.

Facility in Pricing and Costing. Long description of prices is a cumbersome job and fails to satisfy the need of indexing and easy referencing. Referencing also becomes slow in a long descriptive price list. In order to make it easy and more reliable codes are introduced. Costing in a manufacturing units, is to be done in a manner which may make it more reliable and simplified. The codes can do it. As per the requirement the cost may be calculated job wise, for which different cost headings are to be provided. Codes become helpful here, thus, facilitating the costing.

Facility in Locating and Indexing. Again long descriptions are not of much help in locating and indexing the material in the storehouse. The materials in the storehouse have to be kept n an order which may facilities their placement and location. Also Indexing in logical order facilitates the easy operation in the storehouse. For making it less time and energy consuming, items may be arranged according to the codes allotted to the materials. This may, thus facilitates the location and indexing of the materials in the storehouse.

Efficient Inspection. Inspection is checking of the quality and utility of the materials as per the purchase order and requirement of the production unit. This has to be efficient, as on correct and proper inspection of the material depends the standardised and qualitative production, which in turn helps in maintaining the goodwill of the product in the market and ensuring at least the previous margin of profit. Codes and not lengthy description clearly indicate the correct quality of the material assisting in efficiency inspection.

Planned Production. From the above discussion it is evident that codification helps in avoiding lengthy and unwieldy descriptions resulting in accurate identification, achieves standardisation, reduces the varieties, facilitates mechanical recording, simplifies costing and pricing ensures efficiency inspection, takes less time and energy in store-room operation. All these, naturally, help in planned production ; in preparing the production schedule and strict adherence to, it comes within easy reach because of the codification of materials.

Systems of Codification.

There are different systems of codification for nationalised storekeeping. But the best system is that, which also gives The history and sizes of materials along with standard form. Codification should also avoid calling the material by its function or its used name. It should give an objective name which should be versatile in nature. The system is more commendable for adoption where large number of items having large variety are in storage however the following systems for codification are generally used in the modern industrial concern:

* ★ Arbitrary
* ★ Alphabetical
* ★ Numerical
* ★ Alphanumerical
* ★ Mnemonic
* ★ Decimal.

Arbitrary System

Arbitrary system as the word 'arbitrary' denotes, is a system arising from accident rather than from rule. It is based on the serial number under which a material is received in an industry and the same is allotted a code number. This system often adopted in the initial stage of an industry. In this system either one has to follow the arbitrary number allotted to a material upon its receipt or the vocabing department gives the arbitrary number. For example, if vocabing department has so far standardized 9100 items and if now it wants to standardize a lathe machine which has been received, then 9101 code number will be allotted to the lathe machine.

This system has the primary advantage that there is no fixed limit for codifying any number of items. But the disadvantages are that there are changes of duplication of items and one cannot know the history of materials or any specific description of the material but for each item will have to refer to the key which become cumbersome. This is the reason why this system is not popular.

Alphabetical System

Under this system of codification 'alphabets' are the basis and each material is allotted a certain letter from the alphabet which has no relation with number. The first alphabet of the name of the material is invariably the staring point of codification in this system, Subsequently, sub-alphabet is also used depending upon its other characteristics. The following illustration may make the system more clear.

Acid	A
Brass	B
Iron Ore	IN—O
Iron Pig	IN—P
Iron Steel	IN—S
Iron Sheets	IN—ST
Iron Bars	IN—BA

The alphabetical system is not easy to remember and it is not possible to expand it to the desired level because of the limited availability of alphabets. It is difficult to select proper alphabet for each of the items in the store. That is why this system is not so popular and is out-dated. Small concern which have a very limited stock can adopt this system.

Numerical System

This system is based on numbers. The materials are codified with the aid of numbers making provisions for further expansion as well. For instance, number I may be allotted to iron and number 2 to stainless steel and so on. The numerical system like the alphabetical suits only to small stores. Though, this is a simple system, but there are chances of duplication. This system is also very old and not so popular. Hence, it is not so much used in the modern industries.

Mnemonic System

Mnemonic system as the word 'mnemonic' denotes, is a system of assisting memory. Under this system the material is codified by giving letters and figures as codes. It may be decimal or arbitrary. Letters are assigned to different materials according to the sound of the names. For example:

PM	Packing Materials
BS	Brass Screws
SS	Steel Screws.

Under this system one can easily know what a particular item is meant for by just looking the code number. The system is adopted in small concerns where the number of items is not much because it has limitation as one will have to accommodate the code out of 26 alphabets for copying all the items. An ideal system may be to use letters to represent the names of materials and number to represent the size and shape and other specifications of the material. For example:

SS—Represents Steel Screws.

20—The first two numbers, viz., 20 represent the length of the screws in mm.
= 20 mm.

08—The 3rd and 4th numbers, viz. 08 represent the diameter of the
screw in mm.
= 8 mm.

Hence SS2008 means steel screw 20 mm long and 8 mm in diameter. This system is also most common in numbering the vehicles, e.g., DHS—4965. Here DH stands for Delhi, S for scooter and 4965 is the arbitrary registration number.

Alpha-numerical System

This system is same as Mnemonic System. Rather we can say that other name of mnemonic system is alpha-numerical system because in this system the material is codified by giving alphabets and numbers both as codes.

Decimal System

Decimal system of codification is the universal method of codification. It is simple and easy to codify large stores. Under this system items up to 5,00,000 can easily be codified and at the same time each symbol will give the history size, specification and complete picture of the item. To follow this system one has to categorize stores in classes, groups, sub-groups, types and last three digits can be left for sizes or sub-types. Decimal system works from 0 to 9 figures and on adopting 7 to 11 digits. Each digit indicates something or the other. For example:

1st digit	Section
2nd digit	Class
3rd digit	Group
4th and 5th digit	Type of materials
The onwards	Size, part No., specification or any other detail required.

Let us consider that stores of a particular industry can be divided into the following main sections and allotted numeral as shown below:

0	Raw Materials
1	Plants and Machinery
2	Machine and Hard Tools
3	Construction Materials
4	Maintenance, Repair and Operation Materials
5	Spare Parts
6	General Stores
7	Fixtures
8	Paints, Oil and Lubricants
9	Finished Products.

Mild steel bar would be raw material and as such fall under section (0). This raw material could be further sub-divided into classes as follows:

0	Timber	5	Paper
1	Rubber	6	Glass
2	Metals	7	Leather
3	Textile	8	Chemicals
4	Plastics	9	Miscellaneous

Therefore code number of metal would be 02.

Raw Materials

The classification of raw material—metal could again be divided into the following group:

0	Aluminium	5	Brass
1	Zinc	6	Mild Steel
2	Lead	7	Stainless Steel
3	Nickel	8	Alloy Steel
4	Copper	9	Other Metals

Similarly the fourth number could show any of the following type:

0	Ingot	3	Wire
1	Plate	4	Mesh
2	Sheet	5	Bar

Again the fifth digit to indicate the following classification of shape:

0	Square	3	Round
1	Flat	4	Hexagonal
2	Angle		

The sixth, seventh and eight digit could show size—say in millimeters. Thus a round mild steel bar of 25 diameter could be shown as—02653025.

These digits show :

0	2	6	5	3	025
Raw Material	Metal	Mild Steel	Bar	Round	25 mm dia.

With a fully significant coding arrangement like this, the code No. 02553100 would mean 100 mm round brass bar in the raw material stores.

Stores Vocabulary

When all the items of stores are codified according to standard Material Code, these should be compiled in the form of a book in the chronological order. This compiled from of codification will be known as Stores Vocabulary or Stores Catalogue. Stores Vocabulary should be circulated among all the departments concerned for their use. Catalogue should be prepared when there are sufficient items in the stores. However, as and when fresh items are received, these should be codified accordingly and amendment slips should be issued.

Marking of Stores

For identification of materials, marking of stores is another method. These are two types:

★ Colour Marking.

★ Secret Marking.

Colour Marking

Colour marking of materials is often used to distinguish different grades of steel or copper or brass. Alloy steel costs several times more than the mild steel but looks alike to the naked eye. Hence for storekeeper to identify the different sheets and other materials it is necessary to adopt a suitable identification code. Normally, colour bands of three or four colours are adopted and painted at the end of bars, sheet etc. If metals are to be coloured marked the following colour code may be assigned to different metals:

Aluminium	Black	Steel	Blue
Copper	Red	Zinc	White
Iron	Green		

Secret Marking

In order to minimise the risk of pilferage, it is sometimes advisable to mark stock items with the secret code or initials of the firm so that, if anything is stolen, it may subsequently be traced and the mark will serve to prove that the article is the firm's property. Discretion should be exercised in this matter e.g., forging, casting, heavy steel, etc., are not worth the trouble of marking but articles suitable for domestic use, such as hand tolls, soap, towels, clothing, electric lamps etc. will usually repay some attention.

Thus secret marking is a moral check on the staff and theft and pilferage of costly items must be avoided to a greater extent due to secret marking.

Demerits of Codification

The above merits of codification do not lead us to a safe conclusion that it suffers with no disadvantage. There are certain demerits of codification which are enumerated as under:

1. **Mistake in Coding.** The success of codification depends on 'a check-up' line for verification of codes allotted from time to time, place to place and position to position. In normal circumstances this checkup line may not succeed in verifying the codes at all times, in all places and positions. The coding mistake, thus, may crop up and avoidance of which may not be possible, howsoever alert the store management may be. A busy or a new assistant may err somewhere unknowingly and unconsciously, the incidence of which may bring disaster to nay of the department affected by his error.

While filling up a stores requisition slip, an assistant of the production department, if he has wrongly entered the code number, may get the wrong quality of materials from the stores department, resulting in production of higher substandard or high-standard product. Both may be bad from the goodwill point of view. A substandard item may lose the goodwill in the market. In the same way a high standard product may have more cost of production but will not yield the same high price as the product is to be sold out at the same price at which other products of the like standard are being business in the market as a correction will, in future, be made and the same high standard material which came into the market because of error in coding may not come up and the customer might be expecting the same high standard. They may be disappointed and may be desert you by switching their linking for your competitors product out of annoyance and dissatisfaction.

Likewise, a coding mistake in the purchase requisition by the stores department to the purchase department may result in wrong placement of orders to the supplier, consequently wrong supply of materials to the stores department by the supplier. This may be rejected by the inspection department, though the inspection department may be in an awkward position as the materials supplied are conforming to the purchase order but are not confirming to the production plan. They have no moral and legal support for rejection, but they have no moral grounds to accept either. However, if they accept, it may result either in overstocking or may at a later date be declared as obsolete. In both the cases there is a loss, which is due to mistake in coding.

A grave situation may arise when in a certain accounting statement a mistake in coding has crept in. This may lead to wrong decisions, wrong formulation of policies, wrong course of action which may sometimes disturb the business routine.

Though the mistake in coding might have crept in because of mechanical defects, bad handwriting, careless writing, negligence or carelessness, etc., it is enough to say that it may be disastrous, but is unavoidable as human element is involved in the entire store-room. The only things which may be suggested here is that whenever applying codes we have to be cautious and alert.

2. **Detection of Coding Mistake.** It involves long time and labour to find out the mistake which has been committed. Finding out of exact code against a wrong code, is a strenuous task, then, he has to go from page to page. It is, thus, a costlier job with sometimes is not worth undertaking.

Stores Organisation

The stores and receiving jobs are closely related. Therefore, the receiving activity is often placed under Jurisdiction of the storekeeper and stores personnel reporting to materials management department. The following are the two strong views for this organisational arrangement.

(a)   Stores activity is a materials-oriented activity, therefore, it should report to a department whose primary interest likewise lies in materials and supply operations.

(b)   Receiving is the last step in material acquisition process and stores is the last step in materials supply process.

Under this kind of system the relationship between stores, inventory control and purchasing will receive proper attention. In some companies the stores activity are under the Chief of Production. The reason for this set up is as follows:

(a) Production management is responsible for running the production operation smoothly and for meeting product delivery dates. The production department therefore have to control the material supply to ensure smooth delivery of material to the production work stations.

(b) Receiving and stores should not be supervised by the department that buys and authorities payment for the materials they receive and disburse.

Both the above systems of Stores reporting are of equal importance to management. The final decision, therefore, will probably be based on other factors in a particular firm's organisational and operational pattern. A final factor which conditions the decision is the location of the inventory control responsibility. While there is no requirement that stores and inventory management should report to the same department. But if the two departments report to the same department head many mutual control problems can be solved quickly at a lower level before they become serious. In all cases however, the decision should be based on an attempt to fulfill the following objectives as completely and realistically as possible: To keep materials in good condition and to have then immediately available; To develop and order filling and delivery system which satisfies production requirement, and to achieve the activities at a minimum cost.

26
Stores Control Forms and Stores Accounting

Introduction

The mechanised materials control systems are frequently used because of the volume of transactions and number of items involved in industries. The three methods of punched card Materials Control system commonly used are:

(i) Balance-Forward Method
(ii) Unit record Method
(iii) Batch Method

Balance-Forward Method

A card is created for each classification of stock, say for 5000 items; 5000 cards would be required. Information carried on the cards will be; Item; Date; Opening Balance; Minimum Inventory; Re-order Paint; Receipts; Issues; Inventory On hand; On Order; Available. Prepunched cards can be employed to reduce the amount of punching required. The system must be maintained. Stock adjustment due to rotating counts or physical inventory can be made as they are identified. Also discrepancies in counts on issues; receipts etc. can be corrected as they are discovered. Periodic Reports such as shown in fig (26.1) can be generated by the balance-forward cards.

Unit Method

This method requires that a card be punched for each item in stock.

The Batch Method

The batch method permits the most complete machine processing. As issues or sales are made a copy of the issue slip or sales order is processed through key punch. The following information is contained relative to each issue:

(i) Customer Code
(ii) Quantity
(iii) Unit selling or Transfer price.

A Balance-Forward card is created by Batch rather than individually as in Balance-Forward Method described earlier. Fig. 26.2

Repeating Purchase Requisitions

Some companies have developed a repeating purchase requisition for stock material and parts which are ordered on a repetitive basis. This saves making out an individual requisition every time material is required. The heading is typed only when card is originally made up and lists all information regarding that particular part. Alternate Vendors, Price and Price breaks are shown in fig (26.3)

Receipt-and Inspection

In many companies receiving is a function of material control, although sometimes material control takes over after the materials are accepted. The basic receiving functions are unloadings and unpacking materials, checking materials against the purchase order and invoice, having any necessary inspection done, making out receiving and inspection reports, putting materials into standard containers and delivery the materials to store room. One copy of the receiving and inspection report is kept in receiving department and other go to the purchasing department and store room. Fig. 26.4 is a typical receiving report.

The Fig (5) shows the most important forms used during the physical inventory viz Inventory tags and Inventory sheet. The tag method is considered as the best way of inventory materials since it provides a visible check of material which has been inventoried and is a convenient way to work to summary sheets and company records. The summary that form shown in Fig. 26.6 is suitable for all classes of materials when the tags are returned they should be checked in on the summary sheet and any missing tags posted to a sheet such as Fig. 26.7 so that a search can be made for the missing tags. Tags for work in process materials should clearly indicate the stage of completion or progression of that material.

Stores Requisitions. The important information on any materials or store requisition include the following:

   (i)   Description of item wanted.
   (ii)   Quantity wanted and unit of issue.
  (iii)   Point to which material should be delivered.
  (iv)   Date wanted.
   (v)   Charge or order number
  (vi)   Signature of person making out or authorizing requisition.
 (vii)   Date of signing
(viii)   Initials or signature of person receiving material.
  (ix)   Initials of store keeper.

Simpler forms carry the above data, but often additional entires are needed. Fig (26.8) for example has space also for stores location new balance on the bin tag, amount issued, cost data, initials of store record clerk who may apportion material in advances of issue. Forms like there are used most frequently when the items come under different cost charge classifications and items are in different store rooms. But when the bill of material is required for a complete order, it can be conveniently arranged to serve as a group stores issue, as in case of an assembly order, a form similar to Fig. 26.9 This form is prepared as a master form which can be duplicated.

				Reorder Schedule No			
Part No.	Description	In Stock	On Order	Total Available	Required For Schedule	Place Orders For	

Fig. 26.1 **Periodic re-ordering schedule**

Summary Stock Status Report

To ....   Date ....

Item Code	Description	Opening Balance	Actual Transactions			Future Transactions		
			Recd	Issued	On Hand	On Order	Req'd	Not Committed

Fig. 26.2 **Summary Stock Status Report**

Purchase Requisition

To ....   REQ. NO. ....

Date ....

Please purchase the following items

Quantity	Description		

Ship to ....

Via ....

Signed ....

Date Required ....

Approved ....

Fig. 26.3. **Purchase Requisition**

Receiving Report		No	
Received from:		Date	
		P.O. No.	
Quantity	Description	Weight	Dept. Charged

Condition of Shipment:
Order Complete ....   Partial ....

Delivered By   Reed By ....
Freight Charge   Checked By ....

Fig. 26.4. Recieving Report

INVENTORY   4059

Material Symbol ....

....

                    4059
Material Symbol ....

Description ....

....

Operations ....

Performed ....

Quantity .... Unit ....

Location ....

Counter By ....

Remarks ....

Date	Received After Count	issued After Count

Fig. 26.5 Inventory Tag

| | | Date .... |
| | | Supervisor .... |

Issued	Group & Location	Issued			Ret'd Used			Ret'd Unused		
		From	By	To	From	By	To	From	By	To

Fig. 26.6 Issues And Receipts of Inventory Tags.

| | | Date .... |
| | | Supervisor .... |

When Corrected, Cross out and initial in **Red** and post to Form 'A'

Ticket No.	Entry By	Store and Group	Tissue To	Remarks	Block of Tickets Returned	Checked

Fig. 26.7 . Record of Missing Tag.

Material Issue				Order No.
CREDIT OR TRANSFER				Credit
Deliver To	Shop Symbol	Name of Shop	Location	
				Date written
Description (Units/Barrels/Coils) etc.				Date wanted
				Unit wanted
				Quantity wanted
				Approved by

New Bal. Bin Tag	Issued		Price	Stores Values	Handling	Total Cost
	Quantity	Unit				

Apportioned	Storekeeper	Balance Clerk	Cost Clerk	Material Received By

Fig. 26.8 . **Material Issue, Credit or Transfer Slip**

Sheet No.	Bill of Material				Quantity	Assembly No		
Assembly Name								
Part No.	Description	Req'd.	Del. to Deptt.	O.K.	Inventory code No.	Balance in stock		
						Cost	Amount	

Remarks	Total This Sheet
	Total Brought Forward
	Total Material
	Authorised by Date ....
	Filled by Date ....

Fig. 26.9 Assembly Bill of Materials for group Stores issue requisitioning

Stores Accounting

Materials costing is important in terms of the valuation of the cost of materials consumed by the production departments as well as in terms of the estimation of the value of materials help in stock.

Costing of the Receipt of Materials

Costing means classifying, recording and allocating the appropriate expenditure for determining the cost of product or services and the presentation of the suitable arranged data for the purpose of control and guidance of management. The factors that are to be included in the building up of the cost of the materials received are material price; freight charges, insurance and taxes. Price usually refers to the price quoted and accepted in the purchase orders. The freight costs incurred in transporting the goods are usually collected under a separate head. Goods in transit are mostly covered by insurance. All such insurance expenses must be calculated and added to the base coast and transportation cost. Under the miscellaneous head, we need to classify costs incurred by way of custom duties, taxes and packages. In brief, we can say that the coast of the materials received is equal to the price quoted less discount, plus fright, insurance, duties, taxes and packing charges.

Statistical determination of Costs

The process of budget preparation takes into account only one cost-variation factor: Level of operations. Other factors, however, can be of great significance also. Consider the impact on operations and manufacturing cost of introduction of one or more new products into processing during a given month or quarter. Similarly, consider the impact of such factors as the following; changes in size of production orders; changes in rate of labour turnover; Changes in quality of raw materials used, changes in machine process. All these factors reinforce or detract from the cost impact of the current level of operations. The management can improve its forecasts of future coasts and profits by such statistical determination of costs. Its budget will be more accurate and will conform more closely to actual results if operating conditions actually prevailing were correctly anticipated. It has been observed that management's increased cost knowledge will enable it better (i) To evaluate proposals for cost reduction, and (ii) to point out where it should make all efforts for cost reduction.

Costing of the Issues to Production Department

Some of the methods of costing are : First in First Out; Last in First out: Standard cost; Basket stock method; Market price at the time of issue; latest purchase price; Replacement or current cost are some of the methods used in costing the issues to production.

(i)   **FIFO:** The assumption made in First-in First-out costing method is that the oldest stock is depleted first. Therefore the time of issues, the rate pertaining to that will be applied. This is logical in the case of items which deteriorate with time. In FIFO process, the value of the stocks held on hand is the money that has been paid for that amount of stock at latest price levels and hence can straightaway be used in balance sheet reflecting the true value.

(ii) **LIFO:** The assumption made in Last in First out is that the most recent receipts are issued first, we can apply LIFO system in a period of rising price. The latest prices are charged to the issues thereby leading to lower reported profits and hence saving in taxes. However the LIFO systems have the same disadvantages as that of the FIFO systems.

Average Cost Method

In this method, the issues to the production department are split into equal batches from each shipment at stock. It is a realistic method reflecting the price levels and stabilizing the cost figures. As more purchases are made, new average is computed and this average is applied to the subsequent issues.

Market Value Method

This method is also known as replacement rate costing. Here the materials that are issued are costed at the market rate prevailing at the time of issue. This method requires continuous monitoring of the market rates for all materials and hence it is unwieldy.

Standard Costs

In this method a standard rate is determined based on detailed analysis of market price and trends. Efficient use of materials is truly reflected by adopting this method as the accounting is divorced from fluctuations in rates. In this method the standard rate is kept fixed for a definite period of six month or more.

Costing the Closing Stock

In this method the guide line used here is that either the market price or stock at cost, is to be used, which ever is less. The factors which determine the cost of closing stock are price levels, obsolescence and deterioration. While evaluating the closing stock costs, some stock items, such as machinery spares, tools etc. tend to become obsolete earlier than others. In evaluating the cost of stocks, at the close, a provision therefore should be made to account for obsolesence.

27

Transportation and Traffic Management

Introduction

Physical Distribution is concerned with getting goods from the manufacturer to customer. While physical distribution usually involves substantial expenditures for warehousing and internal materials handling, the biggest element in the process is usually "Transportation". Efficient handling of the transportation of materials is of major importance to industry and consumers, who must ultimately pay for the charges involved. In the typical manufacturing company transportation services are the third greatest expenditure, only purchased materials and labour are more important.

Transportation is of particular importance to purchasing personnel for several reasons. The companies organised under the materials management concept often include transportation as a function of that department. However, complex regulations make transportation purchasing sufficiently different from conventional purchasing activities so that in large companies it is usually not carried on in the purchasing department but in a separate traffic department. Because of the complexities traffic management is often placed in a separate department. This department's responsibility include; incoming fright; outgoing fright and internal plant transportation. The traffic matters of greatest concern to purchasing include; Ascertaining rates; Selecting Carriers and routing; Auditing freight charges and preparing claims; Tracing Lost or overdue shipments; Determining cost reduction possibilities by consolidation of shipments and Determining Cost reduction possibilities by consolidation of shipments and many special arrangements available from shipping concerns. The important aspects in traffic and transportation management are choice of mode of transport, route selection, rate verification and auditing, management of claims as well as application of linear programming to minimise transportation costs. Let us consider the various transport avenues that are open to the materials managers viz Shipping; Rail. Road and others.

Transport by Shipping

This is generally resorted to in case of imports. At times method is used for movement of materials from one part to another within the country. Shipping transport is used for carrying such vital commodities as foodgrains, oil; etc. The increased use of shipping calls for more birthing facilities at our ports viz Ports of

Bombay; Calcutta; and Madras. The materials handling facilities at ports also require improvement as at present unloading process is slow due to inferior handling device and labour problems at ports.

Rail Transport

The Railway network is Asia's largest and the world's fourth largest system. There has been a continuous modernising of the track locomotives, signalling equipment and other devices for monitoring and control. However, over the years the percentage of freight traffic handled by road has steadily increased. Railways are ideally suited for long distance over 800 kilometer freight transportation with in the country. Quick Transport Service has been introduced to overcome the consumers' apprehension about undue delay in the transport of goods by introducing a guaranteed Quick Transit Service on payment of a small surcharge which is refundable in case of consignments do not reach destination within the target time. In order to compete with the door-to-door service of road transport, the Railways introduced a scheme of road—cum—rail transport in containers of five tonnes capacity ensuring door to door delivery as catered for by road transport. Recently the Railways have introduced Parcel service for collection of small parcel from specified points through Mobile Booking Service. The introduction of an increased number of parcel expresses for quicker transport of perishable and other consignments at low freight rates in coaching vans is receiving the attention of the Railways.

Road Transport

This is an important mode of transportation of material. There are a large number of transport organisations in the country having a fleet of trucks. They are suited for transporting freight of the other of 5 to 10 tonnes usually over distances of 300 to 500 kms. There are many constraints to the movement of freight through road. Some of them are octroi, interstate permits, check posts. There are other areas of movement of materials viz through inland waterways, pipelines and air etc. Shippers of bulky relatively inexpensive raw materials try to ship by water or pipeline when ever possible in order to keep transportation, costs at a minimum. In fact, chemical, aluminum, steel and power plants often are intentionally located at deep water ports or on inland waterways in order to reduce the cost of inbound raw materials. Water-shipment is slow but the rates always are low.

Routing and Delays

Once the carrier is selected, the traffic manager's job would seem to be complete but it rarely is. The carrier does not necessarily carry out its part of the bargain. Trucking companies do not always have tracks immediately available and, if they are busy, do not maintain their committed schedules.

Tracing

The supplier's traffic department may trace shipments for its customers. The buyers follow up with the supplier's sales department. The sales department may give the appropriate bill of lading or way bill number to the buyer so that his own traffic department could trace the shipment or it might do the tracing for the customer.

Transportation Rates

The basic principles on which freight rates are predicted are quite simple. The rail roads are still a starting point in rate making today. Implicit in this approach is an assumption regarding the price elasticity of demand for the service. If demand is inelastic, rate can be set high without a significant loss in total revenue to the carriers. If demand is elastic, the reverse is true. In fact, charging what the market will bear is only sound a procedure for setting the upper limit on rate determinations. Follow are some of the factors responsible for setting the transportation rates:

(i) **The Value of the Product:** More valuable the product the higher the rate can be.

(ii) **The cost of hauling:** Product density is related to the ability of a carrier to obtain complete loads. Similarly the factors such as Experience of past claims on the commodity; Space required (More the space required higher the rates); Packaging Geographic factors (traffic density at the origin and destination; volume of traffic moved (Quantity discount); Distances of goods must be transported; Special facilities required (such as stopover facility).

(iii) Competitions of the transporters viz. intermodal competitions; Intra-industry competition (Number of trucklines).

(iv) Government Policies and regulations regulating transportation.

Like other prices, tariffs charged by common carriers all influenced by supply, demand and cost of production. They also are influenced by peculiar variables of their own. Like most other sellers of goods and the services, common carriers are not free to set their own rates nor can they operate whenever and wherever they please.

Cost-Reduction Opportunities

As the rate structure is very complex, there are many opportunities for traffic experts to make tremendous savings in buying transportation. Every well managed company audits its freight bill and make enormous savings as a result. The rate structure is so complex that errors are inevitable. Even after freight bills have been audited once, it is still possible for a skilled auditor to detect errors. For this reason every large size company sends bills that its own auditors have checked to an independent auditing firm for second review.

Variations within Established Rate Structures

In addition to differentials existing between carriers, there are also rate variations which can be classified as (1) class and commodity rates, (2) Rates related to specific shipping practices and (3) Rates relating to commodity descriptions.

Commodity Rates

The result of this process, a freight classification is an alphabetical list of articles, together with each commodity's "rating". The rating is the class into which the article is placed for rate purposes. The use of this classification system greatly reduces the number of freight traffics in which rates are published. Tariff is the term used for schedules that contain rates and charges, as well as the conditions governing the movement of products or persons by commercial means.

Rates related to Shipping Practices

Quantity Discounts: A commodity will have a lower transportation rate if the quantity shopped meets truck load weight. The minimum rates vary with the cubic densities of the commodities and the specific modes of shipping.

Rates related to Commodity Description

It has been observed that material changes, description changes or regrouping of commodities can result in transportation economies. When various materials or components are included in a shipment the highest commodity rate will apply for the entire shipment. Lower rates can after be obtained by careful separation of materials.

Carrier Alternatives

There are three types of carriers, common carriers, contract carriers and private carriers. The private carrier provides transportation for his own commodities in company owned or leased equipment while common carriers and contract carriers are regulated by government regulations. However the regulations pertaining be contract carriers are less stringent than those pertaining to the common carriers.

Free-On-Board Terms

The designation of F.O.B. indicates the point of origin from which freight charges are calculated. The routing of shipments has economic and legal significance to buyers, carriers, and shippers. In general, the designated F.O.B. point determines who pays the freight charges and also who has legal titles to the merchandise while it is in the hands of the carriers. If no qualifying phase is added the teem F.O.B. is generally understood to be at the point of origin of the shipment. It is at the F.O.B. point where title changes hands and form which freight charges accrue to either buyer or skipper. For a shipment made F.O.B. destination there responsibilities fall on the seller. As indicated, freight rates are subject to varying degrees of flexibility, depending on many variables. It is imperatives that skippers know enough transportation and traffic management to understand that rate econo-mies can be achieved through careful classification goods.

Packaging and Materials Handing

It is often practical to equip or company trucks with special racks or other materials handling devices to cut packaging and materials handling cots. Since carriers are responsible for damage to goods in transit, they naturally insist that shipments be securely packaged. In some cases, it is possible to ship in lighter, cheaper containers with little increase in damage. Experience has shown that damage losses can be cut by so percent if merchandise is handled more gently in company—operated trucks.

Damage Claims

Common carriers are forced to pay huge amount in damage claims. Even if the skipper is reimbursed in full, damaged shipments cause production delays, and the processing of claims is a huge burden on the typical firm's traffic department. Not all claims are the carrier's fault rather carrier's maintain that most of the claims occur as a result of inadequate packaging. The exact cause of damage is not clear cut in many cases. The shipment would certainly not have been damaged it it had been handled with sufficient care and merchandise been packed more carefully in costlier containers. Traffic departments spend a great deal of time investigating damaged shipments and making claims against carriers. Similarly the traffic department work with company's other departments such as packaging engineers etc. to reduce damages. The procedure for claiming damages is fairly straight forward, where the skipper presents the carrier with claim. Skipper gives him his bill of lading, the paid freight bill and a copy of supplier's invoice or other evidence that establishes the value of the damaged merchandise. The claimant may include photographs of the damage to serve as a basis for the development of ways to prevent future damage. The carriers must include an allowance for damage claims in their tariffs and it is to everyone's interest to help minimise such claims.

Demurrage Charges

Demurrage charges are penalties assessed by carriers on vessels held by or for a consignor or consignee beyond a stipulated free time provided for loading or unloading. Most business concerns operate under what is known as an "average agreement" with respect to demurrage charges. The gist of such an agreement is that a receiver is given a credit for each vessel released before the expiration of the first 24 hours or 48 hours as the case may be as free time. This credit can be used to off set a debit of one or two day's demurrage.

Insurance Buying

The materials manager has to arrange insurance of materials in transit and in stores scientifically, so that the possibility of catastrophic and profit shattering loses either to the organisation or to those dealing with the organisation or to those dealing with the organisation like the suppliers and purchasers is avoided. The responsibility for decision making and management of insurance is handled by an insurance department headed by a specialist in insurance.

Insurance Management

This aspects of insurance management that affect materials management are: (a) Arranging an adequate insurance cover for materials in transit and storage (b) Claims management; and (c) reduction of insurance costs.

Marine Insurance

The insurance department deals with insurance of Transit of incoming material; storage;and transit of outgoing materials including finished goods. Insurance of transit of materials is known as marine insurance and the same holds true for air consignment and inland transit by road or rail or other means of transport. The marine policy for covering ocean voyage is issued in a standard form and the terms and conditions of the standard policy are modified by a set of clauses known as Institute Cargo clauses which have also been standardised by the Institute of London Underwriters. The marine policy is a valued policy; that is the sum at which cargo is to be valued is left at the discretion of the insured. The cardinal principle of insurance law is that a contract of insurance is a contract of indemnity and so the law does not permit the insured to make a profit out of an insurance claim but in case of marine policy, it is possible for insurance to be taken for the invoice value of the cargo plus the amount of import duty and a reasonable amount of profit and incident expenses. All marine policies are issued subject to the transit clause which is part of all Institute cargo clauses. The various clauses impose the following limitation.

(a)   The insurance cover cease as soon as the goods are delivered to the consignee or place of storage at the destination named in the policy.

The cover ceases on expiry of 60 days after completion of discharge of goods from the over seas vessel at the final port of discharge. Some of the important considerations are:

(i)   The sum insured take care of invoice value, duty, incidental expenses and reasonable profits.

(ii)   The scope of the cover should be clearly defined.

(iii)   Where the value of the container is high, it is better to be clear as to whether the cost of the container has to be paid or not even when the contents are not damaged.

(iv)   Where importing is a continuous process it is advantages to take a floating policy or an open policy instead of a specific policy for each consignment.

Inland Transit

The same policy form is used for covering internal transit but the clause attached to the policy are different and the scope of the cover is also different. The term of the cover should be as wide as an "all risks" policy or as restricted as "road risks or rail risks policy".

The nature of packing, the mode of transport and to extent of care exercised in materials movement have a bearing on the probability of losses occurring and on the quantum of such losses. A systematic and periodic, preferably an annual review of the existing insurance arrangement of the organisation, the cost benefits should be undertaken. The insurance cover designed with care today may not remain a perfect fit for the future ever after 2 years.

28

Cost Reduction and Materials Management

Introduction

The Materials Manager should have a program for the department to achieve these goals. In addition, he should require each of his key subordinates to develop programme to achieve similar goals. Authority to develop a program should be delegated to the lowest possible level in the organisation. In doing this materials manager ensures that the efforts of everyone in the organisation will be directed toward goals that are consistent with those of the whole department and of the company. The materials manger's plan is a composite of the plans of his subordinates and would also be submitted annually. If accepted by the top management it would become a standard against which the department's performance would be measured.

There are four basic steps in measuring the performance of any department: Define the limits of the job; Determine the objectives to be achieved within these limits; Develop a program to meet these objectives; comparing progress with the objectives.

The limits of the job are defined by the organisation structure and policies and procedures. The basic objectives of materials organisation are: Low Operating cost. Low prices of purchased materials; Minimum Investment in inventory; Superior supplier performance; Developing materials personnel; Good records. The performance measurements of individual parts of the plan might be made annually; quarterly; monthly; weekly, or even daily. In general, performance in achieving the qualitative goals would be measured less frequently than that in achieving the quantitative goals. The quantitative phases of performance measurement get more attention for several reasons. Among them are the following:

(i) **Quantitative data are more objective.** They are based on operating statistics where as the qualitative data reflect in large part the opinion of the observer. It would be hard for a materials manager to distort performance on cost reduction, where progress is measured in rupees.

(ii) **Quantitative Data Change.** The materials management goals and objectives can only be measured qualitatively are long range in nature. The progress on quantitative goals should be measured frequently so that remedies can be applied immediately should performance fall short of expectations.

(iii) **Quantitative Data Related to Company Performance.** Top management has an immediate interest in materials management goals such as cost reduction and inventory turnover which can be measured quantitatively.

Material Cycle

The material cycle commences from the time when its need is felt in the plant to the time when it is supplied either to stores or to the production service department. At the stage of materials planning, the requirements of all types of materials are estimated for a predetermined period. The industrial engineer, the production engineer, the R & D specialists; the designer and buyer all Joint hands in carrying out an effective value analysis. All repetitive items, their purchase, procurement, replenishment etc may be controlled through the application of ABC analysis. But non-repetitive items need a different treatment as the requirement for them is generally unpredictable. In this case the modern techniques of PERT and CPM analysis can be applied. In order to achieve better results, ABC analysis especially for 'A' and 'B' items, ignoring 'C' items. Application of value analysis in materials planning phases may be considered in two stages : Pre-design value Analysis and Pre-purchase value analysis. The materials management department is in a key position to effect substantial saving in materials cost. An efficient materials department should aim at obtaining: Right Quality; Right Quantity; Right price; Right source; Right delivery. These must also be constant market research and systematic effort for developing new sources of supply which would result in obtaining better quality at lower price, ensures earlier deliveries as also provide a second line of defence in case of a sudden dropping out of a regular source.

Buyer-seller collaboration and standardisation are the two other important aspects. Clear specification ensures right quality. Moreover, more than one manufacturer can supply standardised items, thus providing better availability, better price and better delivery. Since transportation cost constitute a large percentage of total material cost, the area also provides substantial saving if the principles of value analysis are applied here. Similarly for large volume transportation, the techniques of operations research could be applied and possibilities of cost reduction located.

Concept of Cost Reduction

Cost reduction is our prime concern but it should not be at the cost of product quality. Other values like use value, exchange value and esteem value of the product must be given equal importance. Every firm must try to continuously increase its productivity through cost reduction and quality improvement.

Direct costs. Direct cost of material is an important element in the total cost of product. It varies from 50-80% of the total unit cost. In the control of this material cost the method engineers, materials planners, industrial engineers, design engineers; who deal with direct material can play a significant role. They determine standard cost of the product, the Labour Cost; both "direct and indirect is an important element of cost. Performance standards may be set up through "Time and Motion" study. Actual performance should be compared with standard performance and in case of variation. Steps should be taken.

Indirect Costs. Overhead expenses such as indirect Labour, indirect Materials, Administrative and Selling expenses are controlled by budget. For the control of Indirect Materials, norms of consumption must be set up.

Works Cost and Works Overhead. This includes expenses on indirect labour, indirect material, power, rent, repair, depreciation etc. They are all indirect expenses right from the receipt of order to the despatch of finished goods.

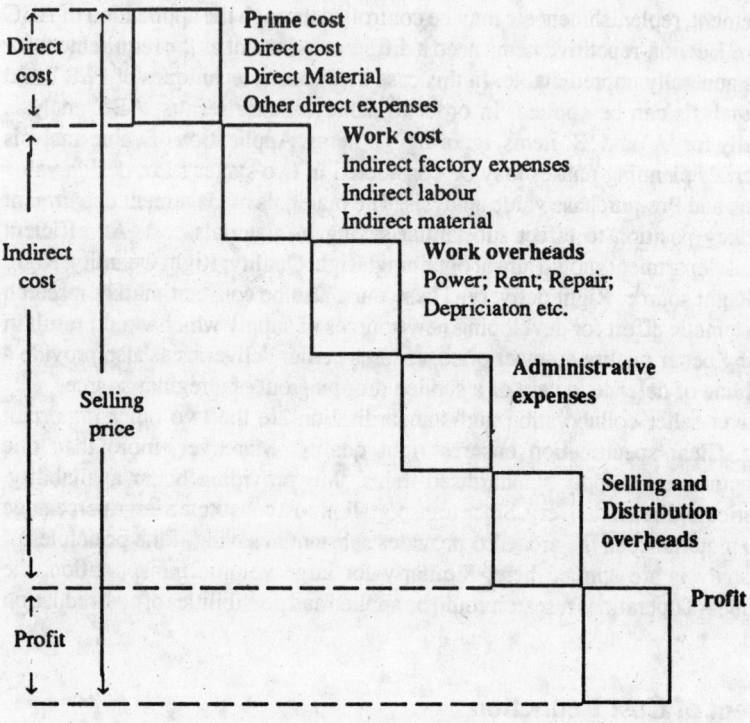

Fig. 28.1

Administrative Overheads. Direction, control and administration of an undertaking, rents and rates of office premises and furnishing if any, office lighting, salaries of office staff, stationary, postage and other expenses are covered under Administrative Overheads.

FACTORS CONTRIBUTING COST REDUCTION
Materials
★ Use of proper size and type of material
★ Vendor evaluation

★ ABC Analysis
★ Reduction in lead time as far as possible
★ Plan for minimum inventory levels to avoid stockouts and maximum inventory levels to control inventory carrying cost.
★ Proper storage to avoid change and ensure quick availability when required.
★ Minimum use of indirect material
★ Dispose of nonmoving inventory and waste to reduce the carrying cost.

Similarly cost reduction can be achieved in the area of Manpower; Machines and Equipment, Energy Conservation, Lands and Buildings and Time Management, which are given as under.

Manpower

★ Optimum utilisation
★ Select right person with right skills.
★ Train in the desired skills to meet the challenges of fast changing techno-economic environments.
★ Recognition of work and proper motivation for best performance and reduced absenteeism.
★ Better working environments.
★ Job redesign; Job enrichment and career planning.

Machines and Equipment

★ Proper maintenance so that (i) Downtime is minimum (ii) Full capacity is available.
★ Proper job on right machines.
★ Replacement at proper time.
★ Use of proper jigs and fixtures to reduce setting time and operating time.

Energy Conservation

Improving fuel consumption in boilers, furnaces and engines, effective plant maintenance, adoption of heat recovery system.

★ Good housekeeping in Industry gets priority.
★ To ensure compliance with energy conservation norms would be particularly worthwhile.

Lands and Buildings

★ Scientific layout to optimise area utilisation.
★ Better vertical space utilisation wherever possible.
★ Design of buildings to make use of natural light and ventilation.
★ Make use of gravity flow/feeding of materials and equipment wherever possible.

Time Management

★ A check list for the top management

★ Are company policies, objectives and mission well defined and made known to staff/employees?

★ Are responsibilities clearly defined and assigned?

★ Is there an established executive development programme?

★ Are vendors rated for services and quality?

★ Is a full organisational analysis made periodically to evaluate competence and redundancy?

Cost reduction embraces unit cost reduction by expenditure reduction in respect of a given volume of output and/or unit cost reduction by the increase in productivity. The another way of looking at cost reduction is to view it as the process whereby permanent savings are made without any reduction in the quality and/or usefulness of the products. There is a need to develop an attitude of mind which challenges all standards with a view to their improvement.

Areas of Cost Reduction

The success of any business depends primarily on the efficient use of those basic cost elements viz electric energy; per unit of production of goods; man-hour of labour; weight of raw material etc. The various constituents of total cost are shown in following diagram.

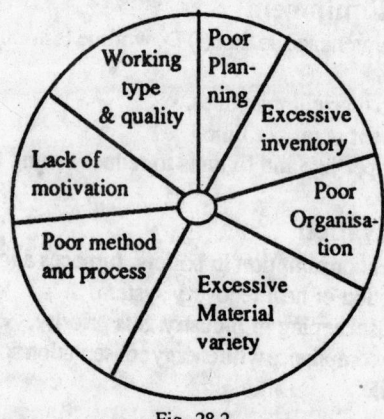

Fig. 28.2

Control of inventories of raw materials, semi finished and finished goods is essential in a well managed business. Speculation in inventory appreciation is hazardous and has no place in manufacturing. Integration of industrial processes under one management can reduce production costs through control of raw materials and selling costs through elimination of middle men.

29

Make or Buy Decisions

Introduction

This is an important policy decision of management and purchase department has a role to play in both cases. After preliminary investigation of cost a detailed study may be needed which can involve engineering, design, production planning, costing as well as purchasing. The final decision on a make-or-buy question is made by the joint efforts of several departments, the purchasing department will always have an important part in the determination. It needs no emphasize that no simple rule can be applied to all cases of make or buy. Some decisions are major, such as a paper manufacturing firm deciding whether to build or buy a pulp mill to supply its own raw materials or a pharmaceutical company to manufacture or buy the basic bulk drug from an outside. Each such case must be decided on its merits, but the important issues may be different in the two cases. In broad terms four relevant considerations are the elements in all purchasing decisions: Quality; Quantity; Cost; and Service aspects.. Make or Buy decisions are taken at the top level and the position of materials manager in the organisation hierarchy determine whether he is to play a key role or supporting role to deal with the problem of Make or Buy. Few organisations have formed committees to analyse and to recommend proposals. Make or Buy decisions arise due to a variety of reasons. Mostly such situations arise due to consistent failures on the part of the suppliers in the supply of items. The management then decides to make the item. Sometimes availability of items from outside suppliers in the supply of items. The management then decides to make the item. Sometimes availability of items form outside suppliers may not be adequate owing to staff competition from many end users.

Importance of Make, Buy or Lease

Every organisation has a choice of following three basic decisions in sourcing a new product:

(i) Purchase the product complete from a contract manufacturer.
(ii) Purchase some components and manufacture the balance.
(iii) Manufacture the product completely starting with the extraction of basic raw materials.

In practice no company can seriously consider the third alternative. Normally large companies rely on outside suppliers for at least some components and materials. Some companies choose the first alternative and obtain a new product completely from an outside supplier. The general rule, however is that an organi-.

sation will make something and buy others. This applies even to service organisations. For manufacturers, make or buy decisions must constantly be made and it is largely a matter of deciding at which stage of fabrication a component should be purchased.

Make or Buy Criterion

Companies will usually buy a finished component from an outside supplier when:

(i) They do not have facilities to make it and there are more profitable opportunities for investing company capital.

(ii) existing facilities can be used more economically to make other components.

(iii) Patent or other legal procedures do not allow the company to make component.

(iv) There is only seasonal demand of the components.

In addition following are some of the considerations which have a say in case for decisions "to buy".

(a) Drawing needs modifications.

(b) Quality to be manufactured.

(c) Fluctuation demand.

(d) Special manufacturing techniques involved.

(e) Frequent changes in the design involved.

(f) Jigs and tools, gauges etc are to loaned.

Companies prefer to do their own fabricating. They buy only raw materials or semi-finished parts.

(i) Components can be made more cheaply by the company than by outside suppliers.

(ii) The fabrication is vital to the company's product and requires extremely close quality control.

(iii) Component can be readily manufactured with the company's existing facilities.

(iv) Fabrication requires extensive investment in facilities that are not available at supplier plants.

(v) Product has a demand that is both relatively large and stable.

In addition following are some of the considerations which have a say in case of decisions "to make"

(a) Prices obtained are the best.

(b) Quantity to fabricate is optimum.

(c) Some other items can be produced along with the product and thus reducing production costs in the whole.

(d) The material to be fabricated is readily available.

Other Important Factors

Some companies, by tradition prefer to make almost every component of their products. Other prefer to buy as much as possible from outside suppliers. In general, an aggressive company in an industry that is expanding very rapidly will prefer to buy many of its components from outside suppliers. In some companies making an item is effectively prevented by its own labour relations policies. If it is extremely generous with wages and fringe benefits for its employees, there will be some industries where it simply is not competitive even if its methods are better than those of potential competitors. When a company decides to make an item that it has been buying, it always eliminates the suppliers sales expenses and usually eliminates some freight cost. If its production costs are identical to those of the suppliers the company also gains the suppliers profit and usually at least part of the fixed overhead expenses incurred by the supplier.

Illusory Savings

Companies often continue to buy the item even though they may be capable of making it. In most cases it is not to their advantages to make the item for a number of reasons, including:

(a) **Start Up Cost.** The company's direct costs may be substantially higher than the direct costs of the supplier. The supplier presumably is one of the most efficient producer of the item and has achieved his existing costs only after management have acquired considerable experience. Assuming that buyer can produce as efficiently as the supplier, his saving will be reduced by **start-up cost.** No company can immediately produce a new item at peak efficiency, costs on initial production runs are always above standard.

(b) **Flexibility.** Company which has invested in equipment to make the item has less flexibility with make items. But if the company has not invested in equipment, it can exploit changes in market conditions on buy items and take advantages of lower prices offered by suppliers eager for new business.

(c) **Savings in Overhead.** Savings in overhead are often illusory. When business expands to capacity operations, fixed overheads increases and must be absorbed by each component that is produced. Similarly if business slumps, fixed overhead is under absorbed and unit cost rises. It has been observed that fixed overheads rarely rises immediately after a company starts using idle facilities to make a part it has previously purchased but eventually it creeps up.

(d) There may be better profit opportunities in other areas. Each make part requires some investment in facilities, inventories, and top management time. Even though this investment is profitable the company may be able to earn even higher by using its resources to expand the business it knows best.

(e) Some companies decide to buy items that they have been making. In some cases they even dispose of plant and equipment used to make the item. Such decisions are almost economic if the suppliers price is lower than that of company's direct cash cost of making the part. Similarly when a company facilities are temporarily taxed beyond capacity, it may decide to buy part of its needs from a outsider supplier.

Factors Governing Make or Buy decisions

Following are the four factors in the context of make-or-buy to show how it might bear on the decision.

Quality Consideration

It has been observed that quality available in an item being purchased differs in no important respect from the quality the company could impart to the same item if it make it. However often there will be sufficient quality differences between the two alternatives to warrant weighing the quality factor in making the decision. If the supplier consistently fail to meet the quality requirements, then the firm may decide to make it. Firstly, it is essential to find out whether such exacting quality requirements are necessary at all. Another important aspect is whether the buyer will be able to make the item to such quality standards when established vendors are unable to do so. A vendor enjoys the benefit of specialisation and probably the item in question is his main product line. Buyers also have the option of changing the supplier and hence there is lot of flexibility.

Quantity Requirements

When the quantum is small, such as specialised jigs and fixtures, it may be worth while to make them especially if the existing facilities and the little amount of sub-contracting is adequate for such manufactures. It is not paying proposition for the supplier to accept such small jobs. Sometimes demand for items may not be certain and it may be advantageous to make it using a flexible production schedule. When the requirements are very large, buying would be desirable.

Cost Aspect

Advantages and disadvantages will have to be quantified and a cost analysis has to be done. The problem of estimation accurately the cost of making an item stems from the nature of overhead costs manufacturing plants. If the make or buy decisions were concerned with an item that was clearly separate and apart from all other items made in the plant, the allocation of overhead would be a simple matter.

Fixed costs do not vary with output and are often referred to as overhead or unavoidable costs. They include such items as management salaries, taxes, depreciation, and insurance. Variable costs fluctuate with output and are often referred to as direct costs. Examples include the direct labour and raw materials required to produce additional units of output. In evaluating the estimated cost of making an item, it should be recognized that the accuracy of estimate depends on the skill and case with which the cost estimates are compiled. The smaller the item in terms of either value or quantity, the more difficult it is to arrive at an accurate cost figure. A plant with excess capacity can increase production by merely adding incremental costs that are variable in nature, where as a plant being fully utilised can increase production only by increasing its capacity, thereby adding incremental costs of both fixed and variable nature.

Similarly cost of transportation; economics of scale and cost of capital must be given a thorough analysis in arriving at a make or buy decision. An adequate return on investment should be considered as a cost of making an item because money invested in raw materials and equipment is tied up and therefore a real cost, since other opportunities must be forgone. Management should ask itself; If as a result of a buy decision, we can free money? what return can we expect from it? On the other hand, if a make decision required an increase in investment, what will be the implicit cost of this money? It must also be remembered that a firm that produces for itself may have only itself as a customer, and therefore its production runs must be large enough to take advantage of economics of scale.

30
Value Analysis and Materials Management

Introduction

Value Engineering and value analysis mean the same thing and are synonymous terms. Value Engineering or value analysis had its birth during the Second World War. Value Engineering comprises a group of techniques aims at the systematic identification of unnecessary costs in a product or service and efficiently eliminating them without detriment to its quality and efficiency. It is a technique which is looking for the function; seeking alternatives; and developing it through practice and test till the cost-wise idea is ready for implementation.

Value analysis involves a creative approach for the systematic identification of unnecessary costs. If these costs are first identified and then systematically eliminated, the overall cost of the product or service can be considerably reduced, and the sales and resulting profit considerably increased. Value analysis is an effective tool for cost reduction and the results accomplished are far greater. It improves the effectiveness of work that has been conventionally performed as it examines into the very purpose, design, method of manufacture etc of the product with a view to pinpointing unnecessary costs, obvious and hidden.

Definition

There are almost as many specific definition of value analysis. Value analysis is the study of the relationship of design; function and cost of any product, material or service with the object of reducing its cost through modification of design or material specification, manufacture by more efficient process, change in source of supply, or possible elimination or incorporation into a related item. Value analysis is the organised, systematic study of the function of a material, part, component, or system to identify areas of unnecessary cost that can be eliminated without impairing the capacity of the item to satisfy its objective. Value analysis focus is on engineering, manufacturing and purchasing with attention to one objective— equivalent performance at a lower cost without affecting functional performance. Value analysis is to buy a function rather than to buy an item. Value analysis techniques are important because of competition; profit squeeze; value of a rupee saved, etc. Value analysis means better value through better purchasing techniques, better suppliers, design, re-evaluation and modification, better manufacturing methods, lower cost office operations, standardisation, substitution, better material handling, better traffic operations and better inventory control.

Applications of Value Analysis

The implementation of a value analysis programme is the selection of products to be analysed. Value analysis makes its impact during the maturity stage of the product. The two approaches for the choice of products for value analysis are through ABC analysis and contribution analysis. It involves the listing of the products in descending order of their annual consumption/production sales. Now select the first 10% of the items accounting for around 70% of the annual turnover of the company. Thus maximum efforts can be directed on these items in order to bring about cost reduction.

In the Contribution Analysis the products are ranked according to their contribution (Sales price-variable Expenses). The product with the least contribution per unit are selected for value analysis as these have a higher cost compared to their sales price. In India several public and private sector undertakings have applied value analysis techniques successfully for cost reduction and import substitution purposes. The use of aluminium in place of copper in the cable manufacture has helped the country to conserve valuable foreign exchange. Similarly aluminising in place of galvanising has helped reduce the imports of zinc.

Vim was packed and sold in market in tin packing, thereafter it was substituted with cardboard packings, saving a lot of money to the company. Value analysis is gold mine for developing country like India which should endeavour to exploit this technique to the fullest extent. This technique originally evolved when Armed forces were faced with shortage of supples and rising cost. It is indeed a valuable technique in the hand of Industrial Management.

Value analysis is neither a cost nor price. It involves some thing more than cost or price. it is relative value, can be defined as 'Lowest price" for a function at the desired time and place with the required essential quantities. It is an inquiry into the functional value or utility of a product for the particular endure. It is a matter of measuring a cost of performance of an item and to determine how a performance can be improved by improving the value by a process of substitution or modification. Value analysis does not necessarily means reducing a price of an *article, at time* higher price articles may be substituted to improve the performance of article, reducing the ultimate cost and thus improving the value. Essential feature of value analysis is the process of identifying unwanted cost and steps to illuminate these costs. Each article purchased has specific function to perform and therefore its utility for performing the required function can be related directly to the price paid for it. Every article has more than one function to perform. These functions can be : Primary functions, Secondary functions and Tertiary functions.

Value analysis is a tool for assessing the value of end products as well as value of incoming materials. It is an important tool of 'Material Management". it is a scientific technique designed to make possible the same or better performance and customer satisfaction at lower cost. Frequently the production is improved by the process. Value analysis may be applied to the design of a product or a component since design changes will often simplify production and make it possible to eliminate some of the processes used.

What is Value?

Value is a broad term often used to denote cost and price. The value of an item is its worth. The various aspects of value are: Use value, Esteem Value, Exchange Value and Cost Value. The emphasis is on use value and at times cost and use value are confused. The value concept possesses tremendous profit maming potential and .epresents and attempt to integrate the commercial skill, manufacturing skill. Design analysis and cost analysis constitutes the two fundamental tools of value analysis.

Value can be expressed as :

$$V = \frac{W}{C}$$

$$Value = \frac{Worth\ of\ an\ item}{Cost\ of\ price\ of\ an\ item}$$

The worth and cost have to be worked out subjectively from the point of view of the buyers or sellers or manufacturers. Worth is determined in terms of functional utility, esteem time, place factor and special advantages if any. There are three ways in which value can be increased by increasing worth, decreasing cost for the same worth and by doing the above two ways.

Value can be classified as:

(a) Use of functional value: The properties and qualities which accomplish a use, work or service.

(b) Esteem Value :The properties, features or attractiveness which cause us to want or to keep it.

(c) Cost Value: The sum of labour, material and various other costs required to produce it.

(d) Exchange Value: Its properties or qualities which enable us to exchange it for something else we want.

Thus value analysis is like a three legged stool with three legs as Performance, Delivery and Cost. All the three legs must be perfectly matched to keep the seat on an even position. The general causes of high costs are: Lack of information, lack of ideas, honest but wrong beliefs, temporary circumstances, habits and attitudes.

Value Analysis and Materials Management

The various analysis process on value develops many questionnaires and ultimately reaching the optimum solution. These analysis are based on its function values. If we analyse the total cost say (100%) it can be composed of following components percentage wise:

    ★ Bought out (Material and Component) Cost   =   40%

    ★ Direct Labour Cost   =   10%

★ Selling and administrative charge cost = 17%
★ Profit = 8%

On the other hand product cost are compared of 75% in design phase and 25% being materials, tooling, processing and factory overheads. Now from this it is very clear that design phase and Materials phase are controlled than 50% cost can be controlled.

Value Analysis Techniques

The most desirable alternatives for a given application depends on the use to which the product will be put to use. The best product is the one that will perform satisfactorily at the lowest cost. Value analysis procedure essentially follow the same pattern in all cases. First there is an analysis of the design of the item and its uses, and then the application of price analysis to the alternatives.

Value analysis or value engineering, as it is called, is one of the several specialised services recently offered as tools of management. *Value Engineering techniques differs from other cost reduction techniques, in the sense that it probes into the economic attributes of value.* The concept is that by continuous process of planned action, it is possible to improve the performance, increase the value in product and thus reduce cost.

Value engineering is defined as the relationship between what function is worth to a buyer and what it costs to him. There are three types of values having relationship with function (a) Use value; —the monetary measures of the properties or quality of an item which contributes to its salability. (b) Esteem value, the. monetary measure of the properties or qualities of an item which contribute to its salability, but not to its performance. It does not include use value (c) Exchange Value—the monetary measure of the properties or qualities of an item which enable it to be exchanged for something else. The above concept can be put in a formula. Use Value + Esteem Value > Exchange Value.

Value analysis techniques can be applied to the following types of materials:

(i) Raw materials   (ii) Semi-finished product
(iii) Component parts   (iv) jigs and fixtures
(v) Production tools   (vi) Packing materials
(vii) Transportation cost

The objectives of value analysis is to increase the value by either increasing the worth for the same cost or decreasing the cost for the same worth or both. It is very important to bear this value concept in mind, in order to increase the productivity. Value analysis concept and techniques have revolutionalised the entire function of purchase department. An equivalent substitute used in place of imported substance may save a tremendous foreign exchange. Example—corn oil use in place of imported oil used in fermentation industries of India. Use of *Hydrol* (indigenous) in place of *Dextrose* which was being imported. The difference in costs of *two* is about Rs. 2500 per ton.

In Indian industry we are normally finding a problem of either to continue the imports of variety of small items or should become creative and look around for substitute. The substitution is an activity which involve high technical skill which a manufacturer himself cannot create. Since the responsibility of procurement is on him, so it is desirable that he should initiate the programme bye creating the technical department to bring about substitution may even be a sheer wastage. It should not be accepted even for a moment that every item of value analysis results in the successful imports substitution. In India value analysis is still a concept. It is high time that this gestation period should be over and enlightened management of various organisations should accept the concept of value analysis with confidence and should be put in practice.

Thus Value Engineering objectives is designed to coordinate the professional skill of design, production and purchasing personnel in an endeavour to minimize all the elements of costs that are within their control, while at the same time retaining the quality, reliability and salability of the product concerned.

Value Analysis places great concentration on the use value but many times cost and use value are confused. Value analysis is very important from the view point of cost accountant and termed as marginal costing. Marginal costing infact means only variable costing and to accountants marginal cost averaged over a large block of units. In practice analysis of marginal costing included direct costs which are related to significant changes in volume and type of output. It is indeed difficult to allow for all the increases in fixed costs which arise from changes in volume and type of output. Accounts do take into consideration the cost of costing, the time available for making calculations, the degree of accuracy required and other special features of a business enterprise. There are following five reasons why unnecessary costs enter into a product or service:

(a) Lack of information
(b) Lack of utilisation of creative minds
(c) Temporary circumstances
(d) Honest wrong beliefs
(e) Habits and attitudes

The techniques of value analysis refers to those techniques which enable to determine the level of value in product or in expenditure. As a result of the value analysis techniques the value of preforming the functions or service comes to 25% or even more below the price currently spent. These techniques are brought into application through:

✯ Organised approach	✯ Getting information and facts
✯ Employing creative thinking	✯ Converting idea into rupee value
✯ Evaluating functions	✯ Using accurate costs
✯ Working on specific matters	✯ Using good human relations
✯ Considering the rupee value of each tolerances	✯ Using ones own judgement
	✯ Using industrial standards

★ Spending company's money as one's own money.

These techniques are, however, used in a framework which is called value analysis job plan which consists of following steps or phases:

(i) Information phase
(ii) Speculative phase
(iii) Analytical phase (Going back and looking at our ideas and converting to rupee value)
(iv) Programme planning
(v) Programme execution
(vi) Conclusion

Design Analysis

The design analysis procedure entails a methodical step by step study of all the phases of design of a given item in relation to the function it performs. Analysis of each component attempts to answer the following four questions :

(i) Can any part be eliminated without impairing the operation of the complete unit?
(ii) Can the design of the part be simplified to reduce its basic cost?
(iii) Can the design of the part be changed to permit the use of simplified and less costly production methods?
(iv) Can less expensive but equally satisfactory materials be used in the part?

The specific manner in which a value analyst approaches the problem of design analysis is highly creative method and different from one analyst to another. Most companies developed some checklist to systematize the analyst's activity. The following is the Value Analysis checklists suggested by **Material Association of Purchasing Manufacturer** to determine the function of the items:

(i) Can the item be eliminated?
(ii) If the item is not standard, can a standard item be used?
(iii) If it is a standard item, does it completely fit the application or is it a misfit?
(iv) Does the items have greater capacity than required?
(v) Can the weight be reduced?
(vi) Is there a similar item in inventory that should be substituted?
(vii) Are closed tolerances specified than are necessary?
(viii) Is unnecessary machine performed on the items?
(ix) Are unnecessarily fine finishes specified?
(x) Is commercial quality specified?
(xi) Can you make the items cheaper in your plant? If you are making it new can you buy it for loss?
(xii) Is the item properly classified for shipping purposes to obtain lowest transportation rates?
(xiii) Can cost of packaging be reduced?
(xiv) Are suppliers being asked for suggestions to reduce costs?

Cost Analysis

Cost analysis involves the investigation of a suppliers probable costs of producing a given product. It helps in negotiating an original price. In recent years cost analysis plays the two following major roles:

    (i) Cost analysis is conducted for currently purchased items whose costs appear excessive. In such cases the informations developed from Cost Analysis is used as a basis for further price negotiations with the suppler :

   (ii) Cost analysis also serves as a means of locating high cost posts which should be subjected to designs analysis.

Steps of Value Analysis

The technique of value analysis can be grouped as follows:

(i) Eliminates	(ii) Retain or keep
(iii) Change	(iv) Modification
(v) Incorporation	(vi) Subdivision
(vii) Substitutions	

    (i) **Eliminate.** If after the examining of the utility of an item it is found that it is not contributing to the functioning of a job than the item can be eliminated.

   (ii) **Retain or Keep.** Retain only those items whose marginal utility for primary, secondary and tertiary function contribute to the productive examples.

  (iii) **Change.** An alteration in the specification can be achieved in the following profiles:

       (1) Modification   (2) Incorporation

       (3) Sub-division   (4) Substitution

  (iv) **Modification.** At times it is advantageous to modify the specifications or design of an item or job if such modification can be achieved by cheaper and better alternatives than those must be introduced.

   (v) **Incorporation.** We may be able to incorporate two or more than two items for the same job. In other words we may also be able to incorporate two or more than two processes in the manufacture of the items into one single process.

  (vi) **Sub-division.** It is similar to the concept of make and buy. It is often possible to manage the requirements more economically by purchasing component than to assemble it.

 (vii) **Substitutions.** Value concept has a greater reliance to the concept of substitutions from both material as well as social point of view. One of the examples which can be taken with precise metal like copper. Sheets at which copper is being consumed in the world is 8% of its total reserve and with the expanding requirements we may be able to consume the total reserve in another 10 years time. Aluminium conductor can replace or substitute the copper conductor. This way we will be rendering a social value to aluminium and at the same time conserving the value and life of the copper by preserving the reserve for longer time. Value analysis concept has revolutionized the entire function of Purchase Department. In Western India in one of the chemical industry depending upon fermentation technique needed a particular type of oil for its fermenter which was not available in India. Thus the import license were required and the oil was getting rare and rarer. Circumstance and hazardous were increasing day by day. The Organisation Production Deptt. collaborated with Purchase Deptt. and after a series of experiments select a particular type of corn oil which replaced the imported oil equally and efficiently. This way a company can save annually a considerable amount of foreign exchange.

In another example an organization which was consuming 100 ton of dextrose annually. The dextrose manufacturing company in the country were finding it difficult to meet the demand as it requires maize as a raw material and maize became a rare commodity due to the economic and political weather as we were depending upon PL 480 programme of USA. As the programme disturbed thus resulted problem for dextrose manufacturing company. This organization substituted dextrose by another chemical by the difference in the cost between the two is about Rs. 1500/ ton so naturally there was a saving of Rs. 1.5 lacs annually.

In Indian industry we are normally finding a problem of either to continue import of variety of small items or should become creative and look around for substitute. The substitute in an activity which involves high skill which manufacturer himself cannot create. Since the responsibility of procurement is on him so its is desirable to initiate the programme. But creating a technical department to bring about substitute may be even a shear wastage. It should not be accepted even for a moment that every items of value analysis results in successful import substitutions.

Application of Value Analysis

There are various applications of value analysis. A few of them are :

(i) **Value analysis of end products.** The end products offered for sale should be value analysed.

(ii) **Value analysis before undertaking of a project.** In one of the project height between floor and ceiling was reduced to 11 feet instead of scheduled 14 feet height. This resulted in tremendous cost reduction without affecting the functional utility.

(iii) **Predesign value analysis.** In developing country like India the importance of predesign value analysis cannot be overlooked. It may bring about a substantial cost reduction and foreign exchange saving by the creative design. Blind copying of foreign design has often led to extra cost in procuring material cost, unwanted expenditure of foreign exchange, high inventory and wasteful methods of production.

(iv) **Value analysis before tendering.** Many firms are now introducing this concept and subjecting their tenders for big jobs or project to value analysis before submission so that they can offer better value to their competitors.

(v) **Prepurchased value analysis.** This is the last stage to use this concept. This is the final chance to save money and to obtain better value before we make over final commitments to prospective suppliers. Example. In ABC analysis, first point is to handle carefully item 'A' whose value is 90%. Thus value analysis in Industry is the latest technique available for producing at a lowest possible cost. It combines price analysis and product analysis.

(vi) **Price analysis.** It may be defined as buyers investigation of suppliers direct and indirect cost of manufacturing the product. This will help buyer to negotiate few prices with the suppliers near the actual (closed) cost of production.

(vii) **Production analysis.** It refers to buyer's investigation to determine the adequate quality of material. Its primary objective is to change the specification of product to correlate the product design with a most economical production sources of supply.

Stages in Value Analysis

The value engineering exercise proceeds systematically through following six stages. These are :

(1) Organisation (2) Evaluation
(3) Speculation (4) Investigation
(5) Recommendation (6) implementation

1. **The organisation stage.** It is composed of *product* selection and its analysis.

2. **Evaluation stage.** This is also called Information stage. It is desirable to design an **evaluation sheet** on which to record all the collected information. The information that is to be recorded on the evaluation sheets includes:

   ★ Design ★ Manufacture
   ★ Purchasing ★ Costing
   ★ Derived information

3. **Speculation stage.** Speculation stage can be regarded as the heart of the matter. The essential technique of this is functional approach. This has the following steps :

   (i) The functional approach to cost reduction.
   (ii) Assessing the value of a function.
   (iii) Alternative actions following functional assessment.
   (iv) Further studies during the speculation stage:
       (a) What are the functions, whether essential or can be eliminated?

(b) What material is used or proposed or can another material be used?

(c) What factors control the amount of material used?

(d) How much of the basic material is wasted during manufacture?

(e) Is the component made from standard raw material, or is the component itself standard?

(f) Through what operations does the component proceed during manufacture?

(g) What labour operations are involved?

(h) What tolerance have been specified and if so, why?

(i) What surface finish is required to meet the function?

(j) What is the "brought out" content of the cost? This division is based on make or buy.

Thus the crux of speculation stage is only *creative thinking*.

4. **Investigation.** It has:

   (i) **Design Contribution.** This include change in design, the raw material used, the dimension and limits etc.

   (ii) **Production Contribution.** Covers process of production, use of standard components, reduction of labour costs etc.

   (iii) **Purchase Contribution.** To reduce the cost of material and parts which is supplied.

   (iv) **Applying the ABC of Value Engineering.** There are 10 points suggested by Miles as long ago as 1947. These heads are:

   $A = Actual$ necessity or otherwise has 4 points.

   $B = Best$ way or not has 3 points.

   $C = Competitive$ cost or not has 3 points.

5. **The Recommendation.** The proposed alteration and comparison of costs before and after all to be submitted for final recommendation by management.

6. **Implementation.** This is the stage at which management revives and decides upon the value engineering suggestions submitted, and take advantage of any "trade offs".

Benefits of Value Engineering

Results are derived on two factors:

(a) Benefits to the company   (b) Benefits to the customer

(a) **Benefits to the company.** It is again divided as tangible and intangible results.

Tangible Results. The results of value engineering will be of two sorts—tangible and intangible. The tangible results are the actual savings which can be expressed in monetary terms, and it is true to say that they seldom does a properly-conducted value-engineering exercise, achieve less than 50% saving in costs. The saving is usually much more than this. Cost reduction on an individual component can be of the order of 25, 50 or even 100%. Example : It is assumed that 5% saving on Rs. 100 selling, manufacturing cost of Rs. 75, and a profit margin of Rs. 8. There will be Rs. 3.75 cost reduction; which increases the profit to 8 + 3.75 = Rs. 11.75

Intangible Results. These are:

   (i)   Increased cooperation.

  (ii)   Enhanced states of the purchasing function.

 (iii)   Simulation of creative thought.

 (iv)   Increased Cost-Consciousness

  (v)   Mutual understanding

 (vi)   Development of latent ability

(b) Benefits to the customer. To live and to prosper, industry must provide the customer with what he wants, when he wants it, and at a price he is willing to pay. Value technique must, therefore consider the customer and his requirements. The crux of value analysis may be on the following factors:

   (i)   If we are in production, consider one of our company's products. What function does it perform for the customer? What values are important to performance, durability, appearance or any combination of these values?

  (ii)   What materials go into the product? What other materials might be substituted for them and what would they cost in comparison with the materials being used at present? If cheaper material could be used for any part of the product, would the values provided for the customer be lessened?

 (iii)   Could the parts be made or the product assembled in some simpler way than at present?

 (iv)   If you are not in production, consider some procedure your department or section is using. What is this procedure designed to accomplish? What does this cost? Is every step in the procedure necessary, or could one or more of the steps be eliminated and the same result accomplished?

Introducing Value Analysis

The first step for implementation of a value analysis programme is the selection of products to be analysed. Value analysis makes its mark during the maturity stage of the product. This means that one has to start thinking along these lines during the later part of the growth stage. What is important here is the switch from. Performance-Oriented thinking to Value Oriented thinking. The two approaches for the choice of products for value analysis are through ABC analysis and contribution analysis. In the ABC analysis, the products are classified according to their sales or consumption value and ranked in descending order. The product which offer the maximum sales or consumption value are selected. This would offer the best result in term of return when analysed. Critical and production holding items are to be subjected to value analysis in order to examine the rigidity of the specification.

Value analysis is concerned with the cost added due to inefficient or unnecessary specification or feathers. It makes its contribution in last stage of product cycle namely the maturity stage. At this stage Research and Development no longer make positive contribution in terms of improving the efficiency of the function of the product or adding new function to it. The prime task of manufacturer becomes one of tackling the value part of the job so well that product leadership will be maintained.

In order to answer the above questions three basic steps are necessary:

1. Identifying the function. Any useful product has some primary function which must be identified, a bulb to give light, refrigerator to preserve food etc. In addition it may have secondary functions such as with standing shock etc.

2. Evaluation of the function by comparison. Value being a relative term, the comparison approach must be used to eliminate functions. The basic question is "Does the function accomplish reliability at the best cost" and can be answered only by comparison.

3. Develop alternatives. Realistic situation must be faced, objection overcome and effective engineering manufacturing and other alternatives develop. In order to develop effective alternatives and identify unnecessary cost the following thirteen value analysis principle must be used:

- ★ Avoid generalities.
- ★ Get all available cost.
- ★ Use information only from the best source.
- ★ Brain storming sustains.
- ★ Blast, create and refine.
- ★ Identify and overcome road block.
- ★ Use industry specialists to extend specialize knowledge.
- ★ Key-tolerance, not to be too light.
- ★ Utilise and pay for vendor skills and techniques.
- ★ Utilise vender available functional products.
- ★ Utilise speciality process.
- ★ Utilise applicable standards.
- ★ Use the criterion "would I spend money this way."

31

Industrial Engineering

History and Development

Industrial engineering had its roots in the Industrial Revolution; it was nourished by individuals who sought to advance organisation and management principles at an early date; it emerged as a separate discipline and was formalised in the late ninteenth and early twentieth centuries and it achieved maturity after World War II.

Industrial engineering is basically a logic attempt to identify and eliminate all redundant activities, and streamline such activities which are unavoidable and are must for achieving the object of the enterprise. Industrial engineer works mainly at the interface of men and machines, but focus his attention and efforts on machines and not on men. It is now high time for industrial engineer to look at the problem of productivity as a human—rather than a technical problem.

Industrial engineering is a service function, consisting of a multidisciplinary approach to bring about changes for improvements in systems relating to men, machines and materials. Industrial engineer is concerned with the cost aspect if the product.

Definition of Industrial Engineering

Industrial engineering is concerned with the design, improvement, and installation of integrated systems of men, materials, and equipments. It draws upon specialised knowledge and skill in the mathematical, physical and social sciences together with the principles and methods of engineering analysis and design to specify, predict, and evaluate the results to be obtained from such systems.

Industrial Engineering Functions

Industrial Engineering Handbook by Morley H. Mathewson summarises the traditional industrial engineering functions as follows :

Methods Engineering—Operations analysis, motion study, material handling, production planning, safety and standardisation.

Work Measurements—Time study, predetermined elemental time standards.

Control Determination—Production control, inventory control, quality control, cost control.

Wage and Job Evaluation—Wage incentives, profit sharing, job evaluation, merit rating, wage and salary administration.

Plant Facilities and Design—Plant layout, equipment procurement and replacement, product design, tool design.

Similarly the American Institute of Industrial Engineers have spelled out the following primary activities as being part of the discipline of industrial engineering :

1. Selection of processes and assembling methods.

2. Selection and design of tools and equipment.

3. Design of facilities, including layout of buildings, machines, and equipment; material handling equipment; raw materials and product storage facilities.

4. Design and/or improvement of planning and control systems for distribution of goods and services; production, inventory, quality, plant maintenance and engineering or any other function.

5. Development of cost control such as budgetory control, cost analysis and standard cost systems.

6. Product Development.

7. Development and installation of wage incentive systems.

8. Development of performance measures and standards.

9. Operation research including such items as mathematical analysis, systems simulation, linear programming, and decision making.

10. Office systems, procedures and policies.

11. Organisational Planning.

12. Evaluation of reliability and performance.

13. Design and installation of Data Processing systems.

14. Design and installation of value engineering and analysis systems.

The principles and methodologies of industrial engineering are being applied in incrcasing measure to consideration of man's environmental problems—social, economic and political in line with the industrial enigneers awareness of the individual worker his motivational requirements. The application of Industrial engineering techniques would meet with only limited success unless there is simultaneously a studied effort for streamlining systems and procedures, at higher levels.

Profile of an Industrial Engineer

Industrial engineering is an umbrella that covers a multitude of functions and encompasses every phase of a conmpany's operations.

In addition, industrial engineers conduct organisation studies, design tools and equipment, make plant location studies, evaluate quality and reliability performance, administer wage and salary functions, solve complex business problems by the use of operations research and develop new products and product applications.

Industrial engineer of today bears little resemblance to the industrial engineer of thirty years ago. Most of the differences are due to the fact that industrial engineering, which is based on an ever-widening body of knowledge, is much changed from what it used to be. Industrial engineers point to the fact that they spend most of their efforts working towards short-range goals, when they should in fact be aiming at long range profits. At present industrial engineers conduct organisation studies, design tools and equipment, make plant location studies, evaluate quality and reliability performance, administer wage and salary functions, solve complex business problems by the use of operations research, and develop new products and product applications.

Objectives of Industrial Engineering Department

The basic objectives of an industrial engineering department is usually two fold : (1) To establish methods for controlling production costs. (2) To develop programs for reducing those costs. Both the methods and the programs are carried out by line management. In most cases the industrial engineering department is ordinarily held fully responsibility for the applicability and accuracy of the programs developed and recommended.

Place of Industrial Engineering Department

The nature of the industrial engineering department, its responsibility, and its working relationships within a compay where it belongs in the organisation structure needs a through study. There can be no hard and fast rule governing the place of industrial engineering in the organisation structure of all companies and plants. The activities of industrial engineering department influence its place in the organisation structure to this extent : it reports to the executive that is responsible for the departments in which it does the bulk of its work. In most cases, however, it provides some service to other departments.

Organisation Chart showing the position of Chief
Industrial Engineer in the organisation.

Organising for industrial engineering is no different from organising for any company function; the same accepted principles of organisation must be applied. In simple out line the organisation of an industrial engineering department demands three steps :

(a) The authority, responsibility, and accountability of the department must be clearly defined.

(b) The department must be integrated in the company plan of organisation.

(c) Every provision must be made so that the department can effectively perform its assigned tasks.

In administrating the department, the chief industrial engineer should of course, strive to have a well knit smoothly functioning unit. His operating plan must also enable him to provide all the services expected of his group.

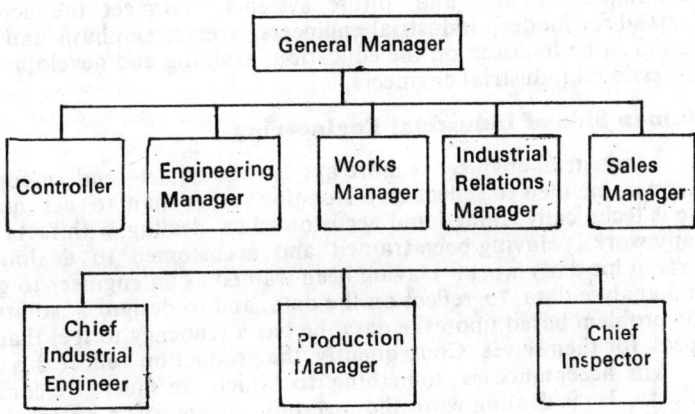

Fig. 31.1 Orgainsation Chart showing the position of Chief Industrial Engineer in the Organisation.

Industrial Engineering Personnel

The same principles of personnel administration that apply to other line and functional departments apply likewise to industrial engineering. The industrial engineering department is staffed by recent college graduates whose major qualification is educational background. These employees lack the stature and maturity to influence action. *It is therefore desirable that the well-staffed department have* adequatly experienced and senior engineers to balance the younger group. An industrial engineering department can provide excellent training to their employees in following specific areas :

(i) The methods and procedures adopted by the department.

(ii) The formal and informal organisation structures of the company.

(iii) The techniques of operating in a service atmosphere and dealing with workmen; supervisors; unions; management and other officials.

(*iv*) Pertinent technical phases of production operations.

(*v*) Company policies on time study; wage administration, areas of participation by the union, craft observance, and the like.

It is to be seen that professional industrial engineer after the training provides solutions to problems before the fact rather than after; Industrial engineer provides more creative assistance in support of dynamic management; and provides factual and unbiased solutions based on the increasing ability to measure understand, stimulate, and manipulate existing and future systems. To meet the increasing demand for modern industrial engineers, greater emphasis and attention must be focussed on the education, training and development of professional industrial engineers.

Human Side of Industrial Engineering

Industrial engineer is more apt to be quiet and modest in manner, inclined to reflect on a situation rather than to act quickly. He is technically trained and accustomed to dealing with facts in his daily work. Having been trained and accustomed to dealing with facts in his daily work. Having been trained as an engineer to gather and analyse data, to reflect on the data, and to design a solution to the problem based upon the data, he has a tendency to feel that facts speak for themselves. Consequently, the production effort necessary to win acceptance is something to which he must give conscious effort. He is dealing with the nervous system of a business in his work which ranges from machines and methods and layout, through standards and systems for payment and control. And it is people who make up the nervous system of the organisation. The industrial engineer must be able to get along with people and win their acceptance, even though he is not inherently sales-minded. Industrial Engineering is concerned with the interface of Man and Resources. In a Man/Machine relationship, the engineer's primary role is with the Machine and Personnel. The industrial engineer's role is with Man/Machine interface which determines the productivity of the system. Man interacts with the resources with the total Mind and Body. Industrial engineer have, therefore been as much concerned with the mental interaction as with physical.

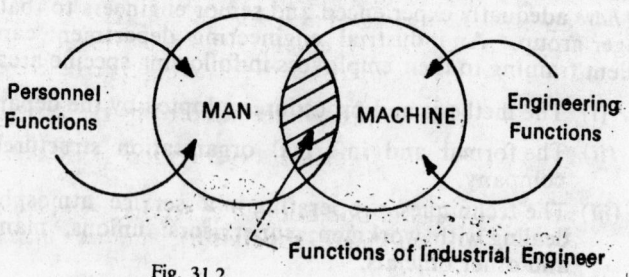

Fig. 31.2

Modern Industrial Engineering

Modern industrial engineering, is a broad discipline encompassing the analysis, design, and improvement of any and all productive elements of an enterprise, indeed of any organised human endeavor. The modern industrial engineering includes the use of the philosophical approach, and techniques of operations researches, systems engineering, and management science, and in concert with other disciplines. It aims towards improvement in the productivity of those production and management systems that provide the society with needed products and services. The industrial engineering functions can be studied by examining the true meanings of the words "Industrial" and "Engineering". "Industrial" is the adjective form of the noun; industry which means skill; cleverness ingenuity; diligence in any employment or pursuit and human exertion for the creation of value. "Engineering" is that branch of applied science which attempts to utilise the resources of nature and of human nature for the benefit of mankind, with regard for the relative scarcity of such resources. *Therefore the industrial engineering is the use of engineering principles* and practices to facilitate the creation of value satisfactions.

Applications of Industrial Engineering

As industrial engineers continue to be attracted to new areas of application many other interesting problems will be recognised and attacked. Opportunities include insurance; communications; Architecture and building construction; hotels and food service; community and civic affairs; cultural pursuits and perhaps the most challenging of all organised religion. Industrial engineering has its application in Agriculture viz. Work Measurement; Methods improvement; Layout; Engineering economy; Systems engineering; Operations Research; Material Handling; etc. The term Industrial is firmly established in manufacturing industry and is becoming increasingly valuable to organisations primarily concerned with the sale of goods or services. More industrial engineers will enter these services oriented organisations. The future for the profession is bright in this segment of the economy; it is limited only by the industrial engineer's ability to produce meaningful results, one such opportunity is in the retailing industry.

32
Work Study

Introduction

Work study is mainly concerned with the examination of human work. In fact planning cannot be done unless one knows how long it will take to do a particular job. It is important in modern time that our lives take cognizance of time. Time is very important to the manufacturer who must keep to promise, to estimate quantities, and to other industrial and business arrangements. The need for this managerial tool arose in the middle of nineteenth century, when the greater use of machinery and increasing size of manufacturing units necessitated a more efficient means of controlling production schedules. It was F.W. Taylor who argued that greater, attention should be paid to the method of timing tasks Taylor advocated the breaking down of a task into what he termed. elements of work and the timing of these elements separately with the aid of a stop watch. He also emphasised the use of a differential piece rate system. In this system a strong incentive was offered to those who reached or surpassed the standard established by Taylor's time study methods.

A contemporary of F.W. Taylor was Gilbreth ; he and his wife, Dr. Lilian Gilbreth were sure that the way in which work was done was far more important than trying to devise timing systems of how long it took to do a set task. Gilbreth often found it necessary to take motion pictures and examine them frame by frame to examine the smallest elements of movement.

Definition of Various Terms

Time Study—means the straight timing of work elements. Work measurement—It implies that human effort is somehow measured and one of its objective is the installation of financial incentive systems of payment. The object of motion study, method study, micro motion study is to devise new and improved working methods. H.B. Maynard suggested that to obtain the maximum effectiveness from the study of work it was necessary to use all the known techniques. He named his co-ordinated and systematic approach to the development of improved methods *"methods engineering"*.

Work Study

Method Study	Work Measurement
It is the study of methods of work for a job to achieve process improvements, improved layout, improved design, better working environments and reduced fatigue.	It is the measure of the work for a job to arrive at the best method of work, improved planning and control.

Work study is term used to embrace techniques of Method Study and work measurement which are used to ensure the best possible use of human and material resources in carrying out a specified activity.

Importance and Scope

Work study is not a theoretical concept but essentially a practical one dealing with human beings who have their own style and attitude. So the success of work study depends upon the existence of good relations between management and workers. Work study involves lot of changes of various kinds of working methods. Since people in general do not like changes but prefer to carry on as they are already doing. There will always be a tendency to resist any new methods suggested by work study officers. But if relations are good and the workers have confidence in the ability, integrity and fair mindedness of work study man, there is a good chance that sound proposals will be accepted willingly by the workers.

Work Study as a Science

Work study have been criticised on the grounds that many of its techniques depend upon human judgement and therefore the result have only limited application. Let us see how far this criticisms hold true. It is true that some of the techniques such as rating as used in work measurement depend upon the estimates of work study man. Thus properly trained and qualified man with correct judgement should be acceptable for all practical purposes. As we know that many times the predictions of weather forecasters are wide of the truth but this by no means that we should ignore all weather forecasts. In any area new concepts and knowledge keep emerging similarly in work study new and more accurate technique will emerge.

Uses

Following are the important uses of work study :

(a) Direct means of raising productivity.

(b) It is systematic, no factor is overlooked.

(c) Most accurate method and yet provide a sound basis for production planning and control and incentives.

(d) It is most important tool of analysis.

(e) Every one concerned with industry benefit from it such as customer, worker and management.

Work Study and Productivity

Productivity increase is the key factor in raising the standard of living. Work study indicates how resources can be effectively utilised and study would help in realising this aim.

Selecting the Work to be Studied

We should select for study the work that is likely to have the longest production run, offers the greatest scope for improvement, and which promises the greatest financial saving for a given outlay.

Rcording Facts

The techniques used in recording the facts may be divided into following three categories :

(a) Process and time,   (b) Diagram and models,

(c) Charts.

Process Charts

It is useful diagrammatic means of presenting information on the major activities associated with particular investigation.

The advantages of process chart are :

(a) It is convenient means of presenting information.

(b) It shows clearly the relationship between several sets of data.

(c) It permits quick analysis of the problem.

(d) It provides a record for future reference.

Charts are like machines which requires the fuels on which it runs. Chart is a means to an end. It cannot solve the problems by itself but it shows up inefficiencies and the way to speedy solution,

Outline Process Charts

It is graphic presentations of the points at which materials are introduced into the process, the sequences of operations and inspections carried out. Charts includes information as the time required the grade of labour, type and location of machine employed. Outline process chart records the following two type of activities :

(1) An operation represented by the symbol O. An operation occurs when information is given or received or when planning or calculation takes place.

(2) An inspection-represented by the symbol □. It occurs when an object is examined for indentification.

Process Charts Symbols

A chart is a diagram or picture or a graph which gives an overall view of process.

Process charts are denoted by symbols. The symbols give a better understanding of the facts. Following are the various symbols to record various events.

Event	Symbol	Description
Operation	○	It represents an action.
Storage	△	It represents a stage when material (raw or finished) awaits an action.
Delay	D	It represents a temporary halt in the process.
Transport	⇦	It represents movement of an item.
Inspection	□	It represents an act of checking.
Combined Activities		
Operation-cum-Transportation	⬤	First activity represents outer part and second activity the inner part.
Inspection-cum-Operation	⊡	

Thus a chart indicates the process with the help of symbols and aids for better understanding with a purpose of improvement for indentification.

Flow Process Chart

It is a graphic representation of the sequence of all operations, transportation, inspections, delays, storage etc. occurring during a process or procedure. This will include in greater details all relevant information for analysis such as time of each activity distance moved, frequency of movement.

The Flow Process Chart will assist in drawing a complete sequence of events occurring in process and keeps us in finding the delays, improper handling frequency of movement and keeps us in finding the tracking etc.

It is a means of recording information and is an extension of outline process chart. In addition to two activities—operation and inspection of outline process chart, it has three further activities :

(a) Transportations—It is represented by the symbol←

(b) Delays—It is represented by the symbol D

(c) Storages—It is represented by the Symbol Δ

Flow process charts can relate to material or individuals.

Multiple Activity Charts

These charts are the pictorial representative of relationship between man-time and machine time. These charts show the relation between two or more separate but selected operation cycles and may also be used to study the work of several operators on a group of operation. Multiple activity charts help us considerable to visualise the sequence and relationship of events.

Travel Charts

This is drawn in conjunction with a flow diagram. In the case of multiproducts and non standardised products this chart helps us to indicate flow between processes, departments of shop areas as shown in Fig. 32·1.

VARIOUS MODELS

The Layout Diagram

It is a floor plant upon which the arrangement of all equipment is marked to scale. Such a diagram help an investigator to record his findings in a form which is simple to visualise.

Flow Diagram

This is a sketch or model to scale of the layout of work places, machines, equipments, floor areas and building particularly showing the location of all activities in a flow process charts. This also indicates the paths and movements followed by men, materials, equipments in executing the activities.

	Departments→	A	B	C	D	E	F	Total
F ↓	A			—	4	1	2	10
O								
R	B			3	2	—	3	9
M	C				3	2	3	8
↓	D		3	3		1	2	9
	E				1		1	2
	F		3	2				5
	Total	9	8	10	4	11		

Fig. 32·1

String Diagram

It is essentially the same scale plan of the layout as used in the flow diagram with the only difference that movements are shown by continuous threads. This diagram will help us in finding out the points of congestion and back tracking. Better routes could be chosen by following different routing of the thread and comparing it with the previous routing.

Micromotion Films

In many cases the high speed movements made by operators cannot be seen by the eye or with stop watch. The techniques of micromotion is used to record rapid movements. It consists essentially of a cine camera which takes a film of the operator's movements. Micro-motion study besides giving a more accurate recording of both movements also provides a permanent record which can be studied at a latter date also. The equipment for micro-motion study is a 16 mm cine camera, a tripod and exposure meter and a timing ock known as microchronometer.

Microchronometers is provided with a self-starting motor with geared mo vement designed to read 1/2000th of a minute.

The Memotion Technique

It is technique named after its inventor Marvin E. Mundel who developed this particular use of the cine camera for recording industrial activities and subsequent analysis of the film. It is mainly used to study the work of dentist, a pharmacist and a group of men in a foundry. This technique is mainly used for long and irregular activities and work of a group of people. The equipment required for memotion is a cine camera capable of carrying a film magazine of at least 30 meters. Memotion is a kin to activity sampling. Activity sampling is quicker and more economical method, but memotion provides a permanent record.

Operation Analysis

Operation analysis may be defined as a systematic procedure, employed to study all the factors which affect the method of performing an operation to achieve maximum over all economy. The factors that surround the simplest process or operation are many and varied. Accordingly small progress will be made toward methods improvement and automation if the job is studied as a whole. The operation analysis procedure is the basis for all the manufacturing research work that is being done in industry today.

Applications of Operation Analysis

Human nature is such that we all feel that "Our work is different". The best intentioned manager with a cost problem consequently feels that the technique will work better in every other situation than it will in his own. However, an industrial engineer will have sufficient experience with the technique to know of the beneficial results that are always obtained as a result of applying it.

Operation Analysis Improves Automation

(*i*) Observe or visualise operation.

(*ii*) Ask questions.

(*iii*) Estimate degree of improvement or automation possible.

(*iv*) Investigate ten approaches to improvement in automation :

    (*a*) Design of Part or Assembly.

    (*b*) Material Specification.

    (*c*) Process of Manufacture.

    (*d*) Purpose of Operation.

    (*e*) Tolerances and Inspection requirements.

    (*f*) Tools and speed, feed and depth of cut.

    (*g*) Equipment analysis.

    (*h*) Work place layout and motion analysis.

    (*i*) Material flow.

    (*j*) Plant Layout.

    (*k*) Compare old and new methods.

The repetitiveness of the operation is another factor that must be considered. The industrial engnieer frequently finds operations that are repetitive from an operation analysis view point even in maintenance work. In this instance when a number of different jobs

Human Side and Work Study

To make the concept of work study successful the human relations must be considered. An organisation can apply work study
successfully only if it maintains a continuous watch over the human
issues. Thus work study is 10 per cent technical and 90 per cent
psychological. In an organised organisation the trade union exist.
The union is concerned with employees hours of work and other
working conditions such as financial conditions governing the
employees. Thus any change brought about in the working conditions of the employees the union must be involved.

Following are likely to be the points in particular upon which
the Union will wish to be satisfied.

(i) There should be adequate consultation before introduction
of any change which affects their member.

(2) There should be definite policy in respect of those people
which may be affected as a result of work study and incentive
schemes.

Management Reaction

(i) One possible source of difficultly is general conservatism
of human being, any change in the existing method and practices is
considered as an interruption of a comfortable situation.

(ii) Work study team is likely to get resistance from department managers.

(iii) It is necessary to get the support of top management
without which any efforts the initiate work study is meaningless.

(iv) Supervisors, Engineer, Foremen, Chargemen should be
given fullest information. This group is closest to the workers and
it is from where the work study group need a day-to-day co-
operation.

Hence any cause or conference intended to introduce work
study to supervision should concentrate on the following points :

(a) Economic necessity of reducing manufacturing cost.

(b) The advantage of systematic method study over occasional
and haphazard attempts made to improve methods.

(c) The advantages of measuring work.

(d) The fairness and advantage to the workers of additional
payments for additional work.

(e) The basis of incentive scheme and the way in which incentives are calculated.

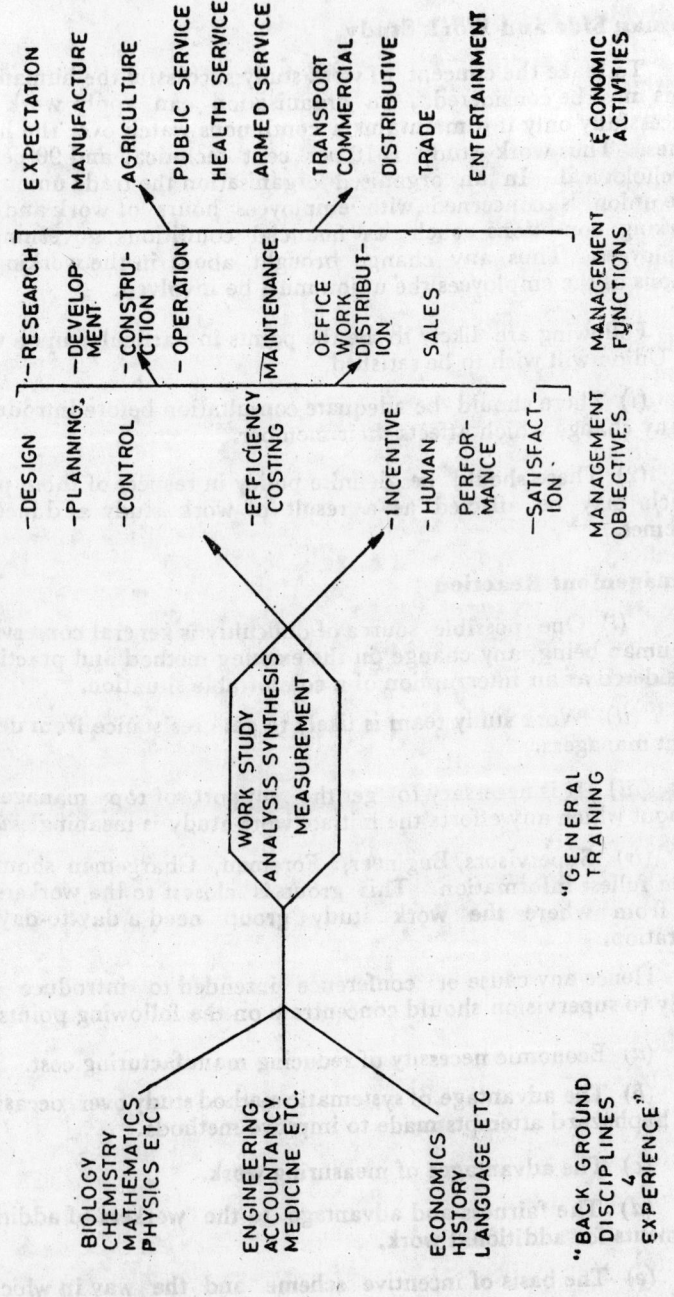

Fig. 32.1

Individual Reaction to Work Study

"What are the goals to which a man works and the fulfilment of which he may regard as successful living ?" Professor Bakke summarized the following points :

(a) **Society and respect of other people.** One of the bases of mutual respect in a working group is that there should be some balance of favourness in the distribution of work and responsibility and that each member should feel that others should attach some meaning and value to his job. This is one of the objective of work study also.

(b) **Degree of creature comforts and economic security.** This scheme is based on work study, demonstrate a clear link between extra efforts and extra award, hence a man may feel that cerature comforts within his reach. As it decides, the number of people to be employed once there is "build in" security also.

Independence in and Control Over his Own affair

Complete independence is not enjoyed by any body at work and perhaps independence for the man at the bench is difficult to visualized. In practice it may be taken as training him to do a job, telling him exectly and what is expected of him and then letting him get on with it. Knowledge of results has been proved to be a great spur to better performance.

Understanding of the Forces and Factors at Work in his World

A worker need to know about the effort of environment to his job and the factors affecting him, the answers is best provided by a work study team.

Integrity

Work study enables a man to do.

"A fair days work for a fair days pay". He feels wholeness in his job.

The figure gives an idea about the application of work study in the service of managment and society at large.

Method Study

It is a systematic recording and critical examination of the factors and resources involved in existing and proposed ways of doing work as a means of developing and applying easier and more effective methods and reducing costs.

This technique is very useful in the following cases :

(a) More valuable in case of repeated operations.

(b) More applicable to people working on incentive system.

(c) More useful in case of automatic machine tools turning out large number of machined parts.

(*d*) More applicable in case of parts to be machined which require high degree of accuracy.

(*e*) More applicable where there is a need to simplify the movements and sequence of operations.

In fact the principles of motion economy are important in all types of operations. It may not be possible to analyse highly skilled man's movements and teach him to improve them but it is almost possible to alter the design of equipment and improve tools layouts to effect economies in production.

Method Study in Relation to Time Study

Motion or method study is necessarily an integral part of time study. Since it would be wasteful and in-effective to carryout an expansive time study on an operation in which the methods had not been previously simplified and improved. Thus time study alone is not enough, because it may only serve to correct inaccuracies in standardized operation times. If the management and the worker and the organisation are to profit in the long run there is a need of economy of motion in all productive operations. To standardise existing production methods may be merely a standardise inefficiency.

Procedure for Method Study

The basic procedure in method study can be classified into following components :

(*a*) Select—The work to be studied.

(*b*) Record—The relevant facts of all present methods.

(*c*) Examine—Facts in sequence.

(*d*) Develop—Economical, practical and effective methods.

(*e*) Install—Method as standard practice.

(*f*) Maintain—Standard practice.

(*g*) Feed Back—To get back the information.

33

Work Measurement

The development of a time standard to perform a task has been and still is controversial. According to Fredrick W. Taylor, the father of time study, to establish an operation standard, it is necessary to sub divide the operation into elements of work, write a description of each element of work, time each element of work with a stopwatch, and add certain allowances to cover unavoidable delays and fatigue.

Methods Time Analysis

A.B. Segur of Oak Park Illinois, was one of the first to establish the relationship between the time element ann the motion itself. His ambition to integrate time with motions led to the developments of his MTA system. When an operation is studied, it is generally discovered that the operation consists of getting something, moving it to some location, processing or assembling it, and then releasing it. For example, the operation of writing with a pen might be motion analysed as follows :

Description	Motion
1. Move hand to pen holder	Transport Empty
2. Grasp pen in pen holder	Grasp
3. Move pen to paper	Transport Loaded
4. Write on paper	Use
5. Move pen to pen holder	Transport Loaded
6. Release pen	Release
7. Move hand back to paper	Transport Empty

In addition to analysing the motions involved, certain additional information must be collected, such as distance moved, the type of grasp, the type of release, and the like. Segur stated that the method must be well defined beiore an attempt is made to time-analyse the motions involved. In the Motion Time Analysis system, motion values are carried out to fifth decimal—0.00150 minute. These base times include no allowance for futigue and delay, which must be provided for as a seprate addition to the base times.

Steps Involved In Establishing Time Standard

Time study is a procedure osed to measure the time required

by a qualified operator working at the normal performance level to perform a given task in accordance with a specified method. In practice, it is difficult to *seprate methods study and time study completely*. The definition of time study states that the task measured is performed with a specified method. The figure (   ) presents a graphic analysis of the steps involved in establishing a time standard. Time study itself begins with "Selection of Operator".

METHODS STUDY/TIME STUDY

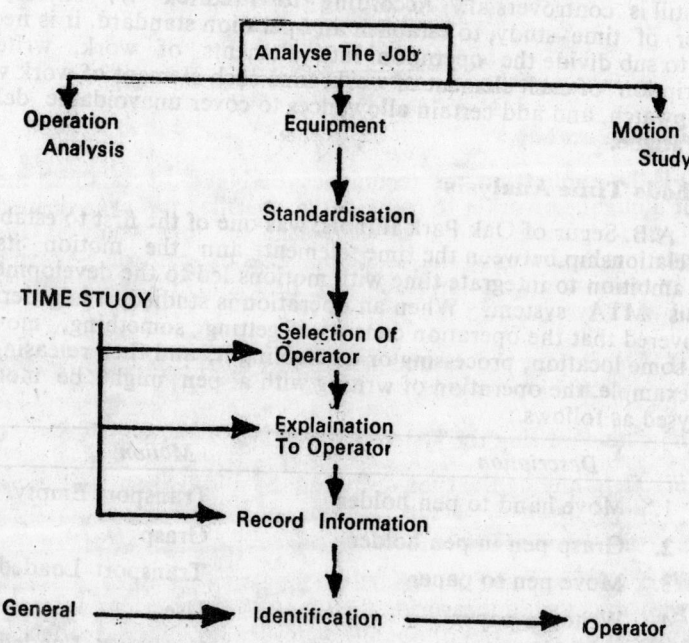

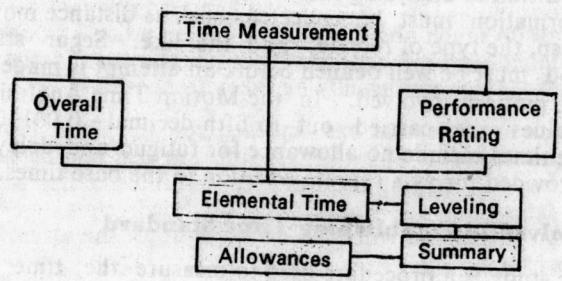

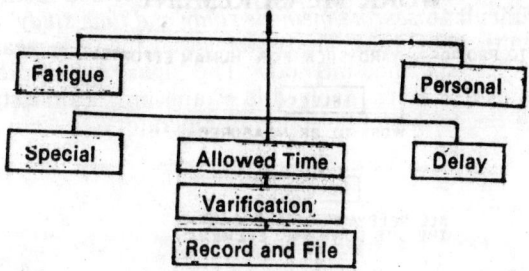

Fig. 33.1

Work Measurement

It is the application of techniques designed to establish the time for a qualified worker to carry out a specified job at a defined level of performance. Work measurement or time study is the technique for determining the standard time to perform a specific task. In other words motion study and time study is the systematic study of work system with the purpose of :

(*i*) Developing the desired system and method usually the one with the lowest cost.

(*ii*) Standardising the system and method.

(*iii*) Determining the time required by a qualified and properly trained person working at a normal pace to do a specific operation.

Work measurement, as the name suggests, provides management, with a means of measuring the time taken in the performance of an application for series of operation in such a way that ineffective time is shown up and can be separated from effective time. In this way, its existence, nature and extent become known where previously they were concealed within the total. One of the surprising thing bout factories where work measurement has never been employed is the amount of ineffective time where very existance is unsuspected or which is accepted as the "usual thing" and something inevitable that no one can do much about built into the process. Once the existence of ineffective time has been revealed and the reasons for it was traced down, steps can usually be taken to deduce it.

Here work measurement has another role to play. Not only can it reveal the existence of ineffective time ; it can also be used to set standard time for carrying the work, so that if an ineffective time does creep in later on it will immediately be shown up as an excess over the standard time and will thus be brought to the attention of the management.

Work measurement may start a chain reaction throughout the organisation.

WORK MEASUREMENT

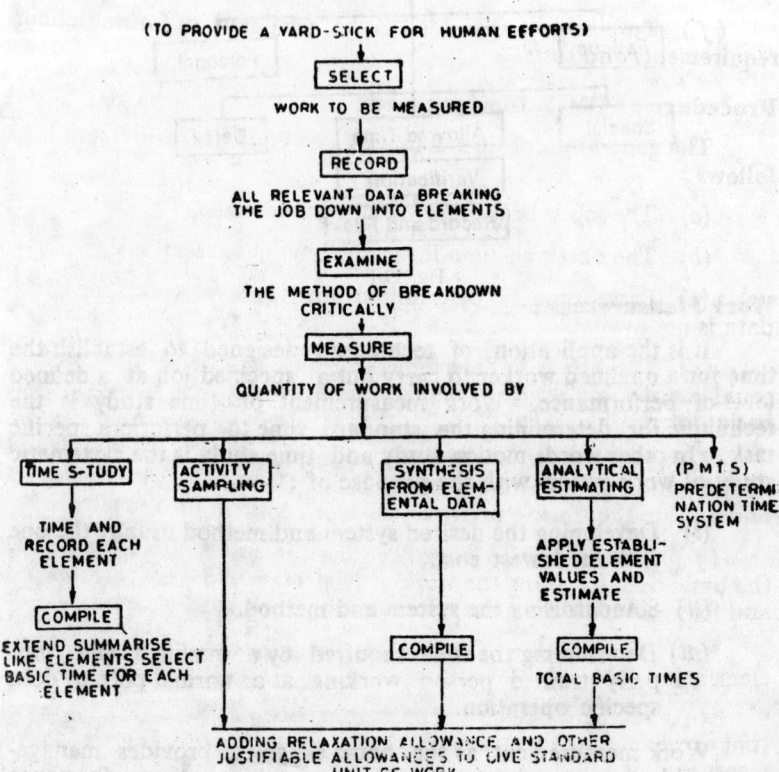

(TO PROVIDE A YARD-STICK FOR HUMAN EFFORTS)

SELECT
WORK TO BE MEASURED

RECORD
ALL RELEVANT DATA BREAKING
THE JOB DOWN INTO ELEMENTS

EXAMINE
THE METHOD OF BREAKDOWN
CRITICALLY

MEASURE
QUANTITY OF WORK INVOLVED BY

TIME S-TUDY — TIME AND RECORD EACH ELEMENT — COMPILE — EXTEND SUMMARISE LIKE ELEMENTS SELECT BASIC TIME FOR EACH ELEMENT

ACTIVITY SAMPLING

SYNTHESIS FROM ELEMENTAL DATA — COMPILE

ANALYTICAL ESTIMATING — APPLY ESTABLISHED ELEMENT VALUES AND ESTIMATE — COMPILE — TOTAL BASIC TIMES

(P M T S) PREDETERMINATION TIME SYSTEM

ADDING RELAXATION ALLOWANCE AND OTHER
JUSTIFIABLE ALLOWANCES TO GIVE STANDARD
UNIT OF WORK

Fig. 33.2

Objectives of Work Measurement

Using as a target the times established for jobs at the defined level of performance, work measurement will be found to have the following uses.

1. To compare the efficiency if alternative methods other conditions being equal, the method which takes the least time will be best method.

2. To enable realistic schedules of work to be prepared by relating reasonably accurate assessment of human work to plant capacity.

3. As the basis of realistic and fair incentive schemes.

4. To assist in the organisation of labour by enabling a daily comparison to be made between actual times and target times.

(e) As a basis for labour budgeting and budgeting control systems.

(f) To enable estimates to be prepared of future labour requirements and costs.

Procedure of Work Measurement

The general procedure followed [in Work Measurement is as follows :

(a) The job is broken down into its elements.

(b) The observed time for each element is recorded.

(c) Basic time is determined for those elements for which data is not available.

(d) The values so determined for any of the elements which could conceivably secure in another job are added to the records of basic times.

(e) Determine the frequency of occurrence of each element in the job, multiply the work content of each element with its frequency and add up the time to arrive at the work content for the job.

(f) The proportion of rest required is assessed and added to the basic time for doing the work at the standard rate of working and for recovering from the effort i.e. the work content.

(g) The addition of the relaxation allowance may be made element by element.

(h) If there are any contingent delays a blanket allowance (not exceeding 5%) may be added since they are not economical to measure.

Basic Time

Basic time is the time for carrying out an element of work at standard rating i.e.,

$$\frac{\text{Observed Time} \times \text{Observed Rating}}{\text{Standard Rating}}$$

Normal Time

This can be computed by the following Formula :

$$\text{Normal Time} \quad \frac{\text{Average observed actual time} \times \text{Performance Rating}}{100}$$

Standard Time

Standard time is the total time in which a job should be computed at standard performance i.e. work contact, contingency allowance for delay, unoccupied time and interference allowance considered.

Observed time+Rating=Process Time Normal Time) Normal time+R.A.% of Normal Time+Contingency allowance for work as % of Work content=work continue (in minutes).

Work content+Continging Allowance for delay (minute) + unoccupied time (minute) + Interference allowance (minute) =Standard time (minute)

Performance Rating

Performance rating will be calculated as follows :

$$\text{Performance Rating} \frac{\text{Normal time}}{\text{Actual time}} \times 100$$

Unit of Work

The Human Work Content of different types of job can be expressed quantitatively in terms of common unit. Allowance for rest, which varies according to mental and physical efforts involved in each job is included in human work, after addition of the relaxation allowance (RA) appropriate to the work, the resultant values are totalled to give a value for complete job in composite units. These composite units are referred to as unit of work.

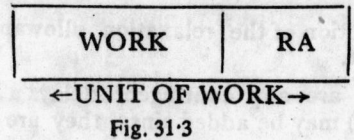

← UNIT OF WORK →

Fig. 31·3

The following are the *principal techniques* by which work measurement is carried out :

Techniques of Work Measurement

(*i*) Time study.

(*ii*) Synthesis from elemental datas.

(*iii*) Predetermined motion-time system.

(*iv*) Analytical estimating.

(*v*) Activity sampling.

Time Study

In is a work measurement technique for recording the times and rates of working for the elements of a specified job carried under specified condition and for analysing the data so as to obtain the time necessary for the carrying out of the job at a defined level of performance.

Usefulness of Time Measurement

It is indeed useful to the management to know how long it should take to carry out vaious kinds of work in a plant. Let us discuss various uses :

(a) It reveals the capacity of each machine and enables the planning department to apportion work correctly.

(b) It increases accuracy in forecasting delivery dates.

(c) It is useful in costing.

(d) It enables reliable forecasts to be made of machine when a new department or factory is planned.

(e) It helps in choosing the best method.

(f) It helps to prevent misuse of manpower in each department.

(g) It fosters the development of equitable incentive payments system.

It is the oldest work measurement technique. It is regarded as the fundamental technique. A systematic method study should be carried out before the time study because it is no good carrying out a time study on any work until the simplest way of carrying out that work has been determined by a systematic method study.

Object of Time Study

Production cost of any commodity is made of three components :

(a) Raw Material Cost.   (b) Overheads.

(c) Labour Cost.

A production manager can easily predict the cost of raw material and with some past experience and judgement overheads can be estimated pretty accurately. To estimate labour cost, we need to know the labour time. The parctical experience of production specialists helps them exercise considerable judgement in estimating direct labour time. However, costly errors can occur if personal judgement is used exclusively. Time studies are conducted to make precise evaluation of direct labour time.

These time standards, in addition to establishing standard labour costs, also establish capacity of productive equipment. The standards are thus used for scheduling production orders.

Another use of time study is that as a result of time study, a time-study engineer inevitably discovers constructive refinements in operation methods, tooling, plant lay-out and materials handling. Thus he is in a position to compare the various alternatives and recommend the best possible method.

Also time standards are used for setting pay incentives. Anybody who can do better than standards by his extra skill or extra labour should be given the suitable award. Thus the time standards are also used for determining pay incentives.

Last but not the least, time standards help in evaluating the productivity of individual employees. Thus line standards help the supervisor in determining fair days work for the individual employees.

To sum up, the objects of time study are :

(a) For cost allocation.

(b) For scheduling production.

(c) For determining alternatives.

(d) For determining pay incentives.

(e) For determining fair days work for individual employees.

Method of Time Study

Any work measurement system must perform the following two tasks :

(i) Measurement of actual observed time, and

(ii) Adjustment of observed time to obtain normal time.

Measurement of actual observed time can be done by the following methods :

(a) **Stop Watch Method**

(i) Decimal-minute stop watch.

(ii) Decimal-hour stop watch.

(b) **Motion Picture Camera**

(i) Slow speed 50/1000 frames/minute.

(ii) Normal speed 960 frames/minute.

(iii) Modified normal speed 1000 frames/minute.

(iv) Sound speed 1440 frames/minute.

(v) High speed 64—128 frames/minute.

(vi) Very high speed 1000—3000 frames.

(c) **Machine using moving Tape and Disk**

(i) Time Study machine-Marsto-Chron.

(ii) Kynograph e.g. Esterline-Angers Operation recorder.

(iii) Series recorder.

(iv) Electronic Timer-Information and time, date punched on tape-IBM Machines.

Conducting Time Study

The time study is not carried out unless the request for the same comes from the right man. The plant manager, chief engineer, production control supervisor, cost accountant, foreman or other members of the organisation may make such requests.

After the request for time study has been received the Industrial Engineering Department will assign the job to a trained and experienced analyst. The analyst should go over the job with the foreman of the department. The time study standard established for a job will not be correct if there is :

(1) Change in the job
(2) Material is not as per specification
(3) Machine speed has changed
(4) Conditions of work are changed.

To have a systematic time study the analyst must ask the following questions :

(a) Can the speed of the machine be increased without affecting the tool or equipment life or the quality of the product ?

(b) Can we bring changes in the tooling made to reduce the cycle time ?

(c) Can we bring the material closer to the work area to reduce the Material Handling time ?

(d) Is the equipment rightly operated ?

(e) Is the operation being safely performed ?

Equipments For Work Measurement

Stop Watch

There are many types of stop watches. One type is controlled by pressure on the winding knob. The second kind is the flying back type which is started and stopped by a slide at the side of the winding knob. Another stop watch may be decimal-minute type. It may be divided into one hundred division called centiminutes. Another type could be decimal-hour. In this the dial is divided into 100 divisions. The larger one makes one revolution in 1/100 of an hour (0·6 min) instead of one minute.

A CENTIMINUTE STOP WATCH

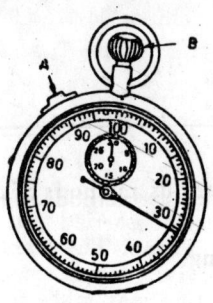

Fig. 33.3

Study Board

This is a sheet, usually of wood slightly larger than the record form to be used. It has a clip to hold the form in place and another clip to hold the stop watch in a convenient position for operation by the left hand.

A SHAPED STUDY BOARD MADE FROM 3 PLY FITTED WITH TERRY'S WATCH CLIP

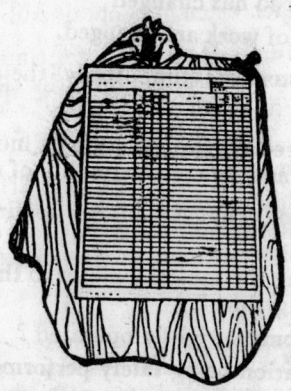

Fig. 33.4. Observation Form.

Study Sheet

It varies with the type of the job under study. In the study sheet brief particulars of the job are entered on the form including the elements which it has been decided are most suitable for study.

(TIME STUDY)

Particulars of job and operator :

Element	Rating	Watch Reading	Observed time	Ineffective or check time	Normal time

Methods of Timing

Following are the various methods of timing :

—Flyback Timing

— Comulative Timing

—Selective Timing

—Differential Timing

Flyback Timing. In this method as soon as the first element starts the winding knob is pressed and the large hand returns to zero. The time elapsed between the check time and the start of first element is noted. Before the start of the next element the hand is returned to zero and the duration of the previous element is entered on the form.

Comulative Timing. In this method the watch is started at the check time and the hand runs without being returned to zero. In the beginning of each element the watch reading is noted on the form. The time of each element is obtained by subtraction.

Selective Timing. It is used to time an isolated element. The watch is started at the beginning of the element till the end and the element is timed as often as necessary. This the most accurate method.

Differential Timing. In this the element is observed and timed with an adjacent longer one, when sufficient times for the combined elements have been taken, the longer element is timed by itself. Time for the short element is then obtained by subtraction.

Production Standards

These are computed and set to answer questions like 'what is a fair day's work'? These standards state how many parts, assemblies should be provided per minute, hour or day. Thus production standards can be expressed in terms of pieces per unit of time or time per piece. Although these standards are designed to determine how much output is expected of an employee, they include more than just work. Actually, production standards include standard allowances for delays that occurs as a part of the job, time for personal needs and where work is heavy, an allowance for physical fatigue.

Standard Time = Normal Time

+Delay Allowance

+Fatigue Allowance.

Application of Time Study in Setting of Production Standards

The first step in setting standards is the determination of normal time for performing a specific task. The actual observed time, however, is not the normal time. In order to arrive at normal time from observed time, rating method is used. Rating is that process during which the time study analyst compares the performance of the operator under observation with the observers own concept of normal performance. The time study analyst must judge the operators' speed while he is making the study, and award the performance rating.

$$\text{Performance Rating} = \frac{\text{Normal Time}}{\text{Actual Time}} \times 100$$

Determination of Standard Time

The actual observed time and hence the standard time can be measured by the following two methods :

1. Stop Watch Method. The following steps are undertaken in work measurement by stop watch method :

(*i*) Standardisations of methods for the operations-standard method for the operation is determined, specifying work place layout, tools, sequence of elements. The resulting standard practice is recorded.

(*ii*) An operator who is trained and experienced in standard methods is selected for study.

(*iii*) For timing purposes, elemental structure of operation is determined. Operation is broken down into elements and elements are separated on the basis of their occurrence (in each cycle, periodically or randomly).

(*iv*) Actual time required for the elements is observed and recorded, making simultaneous performance rating.

(*v*) The number of observations required to yield the desired precision of the result based on the sample data obtained in step (*iv*) is calculated (this step is explained later). More data is collected if necessary.

(*vi*) Normal time is computed now by the formula :

$$\text{Normal time} = \frac{\text{Average observed Actual Time} \times \text{performance rating}}{100}$$

(*vii*) Allowances of personal time, delays and fatigue are determined by the formula.

Standard Time = Normal Time + Allowances.

Number of Cycles to be Timed

Variations in time may result from such things as a difference in the exact position of parts and tools used by the operator, from variations in reading the stop watch, and from possible differences in determining the exact end point at which the watch reading is made. Time study is a sampling process. Consequently the greater the number of cycles timed, more nearly the results will be representative of the activity being measured.

The standard error of the average for each element is given by :

$$\sigma \, \bar{X} = \frac{\sigma'}{\sqrt{N}} \qquad \qquad \ldots (A)$$

$\sigma \bar{X}$ =Standard deviation of the distribution of averages.

σ' =Standard deviation of the universe.

N =Actual number of observations of the element.

A 95% confidence level and ± 5% precision are commonly used in time study.

Let N' is the required number of observation to predict the true value within ± 5% precision and 95% confidence level.

This means there is atleast 95 out of the 100 that the sample mean or average value for element will not be in error more than ± 5%.

Then, $0.05\ X = 2\ \sigma \bar{X}$

$$0.05\ \frac{\Sigma\ X}{[N} = 2\ \sigma\ \bar{X}$$

$$0.05\ \frac{\sigma\ X}{N} = 2 \times \frac{1}{N}\sqrt{\frac{N\ \Sigma\ x^2-(\Sigma x\ 12)}{\sqrt{N'}}}$$

$$N' = \left(\frac{40\sqrt{N\ \Sigma\ x^2-(\Sigma\ x\ 12)^2}}{\Sigma\bar{X}}\right)$$

Rating

It is defined as worker's rate of working relative to the obser-ver's concept of rate corresponding to standard rating. It is infact the assessment cf the speed and effectiveness of a given worker and in order to rate reliably the work study man must be thoroughly familiar with the correct way of carrying out the job, he is studying. It is really an important part of time study.

Rating Standards

The work studying man must have a standard to which he can relate his observation of a particular worker. In other words the rating is that process during which time study analyst compares the actual performance of the operator with the observer's own concept of normal performance. It is a matter of judgment on the part of analyst.

Systems of Ratings

Bedaux introduced a system of rating based on time study. His time standard were expressed in points. Under this system the amount of work one does per minute is known as standard work unit of minute. Thus the basis of Bedaux plan is energy and time. Bedaux point or unit for a work may contain 75% effort and 25% rest or other proportions.

Westing House System of Rating

It is a four factor system for rating operator's performance. The four factors are :

(1) Skill, (2) Effort,
(3) Conditions, (4) Consistency.

Synthetic Rating

It is the method of evaluating an operator's speed from predetermined motion time values. The ratio between the predetermined motion-time standard for the element and the average actual time volume in minutes for the same element is called performance rating factor.

$R = P/A.$

R = Performance Rating factor.

P = Predetermined motion time standard for the element.

A = Average actual time value for the same element in minutes.

Standard Performance

Standard Performance is the rate of output which a qualified worker will naturally achieve without over-exertion as an average over the working day or shift provided they know and adhere to the specified method and provided they are motivated to apply themselves to their work.

This performance is denoted as 100 on the standard rating and performance scales.

Scales of Rating

In order that a comparison between the observed rate of working and the standard rate may be made effectively, it is necessary to have numerical of scale against which we make the assessment.

The rating can then be used as a factor by which the observed time can be multiplied to give the basic time which is the time, it would make the qualified worker, motivated to apply himself to carry out the element at standard rating.

There are several scales of rating where the most common of which are those designated the 100—133 scale, the 60—80 the 75—100 and the British Standard Scale which is 0-100 scale.

Allowances

Various types of allowances may be applied depending upon the type of job. Following are the various allowances :

—Relaxation Allowances.

—Contingency Allowances.

—Interference Allowances.

—Periodic activity Allowances.

—Introductory Allowances.

—Learner Allowances.

—Unusual condition Allowances.

—Unoccupied time Allowances.

Relaxation Allowance. It is the time that an operator is allowed for his personal needs and for fatigue. The personal need allowance is constant for the various operations carried out in a factory. The fatigue allowance will vary with the nature of the job.

Contingency Allowance. Certain elements may occur so in frequently that it would not pay to study enough work cycles to evaluate them accurately. The elements for which the contingency allowance should be made are following :

—Consultation with supervisors

—Sharpening of tools

—Obtaining special materials from stores.

If for a job the total time taken is 50 hours and the time taken for eight odd elements are 15 minutes then contingency allowance would be,

$$\frac{0\cdot25}{50} \times 100 = 0\cdot5 \text{ per cent}$$

Interference Allowance. This allowance is given when two or more elements occur simultaneously. If more than one machine requires attention at the same time and the cycle time of these machines differ; interference will occur more frequently.

The accurate assessment of this allowance requires a thorough studies. The allowance is added as percentage.

Periodic-Activity Allowance. This is the allowance for activities to be carried out once only for example setting up of a lathe machine before production starts. The time for these activities can be noted and the cycle time can be proportionally adjusted.

Introductory Allowance. This is the allowance at times given at the introduction of an incentive scheme. The allowance is strictly temporary and decreases week by week until it disappears say after 4-5 weeks.

Learner Allowance. The introduction of new man into the system may show down the working of the whole team until he became proficient. If no allowance is given, the training of the apprentice or learner will be neglected.

Unusal Conditions Allowance. This allowance covers the additional time required for special case of some specific type of raw material or equipment in handling at loading or unloading points.

Unoccupied-Time Allowance. It is the other name to process allowance. It is the time, the operator has to wait till his machine or his work group is free. The idle time of this nature is undesirable but some allowance is necessary to help the operator to reach his standard performance.

Policy Allowance

It is an allowance given at the discretion of the management due to inherent features of the work. This allowance is not genuine part of time study and should not interfere with calculation of standard as shown in Fig. 33.5.

The examples of policy allowances are the one discussed above that is :

(a) Introductory Allowance

(b) Learner Allowance

(c) Unusual Conditions Allowance.

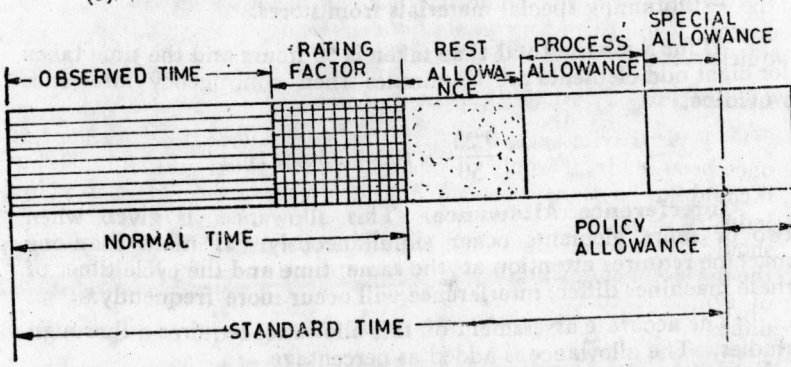

Fig. 33.5

If the policy allowance is also added then it is termed as "allowed time". Standard time is expressed as standard man/machine hour or standard man/machine minutes. Symbolically they are represented by (S.M.H.) and (S.M.M.) respectively.

Production Studies

It does not constitute a new technique but it is specialised form of time study. It is carried out over a specified period of time say for half a day or one shift for the purpose of ascertaining the frequency and duration of activities.

Production study is relatively a continuous study designed to check an existing standard time to obtain additional information affecting the rate of output. The object of this study is to account for everything that occurs during a working period.

The main uses of production studies are :

(a) To check the output of a particular worker or machine.
(b) To find out the idle time and factors affecting further delays.
(c) To secure information to find relaxation, contingency and interference allowance.
(d) To check the accuracy of time arrived at.

Production study is a lengthy and costly business so the activity sampling is used as a substitute.

Synthesis from Elemental Data

It is a work measurement techniques for building up the time for a job at a defined level of performance by totalling element times obtained previously from time studies in other jobs containing the elements concerned. It is seen that many of the timed elements reoccurred in the similar forms in various jobs. This realisation led to the recording and filing of times so that when elements reoccurred the existing records could be utilised in order to save time and energy. This recording would save work and make job times much quicker and cheaper to produce.

Limitation of Synthetic Data

F.W. Taylor suggested that after a detailed time studies had once been made, it would be possible to lift them in handbook but it could not be done because methods vary from plant to plant, industry to industry. However, Gilbreth stated that each therblig should be performed within a certain time range. He could foresee the development of predetermined motion-times. So far no manual of synthetic data is obtained so far because of some particular difficulties.

Predetermined Motion Time Systems (P.M.T.S.)

It is work measurement technique whereby times established for basic human motions are used to build up time for a job at a defined level of performance. This techniques is used for finding the times for basic motions of the therblig. To obtain these times various operations are filmed and then the film is analysed very carefully. The advantage of films is that each element can be observed many times merely by projecting the film repeatedly.

The following diagram gives a quick understanding of the nature of the methods time measurement procedure

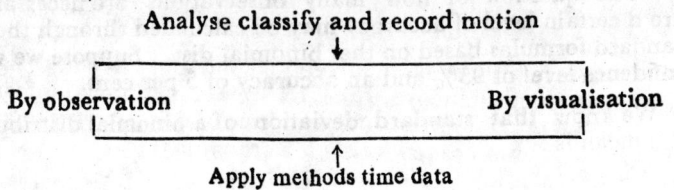

Analyse classify and record motion
↓

By observation   By visualisation

↑
Apply methods time data

Analytical Estimating

It is a work measurement technique whereby time required to carry out elements of a job at a defind level of performance is estimated from knowledge and practical experience of the elements concerned. Analytical estimation is widely used to estimate times in engineering, maintenance and construction work. In this method the estimator should be fully conversant with all the details of work being studied. Analytical estimating can give satisfactory results although these are less accurate than those obtained by time study or by synthesis from elemental date. In this method time is not estimated for the job as a whole but the task is first broken into elements as in time study, but elements are longer. Time for longer elements are worked out by the trained estimator only.

Activity Sampling

It is a work measurement techniques in which a large number of instantaneous observation are made over a period of time of a group of machines, processes or workers.

Such observations indicates :

What is happening at that instant. The concept is that random observations can produce results whose accuracy depends upon the number of observations made. This technique is also known by Random observation, Work Sampling and Ratio-delay.

Work sampling is used in :

(a) Measuring the working and non-working time of man and of machines (for delay allowances or utilization time of M/cs).

(b) To estimate the percentage of time devoted to each of duties included in the jobs performed by such people as maintenance personnel, office workers, managers etc.

(c) To establish time standards.

The work sampling method consists of taking a number of random observations of the activity being studied and from this determining the percent of time devoted to each aspect of the operation.

As it is obvious the procedure is simple. However, the key to the accuracy of the technique is in the number of observations. A greater number of observations provides a higher degree of accuracy. The question of how many observations are necessary to ensure a certain level of accuracy may be calculated through the use of standard formulas based on the binomial dist. Suppose we want a confidence level of 95% and an accuracy of 5 per cent.

We know that standard deviation of a binomial distribution which mean is

$$\sqrt{\frac{P(I-P)}{N}}$$

Since we want 95% confidence level and since we know that 95% area under a normal curve includes two standard deviation, then

$$SP = 2\sqrt{\frac{P(I-P)}{N}}$$

S = desired relative accuracy.
P = % age expressed as a decimal.
(idle or working activity)

Example. Suppose we want to determine the % of idle time of a certain machine. Assume a confidence level of 95% and accuracy of 5%. We want to determine the number of observations to give us the desired results.

Now before we can use the formula.

$$SP = \sqrt{\frac{2P(1-P)}{N}}$$

it is necessary to estimate the value of P. What we do is to conduct a trial study to get a first estimate of the % of idle time (P). Suppose 100 observations were made and in this preliminary study 25 observations showed the machine to be idle.

So

$$P = 0\cdot25$$

$$S = 0\cdot05$$

$$SP = 2\sqrt{\frac{P(1-P)}{N}}$$

$$0\cdot05P^2 = 2\sqrt{\frac{P(1-P)}{N}}$$

$$0\cdot0025P^2 = 4P\frac{(1-P)}{N}$$

$$N = \frac{4P(1-P)^2}{0\cdot0025\,P} = \frac{4(1-P)}{0\cdot0025P}$$

$$= \frac{1600(1-P)}{P} = \frac{1600(1-0\cdot25)}{0\cdot25} = 4800$$

This is only the 1st estimate. As the work sampling study is underway and 500 observations have been made, a new calculation might be made.

Now suppose machine working $=350$ times.

Idle $=150$ times.

$$P=\frac{150}{500}=0 \cdot 30$$

Now $0 \cdot 05 \ (0 \cdot 30) \qquad =2 \ \frac{0 \cdot 30(1-0 \cdot 30)}{N}$

$$N=3633.$$

Actually it is advisable to recalculate N at regular intervals, perhaps at the end of each day.

Advantages of Work Sampling Over Work-study

(a) Eliminates stop watch study.

(b) It is economical in bigger set ups.

(c) Applicable to activities that are impractical to measure by Time Study.

(d) No trained work-measurement analyst required.

(e) Measure can be made with pre-assigned degree of reliability.

(f) Time study evens out the results of day-to-day variations.

(g) A work sampling study may be interpreted any time without affecting the results.

(h) It is easier for the studied people.

(i) It is less resented by workers as no stop watch or other timing device is needed.

(j) A simultaneous work sampling study of several operators or machines can be done by a single observers.

Limitations

(i) Non economical for work at single locations.

(ii) Not much scope for improving work methods.

(iii) Harder to explain to union and management.

(iv) Non-random sampling may bias results.

Example An operator works for eight-hours a day. A work sampling study shows that he was idle 15% of the day and that he worked the remainder of the day at an average performance index of 110%. The record shows that he turned out 420 pieces of acceptable quality during the day. Find the standard per piece. Assume 15% Allowances.

Solution. Total time expended by worker$=8\times60=480$ min.

No. of parts produced $=420$

Working time$=85\%$

Idle time$=15\%$

Average Rating$=110\%$

Allowance$=15\%$

Standard time per piece

$$=\frac{(\text{Total time})\times\%\text{working time}\times\text{Rating}}{\text{Total No. of pieces produced}}+\text{Allowances}$$

$$=\frac{480\times0.85\times1.10}{420}\times\frac{100}{100-.15}=1.26\text{ min.}$$

Example. A furniture making factory contained a group of 13 machines. The management of the factory decided to find the percentage of time spent on the following activities.

1. Working (Processing on timber).
2. Running (Running time of machines without processing).
3. Set up (Being prepared for the next job).

The observations were recorded between 9 A.M. to 12·30 P.M. and from 1·30 P.M. to 5·00 P.M. The total observation of four activities were :

	M/c working	M/c Running	M/c being set-up	M/c idle
Total observation	288	475	83	530
Percentage	20·9%	34·5	6·1	38·5

To determine the limits between which it would be reasonable to assume the true values to lie.

Solution. We know that dispersion of the curve is expressed in terms of the standard deviation represented by the σ (Sigma)

$$\sigma=\pm\sqrt{\frac{p(100-p)}{n}}$$

$p=$It is the percentage of an observed activity

$n=$Total number of observations

$$\sigma=\pm\sqrt{\frac{p(100-p)}{n}}$$

For 68 per cent of chance tolerance when the machine is working the standard deviation is

$$\sigma = \pm\sqrt{\frac{20\cdot9}{1376}\frac{100-20\cdot9)}{}} = \sqrt{1\cdot204} = 1\cdot1$$

Total No. of observations

$$\begin{array}{r} 288 \\ 475 \\ 83 \\ 530 \\ \hline \end{array}$$

1376 Nos.

So limits are $(20\cdot9+2\cdot2)$ and $(20\cdot9-2\cdot2)$

That is $23\cdot1\%$ and $18\cdot7\%$.

The results show that machine is under utilised and something seriously wrong with either production planning or production con--trol mechanism.

Determining time standards by work sampling.

Example. A work sampling study might show that a worker on a drilling machine was idle 15% of the day or 72 minutes in eight hours shift. If he has produced 240 acceptable pieces in the day, standard time can be computed as follows :

Information	Source of data	Data of one day
Total time spent by operator	Time card	480 mts
No. of parts produced	Inspection Deptt.	240
Idle time in %	Work sampling	15%
Working time in %	—do—	85%.
Average performance idex	—do—	110%
Total allowances	Company time study manual.	15%

Standard time per piece,

$$= \frac{(\text{Total time in mts.}) \times (\text{Working time in}\%) \times (\text{Performance index}\%)}{\text{Total No. of pieces produced} + \text{Allowances}}$$

$$= \frac{480 \times 8\cdot5 \times 1\cdot10}{240} \times \frac{100}{100-15} = 2\cdot2 \text{ minutes.}$$

34

Plant Location

Introduction

A new business is always faced with the problem of where to locate. The organizer of a new business seldom consider sites outside of his own community. This is probably because of the more pressing problems involved in establishing a new business. The problem of plant location in established industry receives more attention. A number of conditions can lead to the consideration of the plant location problem. These conditions are : (1) expansion, (2) decentralization, and (3) economic factors, such as a shift of the markets or an inadequate labour supply.

Expansion is the most frequent cause for considering the problem of plant location. Established industries often expand when (1) their facilities become obsolete, (2) market demand grows beyond a capacity of the plant site and facilities, or (3) inadequate services are available to the customer.

The decentralization of industry in this country is increasing at a rapid pace. Many of the industrial giants that historically grew in one location have found that decentralization pays off in a number of respects. The company benefits by the availability of new sources of labour supply, the improved labour relations of smaller plant, a lower absentee rate, and the greater interest in their work shown by citizens of small communities. The worker benefits from the lower cost of living in uncongested town, the elimination of expensive and annoying buying off. The increased opportunities for advancement and better opportunities to utilize his spare time is available in bigger towns. A decentralized market is a cause for decentralization of the industry servicing it. The Coca Cola plant, which are spread or decentralized from coast to coast are good examples of locating in the concentrated market areas (mostly urban).

In the final analysis, the objective of any location is to able to deliver the products to the customer at a cost equal to or less than the competition within the product field. The cost of goods delivered to the customer depends upon

    (a) The cost of gathering the material ;

    (b) The cost of fabricating the material ;

    (c) The cost of distributing the finished product to the consumer.

    Keeping the above cost breakdown in mind it is possible for any manufacturing facility to make a list of those location factors which will tend to effect one or all of the above three costs. The list will vary depending upon the product and production characteristics, the objective and the sizes of the particular installation. There are a number of factors, however, which tend to have an effect in cost plant and location studies e.g. nearness to market and raw material ; availability of power and water ; climate, availability of capital, labour ; transportation facilities ; site availability. There are more factors which will interact with each other.

Definition

    Plant location can be defined as the determination of that location which when considering all factors will provide minimum cost delivered to the customer. Amongst the different problems relating to the management of the production system is that of location of industrial plant. The plant must be located in a place which influences the cost of production. It can be imagined that manufacturing and distribution costs may vary to an extent of 10 per cent simply by virtue of the choice of plant location.

Location Problem

    Two levels of problems must be considered when considering plant location. They are : (1) the selection of the general territory, and (2) the selection of the community and plant site. Although some location factors, discussed later, may be applicable at both levels, there are certain unique considerations when selecting both a general territory and the community with its plant site.

    In Western countries much attention is given now to the problem of dispersal on account of nuclear attacks. This may become a real problem even in this country. Apart from the dispersal needs for unclear attack, there would be a problem of strategic location of plants connected with defence production either as sub-suppliers or main producers. The continuous treat of disruption of trade along the borders automatically prevent any concentration of industry near these areas. Such major factors are generally considered to eliminate (or alternatively, narrow down) the choice of the general territory after which the selection of the community and plant site follows.

Territory Selection

When making the selection of a general territory, obtaining specific information about a given community can be a waste of time. Territory selection calls for information of a more general nature. Location information of a general nature may be obtained from the following sources :

— Ministry of Industrial Development and Internal Trade.

— Institute of Public Opinion.

— Techno-economic surveys conducted by State Governments.

— National Control of Applied Economic Research.

— Indian Federation of Chambers of Commerce.

— State Chambers of Commerce.

— Department of Industry in State Governments.

— Survey of India Maps.

Community and Site Selection

Once the general territory for location has been selected, it becomes necessary to choose community and a site. This choice is based upon consideration of the factors discussion in the next section. It should be noted that some of these location factors must be considered at the community and site level as well as when making the selection of the general territory. An early decision must be made regarding the size of the community in which the plant is to be located. The alternative choices can be classified as : (1) city, (2) sub-urban, and (3) country locations.

City *versus* Sub-urban *versus* Country

The advent of the automobile (and State Road Transport) has brought new mobility to our working force. This is one of the reason for present day industrial rush to the country. Wide open spaces and freedom to expand are probably two of the biggest inducements. The type of manufacturing process may decide the site selection. For example, a country location is desirable for a plant producing explosives. Some of the general conditions leading to the selection of an appropriate type of community might be listed as follows :

I. Conditions suggesting urban location

    (a) Large skilled labour force required.

    (b) Processes heavily dependent upon availability of city utilities.

    (c) Multifloor building desirable.

    (d) Close contact with suppliers demanded.

    (e) Rapid public transportation available.

    (f) Technical and Research requirement.

2. Condition suggesting rural location

(a) Semi-skilled or female labour force required.
(b) Avoidance of heavy city taxes and insurance desired.
(c) Labour force residing close to plan
(d) Plant expansion easier than in city.
(e) Community close to but not in large population centre.

3. Conditions suggesting a country location

(a) Large site required for either present demands or expansion.
(b) Lowest property taxes available desired.
(c) Unskilled labour force required.
(d) Low wages required to meet competition.
(e) Morale of working force improved by country location.
(f) Manufacturing process is dangerous or objectionable.
(g) Processes causing environmental pollution.

One rule of thumb regarding the size of a site is that it should not be less than five times the actual size of the plant itself. This is considered a minimum in order to allow for loading platforms, siding, tractor tailer access, parking facilities, and storage area. Wherever possible, open land is desirable on two or more sides to allow for future expansion.

Unfortunately tempting offers of a fine site or attractive tax promises frequently influence plant location decisions. Objective data is essential to good plant location. Tempting proposal from those with an axe to grind should be considered with caution.

In order to properly select a site, a list of general specification should be prepared as follows :

1. Description of building to be constructed (including sketch).
2. Size of plot.
3. Necessary railways, highway and waterway facilities.
4. Minimum size of water mains, gas line, and power line.
5. Volume of ground water to be utilized.
6. Sewerage and effluent disposal requirements.
7. Safety area for offensive odours, noise, smoke etc.
8. Provisions for spinkler pressure (gravity tank or local waterman).

The maps published by the Geological Survey of India are very useful in selecting a good plant site. These maps show land elevations, water features and works such as roads, dams, buildings, railways, but not power lines.

Influence of Location on Plant Layout

Plant location will determine the proximity of a plant to its source of raw materials and its market area, the distance from the plant to these two areas tends to determine the method of transportation to be used. The type of transportation will in turn, determine whether the layout should provide for railway, truck or water loading and unloading facilities. The arrangement of the shipping and receiving department will vary in the layout according to the type of transportation utilized. A plant location may be determined in part by the fuel requirements of the concern. The plant layout must provide for storage of this fuel, whether it be coal, oil, or gas. Also, the layout must consider the requirements for power generation.

The demands of future expansion on the plant layout are influenced by the location of the plant. When plant expansion in a city location must take place by adding stores to a presently constructed building, the plant layout problems are somewhat different than they would be in a country location, where plant expansion might take place horizontally by adding a wing to a single storey building.

Factors Influencing the Plant Location

Following factors must be kept in mind before final selection of plant location is made :

1. Market. Proximity to market is one of the important factors affecting the choice of plant location because both promptness and cost of service to customers depends upon proximity to the market.

Depending upon the product, market may be concentrated or widely dispersed. When a market is concentrated, the market factor would tend to influence the enterpreneure to choose close to this concentration.

For example, a manufacturer of textiles would opt to locate near Bombay and Ahmedabad or a manufacturerer of automobile parts would be opt to locate near Madras or Bombay.

2. Labour. The business can never bring fruits unless it can obtain and maintain a productive labour force. Not only must the labour force be available in adequate number but also it must contain the skill required in a given manufacturing process.

There may be any development in modern machinery, which in many instances reduces the importance of skilled labour, the key to success in any business or organisation is the human factor ; worker can make or mar any organisation. Wage levels must of course be considered. In general wage and skill are lower in south than in the west or east. Therefore industries requiring many unskilled workers, which traditionally pay low wages are attractive to the south. Lower wages in south are partially explained by the fact that area standards are also low in that area.

3. **Transportation Facilities.** The problem of transportation is of great importance specially when raw materials are bulky and of low value. The more rent of material can consume a very high precentage of the final cost to the customer. An industry tends to be localized at places which have developed means of transport, Faridabad in Haryana is an excellent example. It is well served both by railway and road transport. It lies on the mainline between Bombay and Delhi.

Transportation Media

The following are the transportation media with particular advantages of each.

Rail (all classes of traffic).

Water carries (all classes of traffic).

Highway vehicles (all classes of traffic).

Pipelines (bulk liquids and gases).

Aircrafts (where speed is essential and where access by surface agencies is specially difficult as in Assam).

Pack animals (in difficult terrain).

Belt, cable or rail conveyors of various types (short distances).

Human carriers (short distances and small quantities for example, purchases of compact, finished consumer goods and delivery of letter mailed to final destination).

Self delivery (livestock and self propelled vehicles).

Electric cable (electric energy).

Telecommunication (information, commercial negotiations).

Railways

Indian Railways have very large net work of traffic system. According to the type of the product, Railway carriages are designed for better transportation. This is one of the cheapest mode of transport also.

Roadways

Many transport organisations are existing in India. For hill station and to carry heavy machinery, Road transportation quite good and Rail transportation may not be suitable for this purpose.

Waterways

This mode of transportation is mainly for international transits. Bulk and non perishable materials can be transported to far of places with low charges. But the transportation time is very high. Port

Trust of India, and some Shipping Corporations of India are playing very vital role in this method of transport.

Special Methods

(i) **Pipe lines.** To transport bulk liquids and gases to short distances. This is the economical method, for example the supply of cooking gas to houses near to a Refinary city. Similarly transport of crude oil from a well to a Refinary.

(ii) **Belt and cable convertors.** For internal transport, or from a mine to shipping area, for example Coal after quarring will be transported to Rail or Truck looking places by only bucket converyors.

As you can see, each of these transportation media has its advantages and limitations. In order to select the proper transportation media, the shipper should consider the following :

1. Type and extent of materials handling facilities at origin and destination.
2. The relative costs for the various media.
3. The urgency of the shipment.
4. The demand for special services, e.g., refrigeration.

Transportation costs vary with the type of route used and the type of media selected as well as the length of haul. In general, the cost of moving materials per mile tapers off as the length of haul increases.

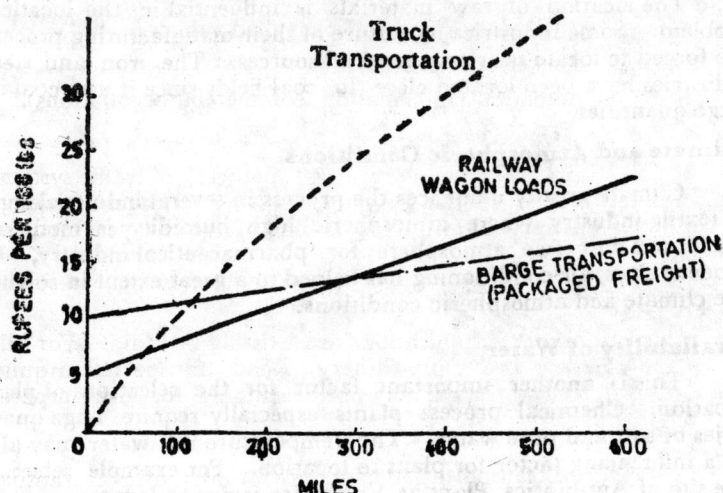

Fig. 34.1. Hypothetical Cost Comparison of Various
Transportation Methods.

We can use as analysis involving the break-even point to select a transportation media (Fig. 25·1). In this case truck transportation appears to be the most economical up to a distance of approximately 40 miles. Transportation by large appear as to be the most economical for long-distance hauling, that is for distances greater than about 280 miles. For hauling between 40 to 280 miles the railway appears to be the most efficient carrier.

The transportation method of linear programming has established itself as a useful tool for considering the transportation problems in locating a plant. The linear assumptions required by this technique are usually quite realistic in this type of problem.

Availability of Power and Fuel

In the last century the industries were located near the coal mines or the locations where coal could be taken cheaply and conveniently. But today all industries require electric power of some sort. There are certain process, that require usually large amounts of electric power. For example heavy water plants or chemical processes of hydrolysis nature are all located in areas where large sources of cheap power are available. In India we have three types of power sources : (i) Thermal power (ii) Hydro electric power (iii) Nuclear power.

For any industry the availability of power should be cheap, continuous and regular. Technologists are exploring the possibility of obtaining power from wind and sea waves.

Availability of Raw Materials

The location of raw materials is influential in the location problem. Some industries by nature of their manufacturing process are forced to locate near raw material sources. The iron and steel industries have been located close to coal fields since it uses coal in large quantities.

Climate and Atmospheric Conditions

Climate greatly influences the process in several industrial such as textile industry where atmospheric high humidity is essential. Clean and dust free atmosphere for pharmaceutical industry. In modern days, air conditioning has helped to a great extent in solving the climate and atmospheric conditions.

Availability of Water

This is another important factor for the selection of plant location. Chemical process plants especially reqnire huge quantities of soft and pure water. The temperature of water may also be a influencing factor for plant to location. For example selecting the site of Antibiotics Plant at Virbhadra is due to low temperature of Ganges water and less bacteria logical load to reduce the refrigreation loads.

Laws and Taxation

State and local laws should be studied when considering various locations. Labour laws and employment regulations vary widely from one location to another. Some of the aspects of industrial operations regulated by laws are hours of work, minimum wage and fringe benefits to workers and working conditions for woman employees.

Special and Recreational Facilities

As social and recreational facilities are the necessities of one's life after the plant hours and on holidays, the employees need the same. Therefore it is essential that Government should provide suitable parks, cooperative stores, community halls, cinemas and education facilities for the employees. All the above facilities shall be for the good health and will therefore create much interest in the work. In certain factories it is seen that the employees are given very cheap subsidised meals.

Waste Disposal

Pollution control act is in force now. The effluents and sewage disposal of a plant should not harm the public. So proper effluents treatment and waste disposal methods should be available. It must be examined that drainage may accumulate water around the plant and create unhygienic conditions. Similarly, if the load bearing capacity of the land is low, a company may not be able to establish sound building foundation or build a heavy building. The installation of heavy machinery may also be difficult.

Technology Know How

The know how techniques should be available near sources; so that delay in consultation will be reduced if any break down or troubles occur in the plant. Thus a plant should also be near to a consultation organisation so that for future expansion problems and present difficulties can be discussed oftenly for smooth running of the plant.

35

Plant Layout

Introduction

Production results from the working of men, materials and machinery. One or more pieces of materials are taken and their form, nature or chemical characteristics are changed by using machinery including tools and equipments to give a product. The product has a greater value than the sum total of all the inputs that goes into the making of it as it serves a need and is functional. The value is created by incurring the cost in the process of production and using of varieties of production facilities. While physical arrangement of production and other facilities have a bearing on the cost due to maintenance, quality control etc. They have also a direct effect on the cost due to the handling of material and movement of men and other facilities. It is estimated that in the cost of production about 30% is the cost of material handling at various stages in various steps. The chapter deals with the study the effect of systematic and good plant layout and material handling system.

Definition

The layout of a factory, workshop or working area means the position of the departments or shops in the factory and of the machines, work places and storage points in the working areas, including officers and staff facilities relative to one another.

A bad layout can add a good deal of time to the total work. Contents of a process which causes unnecessary movement of material and consumes time and energy of the worker without any value addition in completion of a job.

Many establishments have never thought of any change in the layout ever since they were established. Machine, pieces of plant and even whole departments have been added from time to time wherever space could be found. The result has been that material takes often long time and charters a round about journey in the course of process.

The importance of layout to the production of process varies from industry to industry. Equally important variable is the extent to which it is possible to alter the layout once it has been established. These two factors must be borne in mind while studying the flow of material or movement of workers about the plant.

Objectives of Layout

If a finished layout is to present an effective arrangement of related work areas in which goods can be economically produced, it must be planned with objectives of plant layout well placed in mind.

The major objectives are :

(*i*) Facilitate the manufacturing process.

(*ii*) Minimise the material handling.

(*iii*) Maintain flexibility of arrangement of operations.

(*iv*) Maintain high turn over of works in process.

(*v*) Hold down investment in equipment.

(*vi*) Make economical use of floor area.

(*vii*) Promote effective utilisation of labour.

(*viii*) Provide employee convenience and comfort in doing the work.

Manufacturing Process Facilitation

Arrange machines, equipment and work so that material is caused to move as smoothly in a straight line as possible. This does not mean that flow line must be straight, there should be minimum of back tracking and

(*a*) Eliminating all delays possible.

(*b*) Planning the work so that work can be identified and camed with little possibility of mixing with other parts.

Plant maintenance of machines which will maintain quality of work. This may involve capital expenditure. Such air-conditioning of areas or good house keeping to prevent damage of parts.

Minimizing Material Handling

The material should travel from the machine to the other as little as possible, specially the heavy components should flow in line only because excess flow of material require man power as well as equipment for lifting and carrying the material.

Flexibility

The layout should be flexible for the following causes :

(*a*) Design change.

(*b*) Increased production.

(*c*) Adding a new product.

(d) Maintaining high turn over *i.e.* the storage time of the material inside the line should be minimum. Because extra storage of material require more storage areas as well hinders the flow of material.

Economical use of Area

Each square feet area costs money. Unaccepted waste or idl floor is burden on rest of the plant spacing between machines afte necessary spacing required for movement of men and material.

Effective use of Labour

Labour may be wasted through poor layout. Following factors should be considered for effective use of labour utilisation :

I. Reduce manual handling.

II. Minimize walking

III. Balance machine cycles so that man and machine are not idle.

IV. Laying the machine in such a way that supervision is simplified.

Employees Comfort

The layout should be such that the worker is not effected by heat, moisture or dirt. Direction of light should also be considered for laying the machines. Equipment causing more noise should be isolated as far as it is possible. Equipment which vibrates should be cushined or specially mounted to prevent transmission of vibrations to floor or surrounding area.

Plant Layout Problems

Ordinarily when one thinks of plant layout, then comes the problem of planning a complete layout for a new product, that is planning a new plant, starting from scratch. This is usually not the case more frequently problem involves the re-layout of existing plant or alteration of some sort of arrangement. Following are the main problems of plant layout.

Changes in Design

Frequently change in design of a part calls for change in process of operations to be performed. This change may require only alterations of existing layout, or it may cause extensive re-layout programme depending upon nature of change.

Enlargement of a Department

If for one reason or the other it becomes necessary to increase the production of certain part of product, a change in the layout may be called for. This involves the addition of a few machines which call for entirely new layout so that, part need not to travel long.

Reduction of a Department

This problem is a reverse of the Enlargement of a Department. If products demand are reduced drastically and permanently then few machines have to be removed and rest of the machines have to be rearranged.

Adding a new Product

If a new product is to be added to a line then it is similar to the product already being done, the problem is similar to enlarging a department. However if the new product differs considerably from those in production, a different problem presents itself. The present equipment may be used by adding a few machines or completely a new department may be desired.

Moving a Department

Moving a department may or may not present a new problem. If present layout is satisfactory it is only necessary to shift to another alternative position. If, the present layout is not satisfactory, an opportunity presents itself for correction of past mistakes.

Planning a New Plant

This presents a largest problem in plant layout. Building can even be designed to suit the layout. The additional area has already been provided by removing the receiving department from its present position.

Types of Layout

As already noted production results from men, material and machinery together with management. The characteristics are changed to make a product. An analysis of the above gives the various types of layouts. There are three types of layout :

1. **Layout by fixed position or by fixed material location**

Here the material or major component remains stationary. The complete job is done herewith material, men and machinery brought to the location, e g. ship building.

2. **Layout by Process or Layout by Function**

In this case, all operations of the same process or type of process are grouped together, like all welding in one area, all drilling in one area.

3. **Layout by product or line production**

This layout places one operation adjacent to the previous operations. The equipments regardless of process are arranged as per the sequence of operations.

In ideal conditions the raw material comes in at one end of the factory, travels through it in a straight line and comes out at the other as a finished product ready for dispatch.

Ideal conditions are not often possible in real life situation. There is no harm in the work travelling around the factory, provided that it follows a fixed path, that the distances between successive operations are kept as short as possible and that the work moves steadily forward.

In factories where many products are made, or where the products are made up of many different parts, a good layout is much more difficult, especially where batch quantities are small and there is a choice of processes. This situation is generaly found at its worst in the engineering industry. A decision may have to be taken between having a *process* layout and a *product* layout.

A process layout is one in which all machines or processes of the same type are grouped together.

A product layout is one which all machines or processes concerned in the manufacture of the same product or range of product are grouped together.

Most workshops today are a mixture of the two types. The following is a short list of the advantages and disadvantages of both.

Process Layout : Advantages and Disadvantages

(1) The sequence of manufacturing is flexible and a minimum investment in machines is involved. More floor space is usually occupied than in a product layout.

(2) Machines can be kept busy most of the time, low and medium-volume production costs can be kept down.

(3) There are no definite physical channels along which work must flow. The results of this are more handling of materials, larger bank of works in process and a more complicated system of production control than with a product layout.

(4) Workers and supervisors become skilled in the operation of a single type of machine but require longer training to be able to deal with varied jobs. Thus specialisation can be achieved.

(5) Machine breakdowns do not held up a succession of operations ; work can be transferred to other similar machines.

Product Layout : Advantages and Disadvantages

(1) The flow of work over direct physical routes cuts out delays ; there is less material handling, and the definite sequence of operations over adjacent machines simplifies production control and cuts out forms and records.

(2) The total time of production is kept low, less floor space is occupied and the capital tied up in work in process can be kept down. The capital investment in machines may, however, be raised by duplication of the same machine on several lines.

(3) Manufacturing costs are low at high volume production but rise steeply as it drops. If one or more lines are running light there is considerable machine idleness.

(4) It is easier to train workers inexperienced in industry upto a certain level of skill.

(5) A single machine break down may shut down a whole production line.

Which of the layout types to be used

(1) Layout by fixed position—this type of layout is suitable where—

- (a) Operations required hand tools or simple portable machines.

- (b) Making only one or few pieces

- (c) Cost of moving the major pieces is high

- (d) Skill of workmanship lies in the ability of workers or where we wish to fix responsibility for product quality to one workman.

(2) Layout process—this layout is used where—

- (a) Expensive machinery is involved.

- (b) Variety of products are made.

- (c) Operations are unblanced or in other words wide variation in time required in different operations are encountered.

- (d) The demand for the product is small or intermittent.

(3) Layout by-product or line production—this type of layout is used when—

- (a) There is a large quantity of pieces of products to make.

- (b) Design of the product is more or less standardised.

- (c) Demand for the product is fairly steady.

The Fig. 26·1 shown on the next page shows the cost comparison between product and Process type Layouts.

The Combination Layout

The combination method of layout is feasible when a number of products require about the same requence of functional operations but none enjoys sufficient volume to justify individual production lines. The principle of this method lies in the arrangement of functional departments across the building at right angles to flow of product and in the required sequence of operations. Fig. 35·1 shows a typical combination layout.

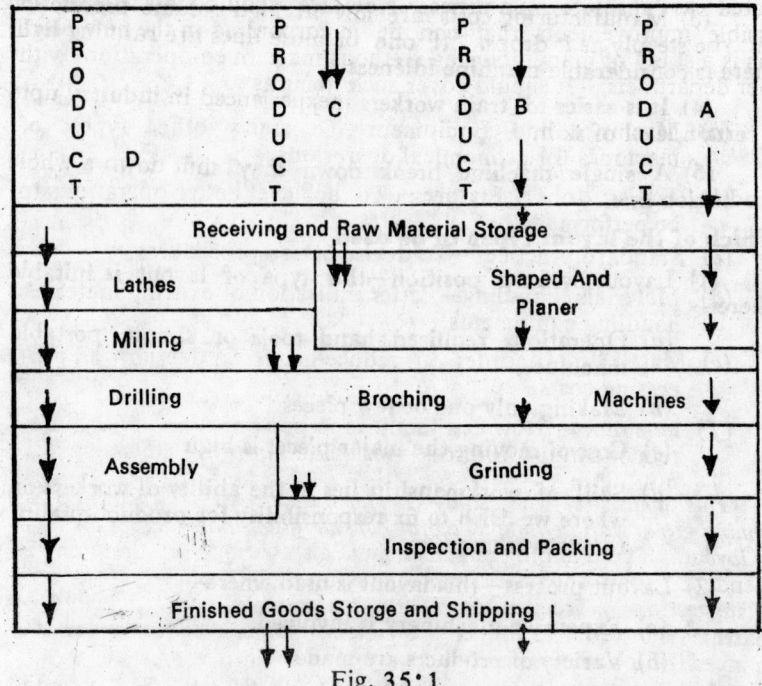

Fig. 35·1

One of the example for Combination Layout is steel office furuiture. The sequence of operations for each component part is generally the same viz Shear or blank stock for part ; Punch and notch blanks ; Bend or form blanks ; Spot weld or gas welding component parts into subassemblies ; Finish (Spray ; bake etc.). Assemble final product ; Instale hardware (Lock ; hinges etc) ; Pack and ship. A company producing several different style of desks-filing cabinets, chairs, cabinets table, and so forth can make profitable use of such a layout.

The Layout of Production Equipment Procedure for Plant Layout

Plant layout is just one limited phase of factory planning. The layout of a factory generally consists of a number of separate problems of arrangement of facilities, and each problem is small enough to be manageable. The more complex the product becomes the more important it is to follow a systematic method to simplify the problem. The procedure should first guide the planner's actions so that the probability of his arriving at an optimum arrangement is maximised, and, second, it should provide him with a basis for judging the relative merits of any one proposed layout versus others. At treated as absolute quantities. Now we should investigate all possible improvements that can be in corporated in the new plan. This is a study of manufacturing methods made in co-operation with other departmets. It should cover such items as :

(a) Machines and Equipment—To study other types of machines for economical operations.

(b) Tools ; Jobs ; Fixtures—To achieve better operations to be performed.

(c) Standard practices—To devise better procedures.

(d) Materials handling—Better utilisation of existing materials handling equipments.

(e) Materials—In order to reduce materials, labour and processing costs.

(f) Personnel—How can facilities to personnel be improved to reduce fatigue, improve morale and improve efficiency.

This phase of the study is most important as it provides the planning group with definite ideas for consideration in arranging for the new layout. The planning engineer must also recognise that he is dependent upon the other departments in the plant for information and suggestions. Thus it pays him to co-operate fully with other departments in order to induce these departments to do as much of this investigation as possible.

The next step is to make the basic decisions and establish alternative methods of arranging the factory. There basic decisions must be treated as tentative, for as the details are worked out it may become necessary to change the decisions partially or totally. *The decisions were made on the basis of their probability of being best, but they can be proved only by actual computation of their results.* Now we will go to prepare and test the several alternative arrangements, choose the best, and prepare the report to management. The report will contain information regarding department boundries ; Arrangement of machines within the department boundries ; analysis of each arrangement to determine cost of production, materials handling and supervision ; Schedule for the installation and a final plan for management's approval. Finally work out a complete instruction for the re-layout and supervise the installation.

Factors affecting layout

Whatever be the type of layout being contemplated, the following factors influence the selection in every case.

(i) Material ; (ii) Machinery

(iii) Man ; (iv) Movement

(v) Waiting ; (vi) Service

(vii) Builing ; (viii) Flexibility.

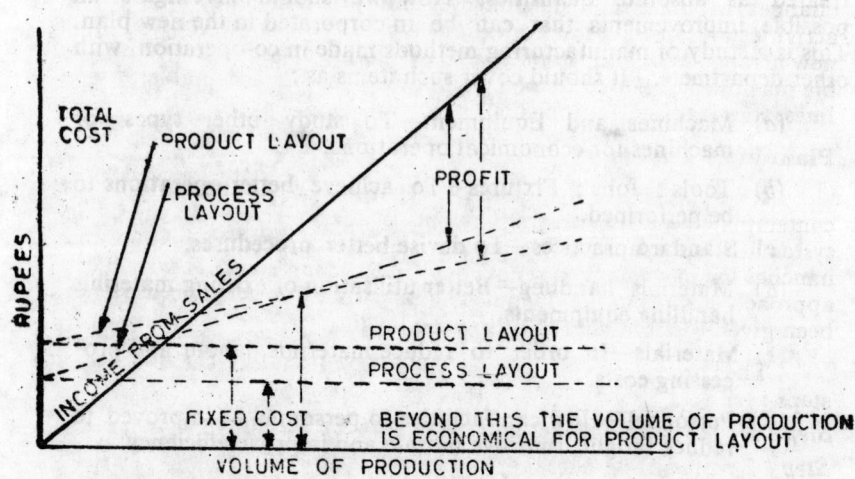

Fig. 35·2.

1. Material factor. This include the design, variety, quantity, necessary operation and their consequence. These are affected by : (a Design and specification, (b) Physical and chemical characteristic, (c), Quantity and variety of products and material and (d) Component parts or material and how they go together.

2. Machinery factor. This include process or method, machinery, tools and equipment, machine utilisation and requirement of machinery and power.

3. Man factor. This include safety and working condition, manpower requirement, man utilisation etc.

4. Movement factor. This includes the flow pattern reduction of unnecessary handling, space for movement and analysis of handling methods and equipment.

5. Waiting factor. This include location of storage or delay points space for waiting, method of storing, safeguard equipment for storing and delay.

6. Service factor. This include services for personnel such as fire protection, lighting, heating and ventilation etc., services for material such as quality control, production control and services for machinery such as maintenance and utilities like, gas steam, water supply etc.

7. Building factor. This includes features like Railway siding, yard areas for parking and storing, shape of building. numbers of floors, facilities of elevation stairs etc.

8. Flexibility factor. This includes consideration due to changes in material, machinery and process, man, supporting activities and installation limitations etc. It means easy changing to new arrangement and is facilitated by mobile or mounting or movable machinery, self-contained equipment, readily acceptable service lines, standardised equipments.

Planning a Layout

A. Systematic Approach. Whether a new layout is being contemplated or an existing layout is to be changed, there is a systematic approach in doing so. Dis-coordinated thinking and hanches for movement do not pay in the long run. A systematic approach ensures to the layout engineer that due consideration has been given to all factors affecting the layout.

The procedure for layout project will entail the following steps :

Step 1 : Study the 'make' and 'buy' items in the material list.

Step 2 : Study the current and future output target for each product. Determine the component quantities to be stocked and manufactured.

Step 3 : Prepare operation planning sheet. Obtain production routine equipment requirements and process times from other departments.

Step 4 : Determine number of equipments of each category required and direct man power requirement.

Step 5 : Determine space requirement at each work spot and make tentative work place layout.

Step 6 : Study the stock policy and find out the storage requirements for raw materials, work in process, 'Buy' items and finished goods.

Step 7 : Obtain packaging informations to determine the quality of packaging materials and the final shape of material for shipment, obtain also the shipping schedule.

Step 8 : Prepare operation process chart, flow process chart and an overall and detailed diagram of the flow.

Step 9 : Make tentative layout by trial and error.

Step 10 : Choose handling procedure. Determine the size of the unit load number of units to be moved and the distance. Also calculate the indirect man power required.

Step 11 : provide adequate gangways for movement.

Step 12 : Discuss with line management and incorporate their suggestions.

36
Material Handling

It is an important area of study with a view to effect cost reduction by suitable material handling system and devices in a plant. The material handling has became more and more complicated due to the development of new methods and techniques. Material handling has now become a specialised function. In the pre-scientific management period the material handling was only man handling and handling was done in most uneconomical may by human labour only.

Definition

Material handling has been rightly defined as an Art and Science involving the moving, packing and storing of substance in any form. Material handling is the preparation placing and positioning of materials to facilitate their movement or storage. The total cost of material handling formed is indicated in the following diagram.

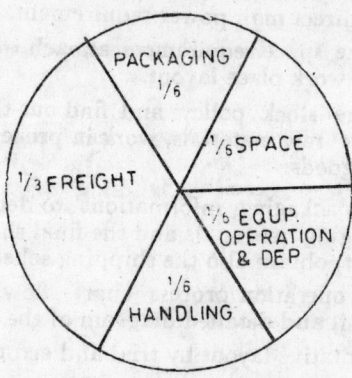

Fig. 36.1

Briefly material handling is the moving of material from the raw storage through production to ultimate consumer with the least expenditure of time and efforts, so as to produce maximum productive efficiency at the lowest material handling cost.

TERMINOLOGY OF MATERIALS HANDLING

The various terms used in the terminology of Materials Handling are :

Materials Handling

Any movement of materials, vertically, horizontally or both, manually or mechanically, in batches or one piece at a time.

Transport

Any movement of a lot or batch of product from one production center or storage area to another production center or storage area. Transport is usually accomplished by "movement" and or materials handling equipment.

Transfer

The movement of pieces, singly or in small quantities from a container to a machine where the operation is performed and back into another container. All materials are handled in either bulk or packaged form. This handling is performed by the machine operator and is not formally charged against the cost of materials handling.

Bulk Materials

Any Materials that are loose and handled in quantities without being contained in bags, boxes barrels, and the like.

The material is usually deposited in a bin, tank, elevator, or other container, from which it is fed into the processing equipment.

Rehandle

A term applied to the *movement of a piece ; package or unit load*. *Rehandle consists of the pick-up ; move and set down*. This is useful in analysing unnecessary or excessive handlings that can be eliminated.

Unit Hendle

The unit load may be made up of a certain number of individual pieces secured to a pallet or deposited in a box. It is used with bulky items in manufacturing items.

Packaged Material

Materials contained in convenient sized packages such as bags, crates, cartons cans, barrels, boxes or other types of containers that can be handled as individual pieces by the materials handling system.

Analysis of Materials Handling Problem

The materials handling problem requires the accumulation of a large amount of factual data and their analysis relative to the particular plant and production conditions. This analysis can best be performed when the collection and analysis of the data follow a logical pattern. Such a pattern will include following items :

(*i*) Plant factors

(*ii*) Methods factors

(*iii*) Products and materials

(*iv*) Present handling methods and equipment.

(*v*) Proposed handling methods and equipment.

(*vi*) Cost data and economic analysis.

Plant Factors

The refers to the conditions of the building and layout as they are proposed to be. The analyst needs to have data on the size and relative location of buildings and interior features such elevators, stairs, and columns ; floor load capacity ; ceiling heights, piping and power circuits etc.

Methods Factors

It encompass all the details of production methods, equipment, processes, sequence of operations, production plan, temporary storages ; volume to be handled, so forth. The methods employed in production impose special restrictions can be used to eliminate whole groups of handling systems thereby simplifying the selection.

Products and Materials

The specific kinds of products and materials to be moved and volumes and distances of each move provide the engineer with additional specifications for the system. Flexibility of the system is a guard against early obsolescence.

Present Handling Methods and Equipment

If the present materials handling problem is in an existing plant, it can be assumed that there is existing handling equipment and that it is either inadequate, expensive to operate, or unsuited to the handling job.

Proposed Handling Methods and Equipment

The fact that existing equipment can be used, apparently with satisfactory results, does not mean that this plan is necessarily the most economical. Only when the engineer has obtained all the data, he is in a position to investigate the available materials handling systems to find those that will satisfy the needs of the plant.

Cost Data and Economic Aanalysis

With the number of alternatives reduced to popular methods used and is the one that is most likely to give the best machine arrangement with the smallest overall cost of planning.

Basis for Material Handling Analysis

The material handling equation may be helpful in inter relating the many factors inherent in the analysis of a handling problem. Fig. 36.2.) gives the checklist for determining the properties of the material to be handled.

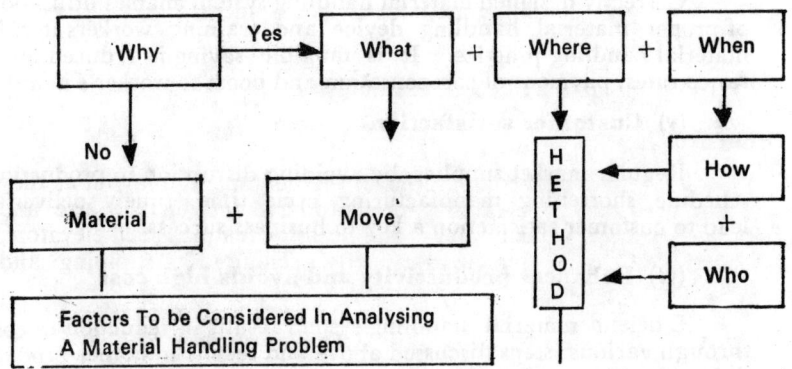

Fig. 36.2 Material Handling Equation

Objective of Material Handling

(*i*) Reduction in wastage of machine time.

(*ii*) Reduction in manufacturing cycle time.

(*iii*) Avoiding disruption in production schedule.

(*iv*) Bring safety to workers and provide for safer working conditions.

(*v*) Bring customer satisfaction.

(*vi*) Enhancing productivity and avoid high costs.

(*i*) Reduction in wastage of machine time

Proper material handling system would ensure regular supply and distribution of raw material and finished product. This will not create the problem of idling of machine and eliminate problem of wastage of machine time.

(ii) Reduction in manufacturing cycle time

Keeping the loading and unloading time of machine at minimum enables the completion of manufacturing cycle at reasonably short time.

(iii) Avoid disruption in production schedule

In mass production system, all parts or materials must come in the assembly line in correct quantity and at precisely the right time to avoid risk of shut down.

(iv) Safety and safe working conditions

Correctly designed material handling system enables utilisation of proper material handling device and training workers in safe material handling practice. It is tangible saving in reduced insurance rates, payment of compensation and boost to worker's moral.

(v) Customer satisfaction

Regular market supplies, by avoiding disruption in production schedule, shortening manufacturing cycle time, timely deliveries lead to customer satisfaction a key to business success.

(vi) Enhances productivity and avoids high cost

Efficient material handling system results in reduction in cost through various steps discussed above and results in higher productivity.

Principles of Material Handling

The various objectives listed above can be achieved by following certain principles. It will be well to remember that these are the general principles only and in practice the individual industrial problems affect the evolution of material handling system. Factors like the production system, types of building and cost of various devices largely effect the system in a particular industry. General principles are :

(i) Materials should be moved over the shortest distance possible. Short movements require less time and cost less money than long movements.

(ii) Do not retrace your path. The aim of material handling system should be to move in forward direction or clockwise direction.

(iii) Terminal time should be kept as short as possible. It is time taken for picking up the finished product.

(iv) Loads should be carried both ways on material handling trips whenever possible.

(v) Partial loads and manual handling should be avoided.

(vi) Utilise gravity as moving force as far as possible.

(vii) Utilise containers and unit loads. The unit load means the product to be moved should be grouped into units of a large and constant size.

(viii) Materials should be already marked to help save time in sorting and avoiding mixing up.

Factors Affecting Materials Handling System

The following four factors affect material handling system.

1. Production system.
2. Product to be handled.
3. Type of building within which material is to be handled.
4. Cost of material handling device.

1. Production System

There are two basic types of production systems. One is intermittent or job-lot production ; the other is continuous-production. Intermittent systems are used where a variety of jobs are being done at the same time on a variety of machines and in a variety of places. Since a great deal of flexibility is necessary it is likely that variable path equipment will be used more than fixed-path equipment. In terms of materials-handling devices lift trucks, pallets, hand trucks and trolleys would be more useful in intermittent production system in continuous production systems. These materials-handling devices would not be so commonly used in continuous-production systems, because fixed-path equipment would be more reliable and more economical in many cases. In mass production or continuous-process systems, one would find extensive use of conveyors, cranes, hoists, pipelines and in modern plants, automatic of transfer machines and tape-controlled indexing machines.

2. Types of Products to be Handled

The type of products being handled greatly affects materials handling decisions. Powders, liquids and gases are well suited to transport by pipelines. Automatic transfer machines should be used in automatic processes where a sufficiently high volume of production is available to offset the high cost of the equipment. Cranes and hoists are best suited to very heavy lifting jobs, where the material to be used cannot be easily lifted from below. Conveyors should be used where high volumes of products move from one fixed point to another. Trucks should be used where flexible destinations are required and where the volume of material to be moved may be low. The characteristics of the products, including weight, size, shape, perishability and whether it is a solid, liquid or gas have significant bearing on the selection of the type of device which should be chosen to move them.

Type of Building

The number of floors has a direct effect on the devices which should be used. Single-storey building, the most common type being built today lend themselves to the use of lift trucks and conveyors. They also eliminate the need for elevators, which often prove to be a bottleneck in materials-handling. Multistoryed building, however, find themselves to the use of gravity flow with pipelines and shuters which is the most economical method of materials handling.

The last major factor is the cost of various devices which are available for use. Cost comparisons among various alternatives are often difficult to make. The initial costs of alternative devices should be considered. The usual life of equipment will also affect the decision. Resale or scrap value is the other factor. Perhaps most important are operating costs which include fuel maintenance, repair,

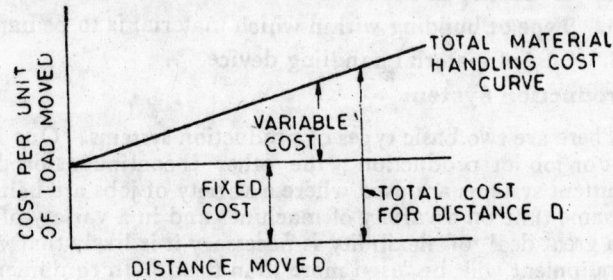

Fig. 36.3.

insurance and labour costs. These costs must be determined for the alternatives before estimates of the value of one device versus another can be determined.

The graphical representation given in Fig. 36.3 shows the relation between cost per unit of load moved and distance moved.

Analysis of Material Handling Problems

Following are the various important factors which need to be studied to analyse the handling problems.

1. Establish the Scope of the Study

(i) What is the degree of thoroughness is expected of the study ?

(ii) Should the study includes related functions such as manufacturing, financing, personal relation etc ? If not what areas should be included ?

2. Pinpoint the Physical Boundries of the Study

(i) What areas in the plant layout should be covered by the study ?

3. Determine Production Forecasts

(*i*) What volume of material is expected to be handled by the equipment ?

(*ii*) What is the expected range of expansion ?

4. Determine the Most-handled Items

(*i*) Which material require the most of the handling work ?

(*ii*) What is the total handling cost of these items in relation of the others ?

5. Compute the Cost of the Present Handling Method

(*i*) What are the functional elements of the system ?

(*ii*) How many man-hours are spent for each element ?

(*iii*) What are the other costs directly related to the handling operations ?

Analysis of Material Handling

Materials handling problems vary from situation to situation. The gravity of the problem varies according to whether production is continuous or intermittent. This in effect means only that a detailed analysis of various aspects of materials movement is necessary.

1. What to Move

Data is collected on what materials are to be moved. The bill of materials coupled with the production schedule would give this information. Additionally all information regarding weight, size, shape, quantity etc., of the materials to be transported is to be collected. The following form could be used.

ITEM SPECIFICATION LIST

ITEM			UNIT				HOW MOVED			
Description	Size	Wt.	Qty.	Qty. per unit	Size	No. of units Reqd	From	To	Dist.	Eqpt.

2. Where to Move

Once the data is available as to what to move, the details of where to move are to be collected. It is necessary that information is collected in the form of distance × weight since this is the most pertinent information. The information could be collected in the form given below :

From \ To	DEPARTMENTS			
	A	B	C	D
A				
B				
C				
D				

3. How to Move

The data required to be collected is (i) the equipment used for movement, (ii) whether material is moved individually or in units like tray, pallets, bundles etc. and (iii) the time taken for the movement. Details of equipments used are to be collected, viz. (i) capacity, (ii) speed, (iii) flexibility and (iv) manoeurability. Hence the factors affecting material handling can be summarised as follows :

What to move	:	Materials-form, characteristics, quantity.
In what	:	Containers
Where	:	Routing
How often	:	Frequency and duration
How far	:	Distance
How fast	:	Speed
On what	:	Paved gangway, curves
By whom	:	Type of labour
By what	:	Equipment.

Material Handling Surveys

These surveys are of mainly three types :

(1) Plant-wide survey

(2) Department survey

(3) Area survey.

1. Plant Wide Survey

It is there to provide a material handling system in the plant. It can be carried out in following steps :

(i) Prepare a plant activity drawings

(ii) Prepare a plant-layout drawings

(iii) Draw the flow of samples of major materials, assemblies and products.

2. Department Survey

After selecting the appropriate department of plant-activity we should prepare a material-handling chart showing methods of transportation, quantities, distances moved, times and frequency of movements.

3. Area Survey

In the area survey we are able to analyse each material handling situation in the department. In this we use the following techniques.

(1) Predetermind times and

(2) Activity sampling.

The survey of material handling reveal—the kind, the type, the volume, the weight and other characteristics during processing, storing and shiping.

In carrying out the survey following information should be available :

1. How often handling occurs.

2. Number of times each type of material is handled.

3. The type of the material to be handled.

4. The form in which the material is received, moved and despatched.

5. The dimension of the material handled.

6. The nature of route the material follows.

7. The probability of plant accident that can be contributed to handling.

8. Cost of material handling at each place.

The material handling equipment will provide three types of services :

1. Material loading and unloading.

2. Inter-departmental movements.

3. Intra-departmental movements.

The choice of the equipment requires a thorough knowledge of all available equipments. The various types of material handling equipments can be grouped into four major heads as shown in the following diagram.

Types of Material Handling Equipments

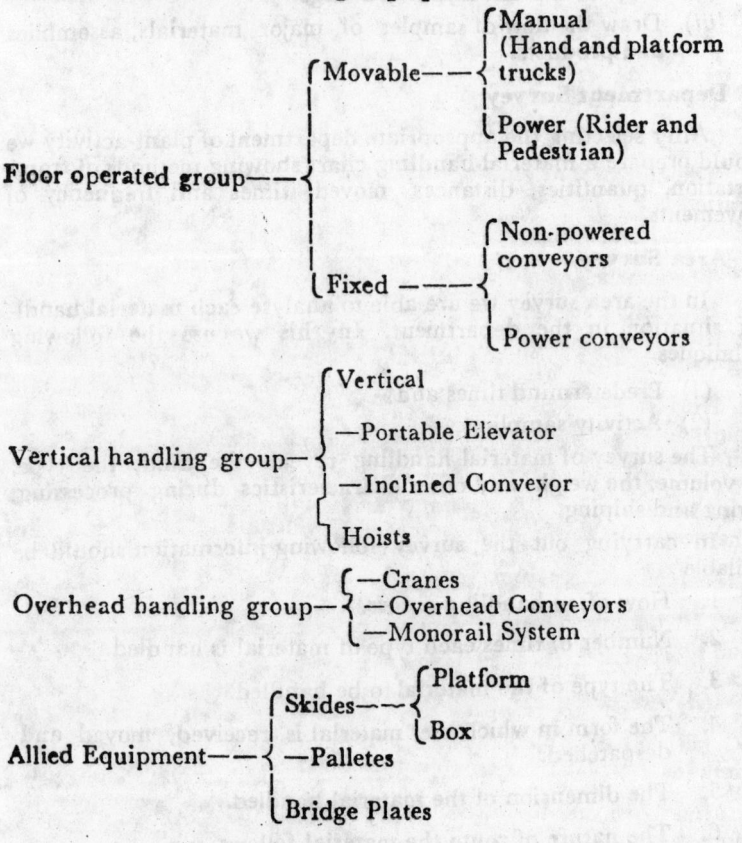

Cost of Material Handling

Following is the formula which helps us in arriving at econo mics of a material handling installation.

The most economical investment $=Z$

$$Z = \frac{(S+L+U-E)X}{A+B+C+D}$$

where $A=$ Interest on investment

$B=$ Percentage allowance for insurance

$C=$ Percentage allowance for maintenance of material handling equipment

$D=$ Percentage allowance for depreciation of equipment

$E=$ Annual power cost to operate the equipment

$S=$ Annual saving in direct labour cost

$L=$ Annual saving in fixed and variable overheads

$U=$ Annual saving or increase in earnings

$X=$ Percentage of year the equipment is used

$I=$ Initial cost of equipment.

The cost of maintaining the equipment (excluding labour) to operation is :

$$Y=I(A+B+C-D)$$

The yearly profit V made from operating the equipment is :

$$(S+L+U-E)X-Y$$

Diagnosis of Material Handling Problems

Normally in a plant there are lot of problems with which an employee is bombarded but, with passage of time he develops a habit to tolerate the inconveniences at that place and accept the problem as inevitable. However, a survey of plant would bring out several facts which need a systematic study. Certain points will show us the need of a study of the Material Handling Problem.

In Receipts and Stores

(i) Does the area where materials are received give a look of bad house keeping ? Are materials being dumped all over the space ?

(ii) Are materials piled up in the receiving yard to be moved ?

(iii) Are the members of group, handling materials are uniformly loaded ? How many men are engaged ?

(iv) Are the materials moved twice in the receiving yard before they are finally sent into the plant ?

(v) Has the storage area been divided into sections for storing different materials ?

(vi) Are the records of materials stored maintained properly so as to locate them properly ?

(vii) Are floor loading capacity and ceiling height are used properly displayed ?

(viii) Are the items of more frequent use kept near the gate ?

(ix) Is the floor properly levelled so that the movement can be made without the danger of accident or obstruction ?

(x) Does the distance of the store from the production department ensure shortest distance moved ?

(xi) Are the packages ready in time to avoid demberrage of wagons ?

(xii) Are there any damages to the materials during handling ?

In Production Areas

(i) Do skilled workers handle material or heavy object ?

(ii) Are there proper demarcation lines for material movements areas ?

(iii) Is the production area clustered with parts and materials for use rejects ?

(iv) Are in process material kept on floor ?

(v) Has the storage area been divided into sections for storing different materials ?

Diagnosis

If the answers to the above questions are observed to be in affirmative then general diagnosis is that there is need for studying in detail the material handling problems.

Developing Better Methods

After the charts have been prepared, following questions must be answered before selecting a handling system.

Materials

(i) What material is handled and why ?

(ii) Can it be handled in any other convenient form ?

Location

(i) What is existing location of the material ?

(ii) Is there a better alternative exists ?

Timing

(i) Why must this material be moved now ?

(ii) Will other operations be effected if this does not reach in time.

Handling

(i) Can the operators be relieved of handling ?

(ii) Does material travel effect the pace of operation ?

Conveyors are of following variety.

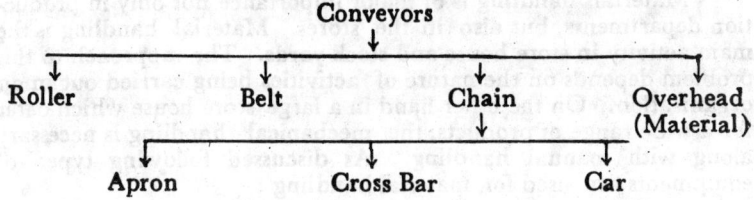

Cranes are of following models :

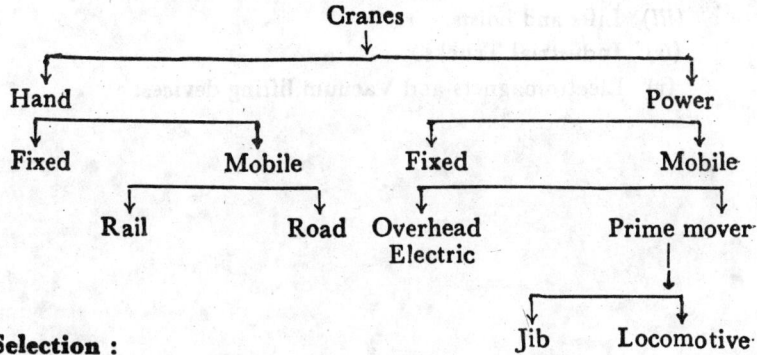

Selection :

The following comparative table can act as a guide in selecting suitable types of equipment.

Table II

Comparison of Various Types of Material Handling Equipments

S. No.	Attributes	Conveyors	Trucks	Cranes
(a)	Flexibility of operation	Move along fixed path	Highest flexibility	Move along fixed path.
(b)	Economic of conveyance and economic distance	Highest	Medium	Least.
(c)	Speed of Travel	Least	Fast	Less fast.
(d)	Cost/Unit work	Cheapest	Less cheap	Costiliest
(e)	Load Bearing	Least	Medium to high	Highest
(f)	Maintenance	Easiest	Not easy	Not easy
(g)	Air Pollution	None	Emit Pollutants except battery driver	None for Electric driven.

Summary

Materials handling is of major importance not only in production departments, but also in the stores. Material handling is the main activity in store house and stock yards. The approach to this problem depends on the nature of activities being carried out in an organisation. On the other hand in a large store house which cater for a wide range of products, the mechanical handling is necessary along with manual handling. As discussed following types of equipments are used for material handling :

(i) Conveyors.

(ii) Cranes.

(iii) Lifts and hoists.

(iv) Industrial Trucks.

(v) Electromagnets and Vacuum lifting devices.

37

Human Engineering

Human factors engineering, in method and concept, has as its principal concern with the design of efficient and safe-man-machine systems. This concern is compatible with and supplements traditional industrial engineering approches to problem of performance and work-space layout. The human factors engineer is ultimately concerned with skilled human performance. Beyond the equipment design and layout factors already there are stress and fatigue effects which should be considered in the total picture.

In fact everything designed is ultimately for the use of or by man. To ensure safe and efficient use of equipment, engineering design and layout should take into account man's capabilities and limitations. Human factors engineering embraces a broad approach with focus upon man's sensory input, motor response output, and information processing characteristics and their interaction with environmental conditions and system function requirements. The human factors engineering task is to define human performance requirements and ensure that they are not compromised by improper equipment design and layout.

The traditional human factors engineering deals with (a) Display and control design (b) Human physical characteristics; (c) Environmental factors and (d) Design for maintenance.

The humam factors engineering is characterised by concern with efficient and safe utilisation of man in Man-Machine system with emphasis on the Selection; Design; and Arrangement of system components so as to take into account both man's capabilities and his limitations. It involves a vast interdisciplinary body of knowledge and methodologies considered by the human factors specialist in his approch to man-machine systems problems.

Human Engineering. The main objective of human engineering or ergonomics is to assist in design and operation of man-machine environment system which will ensure physical and mental ease to the human beings. It is really surprising that equipment designers often overlook the requirements. The details of equipments are most often determined by the engineering require-ments rather than the workers' convenience. It is here that need of the study of human engineering arises. The attempt should always be made to design the machines and equipments in such a manner that not only the users but also those in the vicinity should be protected against dangers of accidents. Guards for the moving parts, barriers for dangerous spots are some example of safety measures that need to be installed in all factories.

What is Human Engineering? Human Engineering means engineering for human use. It can be described as an attitude by which and engineer approaches the problems of desingning macines and equipments to be used by human beings. The human engineer implements seientific experiments and other research methodology to study 'human factor' areas as they pertain to the operation of the man-machine systems and concepts. Human engineering group generally includes engineers, psychologists, mathematicians, anthropologists physicians and specialists from various related fields. Some of the techniques commonly used in human engineering include experiments design mathematical tools, games theory and statistical sampling. The methodology of human engineering research and design relates in many ways to operational research industrial engineering, and systems development engineering.

Human Engineering—Study of People at Work. Human engineering in the study of people at work and of work methods. It includes the study of equipments design, hours of work and environmental conditions or work. Its purpose is to improve productivity and job satisfaction. Beginning in 1940 the term 'human engineering' was associated with equipment design By the mid of fifties several aircraft companies began to utilise human engineering in equipment design and training programmes.

The functions and responsibilities of human engineering have continuously broadened in scope and participation by human factors specialists has also increased. In future also, human engineering will have a wide spread utilisation in civil systems, architecture, heavy equipments agricultural machinery, etc.

To understand the idea of human engineering work clearly for equipment design, Let us study the following problems faced by one aircarft company.

(i)   What should be comfortable fingers pressure for knobs ?

(ii)   How much force should be required for a wheel control ?

(iii) What should be the volume occupied by a man?

(iv)   What should be the angles of vision from a prone position ?

(v)   What should be the height of " no smoking" sign ?

(vi)   What are the effects of drugs on the human body in fight ?

Human engineering should be able to increase the usefulness of many machines and tools. There seems to be enough room for improvements in control indicaters and work space arrangement. To have the insight of the concept of human Engineering let us understand the concept of 'work study'. Work study is comprised of Motion Study and Time Study. In motion study all manual work can be analysed into various elements and consequently studied and rearranged as desired. Motion study and time study were essential parts of the scientific management movement. Time study originated by Frederic W. Taylor is mainly used for establishing time standard and piece rates. Motion study develped by Frank and Gilbreth is employed for the analysis and improvement of work methods. Motion and time study provides a means for perforing work. In other words time and motion study of help to determine a standards day's work and to determine the standard time required to perform a specific task. The various elements of motion study are shown in the table given on next page.

The elements of motion can be used to break down any job to show the details of hand work. The various principles of effective motion can be stated as follows:

(a) Successive movements should be so related that one movement passes easily into that which follows.

(b) The order movements should be so arranged that the mind can attend to the final aim.

(c) The sequence of movements is to be so framed that an easy rhythm can be established.

Elements of Motion Study

Physical Elements	Semi Mental elements
Transport empty	Search
Load	Final
Transport Loaded	Selects
Unload	position
	Mental element
	Plan
change direction	Inspect
Hold	Remember
Assemble	
Disassemble	Delay elements
Use	Avoidable delay
	Un voidable delay
	Normal delay
	Rest for fatigue

(d) The continuous movements are preferable to angular movements involving sudden changes in the direction of movements.

(e) The number of movements should be reduced as far as possible.

(f) Simultaneous use of both hands should be encouraged.

(g) Fixed positions should be provided for tools materals etc.

Motion study Techniques. The tools and techniques of motion study may be divided into three classes
  (a)   Process Analysis (b) Equipment Utilisation (c) Operation Analysis

Process Analysis. It is the process of making a component or performing an activity before undertaking a thorough investigation of specific operations in the process. The device of record a process in a compact manner is the process chart. The chart represents graphically the separate events that occur during the performance of a task . The chart begins with the raw material entering the factory and follows it through every step such as transportation to storage of inspection of the machining operations and assembly till it becomes a finished unit. At times a better picture of the process can be obtained by putting flow lines. Both a process chart and flow diagram are sometimes needed to show clearly the steps in a manufacturing process.

Equipment Utilisation. In order to have a complete breakdown of the process the series of operations are plotted against time scale. This diagram is called as activity chart. In order to eleminate unnecessary waiting time for the operator and for machine it is important to perform the following three steps: (i) Get Ready ; (ii) Do and (iii) Put Away.

Operation Analysis: The overall study of the process should result in a reduction in amount of travel of the operator materials and tools and should bring about orderly and systematic prcedures.

Micromotion Analysis. Most work is done with two hands and all manual work consists of a relatively few fundamental motions which are performed over and over again. Motion pictures are made of the operations. A special clock is placed in the picture to indicate time of the film. After the film has been processed it can be placed in a special analysis projectors and the motion for the hands, arms or other body parts can be determined and the time for each motion can be measured and recorded.

The main objective of motion and time study and other changes in work methods has been to increase production and therefore profits. Demonstration has the spectacular example of loadingpig iron. Through a study of this operation, production was raised and failure reduced. Procedure at the Bethlechem steel company installed by the Taylor, the originator of scientific management increased the amount of pig iron loaded from 12.5 tons, to 47 to 48 tons per man day.

Providing good working conditions is often thought to increase production. This is frequnently justified on the grounds that it leads to improved morale, which in turn is believed by many businessmen to cause higher productivity and profits.

Time Study. In case the time study principle and applications are perfect, we should arrive at the same time for a iob regardless of each worker and the final time should entitle normal amount of effort and which would be same amount of effort for all kinds of jobs studies and result in average earnings each job studies at equitable of the pay levels determined in advance. In other words, if it decided in advance that turret lathe operator should be paid more than capstin lathe operator, the earnings would average out that way and there would be none of the upsets that occasionally occur.

In an experiment it was shown that all the time study men used the same division into elements of motions and an unlimited length of time for each study. It was found that the average deviation or each of time study expert from the average of the whole group was 21 per cent, the average deviation for men from the same organisation was 9 per cent, an indication that training and similar procedure differences. The time study engineers at times over-estimiated considerably the speed of some operators and underestimated the speed of others. Barnes thought that machanical errors of measurement in time study have been reduced by the use of the standard data. The reliability of using standard data has been carried out in various studies.

Acceptance of time study. Time studies in industry have shown that the attitudes of the group was a more powerful motive, Management is now aware that workers are not motivated primarily by the pay cheque. Determination or group attitudes would give management a better ground for understanding the workers.

The Western Electric Company foud that group attitudes also go against to objectives of management. The management and particularly the industrial engineers at Western Electric found that when the task is one that depends on the efforts of the groups, the group will exert pressure on each indvidual to put out as much work as possible so that each member will make the maximum amount of money. The group considers a certain amount of work a proper day's work and the motive of the social group is stronger is the day incentive. Group pressure was exerted on fellow employees not to exceed the group standard and to remain output at a regular rate. On occasions when output actually exceeded the group standard the additional output would not be repcrted but would be held back and reported later. These group attitudes regulated the workers reactions to all outsiders and particularly to supervisors,. Workers did not confine in supervisors, who were looked upon as outsiders. The group attitudes decreed that workers would have to do a certain amount of work but that they should not do too much, even though a group bonus plan was in effect that would have resulted in additional pay for more work. The workers could have done much more than they did without physiological damage. Most efficiency experts, experts, especially a recent years, have sincerely applied motion and time study principles which these sincere men have failed to obtain acceptance, it has probably been because of a failure on their part to be sensitive to the human relations probelms which are inevitable. In the past the details of motion and time studies have been kept secret from the employees. The experts have stressed that the purpose of motion study is to find the one best way of doing work for each job and to train every employee so that he may perform in that one best way. Psychologists have raised the objection that, because of individual differences, the best way of work for one worker is not necessarily the best for others. In fact, the employees should be taught the best way to work, and should be allowed to express their individual initiative after that.

Work Curves. A graph showing the level of performance against time spent at work is called a work curve. Since practices to alleviate fatigue depend to a large measure on the nature of the work curve, its discovery and use are important. No difficulty exists in obtaining such a curve where work is uniform and successive in terms of units produced. However, where work changes are frequent, as in ordinary varied office work, work curves are most frequently obtained by the use of laboratory task tests. The results of these tests are then plotted and used as index of fatigue for such jobs. Motivation might be higher in the laboratory tests than on the jobs. the degree of transfer depends on the similarity of task, environment, attitudes, and neuromuscular processes.

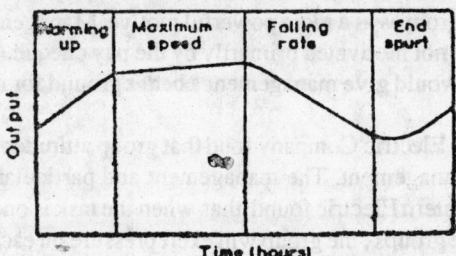

Fig. 37.1 . Simple Activity Curve

The work curves can be typed according to the nature of the task involved, namely, simple muscular, complex muscular and mental, as shown in Figs., 37.1, 37.2 and 37.3.

Simple Muscular Work. The first curve is simple muscular activity curve. It is derived by recording the units and output of a group of muscles against the force of some weight or spring. It can be seen from the drawing that the curve for simple tasks shows a short warm up period followed by a high level of performance. A gradual tapering off then appears. Subsequently, a sudden drop occurs to a point of complete exhaustion. Such work curves have a general resemblance to motor curves, but they involve simpler coordination and fewer muscles. One generally conclude from a study of such curves that the more complex and rapid the task the faster fatigue sets in. Moderate task permits a greater total amount of work before complete exhaustion occurs than doing heavier tasks.

Complex Muscular Work. The second curve depicts the amount of work done per unit of time in analytical industrial situation at a complex muscular task.

This job is assumed to be motor in nature and not monotonous. As interpreted, the initial upward slope of the curve indicates, a warm up period. This is followed by a gradual rise in production until mid-day when a drop occurs. The afternoon curve taken after the mid-day rest is similar in appearance to the morning curve but it does not rise to a high production peak and it falls more rapidly near the end of the day.

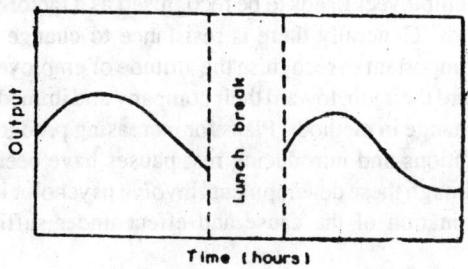

Fig. 37.2 Complex Muscular Activity Curve

Mental Work. The third graph shows a typical mental work curve. As in muscular work, mental work decrement occurs in quantity, speed and accuracy of output. Again large individual differences and the nature of work tend to determine the timing of extent of such declines. The work decrement in mental tasks requiring continuous attention has been attributed to interferences known as "blocks".

Psychological Aspects. There are three main types of psychological contributions to human engineering:

a. Principle and practice of human relations to obtain co-operation in changes.

b. Research on methods to ensure that adequate experimental and statistical controls is there to make findings accurate.

c. Recognition of psychological characteristics of workers such as perceptions, speed and tendency to boredom under certain conditions.

The use of time study as well as of motion study, generally implies that there is one best way of work. The influence of the work group on employee morale is a powerful force. the resistance to the idea of the introduction of improved work method is largely avoided if workers are afforded greater participation in the origin of these improvements. The advantage of workers' participation is

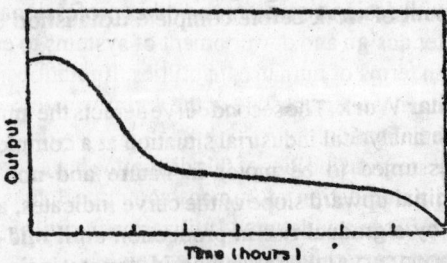

Fig. 37.3. Mental Work Curve

that people are much more willing to do what is forced upon them.

Morale of employees needs to be recognised as a factor in the introduction of new methods. Generally there is resistance to change unless people are changed. It is important to recognise the attitude of employees toward the new methods, toward their job, toward their company and immediate supervisor for promoting a change in methods. Plans for increasing production by improving working conditions and introducing rest pauses have been made mainly by engineers, although these developments involve psychological problems, such as the determination of the cause and effect under sufficiently controlled conditions.

Equipment Design. Human engineering equipment design involves the knowledge of not only the motion and time study analysis but many other areas to fit man over machine. Motion and studies attempt to adopt the man to the machine, whereas human engineering places more stress on changing the machine if necessary.

The design of any tool and equipment should be based on the capabilities of prospective operators. For example, sledge hammers should be light enough to be within the lifting capabilities of an ordinary man. Machines are becoming substitutes for people on the more routine jobs, but people must design, build and maintain the machines.

Duties of a Human Factor Engineer. The system development corporation of America which is probably the largest single group of human factor engineers in that country lists typical duties as follows:

    a.   The human factor engineer designs and implements scientific experiments and other research methodology to study human factor areas as they pertain to the operation of man, machine, weapons and other complex systems and concepts.

    b.   Evaluates existing or proposed man-machine systems and sub-systems in terms of human physiological and psychological requirements so that optimum suitabillity and durability from a human input stand points are established.

c. Consults with design engineers and other professionals prior to, during and after design and development of systems to ensure optimum operation in terms of human capabilities, limitations and variables.

d. Provides the most current state of the art information in his particular field of specialisation.

The functions and responsibilities of this discipline have continually broadened the scope and participation of human factor specialists. The future of human engineering will see a wide spread utilization in civil systems, architecture, heavy equipments, agricultural machinery etc.

Anthropometric Factors

Failure to take into account human body structure and dynamic characteristics in workspace layout can have adverse effects on operator efficiency, health and well being. Both static and dynamic anthropometric data are available in detail in standard guides. *The static anthropometry deals with workspace layout problems.* The authropometric data are further useful in the design of personal equipment viz helmets, goffles ; masks : earphones ; gloves ; work tables ; desks ; vehicles ; furuiture ; appliances ; office equipment, and passenger seats. Determination of the useful visual field is another application. It should be noted that anthropometric data vary with a numbers of factors ; age ; sex ; race and nationality are examples. *The Dynamic anthropometry is ooncerned with the functional range and pattern of body movements and with the operations that can be made by limbs in various positions.* Dynamic data for kneeling, crawling and prove positions have unique application where work is spatially restricted as is the case with mechanics, plumbers or repair-men.

Approaches, Techniques and usefulness of Human Engineering. Some of the wellknown tools and techniques now used in human engineering include the following:

Experimental Design. Mathematical Modelling.

Game Theory. Linear Programming.

Information Theory. Decision and Testing.

Simulation and Testing. Field Testing.

Statistical sampling. Direct cast history studies.

The methodology of human factor engineering research and design relates in many ways to associate technical fields such as operations research; industrial engineering, weapons systems analysis, industrial design; life support engineering and system development engineering. Human engineering should be able to increase the usefulness of many machine and tools. There seems to be enough room for improvements in control indicators and work-space arrangement. Human engineering is the outgrowth of motion study originally developed by Gilbreths. It is broader in concept than motion study and accepts a broader set of criteria such as frequency of error and psychological and physiological costs as well as the older criteria of economy of motion and labour cost.

To have the insight of the concept of human engineering we have to understand the concept of work study. The work study is comprised of motion study and time study. In motion study all manual work can be analysed into various elements and consequently studied and rearranged as desired.

Fatigue. Fatigue is one of the most significant problems before industrial engineers. Fatigue can be defined as negative appetite for an activity. Industrial fatigue affects the worker's muscles, nerves and mind. In nervous fatigue, the nerve fibres terminate at the muscles and the plate stops transmission of the nerve impulses to the muscles which initiate muscular activity. The initiation of nerve impulses in the brain also stops questioning if the organism continues to work for a long period. This stopping or sending the impulses from the brain saves the organism and its muscles from damages. Thus fatigue is defined as a reduction in the activity because of previous work. Different authors have given different definitions to the term fatigue.

—It can be a decrease in the capacity to do work or loss of efficiency.
—A decrease in interest or willingness to work—a feeling of weakness.
—A more or less complete loss of irritability and responsiveness of a tissue.
—A condition of mind resulting from prolonged mental activity.
—A failure to maintain physioiogical or organic equilibrium.
—Not an entity but a convenient word to describe a variety of phenomena.

Nature of Fatigue. The complexity of the fatigue phenomenon can be understood by an interesting experiment carried out at the University of Zagreb in Yugoslavia. In this a group of 25 college students was confined in a lecture room all night without sleep and was kept active by reading, music, dancing, etc. At 6 O'clock in the next morning the students went on a six mile quick walk, after which they were tested for two bours. Another control group of 25 students were given the same tests after a normal night's sleep. Experiment showed that 87.5% of the experimental group rated their condition as "very tired" or exhausted and none reported "no fatigue". Only 12 per cent of the second group rated their condition as "very tired" and none rated it "exhausted", while 28 per cent reported "no fatigue" or slight fatigue".

However, with respect to performance on 12 psychological tests requiring reasoning, spatial judgements, verbal tests, and locating printing errors there were no significant difference between the groups.

It is found that changes in the integration of process account for some of the difficulties in isolating the effects of fatigue in the organism as a whole. The body seems able to compensate for fatigue effects by spreading the work, in the same way a man shifts a suitcase from right to left arm. One doctor has written a book titled. 'The wisdom of the body' in which he points out how certain organs of the body can take over the functions of a deceased organ and serve as second and third lines of defence against loss of life.

Fatigue can be known by studying the feelings of the fatigued person. An individual may 'feel' completely rested, but his work record may show a rapid decline. Under conditions of strong motivation, men may contique to work for long period of time without being aware of fatigue, whereas, under other conditions they may feel fatigued before they go to work. The various signs of fatigue include inefficient eye movements, closing the eyes, looking about, and periods of skimming alternates with series of reading. If in an individual motivation is present, the same indivdual will feel no decrement in reading ability even after six hours continuous reading.

An individual's attitude is an important factor in the ability to do work, but the presence of the attitude cannot be detected by any physioligical measure. The study of fatigue must control not only the actual physical activity but also the indefinite environmental factors which influence a man's outlook and attitude towards the work. It has now become very apparent that emotional stability and mental hygiene cannot be closely related to fatigue. To the extent the fatigue involves the organism as a whole, it is a psychological problem. The work produces chemical and psychological changes in muscles, nerve tissue, and the blood. the problem is the concern of both bioligical chemists and psychologists.

Signs of Fatigue. Mental fatigue expresses in many different ways the phenomena of monotony and boredom. Monotony is a state of mind caused by performing repetitive tasks. Boredom means lack of interest and is generally characterised by depression and a desire for change of activity. Boredom is affected by personality, attitude and interest patterns. Boredom and monotony are different from fatigue because it is a desire for a change in activity rather than for a rest or relief from work altogether.

Prolonged mental work result in an incapacity to evaluate what is being read. Common characteristics of prolonged mental work are increased errors and an increase in the amount of time necessary to assimilate written material or to solve perplexing problems towards the end of the period. In addition to this the physical state of the body is no indication of its mentally satisfied state, although there is a state of tension which expresses itself in both muscular and neutral form during the mental work. The change in the nerve centres that result from prolonged mental activity are uncertain. Consequences of mental overwork in children are disturbances of vision, headache, bleeding from nose, loss of appetite and indigestion, cerebral disorders and nervousness.

Effect of Fatigue. The effects of prolonged work on the body result in excessive muscular activity. These can be grouped as:

(i)   overt changes.
(ii) internal changes.

Overt Changes. If the muscles are strained then the following phenomena occur.

(i)   The amount of work that the muscles can do depends upon the speed with which successive contractions are made.

(ii) Decreased capacity on the part of one set of muscles is accompanied by a decreased capacity on the part of muscles of the body.

(iii) The interference with the removal of waste products or with the supply of blood to the area lessens the working capacity of the muscle, loss of sleep, anaemia, hunger, or dissipation reduces the number and amount of contractions that the muscle can make.

(iv) The rate and extent of contractions diminish rapidly at first, then maintain a constant pace and finally speed up a little just before exhaustion sets in.

Internal Changes. Any muscular activity is accompanied by following changes:

(i)   Waste products are constantly being formed and thrown into the blood stream consistig of lactic acid, potassium phosphate and carbon dioxide. With any accumulation in the active muscles, they exert a poisonous and paralyzing effect on the whole organism.

(ii) The muscles are provided with a store of carbohydrates and are constantly being supplied with other carbohydrates and with a steady supply of oxygen. Normally these serve to balance the effect of the waste products but if constant strain is put on any muscular set, there is a gradual accumulation of lactic acid and carbonic acid, and a steady decrease in the supply of oxygen acidity which expresses itself in the form of the experience of the general fatigue.

(iii) The supply of oxygen is dependent upon the capacity of the organism for transporting oxygen from the lungs to the tissues.

(iv) The carbohydrates supply consists mostly of a simple blood sugar called as glycogen which is manufactured in the liver. During the contraction of muscle activity, the glycogen changes to lactic acid and during the relaxation phase, part of the lactic acid is converted back into glycogen and part is united with oxygen producing carbon dioxide. When that is combined with the water in the tissues, carbonic acid is formed. A great deal of the carbonic acid passes out of the muscles through the cell walls, whereas the lactic acid is removed only by oxygen. There is concentration of sugar in the blood of 0.1 per cent, and in the normal activity of the work the percentage remains fairly constant. In fact, the excess of lactic acid is the sign of fatigue.

Types of Fatigue. Psychological fatigue is used here to designate those factors which cause work decrement. It includes the falling off in efficiency of work commonly referred to as mental fatigue, also known as monotony and boredom. Monotony and boredom are influenced by the way a person views his task from time to time. Causing the output to fluctuate rather than to fall of progressively. The way in which a job is perceived is a individual matter, but certain kinds of tasks and work atmosphere are more likely to induce monotony and boredom than are others. Motivation is an important factor in all forms of fatigue and the rate of fatigue for almost any type of task varies with the intensity of the motivation. When motivation is low, fatigue effects appear very costly but when motivation is high, fatigue may not be apparent until physical exhaustion is manifested. Mental conflicts and frustrations are so commonly associated with man's work that it is to the industrialist's interest to determine methods for reducing their incidence. Motivation influences a man's will to work. The amount of energy a man may have available for a task seems to depend upon the motivating conditions he finds in the situations. It appears that the influence of motivation on work is one of determining the amount of energy which will become available for the task as shown in Fig. 37.4.

Energy is generally rationed and a particular job must have certain priorities if it is to get a good share of the energy. After office work the man may be too tired to work overtime but if game of cards is suggested, plenty of energy becomes available. The man's basic supply of energy is not depleted by his work, but the portion allocated to a given task being expended. If many allotments are made, the total supply is reduced. Rationing then becomes more strict, higher priorities are needed and smaller allocations are made.

Fatigue—Monotony—Boredom

Fatigue. It can be defined as reduction in the ability to do work because of previous work.

Monotony. Monotony can be defined as desired to have change due to similar job. Monotony is the state of mind caused by performing repetitive tasks.

Boredom. It is mental fatigue due to routine job. Attitude and personality are related with boredom or lack of interest and is generally characterised by depression and a desire for change of activity.

Monotony and Boredom can be differentiated from fatigue because the former are desire for change in the activity rather than for a rest or relief from work altogether.

Prolonged mental work results in incapacity to evaluate what is being read. A common characteristic of prolonged mental work are increased errors and increased time to assimilate written material or to solve perplexing problem towards the end of the work. In addition to this, physical state of the body is no indication of its mental tension which expresses itself in both muscular and neural form during the mental work. The changes in the nerve centres that result from prolonged mental activity are of uncertain consequences. The symptoms

Fatigue Charts for Different Tasks

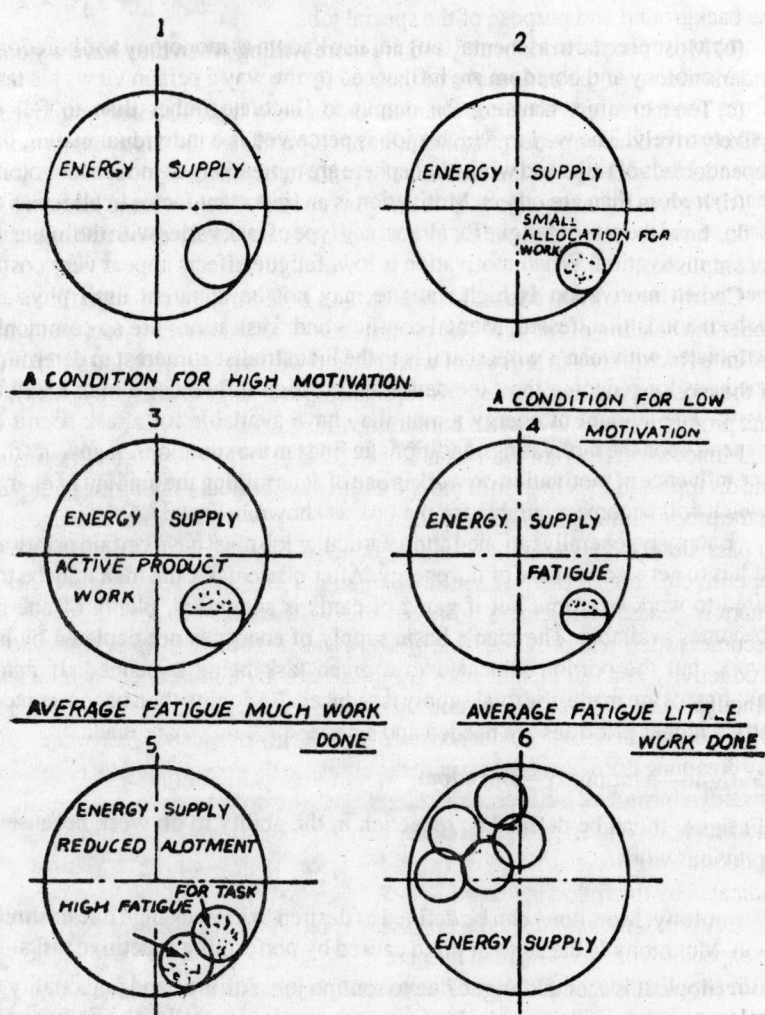

Fig. 37.4. Complete fatigue for task energy allocated to several task.

of mental overwork in the children are disturbances of vision, headache, bleeding from nose, loss of appetite and indigestion, cerebral disorders and nervousness.

Reduction of Monotony and Boredom in Industry. The following steps should be taken by the management in order to reduce monotony and boredom:

(a) When a supervisor gives an unusual job to a worker, he should explain the background and purpose of the special job.

(b) Most people work better, and are more willing when they have a good understanding of the end results of their work.

(c) The increased verbal communication between supervisor and employees is likely to produce a better personal relationship and give the supervisor a more dependable "size up" of the individual members of his group.

(d) It gives lower level jobs a "fourth dimension". To the dimension of what to do, how to do it, and when to do it, it adds why it is done and what it accomplishes.

Conditions of Boredom. Monotony and boredom are the specific areas under the heading of fatigue. The mental state of monotony is associated with definite fluctuations in the rate of working and with a fall in production. In one of the most exhaustive studies this relationship was demonstrated by showing that production was low when boredom was experienced.

Monotony effects are most pronounced during the middle of the work period and disappear in anticipation of the end of the work period, causing the end spurt in the production curve. The individual feels restless and the strain time seems to pass slowly. The extent of the monotony is dependent not only upon the repetitive nature of the task, but also upon the degree of attention required. There is general agreement, however, that workers tend to slow down, talk, become restless and show variable production when bored. In other words both productivity and the mental state are influenced by a variety of factors, and although some of these may be common to both, each is influenced by special ones. Letting the mind wander seems to be one way of escaping monotony. If day dreaming does not interfere with the ability to do good work, it is probably a useful adjustment, but if constant alertness is imperative, it may cause errors and accidents. That a good deal of the loss in production in respective work in due to specific condition of boredom rather than a muscular fiatigue is indicated by the following facts:

1. Afternoon monotony effects do not exceed those of the morning as could be expected from accumulated fatigue.

2. Application of the end of the work period tends to abolish signs of monotony, and

3. Intelligent workers are more subject to monotony effects than are less intelligent ones. these facts indicate that a knowledge of the mental effects of respective work is highly important since such information might suggest methods for eliminating this mental condition.

Fatigue Versus Boredom

Mayo and Lovekin have suggested that what is commonly called "fatigue" had better be called by its right name, "Boredom". It is difficult to eliminate the erm "fatigue" from the scientific discussion. The distinction between the two terms is discussed below:

a. Fatigue is usually associated with physiological depletion while boredom is a concomitant of mental dullness.

b. Fatigue is a decreased capacity for work. Boredom is a decreased interest in work.

c. Fatigue is conscious inability. Boredom is a feeling of incapacity with or without there being a physiological basis for the feeling.

d. Fatigue expresses itself in the form of a gradual decrement in the work curve with a final and spurt indicative of the functioning of second spell. Boredom expresses itself in the form of irregularities in the work curve with intermittent spurts of short duration. There is a sharp drop in the work curve during each work spell and a rapid rise toward the end.

e. Fatigue is due to the work task being temporarily beyond the capacity of the worker while boredom is due to the consciousness of the uniformity of the work task and a temperamental incapacity to subject oneself to that particulr type of routine work. Fatigue is the result of too long continued physical or mental work. Boredom is due to the absence of work challenges.

f. Fatigue is, to some extent, measurable. Boredom is a subjective attitude that defies objective evaluation.

g. Fatigue has phsiological accompaniments which are fairly definite. Boredom has mental and emotional accompaniments that do not lend themselves to ready calculation.

h. Fatigue expresses itself in the desire for rest. Boredom expresses itself in the desire in the desire for change.

Monotony and Boredom. The repetitive work destroys such human values as pride in workmanship and individuality. In fact, boredom and dissatisfaction are common in our present methods of production. Monotony and boredom are described as the undersirable effects of repetitive work. Monotony is described as the state of mind caused by repetitive work. The term boredom is referred as a more inclusive term, taking in the person's unfavourable outlook and feeling tone for the task he is performing. Boredom will be affected more than monotony by the following factors:

(i) The personality of the person' (ii) the attitude and mood of the person; (iii) the perception of the task performed. This means that the individual may not agree on the task that is most boring; individual persons may show more boredom on one day than another.

Work which requires constant alertness and attention is subject to interferences known as Blocking. The phenomenon of mental blocking becomes objectively apparent when we measure the continuous result of mental work. If a person is asked to name a series of colours, give the opposites of a list of words, or add a series of sums, and if his responses are recorded on a revolving drum so that each response makes a mark, it will be found that these marks are irregularly grouped. A few responses occur rapidly, then there is a delay, followed by another set of responses. Such records show that the responses are either very close together or fairly far apart.

The blocks or lapses in performance are associated with the making of errors. There are wide differences among people in the length of their mental blocks, as well as in the frequency with which these occur. These number of blocks range from two or six per minute in different people. Individuals who tend to perform slowly in experimental tests are likely to be the ones who have long or frequent blocks. Blocking probably functions as an automatic method of resting.

Measurement of Fatigue. Fatigue can be measured, if the muscle is taken out from a living organism and put into the laboratory experiment. This is done in the physiological olaboratory by taking a muscle from the living organism and giving electrical impulses rehthmically to the nerve attached to the muscle. The muscle contracts with the electrical impuleses and the contraction of the muscle is recored graphically on a recorded drum by means of a stylus. If the electrical impuse continues to stimulate muscle for certain period, the muscle gradually diminishes its contraction and finally stops contracting. It will not be possible to experiment this in case of an industrial worker. It is a fact that if a muscle remains attached to the living organism, it cannot be made fatigued. The central nerve system of the organism will stop the muscle long before it becomes fatigued in order to save it from damages. so muscular fatigue cannot be measured so long as the muscle is attached to a living organism.

When a muscle works, it undergoes electrical changes. It is possible to amplify the electrical activity of the muscle, so that even the slightest contraction can be detected. This respiration undergoes changes during mental or physical work. The rate of respiration increases in muscular activities, but there is no definite change of the rate of respiration during mental activity. If the individual feels happy, the rate of respiration decreases significantly. The human organism acts as a machine and its activity transforms energy, then a work that it performs can be measured directly in terms of such energy transformation which is technically known as metabolism. In metabolic activity there are two processes going on in any living organism. Metabolism is a destructive process by means of which energy is consumed. Anabolism is a constructive process which assimilates food and converts it into energy for consumption during activity.

Fatigue Tests. Fatigue tests are classified as psychological, physiological and workshop records. The psychological and work-shop tests are valuable indicators of small bodily changes and the physiological tests for gross bodily changes. Studies have shown that unless fatigue has reached the stage in which there is a disturbance of physiological equilibrium, the physiological tests do not show any change. The various tests can be classified as follows:

In this method the usual practice is to isolate as possible the activity of one group of pattern of muscles and infer that as made up of activity representative of the muscles patterns of the whole organism. Following are the representative tests:

Tests of Physical Endurance. In an ergograph, the isolated finger is attached to a set of weights, which the finger repeatedly pulls back and forth, the exhaustion sets in, and the extent of the finger movements is recorded.

The findings of the results from the ergography are:

(a) the speed with which exhaustion takes place if the palter is constantly stimulated.

(b) the ergograph that characterises one muscle pattern, is different from any other pattern, and seems to represent the individual's form of energy expenditure.

(c) the regularity of the work curve seems to depend on the maintenance of a condition of physiological and emotional balance. Characteristics differ between different subjects in their resistance to work.

Ergographic curves seem to fall into the following three clauses:

(i) Those maintaining a high and consistent level for some time, followed

(ii) Those starting with a high level, but quickly falling to a low average.

(iii) Those with a maximal start but with a gradual decrease in activity until the exhaustion point is reached.

Tests of Strength. Various types of strength tests have been used, and in most of the dynamometer tests, norms are provided for the different age groups. The tests can be administered in a very short time, and in these tests, the strength of the hand, the leg, the back, and the shoulders can be determined.

Tests of Motor Speed. Sapping board is an instrument to measure the endurance for such types of work as typing and fitting small mechanical parts together.

Tests of Muscular Precision and Muscular Steadiness. Steadiness tester is used to measure muscular stadiness. It consists of a mental casting with one face at an angle of 45 degrees and pierced with nine holes of different diameters. To one end of the casting is attached a binding post so that it may be put in circuit with a battery and a sounder. A stylus with a connection card is provided, and the task given to the subject is to insert the stylus into the holes without making contact. The accuracy with which this can be done is indicative of muscular control, the degree of control which the subject exercises can be estimated after varying spells of work.

Reaction Time Tests. Many varieties of these tests have been used. The purpose of reaction time test is to form an estimate of the speed and accuracy with which a person can make adjustments under different types of situations.

Tapping Tests. The rate of tapping is calculated by recording the number of tappings and correlating it with time taken in tapping. If an individual continues to tap, his rate of tapping is fast in the beginning and gradually slows down with the lapse of time.

Body Sway Test. Stylus is fitted on the head or on the shoulder of an individual. The individual is asked to stand without moving for a considerable period. The stylus records his body sway on a paper attached to it. It has been observed that the rate of body sway increases with the passage of time.

Steadiness of Hands. The subject places an electric stylus inside a bigger hole first for a period of two minutes without touching the plate. If he touches the plate, an electric bulb or a dazzer indicates the touch. Then he puts in the next hole which is smaller than the first and continues the experiment till he reaches the smallest hole, the steadiness of hand suffers by continuous activity.

Flicker Test. In the flicker test a disc which is partly black and partly white is rotated. At a certain speed, it is seen as a solid grey. At lower speed, it flickers, A fatigued man sees the flicker more. This test is mainly used for the selection of drivers.

Physiological Tests (Skin Tests). Tests of physiological nature have usually been limited to changes that occur in the blood and in metabolism as a result of the expenditure of energy.

Blood Changes. These can be measured in a number of ways like tests of blood pressure, pulse pressure and pulse rate, pulse product.

The Pulse Rate. Normal pulse rate varies in different people. For men the usual estimate is form 68 to 76 beats per minute, and for women 74 to 80. some people, however, may have normal rate as low as 50 or as high as 90. During severe activity the increase in pulse rate is very rapid.

As an indication of a so-called fatigued condition, pulse rate is of value only for the heavier type of activities, in lighter continuous work, *e.g.*, type-writing, no change is noticed even after 8 hours practice.

The pulse rate change may indicate an increased expenditure of energy, it cannot be taken as indicative of the feeling of fatigue.

Blood Pressure. Maximum pressure occurs at the time of contraction of the left verticle, and is termed as diastolic blood pressure. Measuremer ts are taken with a manometer, an instrument consisting of a rubber sleeve with a graph inflated to the point where the flow of blood is cut off, giving scale measurement of blood pressure. Normal systolic blood pressure ranges between 100 and 130 and diastolic between 60 and 65. When the body is under exertion, the systolic pressure usually increases and the diastolic decreases. When the body is in an exhausted state the increase may be as high as 50 to 70 per cent and the decrease from 0 to 50 percent.

Metabolic Changes. Metabolic changes indicate the rate at which body energy is being used. Several devices are in use for measuring the rate of respiratory exchange *i.e.* the ratio of oxygen consumption to carbon dioxide production. Under experimental conditions measurement is made by breathing oxygen from a container through a mouth piece or a gas mask.

During moderate work, the metabolic rate is usually from two to four times the normal rate but is unusual for a person to maintain a level of work that demands more than eight times the basic metabolic rate.

Metabolic rate can be measured as:
(i) Changes in skin temperatures.
(ii) Insensible weight loss, due to evaporation of water from the skin and lungs.
(iii) Decreased skin resistance as a consequence of increased neuromuscular activity.

Indirect Measurement of Fatigue. Psychologists measure industrial fatigue in terms of production, rate of absenteeism, rate of accident, job satisfaction, wastage, etc.

But the environmental conditions such as noise, improper lighting and exterme heat, cold and dust, have shown an increase in fatigue.

An improvement in these conditions contribute to fatigue alleviation which results in higher morale, lower accident rates and less absenteeism.

The factors which affect fatigue are:
(i) Noise and Industrial Music. *(ii)* Illumination.
(iii) Temperature and humidity *(iv)* Ventilation.

Noise. Noise is an unwanted sound and it increases fatigue. If the noise is too much and continuous for a long time it might cause deafness, for example, boiler make deafness, sheet iron worker and blacksmith deafness. The effect of noise depends on the kind of noise and kind of work. Mental work is affected more than manual work by noise. The effect of noise is greater when it is irregular. Sounds of either very high or low tone qualities are more irritating than those in middle zone.

The noise of moderate intensity serves as a psychological inducement to work. One study of reduction of noise concludes that it is an investment which under ordinary conditions will increase productivity by 10 to 15 percent.

Meaningful noise, however, has a great intellectual value. It is very difficult to adopt to a meaningful noise and as such this has a continuous effect on the efficiency of the workers. Meaningful noise with moderate intensity in the form of music increases the output. This is more true in case of monotonous and repetitive work. Some industries such as Glaxo Laboratories, Bata Shoe Company, DCM Chemicals and Telco are using music in packing sections which is giving good results. The effect of noise on learning has been observed more clearly than its effect on fatigue. It is commonly believed that workers produce a great deal more under conditions of quietness. The available evidence suggests that noise has an effect but only slight one on production. But if the individuals are exposed to very high pitch noise such as boiler makers and aeroplane pilots they may suffer permanent deafness.

Industrial Music. Music in factories has become fairly common and therefore, requires an evaluation. Some studies were made in England where researches found the initial interest for dealing with monotony. The effects of musical programmes on simple assembly operations showed production increase upto 6 per cent, when production on days with music was compared to that on days without music. In a radio-tube assembly factory in which music was customarily played, the effectiveness of fast, show and mixed programmes studied. The scrap rate was found to be less when either fast or slow music was played than when there was no music or when fast and slow musical programmes were alternated. Musical programmes also had a beneficial effect on employee morale.

The increase in production varies considerably with the job and the shift. Simple repetitive jobs that can be performed well while talking seem to benefit most be music. The research findings indicate that music seems to encourage young, inexperienced workers to increase production on routine jobs. No significant effects were found for older skilled workers.

The general finding is that employees like music. At least 75 percent strongly favour it and only a small percent oppose it. Slow music seems more desirable than fast music, variety is essential to keep young and older employees satisfied with the programmes. Vocal music may be distracting form some employees, and music with strong beats may disturb or help depending on how it fits into the movements and rhythms of a job. Spaced programmes seem to be favoured. At present, it appears that the music should be for about 12 per cent of the time for the day shift and as much as 50 per cent of the time for the night shift.

Most favourable effect of music is its influence on boredom. It takes the mind from the work as well as frees the brain of the obligation of initiating other activity.

Illumination. Many industrial undertakings take this advantage of natural light during the day time. But it should be remembered that the illumination of the work place by means of artificial modern lights is far better than taking the natural day light in carrying out any work efficiently. Any natural light which the industry uses should be carefully utilized for the working conditions. North light is commonly used in industries because the variation of light is not much during the day time.

Engineers are utilizing modern techniques in the plant-layout, to avoid such changes of illumination during the day in the workshop either by putting the transparent shades on the top of the factory or cutting down lights from all conditions except from the north.

The entire work area should be uniformly illuminated. Areas outside the immediate focus of observation, should also be evenly illuminated otherwise it may cause continual adjustment changes and consequently higher visual fatigue. Since the pupil of the eye must adjust itself when focussing on a bright spot and then on a dark spot, expansion and contraction of the pupil sets up an increased strain and fatigue in the worker. Because of the uniform light the workers do not find any visual difficulty while coming out from a specific work situation which is illuminated brightly.

Indirect lighting is the best method of producing uniformity. Eye muscles tend to pull toward areas of high illumination and a tug-of-war exists between the muscles that are concentrated on the work area and those that veer to the higher illumination.

The human eye adjusts so readily to changes in illunination that the effects of poor lighting are usually not immediately observable to the untrained eye. Errors in adjustment are often found when people's reports on brightness are checked against readings on light meters.

Poor illumination causes fatigue and irritability and is a source of errors and industrial accidents. Money spent to correct improper lighting will not be more than paid as higher output per worker, lower production costs, and as happier, healthier and probably safest work force.

The colour of sorrounding object and surfaces is related to lighting. The use of white or light pastes improves illumination by reflecting a high percentage of light, while deeper pastes have a considerably lower relfective value. The occurrence of glare should however, be avoided. Several studies made on the use of coloured light show that people prefer daylight colour. The use of this so-called white light is said to assist materially in visual efficiency. Various jobs require different intensity of light foot candles.

(i) The severe and prolonged visual tasks, such as fine engraving, and discriminiation or inspection require 50 to 60 foot candles.

(ii) ɪ ne prolonged critical visual task such as drawing, fine assembling, fine machine work, and proof reading, require 50 to 60 foot candles.

(iii) The visual tasks such as skilled bench work, require 10 to 20 foot candles.

Temperature. Indians are accustomed to working in a temperature which will be considered as high. At several places like Calcutta, Madras, Bombay, etc. where the employees work in a moderate climate throughout the year, there the temperature does not rise or drop to the extremes. The efficiency goes down during the extreme temperature conditions.

Humidity. Variations in humidity play very important role in industry. Rise of humidity in low temperature to a moderate extent gives relief to worker from severe dry cold, whereas, rise in humidity further discomfort to the rise of

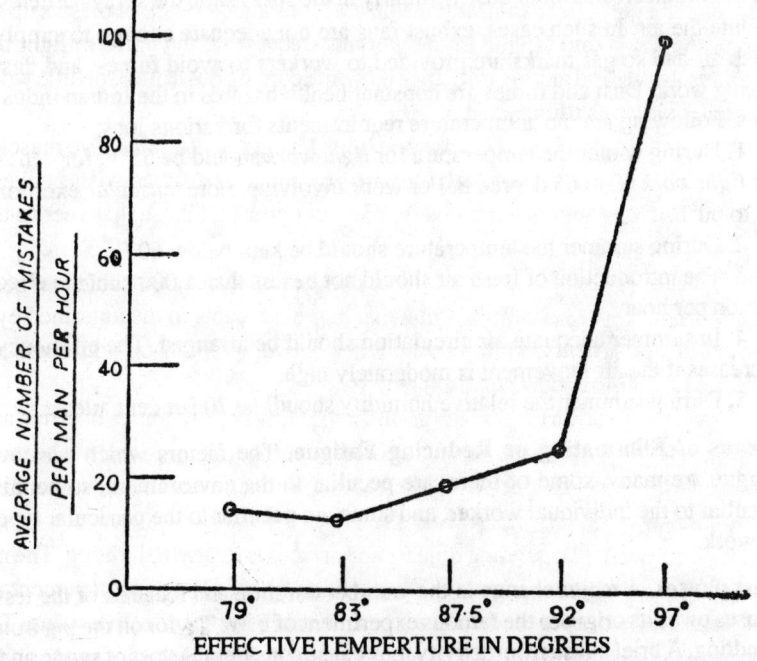

Fig. 37.5 .

temperature. It is more uncomfortable to withstand wet heat and dry cold.

On the other hand, in textile industry high humidity is required. The shinning and weaving departments require as high as 80 to 85% of relative humidity. Prolonged discomfort from heat will no doubt effect efficiency. The effect of temperature on worker's productivity is encountered at much lower temperature levels in physical work. Low temperatures can also affect productivity. Ability to make precise movements with the hands and fingers decreases. Discomfort has been reported where sizable temperature differences existed within room, that is cool air at floor level and warm air at head level. Furthermore, where sizable temperature differences existed within a building and the nature of the work required worker's movement from area to area, discomfort was noted. Individual differences make it possible to set exact temperature and humidity limits within which all workers will be satisfied.

Ventilation. Temperature and humidity can be controlled to a great extent if adequate ventilation by the circulation of fresh air is arranged in the factories. The circulation of fresh air is generally done by fans, ventilators and blowers. Artificial arrangements are also made in chemical industries to avoid polluation of air, through gases and fumes. In cement factory where the stones are crushed the operation creates lot of dust. Similarly in the spray shop the spray particles pollute the air. In such cases, exhust fans are not adequate enough to supply fresh air and so gas marks are provided to workers to avoid fumes and dust during work. Dust and fumes are constant health hazards in the Indian industries. Following are the temperature requirements for various jobs:

1. During winter the temperature for *light work* should be 65°F, for *active yet light work* 60 to 65 degree F. For work involving more *muscular* exertion 55 to 60°F.

2. During summer the temperature should be kept below 60 F.

3. The introduction of fresh air should not be less than 1,000 cubic feet per person per hour.

4. In summer adequate air circulation should be arranged. The efficiency increases if the air movement is moderately high.

5. During summer the relative humidity shoudl be 70 per cent atleast.

Means of Eliminating or Reducing Fatigue. The factors which remove fatigue are many, some of them, are peculiar to the enviornment, some are peculiar to the individual worker, and some are peculiar to the particular type of work.

Rest pauses. A major change in the number duration and location of the test pauses owes its origin to the famous experiment of F.W. Taylor on the pig iron handling. A brief pause from activity may cause considerable loss of swing and only a light recovery from the effect of fatigue. Too long a pause may be equally unfavourable. It may be more than enough to dispel the factors which have been impeading working capability.

It is difficult to tell what is the proper length of a work and rest pauses should be, unless considerable experimental work has preceded the judgement. The length of the pause should vary for different jobs in the same organisation a general rest pause for all employees will not have the same effects.

The fatigue in general is natural. If activity of one set of muscles reduces the ability of other to do work, then recovery must proceed most effectively when the person is completley relaxed. This suggests that work pauses for heavy moscular activity should not only be work stoppages but periods of complete relaxation. Unfortunately, workers who have the most strenuous occupations too often have the least accessible facilities for relaxation.

Various studies have indicated that different amount of rest pauses have a definite effect on production. In one of the study, rest pauses of seven minutes, introduced in both morning and afternoon work periods favourably change the entire shape of the hourly production curve. Favourable results have been found with a five minute rest after each hour of work and even when a two minute rest is alternated with three minutes of work.

The distruction of rest pauses has not been determined for industry but it may be expected to vary considerably with the type of work, length of the work day, length of the work week, individual diferences sex of the worker and the level of motivation. It has been seen that whether or not employees have rest pauses is not entirely under the control of the company. Employees can pace their work so that short rest pasues are continuously taken. Management can control only the rest pauses that are officially given, and these resul in decreased unofficial rest pauses.

Unofficial rest pauses actually reflect the employee's way of budgeting his energy. When men work long days they pace their work differently and work more slowly than when they work short days.

The official rest pauses should be introduced just before production begins to fall off. If an individual is engaged in finishing a unit of work, an externally imposed pause may serve as an irritant because it disrupts the work pattern. Jobs should be carefully studied and the rest pauses adjusted properly. Training individuals to space their rest properly may be the best solution of the problem of distributing rest periods in accordance with variations in jobs and differences in people.

The need for rest pauses varies in different types of work. In many positions there are frequent pauses due to delays of various, kinds and to superimpose additional rest pauses on an already crowded series of undesired rest pauses would be quite illogical. Rest pauses are necessary for the worker under the following conditions.

(a)   When the work is repetitive and monotonous.

(b)   When work involves continuous sitting or standing.

(c)   When workders are exposed to extreme heat and unfavourable ventilating conditions.

From a production stand point the introduction of rest pauses has almost invariably resulted in an increase inproduction rate. A study reports an average increase of 6.2 per cent in 10 different types of work as result of two rest periods daily, and an increase of 2.8 per cent in five other types of work as a result as of 5 to 10 minutes rest per hour or per work spell.

Optimum Work Weeks. Let us consider the effect of reducing working hours below those of the eight hour day. Studies have shown that production for a six hour day was approximately the same as for an eight hours day.

In a more detailed study work weeks of 36, 40, 44 and 48 hours were compared. The hourly output for the 40 hours week was found to be the most efficient, the production being.868 units. With the 44 hours week hourly production fell to 839 units and with the 48 hours week, it dropped to 793.5. For the 36 hours week, production was 734 units which make the work pattern less efficient (per hour) than the 40 hours week but more efficient than the 44 and 48 hours weeks. The infficiency of the shortest work week may have been due to loss in practice and improper work attitude or to any number of other factors.

Apparently, the 45 to 50 hours week is approximately the normal work pattern which yeilds the greatest weekly output, only when the work period exceeds these hours that there is decrease in hourly output sufficiently great to make the longer week productive than the shortest week.

The studies have shown that the work of one day may seriously hamper that of the next. The fact that Monday's production was found to be below that of Tuesday suggests a phenomena analoguous to warming up. The rate of production falls of during the week and is highest on Tuesday and lowest on Saturdays. The persisting fatigue phenomenon is due to incomplete recovery from one day to the next or to a need for recreation without which man is deprived of a necessary stimulus to life.

The Work Shift. The industries are required to operate three shifts in order to give 23 hours of service, to increase the production capacity of the plant. As capital investment or fixed cost per employee increases, it becomes more and more necessary to operate more than one shift to reduce overhead.

The psychological problems created by the shifts are perhaps the most striking. Most employees prefer the day shift, the expressed reasons being health and social life. The investigators found general agreement among the workers that the feeling of fatigue was greater for the night than for the day shift. Failure to get enough sleep at home while working on the night shift is the serious problem.

Answers of workers to the questions of what cause the most trouble in making shift changes are shown in table. The major problems had to do with making adjustments and eating. The family reaction to shift work was negative in 61 and 79 per cent respectively for plants A and B.

Shift rotation is an attempt to distribute the inconvenience to all employees. Although the changes introduce additional problems of adjustment, it does

Area of complaint	View of skift workers mentioning each area in percentage	
	Plant(A) N=22	Plant (B) N=21
Physical and Physiological		
1. Sleeping difficulties	84	67
2. Eating difficulties	45	47
3. Eliminate difficulties (irregular bowel movements, constipations etc.	6	5
4. Staying awake on the job	3	8
Social		
5. Family Life	6	4
6. Social Life recreation	5	7
7. Miscellaneous	5	9

* Percents add upto more than 100 since some respondents mentioned more than one complaint area.

permit each family to live a normal life for a part of the year.

Persons who live out of phase with other also have problems with shoping, recreation etc. When a large number of persons work on a night shift, the possibilities of developing more nearly normal patterns of life should be explored.

38
Productivity and System Concepts

What is a System

According to Mason Richardson, a system is a set of coordinated components. These components can be considered as subsystems. These sub-systems interact with each other to produce an output in relation to the inputs utilised. A system, therefore, has basic objectives to be achieved. The performance or output of a system has to be compared against the objectives. A system approach stipulates to look at performance in totality ; the performance as a result of interaction of components (sub-systems) comprising the system. An organisation can be considered to encompass systems of Production, Marketing, Financial Management, etc.

Productivity of a System

It represents the overall efficiency of the system. It signifies so to how well a system is performing against the objectives or goals. Essentially, therefore, in systems approach, the performance objectives of the system under evaluation form an important aspect in an exercise of productivity measurement. The performance objectives become the pivotal point to start the exercise. These must be laid down and qualified. The performance objectives also help one to know the resources (as inputs) for development for improvement in productivity.

Good Productivity Measurement under Systems Concept

The following essential attributes of Productivity Measurement Model are needed :

(a) Productivity measures should reflect as to how well a system is doing with reference to its performance objectives.

(b) It should take into account all outputs of the system. Outputs are not only the products manufactured or products sold. Outputs are the achieved objectives.

(c) It should take into account all the resources consumed in the process to achieve the output (achieved objectives).

(d) The model should provide information as the deployment or reallocation of sources to obtain better productivity.

(e) If the system has multi-performance objectives, the model should reorganise the relative importance of each of the objectives.

(f) The data, which the model proposes to use should be easily available and comprehensible. It should be management oriented.

What comprises a Production System

A Production System basically is meant to produce marketable goods in right quantity, right quality, at minimum cost and in right time. In the process the production system uses several resources, including men machinery, materials and capital, and through coversion known as manufacturing process converts raw materials or semi-finished goods or components to finished saleable products. The production system, therefore, essentially is an organised entity which uses facilities of man power, machine or plant capacity, materials and capital in a controlled manner to achieve the desired objectives decided well in advance. In manufacturing saleable goods it also takes into account requirements of the customers existing or potential, to meet their demands of quality, quantity and timely supplies. The production system organises (itself essentially in these directions. Besides, product on system as considered above does not vary widely in different types of production operations like the mass production, the batch-size production or the flow production. Only the extent of utilization of different resources vary. The production system to comprise of a number of manufacturing facilities alongwith man-power, direct as well indirect, to convert raw materials to finished goods. The process of conversion calls for expenses on tools, consumables, utilities, maintenance, etc. and it is presumed that the expenditure falls under the perview of the production system. In line with known organizational practices, it is presumed that materials procurement storage, design and development of products do not forms a part of the production system under review.

Performance objectives of a Production System

A Production System can have several objectives. These will vary in consonance with the corporate objectives of the company. However, it is essential that these objectives are known in concrete quantitative terms. A proper modification of objectives helps in planning, organising activities and monitoring of the performance. As a matter of fact, an exercise of productivity measurement is meaningful only if performance objectives are identified and qualitified. For the production function the following task can be considered as important ones for a satisfactory performance :

—Control of Labour Costs.
—Control of Material Costs.

—Optimum use of Plant and Machine capacities.

—Fulfilment of Production Programme on time.

—Improvement of Profitability.

—Achievement of Quality Standards.

—Training of Workmen.

—Control of through-put cycle.

—Control of the use of capital.

—Protecting Company Assets.

—Good House Keeping.

—Reduction of costs.

—Improvement in technology.

—Maintenance expenditure.

In brief the following can be considered as performance objectives of a Production function :

(i) Labour Utilization.

(ii) Capacity Utilization.

(iii) Material Utilization.

(iv) Completion of Schedules.

(v) Capital Utilization.

(vi) Control on Indirect Expenditure.

(vii) Cost Control.

Labour Utilization

The most important function in any production set-up is that the budgeted quantity of work over a period must be achieved. Labour utilization is essentially an objective in that direction. Labour utilization measures as to how work has been planned to go through at various machines/departments or work centres and as to what has been the achievement against the plan. The labour utilization can be higher or lower depending upon several factors or outside the control of the management. Some of the factors can be the availability of work loads, materials, work tools break-downs of machines availability of power, worker efficiency, the quality of basic raw-materials for conversion etc. Performance objectives against labour utilization can be considered as :

—Standard hours equivalent to a minimum of 70% attendance hours of all direct workmen are produced. This objectivated output can be expressed as performance ratio as :

$$= \frac{\text{Standard Hours (Budgeted)}}{\text{Attendance Hours (Budgeted)}}$$

Capacity Utilization

Capacity has been defined in many terms. It can mean almost anything and also virtually nothing. It is, however, safe to define it as the maximum output which can be delivered during a period of time under given conditions. Maximum manufacturing capacity (or simple capacity) is not a constant deterministic quality as it is a function of use and technical design of product manufactured. This can be improved upon through introduction of better manufacturing technology in the form of better cooling, selection of appropriate machine tools, manufacturing processes, automation, etc. To an extent, this measure also reflects on the steps taken by the production system to improve upon the manufacturing technology. Capacity of a facility is also inter-related to other facilities. Measurement of capacity, in case of a production system having multi-product and multi-facility dimensions, is a difficult exercise as a number of cons-traints have to be taken into account to arrive at the figure. One way to define this performance objective is :

—A minimum of 70% of maximum capacity is utilized. Trans-la'ed in terms of a ratio, it would amount to :

$$= \frac{\text{Standard Hours (Budgeted)}}{\text{Capacity Hours (Budgeted)}}.$$

Materials Utilization

In the context of the present case study, procurement, purchase, storage, and the inter-related issues of inventory control, failures in supply, loss of production due to delays in supplies, vendor deve-lopment, etc. are not considered to form a part of the production system. Production system converts materials to finished goods. The material utilization in this context is the utilization or conver-sion of purchased items in the form of raw or semi-finished stages to finished goods against a budgeted norm. Utilization varies depen-ding upon the percentage of rejection, creation of scrap, spoilage an obsolescence, work wastages, etc. The performance objectives against material utilization is the yield of materials used and can be expressed as :

$$= \frac{\text{Useful conversion of material (Budgeted)}}{\text{Total expenditure on material (Budgeted)}}.$$

Completion of Schedules

A customer looks forward to the receipt of goods in time. Delay can result in a bad image of the Company, and consequently there can be a loss in sales, decrease in goodwill etc. For a pro-duction system, therefore, keeping to the delivery schedules, i.e. manufacturing the goods as per the programme is equally an impor-tant performance objective. In the case of a production system manufacturing a single product, it should have been possible to arrive at an acceptable throughout cycle time and co-relate the delay in completion to the same to establish some norms.

Capital Utilization

For any production set up, facilities, machine tools and processing assets of capital nature are essential. Capital utilisation in this context refers to the use of capital for use of capital assets. Decisions are also required as to when to replace fixed assets. Decision to replace early will bring down maintenance costs but will call for capital expenses. On the other hand decisions to replace late will no doubt give a better ratio of production to capital expenditure but simultaneously increase maintenance costs. Somewhere a balance is desirable. The performance objective is to obtain a high ratio of standard hour produced to per unit of capital spent on production assets. Although there are many accounting methods of valuing fixed assets, one such method is written down value. The performance objective to be considered here then becomes :

$$= \frac{\text{Standard Hours (Budgeted)}}{\text{Capitalised cost (Budgeted)}}.$$

A production system incurs a lot of indirect expenditure, which helps the production process. This may comprise of manpower employed in planning, store keeping, record keeping, inspection as well as in the form of labour employed in any indirect operations covering maintenance tool room, tool issue etc. Indirect labour is also employed in materials movement, good house keeping, cleaning operations and all other sundry jobs. Indirect expenditure is also incurred in the form of use of indirect materials like tools, oils, lubricants and other consumable items required in the production of goods.

Performance objective for the control of Indirect expenditure can be considered as :

$$= \frac{\text{Total Expenses (Budgeted)}}{\text{Indirect Expenses (Budgeted)}}.$$

Direct expenditure is incurred in utilities as well. In many an organisation a total of indirect expenditure form a formidable percentage of total expenditure.

Cost Reduction

"What does it Cost" and how to control cost are two important questions in every day transaction of business and the latter one is most important. If costs are not controlled at the point of occurrence, the spiral up and lead to lower productivity. From national angle cost reduction is a vital factor in increasing productivity and standard of living. Lower costs make for lower prices and can facilitate improved quality through value engineering and analysis ; which in turn stimulates consumer demands. These reasons support the view point that cost reduction must be practised much more strongly and on a much wider scale. Cost reduction is the prime concern of today but it should not be at the cost of product quality. Other values like use value, exchange value and esteem value of the product must be given equal importance. Every economy and consequently every business industry must try to continuously increase its productivity through cost reduction and quality improvement. The industry around the world and in India is in many different stages of development ; so action oriented cost reduction approach will go a long way in improving national economy and productivity.

Direct Costs

Certain costs, which can be easily identified with specific products or processes are direct costs. Direct cost of materials is an important element in the total cost of product. It varies from 50 – 80% of the total unit cost. In the control of this material cost, Designers, Method Engineerrs, Material Planners, who determine and deal with Direct Material in various capacities, can play a significant role. They determine standard cost of the product.

Labour Cost, both direct and Indirect, is an other important element of this cost. Performance standards may be set up through Time and Motion Study. Actual performance should be compared with standard performance and in case of various steps should be taken. Money cost of labour is not the only thing.

Indirect Costs

Overhead expenses such as Indirect labour, Indirect material, Administrative and Selling Expenses are controlled by means of budget. For the control of Indirect Materials, norms of consumption must be set up.

Works Costs and Works Overhead

This includes expenses on Indirect labour, indirect material, power, rent, repair, depreciation etc. They are all indirect expenses right from the receipt of order to the despatch of finished goods.

Administrative Overheads

Direction, control and administration of an undertaking, rents and rates of offices, premises and furnishing, if any, office lighting, salaries of office staff, stationary, postage and other miscellaneous expenses are covered under Administrative overheads.

Selling and Distributive Overheads

Cost of soliciting and securing orders for the articles of products and of efforts to find and retain coustomers, publicity and advertising, catelogues, leaflets, and price list, packing and forwarding charges, godown rent etc. includes all the expenditure incurred from the time a product is manufactured until it reaches its destination.

Costing

Costing means classifying, recording and allocating the appropriate expenditure for determining the cost of products or services and the presentation of suitable arranged data for the purpose of control and guidance of management.

Factors Contributing Cost Reduction

—Optimum utilisation.

—Select right person with right skill.

—Training in the desired field to meet the challenges of fastly changing techno-economic environment.

—Recognition of work and proper motivation for best performance and reduced absenteeism.

—Ensure availability of all input at right time.

—Better working environments.

—Job re-design.

—Re-deployment possibilities instead of fresh induction.

—Career Planning.

—Scientific manpower assessment and planning in case of fresh induction.

Machines and Equipment

—Proper maintenance of machines and equipment so that (i) Downtime is minimum, (ii) Full capacity is available.

—Proper machine loading to ensure optimum utilisation and reduced cycle time.

—Proper job on right machines.

—Use of special purpose machines rather than conventional machines e.g. using Numerical Control Machines (Gas Cutting) instead of manual Gas cutting.

—Replacement at prope time.

—Use of proper Jigs and fixtures to reduce setting time and operating time.

Materials

—Use proper size and type of material.

—Make use of Tailoring Charts to ensure optimum utilisation of plates.

—Vendor evaluation.

—ABC Analysis.

—Reduction in lead time as far as possible.

—Proper storage to avoid damage and ensure quick availability when required.

—Minimum use of indirect material.

Energy Conservation

In short term good house keeping in industry gets priority. This involves improving fuel-consumption in boilers, furnaces and engines, effective plant maintenance, adoption of heat recovery systems, re-using waste materials to produce energy etc. The appointment of an energy officer responsible for day to day supervision in each large industrial unit to check machine, idle time, frequent start ups and stops, leaks and spillage and ensure compliance with energy conservation norms would be particularly worthwhile.

Lands and Buildings

—Scientific layout to optimise area utilisation.

—Better vertical space utilisation whenever possible.

—Multistorey buildings for expansion of office areas.

—Design of buildings to make use of natural light and ventiliation.

—Make use of gravity flow/feeding of materials and equipment wherever possible.

—Plan for least material transportation to avoid spillage and damage.

Time Management

Management Effectiveness. A check list for the top management is as under :

—Are responsibilities clearly defined and assigned ?

—Is a full organisational analysis made periodically to evaluate competence and redundancy ?

—Is there an established executive development programmes ?

—Are vendors rated for services and quality ?

Tools of Cost Reduction

Some of the following tools may overlap somewhat, but each may have its purpose in a comprehensive programme.

Cost Accounting and Analysis

Cost Control takes the established cost standards and endeavours to keep operation cost in line with those standards. Cost reduction challenges all cost standards and endeavours to reduce these continuously. Standards are targets in cost control procedures but are suspect in the case of cost reduction.

Cost reduction should be applied in every section of business. It is not dependent upon standards, though target amounts may be set. Cost control seeks to get the best results at the lowest possible cost under existing conditions. Cost reduction recognises no condition as permanent, where a change will secure a lower cost figure. Cost accounting and analysis are most valuable tools for analysing potentialities and possibilities of reducing cost of the product ; the process and operating procedure or the general factory expenses. Cost accounting reports are necessary to give an appraisal of the effectiveness of cost reduction and control.

Budgets and Budgetory Control

Budget plan and Budgetory control figures provide a storehouse of cost reduction possibilities. The operation of Budgtory control shows where operations are not going according to plan. Budgets are often the best source of information. Cost reduction agencies use budgetory control by setting up targets for savings.

Analysis

(a) No organisation is too small or too large to rule out the attractive possibilties of saving through cost reduction.

(b) Cost reduction is a state of mind and should be promoted as such.

(c) Involve as many employees as possible in cost reduction programme.

(d) It should be a continuing programme, for it is never finished. Cost reduction should, therefore, be a permanent, alive and vital part of company's organisation.

(e) Savings can be made in all directions, but priorities must be fixed.

(f) Cost reduction is the real key to national prosperity under a rising standard of living.

(g) No concern can afford to ignore cost reduction.

(h) Cost reduction requires resourcefulness imagination and enthusiasm. Success in effecting savings today merely reduces extra affort tomorrow.

(i) Cost reduction leadership flows from top. It will rarely climb up from the bottom.

Performance Objectives of a Production System

Performance Objectives of a Production System are examined and measurement of productivity demonstrated through the use of Performance Objective Productivity (P-O-P) Approach. The concept lays emphasis on Performance Objectives and corresponding inputs to arrive at an integrated Productivity Index. The emphasis is on production function to be considered as a system.

Conventional Productivity concepts and hence measures relate output to inputs, such as Labour, Materials and Capital. As a variation, some of the productivity measures consider input as an aggregate of all factorial inputs. These measures suffer from several drawbacks as they fail to recognise interactions of various inputs within themselves as well as influences of other factors such as managerial practices, environment, changes in technology and organisational skills. Productivity measurement models have been presented as if the inputs of Labour, Capital, Materials, etc. can be fully identified in isolation to each other without any inter-relationship. The models also suffer from inadequacies of Units of measuring both the outputs and inputs.

Labour Productivity as an Index of Productivity

Often labour productivity is considered as synonym to productivity of a Production system. As a matter of fact it is generally perceived that a level of productivity achieved is what is produced by the labour on machines and work benches. An effort to reach higher productivity is interpreted as an action to achieve more output from labour force. Labour on the other hand is only one of the input resources responsible for output. In many products labour may form only a small percentage of input resources ; the materials may constitute the bulk of the proportion of the inputs. In other cases, technology may play a major role.

Labour Productivity is generally perceived and interpreted as a measure of labour efforts or effectiveness. Higher the labour productivity, it is recognised, higher are the labour efforts having been put in. Labour productivity is a function of many variables and is dependent on factors, such as technology, environment of the plant, the management practices, policies and other organisational characteristics of the plant. These affect production from a workman considerably. It is well known that output per labour hour can be increased through use of better quality of raw materials. Components, with a high degree of fabrication in the form of castings, forgings etc. require less machining hours. In a similar context better manufacturing technology in the form of advanced tooling, jigs, fixtures, tool control systems, machine tools with multiple work and tool stations and higher work parameters, or a high degree of automation and production engineering systems increases the output with perhaps lesser efforts from a workman. Managerial decision to replace manual operations with electronically controlled machine tools and machining centres can result in higher output, and hence a higher labour productivity. In all these instances, labour productivity, as it can be seen, has been determined by factors beyond the reach of a workman. Labour productivity cannot, therefore, be viewed as solely representing the efforts or effectiveness of a worker.

Factorial Productivity as Index of Productivity

Even output cannot be considered as resultant of one output. Several input factors join hands and interact to provide an output. There are trade-offs in between. Savings over one input can correspondingly result in an increase in another input with possible overall result of productivity having been lowered. For example, use of better quality raw-materials or a higher degree of fabricated components would show an increase in labour productivity measured, say in a number of units produced per labour hour, but simultaneously it would also result in increased costs of materials seen as another input and thereby decrease overall productivity measures, say as a number of units produced per unit of money spent together on labour and materials. Similarly, an advanced technology might yield a higher labour productivity, but perhaps in the process lower capital productivity. Besides, output is equally affected by factors, such as environment which might be beyond the controls of the organisation. A factorial productivity, such as capital productivity or Materials productivity measures as ratio of output to factorial inputs of Capital or Material cannot be a representative index of productivity of an organisation. On the other hand, a factorial productivity can lead to wrong perception of performance of the organisation.

Production, Productivity and Quality

The difference between production and productivity is well understood and requires no elaboration. Whereas production is output or performance measures in absolute terms with no reference to its relationships with resources spent. Productivity essentially represents the relationship between the performance (or output) and input. Production, Productivity and Quality cannot be considered in isolation. Both production and productivity have a common dimension in the form of output. It is a fallacy to presume that production as well as productivity can be increased through simplistic means of diluting quality does not carry same economic or use value as foreseen without lowering of specifications. As a safeguard against the tendency of aiming at highest output, and hence productivity, in a product on system at the cost of quality of output, a good productivity measures, should preferably have a built-in provision for accountability of rejection rate defective. workmanship, scrap generation, spoilage, costs of rework, yield of materials used as inputs or similar other characteristics which reflect on the quality of output. As a first step in an exercise of productivity measurement, acceptable quality characteristics of the output must be specified and quantified as performance objectives. Any deviation form the same should have a bearing in the productivity measure.

Productivity and Profitability

It is generally presumed that higher productivity results in higher profits. This is again a fallacy. Profits are essentially related to market conditions, the quality of the product and the marketing strategy. There are instances of increased profits although productivity went down. This happens in cases of monopoly manufacturing when the customer is asked to pay for low productivity of the suppliers. Similarly, increased productivity need not necessarily bring better profitability because of other matching steps having been taken to meet the competition.

Productivity and Good House Keeping

Many an organization spend a lot of money on house-keeping, cleanliness, landscaping, etc. as to present the plant as a show-piece to a visitor. Very often colourful progress reports, charts, graphs, are prominently displayed on notice boards as well as in deptartments. These might be aids to contribute to a good morale or o project a systematic work to a client. To a small extent in this context good housekeeping can influence productivity. But this cannot be considered as an end objective and cannot be a part of the output of a plant. A plant with an extreme good house-keeping outlook cannot be an alternative to an efficient operation. However, it is not a significant input factor to increase output.

39

Management Information System

Concept of Information

We perceive information as paper work, which means nuisance. This feeling is not correct. Information at present is one of the ingredient of industrial operations. In a simple organisation, the amount of information needed to provide efficient utilisation, of men-materials and machinary is not very significant, but in a larger size production unit efficient production depends considerably on systematic flow of information regarding the sales forecasts, materials planning, production scheduling, costs and profitability at various production levels. Information is an action concept. When you want somebody to work, you must tell him what to do and how to do it. Modification for creative and intelligent work come only from job satisfaction and job satisfaction cannot be achieved without understanding the objectives of the job to be achieved which in turn can not be understood without information.

Executives in industries normally complain of :—

(i) Increasing amount of paper work.

(ii) Paucity of information available for decision-making.

The two are contradictory. On one hand, information is pouring in through papers, submerging the executives and on the other hand, there is not enough information for rational decision-making.

The information system is based on the analysis of each aspect of information with respect to different utility for decision making and the cost of getting the information.

The information system reflects the style of management. In many concerns, the management exists only at the top. All the decisions are taken by the top man himself. With this unity style of management, there is a possibility of inefficiency in decision making due to overwork rather than to share power and depend on the loyalty of others. Management normally has two options :

(i) Fire Brigade management and

(ii) Look ahead management.

The Fire Brigade Management can ask for information when required. But look ahead management, believes in formulating carefully its needs for information ahead of time. The information is closely tied to the possible action to eliminating superfluous information. Thus there is a need to devise a system which provides symbolisation and information to substitute for seeing actual work.

The importance of professional managers has considerably increased during last 30 years. Many factors are responsible for this importance such as rapid change of technology increasing size of business, changing socio-political environment in which business have to operate. According to Peter Drucker managing of business goes a way beyond passive reaction and adaption. It implies the responsibility for attempting to shape the economic environment for planning initiating and carrying through changes in the economic environment for constantly pushing back the limitations of economic circumstances and enterprise's freedom of action.

In order to provide this kind of creative leadership in a complex environment, the management needs information. Manager needs the information in order to perform following tasks :

Setting of objectives .

To establish the plan for achieving the objectives.

To execute the plans.

To evaluate the performance of the organisation.

To have the feedback of the plans.

Thus management's number one task is to get the right information in the right form, at the right place and at the right time. The right information is that information which is necessary for the effective functioning of the management. Thus the increased complexity of business has forced management to rely heavily on information system.

Following are the parallel developments which have increased the availability of quantitative and qualitative information.

Standard costing.

Marginal contribution and break-even analysis.

Flexible budgeting.

Electronic data processing.

We know that as a business grow in size and becomes complex it becomes impossible to pay attention to all the aspects of the business. At this stage we need the advice of functional specialists in the area of production planning, production control, accounting, work study, marketing, quantitative techniques etc. In addition to this the changes in the organisational structure had to be made to meet the various changes. The term management information can be defined as an efficient and purposeful system which have the capability to provide all levels of management, a timely, prompt, accurate and reliable information M.I.S. is a nervous system of the organisation and any deficiency in this system is very likely to reflect of the functional and administrative efficiency of the organisation. To understand the concept of management information system. Let us understand each of the following terms :

—Management

—Information

—System.

Management

It is an essential part of any group activity. An organisation cannot survive without management. It is the management which makes people realise the objectives of the organisation and directs the efforts towards their achievement. Management is a multi-purpose organ of an organisation that manages the work and personnel at work. It is a creative and innovative force striving to secure the maximum result by the use of available resources. Management provides new ideas and vision to the work group and integrates its efforts in such a manner as to account for better results. The word Management refers to top, middle and lower level as well as different functional areas of management such as production, operations, marketing, finance, accounts, management services, personnel etc.

Information

It is something which management expects to know at a given time. Information is not data, the data comprises of a host of information, which may be of little or no consequence to management. The information is needed to plan, organise, direct and control the business and suitably designed information pays for itself. The effectiveness of any information is dependent on the timing and content of the information presented and the management action. The need for management information is felt when the managers have to make decisions. The manager will have to rely on his judgement but he must have information on the basis of which he arrives at the decisions. The three important uses of information are :

—Process of management planning (decision making).

—Execution.

—Control.

Once the objective of the business are decided than it is necessary to plan the course of action to be followed in order to achieve these. Such a plan need reliable information so as to obtain best results. The next step is to communicate the plan or decisions to lower levels for execution. This requires information from those higher up as to what action is to be taken all along the line to the lowest levels. The next is the control functions of management. After the plan has been made and executed one must have the information of the results and standards to evaluate whether objectives have been achieved or not. This calls for a regular flow of feedback information which will indicate in what way the results vary from planned targets. By examining these variances management can decide the corrective actions to be taken. Thus planning, executing and controlling the affairs of a business will require information. Information is required by all levels of management free but the same information will not be useful at all levels. At the top management level what is required is a very broad survey relating to areas where results have varied from plans. This is known as management by exceptions. Lower down the organisation structure the same principles will be followed by each executive and will be given more detailed reports directly bearing the responsibility.

Information can be presented by means of graphs, charts, (pie, charts. Bar charts, Histograms, Ideographs and Flow charts). The timing and frequency of reports depend on the nature of the report whether routine or special. In considering the timing and frequency of reports, the financial cost of preparation must be weighed and balanced against the anticipated benefits expected from the use of the reports. Reports normally present information regarding a period of time. Reporting is the physical aspects of manufacturing on its technological aspects and economic aspects. Reporting is a matter of planning and creation of record.

Following are the types of records which are usually created.

(a) Records of basic information.

(b) Records showing what is available.

(c) Historical Records.

(a) **Records of basic information**

(i) Blue prints.

(ii) Bills of materials.

(iii) Production routing.

(b) **Records showing what is available**

(i) Raw materials.

(ii) Work in process.

(iii) Semi-processed stock.

(*iv*) Machinery and equipments.

(*v*) Finished stock.

(*vi*) Tools, jigs, fixture and gauges.

(*c*) **Historical Records**

(*i*) Records of production.

(*ii*) Records of waste and rejects.

(*iii*) Records of machine performance.

(*iv*) Records of sales.

(*v*) Records of absenteeism.

The records exhibit information about actual production, about actual input as compared to pre-conceived standards.

In a small organisation the reports are prepared on the shop floor and the management evaluates the daily report and takes prompt action. In a large plant the same course is followed by the production planning and control on an adequate scale relevant to the operational requirements.

Following are the various type of reports normally used in a large plant.

(*i*)   Daily production reports,

(*ii*) Weekly raw material position,

(*iii*) Monthly cost statements,

(*iv*) Monthly Maintenance Plan,

(*v*)   Monthly Quality Control Report.

Various forms are used for evaluaticn and considering the reports. Management information need not always generate from within the company. It has been realised by now that the top management should have a continuous flow of information. Following are the areas on which routine information is given importance.

—Customers.

—Competitors.

—New areas of Investment.

—Government Policy.

The various types of management information which are commonly required by the general management in most of the business organisation are :

—Budgets.

—Revenue Statements.

—Cost Reports.

—Statistical Information.

—Capital Expenditure.

Budgets

Budgets are normally prepared annually and submitted in advance to the management.

A consolidated budget comprising of detailed budgets is prepared to show the broad aims and plans of the organisation as a whole and give a forecast of the financial effects.

Revenue Statements

The revenue statements are prepared on monthly or quarterly basis. Statistical information is given wherever possible and correlated with the financial information. The revenue statements will give the comparative figures of corresponding and previous periods.

Cost Reports

Costing information is essential for production and service oriented companies in view of its impact on project margins.

Statistical Information

The volume of business is perhaps the most critical single factor having repercussion on net projects arising from the burden of fixed overhead expenses. Statistical information deal separately with each activity usually started in quantitative unit.

Cash Flow

Every good information system have cash under control, amounts locked up in sundry debtors are drawn and so are losses from bad debts.

Every business organisation uses the concept of cash flow. The cash flow statements are the projection for a future period of time. The opening and closing cash balances will give us the surplus or shortfall in cash resources. As borrowed money command high interest rates, it is essential that a proper scheduling of cash resource is achieved.

Capital Expenditure

Capital expenditure statements are used as part of annual budget. This gives us the estimates of expenditure to be incurred during the forthcoming period on capital project to be taken up. The capital expenditure plan are part of the long term forecast.

System

System refers to a group of components which interact to provide management with the information it seeks. The system would have its components as :

1. Inputs.
2. Outputs.
3. Processing devices.

Nature and Design of M.I.S.

Nature. The unique information requirements of different management levels and the functional areas is outlined in Fig. 39.1. The diagram clearly describe the information systems in the corporate hierarchy. Table 39'1 shows the various management areas and the kind of information that might be needed.

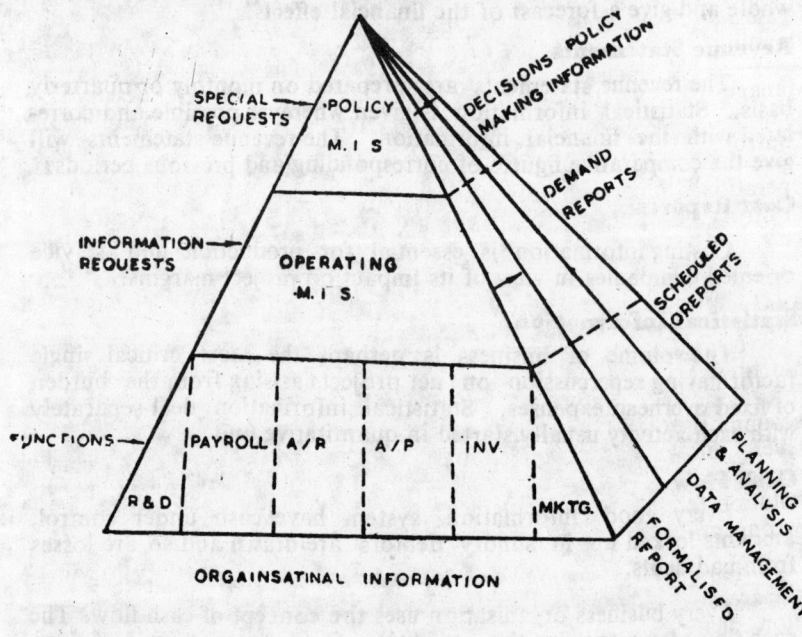

Fig. 39.1

Design

It is indeed a difficult task to design a suitable and efficient information system for the management. This demands a thorough knowledge of the organisational structure working and system. It requires experience, skill and vision of the Information System. Designer who must understand, assimilate and foresee the information needs of the management. A few of the important steps involved in the design of M.I.S. are :

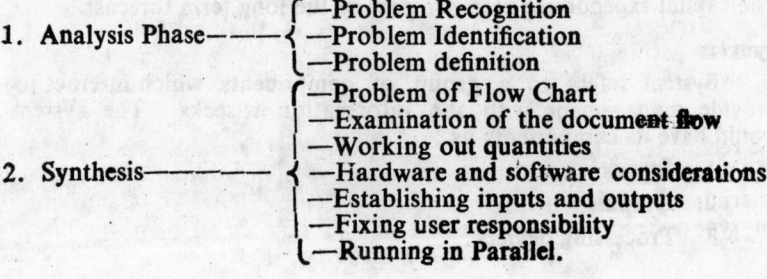

1. Analysis Phase————{ —Problem Recognition
   —Problem Identification
   —Problem definition

2. Synthesis————— { —Problem of Flow Chart
   —Examination of the document flow
   —Working out quantities
   —Hardware and software considerations
   —Establishing inputs and outputs
   —Fixing user responsibility
   —Running in Parallel.

Table 39·1. An Information System and Corporate Hierarchy

Level	Organisation Identity	Activities	Information System	Outputs
Strategic Planning	Top Management	Set objectives	Special "one-time reports"	Goals and Policies
Management Control	Departments heads	Allocate assigned resources	Many regular reports and Formats	Decisions "personal" leadership
Operational Control	Supervisors Foremen Clerks	Use resources to carry out tasks	Formal	Actions

Analysis Phase

(i) *Ploblem recognition.* The designer must recognise the potential areas which need to be served better and supplied with information which is better, prompt and accurate. This has to be accomplished by frequent and free discussion with the various levels of management. Enough time should be spent in the recognition of the problem, because it is preferable to have wrong solution for the right problem rather than the right solution for wrong problem.

(ii) *Problem identification.* This involves a further study into the areas which have been broadly recognised in the problem recognition phase and a closer scrutiny of the selected areas already identified a deficient areas.

(iii) *Problem definition.* This phase is a further detailed study of the problem identification phase. For instance, it will point out the exact discrepancy.

Synthesis

(i) *Preparation of a document flow chart.* The purpose of the flow chart is to present the flow of different documents relating to the discrepancies of the deficient areas.

(ii) *A close examination of the document flow chart.* This is an exercise involving a careful examination of the several deficiencies thrown in by the document flow chart.

(iii) *Working out quantities.* This involves the computation of the record size and the data elements involved in the systems design. The total volume of data to be processed and maintained have to be determined. This will, in turn constitute the basis for indentifying the type of information system that may be necessary and the kind of organisation, hierarchy equipment and procedures which will be required to support such an information system.

(*iv*) *Hardware and software considerations.* Now it is the time for identifying the exact computer configuration, input output terminals, data storage devices, on-line terminals and other peripherals. It is the time when exact software support that will be necessary needs to be analysed, planned for training personnel as an integral part of this phase.

(*v*) *Establishing inputs and outputs.* This calls for the identification of the areas of entry of raw information into the information system including the exact contents of that information. This also involves determination of the kind of outputs which will be required out of the M.I.S.

(*vi*) *Fixing user responsibilities.* It is important to state precisely and clearly the user responsibilities in the matter of data entry including instructions on such vital matters as the formats, the input output times, the devices, the equipment and the location in which the entry will take place. It will also have to be explained to the user the importance of feedback from him to the reporting output in order that discrepancies in the reporting could be rectified without loss of time.

(*vii*) *Running in parallel.* Any newly designed M.I.S. has to be observed through a period before it is finally accepted. It is important that the new system should run in parallel with the existing system. This is referred to as 'running in parallel'. This phase of new system development and testing could last from a few weeks to several months depending upon the feedback received from the user with respect to the report reliability offered by the new system.

Impact of Management Information System on Management

Following points give the impact of M.I.S. on management :

(1) Information is power and a kind of *information or computor man* has grown in the organisation. Information man is consulted by the top management as well as heads of functional groups, in connection with the fulfilment of information needs of the organisation.

(2) M.I.S. growth has resulted in centralisation. The availability of prompt, accurate and timely information about the various activities of the organisation to the management has enhanced their capacity to perform the control function.

(3) M.I.S. one growth of computers has changed the entire philosophy of the management. Certain problems which is used to be considered very complicated and almost impossible to solve are now a matter of fact. Production scheduling, Linear Programming,

Transportation and assignment models problems are now solved in no time by the help of computers.

Accounting Information System

One of the most essential part of a good management information system is the accounting information system. The principles underlying a good accounting information are :

(*i*) Consistency

(*ii*) Flexibility

(*iii*) Comprehensiveness

(*iv*) Applicability.

A clear assignment of responsibility is essential to an effective information system. A good information system gathers data on costs, revenues and profits. The effectiveness of management information is ultimately determined by the quality of its reports. A common defect of most management reports it that they give too detailed information, which may often lose sight of even by the skilled manager. The reports are useful to management if they are designed towards performance measurement. The reports should be confined to exceptions. The reports should be designed to highlight the areas where action is necessary. The reports should be designed to match the organisational structure of the organisation.

Objective of Management Information System

The object is to provide timely and effective information for the management. The management information system can be divided into following 4-sub systems.

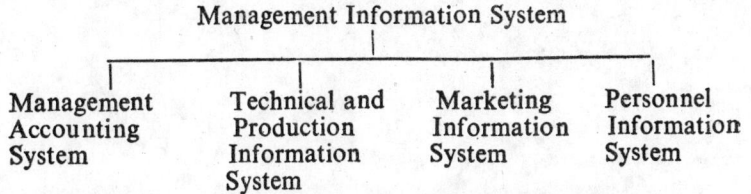

Management Information System

| Management Accounting System | Technical and Production Information System | Marketing Information System | Personnel Information System |

Let us see the line diagram for Marketing Personnel and Production Information Systems.

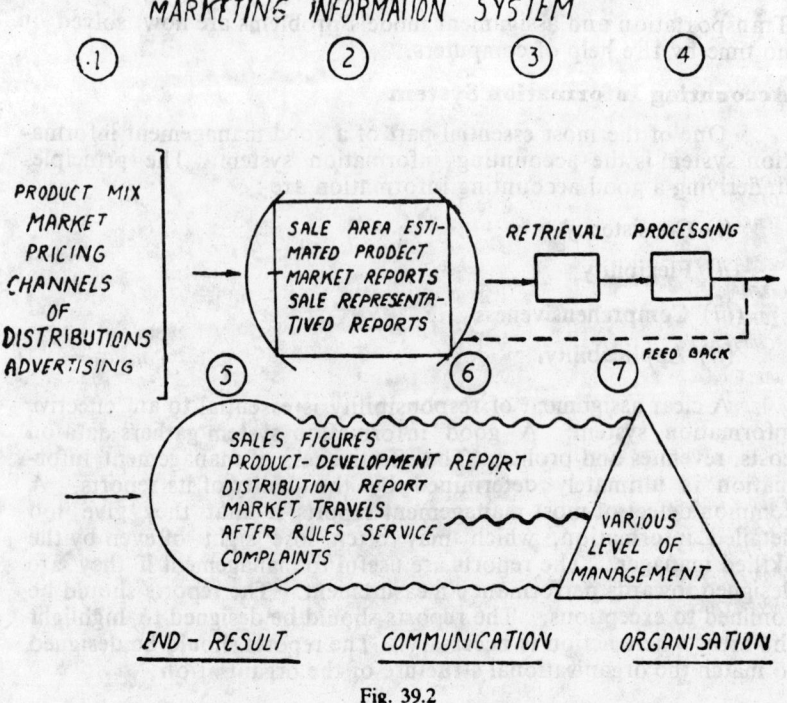

Fig. 39.2

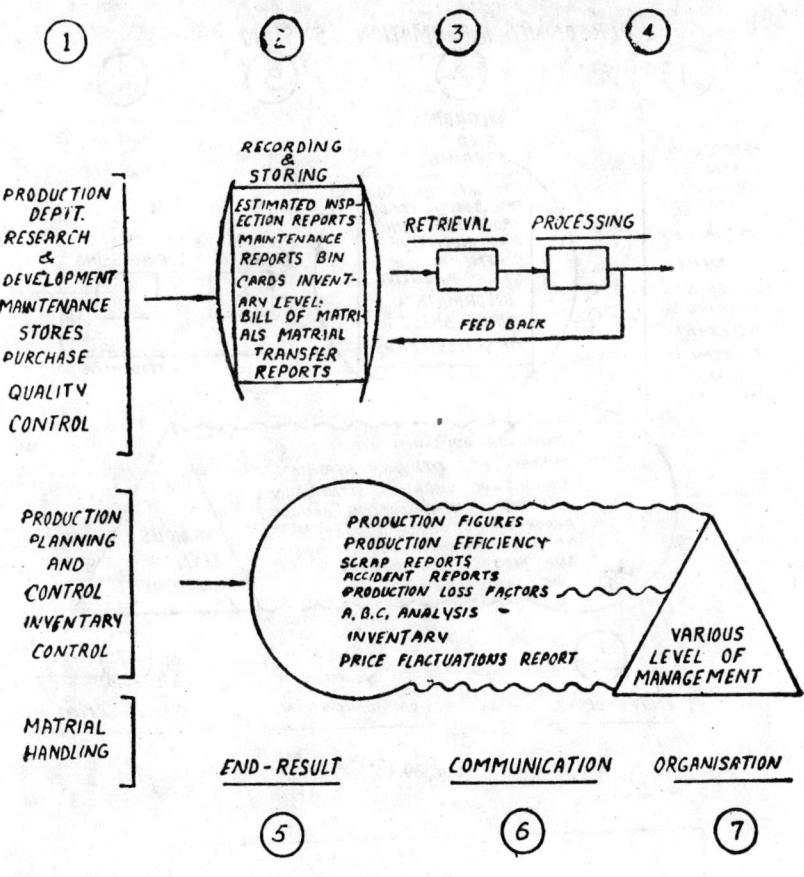

Fig. 39.3

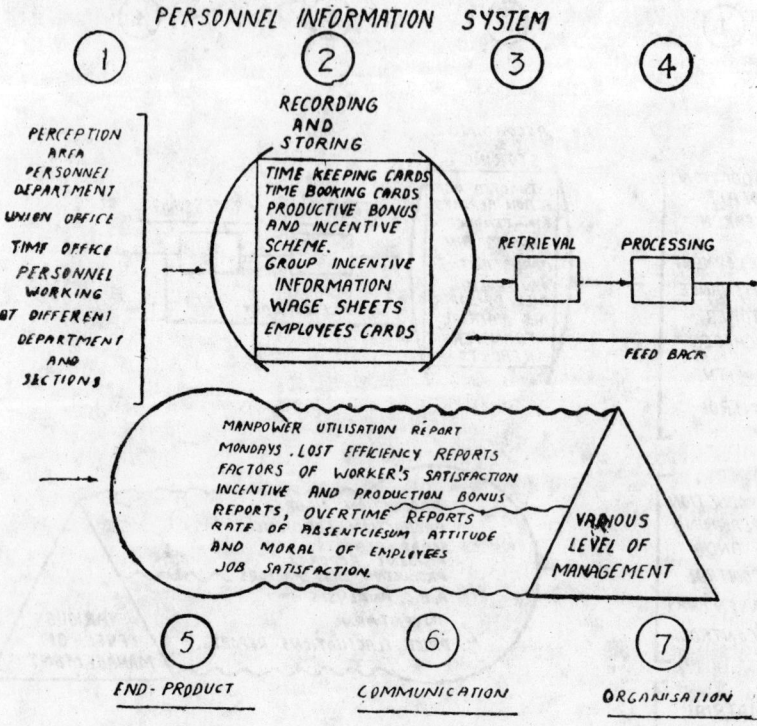

Fig. 39.4.

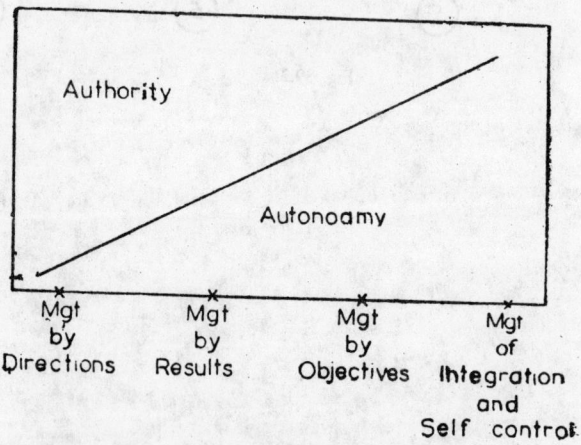

Fig. 39.5

COMPUTOR BASED INFORMATION SYSTEM

Computor Based Information System (C.B.I.S.) is an invaluable aid to the various levels of Management. The computor enables all management philosophy system such as :

(1) Management by direction.

(2) Management by results.

(3) Management by objectives.

(4) Management by integration and self control to operate efficiently and effectively.

Let us see each of these four types of control philosophies of management before studying the impact of (C.B.I.S) on each of these philosophy.

Management by Direction

It is the authoritative and directive form of management. In this approach the top management sets and directs goals, orders, producers for the members of the organisation.

Management by direction organisation uses budget as a control tool. The budget is then issued to the operating manager and he follows these budget. The operating managers operate between the upper and lower tolerances limits thus it limits the creativity on the part of operating managers. In case of problems occurring outside the imposed limits he cannot make, the decision as he seeks the advice of the senior management.

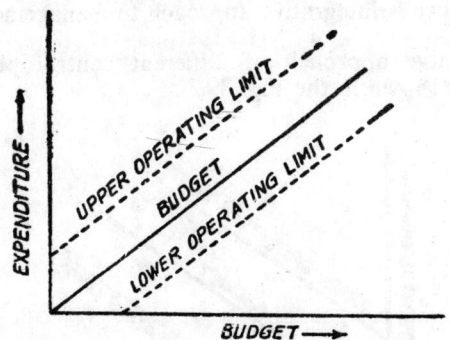

Fig. 39.6.   Influence or C.B.I.S. on management by direction organisation.

Management by Results

It is also similar to directive types approach by management. But the controls are not that rigid as in case of Management by Direction. In this approach the members have same way in the implementation of operations and devising procedures.

Management by Objectives

It relies more on participation style than Management by Results. In Management by objective the goals are set fourthly by the managers and his subordinates. Once the goals are accepted by both parties it becomes a standard for acceptable performance. In an M.B.O. organisation the subordinate has an important role to play.

Management by Integration and Self-control

It is the most participative style. In this approach although the top management is responsible for the achievement of organisational objectives but the management shares the commitment with the work force. To achieve this the organisation through it policy must have enough flexibility to permit, its adaption to the everchanging requirement of its members and changing environments that organisation encounters. The integrative organisation by virtue of its management philosophy has a flat rather than a tall hierarchical structure. The elimination of many layers from the organisation has a number of advantages such as loss of delay, passing the buck and undesirable emphasis on obeying rules. Decisions are made at lower echelons of the organisation. The problems are easily identified, analysed and solved by work groups rather than middle managers.

The integrative approach views self-control as more effective than external imposed control. The approach which is going to make its mark in future is integrative approach to management.

The relative approach of different control philosophies of management is shown in the Fig. 39.7.

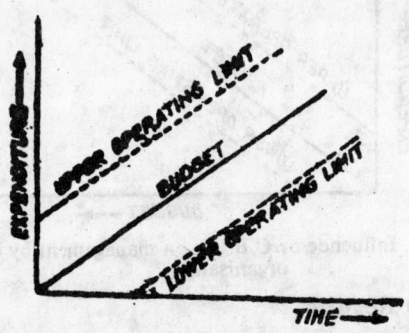

Fig. 39.7

In case of management by integration and self control the senior management has delegated a broad range of authority to the operating managers to make budgetory decisions.

C.B.I.S. suggest that no one type of management philosophy is superior to other but C.B.I.S. can improve the effectiveness of highly centralised directive organisations.

We know that control is both a philosophy and a technique used to guide the organisation in the achievement of its objectives by the use of various skills. The control function emphasises the importance of communications in the achievement of the objectives of the organisation. Control can be viewed as the extension of planning process. The first step in a planning process is the determination of the objectives of the organisation. Management control is the measurement of actual performance *versus* desired results. A properly controlled system enables any manager to quickly spot out the areas of deviation from planned results as shown in Fig. 39.8

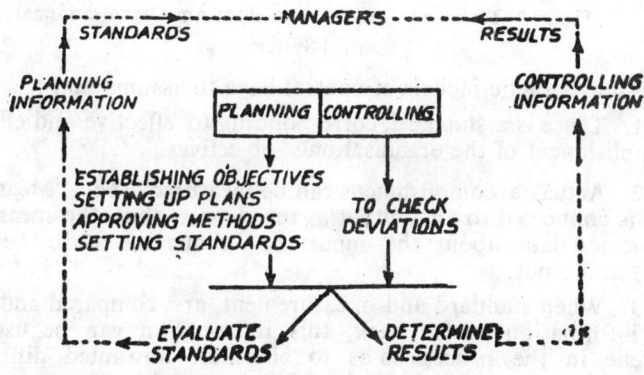

Fig. 39.8

Organisation Control System

The ineffectiveness of many management control system is attributed to the cybernetic philosophy on which are they based. Anthony and Vancil define Management Control as "the process by which managers assure that resources are obtained and used effectively and efficiently in the accomplishement of the Organisation's objectives". Management control is the domain par excellence of formalised systems in organisations, and these systems tend to be designed according to a cybernetic philosophy.

Cebernetics

By "cybernetic" is meant a process which uses the negative feedback loop represented by : setting goals, feeding back information about unwanted variances into the process to be controlled, and correcting the process. This is a much narrower use of the terms "cybernetic" than that advocated by Wiener who coined it to deal with the transfer of messages in the widest sense. In the cybernefic view, a management control process, in ites most simplified form is similar to a technical control process, for example control of the heat of a room by a thermostat. The model in Fig. 39.9 uses only first order feedback.

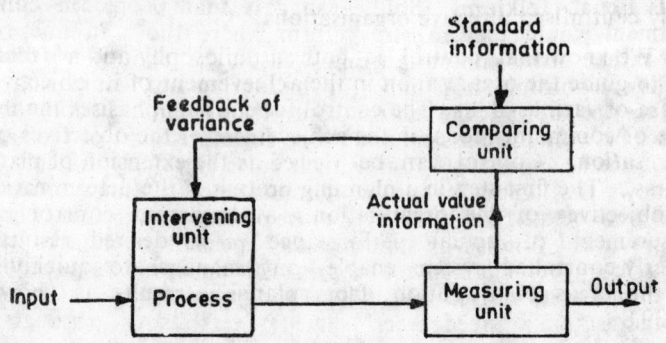

Fig. 39.9 Technical Control Model of An Organisationnal Control System.

All cybernetic models of control have to assume that :

1. There is a standard, corresponding to effective and efficient accomplishment of the organisation's objectives.

2. Actual accomplishment can be measured. The "Measuring Unit" is connected to the output as the process but the measuring may include data about the input, or about the ratio between output and input.

3. When standard and measurement are compared and variance information is fed back, this information can be used to intervene in the process so as to eliminate unwanted difference between measurement and standard for the next round.

Cybernetic Model of Control

There is no doubt that the cybernetic model of control has been eminently successful in the design of machine's electronic circuits, or similar technical systems, but management control in an organisation is a special process in a social or may be socio-technical system. The 'units' in this case are people, or even groups of people. This subjects the use of cybernetic model to severe limitations. The three assumptions of the presence of a standard, the measurability of accomplishment, and the usability of feedback are most justified for routine, industrial-type processes. One remarkable fact about control processes in organisations which has become associated with the cybernetic paradigm is that they are usually tied to a division of labour-different units in the model correspond to different people who are specialised in their tasks. Proper functioning of the control process presupposes commnication : the necessary messages should be sent and correctly received between the various specialised actors; it also presupposes that all will feel motivated to act according to the model. A difference in value between people in controllers' department and line managements is evident to anyone familiar with organisational folklore. Studies suggest that people in control department would tend to stress form where those in line roles would rather stress content. In most cases, the controller's department is responsible not only for measuring and comparing but also for design and the entire controlls system. People at the interface between the classical and the re-structured part of the organsation, such as supervisors found their roles extremely difficult. The case shows that under favourable circumstances, semi-autonomous groups were created and took over most of the management control roles previously fulfilled by superiors and specialists. All tasks within the classical cybernetic control loop-measuring, comparing feedback, intervening—were carried out within the group itself. Its links to the organisation's needs were mainly established through the standards set by others in the organisation for the group's tasks. Like the word "cybernetic", "homeostatic" can be used to mean different things ; the term is used here because of its predominantly biological connotation. Homeostatic processes are composed of cybernetic elements, but without the division of labour between controlling and controlled units ; control is exercised within the system itself. We could also call these processes "self-regulating."

The transfer from cybernetic to homeostatic management control systems will demand a drastically changed control philosophy, especially with regard to the traditional division of labour tied to the cybernetic model. The homeostatic approach needs a new type of controllers. It may also need a new type of information systems.

Non Cybernetic Processes

If we consider the full range of human organisations in which control processes occur, ' those that satisfy the conditions for the cybernetic model tend to be the more structured ones. What we notice in practice when we try to follow a cybernetic approach is : (a) Objectives : may be missing, unclear, or shifting ; (b) Accomplishment may not be measureable ; and/or (c) Feedback information may not be usable. Each of these three conditions is illustrated below :

1. **Objectives are missing, unclear, or shifting**. If there is to be a standard, there should be objectives from which this standard is derived. Setting of standards presupposes clarity about the organisation's objectives. In such cases, decisions, if they are consciously taken at all, are based on processes of negotiation and struggle and cannot be derived from any prior organisational objectives. Objectives may forever remain unclear. This may even be true if someone in the organisation publishes eloquent espoused objectives.

2. **Accomplishment is not measurable**. Even in cases, where objectives are clear to all involved, it is often not possible to translate them into unambiguous; quantitative output standards

against which performance can be measured. How should we measure the output of a police department? One of its final objectives is definitely to prevent crime, so we might consider the decrease of crime rates as on output measure. In such cases, organisations often resort to surrogate measures of performance, measures which are less directly tied to the organisation's objectives but which are more easily measurable. In the case of the police department, the number of people arrested or the amount of fines levied could be such surrogate measures.

For many organisations or activities within organisations, outputs can only be defined in qualitative and vague terms, the only thing really measurable abour such activities is their input—how much money and other resources will be allotted to them. These include most management and indirect activities in industrial organisations, like advertising, personnel departments, control activities in headquarters, research, government services, universities, hospitals and voluntary associations. In all these cases, the sole control of management exists at the time of resource allocation but the criteria for resource allocation to this and not to that activity are judgemental.

Feedback Information is not usable. The cuberneitic model presupposes a recurring cycle of events; variance information is used to correct the present state of affirs to eliminate unwanted variances for the future. A few organisations do sue regular evaluation studies of past investments reviews and claimas as one of their potential effects a "Symbolic use". The Planning/Programming/Budgeting System which became widely known as PPBS a number of objectives, but among these is control which it triies to execute by enforcing the cybernetic model, and in its most ambitious form it claims to apply to any organisation.

MBO is also based on a cybernetic philosophy of objective setting (jointly between the employee, who is often himself a manager, and his superior), performance review and corrective action. However, MBO is also advocated, and applied for indirect jobs, in medical institutions, school systems and government agencies. In these cases, accomplishment is much less measurable, and it is rare to find surrogates acceptable to both parties. If a commonly agreed measurement of accomplishment is lacking, the cybernetic model again does not apply and MBO is simply bound to fail. A second reason why MBO may fail, even if the cybernetic model does not apply, is that MBO is based on simplistic and mechanistic assumption about the relationships among the people involved; it uses a reward—punishment psychology. There is more going on between people than cubernetic objective setting and feedback alone.

The essence of the non-cybernetic situations if that they are political. The decisions are based on negotiation and judgement. Decisions often deal with policies. There is a well-known slogan, "there is no reason for it is just our policy". What this means is that policy is not merely composed of rational elements, its main ingredients are values, which may differ from person to person, group to group and norms which are shared within groups in society but vary over time and form groups to group. It makes little sense to speak of control processes here, at least in the formal sense of which such processes described in cybernetic situations. It does make sense to speak of control structure, taking into account the power positions of the various parties in the negotiations. Within this structure, we may study the control games played by the various actors. Once resources are allocated there is no automatic feedback on the effectiveness of their use, the only controls possible are whether the resources were really spent and if no funds were embazzled. Beyond, that, it is a matter of trust in those in charge of carrying out the programs, the real control takes place through the appointing of a person to task. Activities once decided upon will tend to perpetuate themselves, corrective action in the case of ineffective or inefficient activities are not automatically produced by the control system but ask for a specific evaluation study, deciding upon such a study is in itself a political act which may upset an established balance of power.

40

Computer Technology

Introduction

As a tool of management production, it has profoundly affected society during its relatively short modern history.

History of Compututer Technology

Men has been interested in computing data right from the beginning of human race. Scientists were earlier able to calculate the exact time of solar and lunar eclips through austronomical calculations. It was Pascal in the year 1642 who first invented the mechanical calculating machine to do additions. It was later in the year 1830 that a machine was invented to compute insurance calculations and it was this machine which had the facility of memory. During the Second World War, computers were designed to project with accuracy the trajectory of shells. Computers are used for rail reservations to help the travellers and booking clerks alike. It is used for processing picture taken from satellite which help us to predict fairly accurately the weather position. Computer is used for data processing by which we can handle large volume of data and could process it quickly. Computers are used in printing presses to edit and automatically compose the printing material Computer helps us in processing examination results. Computers are used in chemical industry process control and production planning. Computers have an application in designing machine parts and producing accurate Engineering Drawings. Computers have application in Medical diagnosis. Computers are used for planning health and education and other basic uses in the industry. It can make learning easy and lively. Computers can predict the future more accurately through horoscope. Large variety of computers are used in Government and other service organisations, Research and Development. All the computers are basically machines and are like super-computers. Computers have Input Devices through which data and instructions are communicated. Card Reader is an obsolete means of Input Device. Key Board is an important Input Device and as the key is pressed the computer get the input the own electronic language. Computers are different from calculators as it has a storage capacity. Computers have the Processing Unit which is the brain of the computer. It has arithmetic unit, memory register and control unit. The memory register have memory chips and each chip can have 32000 characters. The chip controls the storage device. A Computer has varieties of outpur *viz*. Text, Drawings,

Graphs, Voice, picture, printers etc. printers give a hard copy. It can handle 30 pages of printing in a minute. Electric typewriters are slow means of printing. Computer screen is a soft copy device. The disc pack has a capacity to store a large amount of data with random assess. This device has really revolutionalised the computer srorage. This fact can store basic data for 2 lac. Villages and data can be retrieved instantaneously. It is like a gramophone recorder where data is recorded on concentric circle. Floppy Drive is another important means of data storage. It can store personal data of 700 employees and any amount of data can be assessed in a few seconds. A hard disc device on a small computer can store 40 million characters, which is around 100 times the data on a floppy. Computer has created an impact on all the segments of the society, still it is a long way to tap the full potential of the computer. Computer revolution has begun in India and computers are manufactured indigenously. The people who write software are called the programmers. The Personal Computers are gaining popularity in the present time. The Systems Analyst analyse a real life problem and develop a computer programmes, design input document and output format report.

Information Technology

The information technology and high speeds computers have a significant impact on the middle and top management of an organisation. The information technology should push upward in the organisation structure crossing the boundry between planning and performance. The availability of high speed computers in conjunction with new mathematical techniques will enable large corporations to move to re-centralisation of the planning, organising and controlling functions. The information technology will draw a line separating top and middle management. The information rechnology will enrich rather than shrink the role of managers at middle level.

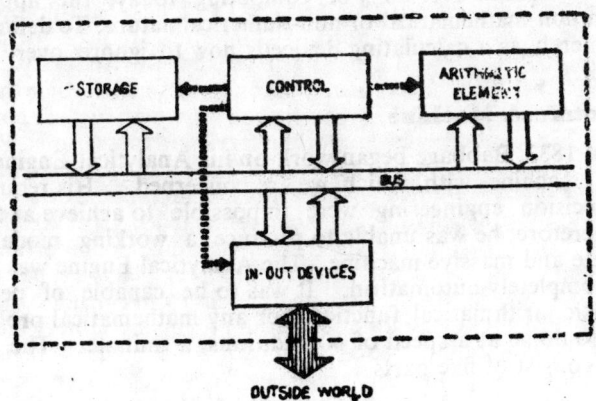

Fig. 40.1. Computer

In 1944, the first automatic digital computer began its operation at Harvard University. By 1954, a few commercially available digital computers had appeared, the UNIVAC I and the IBM 701 and 650. Computer usage has led to savings of 10—25% in clerical costs in many cases, and savings of 10—20% in inventory costs.

What Are Computers ?

There are three essential concepts that we need to examine in order to be able to think sensibly and talk intelligently about computing — What are computers ? What can they do ? How can we communicate with them ? But first of all, What do the terms computer and computing means ? Obviously computing has something to do with reckoning or calculation, but man has been using his brain to do just that for centuries. The Egyptians built the pyramids; whoever built Stonehenge left a calander which can still accurately predict eclipses; the Roman designed and built long straight roads, aqueducts and heating systems ; early explorers navigated the globe, and even radio and television were invented—all without a computer. What is so special about then that we need computers today ? It cannot simply be because they are calculating devices. The problem which early computers had to solve were mostly mathematical. Today computers are used to forecast the weather ; to operate machines ; to cut shapes out of sheet metal, and even to guide spacecraft to the moon. They can set and print newpapers and books. They can be used to help in diagnosing diseases and to find out whether a hospital bed is available for a particular patient. They are used to find obscure documents in archives and elusive criminals on the run. Travel agents around the world have come to rely on them to book seats on air flights or rooms in hotels , either today or a year from now. Companies use them for accounting, stock control and payrolls. The original objective for inventing the computer was to create a fast calculating machine. But in over 80% of computiug today, the applications are of a non-mathematical or non-numerical nature. To define a computer merely as a calculating device is now to ignore over 80% of its work.

Computer—A Machine

In 1833, Babbage began work on his Analytical Engine and it is this machine with which we are concerned. His requirements for precision engineering were impossible to achieve at that time and, therefore, he was unable to produce a working model of the complete and massive machine. The Analytical Engine was intended to be completely automation. It was to be capable of performing the basic arithmatical functions for any mathematical problem and it was to do so at a speed of 60 additions a minute. The machine was to consist of five parts.

—a **Store** in which to hold numbers, *i.e.*, those which were to provide the information (data) for the problems, and those which were to be generated during the course of calculations.

—An **Arithmetic** unit which Babbage called the 'Mill'. This was to be device for performing the arithmetical operations on the numbers which had been stored. All the operations were to have been carried out automatically through rotations of gears and wheels.

—a **Control** unit for seeing that the machine performed the desired operations in the correct sequence and, also by means of a series of gear and sheels, for transferring data between the Mill and the Store.

—an **Input** device to pass into the machine both numbers (the data) and instructions as to which of the arithmetical operations to perform.

—an **Output** device to display the results obtained from the calculations.

Computer Words

Computers store words, generate words, manipulate words, convert words and respond to words. The term "words" is used to convey information ; and information is the commodity that computers accept, store and reproduce. The parts of a computer word are called digits. There are different types of digits in use corresponding to the different types of computer words. The first type is a decimal word which contain only decimal digits (numbers) and a plus or minus sign.

The second is a binary word which contains only binary digits. There are only two binary digits, one and zero. Therefore, a binary word must be a series of 1s and 0s.

The third type is alphanumeric word which contains alphanumeric digits such as letter of alphabet or special symbols.

Most computers work entirely with binary words in their internal operations. A decimal word is a number to the base 10. The decimal word 824 is understood to mean eight 100s plus two 10s plus 1s.

Bytes and Bits

Today many computers are capable of handling variable word lengths upto a certain maximum word length which depends on the specific computer. A byte then is a computer sub-word composed of a number of binary digits or bits.

A binary word is also shown with a word length of 10 binary digits. It takes more digits to represent decimal words in a binary form. There is an important reason for using binary words insid e a computer. It happens to be more efficient to manipulate and store data electronically in binary form than in any other way. Both alphanumeric words and decimal words can be converted to binary words in a computer.

Manipulation and Control of Data

A sequence of instructions called a program. All general-purpose digital computers make use of stored programs which facilitate versatile, high-speed operation. The program instructions are a series of computer words just like the data on which the computer operates. And such data are placed in the computer's internal storage and can also be manipulated by the computer. By developing new programs, the same computer can be made to perform an unlimited number of different operations on whatever input data is provided.

Computer Instructions

We must first understand precisely what the computer will do in response to each instruction. The most fundamental operations are Add, Subtract and Store instructions.

The Clear and Add operation simply brings a single word from a specified storage register to the accumulator.

The Subtract Operation

It is exactly like the Add operation except that the number from register is subtracted from the number in the accumulator.

The Store Operation

It is the reverse of the Clear and Add operation. In this case, the word in the accumulators is placed into the specified register in storage.

The Multiply and Shift Operation

In this case, we must consider the Q-register in different computers, which can best be visualized as an extension to the accumulator. It is capable of holding a standard computer word just like the accumulator. If the number in the accumulator be multiplied by the number "4", which is located in register 300, the computer would be given the instruction MPY 300. After completing the multiplication, the computer would then hold the answer, a double size word, in its accumulator and Q-register just as if they made up one long register. In a computer program it is often desirable to report intermediate results. This printing operation is accomplished by a rather standard sequence of instructions, referred to as a sub-routine. The sub-routine will occupy a block of registers in computer storage. The sequence need be stored only once and a TRU instruction can be used wherever a printout is required.

Storage

It is a warehouse for words. A useful way to visualise the internal storage of a computer is to think of a bank of pigeonholes such as you might find in a post office. Each slot can hold one computer word and there is one register associated with it. We say that the computer can read or write in any one of these slots. It reads by taking a word from the pigeonhole and using it somewhere else. Another important characteristic of computer storage is that the main storage is usually addressable.

Stored Program

A digital computer program is the sequence of instructions which the arithmetic element or working part executes while the computer is operating. Every stored program computer is potentially a general-purpose computer. Computer words represent an instruction or command for the computer. Each of these words is stored inside the computer, and the computer executes them one at a time. Programs usually contain thousands of individual instructions and each instruction is usually a computer word.

Capacity of a Computer

We speak of small, medium and large-scale computers; the price of general-purpose digital computers various from about 8,000 dollars to several million dollars. The three most important things in determining the capacity of a digital computer are speed, storage volume and input capability. Speed refers to tl e speed of operation of the control and arithmetic elements. It's the sum total of all these capabilities that represents the overall capacity of a computer.

Input-Output Limitations

The input devices do require a machinable medium; that is, the words to be inserted must be in a standardized form which the computer should be able to accept these new words at a reasonable high rate speed. Similarly, many of the output devices produce information in a machinable form, since it may be necessary to use these results in another computer or reuse them at a later date. Hence, data in a machinable form can usually be moved from place to place independent, of the computer, and the data can be stored more or less permanently when desired.

One of the most frequently used forms is the punched card, which was widely used long before the advent of electronic digital computers. Small holes are punched in the card by a device called a key punch according to one of several codes which the input devices can interpret. Decks of these cards can be stored, duplicated, or used over and over again. Some or all of the information can be printed on the cards also so that they can be visually checked or interpreted by the operator.

A very similar and widely used form is punched paper tape. In this case a series of holes is punched across the 'ape to represent each character. This is accomplished by a specia. electric type-written record at the same time.

Also widely used and stored on a reel is digital magnetic tape. The technique is similar to the paper tape except that the words are stored as magnetized spots on the tape which are not visible to the human eye. The magnetic tape, however, is a much more efficient medium from the computer's point of view. Much more information can be stored on a single reel, and data can be transferred much faster between magnetic tape and the computer.

TYPES AND USES OF STORAGE

Magnetic Core Storage

The type of storage has an important effect on a computer's capability. It can limit the maximum speed of a computer, and it can represent a significant portion of the computer cost. The computer can only work with information that can be converted to computer words and stored internally.

Magnetic core storage is most analogous to the pigeonhole concept. It is used in almost every computer today. It is the fastest internal storage in many medium and large-scale computers. Magnetic core storage has been one of the most significant component developments in the short history of electronic computer.

A single magnetic core looks like a small doughnut and is made of a ferrite materials. It is a tiny magnet which can be magnetized in one of 2 different ways. It is so tiny that the little loop would not fit on the end of a very sharp pencil. Typically each core has 3 or 4 wires going through it and thousangs of these cores are arranged in what is called a plane. Each core is able to store one bit of information. It must be possible for the computer to write a 1 or 0 in one particular core out of the entire plane. When writing new information in the memory, the old information will be destroyed. When reading information from the memory, it is important not to destroy the existing information.

Magnetic Drum Storage

Somewhat similar to the magnetic core is the magnetic drum storage. We can still keep the pigeonhole analogy here. The magnetic drum is mounted on a shaft which is driven by a motor at a very high speed. The face of the drum is coated with a metallic oxide which can be magnetized, in one of the two different ways. Now, consider the surface of the drum, and divide it into many small squares or spots. This becomes set of pigeonholes registers. Each register has an address based on its position on the drum. The drum is rotating at a very high speed and is connected to the computer through READ-WRITE heads which are in a fixed position

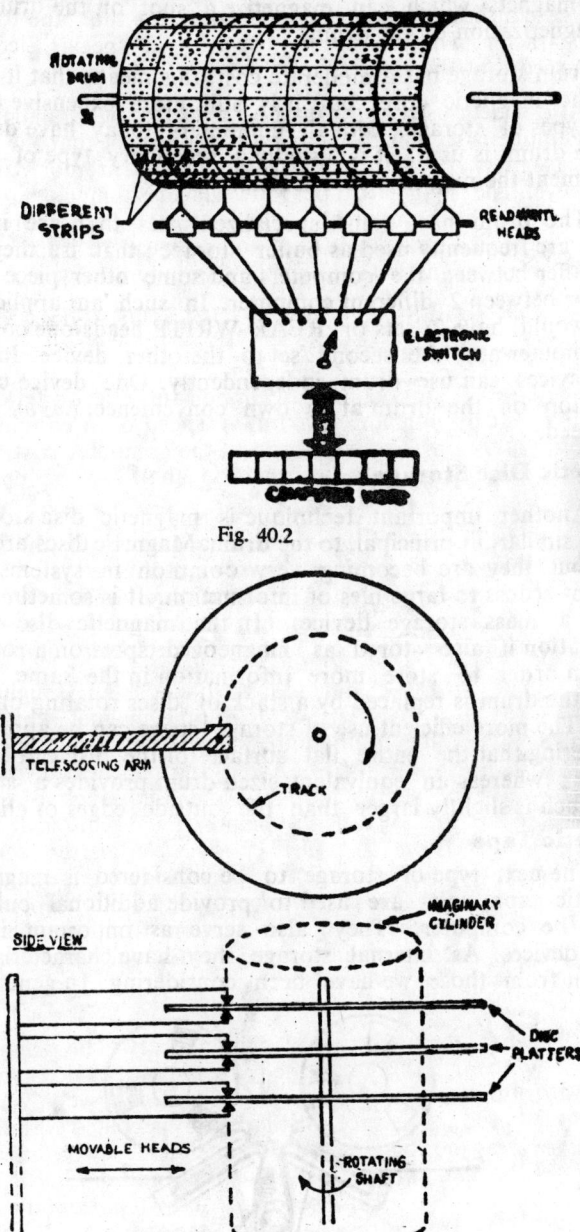

Fig. 40.2

Fig. 40.3

relative to the rotating drum. The READ-WRITE heads are small electromagnets which can magnetize a spot on the drum or sense the magnetization on the drum.

Drum storage has a moderate cost, this means that it is cheaper than the magnetic core, but it is still more expensive than some other types of storage. Not all computers today have drums. But when a drum is used, it is usually a secondary type of storage to supplement the magnetic core second.

The drum has some specialized uses that are inertesting. Drums are frequenlty used as buffer storage ; that is, they are used as a buffer between the computers and some other piece of equipment or between 2 different computer. In such an application, the drum would have 2 sets or READ-WRITE heads-one connected to the computer and the second set to the other device. In this way both devices can use drum independently. One device can put information on the drum at its own convenience, *i.e.* at a slow or fast speed.

Magnetic Disc Storage

Another important technique is magnetic disc storage. This is very similar, in principal, to the drum. Magnetic discs are relatively new, but they are becoming very common in systems requiring frequent access to large files of information. It is sometimes referred to as a mass storage device. In the magnetic disc technique, information it also stored as magnetized spots on a rotating surface. In order to store more information in the same volume of space, the drum is replaced by a stack of discs rotating on the same shaft. The more efficient use of storage space can be appreciated by considering that the entire flat surface of the disc can be used for storage ; whereas an equivalent sized drum provides a storage surface which is slightly larger than the outside edges of all the discs.

Magnetic Tape

The next type of storage to be considered is magnetic tape. Magnetic tape units are used to provide additional bulk storage inside the computer. They also serve as important input and output devices. As internal storage they have characteristics quite different from those we have been considering. In general, words

Fig. 40.4

stored on tape do not have an address, which means that the computer may have to search for the words it wants. It is cheaper, per digit than the other methods of storage, and it is often possible to organize the bulk data and its processing so that random retrieval is seldom required.

The tape itself is similar to that utilized in tape records but is manufactured to more rigid specifications. A reel contains several thousand feet of tape and can be removed from the tape unit and stored in a tape library.

Magnetic tapes are used as input-output devices. Compared to other machinable media such as punched cards or perforated tape, magnetic tape has such higher data transfer rate and considerably more information can be stored per cubic feet. One reel of magnetic tape can store on the order of 10 milion bytes.

Utilizing Storage

A computer system is usually available with several options in regard to the total internal storage available. But more storage capacity will add significantly to the equipment cost or rental.

Input Output Devices

The computer performs the functions and store results in its internal memory. The results are than transferred out of the computer and presented to the person who wants to use it.

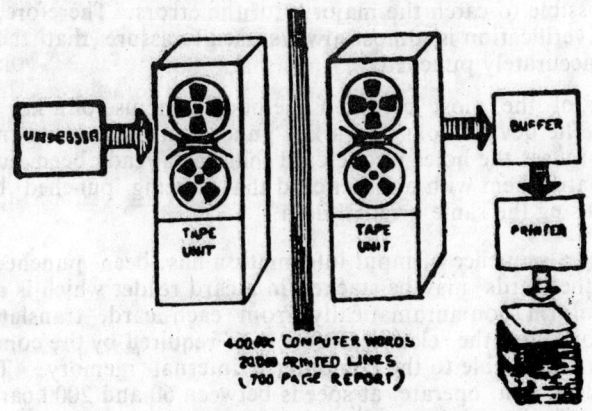

Fig. 40.5. Input Output Device.

Input-output devices, fulfil many needs in a computer system. There is a need to originate and convert data into a machinable form, to match and synchronize the computer speed with other data handling devices, to provide for off-line operations, and to permit more direct man-machine communication.

Punched Cards

International Business Machines Corporation utilizes an 80-column card in which information is punched with rectangular holes.

In the common use of the IBM card, one vertical column of the card is used to represent a single character and this character might be a letter of the alphabet, a number, or a symbol such as a rupee sign. However, it is only necessary that 2 or 3 holes be punched in a column in order to indicate each of the possible characters.

The punched card not only be used to present information to a computer but also serves occasionally as an output form which can be employed by other computers or automatic processing devices Frequently the information punched on a card is also printed on the card so that it can be used as a document or as part of a manual file system.

The first step in entering punched card information into a computer is that of punching the card. This is most commonly done by means of a key punch, a device which resembles an electric typewriter in appearance and operation. Cards are fed automatically into the device one at a time, and the information is punched one column at a time by repressing the appropriate key on the keyboard. It is possible to use a previously prepared program card which serves as a master card and facilitates the punching operation when many cards with the same format with repeated information are punched.

Once a computer accepts false information, it is usually extremely difficult to detect and remedy the error. Most computer programs do include some simple checks on the input data, but it is seldom possible to catch the majority of the errors. Therefore, some system of verification is almost always used to assure that the data has been accurately punched.

One of the most common methods is the use of a key punch verifier. The device looks like a key punch but includes a mechanism that senses the holes in one card that has already been punched and compares them with another card that is being punched by the operator using the same original data.

Once a sequence of input information has been punched and verified, the cards may be stacked in a card reader which is able to read the information automatically from each card, translate this information into the electrical form as required by the computer, and make it available to the computer's internal memory. Typical card readers can operate at speeds between 60 and 2000 cards per minute, while the card punch operates at 100 to 400 cards per minute.

Punched Card Processing

The operations of sorting merging, matching can also be performed by the computer without additional equipment.

Because of the relatively slow speed of card input to the computer cards are seldom used as a direct input to large fast computers.

It is frequently more economical to convert the cards to magnetic tape before entering the data in the computer.

Punched Paper Tape

A punched paper tape is sometimes produced as a by-product of other recording operations such as the typing of a sales order or an inventory list. If a computer accepts a paper tape input through a paper tape reader, it is usually capable of punching paper tape as one form of output. Typical type reading speeds are about 100 to 1800 characters per second, and typical tape punching speeds are about 60 to 300 characters per second.

Limitation of Punched Hole Equipment

The punched hole records have some serious limitations. Because of the hole size required, the data density is low. This is particulary true of punched cards and it means that large files involving hundreds of millions of characters require a considerable amount of space for physical storage.

From the computer's viewpoint, punched holes imply a relatively slow operation. With a fact computer and voluminous input and output data on punched cards or tape, the total operating time might be completely determined by input and output speed with the computer sitting by idly during most of the time.

Character Reading

The idea of character reading devices as a computer input is certainly attractive. Many types of input data are available in printed form, but not in a truly machinable form. The cost of converting the data to punched cards or punched tape is manual job which is inherently costly, time consuming and likely to introduce

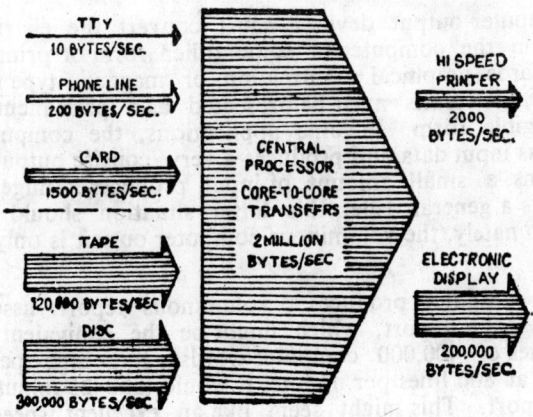

Fig. 40.6. Input-Output Device.

errors. Since most computers can produce a printed output auto-matically, we except them to be able to accept a printed input. The output in under computer control, and there are definite limitations on the size and selection of characters and on the possible formats that can be uséd. On the other hand, printed input information can seldom be so closely controlled.

Speed Problem

One of the keys to understanding input-output devices is the problem of speed. Speed is very important in computers ; in fact, the utility of digital computer hinges on their tremendous speed, and it is difficult to communicate with the outside world at such high speeds. We know that the assumed internal transfer rate is 2 magabytes per second. This assumes that the processor is simply transferring information from one place to another within its internal (core) memory. However, a teleprinter terminal operates at its maxi-mum speed of 100—150 words per minute. This data rate is about 10 bytes per seconed. Some of the more conventional devices—punched cards, punched paper tape—have a rate something like 1500 bytes per second or less. Most of the input devices just mentioned can also function as output devices at approximately the same rates. In some cases it may be a separate but complementary device—for instance, the computer uses a punched card reader for input and a card punch to produce the equivalent output.

An in-put processor is employed to make more efficient use of the high speeds of the internal computer and to be able to match the many different speed of these input-output devices. Such an approach may not be justified in moderate size installations that do-not require extensive inputoutput operation.

Output and Off-Line Operations

Computer output devices, must convert the eletrical signals used within the computer to some other form of printed words, punched cards, graphical information, or magnetic type recordings. Ordinarily, the results must be presented as a permanent record in some tangible form. In some applications, the computer accepts voluminous input data and produces a very concise output. In other applications, a small amount of input produces a huge volume of output. As a general rule, this latter situation should be avoided but unfortunately, the voluminous computer output is only too often the case.

If a computer produces a voluminous report, assuming that it is a 700 page report, which might be the equivalent of 20,000 printed lines of 400,000 computer words. If a high-speed printer were used at 800 lines per minute, it would take 25 minutes to print out the report. This might seem like an excellent speed until you consider that the computer could transfer all that information to a tape unit in half a minute. If the computer were to control the high-speed printer while it is printing out this information, it would tie up the computer for 25 minutes. However, if a suitable off-line equipment were available, the computer could transfer the complete report to a magnetic tape unit in about 30 seconds, then move on to other work. This off-line unit is connected to an off-line printer through a buffer unit. Although it would still require 25 minutes to print the report but the computer would not be involved now.

6. Equipment and Installations. Different computer generations can be reasonably distinguished by important differences in both hardware and software. As the industry continues to develop, it is unlikely that any clear distinction will be possible for future generations. It is, however, helpful in looking at past history. The first generation itself began with the delivery of the UNIVACI on June 14, 1951. The first generation computers used the stored program concept, programming was a difficult and tedious process. However, the mid-1950 is the development of improved software.

The second generation began about 1959 with the delivery of the first commercially available, fully transistorized computers.

During the second generation, batch-type operations became quite common, particularly at university computer centres where many small jobs were handled per day. The year 1965 is generally considered to be the dividing line between the second and third generation. Most computers that continued to be marketed after 1965 are considered at least by their manufacturers to be third generation machines.

The characteristics of the third generation vary from one manufacturer to another. One major technical development is the use of integrated circuits. These represent a further development of solid state technology which permits more circuits to be packaged in a smaller space and ultimately promises lower costs and higher reliability.

By the beginning of the 1970s, the generation concept becomes confused. Another phenomenon of the third generation is the emergency of the mini-computers. They have a small word size and limited peripheral equipment. Mini-computers are designed so that they can function as a satellite processor connected remotely to a larger computing installation.

Economic Aspects

Computers equipment costs represent about one-half or one-third of annual operating costs. Other significant cost are associated with computer programming, planning procedural changes and training or retraining of personnel. The purchase price of the computer will normally be 30 to 50 times the monthly rental cost. The cost of maintenance service is typically 5 to 15 per cent of the monthly rental charge.

The decision to rent or buy depends to a greater extent on such factors as government policy, interest rates and availability of money, hours per day of expected usage, ratio of purchase price to rental price for the specific equipment selected.

The other cost consideration is the selection of optional or peripheral equipment. In large installations the cost of peripheral equipment frequently exceeds the cost of the basic computer. The preparation of the physical installation is frequently an important item of cost.

Adequate provision must be provided not only for the equipment itself but maintenance personnel and parts, for storage of tapes and other supplies, for off-line equipment and for possible expansion of the configuration. Special electrical power is usually provided because most medium and large computer installations require 10 to 100 kW or more of electrical power.

The computer's program determines what the computer will do at any particular time. When that program is replaced by a different program, the same computer performs entirely different functions. Computer installations, therefore, are operated in a manner to make efficient and economical use of the equipment to insure that vital jobs are completed on schedule and to provide some flexibility for the immediate users.

The Analog Computer

In the first place, it will clarify our understanding of the digital computer and its capabilities and limitations if we contrast it with the analog computer. Secondly, many digital compute applications involve the use of analog data which pose special equipment problems. Third, there is a growing group of applications that can best be performed on a hybrid analog-digital computer, which actually means 2 computers linked intimately together.

The analog computer is not as widely used as the digital computer, although it appeared on the scene first. But the characteristics of analog computers restrict their use to certain types of mathematical problems.

The sharp contrasts between analog and the digital computer are the differences in data, in computational technique and in programming. These differences result in certain advantages and disadvantages for the analog computer as compared with its digital counterpart.

The digital computer works with computer words. It accepts words as its raw data and it produces its results as words. These words might be numbers, coded information, or readable text. In an analog computer, data are frequently represented as electrical voltages which can vary continuously in amplitude and time.

Since digital data is represented by words, it is inherently discrete or intermittent. If the temperature is changing, it can't be represented by a single word.

Datas like street addresses or dates are essentially digital in form. Other types of data such as speed, pressure, flow rate and temperature are essentially analog in form. Many digital computer applications do use data which were originally analog data but have been "digitilized" converted to a digital form.

In an analog computer computations are performed simultaneously by many different physical elements. In the digital computer the step by-step computation is specified.

The real difference between the two types of computers hinges on accuracy and speed. Generally, the analog computer, because of its parallel operation has the speed advantage. This advantage, is being crowed by the increasing speed of digital computers. The digital computer has the advantage of accuracy or precision in computations. Typical analog computers are limited to an accuracy of about one part in 10,000. In the digital computer, accuracy of computation is determined by word length and by the approximations that might be used. Computational accuracies of one part in a million or better are quite normal in scientific applications or digital computers.

Some digital computer applications require the use of analog type data, it is important to consider some of the data conversion pit falls. Both analog-to-digital and digital-to-analog converters are available with wide variation in capabilities and cost.

Characteristics of Computers

It is important to appreciate that once a computer has been given a piece of information, it is always capable of remembering it any time. In this sense it is far superior to human being. The computer can always access any piece if imformation that it has in store, and this data can be reproduced in years to came exactly in its original form. A computer can be programmed to generate poetry or music, but it can not appraise it or judge its quality. A human brain subconsciously acquires certain priorities, a computer can not. A computer can help mankind but it do not threaten it.

The various characteristics of Computers can be summarised as follows :

Speed. The computer was invented as a high-speed calculator. This has led to many scientific projects which were previously impossible. The control of the moon landing would not have been feasible without computers and neither would today's more scientific approach to weather prediction. If we want tomarrow's forecase today, meteorologists can use the computer to perform quickly the necessary calculations and analysis. Electrical pulses travel at incredible speeds and, because the computer is electronic its internal speed is virtually instantaneous. We do not talk in terms of seconds or even, today of milliseconds (thousandths of a second). Our units of the speed are the microsecond (millionths), the nanosecond (thousand-millionths) and latterly even the picosecond million-millionths). A powerful computer is capable of adding together two 18 digit number in 300 to 400 nanoseconds (*i.e.*, about 3 million calculations per second).

Storage. The speed with which computers can process large quantities of information has led to the generation of new information on a vast scale, in other words, the computer has compounded the information 'explosion'. In computers, the internal memory of the CPU is only large enough to retain a cerain amount of information. Small sections of the total data can be assessed very quickly by the CPU and brought into the main, internal memory, as and when required for processing. The internal memory (in the CPU) is built up in 1 K or K modules where K equals 1024 storage locations. Small micro-computers have an 8 K or 16 K store while super computer may have storage capacity upto 1024 K stores.

Accuracy. The computer's accuracy is consistently high. Errors in the machinery can occur but, due to increased efficiency in error-detecting techniques, these seldom lead to false results. Almost without exception, the errors in computing are due to human rather than technological weaknesses.

Versatility. Computers seem capable of performing almost any task, provied that the task can be reduced to a series of logical steps. Yet the computer has itself only limited ability and, in the final analysis, actually performs only for following basic operations.

— it exchanges infomation with the outside world via I/O devices ;

— it transfers data internally within the CPU

— it performs the basic arithmetical operations

— it performs operations of comparison.

In one sense, then, the computer is not versatile because it is limited to four basic functions. Yet because so many daily activities can be reduced to an interplay between these functions, it appears are highly ingenious.

Automation. A computer is much more than an adding machine, calculator or check-out till, all of which require human operators to press the necessary keys for the operations to be performed. Once a programme is in the computer memory, the individual instructions are then transferred, one after the other to the control unit for execution.

Diligence. A computer does not suffer from the human traits of tiiedness and lack of concentration. If 3 million calculations have to be performed, it will perform the 3 million with exactly the same accuracy and speed as the first. This factor may cause those whose jobs are highly repetitive to regard the computer as a threat.

THE MICROCOMPUTER

Computers are classified as Super, Maniframe, Mini and micro computers, principally on computing power. The difference between a mini and micro is not always clearcut as mini computer may contain the same components as a micro computer.

In order to understand micro computer in detail there is need to understand basic structure. This is because of the application of microelectronic technology to the design of computers. The structure is a little different but it is important to appreciate that the functions performed are the same. A program which runs or works on a microcomputer could well run on a large mainframe computer without any alteration. The design differences do not therefore affect the computer process.

Micro processor Unit. In 1971 the first microprocessor was produced on single chip of silicon. Today, many thousands of components can be built on to a single chip some 5 mm square and no more than 1 mm thick and the destiny of circuits is increasing all the time. A singal chip can contain all the circuitry perform the combined functions of the control unit and arithmetic/logic unit of the traditionally structured computer. Such a chip is called a microprocessor unit. It is not a complete computer as it lacks memory and input/output capability.

Memory. Memory can also be contained on a chip which like the microprocessor chip is very small and cheap to manufacture when produced in large volumes. Memory may also be assembled on the same chip as the MPU. There are various types of memory chips and a single microcomputer might utilize more than one type. The most common is random access memory. When switched off, any information stored in the memory is lost. Another type of microcomputer memory is read only memory. Information is 'burnt' into the ROM chip at manufacturing time. It cannot be altered and fresh information cannot be 'written' into a ROM. ROMs are used for applications in which it is known that the information never needs to be altered, for example the operating system software which controls the use of computer microcomputer system or a monitor program for controlling a washing machine.

A variation of the ROM chip is programmable read only memory. PROM can be programmed to record information using a facility know as a prom-programmer. However, once the chip has been programmed the recorded information cannot be changed, *i.e.*, the PROM becomes a ROM and the stored information can only be 'read'. A fourth type of memory is erasable programmable ready memory. As the name suggests, information can be erased and the chip programmed anew to record different information using a special prom-programmer facility. Erasure is achieved by exposing the chip to ultraviolet light. When an EPROM is in use information can only be 'read' and the information remains on the chip unit until it is erased.

MICROCOMPUTER SYSTEMS

The present generation of microcomputers is based on the use of 8-bit microprocessors, *i.e.*, information is handled within the MPU in 8-bit registers, accumulators, etc. There is also a rapidly evolving new generation built around 16 bit microprocessors. These more powerful microcomputers provide minicomputer capability at micro prices. A microcomputer is often designed that it can be configured with different amounts of RAM. Typical ranges are 4 K bytes to 64 K bytes and 32 K to 512 K bytes, expandable normally in add on steps of 4 K and 16 K or multiples of same. A micro-computer is usually equipped with a typewriter keyboard for input and can normally be linked to a CRT monitor device display of input and output. A terminal printer device may also be connected, to provide printed copy of output on paper. A microcomputer system is given in Fig. 40.7.

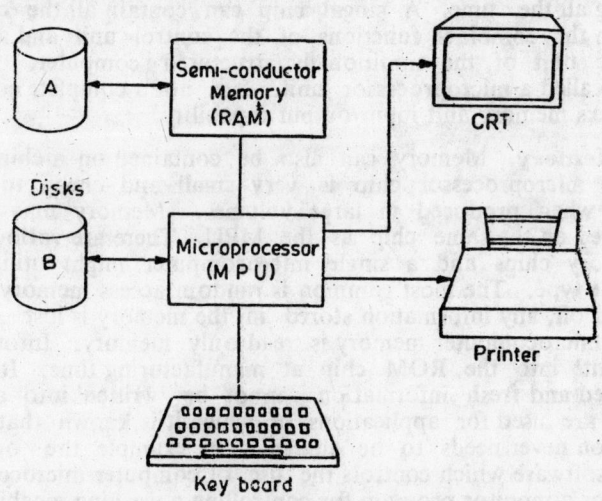

Fig. 40.7: Microcomputer System

Summary

The capabilities of computers have revolutionised the thinking and approach of the people. The computers have accelerated the research work. Today computers finds its applications in Business; Industry ; R and D ; Education.; Training and Development and even in personal life. The computer is used to assist man in business, in research and in many other walks of life. However, versatility has been provided by man's ability to reduce what are often highly complicated peoblems to the simple level at which the computer can be used, and to design and implement ingenious computer systems which can provide a myriad interplay of the basically simple tasks teat the computer can handle. In science, the advent of computers has meant that calculations which were previously beyond contemplation, because of the time span in carrying them out, have now become possible. This has greatly accelerated and expanded research in such sciences as Physics, Chemistry, Astronomy and Genetics. More recently there has been an increasing use of computers for research and data analysis in less mathematical areas such as Medicine, the Social Sciences and even the Humanities. Computers are now a standard feature of life in universities and industrial laboratories.

41

Utilisation of Computers in Production

The degree of utilisation of computers in manufacturing depends on the state of (a) Hardware technology (b) Relevant problem solving techniques, and (c) the risk taking activity of an organisation.

Althogh hardware technology is reasonably uniform throghout the world, an organisation's awareness of relevant problem solving techniques will vary from industry to industry. Siṁilary the individuals who have acquired combination of technological competence, mana gement orientation and organisational influence are somewhat rare. The state of risk taking proclivity of an organisation varies from industry to industry, but competition enforces a good deal of uniformity within an industry. *There are four stages to represent for increasing degree of complexity and scope in pursuing a computer-based technology by a company. These stages are viz Basic computer applications ; Individual computer applicatitions ; Interdivisional computer applications and Advanced computer applications.*

Automated Design and Production

One of the more advanced phases in the development of computer automation is the integration of design and production, particularly in areas where product homogenity and standardisation have reached a high level. The automated design process is one example of a growing tendency toward man-machine or interactive computer systems which are intended to solve very complex problems. The value of interactive computer graphics lies in its capacity to augment human resources through :

1. Visual information displays—the form which is most readily comprehended by human.

2. The availability of large store of information.

3. The elimination of tedious and repetitions activity.

4. The enforcement of constraints in a particular process or activity.

Because a complete design results in complete products specifications, *it is possible through numercial control techniques to command the computer to produce instructions for parts fabrication in certain metal products or to generate a mold in case of molded goods.*

These design manufacturing systems can be readily extended to other computerised subsystems such as production control and purchasing.

Advanced Process Control

Another application in applying computer technology is the development of more advanced process control. The control action consists of a computer initiated signal that changes the set point of a current controller. The value of the set point either can be calculated by the processor or can consist of a time variant value such a predetermined temprature profile. The set point change is accomplished by means of a signal output applied to and compatible with the process instrumentation. *Direct Digital Control* of one or several closed loops involves the replacement of individual hardware elements—mostly the controller—with components of the digital control computer.

Systems Engineering

The progress of a manufacturing company toward computerisation can be considerably strengthened by the concept of system engineering.

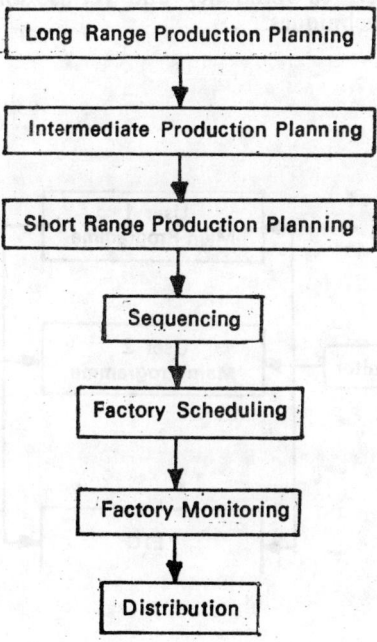

Fig. 41.1

In general, systems engineering is the activity which attempts to arrive at the design of large-scale complex systems. The system engineering concepts have evolved from control theory, many of the terms and techniques in the field stem from the latter activity. In a large manufacturing enterprise following three distinct systems ngineering activities can arise :

(a) Design of complex electronic gear such as radar systems, computers or aerospace missile guidence systems.

(b) Direct control of manufacturing equipment such as chemical processes or machine tools.

(c) Optimal conduct of the business activity.

Sometimes all these classes of systems activity are merged, but more often they are separate efforts employing persons of different educational backgrounds. The computerised fystem, disgned throughout by means of systems engineering principles conducts the production planning and distribution decisions of a large company almost without human intervention as shown in Fig. 41.1.

The fig. (41-1) explains how it is possible to achieve high degree of approach to optimality and yet use somewhat standard optimum seeking techniques.

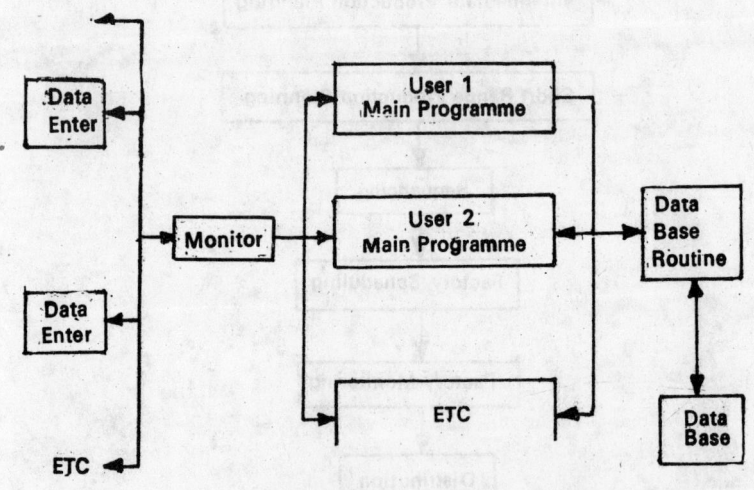

Fig. 41.2

Business Applications

The first non-scientific use of computers concerned with routine clerical work. Computer applications to business and commerce data from the middle of 1950's a decades of vigrous recovery from world war II which led to economic expansion and rapid technological development throughout the industrialised world. Today most large and medium sized companies are almost totally dependent on their computers for numerous administration functions.

Information Systems

Information system packages are available from major computer manufacturers. but in certain cases, these offering may contain an excessive amount of computing overhead. There is also a great amount of effort left to the user in adopting such generalised system design to his own operations. An information system is an interactive process of human judgment alternating with computer actions. The management of information flow by a monitoring programme is given in Fig (41.2).

Computer Attributes

The computer provides such functions as :

- (a) Rapid calculation, routine processing, and fast editing.
- (b) Fast regarding of certain types of documents ; high-speed printing and quick preparation of documents.
- (c) Filing and maintenance of vast bulks of information and retrival of such information on displays and other terminals.
- (d) Scanning, sorting, and searching for specific facts.
- (e) Collecting information rapidly from many sources by either local or remote communication links.
- (f) Giving immediate answers to routine inquiries.

Human Attributes

Following tasks still can be best done by man :
- (a) Handling unforeseen events.
- (b) Selecting goals and criteria.
- (c) Selecting aqproaches to problems.
- (d) Formulating questions and hypothesis.
- (e) Producing new ideas ; planning new approaches ; products, and techniques.

Word processors are already influencing the office practices. Word Processor are used for preparing reports that may need to go through one or more revisions and for producing standard letters and documents. Computers are capable of inserting or deleting words or lines or paragraphs and giving final letters and reports at a very fast rate. Microelectronic technology is enabling officies offices to function faster and more efficiently, and the cheapness of the technology means computing is within the grasp of even the smallest business. Word processors are particularly useful for preparing reports that may need to go through one or more revisions and for producing standard letters and documents. They provide the capability to insert or delete words, lines or paragraphs, and to print out drafts and final copies at speed. No matter how many revisions are required the full text is only typed in once i.e., the initial draft. Systems may include special purpose packages for specific applications such as merging address lists.

Payroll and Personnel Records. Payroll accounting was the first commercial area to become widely computerised. The calculation of wages or salaries involves a number of variable but common factors which relate to the personal details of each employee such as gross pay or rate for the job, tax code, national insurance, etc. The program contains formulae for calculating all the deductions to arrive at net pay. It will also build up records of tax-to-date, pay to-date, etc. It computes these details, prepares a wage slip for the employee and writes up-dated information on the backing store for future use before moving on to consider the next employee. The overall payroll system with attendant personnel details and records may comprise tens or even several hundreds of different though related programs to be produced entirely within the business concern.

Stock Control and Sales. When handling sales orders by computer, the present volume of sales is known as this information, together with past records, and perhaps statistics obtained from market research, forms some of the data needed for sales forecasting, which in turn influences stock control. Basic information for billing is obtained from sales orders and the computer can be used to maintain customer accounts and to print invoices. All these various applications make effective use of the machine's ability to store large quantities of information and to retrive items at speed, and they rely on program to ensure that information is up dated promptly, accurately and usefully. The computer is being used more and more everyday cash transactions.

Personnel Cordex System. Computerisation of personnel cardex system has enabled the organisations
  —for clearcut manpower planning
  —for ready made succession plans
  —to identify the training needs based on the patterns of skill and educational level and performance reports
  —to frame vivid pramotional and career planning programmes etc.

Purchase and Stores. Computers are used in framing the inventory control policies and also for monitoring the procurement so as to affect the atmost economy in procurement and storage of the material. The basic documents like material requisition/order, material receipt and with drawal documents are computerised. With these the material manager will be able to.

—get the consumptional analysis. (*ABC*), stocks analysis (*XYZ*) and movement analysis of stocks (*SFN*) even where lakhs of items are in use.

—get the analysis of lead time of procurement of each material supplier-wise

—decide the procurement and inventory policies.

A similar analysis by computers will help the sales departments in their ware-houses.

Production and Maintenance. Scheduling is a big problem particularly in Jobbing industry. Different machines have different capacities with varied speeds. Computer with greatly help in optimal scheduling taking into account the delivery date, available machines manpower. Quiet often marketing wants all orders whether smaller or bigger to be honoured. Production department will complain the insufficiency of the resources to meet the orders and even the orders are completed, no one is sure of the profitability. For an optimal product mix either to minimise costs are maximise the margin, the linear programming an *OR* Technique is used and hundreds of simult aneous equations are formed and the solution is obtained only by resorting to computer. Similarly a number of waiting problems in maintenance are solved by computer through simulation techniques.

Banking. Banks were among the first large organisations to invest heavily in computing, and today banking is almost totally dependent on the computer. In the past, a large but manageable amount of book-keeping was handled manually. The computer is necessary because there is no other way to deal with the problem. In most instances the computer is sited centrally. Branches are equipped with terminals, giving them an on-line accounting facility and enabling them to interrogate the central system for information on such things as current balances, deposits, overdrafts, interest charages, shares and trustee records. The computer also provide each branch, and its customers, with promot access to information from a much wider financial word than would be possible under a manual local system. New developments such as the automatic cash dispenser, are making in ever easier for customers to deal with banks. Leading international banks and financial insitutions are able to obtain up-to-date news on foreign currency rates from the world's money markets using an on-line world wide information retrieval service.

Insurance and Stockbroking. Insurance Companies, finance houses and stock-broking firms also make use of computers.

Here conditions and requirements are similar to those in banking. In the buying and selling of stocks and shares various calculations have to be made, contract notes drawn up and files consulted and amended. The computer is also beginning to be used as an aid to more efficient investment management. Armed with this information, the stockbroker is in a better position to use his judgement and to take decision.

An Aid to Management. The computer can also be used as a management toll to assist in solving business problem. In operational research, which is the name given to the application of scientific procedures to decision-making certain techniques are used which required the calculating and storage abilities which the computer can provide. Critical path analysis and linear programming are two methods of analysis which are used. In assisting in the decision making process the computer uses rather than provides information.

Industrial Applications

In industry, production may be planned, co-ordinated and controlled with the aid of a computer. The computer may also be used to direct the operation of individual machine tools. The control of a chemical plant by computer can be much safer and more efficient method than by manual control, since changes in conditions which occur during a process can be detected and compensated for immediately. Oil refining, the separation of crude oil into its many component oils, is a continuous process and it depends on the maintainance of certain conditions thoroughtout the process. These two factors make refining a suitable application for computer control Instruments measure such variable as temperature, flow and pressure. Any deviation from the standard is detected and regulating devices are adjusted to bring the process back into line.

Electricity. The computer is used by electricity authorities for load control. Demand for electricity is not constant throughout the day nor throughout the year. Generators have to be phased in and out to meet changing situations. Because of the time lage required to build up the necessary power, fluctuations in the load have to be anticipated in advance. Under computer control past records stored in the system, relating to changing hourly demands under various weather conditions, are scanned and compared with the actual present load in different parts of the supply network. This ensures that extra power is transferred to those areas where it is most needed at peak periods.

Steel. Process control applied to certain parts of steel production has increased efficiency in the industry. One example is in the cutting of the steel into lengths to match the firm's order book. In the rolling mills, which run at great speed, red-hot steel billets are rolled out into strips.

Before the use of computers, the mill would cut the sheets or rods, of varying lengths, into standard sizes or a particular size for one order. The lengths of steel left over would be scrap which would have to be re-smelted resulting in a lower grade steel. With the advent of computers the amount of scrap was reduced to a minimum, for it became possible to calculate the lengths that the billets would make whilst still red hot and being rolled out.

Printing and Paper

Computers are used by the printing trade where they are particularly useful in the production of newspapers and magazines where strict deadlines have to be met and time is short. Articles can be transposed to megnetic tape and then rapidly typeset, under computer control, in several type sizes, widths, and depths as necessary. Complete texts may be retained on tape, enabling amendments to be incorporated easily when reprinting. Computers are also used to update the listings in telephone directories, catelogues, parts and price lists so that they can quickly typeset whenever require. Computer assist the paper mills with process control in ways similar to those described in the chemical, oil refining and steel industries.

Engineering Design

Computers can help in calculating that all parts of a proposed design are satisfactory. This means a great saving in time and elimination of technical faults and human error before a design is further developed. Computer can also be used in calculations of space and layout as well as strength requirements. On metorway construction, the computer can calculate the amount of soil needed to raise an embankment, or the amount of rock to be removed in cutting through a hill, and it can work out the most efficient movement of such materials. The computer can provide graphical and perspective views to show the shape of a proposed aircraft wing or car body the slope of a curve for a new road, the visibility that the pilot or motorist will have or the accessibility of the instruments that he might have to operate. Computers are also used as an aid to electronic circuit design even assisting engineers in designing circuits for other computers.

Computers are pointing the way to an automated society. The development of robots to work in certain hostile situations, such as fire fighting, mines, security of sensitive establishments and space exploration, could bring considerable benefits. Industrial robots, which have the capacity to change industrial practice radically and to bring about automation on a scale not yet envisaged or understood, may not be so socially acceptable. We have to make sure that the computers themselves are protected so that they can provide the benefits, the power and knowledge which they offer us. The greatest pay off, however will be in sophisticated total management information systems, employing advance management techniques such as operations research and linear programming. Computers are used in organisations to improve the quality and accelerate the flow of information and thus to speed up and improve the performance of planning, decisionsm-aking and control activities. Computers are used by many government agencies for control purposes. Computers are used for environmental control purposes. Computers are also helping to conserve natural resources.

Information Processing and Computer.

We have seen that Computers can be used to store, up-date and analyse information, to direct and control industrial processes, and to aid and encourage scientific research. These, and any other applications, have brought about many changes in the organisation and quality of life which affect us directly as individuals or indirectly as members of society. To complicate matters, the speed of change is accelerating and the changes that are to come are likely to be far more radical than those we have already experienced. Computers have helped to reduce the intolerable burden of handling the ever-increasing amount of information with which government depart-ments, public services and business concerns are expected to content.

Improving Productivity

Computers are sophisficated that can significantly improve productivity. And productivity gains in turn, can lead to stronger competitive position in the world.

If we equate the efficiency of a process in terms of its cost, speed and accuracy of output, there is ample evidence in industrial applications that the introduction of the computer has brought positive results. In process control, the use of computer frequently leads to increased efficiency in terms of minimis-ing wastage and/or raising the quality of output which more than compensates for what many indeed be very high computing costs. It may also raise the safety level of a process, a benefit which is beyond measure in terms of money. In business and commerce the efficiency claim is well founded. A large part of business operations involves non-creative, routine clerical tasks and the computer can handle a great column of such tasks more quickly, more accurately and more reliably than is possible manually. Organisations, large or small, do not thrive or fail because of the presence or absence of a computer, but because of the quality of the people who run them. The computer can provide many advantages but we must insure that those advantages are not abused. Any computer system must be so conceived and operated that it is man who runs the machine and can adapt it to his changing needs. The machiner cannot be allowed to determine the patterns of our lives, and those of us who enjoy its benefits automatically acquire the responsibility for ensuring that it is used as wisely and as humanely as possible.

Automation

The object of bringing in a computer, or of extending existing computer facilities, may be to increase productivity, to attain a consistently high quality of product, to improve a service/or to gain knowledge. It may be to extend an existing manual system, perhaps to increase the speed and efficiency of operation, or to handle more work without increasing staff or accommodation, or to provide management with information more quickly.

In business and commerce the computer has complemented rather than replaced the workforce. In industrial situations the computer has taken over many routine, semi-skilled jobs, some of which were also hazardous, and it has offered in their place the possibility of more regarding jobs. The use of computer technology has also helped to create new products and services, and hence new jobs. The retraining of personnel for new, more highly skilled jobs, or for jobs in another industry, is a very important issue. So far computerization has altered work patterns rather than caused people to be out of work.

Security

The threat to security of information, and the threat to individual privacy, are computer arguments for making information systems too widely interchangeable. But if security and privacy are valuable, they are, in computing as elsewhere, qualities which are expensive achieve and they require continuous maintenance.

Metrology

Predicting the weather has long been considered something of a mystique based on country signs and folklore. The problem in meteorology has always been to obtain sufficient data, and to analyse that date quickly enough so that predictions can be made. Satellites linked directly with computers now provide the meteorologist with information, in addition to the data obtained from weather stations on land at sea. Armed with more information, the relationship between the variable factors which constitute weather can be analysed in greater depth, and a more accurate prediction of future behaviour can then be made. The computer system is also able to analyse vast quantities of past measurements to test for weather patterns and, based partly on these results, long-range forecasts can be made.

Space Technology

The development of space technology which culminated in the first moon landing was only possible because of the calculating powers and speed of the computer. Computers were used at the design stage of the project and in all phases of development right through to flight control. The collection of information is the main reason for putting satellites into orbit. Computer-linked space satellites provide previously unavailable information about the universe around us. This information is not merely of interest in scientific research but it increases our knowledge of our own planet and it is of immediate practical value, for example, in meteorology.

Satellites are also used to provide worldwide degital communication links for data transmission. The computer has played a leading role n, and continues to be an integral part of these new wonders of the second half of the twentieth century.

Communication

Air Travel. Air traffic control, which is responsible for organising the safe movement of our crowed airlines, depends on a significant amount of computer support. As flying speeds increase, control decisions have to be taken more quickly. These provide the pilot and the flight engineer with the information they need to control the flight and to make navigational calculations. It takes time, however, to scan the instruments and to assimilate the information and time can be critically precious at the speed at which the plane is flying. Besides the many-in-flight uses, the computer plays an increasing vital role in the training of pilots. A flight simulator provides an exact replica of the flight deck and performance of an aircraft, enabling the equivalent of many hours of flying to be undertaken without leaving the ground. The computer resolves the tasks, monitors and controls the pilot's action, and maintains a record of the pilot's performance. At ground level, the information needs of a busy airline are extensive. Computers are used for the efficient handling of seat reservations, crew schedules, timetables, tariffs, cargoes, maintainance schedules, personnel records, accounting, and stock control. Computer controlled seat reservation brings benefit to customers and to the airlines. It is an economic necessity that airlines operate as near to capacity as possible. To avoid over-booking, a complete list of all bookings needs to be maintained and be available for immediate interrogation. Communication between continent is established by using transoceanic cables and satellites. A travel agent is able to find out the current status of any flight and can book seats on the basis of the informotion obtained. This is all done in a matter of seconds and when a seat is reserved, the information system is automatically updated. Computer controlled reservaiions are also applicable to hotel rooms, theatre seats, sporting events at large stadiums and package tour holidays.

Traffic Control

Maintaining the flow of automobile traffic in congested areas is of paramount importance and in ever-increasing problem. It is not unusual for television cameras to be installed at major traffic junctions so that potential trouble sports can be kept under surveillance.

Transportation

Other transport facilities are making increasing use of computers. Railways prepare timetables, scheduling at the same time the distribution of the rolling stock to operate their services, and they control busy stretches of track with computer assistance.

Computer controlled ticket machines facilitate the automatic checking of tickets are likely to be introduced on busy main line stations and by London Transport on the Underground system.

Service Areas

Computerization is also encouraging the centralization of information, for greater convenience and accessibility, in large data banks. The computer is used extensively to carry out routine clerical functions, in general supplementing rather than replacing clerical staff, undertaking for example electricity and telephone bills. The computer is also playing an increasingly prominent part in the organisation and running of the public health and social services, in the maintenance of law and order, and as an aid to education.

Telephones

Computerized telephone exchanges handle an ever-increasing volume of calls. Cross-country, and even overseas, calls which previously meant a slow link-up through several switchboards and/or operators, can now be made directly and quickly. By way of satellites calls can also be transmitted at faster speeds than through conventional networks. The computer can also maintain a log of calls for subsequent billing.

Medicine

Computer being increasingly used in hospital administration for such tasks as maintaining inventories of drugs, surgical equipment, and linen ; for payroll ; hospital accounting ; and for bed allocation. Information on the condition of patients, details of rests and clinical reports may be stored on a computer system. In intensive care units the computer can be used to monitor a patient's condition. Scannint instruments attached to the patient are linked on-line to the system that the nursing staff can be notified as the patient's condition changes. The computer may print out or display a long of the patient's condition drawing attention to measurements that all outside the critical limits set by the doctor, or the computer itselt may tigger directly the necessary corrective action.

In some clinics, the computer is used, albeit in an experimental way to interview patients before or after they see a doctor in order to collect information for the patient's records and even to assist with the diagnositic process. The computer may assist in medical diagnosis, for example programs exist which can carry out electrocardiogram analysis to determine both normal and abnormal heart conditions. The computer system can act as a vast encyclopaedia of medical knowledge, providing doctor with access to an ever-increasing quantity of information which he could not possibly hope to carry in his head. The computer may assist in prescribing the correct dosage and pattern of treatment, for example, in treating cancer by radiotherapy where it is vital that the correct dosage of radium is administered and only to the exact area required. The computer has an important part to play in medical research and in the teaching of doctors and nursing staff. It is also possible to use computer program to test the effect that a form of treatment might have on a patient before it is administered.

Law and Order

Police forces make extensive use of information retrieval capability of computer systems for this purpose. Records are maintained concerning accidents, vehicle owners, disqualified drivers, traffic tickets, stolen vehicles, fingerprints, criminals, wanted and missing persons, stolen property and drugs. Computers are finding their way into solicitors and lawyers offices for the purpose of legal data retrieval as well as for conventional accounting and record keeping.

Libraries and Museums. Increasing use is now being made of computers in library organisation. Lists of borrowed books are maintained by the system and reminders for those which are overdue can be generated by computer output. Some universities libraries in North Amercia are linked to a network so that an abscure document in a distant archive can be quickly located and, in some cases, photocopies can be telexed over a wide area. Museums are also making use of computers to help with cataloguing and indexing. Information about the exhibits in a museum's collection is assembled and retained as a large data base, and the information retrieval capability of the computer can then be exploited to the benefit of staff and visitors.

Artificial Intelligence

Eversince the integrated circuits and silliconchip have been made use, the computer has become smaller and smaller in its size and increased in its capabilities and reduced its costs. This happened in four stages. Now, the computers are in the 5th generation stage. Instead of going further, miniaturisation specialists are going for enhancing the capabilities of the computers in the fifth generation stage. Artificial intelligence is such one thing which comes under fifth generation stage. As per the dictionary, meaning of artificial is contrievance/equipment and intelligence means learning to cope-up. Even the crude and elementary mechanical means like pressure or safety valves, thermostat etc. do exhibit some intelligence. Because as and when, the parameter goes beyond the range, the mechanical means immediately start functioning based on the sensation. A small baby can generally be said that it does not have the intelligence, because irrespective of the nature of the objectives it catches and it may get hurt. It is only as the time advances bit by bit data, instructions are stored in the brain. Subsequently, whenever it comes across with the object, it tries to compare the data/instructions stored in the brain and if the object under the consideration matches with the data/instructions it receives, otherwise an extraneous attitude is resulted. Whenever, the data and instructions are given by various people or near by will be stored in the brain and used for recognising the objects. Similarly, lot of data and instructions need to be stored in the machine and programme for all possible alternative course of actions, that it can display. This can amount to millions of data points and instructions. For example, where safety and security is par amount, the restriction of the man entering the particular zone is imposed. For this purpose, if a door with electro magnetic signal is provided, it should serve the purpose. But this alone cannot be sufficient because any object that approaches near by the door can pass through. Therefore, a number of logical instructions and sensors are provided so that only a man can pass through the door. As soon as some one reaches the door based on the instructions stored, first it matches the weight of the object. If the weight is within the range of say 50 to 70 kg, the machine goes to search for the next instruction like range of temperature. If the temperature sensed by the machine is within the range say 94 to 102°F it comes to the next instruction to screen out further. It can compare with the patterns of various animate things say birds, animals, man etc. If the object matches with the pattern of the man stored, in the memory, then it asks for the ticket to be inserted in the slot. That means it has recognised that a human being has entered the door. Once the ticket is placed in the place provided for it, the door will get opened. This is how the machine can derive a number of logical inferences. The logical inferences is unique in the artificial intelligence.

Quite often it happens that two specialists will conclude the case with the same symptoms, in different way. Out of the two one can be correct. Though the symptoms exhibited and the tests carried out etc. are the same, the conclusion drawn by two specialists could be different. The specialists who failed to draw correct conclusion as well as a specialist who could succeed in diagnosing correctly will not be capable of understanding the mechanism with which they could arrive to these conclusions. Actually very fast thinking process takes place in the brain but it is difficult to perceive it and link up with logic. Only few will be able to perceive it and build the logic in their decision. It is this logical decision that finally leads to the detailed programming and passing it on to the machine. Today the atrificial intelligence have developed to such an extent that it has surpassed in decision making even the combined decision of the multi specialists. The reason is that the machine cannot have any subjective liking or error. Any number of times and any number of instructions it meticulously follows without any tiresome. Whereas a man cannot have the same capability, therefore it surpasses even the capacity of the man. A new branch of Science called knowledge engineering has been developed. Irrespective of nature of capacity the knowledge engineering is capable of building up the logical questioning and answering and making inferences and instructions. Certain shells have been developed in advanced countries like Japan and states, wherein these shells can be made use for forming the instructions to enable the machine with artificial intelligence. In India, Indian Institute of Science, Bangalore, IIT Delhi, IIT Madras are persuing the research in obtaining the shell which can be a powerful tool in making the machines with artificial intelligence.

Computers Application in Personnel Management

A survey was sponsored by Personnel Journal in America in September 1984 which revealed that 99·7% of the respondents to a *Personnel Journal* agreed that computers are an integral part of personnel management and that they are all using computers in one capacity or another. These results cut across company size and industry. The use of some type of computer system—either in a mini, micro, or mainframe or through an outside service bureau and/or time sharing system—is rapidly becoming common place.

A sample of 1,000 *Personnel Journal* subscribers was selected on basis from the list of US subscribers. This sample was used for the survey. The mailing consisted of a cover letter, a six-page questionnaire, a postage-paid return envelope and a $1 incentive. The initial mailing was done in September, 1984. In November 1984, a second mailing (without an incentive) was mailed to all nonrespondents. Almost half of the subscribers (49·9%), in fact, work for organisation with 1,000 or more employees. Actual titles of the respondents includes Owner/Partner/Corporate Officer/Vice president (15·5%), Director (22·9%); Manager (39·5%); Administration/Supervisor/Officer (14·3%) ; and Specialist/Analyst or Consultant (8·3%). All returns were received directly by Globe Research Corporation in Melville.

Survey revealed that companies are moving toward integration of payroll and personnel systems, according to the survey. Two in three (65·4%) organisations now have systems with some degree of integration, ranging from duplicate information in separate systems (26·9%) to total integration (24·5%). Organisational problems present the major obstacles to integration : priorities (43·4%), budget (31·1%) ; management philosophy (29·8%), and internal politics (26·4%).

Obstacles to computerisation still abound, but personnel professionnal see the automation of personnel systems as inevitable and, in many cases, an accomplished fact. The computer age has arrived, and personnel managers are embracing it. Today top management in USA considers computerised information systems for personnel management more important than they were two years ago. Two-thirds (69·8%) of the survey respondents indicated that top management considers information systems more important. One-fourth (25·3%) say it is neither more nor less important. Automation in the personnel department to date has focused on employee records, payroll, and compensation and benefits administration. Personnel management is obviously still primarily a people management profession. The survey results suggest, however, that more than ever it is also a machine management profession. The study reveals that although everyone has some form of automation, the type of automation and the functions automated varied widely. Almost three-quarters of the survey respondents use personnel software developed in-house for their organisation's mainframe computer. Four out of 10 of the respondents use purchased software packages. Another four in 10 use an outside service bureau of external timesharing.

Micro Computers in Personnel Department

Almost half (43·3%) of the personnel departments responding to the survey indicated that they plan to begin using micro-based systems within the next 12 months. Of those planning to use micro-based systems, opproximately half (51·1%) plan to purchase their software, and four in 10 (42·3%) plan to develop their software in-house. Regarding micro-based system, almost 80% plan to use it for administration and record-keeping. In addition, two-thirds plan to use it for compensation administration. More than half plan to use it for benefits administration, human resources planning and employment or recruitment, and 17·0% plan to use their micro system for labour relations.

One-third plan to convert their manual system to computerised one, and one-tenth plan to replace their current computerised system or a significent element of it.

Anticipated changes in the computerization of the Department

Upgrade or expand current computerized system	49·5%
Begin to use micro-based personal computer system	43·4%
Convert manual system to a computerized one	34·5%
Replace current computerized system or significant element of it	9·8%

Meeting future software needs

Purchase software	58·4%
Purchase for mainframe	22·3%
Purchase for mini	14·2%
Purchase for micro	31·6%
Develop software in-house	44·5%
Purchase external timesharing or use outside service bureau	5·8%

The pay roll issue are still the predominant priorities. However, employment/recruitment and human resources planning are quickly gaining status as equal priorities. Other plans include: pay roll (33·5%), training and development (35·7%) OSHA compliance and reporting (24·0%), and labour relations (14·5%).

Problems in Computerisation

Although top management theoretically supports a computerized personnel system, actually implementing a system inevitably involves overcoming a few obstacles. These problems were closely followed by : (1) time and staff for data entry process and the time consuming conversion process   (16·3%) ; and (2) getting MIS (management information systems) and DP (data processing) to understand the computer needs of personnel, to develop programs for personnel and to give priority to human resources (13·0%). Another 8·2% indicated personnel had a low priority as compared with other departments. Major difficulties are training staff to use the system and having people available who understand the system.   Almost 11·5% of the respondents said that lack of training   and lack of knowledgeable staff were major obstacles to computerizing the personnel department. Systems issues included : finding appropriate software for specific needs ; the need to customize purchase software packages ; going from decentralized to centralized records ; intergrating personnel/payroll/benefits systems ; and interfacing with corporate head-quarters.

Problems in Computerisation

28·9%	Time and staff
18·7%	Cost/Budget
16·3%	Priority
13·0%	Time
11·5%	MIS/DP interface
11·5%	Understanding Technology
6·3%	Appropriate Software
5·7%	Data base problems
5·7%	Input problems
5·3%	Data Input
3·4%	Payroll/Personnel Integration Defining Needs

Application of Micro Computers

One of the example of Micro Computers application is BBC Television consisting of two national networks, BBC1 and BBC2. Together they produce about 200 hours of programmes a week, some of them in the regions but most of them in London. The current annual budget is £500 million and about 9,000 staff are employed. The staff are the responsibility of BBC Television. Personnel Group which consists of 45 personnel officers and senior personnel managers, with support from secretaries and record clerks. The system they operate has been available from a central ICL 2900 mainframe. This holds management data like salary scales, establishment levels, and other information. Nevertheless, the day-to-day work of the Personnel Group entailed regular reference to and handling of individual staff files and their associated paper work. A feasibility study was undertaken in 1981, as a result of which it was decided to install a microcomputer to handle part of the process. The first network was installed for testing in June 1983 and the complete installation was in full operation by April 1984. The Television Personnel Group is organised in eight units, six of which are participating in the project. Three of them are based in various parts of the main Television Centre site in Wood Lane. Each unit is staffed by one or more senior personnel officers, a number of personnel officers, secretaries and record clerks. The work of each unit, carried out within the context of the BBC's overall personnel policies is largely self-contained.

While some aspects of BBC peronnel work are handled centrally, e.g. job evaluation and pay policy, the local officer is responsible for providing virtually the whole range of personnel services to one or more line manager and staff. Typically, these amount to a group of 250—300 people covering a wide range of skills and grades. Among them are professionally qualified engineers creative staff like directors and designers, support staff like scenic craftsmen, and industrial categories like electricians, cleaners and catering employees.

Each of the six units taking part in the project has a local area network which supports number of Racal 6000 workstations, a 20MB Winchester disc unit and two printers, usually one letter quality and one document quality. Some of the larger networks have additional local printers. The Racal 6000 is an 8-bit Intel 8085-based workstation with two floppy disc drives as standard. The CP/M operating system is used and Racal supply a range of Micropro software. Switching into and out of software packages is a simple operation that benefits. Each network supports multiple-access files, print spooling electronic mail, archiving/and a two-tier file security system—essential for confidentiality in personnel work. The principle uses of the networks are for word processing and the handling of data files. Wordprocessing involves standardised personnel material, stored centrally for multiple access, and non-standard letters, minutes etc. many of which go through several drafts. In addition, specific routines incorporating mail-merge techniques, is used for processing correspondence associated with recruitment campaigns, removal and return of files from staff records areas, and other personnel functions. Data files, besides providing up-to-date information for immediate processing can be used via the integrated software to develop management information in the form of listings on subsidiary files. Nearly all the facilities have been developed by personnel staff using the software. Programs in BASIC are written professionally for two tasks : the conversion of the mainframe files for local use and the provision of a fast look-up to an abbreviated version of the main data file. The project is formally managed by the

Television Service's computer services department while a senior personnel officer has been released full-time to act as project manager. A working party of personnel staff from all the networks, together with representatives of the other departments involved in developing the system, meets, regularly to monitor progress and discuss new proposals.

Mail Merge is a means of creating personalised documents quickly by merging standard text with personal information e.g. the recipients' name and address. Simultaneous Peripheral Operation on Line enables data to be transferred rapidly to disc and from there fed to a slow printer, freeing the processor for other tasks. Local Area Network is a means of connecting a number of user devices so that they can communicate with each other and share local and remote resources.

Computers in Health Care

In developed countries the computers are used for following areas in the health fields :

—By relieving scarce people of routine tasks, computers can help increase their effectiveness.

—Computers can help the health scientists to conduct research that will extend the frontiers of medical knowledge.

—Computers can help to improve the quality of physician's diagnoses on the one hand and can help to improve the control of important medical processes on the other.

—Medical knowledge is advancing rapidly computers are needed to retrieve relevant information rapidly.

Computers are also being used for such diagnostic purposes as (i) displaying heart function on a terminal screen from motion—picture X-rays and calculating the volume and width of the patient's left ventricle—the heart pump. (ii) determining by means of a computer aided tomography (CAT) scanner. Computer is also a research tool that's providing insights into :

(a) Causes and prevention of stroke

(b) Patterns of drug addiction.

Computers are also used for Medical History and Retrieval. Once the patient history and medical records are available in machine accessible form, they may be retrieved by the doctor as necessary for review and updating. These data could include type of ailment, level of severity, results of most recent examination etc. Control of the physiological status of patients and laboratory tests are among the many applications of computers in medicine.

Computers in Education

The introduction of millions of personal computers in developed countries into elementary and high schools in recent years has enabled those schools to also offer computer courses. Computer-managed instruction is a name sometimes given to this use of machines. A properly programmed computer can help teachers manage a student's schedule of activities as the student progresses through a program of instruction. Computer-Assisted instruction is a term that refers to a learning situation in which the student interacts with, and is guided by, a computer through a course of study aimed at achieving certain instructional goals. Computer is increasingly being made of computer as a resource in teaching and learning at all levels of education. Instructional material can be prepared and stored within the computer system in the form of programs which are carefully structured to teach specific lessons.

A programme can be used by many students thus freeing the teacher to spend the time on more personal tuition. The computer can also be used time and again with different sets of data so that a variety of conditions can be studied.

The computer can relieve the teacher of some administrative duties, giving him more time to concentrate on teaching. For example, computer can be used to assist in constructing time-tables, to monitor and schedule teaching resources, to build up and maintain comprehensive student records, to provide a complete student profile, and to accumulate information, internal and external, for assistance in careers guidance.

Computer Applications in Manpower Planning

The Computer is an accepted tool in many Organisations today. Quite often the Manpower Planner will be facing the challenge of computiersing Manpower Data.

Computer applications in Manpower Planning helps us to understand questions such as : How many employees can speak/read/write Russian language or German Language in the organisation so that foreign assignments can be taken up easily or the personnel can be trained in the required field immediately ? How many people in the organisation are stagnating more than 8 years without promotions. What will be the status of Stagnation and Promotion rates in another 10 years with the present growth rate of the organisation etc. These questions will present difficult problems to Manpower Planners, but if the department uses a computerised system, the answers could be furnished easily and quickly. Complete personal records relating to education, training, experience, age, promotion performance, and other aspects can be brought into the computer file. Changes like fresh appointments, resignations, promotions, training, can be continually fed into the computer file. This helps the Manpower Planner in maintaining accurate and up-to-date personal files for all employees and producing, all reports and returns easily, reliably in time. If the records are properly designed, and computerised, from the same records, a lot of analysis can be done to enable the Manpower Planner to plan much effectively. The age-group wise analysis, with in a cadre can be used to high light the management succession to avoid a managerial skill vaccum at a later date, due to retirement, resignations etc. A comparison of prospects at various grades of pay in different "Cadres" can be reflected which can help the management in maintaining a "horizontal parity" in promotion rates in different cadres. The stagnation levels analysis in various grades can be easily shown where and to what extent upward movement would be needed to control the situation. The computer could be used to generate reports on labour turnover-the rate of labour turnover for the entire firm, by departments, by jobs and by other categories which will help in Manpower Planning. Scores made by applicants on centain employment tests could be corrected with turnover data to see whether short term employees could be predicted. Another report which would be useful is the evaluation of recruitment methods and sources of applicants. But it is difficult to maintain data regarding sources of probable employees of specific skills without the computers' ability to manipulate data extremely fast. Similarly computer provide information regarding analysis of rate of labour turnover like normal or early retirement, death, dismissal, redundancy etc. There could be a host of personal reasons for leaving such as supervision low salary etc. which can be used for the corrective action. Similarly the analysis of the occupational diseases, short lists for certain vacancies etc, can be prepared direct from the personnel records by instruction to the computer to search for individuals with certain characteristics, one of which might be qualifications of 'degree-level'.

Models and Simulation on Manpower Planning

Although using the computers ability to produce reports is a step in the direction of full utilisation, more sophisticated uses involve models and simulation. A model is an abstract representation of some thing, usually expressed in mathematical terms. An example of non-mathematical model is the diagram used to describe the forces of supply and demand in the market place. The formula used by economists to describe equilibrium is a mathematical model. Simulation involves building a model of a real system in such a way that when one part of the model is changed, the rest of the model will be changed, and reflected or simulate how the real system would behave. The most frequent use of models in the personnel function involves Manpower Planning for forecasting. For example, a firm could develop a model which describe the normal flow of manpower through the firm. By inserting factors such as the normal rate of growth the expected growth rate, the model should provide an estimate of the number of tires, which must be made up to maintain a fully staffed organisation.

Alan Patz has developed a Linear programming model which can be used to evaluate various personnel, policies, such as promotion and retention etc. While the model is limited, it assumes all entries into the organisation are made at the lowest level, it is still very useful because few firms recruit a large number of outsiders for the top levels. Another limitation in the present model is the assumption that management goals remain constant over a time period but work is under way to incorporate variable goals. Given a set of inputs, the model will provide answers such as the promotion rate, retention rate and over ages and under ages, of personnel in each level of the organisation, as people move into, through and out of the organisation.

Simulation

A Computer simulation programme can be used also to explain and predict behaviour, and since it uses concepts which can be measured in an actual situation, it is possible to validate the model. Included among the inputs and outputs of the programme are items relevant to the finance, marketing and production functions of the organisation. Thus the programme can be integrated into the over all planning process.

It has been observed that problem in any industry is not of "poor workers" but of "poor officers". It is therefore, of utmost importance to develop a competent devoted team of officers. The computer services are used to prepare a detailed analysis of stagnation patterns, retirement statistics employment stability indicies, a growth prospects of different cadres of different cadres of officers and supervisors employed, to enable the preparation of career plans.

Various mathematical models on all possible management decision have been developed. Depending on the specific position, certain parameters on the mathematical models are fed and simulation of the various alternatives, are carried out. The outputs from the computer clearly indicate the immediate and long-term effects of a policy decision. The changes in policy relating to promotion, recruitment, and allowing officers to seek outside employment showed what was likely to be the pattern of the number of officers stagnating in one cader.

The arguments about the appropriate location of a centralised manpower planning are fairly, evenly balanced. If planning is at a highly centralised activity so that the major part of the planning of marketing, production or finance are carried out in a central planning division, responsible direct to the board, then the appropriate place for manpower, planning is also in the planning division, because it is there that the work on corporate, objectives is done. Some manpower planning, however needs to be undertaken at the level of the firm as a whole, certain environment will effect the use of some types of manpower irrespective of the various parts of the organisation which have a demand for them.

Achieving corporate objectives may lead to surplus manpower in some departments which can be re-employed elsewhere. · Strategies, such as, carrier, planning and management development may demand that individuals are seen in the context of the total organisation. Most Importantly, a board of directors has to set and achieve objectives for the organisation as a whole. In order to do this, it has to balance departmental interest. Deciding what the proper balance is, requires specialist help diversed from these interests. If the Planning activity is not highly centralised, it may be better to locate manpower planning within the personnel department. In many respects it is the planning function of the personnel activity and must, whatever its position, maintain a close contact with the other personnel activities whose objectives (recruitment or training) it will to establish. The major source of information for manpower planning and several of these other activities (such as a salary admistration) will be a central personnel record, and there is a common interest in seeing that it is kept accurate. It is also true that while a proportion of the manpower planning problems are simply economic, many of the others (possibly more important in the long run) are socialogical investigation from within the personnel department than outside it.

Further, personnel department being the the source of complete information of personnel, the personnel manager will be in an advantageous position in implementing manpower planning. The very status of the personnel manager, in an organisation, itself has the advantage to convince the top level management and negotiate with the union for the affective implementation of manpower, planning.

Similarly through computerisation of Manpower Planning, the various occupations can be systematically classified as : Proprietors, Managers, Admintstrative and Executive Staff, Clerical and allied occupations, Professional and technical occupations (higher), Professional and Technical occupation (lower) Skilled mannual occupations, Semi-skilled manual occupations, Labour and Unskilled-occupations etc. Old skills are displaced, but there may also be some introduction of newer skills to operate the new investments. Maintenance manpower tends to increase at the expense of operating manpower. Other technical innovation would occur irrespective of the change in relative costs, as a result of the normal replacement or growth occurring in the economy as a whole. This, too, leads to a requirement for few skills and for different proportions of the older skills. Similar changes also result as new markets are developed and old ones decline. Market changes provide many of the occassions for new investment and technology, but over a span of time they also cause a movement of accupations and skills from one industry to the other. Computerisation of Manpower Planning helps us to meet requirements for new skills, create requirements for new occupations, identifying new training skills and new education system. Occupational change is both a cause and an effect of long-range manpower planning.

Similarly, the largest proportional change in computerisation is the increase in administrative, technical and clerical occupations. Within this category, scientists, engineers and technologists show the greatest increase, followed by industrial technicians, other higher professional and Financial occupations ; directors, managers, administrative and executive staff ; and other lower professional and technical occupations. Those changes are partly due to increasing productivity. They represent the continuation of a trend which has been apparent for almost fifty years. Many of the other changes are partly due to the effects of different processes and technologies in particular automation. Such as explanation is impossible to quantify precisely, but seems likely to apply to most of the manual occupations with the possible exception of these skilled grades.

42

Computer Languages

Natural languages are highly developed and with them we can express not only facts but also abstract ideas, and we can convey shades of meaning or suggest subtle feelings and sensations. But in order to do so, a large vocabulary is required. A programming language, therefore, uses a limited or restricted vacobulary. Indeed a programming language by its very nature and purpose does not need to say very much. People can use natural languages incorrectly and still make themselves understood. Computers, however, are not yet able to correct and deduce meaning from incorrect instructions. Computer languages are smaller and simpler than natural languages but they have to be used with great precision. Unless a programmer adheres exactly to the 'grammer' of a programming language even down to the correct 'punctuation' his commands will not be understood by the computer.

For processing the data, computer programme is required. Computer programming is a set of instructions explaining the computer step by step procedure to be followed by it in processing the data or in solving a problem. Computer instructions are given in the coded language using the *machine language* or the *higher level language*. The computer operates with absolute machine language instructions and therefore all other higher level languages should be converted to the machine language. The computer inverts higher level language machine language. Each machine has its own set of instructions based on its circuitry.

The machine language is generally 'Binary' language. It is composed of only '0' and '1'.

The higher level languages are machine independent languages. They are basically procedure oriented or problem oriented. The main advantages of higher level languages are that they are easy to learn and are independent of machine.

Higher Level Languages

The higher level languages can be classified into :

—Algebraic formula type processing language

—Data processing languages
—String and list processing languages
—Multipurpose languages
—Simulation languages
—Specific Application languages.

Processing Languages

Examples of such languages are FORTRAN, PL/1, ALGOL. These languages are oriented towards computational procedures for solving mathematical and statistical problems. These languages have poor facilities for describing complex input and output of non numeric data and handling files.

Data Processing Languages

These languages like COBOL, PL/1 are useful for handling data files which requires considerable manipulation of non numeric data and high volume of input and output data. Only single addition, subtraction, multiplication and division can be carried out by these languages.

Therefore in most of the business organisations, for payroll, inventory accounting, receivables, cash balances etc. these languages are used.

String and List Processing Languages

These are the languages specifically useful for handling of text material. SNOBOL and LISP are the high level languages under such type of category.

Multi Purpose Languages

PASCAL and PL/1 are multi-purpose languages. These languages are used for mathematical applications as well as for business applications or data-processing.

Simulation Languages

Development of simulation languages has simplified the task of writing simulation programmes for different types of models and systems. Simulation languages are GPSS, SIMSCRIPT GASP, SIMPAC, DYNAMO and SINTLATE.

The objectives of these simulation languages are to furnish a generalised structure for designing simulation models and to speed the conversion of a simulation model to a computer program. The reason for several simulation languages is that they can be applied to different types of problems which make their simulation procedures more automatic for the user. For example DYNAMO and SIMLATE were designed primarily for simulating large scale economic systems that have been formulated as economic models, consisting of large set of equations. GPSS, SIMSCRIPT and GASP are well suited for scheduling and waiting line problems.

Specific Applications Languages

A.P.T. (Automatically Programming Tools) used for a machine tools and COGO used for Civil Engineering survey purpose. They have been developed for specific application purpose.

It should be appreciated that a given machine is designed to react to or obey only one language or code. Since the inception of computers, three types of computer language have evolved :

(a) Machine Codes ;   (b) Assembly Codes ;
(c) High-level Languages.

Machine Codes

The set of instruction codes (whether in binary or decimal) created in conjunction with the computer designer, is called a machine code or machine language. It will be determined by the actual design or construction of the ALU, the control unit and the size, as well as the word length, of the memory unit. Clearly because of the dependence of the language on the particular machine, such languages are called machine dependent languages.

Assembly Codes

The MLT has to be translated into the binary pattern '1001' before the machine can 'understand' the operation intended. The act of translating is carried out by a special pre-stored program called assembler. It translates the program written by the programmer into that version on which the machine recognises and responds to, and assembles it in the main memory ready for execution hence the term assembly codes. Machine and assembly codes orientated towards the basic design of computers, are referred to as low level languages.

High Level Languages

The commercial viability and wider use of computers led, by the mid 1950s, to the necessity for and development of high level languages. These languages instead of being machine based, are orientated more towards the problem to be solved. Such problem-oriented languages enable the programmer to write instructions using certain English words and conventional mathematical notions, therefore making it easier for him to think about his problem.

Compilation Process. By using a high level language, the programmer saves himself a great deal of time and effort. To describe the above mathematical expression in a low level language the programmer would have to write perhaps five instructions instead of one. The computer, of course, cannot directly understand a high-level language. A translation stage is necessary. Assembly languages use an assembler to perform this conversion process. High level languages use instead a compiler a more complicated pre-stored program to translate the programmer's instructions, into its machine level counterpart which the particular machine has been constructed to obey. Compilers are large programs which reside permanently on secondary storage. When required they are copied into the main memory. The processes of compilation and execution take places as one followed through operation. The compiler being a program is executed in the CPU. Its data is the source program statements, or instructions, each of which is converted into the appropriate machine instructions in binary, and then stored in the main memory of CPU as the object program. The compiler, having performed its work is no longer required. The space it has taken up during compilation is then used by the object program for manipulation of the data, the result of which then be output.

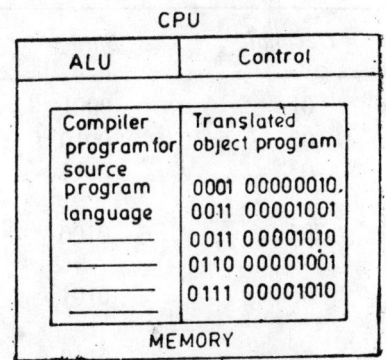

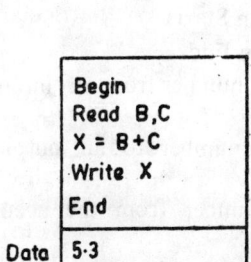

Fig. 42·2. Compiling the Program.

Statement	Meaning
(i) Begin	Inform the Com
(ii) Read B, C	Read two numbers (from a card) associating the first with location B and second with location C.
(iii) $X=B+C$	Add the numbers stored in locations B and C, place their sum in location X.
(iv) Write X	Output the contents of location X (To the line printers).
(v) End	Inform the computer control unit.

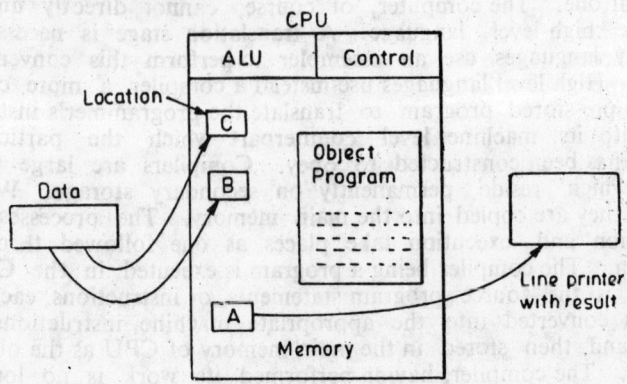

Fig. 42·2. Execution of Object Program.

Machine Level Version

Decimal	Binary	Meaning
01	0001	Program Starts
02	0010	Program Ends
03	0011	Read a number from an input device
04	0100	Write a number to an output device
05	0101	Store number from the accumulator
06	0110	Load a number into the accumulator
07	0111	Add
08	1000	Subtruct
09	1001	Multiply
10	1010	Divide

The programmer does not need to know anything about the accumulator other than that it performs the arithmetic which he has commanded. Although in the original statement locations *B* and *C* and *X* were used, the machine has to equate these 'names' with actual addresses, and this is something which the operating system itself will do.

In most cases, programmers mainly work in a high-level languages, but they can be grouped into four broad applications areas; scientific, business, special purpose and interactive. The language chosen will be largely determined by two factors ; the first being the application in hand and the second, the choice offered by the computer installation *i.e.* the range of compilers which are available. It is normal when using high level languages to use menmonics for addresses in store rather than actual numbered locations. The letters B, C and X refer to three locations in the memory of the CPU. We know that a high level language cannot be understood by the computer and a translating process has to take place. Statements will have to undergo compilation so that they can be reduced to machine level instructions. A copy of the compiler for our pseudo language has already been brought into memory and is ready to compile the source program.

Development of a Computer Program

Before a programmer can begin his task, he will require two specifications. The first will detail the precise problem to be solved, and the type of information which is required. The second will itemize the computer configuration which will handle the job *i.e.* the available equipment. We now summarize the seven stages with which a programmer is concerned in the development of a program, from the specification of the problem to its successful completion.

Algorithm. The construction of the algorithm is the stage which requires creative thinking. The algorithm is the sequence of steps of computer operations which collectively solve a given problem.

Flowchart. The next step is to record the algorithm in a flowchart form.

Code into a high-level language. The sequence of operations outlined by the flowchart is transposed into a programming language.

Input preparation. The instructions must now be prepared in some form suitable for the computer to receive them, *i.e.* punched cards, paper tape, or directly input at a VDU.

Compilation. The source program is fed into the computer and the compiler converts it into a machine version. At the same time, the compiler scan the source program for any syntactical errors, *i.e.* mistakes in the use of the language.

Corrections. Syntax errors are corrected and the source program goes through the compilation process again.

Testing process. The compiler can only detect errors in the syntax and not in the logic of the program. It is the programmer's task to create instructions which are logically sound and in the correct sequence. Any faulty logic that still remains can only be detected by examining the output. Programmers are usually concerned with a general problem rather than a particular problem, *i.e.*

add any two numbers together rather than two specific numbers. In some cases, programmers may be asked to solve a particular problem when it is very large one, *e.g.* to monitor the moon landing.

Personal Computers

Personal computers may be classified as general interest systems designed either for home use or for use in the offices of organisations and professionals. Personal computers are used to entertain, educate and increase personal productivity. Some of the initial manufacturers of Personal Computers are Apple, Atari, Coleco, Commodore, I.B.M. and Radio Shack. The least expensive models have limited keyboards, and the user's TV set is generally used to display output. Casette tape is often used for offline secondry storage of programmes and data. Personal Computers today came in a variety of configuration from self contained units to central processors. Personal Computers contain all the elements found in any larger system and individuals in developed countries are taking advantage of the technology in every imaginable way. Personal Computers can be manipulated in countless ways to entertain educate and serve people and there is virtually no limit to the number of possible applications that can be developed. The Personal Computers can be used to : Entertain with hundreds of challenging games; Balancing one's budget; To marit or one's have energy usage ; To accumulate income data by categories ; Compute and keep track of one's installment payments; To maintain and easy retrieval and reference to such information as names and addresses ; telephone numbers, and dates of birthdays and anniverseries, insurance policies ; tax returns etc.

In the present time thousands of business and accounting packages are available for use on p.c. systems.

The functions performed in offices consist of such data processing activities as originating-recording, classifying, forting, calculating, summarising, communicating, storing and retrieving. Traditional office tools and techniques are still often followed to perform these tasks. But the relatively low productivity of office workers, rising costs are resulting in the introduction of new electronic technology into offices.

The p.c. system can also be used in an office to : compute payrolls and process orders, keep track of appointments, schedule meetings, plan daily activities and maintain an outline alphabetical and subject file showing office locations of original letters and documents ; schedule production ; analyse production costs and control machine tools and other productive equipment; control inventory levels of thousands of different items ; produce personalised letters, mailing labels and other documents, Monitor and compare costs of fuel, oil and repairs of many vehicles. Prepare maintenance schedule, prepare the graphic images etc.

The coming decade will likely bring the marriage of personal computers with communication technology, producing the health off spring of electrical mail. Communications capabilities can be added to a personal computer today. The personal computer driven electric mail is likely to flourish. Word processing and electronic mail/message systems are two of the basic building blocks for the automated office. Since a computer is a fast electronic symbol manipulating device, and since office work basically involves symbol manipulation, it is natural that organisations have turned to computers and other technology help. Organisations have turned to computers due to following reasons :

(i) Low productivity of office workers relative to other groups.

(ii) Rising costs in offices.

(iii) Misfiling of documents.

In most organisations, the change from the traditional office setting to an electronic office environment is likely to evolve through several development stages.

The traditional office setting to electronic office environment is evolved through one of the following stages :

(i) **Word Processing.** Efficient creation, editing and printing of documents.

(ii) **Message Distribution.** Electronic mail/message systems accept messages addressed to one or more persons from a sending station and transmit them over communications channels to the stations of the recipients.

(iii) **Document transmission and reproduction.** A device which scans documents, graphs, pictures etc.

(iv) **Computer conferencing.** People at different locations are using their workstations to attend conferences. Computer conferencing and tele-conferencing are both used to link people who are geographically separated but computer conferencing permits people to participate at different times while tele-conferencing requires participants to be online at the same time.

Word Processing

It is the back bone of office automation. It is the key that opens office doors to the introduction of automated equipment. Word process describes the use of electronic equipment to create, edit manipulate transmit, store retrieve, and print text material. Once a Word Processor program diskette is loaded into a pc system, the computer effectively becomes a word processor. A modern WP system is an amazingly flexible tool. A text is keyed into the system it is displayed on the screen and can easily be recorded on a storage medium such as a floppy disk. So in brief word processing is the rapidly growing use of electronic equipment to create, view, edit, manipulate transmit, store, retrieve and print text material. Word processors are basically computers that are programmed to process text material. Some word processors are teamed with optical character readers so that existing office type writers can talk directly to the WP Equipment. The original text draft is prepared at a type writer. Then OCR reader scans this text automatically enters it into the WP system. The WP system is then used to edit, correct, and print the stored text as shown in the following diagram.

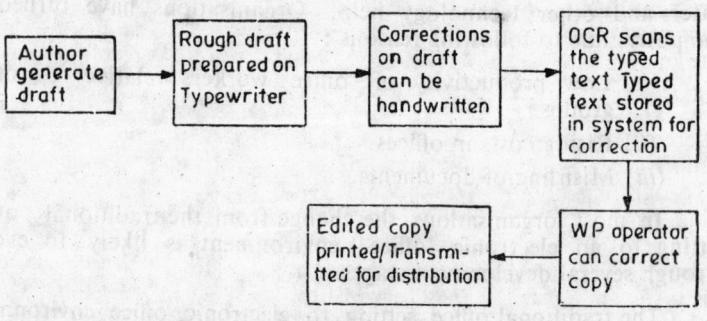

Fig. 42·3

Electronic Main/Message Systems

In United States most large business houses have the advanced Electronic Mail/Message Systems. (EMMS). Electronic mail/ message system is one that can store and deliver, by electronic means, messages that would otherwise probably be forwarded through the postal service or sent verbally over telephone lines. EMMS concepts have been developed in response to the short-comings in other message delivery systems.

An EMMS can provide message distribution services, a means of transmitting copies of documents and pictures computerised conference functions and other services. A message distribution system allows message to be sent at any time to specified individuals/ groups. Recipients periodically review stored voice or keyed text messages at convenient times. Interruptions are reduced, and communication between times zones is facilitated. In brief EMMS can perform a number of functions such as : Message distribution services ; Transmission of documents and pictures ; computerised conferences ; Follow up services.

In the message distribution services a keyed or spoken message is sent on the first try and at any time of the day or night, to an individual who has a storage mail box in the message system. The receiver can periodically review stored message at a time that is convenient. Some EMMSs handle spoken messages, while others are designed to manage keyed text. In the spoken message the sound waves produced by the sender are converted into digital pules and then stored on a hard disk for later retrieval. When message recipients call the voice store-and-forword system, they are notified that messages are waiting. After listening to the messages produced as reconstituted speech, they can then delete them or save them for future reference or pass on to other parties. Teleprinter and visual display terminals are generally used to enter and receive keyed messages. In the transmission of documents and pictures an original document can be placed in a sending facsimile or fax machine. A communication link up is then made with a receiving fax device at another location. As the sending machine scans the document, the receiving device reproduces the scanned images. Thus when the transmission is completed the receiving device has produced a duplicate or facsimile of the original.

The computer conferencing permit people to participate at different times. The alternative to computer conferencing is teleconferencing. The teleconferencing refers to the electronic linking of geographically ocattered people who are all participating at the same time. In the Follow up Services of EMMS function, the message system provides additional services to help the recipient take appropriate action after the message has been delivered. In this system the receiver may forword the message to others with or without comments store the message in a "personal attention needed" electronic file ; store the message in a subordinate file with instructions for the subordinate to take necessary action.

Operating Systems for Personal Computers

Operating systems direct the flow of instructions, data and results from one part of a computer system to another. A system software package is a collection of programs designed to operate, control and extend the processing capabilities of the computer itself. An important system software package found in almost all computer installations is the operating system. An operating system is an integrated set of specialised programs that's used to manage the resources and overall operations of a computer. It permits the computer to supervise its own operations by automatically calling in the applications programs, translating any other special service programs and managing the data needed to produce the output desired by users. The computer systems running under operating systems control today range in size from small personal computers to largest mainframes and super computers. The Mini computers and their operating systems also came along in 1960's and the first personal computers equipped with floppy disc drives were introduced (in 1970s. But most of the pc makers and manufacturers of large systems have developed proprietary operating systems that are indeed for use only with their machines.

The operating systems of 1950s and early 1960s in developed countries reduced the idle time by allowing job to be stacked up in a waiting lines.

System Scheduling

Whenever possible, multiple jobs are scheduled to balance Input/Output and processing requirement. This often involves overlapping Input/Output and processing operations. The various channels as shown in figure are used to facilitate overlapped proessing. A channel consists of hardware, monitoring and connecting elements, controls and path for the movement of data between the Input/Output put devices and high speed CPO. A channel can be a physical part of the CPO which is accessible to both Input/Output devices and other CPU elements. Once a channel has received appropriate instruction signals from the Operating Systems it can independently while the CPU is engaged in performing computations.

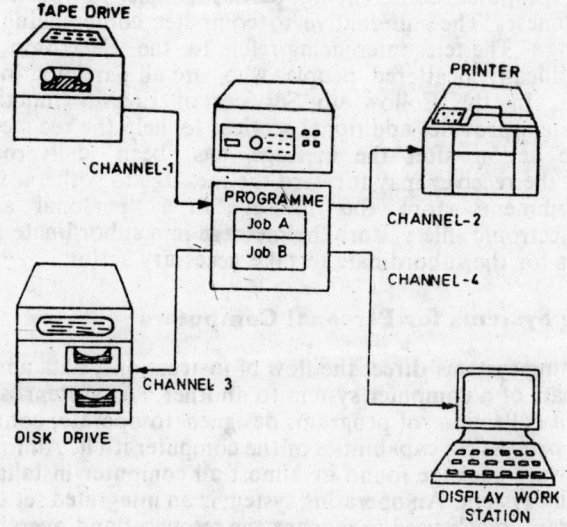

Fig. 42.4.   Channels Control the Execution of I/O Instructions.

In addition to channels small high speed storage elements called buffers also play an important role in overlapping input, processing and output operations. Data from input devices are fed under channel control into an input buffer. This input buffer has an important characteristic of accepting data at slow input speeds and release them at electronic CPU speeds. The output buffer accepts data from the CPU at electronic speeds and releases them at the slower operating speeds of output devices.

Multiprogramming

Multiprogramming is the interleaved execution of two or more different and independent programes by the same computer. In multiprogramming there are a number of programmes available to the CPU and that a portion of one is executed, then a portion of another and so on. and the operating systems switches control from one programme to another almost instantly.

The CPU can thus keep busy while channels and buffers are occupied with the job of bringing in data and writing out information.

Multiprocessing

It is used to describe a processing approch in which two or more independent CPU's are linked together in a co-ordinated system. In multiprocessing the CPU's may simultaneously execute different instructions from the same programme. However it is the job of the operating systems to schedule and balance the input, output and processing capabilities of these systems. The large CPU's in a multiprocessing system may have separate primary storage sections, they may show a common primary storage unit or they may have access to both separate and common memories. A common Operating System may control all or part of the operations of each CPU. Each CPU may be dedicated to specific types of applications. It is also possible for one CPU to be able to take over the workload of another malfunctioning machine untill repairs are made.

Data Base Management Systems

Data base is a collection of logically related data elements that may be structured in various ways to meet the multiple processing and retrieval needs of organisations and individuals. The data bases are commonly recorded on magnetizeable media and computer programs are required to perform the necessary storage and retrieval operations.

Data Processing activities in organisations have traditionally been grouped by departments and by applications. A Data Base Management System can organise, process and present selected data elements from the data base. The capability enables decision makers to search, probe and query data base contents in order to extract answers to non-recurring and unplanned questions that are not available in regular reports. The DBMS will manage the stored data items and assemble the needed items from the common data base in response to the queries of those who are not programmers. The availability of a DBMS, however offers users a much faster alternative communications path as given under in Fig. 50·5. Similarly Personal Computer Operators can use a DBMS package to create and maintain files and records. The DBMS is able to integrate data elements from several files to answer specific user in quiries for information. In other words DBMS is able to access and retrieve data from non-key record fields.

To summarise a data base system is designed around an integrated data file that emphasised the independence of data and programmes. DBMS package can organise, present the process selected data elements. from the data base in response to queries

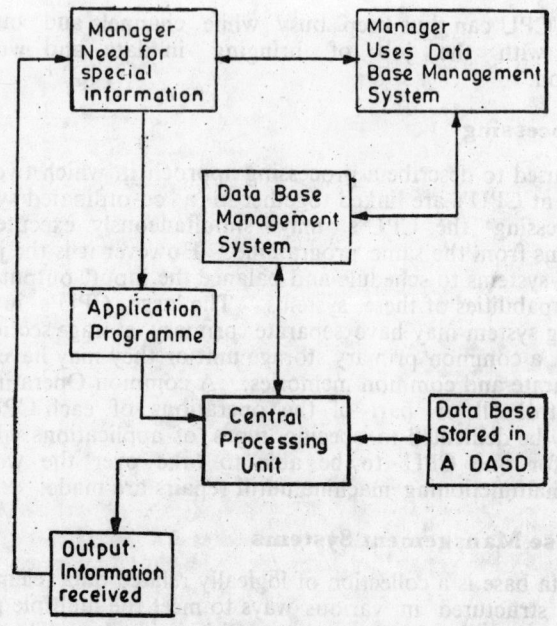

Fig. 42·5. Data Base Management System.

from users. A DBMS employ list, hierarchical network or relational structuring techniques to perform its functions.

Potential Dangers from Computers

There are tremendous benefits to an organisation and its people using computers. The people who initiate changes in computer-based information systems often move forword with the belief that such changes. will increase employee productivity and improve organisation's efficiency. But at times they fail to appreciate the effects their decisions have on social groups, on human skills and feelings and on existing jobs. Some of the problems are : Employment Problems Resistance to change ; Displacement and Unemployment. The reduction in social satisfaction ; The reduction in self-esteem and reputation. Some of the problems have been discussed as under.

Resistance to change. People resist change in different ways. At one extreme, they may temporarily feel threatened by a change, but after a brief adjustment period they resume their previous behavior. Between these extremes are a number of other symptoms including with holding facts, providing inaccurate data and displaying an attitude of in difference. Resistance is often rather than the exception because the changes may appear to others to be a threat—a threat that may take one or more of following paths.

(i) *The threat to job security.* Computers have a reputation for replacing people. Thus there is an understandable fear of loss of employment and/or of reduction in salary.

(ii) *The reduction in social satisfaction* The introduction of computer system often calls for re-organisation of departments and work groups.

(iii) *The reduction in self-esteem.* Many department managers oppose a change because to admit that the change is needed may imply that they have tolerated in efficiency which in their point of view may effect their reputations.

In order to increase the possibility of change following points need to be considered.

(i) *Keep People Informed.* Information relating to the effects the changes on their job.

(ii) *Seek Employee Participation.* The participation of know-ledgeable people yields valuable information. Participation also helps these people satisfy ego and salf fulfillment needs. The computer systems are now performing vital functions in society. In the present time the advancement in scientific and engineering knowledge makes it increasingly difficult for scientists and engineers to keep abreast of their fields. Engineers and scientists therefore must have the ability and willingness to learn about computing, adopt new techniques to retain marketable skills. The possible suffering and anxiety associated with the conviction that technical obsolescence is likely in a relatively short-time is thus something that certain professionals may have to learn to live with. We should not forget that humans still remain for superior to computers in this area of intellactual work. The potential of an alliance between humans and computers can not be restricted.

Future Trends

In future small and medium scale computer systems will gain greater precedence than large and costlier machines. The future buyers of computers will increasingly ask for integrated systems, common standards in hardware and software and high level of service and support. The old and large-size computer makers are growing at less than 15 per cent per annum, against over 50 per cent for medium and small computer manufacturers. It is estimated during one of the studies conducted in U.S.A. (1988) that hardware market would grow between 6 to 9% with PCs jumping to 12 to 16 per cent during 1988. The Minis are expected to grow at 4 to 8%.

and main frames growth will be zero to 4%. The present share of computers manufactured as percentage of India's GNP has been a small 0˙13% during the year 1988 and is expected to be 1% by 2000 AD.

43

Quantitative Methods in Management

Introduction

The growing complexity of problems of management calls for development of sound methods and techniques for their solution. Most of these techniques and approaches owe their origin to military or industrial spheres of activity. Some of these are quantitative in character while others are qualitative. Neither of these two sets of approaches is adequate itself for optimum solutions to all the problems we face. There are problem situations which warrant use of qualitative techniques alone, while others deserve a highly quantitative approach. Most real-life problems of management are such as would readily conform to specifications of a standard technique. However, most problems in a real life require a more detailed knowledge of the methods. The decision maker may rely on professional help for this purpose, now that he knows what types of a consultant to call upon or he may like one of his deputies to acquire additional knowledge or skills by undergoing a training programme, or though sel-study. What is most important in this context is that the problem is correctly identified, or the decision making situation is rightly understood.

Application Of Management Techniques

Since most of the concepts and tools of management sciences owe their origin to industry or military situations. We know that all systems operate within certain constraints and should seek achievement within the limits, set by constraints. The industry aims at profit maximisation while the service sector struggles to maximise the benefits to the people. Both the sectors are dealing with five M's— Man ; Machines, Materials ; Money ; Minutes, (time).

Quantitative Methods in Management

The methods and approaches for handling these in Industrial Sector and Service Sector can not be different. Some of the quantitative Techniques are discussed as under !

Operations Research (OR) : OR is a body of techniques and approched heavily relying on mathematical and statistical applications which enable a decision maker to choose among a number of alternatives. In real life situations the alternative course of action may not at all be clear. It is often a task by itself to indentify the correct alternatives. Once the alternatives are identified the OR approach seeks to determine a measure of effectiveness by which to weigh the alternatives. The decision maker should select the alternatives which maximises the measure of effectiveness. This task is not as simple as it may appear to be. It is often very difficult to agree on a measure of effectiveness, or having agreed on it conceptually, to measure it using available data. The decision making situation itself may be quite complicated in the sense that it may be decision in a wider frame work comprising a large number of interrelated decision situations. The OR approach attempts to develop a model—usually a mathematical one—to describe the decision making problem. In trying to formulate the model the decision maker makes some assumptions some of which may not hold good in real life. And, if the problem fits into any of the available models he just chooses the appropriate techniques for solviny the problem. The major sub-fields of OR representing its various techniques are broadly the following :

The Technique	*What it does*
Optimisation Techniques Linear Programming (LP), Non-linear Programming, Dynamic Programming.	These techniques help one to identify the optimum courses of action which a decision maker has to select. These are applicable to problems of resource allocation, transportation, personnel assignment to jobs, and a multitude of other problems.
Inventory Control Models	These models are useful in deciding on ordering points and ordering quantities for materials stocked.
Project Management Techniques—Network Analysis (PERT/CPM)	These techniques help the programme manager to plan a project in minutes details in terms of time and resources required for each activity. These enable the programme manager to monitor the implementation of the project.
Waiting Line Models	These models help planning of facilities which require waiting for service according to certain rules governing arrival times, service times and the order of rendering services.

The Technique	What it does
Replacemant Models	These help deciding on time to replace equipment so that an overall measure of effectiveness is maximised.
Sequencing Models	These models enable scheduling of a few jobs on a few machines, so that a measure of effectiveness is maximised subject to technological constraints.
Simulation Models	Certain decision situations are too complicated to be modelled adequately. Simulation techniques help seeking the solution in such contexts.

Apart from applications of techniques which can be broadly classified as OR, industrial engineering/productivity techniques such as Method Study, Work Measurement and Utilisation study, statistical tools, survey approach. Management Information systems, monitoring and evaluation systems, etc. are often good aids to management. Cost Benefit Analysis helps selection of a strategy of operation for management and this requires quantification of a series of benefits some of which are qualitative in character. It is possible to introduce a greater degree of quantification and modelling approach into financial management than is normally attempted, with a view to obtaining greater returns to investment of money. Cost-effectiveness studies are useful in making assessment of cost performance in comparable projects. Mathematical modelling of observed phenomena leads to clearer understanding of observed phenomena for developing suitable control mechanisms. Use of statistical techniques in resolving problems of inference based on data is of great utility. Much of the terminal evaluation of projects depends on sample surveys which help quantifying results of activities. This is applicable equally to marketing of industrial products, consumer goods or social services such as family planning.

Role of Operations Research

The management processess essentially involves decision-making relative to finance, production, marketing, distribution etc. The present-day science of management helps in the decision-making task by developing and recommending logical basis for choosing from among available operational alternatives. The decisions can, thus, be made on certain accepted quantitative decision rules instead of an institution, experience and subjective judgement alone. Operations research was first applied to military decision-making during world war II, in U.K. Operation research groups, which included mathematicians, engineers, phycologists, physicists and others, were constituted and attached to operational commands at various levels to recommend policies that would improve military offectiveness.

After the war, O.R. found its application in the field of industry and was directed to dead with problems relating to the product development and marketing. In this context O.R. may be considered as an application of scientific methods to problems arising from operations involving integrated systems of men, machines and materials. Operational Research is the application of scientific method to management decision making. It is being used to tackle a wide variety of problems ranging from space exploration to the control of supermarket stocks. Operational Research is concerned with problems which exists in the real words of business, Government, the Armed Forces, etc. One of the features if this reality is the existence of uncertainty. The break-down of machines, the receipt of customers orders at a factory and the arrival of ships at a port are all, to some extent, unpredictable. A company cannot forecast its sales exactly as there is always an element of uncestainty. In dealing with such problems, it is just as essential to obtain a measure of the uncertainty as it evaluate the expected outcomes.

Operations Management

Operations Management can be defined as the management of systems for the provision of goods or services. It is concerned with the design and operation of systems for the manufacture, transport, supply or service. Since the nature of certain of the problems which face operations management is influenced by system structure, the nature or role of operations management is in part influenced by the structure of the operating system. Similarly the role of operations management is influenced by the objectives which are adopted by or prescribed for operations management. The problem characteristics of the system and objectives together necessitate the use of particular operation management strategies as given in following line diagram. The above diagram outlines the relationships between system structure problems, objectives and strategies and the role of operations management. As indicated earlier the two factors which influence the nature of operations management are :

(i) Objectives of operating systems and
(ii) Problem areas of operation management.

Operations Management Objectives

The objective of operating systems is the conversion of inputs for the satisfaction of customer wants. Customer satisfaction is therefore a key objective of operations management. The need for operations management arise from the fact that operating

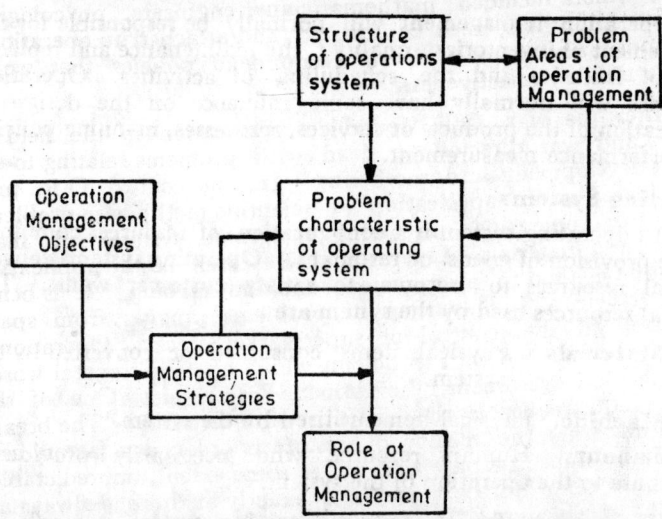

system must satisfy multiple objectives customer service must be provided simultaneously with the achievement of efficient operation. *i.e.*, effective or efficient utilisation of resources. Either inefficient use of resources or in adequate customer service is sufficient to give rise to the commercial failure of in operating system. Operations management is concerned essentially with the utilisation of resources *i.e.* obtaining maximum effect from resources or minimising their loss, under utilisation or waste. The operations management is concerned with the achievement of both satisfactory customer service and resource productivity. However an improvement in one will often give rise to determination in the other. Often both cannot be maximised, hence a satifactory performance must be achieved to both and sub-optimisation must be avoided. All the activities of operation management must be tackled with these twin objectives in mind. The scope of operations management can be adequately indicated by the following fields of activity :

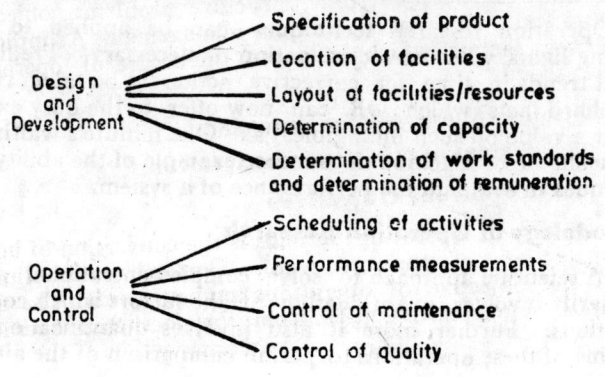

Operation management will normally be responsible for the management of inventories, quality ; the maintenance and replacement of facilities and the scheduling of activities. Operation managers will normally have some influence on the design or specification of the product or services, processes, manning policies and performance measurement.

Operating Systems

An operating system is a configuration of resources combined for the provision of goods or resources. Operating systems convert physical resources to outputs to satisfy customer wants. The physical resources used by the system are :

Materials. Physical items consumed or converted by the system.

Machine. Physical items utilised by the system.

Labour. Human resource who necessarily provide of contribute to the operation of the system.

The primary function often operating system is a reflection of the purpose it serves for its customer that is the utility of its output to the customer. The various functions of operating systems are categorised as follows in Table I.

Principal function	Common characteristics	Type of organisations
Manufacture	Change in form utility	Builder Coal Mines
Transport	Change in location or place utility	Vehicle or Train Services
Supply	Change in the possession utility of resources	Retail shop Petrol Station
Service	Change in state utility of resources	Dentist Hotels

Operation research techniques can be applied to operate 'warning lights' when corrective action is necessary. Prediction of market trends in time for corrective action to be taken is typical of the hard facts which OR can now offer to the busy executive. To put a value on such intangibles as 'One minute's waiting time' of a mechanic or machine is another example of the ability of OR techniques to evaluate the performance of a system.

Methodology of Operation Research

A scientific approach to solve complex operational problems necessarily involves quantification of the factors which control the operations. Further more it also involves quantification of the outcome of these operations to permit comparison of the alternative courses open to achieve the objective. The course adopted by these scientists, working in groups of three or four (called operational Research Groups) can be summarised in the following steps :

(i) Formulation of the problem

(ii) Framing a mathematical model to represent the system or problem under the study

(iii) Generate alternatives and find an optimal course of action

(iv) Test for the optimality

(v) Identify the perameters that control the system

(vi) Impliment of the section found.

Mathematical Models

The most important step in operation and research is constructing the mathematical models of the operations under study. Models are analogies and abstrations and all abstractions and analogies in mathematical model have their limitations. But these models help us to manipulate the various quantifiable factors involved and study and understand the significance of the various factors in relation to the outcome. In any group of operations, which constitutes a sub-system, there are certain factors which can be controlled. An executive incharge of these operations can control these factors within certain limits. The outcome, however, depends on certain factors which are not under the control of the executive. The object of operations research is to understand how the uncontrollable factors affect the outcome and how to assign value to the controllable factors so as to optimize the outcome. Since the outcome depends on certain uncontrollable factors all that an operations researcher can do is to work out the risk involved in adopting a certain course of action and then work out that particular course of action which minimize the risk. In this he uses the well-established theory of probability and statistics.

Areas of Study

One of the advantages which mathematical modelling has led to is the recognition of similarity of underlying structures of many problems in different fields. The tools which are mostly used in allocation problems are linear programming, transportation and assignment models. The possible areas of study of Operations Research are :

(i) *Allocation*. Allocation problems involve the allocation of available resources to jobs that need to be done. They occur when the available resources are not sufficient to allow each jobs to be carried out in the most efficient manner. Hence the objective to allot the resources to jobs in such a way as to either minimize the total cost or maximize the total return.

    (*ii*) *Inventory Problems.* An inventory consists of usable but idle resources. Inventory problems we have to find an answer to the two basic questions *viz.* :

    (*a*) How much to buy ?

    (*b*) When to buy ?

    These two questions arise because of certain costs associated. These costs are the cost of ordering, the cost of holding and the cost of shortage. During any planning horizon we have to minimize the sum total of these costs.

    (*iii*) *Replacement Problems.* Almost all industrial equipment deteriorates with age or usage, unless action is taken to maintain it. At a certain stage it may be more economical to replace the equipment rather than to maintain it. It often happens that items are replaced not because they no longer perform to their designed standards, but because more modern equipment performs to higher standards. Hence a decision has to be made about replacement and maintenance.

    (*iv*) *Queueing Problems.* Problems of congestion are common. Very often we have to tolerate such situations. However, occasions arise when undue congestion has to be eliminated. Basically the problems arises when units arrive at a servicing facility. If the facility is unable to cope with the number of units arriving the units have to wait in a queue. If we can attach a cost to the time spent by a unit in the queue and the cost of providing service at a given rate the problem to decide is the size of the service facility in order to minimize the total cost. The types of problems that can be studied are :

    (*a*) How many counters to provide in a ship ?

    (*b*) How many runways to provide at an air port ?

    (*c*) How many berths for ships at a port ?

    (*d*) How many maintenance crews in a factory ?

    (*e*) How many doctors in a clinic ?

    (*f*) How many doctors in a hospital ?

    (*g*) How to schedule trains, air lines flights etc.

    (*h*) What should be the size of a telephone exchange ?

    (*v*) *Sequencing and Co-ordination Problems.* In these problems the object is to determine the sequence in which certain jobs have to be done and the way in which the performance of these jobs can be co-ordinated in order to accomplish a project of which the jobs are its components. There are essentially two basic techniques, *viz* PERT and CPM, which have now become standard tools in accomplishing all projects.

    (*vi*) *Competitive Problems.* All modern business enterprises operate in a competitive environment. The object of each of the competitions is to minimize his profit (or minimize his loss) without being aware of what his competitor is likely to do. Certain types of games of skill provide elements which are typical of competitive situations. The Mathematical Theory of Games attempts to explain what should constitute the rational behaviour of the individual competitors in a given situation. The theory has been used in :

(a) Advertising campaigns ;

(b) Allocation of search effort in locating minerals and

(c) In military operations.

(vii) *Search Problems.* In these problems acquisition of information is a part. An audit is a search problem. The design of inspection and quality control procedures are search problems. Information storage and retrival are also search problems. The object is to gain the maximum information at a minimum cost.

In an operation research study we encounter several of the above prototype problems. An allocation problem is interlinked with an inventory control problem which in turn is interlinked with production schedulir g problem.

Definition of Operation Research

There are various definitions of operations research. The earliest definition by Morse and Kimball is :

"Operations Research is a scientific method of providing operations quantitatives basis for executives to take decisions on the under their control".

Miller and Starr defined "OR is applied decimal theory. Operation research uses an scientific, mathematical, or logical means to attempt to cope with the problems that confront the executive when he tries to achieve the objective going rationality in dealing with his decision problems".

The operations Research Society of America defined "Operations research is an experimental and applied science devoted to observing, understanding and predicting to behaviour of purposeful man-machine systems, and operation research workers are actively engaged in applying this knowledge to practical problems in Business, Government and Society."

The common ideas emerged out of all such definitions are

—the use of the scientific method

—study of the complex relationships

—provision of a basis for decision making

By combining the above three main attributes it can be defined as "operations research utilises the planned approach (scientific method) and an inter disciplinary term in order to represent complex functional relationships are mathematical models for the purpose of providing a quantitative basis for decision making and uncovering new problems for quantitative analysis".

Types of Models

A model is defined as a representation of an actual situation. It exhibits relationship between cause and effect. The purpose of developing a model is to find out the variables that affect a given situation significantly so that an executive can exercise control on such variables for the desired level of the result.

Basically there are three types of models *viz* physical models, Analogies models and symbolic models.

S. No.	Type of models	Main features	Examples
(i)	Physical models	1. Describes events at a particular time 2. It cannot accommodate modifications so easily 3. More than three dimensions can not be represented.	Photograph Blue print maps, three dimension models
(ii)	Analogies models	1. It represents Dynamic situation 2. It represents quantitative relationships 3. It can be used to study the effect of certain changes in the parameters.	Frequency distribution curves, Demand curves, Flow charts
(iii)	Symbolic models (Mathematical models)	It is superior to any of the above models and a ful representation of a real situation.	Mathematical equation with rotations or symbols.

The mathematical models can further be classified and are of concern for operations research.

—Quantitative and Quantitative model.
--Tailor made and Standard models
—Probabilistic and Deterministic models
—Descriptive and optimizing models
—Static and Dynamic models
—Simulation and non-simulation models

The application of mathematical models has been discussed further under the heading Operations Research Approach to problem solving.

Operations Researce and Systems Theory

The term operations research and systems theory have often been used interchangeably although many people suspect that there is a distinction between the two. Traditionally, systems theory has been closely related to hardware systems, while OR is identified with the software. Operation research tends to be concerned with problems which can be represented by mathematical models, which in turn can be analytically optimised. Systems theory, on the other hand, although formal in nature, is concerned with problems of greater complexity, and is more global and abstract in its approach.

A system is any cohesive collection of items that are dynamically related. Every system does something, and what it does can be regarded as its purpose. Systems theory is primarily concerned with the discovery of the mechanisms by which such purpose is achieved and by which equilibrium or self regulation is maintained.

Application of Linear Programming

According to Robert W. Metzger following are the applications of Linear Programming :

(i) **Product Allocation :** With a number of jobs that can run on a number of different machines, it is possible to determine how best to allocate the work to the machines so as to minimise either the total time or total cost to produce the entire work load.

(ii) **Distribution and Shipping :** With the supply of products at several warehouses, it is possible to determine which ware house should ship how much product to which customer so that the total distribution costs are at minimum.

(iii) **Market Research :** It is possible to determine the best of several possible warehouse or factory or outlet locations, from various facts about each location.

(iv) **Job and Salary Evaluation :** The mathematical programming is used in place of multi correlation analysis to determine the relative weights of the factors considered.

(v) **Blending :** It is applied to blending oils, gasolines, alloy elements, etc. It is possible to determine how to blend available ingredients or what ingredients to obtain to meet at minimum cost, a specific end, product demand.

(vi) **Material Handling :** It is a new area of application. This presents an approach that can increase hand or non automated material handling utilisation upward to 80%.

(vii) **Production Planning :** It is possible to develop the lowest cost producing plan starting from forecast, available plant capacity and the tangible cost factors.

Purpose of Operations Research

The name 'Operational Research' was coined to cover the research activity undertaken in the operational areas of armed forces. The purpose of operational research is the utilisation of modern

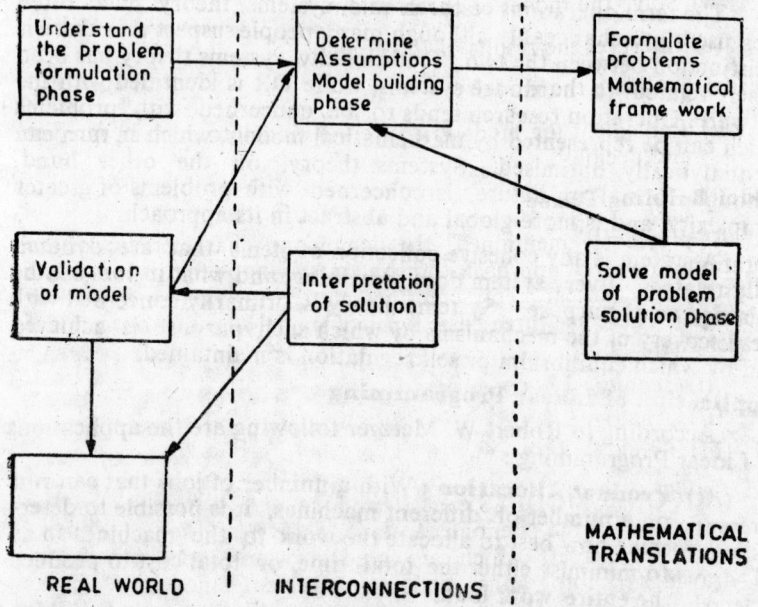

Fig. 43.1.   Major stages in the Operational Research Approach to Problem Solving.

science on complx problem arising in the direction and management of men, machines, materials and money in industry, business, government and defence. After the war some of the techniques that had been applied to tHe complex problems of war were successfully transferred and used in the inJustrialized sector of business.   The object of operation research is to determine their strategic and tactical decisions more scientifically.

Operations Research Approach to Problem Solving

The demoniant characteristic of OR is that it constructs and uses mathematical representations-models-to analyse and study problems. This distinctive approach is an adaptation of the scientific method used by the other physical sciences. The scientific method translates a real problem from its initial context into a mathematical representation of the problem.   This mathematical representation is solved and then translated back into the original context. The OR approach to problem solving can be considered to contain the following seven stages :

(*i*) Understand and formulate the problem.

(*ii*) Determine assumptions—model building phase.

(*iii*) Formulate the problem in a mathematical frame work.

(*iv*) Solve the model of the problem.

(*v*) Interpret the results of the model.

(*vi*) Validate the model.

(*vii*) Implement the model, *i.e.* use the model to explain, predict and decide.

Model Building Phase

As previously mentioned, the demoniant characteristic of the operational research approach to problem-solving is that it constructs and uses mathematical representations-models-to analyse and study problems. There are three basic types of model used in operational research, namely :

(*i*) Iconic.

(*ii*) Analogue.

(*iii*) Symbolic.

(*i*) ICONIC MODELS can be either smaller or larger physical representations of the phenomena under analysis. Common examples are model aircraft which are used in wind-tunnel experiment and drawings, photographs and reproduction models such as planetariums. Typically, iconic models are built with a specific purpose in mind and are not suitable for manipulation or experimentation.

(*ii*) ANALOGUE MODELS use one set of properties to represent another set of properties. For example, an economic system can be modelled by the use of a hydraulic system, where the flows represent the dynamic nature of the real system. Analogue models are preferred to iconic models because they permit a degree of manipulation and experimentation.

Symbolic models or abstract models, as they are sometimes termed, make use of mathematical and statical notation to represent the problem under analysis. For example, the time required to undertake the journey from home to the office could be simply modelled as :

$$T = a + b + c$$

where a = remove car from garge + initial waiting time if any

b = the actual journey time from home to office car park.

c = time to walk from car park to office.

Symbolic models take the form of mathematical relationships which reflects the problem structure. While it is not unusual for all three model types to be used in sequence, symbolic models are generally used wherever possible. Their main advantage is the case with which they can be manipulated and the fact that they yield more accurate results.

Advantage of using models

Some of the major advantages to be gained from using models are :

(a) They assist in the understanding of the problem.

(b) Models can be more easily manipulated and analysed than the real system.

(c) They provide an indication of the information flow-and the gaps therein.

(d) Model building is an aid to decision-making-a tool to help managers to decide what questions to ask themselves.

The additional requirement could be met by structuring the problem using the Decision Tree. Decision Trees allow for an overall picture to be presented of both the problems and the outcomes associated with the various alternatives. The measurement of risk could be included by expressing the various outcomes in the trees in terms of probabilities. The speed at which the model building process can proceed will invariably depend upon the relationship that has developed between the model builder and the manager.

Interpretation of results and model validation

The edge or trim waste problem, which we have detailed, was initially modelled in practice, by the use of a linked. Linear and Dynamic programme. The major variables, some of which are formulated into an objective function which gave greatest weight to the material costs ; in effect an optimization model. At the model validation stage, however, the following emerged.

The important factor to be remembered is that before any solution is implemented the model must be thoroughly tested to ensure that it is a correct description of the problem. As can be seen from the above, significant changes were required to the model before moving onto the final stage of implementation.

Implementation

Although the techniques and model patterns utilized by operational research can be easily detailed in mathematical terms, such statements take almost no account of the human aspects involved such as :

The resistance to change.

The fear of redundancy.

The desire to be consulted and informed.

Problems related to attitudes and motivation.

For example, the implementation of the trim model, in a furniture company, affects a wide range of people and physical systems, which require skills in handling complex qualitative data and interpersonal issues. While all personnel may publicly ascribe to the importance of getting the wastage figures down, for each it will raise quite different fears. For the production manager there is the prospect of more complex scheduling : for the mill manager there will be problems associated with the sorting and storage of the panels ; for the sawyers there will be more difficult cutting patterns, and perhaps less opportunity to earn a good bonus. Tackling these issues then is just as important the effort and ingenuity given to the previous stages in model development. Certainly, a solution that achieves 60-70 per cent of the theoretical maximum benefit and which can be implemented, is preferred to one which could achieve 95 per cent but which cannot be implemented.

A temptation in model construction is to identify the data and structure the variables so that they fit neatly into a specific operational research problem type. An explanation of this phenomena may be found in the fact that the majority of problems are simplified so that they can be modelled from a known set of problem types.

Operational Research Some Inter Connections

Statistics

Many people regard the world as being something fixed and unchanging, or at least something which changes in a regular, ordered way. Any kind of behaviour which upsets this regular, ordered pattern is seen to be undesirable and something which is to be avoided. However, in reality life is full of change, of deviations from the preferred stable equilibrium. Man's effort to understand, control and impose order on such phenomena has led to the study of variability—the science of statistics. The study of variability helps us to understand the variation inherent within our reality, as the all important first step of understanding, controlling and imposing our will on it.

Statistical Data

Data are the essential building blocks of statistics. Statistical data can be classified as primary and secondary. Primary data are the result of the direct collection of raw material, while secondary data consist of material which has been collated from other sources of information.

Collection of Data

There are numerous ways of collecting data for statistical enquiries :

(i) *Primary data.*

(a) Direct personal observation due to the high cost of collection, this method is restricted to use in laboratories and in scientific experimentation.

(b) Personal interview approach. This method is often used by market research companies to ascertain specific product information. (See Marketing in this series for additional details on data collection).

(c) By forms and questionnaires (which are sent out through the post). This approach, while being less costly than (a) above, has the disadvantage that a high percentage of the targeted population often fail to complete and return the forms.

(ii) *Secondary data.* This type of data is collected from information previously collected by other e.g. :

(a) Government departments.

(b) Royal commissions.

(c) Trade associations and research bodies.

(d) Technical and trade journals.

Statistical Measurement

As can be seen from the previous paragraphs on statistical data, statistics implies counting or measuring. In certain cases, direct counting may be possible and is used for objects which can easily be separated, e.g. houses, men, cattle, etc. In other cases the measurement must be made by reference to some acceptable unit, e.g. in the Gulf Oil States, production of oil is not counted in litres but is measured in barrels. Statistical units of measurement are derived from a variety of sources and, for this reason, some caution has to be exercised in how they are applied. The method of measurement adopted depends upon the exact nature and purpose of the investigation. However, the following precautions should be observed :

(a) The unit of measurement must be definite and specific. Any special meaning implied by scientific, trade or customary usage should be identified before the classification of data takes place.

(b) Homogeneity and uniformity. A given choice of unit is unsatisfactory if it implies different properties or characteristics on different occasions.

(c) The unit of measurement must be stable. For example, the fluctuations in rupees make it an unsuitable statistical measure of value. Hence, the practice has arisen of converting current £ sterling values into £ sterling values of the current year by means of suitable coefficients.

Statistical Variates

It can be seen from the above that any set of facts may form the subject of statistical measurement, provided that the individual facts can be specified and identified. Basically this requires :

(a) the grading or classifying of the various kinds of facts according to their relevant characteristics.

(b) the counting or measuring of the items contained in each class or grade.

The name given to the set of characteristics in (a) above is the variate, whereas the counting or measuring is commonly referred to as the frequency or measurement. The variate—the collection of facts may be either quantitative on qualitative. A variate is said to be quantitative when it can be measured numerically, e.g., age, height, income, and qualitative when it can only be described verbally, e.g. hair colour, geographical location. In addition, if the variate is quantitative then it may either be continuous or discrete. A variate is said to be continuous when it passes from one value to the next by infinitely small steps (e.g. height) and discrete when there are gaps between one value and the next (e.g. children per family). This distinction between quantitative and qualitative variates is statistically important, since the former lend themselves to mathematical treatment, whereas the latter do not.

Classification

In the process of statistical enquiry, it is necessary to characterize or classify the subject matter as a preliminary to enumeration. Classification is the process of arranging things in groups or classes according to their resemblances and affinities. For example, coins, bank notes and cheques are different varieties of monetary instruments with different characteristics, but since they share the ability to settle debts, they can all be classified under the one heading 'currency'. Classification, by distributing a definite population, or sample from that population, among the groups/ classes so that nothing is left over, sets an enquiry on a numerical basis and more importantly, assigns to each class its proper importance in relationship to the whole. In addition, classification also facilities the following :

(a) It displays points of similarity and dissimilarity.

(b) It reduces mental effort by the systematic suppression of irrelevant detail.

(c) It suggests bases for comparison and inferences.

Bases of Classification

Variates have already been distinguished as either quantitative or qualitative. However, in discussing bases of classification, it is necessary to improve upon that distinction. Logically speaking, classification may be performed on any of the following four distinct bases.

(i) *Quantitative. i.e.* the basis of distinction rests upon differences in quantity. For example, the analysis of furniture sales according to differences in volume (m³) or value (£) would be a quantitative classification.

(ii) *Temporal, i.e.* referring to the time at which the subject matter in question occured. For example, the analysis of annual furniture sales by weeks, months or quarters would be a temporal classification.

(iii) *Spatial, i.e.* referring to the distribution of items in space. For example, the classification of annual furniture sales by geographical area would be a spatial classification.

(iv) *Qualitative i.e.* involves the necessity to distinguish between differences in quality or condition. For example, an analysis of furniture sales by product group-kitchen, bedroom, dining room, lounge-involves qualitative distinctions.

Tabulation

Tabulation involves the orderly and symatic presentation of numerical data, in a format designed to and assist the user to understand the problem under analysis. There are two stages in the operation of tabulation, as given below :

(i) The first stage requires that the working sheets are summarized. It is these summary sheets of information which form the subject matter for the actual statistical investigation.

(ii) The second and final stage requires the preparation of the final tables, in as much detail as is necessaay. As an added note, the tables should state whether they are original or derivative. Original tables present information in the same form as it was collected, while derivative table imply some process of manipulation, for example, grouping, totalling or averaging.

Uniform Approach to Tabulation

Although tabulation is a routine process the student must take care to ensure that a uniform approach is adopted throughout. No general rules can be given, but the following suggestions, may be of use.

(a) Too much should not be attempted at once : the data should be analysed stage by stage.

(b) Accuracy and consistency of approach are essential. The data should be checked and cross-checked at each stage before starting the next stage.

(c) Neatness of presentation, proper titles and columns that are not overcrowded make tables more readable and understandable.

Pictorial and Graphical Representation of Data

Statistical data can be represented by the following pictorial or graphical methods :

Frequency table. The first requirement, after completion of data collection, is to reduce the data into suitable groups. A frequency table is formed by simply counting how many occurrences are in each group and producing a table of those frequencies. (see Table 1). These are usually referred to as classes.

Table 1

(a) Frequency of productive outputs

Productive output	Frequency of occurrence	Productive output (units/hour)	Frequency of occurrence
200	1	222	7
202	1	224	7
204	3	226	9
206	2	227	8
208	2	229	10
209	3	231	14
211	2	233	9
213	4	235	6
214	4	236	5
215	3	238	4
217	4	240	2
218	5	243	2
220	8		

(b) Grouped frequency table

Productive output	Number of operatives	Productive output	Number of operatives
200—4	5	225—29	27
205—9	7	230—34	23
210—14	10	235—39	15
215—19	12	240—44	4
220—24	22		

The frequency table or grouped frequency distribution as it is sometimes called is the most common form used when presenting numerical data.

Airthmetic Mean

The airthmetic mean is found by adding the sum of values with in a group.

Simple Arithmetic mean is given by :

$$a = \frac{1}{n} (x_1 + x_2 + \ldots x_n)$$

where x is a variate, x_1, x_2, x_n are the individual values of x

$$a = \sum \frac{x}{n}.$$

Geometric Mean

When the percentage change rather than the actual change is the important factor, it is necessary to calculate the average using the geometric mean. For example, the Real Price Index, which is concerned with the percentage change within a period, is calculated using the geometric mean. Another example is the Share Price Index in the Financial Times.

The geometric mean g, can be expressed as the nth root of the product of the n quantities comprising the group.

Symbolically, $g = \sqrt[n]{x_1 . x_2 \ldots x_n}.$

The Median

The median is that value of the variable which divides the group into two equal parts, one part comparising all values greater than the median and the other all values less than the median. The task of finding the median is sometimes extremely simple. All that is required is to arrange the individual items in order of magnitude. The median, which is the central value, can often be located by inspection. For example, in the following series 4, 6, 8, 10, 12, 14, 16, 18 and 20, the median is 12—the central value which has four figures greater than its value and four figures less than its values, *i.e.* four figures on either side of it. When the data are given as ungrouped data, the median may be calculated using :

$$m = \frac{N+1}{2}$$

where N represents the number of items in the series.

Bar Charts

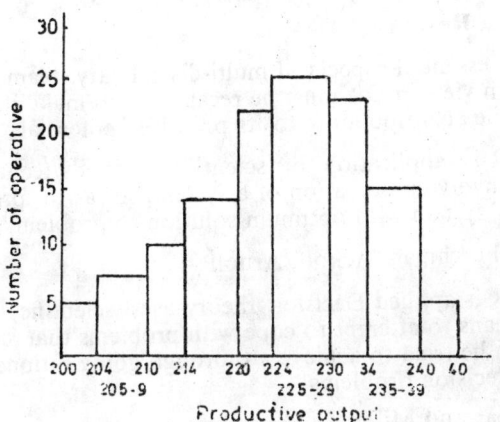

Fig. 43.2. Productive output bar chart.

Ideograph

Productive output 200-4	Productive output 205-9	Productive pouput 210-14	Productive output 215-19	Productive output 200-24

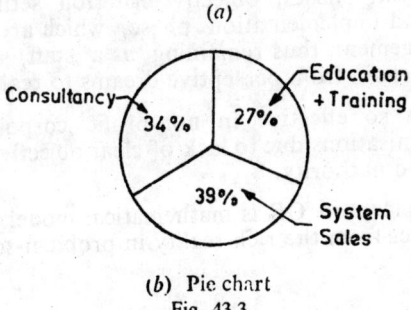

= 5 Operatives

(a)

(b) Pie chart
Fig. 43.3

Concepts, Scope and Relevance in Indian Context

Origin and Development

Multi-disciplinary teams, with fairly clear cut, defensive/ offensive objectives in war, took scientific 'system' approach and recommended plans, which lukcily got implemented resulting in unambiguous improvements in effectiveness and efficiency of system vis-a-vis objectives giving birth to Operations Research.

Operation Research

The essential aspects of multi-disciplinary team and scientific and system view in OR must be recaptured to make its contribution worth while corresponding to its promised potential.

"OR is application of scientific methods/techniques/tools to problem involving operation of a system, so as to provide these in control of system with optimum solution to problem".

—Churchman, Ackoff, Arnoff

"OR is applied Decision theory-uses scientific, mathematical, logical means to attempt to cope with problems that confront executive, when he tried to achieve a thorough going rationality in dealing with his decision problems".

—Star and Miller

"OR is a scientific approach to problem solving for Executive Management".

"OR is application of scientific (quantitative) method to managerial problems-to extend and refine analytical approach to decision making".

However, quality of decision making, and methodologically speaking, 'model making' are central aspects of OR approach, process and reason centre, on which we must deliberate sufficiently in next section.

OR in theory and practice has remained primarily concerned with pre-calculating consequences of alternative courses of action, thus indicating the 'best' through its models, without directly being concerned with vital phases, objective/criterion setting, generation of alternatives and implementation phase, which are left to broader process of management, thus remaining as a staff, advisory aid at one point in the chain from perceptive dreams to reality.

OR is not so effective in non-profit corporations and in government organisations due to lack of clear objectives, and diffused responsibility and authority.

The key concept in OR is mathematical-model building to represent in abstract from the rich reality in problem-solving situation and then solving the models, to predict and control consequences of alternative actions before committing resources to a course of action. Concept of model making is not new-engineers and Scientists have made them extensively-physical, structural and mathematical ones. However, newness in MS models are in terms of complexity, dynamic situations, in extent, magnitude, consequences, inputs, outputs, a simultaneous consideration of objectives and constrainsts in its search for optimality and often only for a feasible and workable solution—satisfying if not optimising—involving varieties of variables-physical organisational, behaviural limitations of all models in simplifying complexity is well known.

Essence of OR is 'Model making' capability, scientific temper, group effort and system view. It is not necessarily a theoretical appreciation of underlying mathematical-logical basis of available standard models except for their academic training value. ,

The most elegant part of QR are the new very well known standard models for evaluation, prediction and optimisation such as: Linear and Non-linear programming, New-work, Dynamic programming, Goal programming, geometric programming, Stockhastic programming, Inventory, Replacement, Queueing, Simulation, Decision trees, Forecasting, System Dynamics Markov analysis. Basis for all these models are : Linear Algebra, Finite differences, Matrices, Vectors, Probability and Statistics, Utility and Information theory, discounted cash flow, economics and accounting, Invariably, solution of all models.

Operations Research in India : Relevance and Potentialities

In India history of OR is nearly as long as it is in most developing countries. In late fifties, we had associations and formal college level courses and beginning of MS function in some of our industry. Professor Mahalonobis association with Indian planning for 1951 and use of quantitative model is well known An aggregate 'thumbnail' picture of achievements and growth in OR will include : Several University level programmes in OR ; limited Ph.D research in OR field ; increasing OR content in all collegiate level. Management Education ; great interest in short term courses, appreciation level in OR for Managers and Public Administration ; Management Services Departments, OR cells, Industrial Engineering Departments, Computer Centres, increasingly coming into being for OR type work in large corporate sectors in business/Industry- Public and Private. in transportation sector—Rail/Airways ; Power generation and distribution Planning and Control Problems in OR application is donated in forecasing, planning and control problems in Operational Production function such as Production Planning and Control Inventory Management replacement investment decisions, Project Management through PERT/CPM. Finance and Marketing functions follow in order of extensiveness of Indian applications.

Future scope for OR/MS in India, viewed in the context of the limited achievement, is infinite in its range and variety. In fact, in our State of Development, the scope for any professional activity is tremendous. All our intellectual and physical faculties. will have to be strained-with a spirit of dedication and service.

Essential Characteristics of Operations Research

The essential characteristics of OR are :

(i) Functional relationships from a systems overview.

(ii) Mised-team approach.

(iii) Planned approach (updated scientific method)

(iv) Uncover new problems.

(i) **Functional Relationships from a Systems Overview.** The first characteristic of operation research means that the activity of any function or part of a firm has some effect on the activity of every other function or part. In order to evaluate any decision or action in an organization, it is necessary to identify all the important interactions and to determine their impact on the whole organization. versus the function originally involved. Initially, the functional relationships in a project are deliberately expanded so that all the significantly interacting functions and their related components are contained in a statement of the problem. Thus, when dealing with an OR problem, the functions and their related components must be examined and selected in light of the significance and the measur-ability of their interactions. A systems overview consists of surveying the entire area under the manager's control instead of one specialized area. This approach provides a basis for initiating inquiries into problems that seem to be affecting performance at a higher or lower level or at the same level. The OR group should analyze the problem with painstaking care, examining all elements in each department affected. These elements might include : the cost of material procurement ; manufacturing set-up, and clean-up costs ; competitive forces and prices ; and the costs of holding inventories and stockout costs. When all factors affecting the system are known, a mathematical model can be started.

(ii) **Mixed Team Approach.** During the early years of operations research, there was a great shortage of scientists (mathematicians, physicists, chemists, engineers and statisticians). The military operations research groups had to be build up their staffs, not by selection but by acquisition. One of the main reasons for the existence of operations research groups is that they bring the latest scientific know-how to bear on the problem. Just as important is their ability to develop new methods, procedures, and systems which are more effective in approaching the problem than any that are presently available. The interdiscipline approach has this added advantage—it recognizes that most business problems have accounting, biological, economic, engineering, mathematical, physical psychological, sociological and statistical aspects. It stand to reason that the individual phases of a problem can best be understood and analyzed by those trained in the appropriate fields.

Planned Approach

Operations research, like most other disciplines, makes use of the scientific method which has been updated to reflect technological advances, such as the computer. OR problems are generally different and sometimes impossible to manipulate and control in their own environment for purposes of experimentation. The operations researcher can change certain variables while holding others constant in an effort to find out how the system would be affected. Hence it is possible to simulate the real world and experiment with it in abstract terms.

The mathematical models may be very difficult to construct and may turn out to be complex mathematical expressions. Business models, ultimately, take form of a system of equations. Underlying these mathematical relationships is an equation which has a relatively simple structure. Some term is equated to some relationship between a group of controlled variables and uncontrolled variables.

The basic approach of the operations research is the scientific method. Its basic steps are : observation, definition of the real problem, development of alternative solutions, selection of optimum solution using experimentation, and verification of optimum solution through implementation.

Uncover New Problems for Study

All interrelated problems uncovered by the OR approach do not have to be solved at the same time. However, each must be solved with consideration for other problems if maximum benefits are to be obtained. It can be said that operations research is not effectively used if it is restricted to one-shot projects.

Based on the interrelationships within a problem, the project may appear to be endless. However, the project can end where the limits of the control exercised by the manager. It should be observed that the dynamics of business today necessitates going back and reviewing projects undertaken some time ago, say two to three years. Subsequent findings may require adjusting previous solutions through the use of revised mathematical models, new input data, and like items. The operations research is a scientific method of providing executive departments with a quantitative basis for decisions regarding the operations under their control. "OR in the most general sense can be characterized as the application of scientific method, techniques, and tools to problems involving the operations of systems so as to provide those in control of the operations with optimum solutions to the problems". Miller and Starr, view operations research as being applied to executive type problems. They define operations research as follows : "OR is applied decision theory. Operations research uses any scientific, mathematical, or logical means to attempt to cope with the problems that confront the executive when he tries to achieve a thoroughgoing rationality in dealing with his decisions problems. Harvey M. Wagner stated : "For convenience, and with reasonable accuracy, you can simply define operations research as a scientific approach to problem-solving for executive management.

Upon examination of the foregoing definitions, certain common ideas seem to emerge : the use of the scientific method, study of complex relationships ; and provision of a basis for decision making. These attributes relate to the essential characteristics of operations research. Operations research, then, can best be defined in terms of its basic characteristics as follows : "Operations Research utilizes the planned approach and an interdisciplinary team in order to represent complex functional relationships as mathematical models for the purpose of providing a quantitative basis for decision making and uncovering new problems for quantitative basis for decision making and uncovering new problems for quantitative analysis".

The OR techniques provide a range of feasible solutions for management. Operations research, then includes more than just developing models for specific problems. Its important contribution is the application of its output for decision making at the lower, middle and top management levels.

Model Defined

One of the basic reasons for developing models is to discover which variables are the important or pertinent ones. The discovery of the pertinent variables is closely associated with the investigation of the relationships that exist among the variables. Quatitative techniques such as statistics and simulation are utilized to investigate the relationships that exist among the many variables in a model.

History of Operations Research

It is difficult to mark the official beginning of operations research. As early as 1914. F.W. Manchester, in England, published. papers on the theorteical relationships between victory and superiority in manpower and firepower. In the United States, Thomas Edison, as early as World War. I was given the task of finding the maneuvers of merchant ships that would be most effective in minimizing shipping losses to enemy submarines.

In the 1930s, Horace C. Leveson applied sophisticated mathematical models to large amounts of data which would otherwise have been totally unmanageable. Starting in 1937 British scientists were increasingly asked to help the military learn how to use the newly developed radar in locating enemy aircraft. Sir Robert Watson-Watt, who claims to have launched the first two OR studies in 1937, recommended that operations research be introduced into the departments of the Secretary of War and the Secretary of Navy in the United States. When World War II terminated, new types of management problems, created by the nationalization of industry and the need to rebuild large segments of the nation's industrial facilities, called for a new approach in Britain. This 'call was answered by the operational research workers who had moved into work on government and industrial problems. The United Steel Companies group has over 100 people, the National Coal Board about 100. The British Iron and Steel Research Associates, British Petroleum, and Richard Thomas and Baldwin all have more than 50 OR personnel.

In the United States, "Operations Research" took on a somewhat different direction. Military research increased at the end of the war, resulting in OR men being retained by the military. In fact, many more were added. Industry and government were subjected to the same stimulation as their counter part in Britain. Initially, industry and government were somewhat in different to operations research. It was not until 1950 that OR began to be taken seriously by American industry.

During this period, linear programming gave industrial operations research a major boost. This technique, basically the application of linear algebra to resource allocation, had applications in many industries. It gave OR personnel a foot in the door of many industrial firms. Many techniques known only to operations researchers, such as PERT and simulation, are used widely today. Probability and statistics, basic to any work in operations research, introduced the notions of confidence limits and probability of occurrence in place of simple averages. Today, operations research is practised and taught extensively not only in the United States and United Kingdom, but also in Europe, Australia, India, Japan and Israel.

Organizing for Operations Research

OR studies can benefit every firm, the only difference being one of size and not of substance. The initiation of OR activity in a firm can be approached in four ways : internally training people from the organization, hiring experienced OR specialists, hiring an external consulting group, and using a combination of company personnel and outside consultants. The cost involved in setting up an operations research department is high, which precludes its use by small firms. This is where outside consulting firms can be extremely useful to small business. The task OR specialists is to provide management with appropriate quantitative methods necessary in decision making.

Internal Operations Research Group

Starting with internal personnel for an operations research group in a medium sized firm, there are a number of important factors that should be considered. The first is that atleast two men should be selected in order to provide a fruitful interaction and exchange of ideas. This is in line with the interdisciplinary approach to operations research. Since operations research makes great use of mathematical modelling, each person should have atleast a degree in mathematics, engineering, or science. Advanced degrees are helpful, but more important is a knowledge of probability, statistics, and mathematics as well as an understanding of the standard OR techniques. In addition, the individual should enjoy working on real world problems that are simple or complex and should be interested in solving problems rather than just applying techniques. The task of operations research is to build models for existing real world problems rather than to search for problems to fit existing techniques. The two men should be able to communicate with all levels in the firm-management and operating personnel alike. The marriage of management and operations research, then, will provide more effective control of firms where complexity is the norm rather than exception. The most effective and efficient way of initiating OR projects is to use the outside services of a consulting firm that has the capability of working with the internal group of two men described previously. Outside guidance may be needed for a period of approximately two years. Quantitative methods of OR are not aimed at replacing or reducing the manager's role in the decision. Rather they are adding a new dimension to managements decisions.

As in most cases, experience is an excellent teacher. The second project always seems to go more smoothly because the team builts knowledge from the lessons learned in its first encounter. Success in initial projects will further management's confidence in the OR group undertaking problems of greater complexity. As times goes on, the OR group will be able to direct the attention of management to important future problems. Not only is the OR team solving problems for management—the team also is assisting management is identifying possible future crises which are, many times, wide in scope and of a long range nature. This can be a major contribution of operations research.

An OR study can raise more questions than it can answer. This can lead to a more penetrating inquiry into the operations of the system and what it is intended to do. The final result is a greater insight and overview of the system with far-reaching benefits and improvements to the firm.

On the other hand, the glowing benefits just set forth can be lost forever if the initial projects are too broad in scope too complex, too slow, or nonexistant in results. Too often OR groups make glib promises of large savings for a small cost. These optimists can be spotted by a simple test. It will take two good, experienced men several months to get familiar with the problem, collect existing data, and perform the necessary steps (except implementation) of the planned approach on the simplest procedures, forms, people, machines, and working out the inevitable 'bugs' that are part of the real world in the implementation phase. The important point is that operations research is basically "research", which takes time to do properly, No one can ask realistically that a project be completed in a few weeks, nor can anyone in operations research promise this.

Mixture of Operations Research Team

As the OR group grows in members, new disciplines are brought to bear on problems. The participants of the team should be chosen according to the kinds of problems that must be analysed. For example, military problems may benefit more from certain disciplines than those of private industry. The firm's team might include personnel trained in mathematics, statistics, a physical science, engineering, economics, sociology, and psychology. Generally, a well balanced OR group would be roughly one-third mathematicians and statisticians, one third physical scientists and engineers, and one-third economists and behavioural scientisis. Courses for the team may be organised within the firm with participation and supervision by operations research consultants.

For most operations research projects, it recommended that the OR Task Force consist of the following :

(*i*) One or more management scientists (operations researchers) --persons with skills and experience in applying operations research techniques to real management decision problems.

(*ii*) One or more computer scientists-individuals whose primary orientation is toward the use of computing power to manipulate data and develop information. (In some cases, the computer scientists and management scientists may be the same persons).

(*iii*) One or more well-rounded, non-technical. personnel-employees who know the organisation better than the computer scientists or management scientists, know where to obtain information, facilitate the cooperation of operating personnel, and finally are better equipped to help the Task Force, "sell" the project solution (s) than anyone else.

(*iv*) One or more representatives from the problem area--personnel who are concerned with the operating functions being studied and who perhaps know the data led exceptions and problems better than anyone else:

The foregoing composition will vary in size based on the complexity of the problem and the time factor involved in the project.

The OR Task Force is responsible to a Management Steering Committee composed of these executives whose functions (departments) are affected by the study. The purpose of this "interpartic ipative" project effort is to allow management the opportunity of providing judgemental inputs and participating in the discussion of difficult problem areas. The Steering Committee, in turn, reports to top management on the project's progress. This apparently unwieldly procedure of committees is beneficial in the sense that it serves to filter the research and help prevent errors. It helps pinpoint all the relationships between processes and departments that must enter into a successful model. The committees also prepare the way for OR personnel, in the sense that they enlist the cooperation of operating personnel and boost their chance of acceptance.

Successful Operations Research Areas

Unfortunately, most managers normally do not think of problems in terms of basic forms or underlying structures, but more in terms of their content—the area of business in which these problems arise. Thus the wide range of problems where operations research has been successfully applied in business and industry will be better understood by discussing problems in terms of their content (functional areas of the firm). Some of these functional OR projects may well involve the business as a whole.

Marketing and Physical Distribution

Operations research is being applied increasingly to marketing. It has been used to determine how much of the total marketing budget should be sent on personal selling, advertising, and sales promotion. In terms of personal selling, OR studies have determined the number of salesman, number of accounts to be assigned for each salesman, and how often they should call on a particular puchasing agent. Similarly, operations research has been used to determine the optimum size of sales territories, the number and distribution of sales offices, and the allocation of sales effort to the different territories. Advertising has been approached in terms of optimizing the frequency of exposures and messages to the particular market segment at the lowest cost. The effect of promotional activity over the firm's many accounts has been studied.

Operations research has helped in the development and introduction of many new products. It has helped in product selection, timing of new products, demand forecasting, and in forecasting the competitive actions of other firms. OR has determined what package sizes, models, colours, and shape the final product should have in order to be successful in the market place. It has determined the guarantee policy and period as well as the service that should be provided.

Purchasing

Operations research has undertaken many studies directed at determining whether to produce assembled parts or to manufacture the items. Factors to be considered are : how often the item is purchased, purchasing prices, fixed and variable costs of manufacturing, present plant loading and delivery time. This is basically a breakeven problem.

OR studies are concerned also with the optimum purchasing of fixed assets in terms of machine models and specific manufacturers. Models have been developed for optimum replacement of equipment after so many years. Studies have been made to determine when new or used equipment should be purchased or whether equipment should be rented.

Manufacturing

Operations research has been used in the selection, of plant sites and sizes (and related warehouses), the optimum manufacturing mixes, and the type and amount of equipment that should be installed. When there is a multiplant operation, OR models have been developed to determine what plants should be shut down, under what circumstances, and how the work load of the closed plants should be reallocated to the remaining operating plants. Quality control systems has been studies by developing a criterion of acceptance for products based upon balancing quality with cost factors. Maintenance in terms of prevention and correction is another fertile area for operations research. Quantitative methods have determined maintenance crew requirements as well as how these requirements relate to the timing of equipment replacement. The effects· of maintenance policies on the smoothness of manufacturing facilities and labour utilization have been determined. Material handling and traffic (internal and external) problems have been solved. The problem of how many machines (varying tapes) has been solved in terms of a single operator to produce the best balance of material flow and operating costs.

Accounting and Finance

Operations research studies have been employed in developing automated data processing and accounting procedures that minimize office costs and at the same time, are conducive to good internal control. Such studies have been used in the development of computer sampling techniques in order to provide greater audit reliability. Sampling has also been used in providing optimal procedures for dealing with claims and complaints. Pilferage problems have been examined and solved.

Cash flow analysis, long-range capital requirements, alternative investments, source of capital, and dividend policies are common finance problems solved. Operations research has been applied to the design of portfolio in order to maintain their value under changing conditions. Credit policies, credit risks, and delinquent account procedures have been studies with promising results.

Personnel

The impact of automation and the feasibility of accelerating automation to reduce manufacturing costs without causing hardships to factory employees is a problem for OR study and solution. Past studies have determined with mixture of age and skills is best under certain conditions. Causes of accidents, problems of labour turnover and absenteeism are examined so they can be minimized for the firm. The methods of operations research have been utilized to recruit [personnel, effectively classify them, and assign jobs where they will have the highest performance. Evaluation of incentive pay, in terms of increasing production, has also been explored and solved in the area of personnel.

Research and Development

Operations research has been used in determining areas where R & D should be concentrated based upon assumed or actual results, in developing duielines for evaluating different designs ; and in determining the life expactancy and reliability that should be designed into a product. With the pressure on profit these days, it is being used more and more to schedule and control the development of R & D projects in order to reduce their time and cost requirements.

Planning

Long-range plans for enlarging the firm through product diversification and mergers are common problems handled by more established OR groups.

In many extremely complex situations, such as R & D projects construction of a new plant, and launching of a new product, the total effort can be broken down in a large number of activities. Some of the activities must wait for others to be completed before they can be started, whereas others can be carried out in parallel. Once the sequence of activities has been logically defined, the critical activities can be determined to spot-light future bottlenecks. Planning methods (PERT) have been devised to control and reduce the time and related costs where possible.

Computer systems, especially those that are capable of realtime responses, have provided the needed thrust for greater application of OR models. In essence, real-time systems that have the ability to feed back information in sufficient time to control their environment have made it possible to bring decision makers and operations researchers into a much more intimate contact for timely decisions.

44

Queuing Theory Analysis

Introduction

Waiting lines or queues are a common feature in everyday life. Waiting lines can be of human beings demanding a service as for example before a railway ticket counter, or of machines as for example cars waiting to be serviced or repaired. The basic ingredients of any waiting line are : arrivals, servers or processors, and waiting lines or queues. Arrivals constitute the input process while the manner in which the inputs are handled constitute the service mechanism. The waiting line or queues depend obviously on the time taken in the service process. It also depends on what is commonly referred to as queue discipline. The queue discipline determines the order in which the items in the queue will be served (first-come-first served or some priority rule, etc.). It can be easily seen that most government offices have all the ingredients outlined above. Papers are received daily in such offices. They are processed by dealing hands, moving vertically up the official hierarchy and finally the applicant is informed of the government's decision thereby ending the process. For all the applicants, there is a waiting time involved from the date of filing of the application and the receipt of government orders on it. Also at any given time in a government office there are papers in various stages of process which can be said to constitute the waiting line or queue.

Waiting Line Process

In any government office work is usually distributed subjectwise among the personnel manning it. Thus a particular subject or compilation is allotted to one individual, who in turn is supervised by a more senior official and so on up the hierarchy. Thus an application or paper pertaining to a particular subject in most government offices travels through the same dealing hands vertically till it reaches the level of decision making. The level official decision making will obviously depend on the nature and extent of the demands contained in an application and the pattern of deligation of powers in the Government department concerned. In a business type situation in mind, waiting line processes are generally categorised into four basic types depending on the nature of the service facilities required. They are (i) single channel, single phase ; (ii) multiple channel, single phase : (iii) single channel, multiple phase ; and (*iv*) multiple channel, multiple phase. The number of channels in a queuing process indicates the number of parallel servers available for handling th inputs while the number of phases indicates the number of sequential stages which each arrival must go through for the service to be completed. It will be seen that paper handling in government offices is best characterised by the single channel multiple phase category mentioned above.

Service and Waiting Costs

In traditional queuing theory analysis as applied to business organisations, the usual practice is to try and reach an optimal level of service keeping in mind (*i*) the service costs and (*ii*) the waiting costs. Service costs are obviously a function of the number of channels provided to handle the inputs. For example, if the Railways decide to open more windows to sell the same class of tickets, they will have to incur more expenditure. Waiting costs for business organisation usually constitute the cost of lost business due to customers becoming impatient while waiting in queues and switching their clientele to rival firms providing the same or similar services. Though the traditional concept of waiting costsis not relevant for our purposes, it would be wrong to assume that there is not cost of waiting involved in the disposal of papers in government offices. Beyond a reasonable point, waiting may involve public criticism and public dissatisfaction. Often, of course, the delay in the disposal of a paper may involve tangible and measurable money costs. For example, a delay in the sanction of a loan, or the allotment of land, or the grant of an electric connection may delay the start of an industrial project, the cost of which consequence could conceivably be measurable. Again the failure to dispose of a stock of perishable commodity in time may cause the government considerable financial loss. Such examples could be multiplied and yet to any one familiar with the normal government functioning it would be evident that such instances can hardly be categorised as remote occurrences. The fact that delay often involves a tangible economic cost is yet to find wide appreciation among government departments. However, the development of such a consciousness is an essential ingredient for a better functioning of such offices.

Queue Discipline

The concept of queue discipline lies at the very heart of the problem of paper disposal in government offices. In traditional queuing theory, models are constructed on the assumption that inputs would be served on a first come first served basis. In government offices while this assumption still holds on a theoretical plane, it breaks down in practice. A characteristic of paper disposal in government offices is the virtual absence of any queue discipline. In actual practice queue jumping takes place in an indiscriminate manner with the result that there is more often than not little correlation between the position of an application on the waiting line and its final disposal in the office. It may be mentioned here that even if there existed perfect queue discipline there would be different waiting times involved for different applications depending among other things upon (a) the complexity of details, (b) scale of demands and (c) level of decision making necessary for disposal. Queue jumping, however, leads to its own set of problems and it would be necessary to analyse the phenomena in greater detail. Queue jumping can be broadly categorised as belonging to two main types (a) legitimate queue jumping and (b) illegitimate queue jumping. Legitimate queue jumping occurs when matters demanding immediate attention are processed out of turn. Which matters are to be given such priority is often determined by the government itself, so that any particular government office has little discretion in this except perhaps pertaining to those issues which it is competent to dispose of within the framework of delegated powers. The usual devices of legitimate queue jumping in any government office can be seen in the form of papers/files marked "Top Priority". 'Immediate', 'Urgent', etc. To the extent such papers get a priority in disposal and processing, others in the waiting line ahead have to wait longer before being disposed of.

Queue Jumping

In a rational system it may be argued that queue jumping if at all it is to occur should bear some relationship to the waiting costs. To elaborate this a little further, a matter where the consequences of delay can be serious should get priority over another where this is not the case. However, it is quite impossible to quantify the waiting costs of large number of papers passing through most government departments. As any government official would know, a starred question of Parliament or a Legislative Assembly is always given top priority in disposal. In a democratic set up no body would even remotely argue that it should be otherwise. However, it would be extremely difficult to quantify the waiting costs in such cases. Such queue jumping is structural since it emanates from the very nature of the political and administrative system itself. However, there can be little quarrel with a principle which asserts that legitimate queue jumping must bear some relationship with a rationally conceived waiting cost.

Since in government a lot of stress is laid on the quantum of disposal as against quality, there is an inbuilt tendency to tackle the least difficult matters first in order to show a large quantity of disposal. This is perhaps one of the lesser appreciated aspects of the well publicised paper disposal drives indulged in by most government departments. The consequences of such a practice can, however, be far-reaching. Important matters which perhaps require greater application of mind and study usually tend to be taken up last unless one of the other imperatives described earlier impinge on the situation making for speedy disposal. To place paper disposal on a more rational basis, it would be necessary to radically curb the practice of illegitimate queue jumping. As would be amply clear, queue jumping is not an undesirable practice so long as it bears some relationship to a rationally conceived waiting cost. This would naturally imply certain fundamental modifications in the existing office procedure.

Setting up a Theoretical Framework

Once we accept the principle that queue is not only necessary, but definitely desirable in certain cases, the next step would be to ask ourselves the question : to what extent such que jumping would be desirable from a theoretical view point ? The basic rationale for queue jumping as we have seen is that it reduces waiting time and hence waiting cost. In matters where waiting costs are high, priority is very much a necessity over other matters involving lower waiting costs. However, such queue jumping can only be carried on upto a point for the obvious reason that for every instance of saving in waiting time and hence waiting cost for the items (papers) overtaken in the queue. Queue jumping will thus be profitable only so long as the saving in waiting cost as a consequence of the queue jumping is greater than the increase in waiting cost of the items overtaken as a result of such queue jumping. There will obviously by a stage where further queue jumping would prove counter productive since saving in waiting cost would be more than offset by the increase in waiting cost of the overtaken items. In Symbolic terms, if Q_1, Q_2, Q_3.....................Q_n denote queue jumping events, and WC_1, WC_2, WC_3........WC_n denote the saving in waiting cost as a consequence of these queue jumping events, and

wc_1, wc_2, wc_3.........wc_n denote the increase in waiting cost of the overtaken items as a consequence of these events.

But a particular queue jumping event Q_1 will be productive only if,

$$WC_1 > wc_i$$

and an optimum situation would be reached when

$$WC_1 = wc_i$$

Any queue jumping beyond this point would not save waiting costs at all in fact it may even prove to be counter productive. Mathematically the total waiting costs at this point would be minimum for a given queue at a given point of time. In other words, the object of the queue jumping is to emphasise the following :

(a) there is a waiting cost involved in respect of every paper pending in an office ;

(b) queue jumping is both necessary and desirable provided it bears a relationship to waiting cost ;

(c) every office should strive to minimise the total waiting costs of the pending papers at any given point of time through the mechanism of legitimate queue jumping.

Queuing Theory

Queuing theory deals with the anylysis of queues and queuing behaviour. A queue forms and grows when the rate of customers arriving is greater than the number of customers being served and leaving the queue. If the rate of arrival is less than the rate of the service, then the queue will be reduced and may even disappear altogether. As and aid to further discussion, a queue can be thought of as a buffer between an input process and an output process as shown in Figure.

Minimize the Total Cost

A queue is a dynamic phenomena. Either the customer is waiting for service or the service facilities are waiting for customers. Clearly if both the arrival and departure of customers to the system is constant, then no queue will be formed. However, given that it costs money to have service facilities waiting for customers and it also costs money, in certain circumstances, to have customers waiting for service, then in practice it is not surprising to discover an out of balance situation between the input and output processes. The aim of queuing theory then is to minimize the total cost of idle facilities and the time lost waiting for them.

Basic Queue Characteristics

All queues display a number of basic characteristics, namely :

(i) The arrival pattern
(ii) The service pattern
(iii) The number of service channels
(iv) The traffic intensity or utilization
(v) The queue discipline.

These characteristics are outlined below.

(i) **The Arrival Pattern.** The customers can arrive at a queue in a variety of different ways. They can arrive in small or large groups, and at regular or irregular, steady or random intervals of time. Hence, the arrival pattern of a queue can exhibit great variability. This variability can be observed and analysed in the following manner. Given that customers are arriving at random, on unit time scale, then arrivals can simply be represented as shown in Figure.

The vertical lines represent the points in time at which there is a customer arrival. If the arrivals are truly random, then by grouping the data into equal unit time intervals a Poisson distribution should be obtained.

From statistical theory it is known that if the arrivals are random, then the interval between arrivals will follow an exponential distribution. The inter-arrival times for the previously collected data are shown diagrammatically in Fig. 44.1 and are also summerized as a frequency table below.

Interval between arrivals (mins)	Frequency
0—9	38
10—19	26
20—29	13
30—39	9
40—49	5
50—59	2
60—69	2
70—79	2
80—89	1

Simple Test for Randomness

A simple test to check that the collected data follows that Poisson distribution, *i.e.* are truly random, is to substitute the mean of one of the distribution into the following equation :

$$\text{Mean of exponential} = \frac{1}{\text{Mean of Poisson}} \qquad \ldots(33)$$

Given that the mean number of arrivals per hour (Figure 44.2) is three, then the equivalent mean of the exponential ditribution is given by :

$$\text{Mean of exponential} = \frac{1}{\text{Mean of Poisson}}$$

Mean of the exponential equals **20 minutes**

$$\frac{1}{3} (= 20 \text{ mins})$$

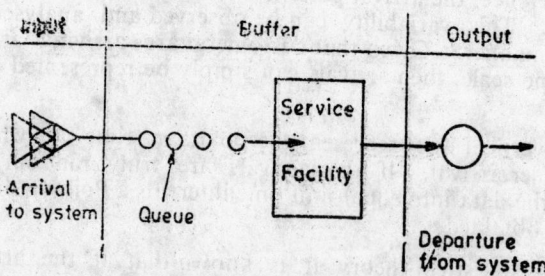

Fig. 44.1.   A Queue with one server.

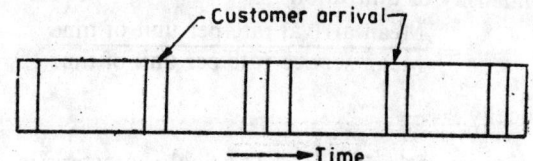

Fig. 44.2. Random Arrival Pattern.

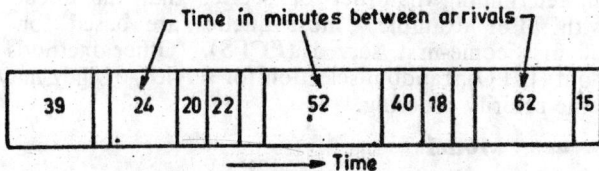

Fig. 44.3. Inter Arrival Times.

Average Arrival Rate

The average or mean arrival rate of customers arriving is calculated by dividing the total number of arrivals by total units of time.

(*ii*) **Service Pattern.** The arrival patterns are random in many situations and can be satisfactorily modelled by using Poisson distribution. But for simplicity and in order to reduce the necessity of complex mathematics, simple queuing theory assumes that service patterns can be represented by treating them as being exponential. The mean service rate denoted by '*s*' is simply the mean number of services per unit of time.

(*iii*) **Number of Service Channels.** In real life situation there are numerous service channels or facilities available to process the customer. Some of the examples of multi-channels queuing are Post Offices, Supermarkets, Banks and Airports terminals. The simple queing models are concerned with only one service channel *i.e.* only one service channel facility exists to serve the queue for example car service or car wash at a garage. The best known example of a limited input queue is that of a single operator tending a number of machines *e.g.* a setter. Operator tending a number of machines say Capston Lathes.

(*iv*) **Traffic Intensity or Utilisation.** The term 'traffic intensity' is derived from the pioneering work on stochastic processes carried out by Erlang, a Danish telephone engineer, in the early part of this century. Traffic intensity, symbolized as *i*, indicates the likelihood and extent of the queue and is the ratio of the mean arrival rate to the mean service-rate *i.e.*

Traffic intensity of utilization

$$= \frac{\text{Mean arrival rate per unit of time}}{\text{Mean service rate per unit of time}}$$

Symbolically $i = \frac{a}{s}$.

(v) **The Queue Discipline.** The final characteristic required to enable a queuing situation to be specified completely is the queue discipline. The term 'queue discipline' refers to the method or set of rules used in determining the order of service that the queue will be dealt with. For example, most queues are based on the principle of first come-first served (FCFS). Other methods are last in-first out (LIFO), random selection for service, or they may be based on some priority ordering.

Simple Queues : Model

For example, a simple queuing model can be described as having the following characteristics :

*Poisson arrival rate.

*Exponential service time.

*Single channel server facility.

*First come first served queue discipline.

Given that the simple queue has reached a steady state, the following relationships may be stated.

*Traffic intensity or utilization $= a/s$.

*Average number of customers in queue $= i^2/l-i$.

*Average number of customers in system $= i/l-i$.

*Average number of customers in queue when a queue exists. $= l/l-i$.

*Average time a customer is in the system $= l/s-a$.

*Average time a customer is in the queue $= i/s-a$.

*Probability of more than n customers in queue $= (a/s)^{n+2}$.

where a is defined as the mean arrival rate.

s is defined as the mean service rate.

i is defined as the probability that a given customer has to wait for service. Hence the probability of service can be calculated as $(l-i)$.

Queuing Theory Concepts

A Danish telephone Engineer A.K. Erbeng started developing he Queuing theory which is also called waiting lines, in 1905. After World War II the theory was applied in various other areas. The Queuing theory deals with problems of congestion and delay which occurs in many situations like :

—Dock yard where ships arrive for loading or unloading.

—Clinics where the patients arrive and are attended by the doctors.

—Manual assembly lines where parts are assembled by a number of different people.

—Booking office where the customers come either for the purchase of a ticket or for booking reservation.

—Bank counter where the public comes for withdrawal or depositing of money.

Common Phenomena

In all such areas the common phenomena, one observes in that

(*i*) Customers arrive to avail a facility.

(*ii*) The customers are attended by a service facility.

(*iii*) Either the customer will be waiting for availing service or the service facility will be waiting for the arrival of the customers.

Minimising the idle time or waiting time of the customer is the most desirable situation. For this purpose the parameters such as service rate, arrival rate of customers, idle time of service facility, idle time of customers, queue length of the customer etc. are found out. Based on this information and cost data, an optimal solution is worked out in order to minimize the idle time of customer.

Types of Waiting Lines

Generally two types of waiting lines situations are found.

(*i*) Single Queue and single service (channel)

(*ii*) Single Queue multi channel.

Again based on the rate of arrival and rate of service, four types of problems are encountered.

(*i*) Constant service time and random arrival of customers.

(*ii*) Random service time and constant arrivals.

(*iii*) Random service time and random arrivals.

(*iv*) Constant service time and constant arrivals.

In a real life situation, random service time and random arrivals are more prevalent. The other three are very rare and generally considered as hypothetical. The arrivals are independent of service facilities. Therefore, normally constant arrivals are not expected. Similarly service time differs with the nature of the job that is attended. Jobs do differ and hence service time also differs and very rarely it can be observed as constant unless a uniform job is attend with automatic devices.

Example 1. Following are the arrival and service times at a tool room where tools are issued to the fitters who arrive at the tool room. Here the service time is taken as constant and arrival is at random to explain the waiting situation given in Table 1.

The calculations of idle time, waiting time, length of queue etc. are given in the table.

No. of fitters arrived to tool room=20

Average rate of arrival per hour. $\Big\}$; 11·15—9·30=105 mts.

$$\frac{60}{105} \times 20 = \frac{80}{7} = 11·4 \cong 12$$

The rate of arrival of fitters to the tool room

=12 per hour

The rate of service is issuing the tools to the fitters $\Big\} = \frac{60}{3} = 20$ per hour

Per hour 12 fitters will arrive and the tool keeper will be capable of issuing tools to 20 fitters per hour.

Since the arrival of fitters is at random, there is bound to be idle time on the part of the fitters (not being attended having reached the tool room) as well as on the part of the tool-keeper.

The waiting time of the fitters=30 mts.

Idle time of the tool-keeper =49 mts.

Average waiting time $= \frac{30}{15} = 2$ mts. per fitter.

The idle time has been calculated based on the sample study of 105 minutes of working hours per day the idle time of

the fitters =137 minutes a day $\Big(\frac{480}{105} \times 30 \Big)$

tool keeper =224 minutes a day $\Big(\frac{480}{105} \times 49 \Big)$

If the idle time of the fitters is to be reduced one more tool keeper has to be added. But by adding the tool keeper the second tool keeper will not have sufficient work and the idle time of the tool-keeper will increase.

Poisson Arrival Rates

In real situation, the rate of arrivals follow Poisson distribution. The frequency of arrival of job. Orders at machines if measured for a long period, it exhibits Poisson distribution. Orders received at stores for material issue, receipts of orders for rectification of

Table 1

S. No.	Time of arrival of fitter	Issue of tools begins at	Issue of tools completes at	Elapsed time between arrivals in minutes	Issue time of tools in Mts (4)—(3)	Idle time of tools keeper in mts (3)ₙ—Aₙ₋₁	Waiting time of Fitters in mts (3)—(2)	Length of queue excluding fitter receiving tools (3)—(2)
1.	09·30	09·30	09·33	—	3	—	0	0
2.	09·39	09·39	09·42	9	3	6	0	0
3.	09·43	09·43	09·46	4	3	1	0	0
4.	09·49	09·49	09·52	6	3	3	0	0
5.	10·04	10·04	10·07	15	3	12	0	0
6.	10·06	10·07	10·10	2	3	0	1	1
7.	10·07	10·10	10·13	1	3	0	3	1
8.	10·08	10·13	10·16	1	3	0	5	2
9.	10·09	10·16	10·19	1	3	0	7	3
10.	10·12	10·19	10·22	3	3	0	7	3
11.	10·32	10·32	10·35	20	3	10	0	0
12.	10·33	10·35	10·38	1	3	0	2	1
13.	10·35	10·38	10·41	2	3	0	3	1
14.	10·41	10·41	10·44	6	3	0	0	0
15.	10·45	10·45	10·48	4	3	1	0	0
16.	10·55	10·55	10·58	10	3	7	0	0
17.	10·59	10·59	11·02	4	3	1	0	0
18.	11·10	11·10	11·13	11	3	8	0	0
19.	11·12	11·13	11·16	2	3	0	1	1
20.	11·15	11·16	11·19	3	3	0	1	1

broken down machine etc. all follow Poisson distribution because the arrivals in all such cases are random and Poisson distribution corresponds to completely random arrivals since it is assumed that an arrival is completely independent of other arrivals.

The Poisson distribution is a single parameter distribution. If can be completely described by one parameter the MEAN. For Poisson distribution the standard deviation in the square root of the mean. This is the mathematical characteristic of Poisson distribution.

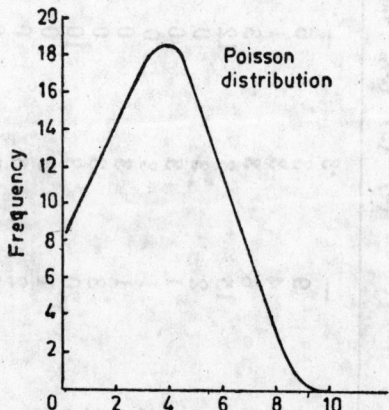

Fig. 44.4 Arrival rates of Jobs orders/men

In waiting line models the average arrival rate is represented by a Greek letter lamda (λ) representing number of arrivals per unit time.

When the arrival rates follows a Poisson distribution (with mean arrival rate of λ), the time between arrivals follows a negative exponential distribution, with mean time between arrivals of I/λ. This relationship between mean arrival rate (λ) and meantime between arrivals does not necessarily hold for other distributions. Therefore the negative exponential distribution is also representative of a Poisson distribution.

Exponential Service Times

The distribution service times follow the negative exponential distribution. The negative exponential distribution also is a single parameter distribution and it can be compeletly described by the mean which is also equal to standard deviation.

Normally Greek letter mu (μ) is used for representing the service rate in waiting line models. μ = number of services completed per unit time.

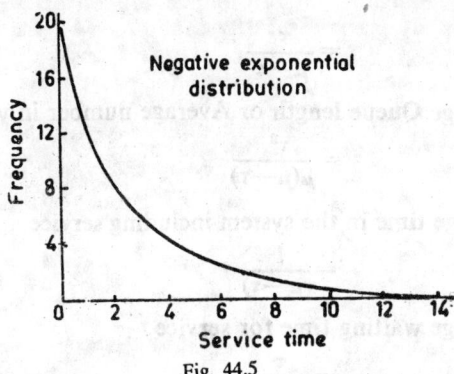

Fig. 44.5

Service rates follow a Poisson distribution with mean service rate μ while the distribution of service times follow the negative exponential distribution with mean service time $1/\mu$.

Thus the rates of arrivals and service follow Poisson distribution while arrival times and service times follow the negative exponential distribution.

With the largest mean processing time or service time, the service time in case of number of machines, issue of materials, etc. follow the negative exponential distribution.

It is based on these distributions that the waiting time mathematical models have been developed.

WAITING LINES MODELS

Because of the mathematical properties of Poisson and negative exponential distribution which represent the arrival rates and service times respectively, it has been possible to develop waiting line models. Though the waiting line models have been developed for single as well multi channel systems, only single channel, system is dealt here.

Single Channel Queue. Following are the various assumptions in Single Channel Queue.

(i) Rate of arrivals follow Poisson distribution.

(ii) Service times follow negative exponential distribution

(iii) First come-first served

(iv) The mean service rate (μ) is greater than mean arrival rate (τ).

The waiting line models for single channel Queue are given below.

(N) Average number in the system including the one being served

$$= \frac{\tau}{\mu - \tau}$$

(L) Average Queue length or Average number in waiting line

$$= \frac{\tau^2}{\mu(\mu - \tau)}$$

(T) Average time in the system including service

$$= \frac{1}{(\mu - \tau)}$$

(W) Average waiting time for service

$$= \frac{\tau}{\mu(\mu - \tau)}$$

(P) Probability of n units in the system

$$= \left(1 - \frac{\tau}{\mu}\right)\left(\frac{\tau}{\mu}\right)^n$$

(ρ) Utilisation of service facility

$$= \left(\frac{\tau}{\mu}\right)$$

Utilisation of service facility. The ratio of mean arrival rate to mean service rate is simply called utilisation factor.

From the model $L = \dfrac{\tau^2}{\mu(\mu - \tau)}$, if a graph is drawn between L the average number in the waiting line and (τ/μ) utilisation factor, L approaches to infinity as the utilisation factor (τ/μ) approaches to unity i.e. $(\tau/\mu) = 1$.

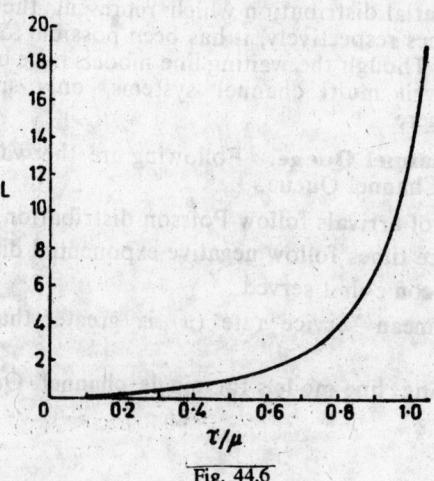

Fig. 44.6

Generally such a situation will not happen since :
- —new arrivals (men) are not expected by seeing long Queue
- —working hours will restrict the people coming into the waiting line
- —It a long Queue is there people at service counter will react to it and speed up the work.

Limitations of waiting line

If actual distributions of arrival rates and service times do not follow standard Poisson and negative exponential distributions, the formula will not hold good correctly.

Where the arrivals service rates do not follow the standard distributions, SIMULATION by MONTE CARLO method is the only solution.

Example 2. The customers arrive in a single Queue to booking office at random. The average arrival rate of the customer is one for every 5 minutes. The booking clerk takes on an average 3 minutes to dispose the customer. Find out the following :

(*i*)　The average number of people in the system

(*ii*)　Average Que length

(*iii*)　Average time in the system including service

(*iv*)　Average waiting time for service

(*v*)　Probability of zero units in the system

(*vi*)　Utilisation of service facility

$$\tau = \text{Arrival rate} = \frac{60}{5} = 12 \text{ customers per hour}$$

$$\mu = \text{Service rate} = \frac{60}{3} = 20 \text{ customers per hour}$$

Based on τ and μ values all the values can be calculated by using the f models

$$\tau = 12, \mu = 20$$

(*i*) Average number of people in the system

$$N = \frac{\tau}{\mu - \tau} = \frac{12}{20 - 12} = \frac{12}{8} = \frac{3}{2} = 1 \cdot 5$$

(*ii*) Average Queue length

$$L = \frac{\tau^2}{\mu \, (\mu - \tau)} = \frac{12^2}{20(20 - 12)}$$

$$= \frac{144}{20(8)} = \frac{144}{160} = 0 \cdot 9$$

(iii) Average time that is taken in the system including service

$$T = \frac{1}{(\mu - \tau)} = \frac{1}{(20 - 12)} = 0.125 \text{ hrs.}$$

$$= 7.5 \text{ mts.}$$

(iv) Average waiting time for the service

$$W = \frac{\tau}{\mu(\mu - \tau)}$$

$$= \frac{12}{20(20 - 12)} = \frac{12}{20 \times 6}$$

$$= \frac{12}{160} = 0.075 \text{ hrs} = 4.5 \text{ mts.}$$

$$T - W = 7.5 - 4.5$$

$$= 3 \text{ mts. service time confirms.}$$

(v) Probability of 'zero' units in the system

$$P_0 = \left(1 - \frac{\tau}{\mu}\right)\left(\frac{\tau}{\mu}\right)^0$$

$$= \left(1 - \frac{12}{20}\right)\left(\frac{12}{20}\right)^0 = \left(1 - \frac{12}{20}\right)$$

$$P_0 = (1 - 0.6) = 0.4$$

(v) Utilisation of service

$$\frac{\tau}{\mu} = \frac{12}{20} = 0.6$$

INFINITE WAITING LINE MODELS

Waiting line models are among the most used of quantitative methods today. The common assumptions of Poisson arrivals and exponential holding times are reasonable in quite a number of instances. In many cases where they are not appropriate, we have a non-homogenous situation where two or more basic conditions are mixed and in a printing press that handle both repair work and production work. In such a situation the resulting distribution of service times really reflects two basic distributions :

In practice we would apply infinite waiting line models where the number in line could grow very large. The high average rate of arrivals of cars at the enterance of the Bridge does imply the potential of the very large waiting line. In contrast, consider the ease of maintaining a bank of 5 computers machines. The break down of machines is considered to be an arrival. Obviously, the maximum possible waiting line of machines to be served is five. It is finite and required a different analysis resulting in a different model compared to the infinite waiting line problem.

We assume that the following conditions are valid :

(i) Poisson arrival rates.

(ii) Exponental service times.

(iii) First come—first served queue discipline.

(iv) The mean service rate μ is greater than the arrival rate λ.

Under these conditions, the following equations apply :

Mean time in system, including service

$$W = \frac{L}{\gamma}$$

Where $L = Lq + \frac{\lambda}{\mu}$

$\quad$ = Mean number in system including the one being serviced.

Mean Waiting time

$$Wq = \frac{Lq}{\lambda}$$

Probability of n units in the system

$$Pn = \left(1 - \frac{\lambda}{\mu}\right) \left(\frac{\lambda}{\mu}\right)^n$$

At first glance, it might seen that the difference between the mean number in the system and mean number in the waiting line $(L-Lq)$ should be 1, the unit being served, but on checking it is found to be less than 1, Similarly the difference between the mean time in the system and the mean waiting time $(W-Wq)$ is simply the average time for service. If we take the ratio between the mean arrival rate and the mean service rate, we have an index of the utilisation of service facility. The ratio is commonly called the utilisation factor and is denoted by the Greek letter "rho" (ρ). If the two rates are equal, $\rho = 1$ and theortically the service facility could be use 100 per cent of the time.

When conditions of arrival and service rate are such that the average length of the waiting line increases, one can reason that the probability that the service facility will be idle must become rather small. The probability of zero units in the system (service facility idle), P_0 is :

$$P_0 \left(\frac{1-\lambda}{\mu}\right) \left(\frac{\lambda}{\mu}\right)^0$$

and since $\left(\dfrac{\lambda}{\mu}\right)^0 = 1$.

$$P_0\left(\frac{1-\lambda}{\mu}\right) = \left(1-\rho\right)$$

This is the probability that the service facility will be idle and check our intution that utilisation of the facility plus idleness should total 100 percent.

Constant Service Times

It modifies the mean number in the waiting line and mean waiting time only slightly. The formulas are :

Mean number in waiting line

$$Lq = \frac{\lambda^2}{2\mu(\mu-\lambda)}$$

Mean Waiting Time

$$Wq - \frac{\lambda}{2\mu\,(\mu-\lambda)}$$

The constant service time does not represent a large number of real situations, is is reasonable in $\cos\theta$ where a machine process arriving item by a afixed cycle.

Example :

The usefulness of the basic equations in a typical decision problem by assuming average time between arrival and service rates per minute are :

$$\lambda = \frac{1}{60} \times 60 = 1 \text{ arrival per minute}$$

$$\mu = \frac{1}{50} \times 60 = 1\cdot2 \text{ service per minute}$$

We can determine the waiting line lengths; Waiting Time; and the percent of the idle time of the attendent using various equations discussed above :

$$Lq = \frac{\lambda^2}{\mu\cdot(\mu-\lambda)} = \frac{1^2}{1\cdot2\,(1\cdot2-1\cdot0)}$$
$$= 4\cdot17 \text{ Mechanics in line.}$$

$$L = \frac{\lambda}{\mu - \lambda} = \frac{1}{1 \cdot 2 - 1 \cdot 0}$$

= 5 mechanics in line, including the mechanics being served.

$$Wq = \frac{\lambda}{\mu \, (\mu - \lambda)} = \frac{1}{1 \cdot 2 \, (1 \cdot 2 - 1 \cdot 0)} = \frac{1}{1 \cdot 2 \times 0 \cdot 2}$$

= 4·17 minutes per mechanic.

$$W = \frac{1}{\mu - \lambda} = \frac{1}{1 \cdot 2 - 1 \cdot 0}$$

= 5 minutes waiting time including service.

$$\text{Idle time} = 1 - \rho = 1 - \frac{1}{1 \cdot 2}$$

= 0.1667 or 16·67% Idle time of attendent.

If attendent is paid Rs. 35/- per day and mechanics is paid Rs. 70/- per day of eight hours work what policy or service should be established ? What cost function do we wish to minimise ? If we add more attendents, the waiting line and mechanics, waighting time will be reuduced, but the idle time of the attendents will be increased.

Finite Waiting Line Models

Many practical waiting line problems which occur in production srstems have the characteristics of finite waiting line models. This is true whenever the population of machines, men or items which may arrive for service is limited to a relatively small finite number. The result is that we must express arrivals in terms of a unit of the population rather than as an average rate. Although there is no definite number that we can point to as a dividing line between finite and infinite applications, the finit queuing table have date for population from 4 to 250 and may be taken as a general guide. In real life situations a wide variety of seemingly diverse problem situations are now being recognised. The time between the arrival of the individual inputs at the service facility is commonly a random variables, the following Table shows the waiting line model elements for a number of common situations.

Table   Elements of Waiting Line Model

S.No.	Situations	Unit Arriving	Service or Processing Facility	Service or Process Being Performed
1.	Doctor's Clinic	Patients	Doctor, Clinic Staff and other Facilities.	Medical care.
2.	Vehicular Traffic at the intersection or at the enterance of Bridge.	Vehicle	Control Points such as Traffic Lights.	Passage through intersection or Bridge.
3.	Purcnase of items at a Super market.	Customers	Check out counter	Tabulation of Bill, Receipt of Payment.
4.	Inventory of various Items in a Store.	Order for withdrawl	Warehouse	Replenishment of Inventory.
5.	Ships entring a Port	Ships	Docks	Unloading and Loading.

Example. Application of Monte Carlo Simulation. This method utilises random numbers for originating the data itself. The random numbers are used in creating a hypothetical data for a problem whose behaviour is known from the past experience (frequency distributions). If no past data is available, the individual must decide whether the variables in the situation under study can be assumed to act at random. The use of cumulative probability distri-

Frequency distributions of

	Arrivals	Service times
S. No.	Atrival time in mts.	Frequency
1.	3·5—4·5	5%
2.	4·5—5·5	20%
3.	5·5—6·5	35%
4.	6·5—7·5	25%
5.	7·5—8·5	10%
6.	8·5—9·5	5%

bution for arrival and service time should closely resemble to the real world.

The frequency distribution of arrivals and service times are given below based on the past record from a tool room wherein the fitters come to take tools from tool keeper. The Queue starts from 10 to 12 o'clock.

S. No.	Service time in mts.	Frequency
1.	3·5—4·5	10%
2.	4·5—5·5	20%
3.	5·5—6·5	40%
4.	6·5—7·5	20%
5.	7·5—8·5	10%

Only from these distribution and random numbers the entire process can be treated.

From the frequency distributions, the cumulative frequency, mid point are calculated and random numbers are indicated as shown below :

Frequency Distribution of Arrivals

S. No.	Mid point of arrival time mts.	Frequency	Cumulative frequency	Random Numbers Range
1.	4	5%	5%	1—4
2.	5	20%	25%	5—24
3.	6	35%	60%.	25—59
4.	7	25%	85%	60—84
5.	8	10%	95%	85—94
6.	9	5%	100%	95—99

Frequency Distributions of Service Times

S. No.	Mid point of service times in mts.	Frequency	Cumulative frequency	Random Number Range
1.	4	10%	10%	1—9
2.	5	20%	30%	10—29
3.	6	40%	70%	30—69
4.	7	20%	90%	70—89
5.	8	10%	100%	90—99

Simulation by Monte Carlo Method

S. No.	Random Number for arrival	Time interval between arrivals mts.	Time of arrival	Service begins	Random Number for service	Service time in mts.	Service cards at	Waiting time of Fitters mts.	Tool mts.	Length of waiting line
1.	15	5	10'05	10'05	20	5	10'10	0	0	0
2.	09	5	10'10	10'10	72	7	10'17	0	0	0
3.	41	6	10'16	10'17	34	6	10'23	1	0	1
4.	74	7	10'23	10'23	54	6	10'29	0	1	0
5.	72	7	10'30	10'30	30	6	10'36	0	1	0
6.	67	7	10'37	10'37	22	5	10'42	0	1	0
7.	55	6	10'43	10'43	48	6	10'49	0	1	0
8.	71	7	10'50	10'50	74	7	10'57	0	0	1
9.	35	6	10'56	10'57	76	7	11'04	1	0	0
10.	41	6	11'02	11'04	02	4	11'08	2	3	1
11.	96	9	11'11	11'11	07	4	11'15	0	1	0
12.	20	5	11'16	11'16	64	6	11'22	0	0	0
13.	45	6	11'22	11'22	95	8	11'30	0	0	0
14.	38	6	11'28	11'30	23	5	11'35	2	0	0
15.	01	4	11'32	11'35	91	8	11'43	3	0	0
16.	67	7	11'39	11'43	48	6	11'49	4	0	0
17.	63	7	11'46	11'49	55	6	11'55	3	0	0
18.	39	6	11'52	11'55	91	8	12'03	3	0	0
19.	55	6						3	0	0
						110		19	8	3

Monte Carlo Method

Take the random number from random table and put them at column 2 in the table for arrivals. The first random number is 15. From frequency distribulion of arrivals, read the next arrival time in minutes against the range of random from 15. The next arrival time is 5 minutes. Write this 5 minutes at column number 3 in the table. Since, the process starts at 10'0 clock, the fitters arrive at tool room by 10'05. Therefore, time 10'05 will be written at column 4 *i.e.* time of arrival. Since the issue of tools begins immediately at column No. 5 in the table, 10'05 will be written as service begins.

Pick up random numbers from random table and put them at column 6. Provide for random number frequency distribution of service times read-service time in minutes against random number 20. The service time is 5 minutes and note this time at column 7. Since 10'05 was the beginning of the service time, 10'05+5=10'10 is the completion of the service time and it is written at column 8.

Waiting time of the fitter is the difference between the time of arrival and time of the service begins *i.e.* column number 4 and 5. This is zero in this case, which is written at column number 9. Similarly, waiting time for tool keeper is the difference between the column 5 and 4 which is zero and written at column 10. This entire process is repeated till 18th arrival when it completes two hours.

Once the events are simulated for a period of 2 hours, the following type of questions of waiting line can be answered.

(*i*) What is the length of the waiting line ($L=?$).

(*ii*) What is the average time a customer waits before he is served. ($W=?$)

(*iii*) What is the average time a customer spends in the system ($T=?$)

(*i*) $L=\dfrac{\text{Number of customers in Queue for 2 hours}}{\text{Number of arrivals}}$

$=\dfrac{3}{18}=0'16$ people (Length of waiting Line)

(*ii*) $W=\dfrac{19}{18}=1'05$ Mts. (Average time a customer waits)

(*iii*) $T=$ Average waiting time + Average service time

$=1'05 \text{ Mts.} + \dfrac{110}{18}=1'05+6'10$

$T=7'15$ Mts. (Average Time Customer Spends in system)

Example. Find out the average member of customers in the system. Average number of customers in the Queue (including occasions when Queue length is zero) if the utilisation factor of the service system is 0·5, 0·7, 0·9 and 0·95.

Utilisation factor $\rho = \dfrac{\tau}{\mu}$

Since the value of ρ only is given, formula for the above questions be converted in the form of ρ.

Average number of curtomers in the system

$$N = \frac{\tau}{\mu - \tau}$$

where τ is the arrival rate and μ is the service rate

$$N = \frac{\tau}{\mu - \tau} = \frac{1}{\dfrac{\mu - \tau}{\tau}} = \frac{1}{\dfrac{\mu}{\tau} - \dfrac{\tau}{\tau}}$$

Since $\dfrac{\tau}{\mu} = \rho$

and $\dfrac{\mu}{\tau} = \dfrac{1}{\rho}$

$$N = \frac{1}{\left(\dfrac{1}{\rho}\right) - 1} = \frac{1}{\left(\dfrac{1-\rho}{\rho}\right)} = \frac{\rho}{1-\rho}$$

Similarly average number of customers in Queue or Average Queue length (L)

$$= \frac{\tau^2}{\mu(\mu - \tau)}$$

$$L = \frac{\tau^2}{\mu(\mu - \tau)} = \frac{\tau}{\mu}\left[\frac{\tau}{(\mu - \tau)}\right]$$

Since $\dfrac{\tau}{\mu - \tau} = \dfrac{\rho}{1 - \rho}$

as has been already shown

and $\rho = \dfrac{\tau}{\mu}$

$$L = \frac{\tau}{\mu}\left[\frac{\tau}{(\mu - \tau)}\right] = \rho\left[\frac{\rho}{1-\rho}\right] = \left[\frac{\rho^2}{1-\rho}\right]$$

Now N and L both are available in terms of ρ

S.N.	ρ	ρ^2	$(1-\rho)$	$L=\dfrac{\rho^2}{1-\rho}$	$N=\dfrac{\rho}{1-\rho}$
1.	0·50	0·25	0·50	0·50	1·00
2.	0·70	0·49	0·30	1·63	2·33
3.	0·90	0·81	0·10	8·10	9·00
4.	0·95	0·9025	0·0500	18·05	19·00

Thus the average number of customers in the queue including the one being serviced and the queue length both will increase as the utilisation of service facility increases. Therefore in order to keep the queue length low, it is essential that utilisation of service facility be low since the customer service is more precious.

Example. If the average service rate is 5 per hour, what is the average time of a customer in the queue for the utilisation of service facilities at 0·5, 0·7, 0·9 and 0·95 in the service system.

Average time of a customers in queue

= Average waiting time of the customer for service in queue

$$W=\frac{\tau}{\mu/\mu-\tau}=\frac{1}{\mu}=\left[\frac{\tau}{(\mu-\tau)}\right]$$

$$W=\frac{1}{\mu}\left[\frac{\rho}{1-\rho}\right]=\frac{\rho}{\mu(1-\rho)}$$

Now values of ρ and μ are available

$$\rho=0·5, 0·7, 0·9 \text{ and } 0·95$$

whereas μ=Service rate=5. The waiting time is calculated as below :

S.No.	ρ	$(1-\rho)$	$\left(\dfrac{\rho}{1+\rho}\right)$	μ	$W=\dfrac{\rho}{\mu(1-\rho)}$ hours
1.	0·50	0·50	1·00	5	0·20 hour
2.	0·70	0·30	2·33	5	0·47 hours
3.	0·90	0·10	9·00	5	1·80 hours
4.	0·95	0·05	19·00	5	3·80 hours

Rate of service is constant at 5 per hour. Therefore value of ρ increases since $\rho = \dfrac{\tau}{\mu}$

Since the service rate is 5 per hour and as the arrival rate of customers increases ; the waiting time of the customer increases.

45

Linear Programming

Linear Programming is an extremely useful technique used for the purpose of maximising some objective such as profit or minimising, some objective such as costs in order to achieve the objectives of the organisation.

Linear programming is now used increasingly to solve management problems. Linear programming is used in situations where numerous activities are competing for limited resources. The person responsible for developing linear programming was George B. Dantzig, who in 1947 developed the simplex method of Linear Programming. Since World War II the linear programming models have been used increasingly to solve management problems. The term linear reveals that the technique is mathematical and involves linear functions and the term programming refers to certain mathematical procedures to get the best solution to a problem involving limited resources. Linear Programming is a technique of numbers and the application of the numbers to arrive at a better decision. It is a tool in the hands of managers to take better decision among the number of alternatives.

Definition

Linear Programming is a technique of selecting the best possible (optimal) strategy among a number of alternatives :

(*i*) Linear Programming, is an important tool used by the management, involved in choosing most appropriate (best possible) alternative among a number of alternative action.

(*ii*) Linear Programming helps the management discover whether the maximum profit or the least production programmes open to it.

(*iii*) It is applied to problems which involve interaction between alternatives.

Uses and Applications of Linear Programming

(*i*) Linear Programming helps in determining the optimum combinations of several variables with given constraints.

(*ii*) It provides additional information for proper planning and control over various operations in the organisation.

(*iii*) It also helps the managers to have better understanding about the phenomenon and various activities of the organisation for the organisation construction of suitable mathematical model visualizing the relationship between variables if any and making improvement over them.

(*iv*) It also contributes the development of executives through the technique of model building and corresponding interpreta‚ tions.

(*v*) It provides a clear specification of the management problems by defining :

(*a*) the objectives to be persued by the organisation.

(*b*) the various restrictions imposed by the system.

(*c*) the various alternatives and relationship between various variables.

(*d*) the contribution made by each alternative to the objectives of the organisation.

(*vi*) Linear Programming has found useful avenues in the fields of agricultural, industrial, military, space and research etc.

(*vii*) Specific areas where it can be fruitfully applied are :

(*a*) Production scheduling and inventory control satisfying the demand and minimising the cost of production and inventory cost.

(*b*) Blending operations and problems—where basic compo‚ nents are combined to produce a product that has definite set of specifications and one may be interested in calculating best possible combination of these components which maximises the profit or minimises the cost.

(*viii*) Other important applications of linear programming techniques can be in the area of :

> purchasing and materials management.
> routing and scheduling.
> assignment problems.
> marketing.
> space exploration.

The problem of linear programming can be formulated with the help of the following terms.

Objective Function

An objective function is some sort of mathematical relationship between the variables under consideration. In the case of Linear

Programming this relationship is always to be taken as linear. The construction of objective function is a process whereby most important features of a system are considered. Given variables $x_1, x_2 \cdots,$ $x_3 \cdots x_n$. An objective function can be of the form

$$Z = A_1 x_1 + A_2 x_2 + A_3 x_3 + \cdots A_{11} x_{11}$$

where A_1, A_2, A_3 are certain constants.

The problem of linear programming is to find out some suitable combination of $x_1, x_2, x_3 \cdots x_n$ etc. which optimises (*i.e.* maximises or minimises) the objective function. The objective function is always non-negative.

The objective function is to be optimised under certain restraints imposed on the variables occurring in the objective function. These restrictions in most of the cases are never exact, had these been exact the objective function could be easily optimised by the use of differential calculus. These restrictions must be known and should be expressed in algebraic expressions.

A set of values of x_1, x_2 which satisfies constraints and the non-negativity restrictions is called a Feasible Solution. A feasible solution which optimises the objective function is known as a *Optimal Solution.*

Methods to Determine Optimal Solution

There are a number of ways to determine optimal solution for a given linear programming problem. Following three methods are generally used :

(*i*) Graphie Method

(*ii*) Simplex Method

(*iii*) Transportation Method.

Graphic Method

The problems involving two variables only can be effectively and easily solved by the method of graphs. The graphical method has only limited applications to industrial problems due to its limitations of handling only two or three variables at a time. In practice it is convenient for two variable only because the geometry of 3 dimensional case becomes so complicated that accurate results are difficult to obtain. The problems containing more than 3 variables are difficult to be solved.

The graphical method consists of following three main steps which are as follows :

(*i*) The constraints of the model are plotted on the graph paper by considering them as equalities.

(*ii*) Lines are drawn in two dimensional plane corresponding to given equations.

(*iii*) These lines refers to the region of permissible values.

(*iv*) The objective function is also plotted on the same graph.

(*v*) Using the lines in steps (*i*), (*ii*) and (*iii*) the optimal solution that satisfies the constraints and optimises the objective function is determined by trial and error method.

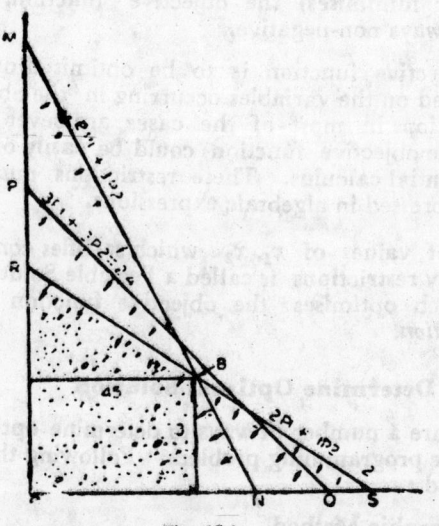

Fig. 45.1

From the graph we find that point 'B' represents the optimum point in the feasible area ORBN which satisfied the equations (*i*), (*ii*) and (*iii*). Point B gives the optimum value.

PROBLEMS

Problem 1. Mr '*X*' can purchase his tomatoes of three varieties (*A*, *B* and *C*) from two sources which differ in yields of various sizes and quality. The yield characteristics are given in table with purchase limitations. We find that from source 1, there is 20% yield of product '*A*', 20% yield of product '*B*' and 30% yield of product '*C*' remaining 30% is waste. From source '2' for product '*C*' and '*A*' the yield is 30% for product '*B*' the yield is 10% and waste 30%.

From which source the customer should buy tomatoes and how much in order to maximise profit.

Table

Product variety	Source 1	Source 2	Purchase limitations tons
A	0·2	0·3	1·8
B	0 2	0·1	1·2
C	0·3	0·3	2·4
Waste	0·3	0·3	
Relative Profit	Rs. 5	Rs.6	

Solution (Graphical Method)

Let P_1 be the tomato purchased from Source '1' and P_2 be the tomato purchased from source '2'

then $\longrightarrow$

$$Z = 5P_1 + 6P_2$$

$$.2P_1 + .3P_2 \geqslant 1·8 \qquad \qquad ...(1)$$

$$·2P_1 + ·1P_2 \geqslant 1·2 \qquad \qquad ..(2)$$

$$·3P_1 + ·3P_2 \geqslant 2·4 \qquad \qquad ...(3)$$

Changing the above equations in the form $\dfrac{x}{a} + \dfrac{y}{b} = 1$ where $a + b$ are the intercepts on x and y axis respectively.

$$\frac{P_1}{9} + \frac{P_2}{6} = 1 \qquad \qquad ...(4)$$

$$\frac{P_1}{6} + \frac{P_2}{12} = 1 \qquad \qquad ...(5)$$

$$\frac{P_1}{8} + \frac{P_2}{8} = 1 \qquad \qquad ...(6)$$

Problem 1. A small manufacturing firm produces two types of gadgets, A and B which are first processed in the foundry, then sent to machine for finishing. The number of man hours of labour required in each shop for the production of each unit of 'A' and 'B' and the number of man hours the firm has available per week are as follows :

	Foundry	Machine Shop
Gadget A	10 hrs.	5 hrs.
Gadget B	6 ,,	4 ,,
Firm's capacity per week	1000 ,,	600 ,,

(i) Construct the objective function and the corresponding restraints for calculating that how many units should be produced per week so that the profit is maximum. The profit on the sale of 'A' is Rs. 30/unit as compared with 'B' Rs. 20/unit.

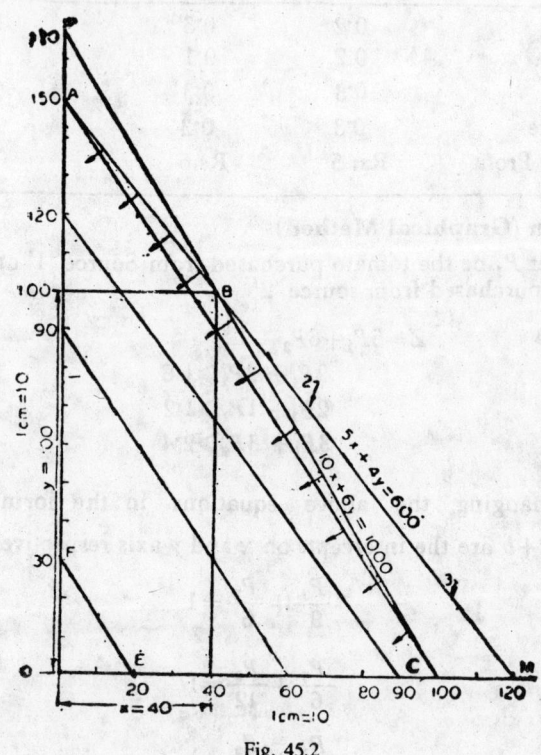

Fig. 45.2

(ii) How many units of A and B should be produced to maximise the profit.

Solution. Let the firm manufacture x units of A and y units of B then,

The objective function and the constraints for the given problem are :

Max. $\qquad Z = 30x + 20y$ $\qquad$ ...(1)

$$10x + 6y \geqslant 1000 \qquad ...(2)$$

$$5x + 4y \geqslant 600 \qquad ...(3)$$

$$x, y \geqslant 0 \qquad ...(4)$$

the above equations are explained below :

The firm's capacity per week in foundry is 1000 hrs and x unit of A will require $10x$ man hours per week in foundry whereas y units of B will require $6y$ hrs. Hence one constraint is

$$10x + 6y < 1000$$

Similarly other constraint is

$$5x + 4y \leqslant 600$$

The last constraint is that both $x + y$ should be positive $x, y \geqslant 0$.

For graphical solution of the above problem, the inequalities are changed into equalities and then the equations thus obtained are plotted on the graph paper. Here equations to be plotted as :

$$10x + 6y = 1000 \qquad \ldots(1)$$
$$5x + 4y = 600 \qquad \ldots(2)$$
$$x = 0 \qquad \ldots(3)$$
$$y = 0 \qquad \ldots(4)$$

Now the line $x = 0$ is y-axis and $y = 0$ is x-axis for plotting the remaining two lines.

Converting the above lines in the form $\dfrac{x}{a} + \dfrac{y}{b} = 1$ we get,

$$\frac{x}{100} + \frac{y}{167} = 1$$

$$\frac{x}{120} + \frac{y}{150} = 1$$

Plotting these lines on the graph, we get two lines PC and AM which intersect at B.

Area covered between the two axes and the line PC contains at those points which satisfy $10x + 6y < 100$. Similarly area below the line AM and between two axes satisfies the inequality $5x + 4y < 600$. The area $OABC$ satisfies all the constraints of the given problem. It is noted that area below the lines PC and AM as the constraints for these equations have $<$ sign. The region $OABC$ is the required space for feasible solutions and from this region we want to find X and Y for which Z is maximum. For this we plot the objective function $30x + 20y$ for some numerical value of Z, and move this line parallel to itself till it contains a point of the feasible region, say line EF is plotted. The line EF moves upwards until EF cuts most remote corner, i.e., the point B of the polygon $OABC$. This is the maximum value that Z can obtain, xy still remaining in the feasible region.

This co-ordinates of the point B ($x = 40, y = 100$) gives the required solution. The corresponding profit is

$$Z = 30 \times 40 + 20 \times 100$$
$$= 1200 + 2000 = \text{Rs. } 3,200.$$

It should be noted that other lines parallel to *EF* will not contain even a single point of the feasible region. The extreme point of the feasible region will be the optimum value. Hence optimum value of *Z* should be tried at the corners of the polygon only.

Problem 2. A manufacturer can produce two different products *A* and *B*, during a given time period. Each of these products requires four different manufacturing operations—grinding, turning, assembly and testing. The manufacturing requirements in hours per unit of product are given below for *A* and *B*.

Operations	Product	
	A	*B*
Grinding	1	2
Turning	3	1
Assembly	6	3
Testing	5	4

The available capacities of these operations in hours for the given time period are grinding 30, turning 60, assembly 200, and

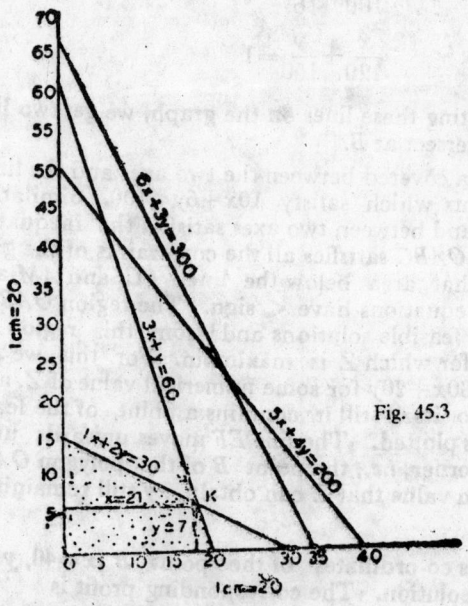

Fig. 45.3

testing 200. The contribution to overhead and profit is Rs. 2 for each unit of *A* and Rs. 3 for each unit of *B*. They can sell all that can be produced at the prevailing market price. Determine by graphical method the optimum amount of A and B to produce during the given time period.

Solution. Let the manufacturer produce x units of A and y units of B. Then the objective function to be optimised is

$$Z = 2x + 3y$$

The constraints are

$$x + 2y \leqslant 30 \text{ for grinding}$$
$$-3x + y \leqslant 60 \text{ for turning}$$
$$6x + 3y \leqslant 200 \text{ for assembly}$$
$$5x + 4y \leqslant 200 \text{ for testing}$$

and

$$x, y \geqslant 0$$

Considering the constraints as equalities we plot the equations as shown in the graph. We find from the graph that limiting factor is the capacity of the grinding section and the point K determining the maximum profit under the given circumstance.

Problem 3. Solve the following linear programming problem graphically. Minimize

$$Z = 3x_1 + 5x_2 \qquad x_1 \leqslant 4$$
$$-3x_1 + 4x_2 \leqslant 12 \qquad x_2 \geqslant 2$$
$$2x_1 - x_2 \leqslant -2 \qquad x_1, x_2 \geqslant 0$$
$$2x_1 + 3x_2 \leqslant 12$$

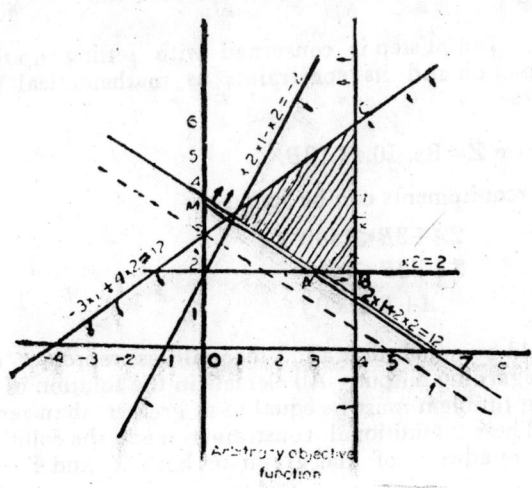

Fig. 45.4

..gure of the given constrained equations is drawn in the graph for plotting the objective function for any value of Z, take

$$Z = 15, \text{ so that}$$
$$Z = 3x_1 + 5x_2 = 15$$

Now draw lines parallel to PQ, the line MN is the first to touch feasible region at the point A. Hence the co-ordinates of A give the optimal solution. At the point A $x_1 = 3$, $x_2 = 2$ and the value of

$$Z = 3 \times 3 + 5 \times 2$$
$$= 9 + 10 = 19$$

In this case the feasible region lies above the line MN as the inequality is $\geqslant 0$.

Problem 4

Example 1

Deptts.	Hours Product A	Required Product B	Available hours in Month
1	2	3	1500
2	3	2	1500
3	1	1	600

There are 4 basic steps used in the graphical method of linear programming.

Step 1. Initial step is concerned with setting up the firm's objective function and its constraints as mathematical equation or inequalities.

Maximize $Z = $ Rs. $10A + 12B$.

There requirements can be stated as

$$2A + 3B < 1500$$
$$3A + 2B < 1500$$
$$A + B < 600$$

It should be noted that all 3 inequalities represent capacity restrictions regarding output. All elemets in the solution of a linear programming problem must be equal to or greater than zero ($A \geqslant 0$ and $B \geqslant 0$). These 2 additional constraints mean the solution must lie in the $+$ quadrant of the graph or both X and Y must be positive.

Step 2. Next we graph the constraint inequalities. If in Dept. one, units of B are produced then

$$2A+[0\times 3B]=1500$$

or
$$A=750$$

The first point (750) units of product A and zero unit of product B. 2nd point is computed in the same manner, only this time all the hours available are used in making the maximum number of units (500) of product B and 0 units of product A.

$$2(0)+3B=1500$$

$$B=500$$

Plot this A B line on X, Y axis.

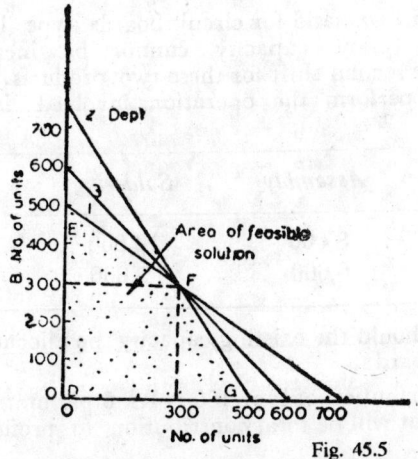

Fig. 45.5

The similar calculation is made for other equation, *i.e.*

Dept. 1 $2A+3B \leqslant 1500$

2      $3A+2B \leqslant 1500$

3      $A+B \leqslant 600.$

or

1      $B=750,\ B=500$

2      $3(0)+2B=1500$
$$B=750$$
$$3A+2(0)=500$$

3      $A+(0)=600$
$$B=600.$$

Diagram is given in graph

Step 3. This step is to plot the objective [function n, which is given as :

$$Z=\text{Rs. } 10A+12B$$

Step 4. This step is to test the corners, *i.e.,* point $D, E, F, G,$ The stripped area to determine which yields the highest contribution

AB

Point D (0, 0) $=$ Rs. 0.

Point E (0, 500) $= (10 \times 0) + 12(500) = 6000$

Point F 300,300 $= 300 \times 10 = $ Rs. 3000

$= 300 \times 12 = $ Rs. 3600

$= $ Rs. 6600

Point G 500$+$0 $=$ Rs. $500 \times 10 = $ Rs. 5000.

The maximum profit is yield on F point.

Problem 5. Demand for circuit boards type 100 and 101 is unlimited. The plant capacity cannot be increased. It is decided to go for second shift for these two products. The capacity of this shift to perform the operations involved is set forth as follows :

Circuit type	Assembly	Soldering	Inspection
100	8,000	12,000	7,000
101	6,000	5,000	14,000

(a) How should the existing capacity be allocated to the two type of circuit boards.

(b) If contribution of type 100 is Rs. 6 per unit and Rs. 4·50 for type 101, what will be total contribution to profit in proposed plan.

Solution.

Assumptions. Let us say we produce x_1 units of 100 type board x_2 units of 101 type board.

Utilization of assembly shop will be :

$$x_2 = (8000 - x) \times \frac{6000}{8000}$$

$$x_2 = (8000 - x_1) \times \frac{3}{4}$$

or $4x_2 = 8000 \times 3 - 3x_1$

or $3x_1 + 4x_2 = 24000$ ... (1)

Soldering Shop

$$x_2 = (12000 - x_1) \times \frac{5000}{12000}$$

or $12x_2 = 5 \times 12000 - 5x_1$

or $5x_1 + 12x_2 = 60,000$ ...(2)

Inspection

$$x_2 = (7000 - x_1) \times \frac{14000}{7000}$$

or $$7x_2 = 14 \times 7000 - 14x_1$$

or $$14x_1 + 7x_2 = 98,000$$

or $$2x_1 + x_2 = 14000 \qquad \qquad ...(3)$$

$$x_1 \geqslant 0 \qquad \qquad ...(4)$$

$$x_2 \geqslant 0 \qquad \qquad ...(5)$$

From equation (1)

$$\frac{x_1}{8000} + \frac{x_2}{6000} = 1$$

$$x_1 = 8000 \qquad x_2 = 0$$
$$x_1 = 0 \qquad x_2 = 6000$$

From Equation (2)

$$\frac{x_1}{12000} + \frac{x_2}{5000} = 1$$

$$x_1 = 12000 \qquad x_1 = 0$$
$$x_1 = 0 \qquad x_1 = 5000$$

From Equation (3)

$$\frac{x_1}{7000} + \frac{x_2}{14000} = 1$$

$$x_1 = 7000 \qquad x_2 = 0$$
$$x_1 = 0 \qquad x_2 = 14000$$

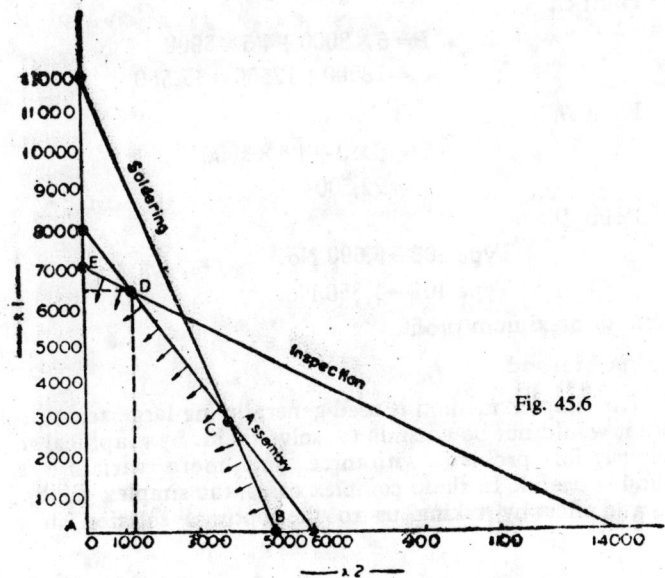

Fig. 45.6

If there is no solution to enforce the capacity of assembly shop to point F. The optimum utilization will at points E, D, C and P.

i.e.

Point	Type 100 x_1	Type 101 x_2
E	7000	0
D	6600	13500
C	3000	3900
B	0	5000

Optimum utilization of the three shops is possible in the above four situations for these two product lines.

For (*b*) part of the question, we have to optimise the profit.

For the graphical we have to calculate the profit for all the four points.

Point E,

$$P = 6x_1 + 4\cdot5x_2$$
$$= 6 \times 7000 + 4\cdot5 \times 0 = 42,000$$

Point D,

$$P = 6 \times 6600 + 4\cdot5 \times 1350$$
$$= 39600 + 6075 = 45,675$$

Point C,

$$P = 6 \times 3000 + 4\cdot5 \times 3900$$
$$= 18000 + 17550 = 35,550$$

Point B,

$$P = 6 \times 0 + 4\cdot5 \times 5000$$
$$= 22,500$$

Point D,

Type 100→6,600 No.

Type 101→1,350 No.

is point of maximum profit.

Simplex Method

The simplex method is used generally for large scale problems where it would not be possible to solve them by graphical method. Obviously for problems with three and more variables simplex method is useful. In these complex cases, the simplex method saves time and effort by taking us to the optimal solution in a finite number of steps, thereby each step bringing us closer to the optimum.

To apply the procedure to complex problems it will be useful to reduce it to a set of rigorous rules which can be applied rather mechanically to save time and effort to ʻreduce the computation required to a minimum.

To start with let us try to understand the terminology in simplex method.

Let the objective function be represented by

$$60x_1 + 50x_2 = \text{Maximisation}$$

Subject to the constraint equations

$$2x_1 + 4x_2 \leqslant 80$$
$$3x_1 + 2x_2 < 60$$

To remove inequalities introduce slack variables x_3 and x_4 as shown below

$$\left.\begin{array}{l} 2x_1 + 4x_2 + x_3 = 80 \\ 3x_1 + 2x_2 + x_4 = 60 \end{array}\right\} \text{constraint equations}$$

$$60x_1 + 50x_2 + (0)x_3 + (0)x_4 = \text{Maximisation objective function}$$

Since slack variables do not contribute any thing to objective function, their coefficient is taken as zero. The above three equations can be represented in the form of matrix as below :

Solution Step			60 x_1	50 x_2	0 x_3	0 x_4	Objective Row Variable Row
0	x_3	80	2	4	1	0	} constrainst
0	x_4	60	3	2	0	1	} equations

Body — under x_1, x_2
Identity — under x_3, x_4
constant column
variable column
objective column

Procedure Summary

1. Formulate the problem into linear equations with ʻslack variables and the objective function.

2. Develop the initial simplex matrix from the above equations including the initial trival solution and the index row

numbers. The index row numbers in the initial matrix are calcu-lated by the formula.

Index number

$$= \sum \left[\binom{\text{Numbers}}{\text{in column}} \times \binom{\text{Corresponding number}}{\text{in objective column}} \right] \\ - \left[\begin{array}{c} \text{Number in objective row} \\ \text{at head of column} \end{array} \right]$$

3. In the above matrix mark the key column, the column with the most nagative index number in the body or the identity.

4. Mark the key row, the row with the smallest non-negative quotient obtained by dividing each number of the constant column by the corresponding positive, non-zero number in the key column.

5. The key number is at the intersection of the key row and key column.

6. Develop the main row of the new table.

$$\text{Main row} = \frac{\text{key row of preceding table}}{\text{key number}}$$

The main row appears in the new table in the same relative position as the key row of the preceding table.

7. Develop the balance of the new table.

(a) The variable and its objective number at the head of the key column are entered in the stub of the new table to the left of the main row, replacing the variable and objective number from the key row of the preceding table.

(b) The remainder of the variable and objective columns are reproduced in the new table exactly as they were in the preceding table.

(c) The balance of the coefficients for the new table are cal-culated by the formula :

New nember

$$= (\text{old number}) - \frac{\binom{\text{corresponding}}{\text{number of key row}} \times \binom{\text{Corresponding}}{\text{number of key column}}}{(\text{key number})}$$

8. Repeat (iterate) steps 3 through 7C until all the index numbers (not including the constant column) are positive. An optimal solution then results.

9. The interpretation of the resulting optimum solution is as follows. The solution appears in the stub. The variables shown

in the variable column have values shown in the corresponding rows of the constant column. The value of the objective function is shown in the constant column, index row. All variables not shown in the stub are equal to zero :

Example. Maximise the objective function

$$P = 6x_1 + 4{\cdot}5x_2 \qquad \qquad \text{... (1)}$$

Subject to the constraint equations

$$3x_1 + 4x_2 < 24{,}000 \qquad \qquad \text{... (2)}$$
$$5x_1 + 12x_2 < 60{,}000 \qquad \qquad \text{... (3)}$$
$$2x_1 + x_2 < 14{,}000 \qquad \qquad \text{... (4)}$$

Step 1. Formulation of problem. By introducing the slack variables in equations (2), (3) and (4), the inequalities can be removed and written as follows

$$3x_1 + 4x_2 + x_3 = 24{,}000 \qquad \qquad \text{... (5)}$$
$$5x + 12x_2 + x_4 = 60{,}000 \qquad \qquad \text{... (6)}$$
$$2x_1 + x_2 + x_5 = 14{,}000 \qquad \qquad \text{... (7)}$$

and let the contribution of each x_3, x_4 and x_5 be zero.

The objective function equation (1) can be written as

$$P = 6x_1 + 4{\cdot}5x_2 + (0)x_3 + (0)x_4 + (0)x_5 \qquad \text{... (8)}$$

Now these equations (5) to (8) are written in the matrix form and solved by simplex method as per the procedure.

Step 2. Initial Simplex Matrix

Table 1

Solution Step			6 x_1	4·5 x_2	0 x_3	0 x_4	0 x_5	Coefficient of objective function called Objective Row
0	x_3	24,000	3	4	1	0	0	Constraint equations
0	x_4	60,000	5	12	0	1	0	
0	x_5	14,000	2	1	0	0	1	

Body — Identity
constant column
variable column
objective column

Calculate index row number and introduce Index Row in the matrix.

Index number

$$-\sum\left[\left(\begin{array}{c}\text{numbers}\\\text{in column}\end{array}\right)\times\left(\begin{array}{c}\text{corresponding number}\\\text{objective column}\end{array}\right)\right]$$
$$-[\text{Number in objective row at head column}]$$

1. Index number of constant column
$$=(24000\times0+60000\times0+14000\times0)-(0)=0$$

2. Index number of 1st column of body
$$=(3\times0+5\times0+2\times0)-(6)=-6$$

3. Index number of 2nd column of body
$$=(4\times0+12\times0+1\times0)-(4\cdot5)=-4\cdot5$$

4. Index number of 1st column of identity
$$=(1\times0+0\times0+0\times0)-(0)=0$$

5. Index number of 2nd column of identity
$$=(0\times0+1\times0+0\times0)-(0)=0$$

6. Index number of 3rd column of identity
$$=(0\times0+0\times0+1\times0)-(0)=0$$

Table 2. Initial Simplex Matrix with Index Row

			6 x_1	$4\cdot5$ x_2	0 x_3	0 x_4	0 x_5	
0	x_3	24000	3	4	1	0	0	
0	x_4	60000	5	12	0	1	0	
0	x_5	14000	2	1	0	0	1	
		0	-6	$-4\cdot5$	0	0	0	Index Row

Step 3. Select Key Column. The column containing -6 the most negative number in index row is marked as *key column*.

Step 4. Select Key Row. Divide each number of constant column by the corresponding positive non-zero number in key column. The row with the smallest non-negative quotient is called *key row*.

(1) Number of key column   3   5   2

(2) Number of const. column   24000   60000   14000

(3) quotient $=(2)\div(1)$ 8000   12000   7000

                                   Smallest non negative instient

$\therefore$ Key Row is $|\,0\quad x_5\quad 14000\quad 2\quad 1\quad 0\quad 0\quad 1\,|$

Step 5. Select Key Number. At the inter section of key columns and key row :

All the above three steps are incorporated in the initial simplex matrix as below.

Table 3. Initial Simplex Matrix with Key col. Key Row and Key Number

			6	4·5	0	0	0		
			x_1	x_2	x_3	x_4	x_5		
0	x_3	24000	3	4	1	0	0		
0	x_4	60000	5	12	0	1	0		
0	x_5	14000	2	1	0	0	1	←Key Row
		0	−6	−4·5	0	0	0	Index row

Key column (pointing to x_1)

Step 6.

$$\text{Main Row} = \frac{\text{Key Row of previous table}}{\text{Key Number}}$$

Main Row $= (14000 \quad 25 \quad 1 \quad 0 \quad 0 \quad 2) \div 2$

$\qquad\qquad 7000 \quad 1 \quad \frac{1}{2} \quad 0 \quad 0 \quad \frac{1}{2}$

Write the main row in the new table in the same position as the key row of the preceeding table.

Step 7. Develop the balance of the new table 4 :

(a) Introduce variable x_1 and objective coefficient 6 in the main row in the stub replacing x_5 and 0 i.e.

6 $\quad x_1 \quad$ 7000 $\quad$ 1 $\quad \frac{1}{2} \quad$ 0 $\quad$ 0 $\quad \frac{1}{2} \quad$ Main row

(b) The remainder of the variable and objective columns are reproduced in the new table −4 exactly as they west in Table −3. i.e.

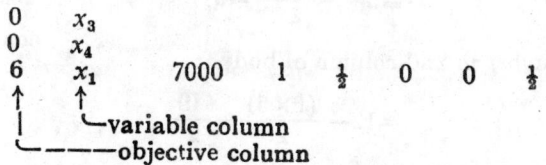

0 $\quad x_3$
0 $\quad x_4$
6 $\quad x_1 \quad$ 7000 $\quad$ 1 $\quad \frac{1}{2} \quad$ 0 $\quad$ 0 $\quad \frac{1}{2}$

└variable column
└────objective column

(c) All the remaining rows numbers are calucated by

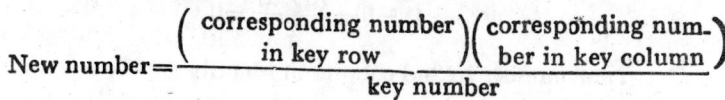

$$\text{New number} = \frac{\left(\begin{array}{c}\text{corresponding number}\\\text{in key row}\end{array}\right)\left(\begin{array}{c}\text{corresponding num-}\\\text{ber in key column}\end{array}\right)}{\text{key number}}$$

First row of Table 3

New number in constant column

$$=24000-\frac{(14000)(3)}{2}=3000$$

New number in 1st column of body

$$=3-\frac{(2)(3)}{2}=0$$

New number in 2nd column of body

$$=4-\frac{(1\times3)}{2}=\frac{5}{2}$$

New number in 1st column of identity

$$=1-\frac{(0\times3)}{2}=1$$

New number in 2nd column identity

$$=0-\frac{(0\times3)}{2}=0$$

New number in 3rd column of identity

$$=0-\frac{(1\times3)}{2}=-\frac{3}{2}$$

Second row of Table 3

New number in constant column

$$=60000-\frac{(14000\times3)}{2}=39000$$

New number in 1st column of body

$$=5-\frac{(2\times5)}{2}=0$$

New number in 2nd column of body

$$=12-\frac{(1\times5)}{2}=\frac{19}{2}$$

New number in 1st column of indentity

$$=0-\frac{(0\times5)}{2}=0$$

New number in 2nd column of identity

$$=1-\frac{(0\times5)}{2}=1$$

New number in 3rd column of identity

$$=0-\frac{(1\times 5)}{2}=-\frac{5}{2}$$

Index row of Table 3

New number in 1st column of body

$$=-6-\frac{(2\times-6)}{2}=0$$

New number in 2nd column of body

$$=-4\cdot5-\frac{(1\times-6)}{6}=-1\cdot5$$

New number in 1st column of identity

$$=0-\frac{(0\times-6)}{2}=0$$

New number in 2nd column of identity

$$=0-\frac{(0\times-6)}{2}=0$$

New number in 3rd column of identity

$$=0-\frac{(1\times-6)}{2}=3$$

All these above numbers are introduced in the matrix 7(*b*) and written at table 4 as shown below.

Table 4. Simplex Matrix Ready for Second Iteration

			6	4·5	0	0	0	
			x_1	x_2	x_3	x_4	x_5	
0	x_3	3000	0	5/2	1	0	−3/2	
0	x_4	39000	0	19/2	0	1	−5/2	
6	x_1	7000	1	1/2	0	0	1/2	Main Row
			0	−1·5	0	0	3	Index Row

Step 8 : Repeat steps 3 to 7 *c*

Table 5. 2nd Iteration

			6	4·5	0	0	0	
			x_1	x_2	x_3	x_4	x_5	
0	x_3	3000	0	5/2	1	0	−3/2	Key Row
0	x_4	39000	0	19/2	0	1	−5/2	
6	x_1	7000	1	1/2.	0	0	1/2	
			0	−1·5	0	0	3	

↑
Key column

Since $-1\cdot5$ in the most negative number in index row, the corresponding column is the key row

(1) Number of key column   5/2   19/2   1/2

(2) Number of constant column   3000   39000   7000

(3) Quotient $=(2)\div(1)$ 1200   4104   14000

                         ↑
                       └─Smallest running
                           Quotient

Therefore,

0   x_3 3000   0   5/2   1   0 $-3/2$ in the key row.

$[5/2]$ the intersection of key row and key column is the key number.

Main Row $[3000 \quad 0 \quad 5/2 \quad 1 \quad 0 \quad -3/2] \div 5/2$

           $=1200 \quad 0 \quad 1 \quad 2/5 \quad 0 \quad -3/5$

Introduce variable x_2 and objective coefficient $4\cdot5$ in place of x_3 and '0' respectively in the main row and write the main row in place of Key Row of previous table. Remaining rows and numbers are calculated as per the formula of new number and written in Table 6.

Table 6. Simplex Matrix 2nd Iteration Completed.

			6	4·5	0	0	0
			x_1	x_2	x_3	x_4	x_5
4·5	x_2	1200	0	1	2/5	0	$-3/5$
0	x_4	27600	0	0	$-19/5$	1	26
6	x_1	6400	1	0	$-1/5$	0	4/5
			0	0	3/5	0	21/10 Index row

Since all the numbers in the index row are positive, the solution is optimal.

Therefore, $x_1 = 6400$

               $x_2 = 1200$

By substituting the values of $x_1 = 6400$ and $x_2 = 1200$ in objective function

$$P = 6\cdot0x_1 + 4\cdot5x_2$$

$$= 6 \times 6400 + 4\cdot5 \times 1200$$

$$= 38400 + 5400\cdot0$$

$$P = 43800 \text{ maximum.}$$

Problem 6. A lock manufacturer can sell all it can produce. It is making two types of locks; Type 1 fetches a profit 0·30 per lock and Type 2 ; 0·34 per lock and total capacity of the shops are as follows :

	Lock Type 1	Lock Type 2
Forging	1,000 No.	1,000 No.
Milling	1,500 No.	800 No.
Stamping	900 No.	700 No.
Assembly	1,000 No.	1,100 No.

Advice the product mix.

Solution.

Let $x_1 =$ Quantity of Lock type 1 produced

$x_2 =$ Quantity of Lock type 2 produced.

Forging shop utilisation

$$x_2 = (1,000 - x_1) \times \frac{1000}{1000}$$

·or $\qquad x_1 + x_2 = 1000 \qquad\qquad\qquad ...(1)$

Milling shop utilisation

$$x_2 = (1500 - x_1) \times \frac{800}{1500}$$

·or $\qquad 15x_2 = 12,000 - 8x_1$

·or $\qquad 8x_1 + 15x_2 = 12,000 \qquad\qquad ...(2)$

Stamping shop utilisation

$$x_2 = (900 - x_1) \times \frac{700}{900}$$

$$9x_2 = 6300 - 7x_1$$

·or $\qquad 7x_1 + 9x_2 = 6300 \qquad\qquad\qquad ...(3)$

Assembly shop utilisation

$$x_2 = (1000 - x_1) \times \frac{1100}{1000}$$

or $\qquad 10 x_2 = 11,000 - 11x_1$

or $\qquad 11x_1 - 10x_2 = 11,000 \qquad\qquad ...(4)$

In this case limiting shop is the stamping one, the combination of the mix will fall on stamping line. The optimum mix could be found by taking out the profit. The point which gives maximum profit will be the mix

$$x_2 = 0, \qquad \text{and } x_1 = 900$$

Putting the value in profit equation, we get

$$P = 0·30 \times 900 + 0·34 \times 0$$
$$= \text{Rs. } 270.$$

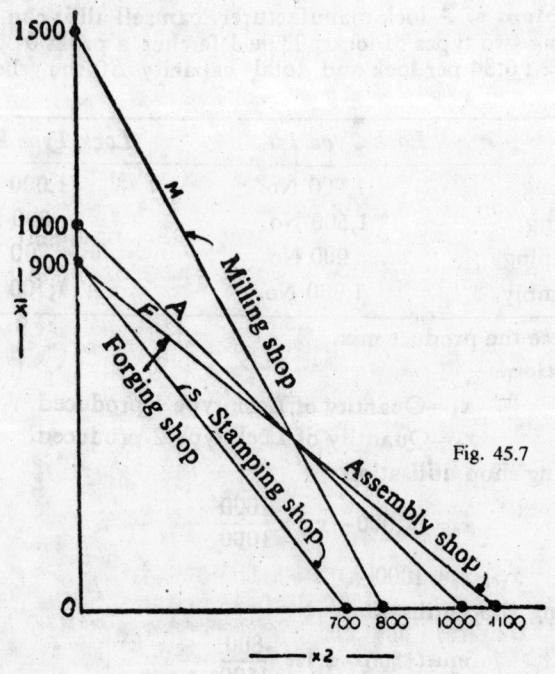

Fig. 45.7

Problem

Cost reduction. A company combines factor A and B to form a product which must weigh 50 kg. At least 20 kg of A and no more than 40 kg of B can be used. The cost of A Rs. 10 per kg and that B Rs. 25 per kg.

Find out quantity of A and B which should be used.

Solution.

If x_1 is kg of A used

x_2 is kg of B used

$$x_1 > 20$$

$$x_2 < 40$$

$$x_1 + x_2 = 50$$

$$C = 10x_1 + 25x_2$$

$$x_1 + x_2 + x_3 = 50$$

$$x_2 - x_4 + x_5 = 20$$

$$x_2 + x_6 = 40$$

The cost function

$$C = 10x_1 + 25x_2 + Mx_3 + 0x_4 + Mx_5 + 0x_6$$

here,

$$\left.\begin{array}{c} x_3 \\ x_5 \end{array}\right\} \text{ are artificial variablds}$$

$$\left.\begin{array}{c} x_4 \\ x_6 \end{array}\right\} \text{ slack variables}$$

By adding variables to all the equations

$$x_1 + x_2 + x_3 + 0x_4 + 0x_5 + 0x_6 = 50$$
$$x_1 + 0x_2 + 0x_3 - x_4 + x_5 + 0X_6 = 20$$
$$0x_1 + x_2 + 0x_3 + 0x_4 + 0x_5 + X_6 = 40$$

Simplex Table 1

C_j	Sol. max.	Sol. value	M x_3	0 x_4	M x_5	0 x_6	10 x_1	25 x_2
M	x_3	50	1	0	0	0	1	1
M	x_5	20	0	-1	1	0	1	0 ←—
0	x_6	40	0	0	0	1	0	1
	Z_j	$70M$	M	$-M$	M	0	$2M$	M
	$C_j - Z_j$		0	$+M$	0	0	$10 - 2M$	$25 - M$

↑

Note : We must try to drive out artificial variable first.

Simplex Table 2

C_j	Sol. max.	Sol. value	M x_3	0 x_4	M x_5	0 x_6	10 x_1	25 x_2
→M	x_3	30	1	1	-1	0	0	1
10	x_1	20	0	-1	1	0	1	0
0	x_6	40	0	0	0	1	0	1
	Z_j		M	$M - 10$	$10 - m$	0	10	M
	$C_j - Z_j$		0	$-m + 10$	$-10 + 2m$	0	0	$25 - m$

↑

Simplex Table 3

C_j	Sol. max.	Sol. value	M x_3	0 x_4	M x_5	0 x_6	10 x_1	25 x_2
0	x_4	30	1	1	-1	0	0	1 ←—
10	x_1	50	1	0	0	0	1	1
0	x_6	40	0	0	0	1	0	1
	Z_j	500	10	0	0	0	10	10
			$M - 10$	0	M	0	0	15

↑

$$\begin{array}{cc} x_1 & x_6 \\ 20-(-1\times30)=50 & 40-0 \\ 0-(-1\times1)=+1 & 0-0 \\ -1-(-1\times1)=0 & 0-0 \\ 1-(-1\times-1)=0 & 0-0 \\ 0-(-1\times0)=0 & 1-0 \\ 1-(-1\times0)=1 & 0-0 \\ 0-(-1\times1)=1 & 1-0 \end{array}$$

Simplex Table 4

C_f	Sol. max.	Sol. value	M x_3	0 x_4	M x_5	0 x_6	10 x_1	25 x_1
25	x_2	30	1	1	-1	0	0	1
10	x_1	20	0	-1	1	0	1	0
0	x_6	10	-1	-1	1	1	0	0
	Z_j	950	25	15	-15	0	10	25
	C_j-Z_j		$M-25$	-15	$M-15$	0	0	0

$$\begin{array}{cc} x_1 & x_6 \\ 50-1\times30=20 & 40-1\times30=10 \\ 1-1\times1=0 & 0-1\times1=-1 \\ 0-1\times1=-1 & 1-1\times1=-1 \\ 0-1\times0-1=0 & 0-1\times-1=1 \\ 0-1\times0=0 & 1-1\times0=1 \\ 1-1\times0=1 & 0-1\times0=0 \\ 1-1\times1=0 & 1-1\times1=0 \end{array}$$

C_j-Z_j should be 0 or possitive for minimization and 0 is obtained in this case for x_1 and x_2 hence solution is reached

$$x_1=20$$
$$x_2=30$$
$$C=950.$$

Problem. A carpenter has 90, 80 and 50 running feet of plywood, pine and birch respectively. Product A requires 2, 1 and 1 running feet of plywood, pine and birch respectively, and product B requires 1, 2 and 1 running feet plywood, pine and birch respectively. If A sells for Rs. 12 and B sells for Rs. 10, how many of each should he make to obtain the maximum gross income ?

Maximum Gross Income

Solution. To solve this problem first we formulate the problem

(i) Let x_1 be the unit of product A and x_2 be the product B.

So

		Plywood	Pine	Birch
	A	2	1	1
	B	1	2	1

Maximum quantity available

	90	80	50

Selling price per unit Rs. 12 for A and Rs. 10 for B.

So
$$Z = 12x_1 + 10x_2$$
$$2x_1 + x_2 < 90$$
$$x_1 + 2x_2 \leqslant 80$$
$$x_1 + x_2 < 50$$

(ii) Now next step is the Development of Equation from the inequalities adding the slack variables. Now the new equation will be :

$$2x_1 + x_2 + x_3 + (0)x_4 + (0)x_5 = 90$$
$$x_1 + 2x_2 + (0)x_3 + x_4 + (0)x_5 = 80$$
$$x_1 + x_2 + (0)x_3 + (0)x_4 + x_5 = 50$$
$$Z = 12x_1 + 10x_2 + (0)x_3 + (0)x_4 + x_5$$

(iii) Let us now develop the initial Simplex Table 1.

Table 1

			0	0	0	12	10	
C_j	Sol. variables	Sol. value	x_3	x_4	x_5	x_1	x_2	
0	x_3	90	1	0	0	2	1	←key row
0	x_4	80	0	1	0	1	2	
0	x_5	50	0	0	1	1	1	
	$C_j - Z_j$		0	0	0	12	01	

Key column

Table 2

key column
↓

C_j	Sol. veriables	Sol. value	x_3	x_4	x_5	x_1	x_2
			0	0	0	12	10
12	x_1	45	1/2	0	0	1	1/2
0	x_4	35	—1/2	0	—1	0	3/2
0	x_5	5	—1/2	0	1	0	1/2	←KR
		$Z_j =$ 6		0	0	12	6
		$C_j - Z_j =$ —6		0	0	0	4

Table 3

C_i	Solution variables	Solution values	x_3	x_4	x_5	x_1	x_2
			0	0	0	12	10
12	x_1	40	1	1	—1	1	00
0	x_4	20	1	3	—4	0	0
10	x_2	10	—1	0	2	0	1
	$C_j - Z_j =$		—2	—12	—8	0	0

Here the value $C_j - Z_j$ for x_1 and x_2 are zero and for others is in negative so we can't move further. Hence profit maximisation will be at that time when we sell the units of product A and B in the following manner :

$$x_1 = 40 \times 12 = 480$$
$$x_2 = 10 \times 10 = 100$$

Total income would be

$$= Rs. 580. \qquad \textbf{Ans.}$$

Problem. In a metal shop. two articles are produced, *viz.* bucket and pressed metal handles. The lids takes 20 sec to stamp, 30 seconds to form and 20 seconds to paint. The handles takes 40 sec to stamp, 10 seconds to form and 16 seconds to paints. The gross profit margins on lids and handles are 6P. and 9P. respectively. The time available per day on each process is 8 hours and 20 minutes, *i.e.* 30,000 seconds.

How many lids and buckets should be produced to obtain maximum gross profit ? What will be the maximum gross profit ?

Solution. *First step.* First step is to formulate the linear programming problem,

The objective is to maximise profit. The function will be mathematically stated as under :

$$Z=6x_1+9x_2$$

Capacity retains are

$$20x_1+40x_2<30,000$$
$$30x_1+10x_2<30,000$$
$$20x_1+16x_2<30,000$$

Second Step. Develop equations from inequalities.

$$Z=6x_1+9x_2+(0)x_3+(0)x_4+(0)x_5$$
$$20x_1+40x_2+x_3+(0)x_4+(0)x_5=30,000$$
$$30x_1+10x_2+(0)x_3+x_4+(0)x_5=30,000$$
$$20x_1+16x_2+(0)x_3+(0)x_4+x_5=30,000$$

Table 1

			0	0	0	6	9	
C_j	Solution variable	Solution value	x_3	x_4	x_2	x_1		
0	x_5	30,000	0	0	1	20	16	
0	x_4	30,000	0	1	0	30	10	
0	x_3	30,000	1	0	0	20	40	←K.R.
	Z_j		0	0	0	0	0	
	C_j-Z_j		0	0	0	6	9	

Table 2

			0	0	0	6	9
C_j	Solution variable	Solution value	x_3	x_4	x_5	x_1	x_2
0	x_5	18,000	−2/5	0	1	12	0
0	x_4	22,500	−1/4	1	0	25	0
9	x_2	750	−1/40	0	0	1/2	1
	Z_j		0	3	0	9/2	9
	$C_j-Z_j=$		9/40	0	0	3/2	0

Table 3

C_j	Production Variable	Solution value	x_3	x_4	x_5	x_1	x_6
			0	0	0	6	9
0	x_5	7200	373/50	−12/25	1	0	0
6	x_1	900	−25/4	1/25	0	1	0
9	x_2	300	126/40	1/50	0	0	1
	Z_j		888/4	3/50	0	6	9
	$C_j - Z_j =$		−221	−3/50	0	0	0

Profit Maximisation

$$= x_1 = 900 \times 6 = 5400$$
$$x_2 = 300 \times 9 = 2700$$

Total profit $= 8100$

Rs. 81·00   **Ans.**

Problem. Maximise Profit $4x_1 + 3x_2$

$$4x_1 + 2x_2 < 10$$
$$2x_1 + 8/3x_2 < 8$$
$$x_1 > 0$$
$$x_2 \geqslant 1·8$$

Removal of inequalities.

$$4x_1 + 2x_2 + x_3 + (0)x_4 + (0)_5 + (0)x_6 = 10$$
$$2x_1 + 8/3x_2 + (0)x_3 + x_4 + (0)x_5 + (0)x_6 = 8$$
$$(0)x_1 + x_2 - x_5 + x_6 + (0)x_3 + (0)x_4 = 1·8$$
Profit $= 4x_1 + 3x_2 + (0)x_3 + (0)x_4 + (0)x_5 - Mx_6$.

Table 1

C_j	Solution variable	Solution value	x_6	x_5	x_4	x_3	x_2	x_1
			−M	0	0	0	3	4
0	x_3	10	0	0	0	1	2	4
0	x_4	8	0	0	1	0	8/3	2
−M	x_6	1·8	1	−1	0	0	1	0
			−M	M	0	0	−M	0
	Z_j		−M	M	0	0	−M	0
	$C_j - Z =$		0	−M	0	0	3+M	4

Table 2

C_j	Solution variable	Solution value	$-M$	0	0	0	3	$\lvert 4 \rvert$
			x_6	x_5	x_4	x_3	x_3	$\lvert x_2 \rvert$
0	x_3	32/5	$\lvert -2$	2	0	1	00	$\lvert 4 \rvert$
0	x_4	16/5	$-8/3$	8/3	1	0	1	$\lvert 2 \rvert$
3	x_2	1·8	1	-1	0	0	1	$\lvert 9 \rvert$
	Z_j		3	-3	0	0	3	$\lvert 0 \rvert$
	$C_j - Z_j$		$-(M+3)+3$	0	0	0		$\lvert 4 \rvert$

Table 3

C_j	Solution variable	Solution value	$-M$	0	0	0	3	4
			x_6	x_5	x_4	x_3	x_2	x_1
4	x_1	8/5	$-1/2$	1/2	0	1/4	0	1
0	x_4	0	$-5/3$	5/3	1	$-1/2$	0	0
3	x_2	1·8	-1	-1	0	0	1	0
	C_j		1	-1	0	1	3	4
	$C_j - Z_j$		$-(M+1)+1$	0	-1	0	0	

Profit maximisation

$$x_1 = 8/5 \times 4 = 6·4$$
$$x_2 = 1·8 \times 3 = 5·4$$
$$\text{Total profit} = 6·4 + 5·4 = 11·8$$

Problem. Find out the maximum profit, subject to following conditions.

$$P = 2x_1 + 4x_2$$
$$x_1 + 4x_2 < 24$$
$$3x_1 + x_2 < 21$$
$$x_2 + x_4 < 8$$
$$x_1 > 0$$
$$x_2 > 0$$

Solution. Removal of Inequalities.

$$P = 2x_1 + 4x_2 + (0)x_3 + (0)x_4 + (0)x_5$$
$$x_1 + 4x_2 + x_3 + (0)x_4 + (0)x_5 = 24$$
$$3x_1 + x_2 + (0)x_3 + x_4 + (0)x_5 = 21$$
$$x_1 + x_2 + (0)x_3 + (0)x_4 + x_5 = 8$$

Table 1

C_j	Solution variable	Solution value	0 x_5	0 x_4	0 x_3	4 x_2	2 x_1
0	x_5	24	0	0	1	4	1
0	x_4	21	0	1	0	1	3
0	x_5	8	1	0	0	1	1
	C_j	0	0	0	0	0	0
	$C_j-Z_j=$	0	0	0	0	4	2

Table 2

C_j	Solution variable	Solution value	0 x_5	0 x_4	0 x_3	4 x_2	2 x_1
4	x_2	6	0	0	1/4	1	1/4
0	x_4	15	0	1	−1/4	0	1/4
0	x_5	2	1	0	−1/4	0	3/4
	C_j	0	0	0	0	0	1
	$C_j-Z_j=$	0	0	−0	0	0	1

Table 3

C_j	Solution variable	Solution value	0 x_5	0 x_4	0 x_3	4 x_2	2 x_1
4	x_2	16/3	−1/3	0	1/3	1	0
0	x_4	23/3	−11/3	1	2/3	0	0
2	x_1	8/3	4/3	0	−1/3	0	1
	Z_j		4/3	0	2/3	4	2
	$C_j-Z_j=$		4/3	0	2/3	0	0

Hence maximum, profit will be

$$x_1=8/3\times2=16/3=5{\cdot}33$$
$$x_2=16/3\times4=64/3=21{\cdot}33$$
Total Profit$=26{\cdot}66.$ **Ans.**

Problem.   The Meerut Manufacturing Company makes 3 products x_1, x_2, x_3. Their contributions towards profits are Rs. 2, Rs. 4, and Rs. 3 per unit respectively. Each of these products passes through 3 manufacturing centres as a part of the production process. Time required in each to produce one unit of each of the 3 products as well as the total time available at each centre for next week in hours are as follows :

Product	Centre 1	Centre 2	Centre 3
x_1	3	2	1
x_2	4	1	3
x_3	2	2	2
Total time available	60	40	80

Determine the optimum product plan for the next week's production schedule

Solution. Objective function

$$z = 2x_1 + 4x_2 + 3x_3$$

Subject to the constraints

$$3x_1 + 4x_2 + 2x_3 < 60$$

$$2x_1 + x_2 + 2x_3 < 40$$

$$x_1 + 3x_2 + 2x_3 < 30$$

Here z is to be maximised.

Introducing slack variables to change the constraints into equality

$$3x_1 + 4x_2 + 2x_3 + S_1 + 0S_2 + 0S_3 = 60$$

$$2x_1 + x_2 + 2x_3 + 0S_1 + S_2 + 0S_3 = 40$$

$$x_1 + 3x_2 + 2x_3 + 0S_1 + 0S_2 + S_3 = 80$$

and the objective function becomes

$$Z = 2x_1 + 4x_2 + 3x_3 + 0S_1 + 0S_2 + 0S_3$$

Table 1

C_j			0	0	0	2	4	3		
	Sol. variable	Sol. value	S_1	S_2	S_3	x_1	x_2	x^0	Ratio	
0	S_1	60	1	0	0	3	4	2	15	←KR
0	S_2	40	0	1	0	2	1	2	60	
0	S_3	80	0	0	1	1	3	2	26·6	
	Z_j	0	0	0	0	0	0	0		
	$C_j - Z_j$		0	0	0	2	4	3		

Key Column

Replacing S_1 by x_2 in table 2

Table 2

C_j			0	0	0	2	4	3	
	Sol. variable	Sol. value	S_1	S_2	S_3	x_1	x_2	x_3	
4	x_2	15	1/4	0	0	3/4	1	1/2	Rotio 30
0	S_2	25	−1/4	1	0	5/4	0	3/2	50/3 ←KR
0	S_3	35	−3/4	0	0	−5/4	0	1/2	70/
	Z_j	60	1	0	0	3	4	2	
	$C_j−Z_j$		−1	0	0	−1	0	1	

Key Column

Replacing S_2 by x_3 in table 3

Table 3

C_j			0	0	0	2	4	3
	Sol. variable	Sol. value	S_1	S_2	S_3	x_1	x_2	x_3
4	x_2	20/3	1/3	−1/3	0	1/3	1	0
3	x_3	50/3	−1/6	2/3	0	5/6	0	1
0	S_3	80/3	−2/3	−1/2	1	−5/3	0	0
	Z_j	230/3	5/6	2/3	0	23/6	4	3
	$C_j−Z_j$		−5/6	−2/3	0	−11/12	0	3

Since in column $C_j−Z_j$ under, x_2 and x_3 has become negative or zero. Therefore $x_2=20/3=6{\cdot}67$ $x_3=50/3=1666$.

Maximum profit$=230/3=76{\cdot}66$

or, $Z=4 \times 20/3 \times 3 \times 50/3=230/3=76{\cdot}66$

Problem. Find the maximum value of $z=8x_1+9x_2$.
$$x_1, x_2 \geqslant 0$$

Subject to the following conditions
$$5x_1 + 4x_2 < 40$$
$$2x_1 + 3x_2 < 24$$
$$x_1 + 2x_2 < 14.$$

Solution. Introducing the slack variables and changing inequalities into equalities we get the following :

$$5x_1 + 4x_2 + S_1 + 0S_2 + 0S_3 = 40$$
$$2x_1 + 3x_2 + 0S_1 + S_2 + 0S_3 = 24$$
$$x_1 + 2x_2 + 0S_1 + S_2 + S_3 = 14$$

and $z = 8x_1 + 9x_2 + 0S_1 + 0S_2 + 0S_3$

Table 1

C_j		0	0	0	0	8	9	
	Sol. variable	Sol. value	S_1	S_2	S_3	x_1	x_2	Ratio
0	S_1	40	1	0	0	5	4	10
0	S_2	24	0	1	0	2	3	8
0	S_3	14	0	0	1	1	2 7	←K.R.
	Z_j		0	0	0	0	0	0	
	$C_j - Z_j$		0	0	0	0	8	9	

↑ Key Column

S_3 will be replaced by x_2 in table 2.

Table 2

C_j			0	0	0	8	9	Ratio
	Sol. variable	Sol. value	S_1	S_2	S_3	x_1	x_2	
0	S_1	2	1	0	−2	3 0 4		←K.R.
0	S_2	3	0	1	−3/2	+1/2 0 +6		
9	x_2	7	0	0	1/2	1/2 1 7½=14		
	Z_j	73	0	0	9/2	9/2	9	
	$C_j - Z_j$		0	0	9/2	7/2	0	

↑ Key Column

Calculation

$$S_1 = 40 - 4 \times 7 = 12$$
$$1 - 4 \times 0 = 1$$
$$0 - 4 \times 0 = 0$$
$$0 - 4 \times 1/2 = -2$$
$$5 - 4 \times 1/2 = 3$$
$$4 - 4 \times 1 = 0$$
$$S_2 = 24 - 3 \times 7 = 3$$
$$0 - 3 \times 0 = 0$$
$$1 - 3 \times 0 = 1$$
$$0 - 3 \times 1/2 = -3/2$$
$$2 - 3 \times 1/2 = 1/2$$
$$3 - \times 1 = 0$$

Table 3
S_1 to be replaced by x_1

C_j	Sol. variable	Sol. value	0	0	0	8	9	
			S_1	S_2	S_3	x_1	x_2	Ratio
8	x_1	4	1/3	0	$-\frac{2}{3}$	1	0	
0	S_2	1	$-\frac{1}{6}$	1	$-7/6$	0	0	
9	x_2	5	$-\frac{1}{6}$	0	5/6	0	1	
	Z_j	77	7/6	0	13/6	8	9	
	$C_j - Z_j$		$-7/6$	0	$-13/6$	0	0	

Calculation

$$S_2$$
$$3-\tfrac{1}{2}\times 4=1$$
$$0-\tfrac{1}{2}\times\tfrac{1}{3}=-\tfrac{1}{6}$$
$$1-\tfrac{1}{2}\times 0=1$$
$$-9/2-\tfrac{1}{2}\times\tfrac{2}{3}=7/6$$
$$\tfrac{1}{2}-\tfrac{1}{2}\times 1=0$$
$$0-\tfrac{1}{2}\times 0=0$$

Calculation

$$x_2$$
$$7-\tfrac{1}{2}\times 4=5$$
$$0-\tfrac{1}{2}\times\tfrac{1}{3}=-\tfrac{1}{6}$$
$$0-\tfrac{1}{2}\times 0=0$$
$$\tfrac{1}{2}-\tfrac{1}{2}\times -\tfrac{2}{3}=5/6$$
$$\tfrac{1}{2}-\tfrac{1}{2}\times 1=0$$
$$1-\tfrac{1}{2}\times 0=1$$
$$x_1=4$$
$$x_2=5$$

Therefore, maximum profit

$$=8x_1+9x_2$$
$$=8\times 4+9+5$$
$$=32+45$$
$$=77.$$

Problem. A manufacturer of T.V. sets makes following four models :

1. **Monarc**
2. **Monitor**
3. **Wintage**
4. **Koral**

Each set requires time to assemble and test. The assembly and testing requirement of four models are :

	I	II	III	IV	Total available time
Assembly time (hr.)	8	10	12	15	2000
Testing time (hr.)	2	2	4	5	500
Profit (Rs.)	40	60	80	100	

The supplier of the picture tube indicated that he would not be able to supply more than 180 tubes in the next month of which not more than 100 could be coloured one.

(1) What is the optimum production schedule for Wintage T.V. manufacturer.

(2) What is the marginal value of an additional hour of assembly time.

Solution. Let x_1, x_2, x_3, x_4 be the number T.V. sets of 4 types.

Objective function. $Z = 40x_1 + 60x_2 + 80x_3 + 100x_4$

Constrained equations :

$$8x_1 + 10x_2 + 12x_3 + 15x_4 \leqslant 2000 \qquad \ldots(1)$$
$$2x_1 + 2x_2 + 4x_3 + 5x_4 < 500 \qquad \ldots(2)$$
$$x_1 + x_2 + x_3 + x_4 < 180 \qquad \ldots(3)$$
$$x_3 + x_4 < 100 \qquad \ldots(4)$$

Adding slack variables and removing inequalities.

Objective function

$$Z = 40x_1 + 60x_2 + 80x_3 + 100x_4 + 0x_5 + 0x_6 + 0x_7 + 0x_8$$
$$8x_1 + 10x_2 + 11x_3 + 15x_4 + 1x_5 + 0x_6 + 0x_7 + 0x_8 = 2000$$
$$2x_1 + 2x_2 + 4x_3 + 5x_4 + 0x_5 + 1x_6 + 0x_7 + 0x_8 = 500$$
$$x_1 + x_2 + x_3 + x_4 + 0x_5 + 0x_6 + 1x_7 + 0x_8 = 180$$
$$0x_1 + 0x_2 + x_3 + x_4 + 0x_5 + 0x_6 + 0x_7 + 1x_8 = 100$$

Initial Simplex Table 1

	C_j		0	0	0	0	0	40	60	80	100	
	Sol. variable	Sol. value	x_5	x_6	x_7	x_8	x_1	x_2	x_3	x_4	Ratio
0	x_5	2000	1	0	0	0	8	10	12	15	133·3
K.R.→0	x_6	500	0	1	0	0	2	2	4	5	100
0	x_7	180	0	0	1	0	1	1	1	1	180
0	x_8	100	0	0	0	1	0	0	1	1	100
	Z_j	0	0	0	0	0	0	0	0	0	
	$C_j - Z_j$	0	0	0	0	0	40	60	80	100	

↑
Key Column

Replacing x_6 by x_4 in table 2

Table 2

C_j			0	0	0	0	40	60	80	100	
	Sol. variable	Sol. value	x_5	x_6	x_7	x_8	x_1	x_2	x_3	x_4	Ratio
K.R.→ 0	x_5	500	1	−3	0	0	2	4	0	0	125
100	x_4	100	0	1/5	0	0	2/5	2/5	4/5	1	250
0	x_7	80	0	−1/5	1	0	3/5	3/5	1/5	0	133·5
0	x_8	0	0	−1/5	0	1	−2/5	2/5	1/5	0	0
Z_j		10000	0	20	0	0	40	40	80	100	
$C_j - Z_j$		−100000	0	−20	0	0	0	20	0	0	

Key Column

x_5 is to be replaced by x_2 in table 3

Table 3

C_j			0	0	0	0	0	40	6	80	100
	Sol. variable	Sol. value	x_5	x_6	x_7	x_8	x_1	x_2	x_3	x_4	
60	x_2	125	1/4	−3/4	0	0	1/2	1	0	0	
100	x_4	50	−1/10	1/2	0	0	1/5	1	4/5	0	
0	x_7	5	−3/20	1/4	1	0	3/10	0	1/5	0	
0	x_8	50	1/10	−1/2	0	1	−1/5	0	1/5	0	
Z_j		12500	5	5	0	0	50	60	80	100	
$C_j - Z_j$		−12500	−5	−5	0	0	−10	0	0	0	

$$x_2 = 125$$
$$x_4 = 59$$

Therefore, maximum profit $= 60 \times 125 + 100 \times 50$
$$= Rs.\ 12500$$

Problem. Maximise $P = 6x_1 + 7x_2$
with given constraints as

$$2x_1 + 3x_2 \leqslant 12$$
$$2x_1 + x_2 \leqslant 8$$

Solution.

Introducing slack variables and changing the inequalities into equalities, we get

$$2x_1 + 3x_2 + S_1 + 0S_2 = 12$$
$$2x_1 + x_2 + 0S_1 + S_2 = 8$$
Also $$P = 6x_1 + 7x_2 + 0S_1 + 0S_2$$

Initial Simplex Table 1

C_j			6	7	0	0	
	Sol. variable	Sol. value	x_1	x_2	S_1	S_2	Ratio
0	S_1	12	2	3	1	0	4 ←K.R.
0	S_2	8	2	1	0	1	8
	Z_j	0	0	0	0	0	
	C_j-Z_j		6	7	0	0	

↑
Key Column

Replacing S_1 by x_2 in table 2

Table 2

C_j			6	7	0	0	
	Sol. variable	Sol. value	x_1	x_2	S_1	S_2	Ratio
7	x_2	4	2/3	1	1/3	0	6
0	S_2	4	4/3	0	−1/3	0	3 ←K.R.
	Z_j	28	14/3	7	7/3	0	
	C_j-Z_j		4/3	0	−7/3	0	

↑
key Column

S_2 to be replaced by x_1.

Table 3

C_j			6	7	0	0
	Sol. variable	Sol. value	x_1	x_2	S_1	S_2
7	x_2	2	0	1	1/2	−1/2
6	x_1	3	1	0	−1/4	3/4
	C_j	32	6	7	2	i
	C_j-Z_j		0	0	−2	−1

Optimal solution $x_1=3$

$x_2=2$

Maximum profit $=P=6x_1+7x_2$

$=6x_2+7x_2$

$=18+14=32.$

Problem. The management of a garment factory hires to recommend the most profitable mix of the three types of garments A, B, C, it is desirous to include in its future plans. The process ng takes place in three sections x, y, z. The production of one unit in the shops (in minutes) and the total available capacity of the shops is compiled in the table. Table also contains the profit contribution for unit of each product. Recommend the most profitable mix to the management.

Table

Departments	Type of garment			Available Capacity
	A	B	C	
Cutting (x)	10·7	5·0	2·0	2705
Stitching (y)	5·4	10·0	4·0	2210
Finishing (z)	0·7	1·0	2·0	445

Profit per unit   Rs. 10·00   Rs. 15·00   Rs. 20·00

Solution. Let x_1, x_2 and x_3 be the number of garments A, B and C respectively.

Framing Problem

$Z = 10x_1 + 15x_2 + 20x_3$ to be maximised :

Express the constraints

$$10·7x_1 + 5x_2 + 2x_3 \leqslant 2705$$
$$5·4x_1 + 10x_2 + 4x_3 \leqslant 2210$$
$$0·7x_1 + 1·0x_2 + 2x_3 \leqslant 445$$
$$x_1, x_2 \text{ and } x_3 \geqslant 0$$

Convert the inequalities by adding slack variables :

$$10·7x_1 + 5x_2 + 2x_3 + 1S_1 + 0S_2 + 0S_3 = 2705$$
$$5·4x_1 + 10x_2 + 4x_3 + 0S_1 + 1S_1 + 0S_3 = 2210$$
$$0·7x_1 + 1x_2 + 2x_3 + 0S_1 + 0S_2 + 1S_3 = 445$$
$$Z = 10x_1 + 15x_2 + 20x_3 + S_1 + 0S_2 + 0S_0$$

Simplex Table :

Table 1

Profit per unit C_j	$C_j \rightarrow$		0	0	0	10	15	20	
C_j	Solution variable	Solution values	S_1	S_2	S_3	X_1	X_2	X_3	Ratio
0	S_1	2705	1	0	0	10·7	5	2	2705/2 = 1352·5
0	S_2	2210	0	1	0	5·4	10	4	2210/4 = 552·5
0	S_3	445	0	0	1	0·7	1	2	445/2 = 222·5
0	Z_j (Contribution lost)	0	0	0	0	0	0	0	
PTP	$C_j - Z_j$ (Net contribution)	0	0	0	0	10	15	20	

Key row ← (at S_3 row)   ↑ Key column (at X_3)

x_3 enters in place of S_3

x_3 enters in place of S_3

Table 2

Cj↓	Sol. variable	Sol. value	0 S1	0 S2	0 S3	10 X1	15 X2	20 X3	
0	S1	2260	1	0	-1	10	4	0	2260/4=565
0	S2	1320	0	1	-2	4	[8]	0	320/8=165
20	X3	222.5	0	0	1/2	7/20	1/2	1	$\frac{222.5}{1/2}=445$
	Zj	445	0	0	10	7	10	20	
	Cj-Zj		0	0	-10	3	5	0	

X2 enters in place of S2

x_3 enters in place of S_2

Calculations :

S_1
$$2705-(2\times222.5)=2260$$
$$1-(2\times0)=1$$
$$0-(2\times1/2)=-1$$
$$10.7-(2\times7/20)=10$$
$$5-(2\times1/2)=4$$
$$2-(0\times1)=0$$
S_2
$$2210-(4\times222.5)=1320$$
$$0-(4\times0)=0$$
$$1-(4\times0)=1$$
$$0-(4\times\tfrac{1}{2})=-2$$
$$54-(4\times7/20)=4$$
$$10-(4\times\tfrac{1}{2})=8$$
$$4-(4\times1)=0$$

Table 3

Cj↓	Sol. variable	Sol. volume	0 S1	0 S2	0 S3	10 X1	15 X2	20 X3	Ratio
0	S1	1600	1	-1/2	0	[8]	0	0	-200
15	X2	165	0	1/8	-1/4	1/2	1	0	330
20	X3	140	0	-1/16	5/8	1/10	0	1	1400
	Zj	5275	0	5/8	70/8	9/2	15	20	
	Cj-Zj		0	-5/8	-70/8	11/2	0	0	

S1 is replaced by X1

[Calculations :

S_1
$$2260-(4\times165)=1600$$
$$1-(4\times0)=1$$

$$0-(4\times 1/8)=1/2$$
$$-1-(4\times -1/4)=0$$
$$10(-4\times 1/2)=8$$
$$4-(4\times 1)=0$$
$$0-(4\times 0)=0$$

Table 4

10	x_1	200	1/8	-1/16	0	1	0	0
15	x_2	65	-1/16	5/32	1/4	0	1	0
20	x_3	120	-1/80 $\frac{9}{100}$	5/8	0	0	1
	Z_j	5375	1/16	39/32	35/4	10	15	20
	C_j-Z_j		-1/16	-39/32	-35/4	0	0	0

$$x_2$$
$$225\cdot5-(1/2\times 165)=140$$
$$0-(1/2\times 0)=0$$
$$0-(1/2\times 1/8)=1/16$$
$$1/2-(1/2\times 1/4)=5/8$$
$$7/20-(1/2\times 1/2)=1/10$$
$$1/2-(1/2\times 1)=0$$
$$1-(1/2\times 0)=1$$

$$X_2$$
$$165-(1/2\times 200)=65$$
$$0-(1/2\times 1/8)=-1/16$$
$$1/8-(1/2\times -1/16)=5/32$$
$$-1/4-(1/2\times 1)=0$$
$$1/2-(1/2\times 1)=0$$
$$1-(1/2\times 0)=1$$
$$0-(1/2\times 0)=0$$

$$x_3$$
$$140-(1/10\times 200)=120$$
$$0-(1/10\times 1/8)=-1/80$$
$$-1/6-(1/10\times -1/16)=9/14$$
$$5/8-(1/10\times 0)=3/8$$
$$1/10-(1/10\times 1)=0$$
$$0-(1/10\times 0)=0$$
$$1-(10\times 0)=1$$

Since all the values of C_j-Z_j are either negative or zero, hence optimal result is given by task $4x_1=200$, $x_2=65$, $x_3=120$

∴ Quantity of product A should be produced $=200$

Quantity of product B should be produced $=65$

Quantity of product C should be produced $=120$

Therefore $Z=10\times 200+15+65+20\times 120$

$$=\text{Rs. }5375\textbf{ Ans.}$$

Problem. An animal feed company must produce 200 lb of a mixture containing the ingradients x_1 and x_2, x_1 costs Rs. 3 per lb and x_2 costs Rs. 8 per lb. Not more than 80 lb of x_1 can be used and minimum quantity to be used for x_2 is 60 lb. Find how much of each ingredient should be used if company wants to minimise the cost.

Solution. The objective function to be minimised is

$$Z = 3x_1 + 8x_2$$

Subject to the constraints

$$x_1 + x_2 = 200 \qquad \qquad ...(i)$$
$$x_1 \leqslant 80 \qquad \qquad ...(ii)$$
$$x_2 \geqslant 60 \qquad \qquad ...(iii)$$

Solution.

We introduce a slack variable S_1 in constraint (ii) to get

$$x_1 + S_1 = 80$$

In constraint (i) only one artificial variable A_1 is introduced

$$x_1 + x_2 + A_1 = 200$$

and in constraint (iii) one slack variable S_2 and an artificial variable A_2 is introduced to get

$$x_2 - S_2 + A_2 = 60$$

The objective function becomes :

$$Z^* = -Z = -3x_1 - 8x_2 + 0 \cdot S_1 - 0 \cdot S_2 - MA_1 - MA_2$$

Here Z^* is to be maximised, hence the penalty for A_1 and A_2 n the objective function will be $-M$ which has a very large value.

Various simplex tables are now framed as follows :

Unit profit, C_j C_j	Solution variable	Sol. value	S_1	S_2	A_1	A_2	X_1	X_2	Ratio
			0	0	$-M$	$-M$	-3	-8	
0	S_1	80	1	0	0	0	1	0	80
0	S_2	60	0	-1	0	1	0	1	80
$-M$	A_1	200	0	0	1	0	1	1	200
	Z_j	—	0	0	$-M$	0	$-M$	$-M$	
	$C_j - Z_j$	—	0	0	0	$-M$	$M-3$	$M-8$	

↑ Key Column

S_1 is replaced by x_1

Table 2

C_j	Sol. variable	Sol. value	S_1	S_2	A_1	A_2	x_1	x_2	Ratio
			0	0	$-M$	$-M$	-3	-8	
-3	x_1	80	1	0	0	0	1	8	0
0	S_2	60	0	-1	0	1	0	1	60
$-M$	A_1	120	-1	0	1	0	0	1	120
	Z_j		$M-3$	0	$-M$	0	-3	$-M$	
	$C_j - Z_j$		$3-M$	0	0	$-M$	0	$M-8$	

↑ Key Column

S_2 is replaced by x_2

Table 3

−3	x_1	80	1	0	0	$\|$ 0 $\|$	1	1	∞
−8	x_2	60	0	−1.	0	$\|$ 1 $\|$	0	1	−60
$\|-M$	A_1	60	−1	1	1	$\|$ −1 $\|$	0	0	60 $\|$
	Z_j		−3+M	8−M	−M	$\|$ M−8 $\|$	−3	−8	
	C_j-Z_j		3−M	M−8	0	$\|$ 8−2M $\|$	0	0	

↑
Key Column

S_2 replaces A_1

Table 4

−3	x_1	80	1	0	0	0	1	0
−8	x_2	120	−1	0	1	0	0	1
0	S_2	60	−1	1	1	−1	0	0
	Z_j	−1200	5	0	−8	0	−3	−8
	C_j-Z_j	—	−5	0	8−M	−M	0	0

Since all the values of C_j-Z_j are either negative or zero hence table gives the values of $x_1=80$ and $x_2=120$ lbs.

and $Z^*=3\times80+8\times120=240+960$

∴ Minimum value$=1200$ **Ans.**

Besides profit maximization problem, the simplex method can be used in cost minimization problems.

Problem. A cattle feed company must produce 400 quintals of a mixture consisting of ingredients x_1 and x_2. x_1 cost Rs. 6 per quintal and x_2 Rs. 16 per quintal. No more than 160 quintals of x_1 can be employed and a minimum of 120 quintals of x_2 must be used. The problem is to ascertain how much of each ingredient should be employed if the company is to minimise cost.

The cost function can be written as :

$$Cost=6x_1+16x_2 \qquad \ldots(i)$$
$$x_1+x_2=400 \qquad \ldots(ii)$$
$$x_1<160 \qquad \ldots(iii)$$
$$x_2\geqslant120 \qquad \ldots(iv)$$

I. Artificial Variables. As case of maximization problems, stock variables are to be introduced, in this case *Artificial variables* are used for the treatment of 2 types of restrictions :

(*i*) for equality

(ii) For greater than or equal to type, artificial variable is used.

Now the equation will be

$$x_1 + x_2 + A_1 = 400 \qquad A_1 = [\text{Artificial variable 1}]$$

With stock variable

$$x_1 + x_2 + A_1 + 0S_1 + 0S_2 + 0A_2 = 400 \qquad \ldots(1)$$
$$x_1 + 0x_2 + 10A_1 + 1S_1 + 0S_2 + 0A_2 = 160 \qquad \ldots(2)$$
$$0x_1 + x_2 + 0A_1 + 0S_1 - S_2 + A = 120 \qquad \ldots(3)$$

Cost problem $6x_1 + 16x_2 + MA_1 + 0S_1 + 0S_2 + MA_1$

Table 1

			M	0	0	M	0	16	
C_j	Solution variable	Solution value	A_1	S_1	S_2	A_2	x_1	x_2	
M	A_1	400	1	0	0	0	1	1	
0	S_1	160	0	1	0	0	1	0	
M	A_2	120	0	0	-1	1	1	1	K.R.
		$400M$	M	0	0	0	M	M	
		$120M$	0	0	$-M$	M	0	M	
	Z_j	$520M$	M	0	$-M$	M	M	$2M$	
	$i_j - Z_j$		0	0	$-M$	0	$6-M$	$16-2M$	

K.C.

II. **Optimum Column.** Which has the largest negative value would be a key column in $C_j - Z_j$.

We have to remember here that the profit maximization, key column is one which has the highest positive value.

Table 2

			M	0	0	M	6	16
C_j	Solution variable	Solution value	A_1	S_1	S_2	A_1	x_1	
16	x_2	120	0	0	-1	1	0	0
M	A_1	280	1	0	1	-1	1	0
0	S_1	160	0	1	0	0	1	0

K.R.

		1920	0	0	-16	16	0	16
		$+$ 280M	0	0	M	$-M$	M	0
	C_j		M	0	$M-16$	$16-M$	M	16
	$C_j - Z_j$		0	0	$16-M$	$2M-16$	$6-M$	0

Key Column

Table 3

C_j	Solution variable	Solution value	A_1	S_1	S_2	A_2	x_1	x_2	
		M	0	0	M	6	16		
6	x_1	160	0	1	0	0	1	0	
16	x_2	120	0	0	−1	1	0	1	
M	A_1	120	1	−1	1	−1	0	0	K.R.
		720	0	0	6	0	6	0	
		1120	0	0	0	−16	16	16	
		120M		M	−M	M	−M	0	
Z_j (Total)			M	6−M	M−16	16−M	16		
C_j−Z_j			0	M−6	16−M	2M−16	0		

↑
Key Column

Table 4

C_j	Solution variable	Solution value	A_1	S_1	S_2	A_2	x_1	x_2
			M	0	0	M	6	16
0	S_2	120	1	−1	1	−1	0	0
6	x_1	160	0	1	0	0	1	0
16	x_2	240	0	−1	0	0	0	1
		960	0	6	0	0	6	0
		3840	16	−16	0	0	0	16
	Total Z_j		10	−10	0	0	6	16
	C_j−Z_j		M−16	+10	0	M	0	0

Step III. This is the last step, when in (table 4) C_j−Z_j column, no negative value remain the optimum solution is reached.

Thus we should use 160 quintals of x_1 and 240 quintals of x_2, with this the cost comes to Rs. 4800.

Minimising an Objective Function

In case we have to deal with the costs, the minimisation of costs is often a desired objective. Then it can be expressed as :

Objective function x_1+x_2=minimum.

The procedure given earlier for maximisation will hold good provided the above objective function x_1+x_2=minimum is modified. Therefore, multiply the objective function throughout by −1.

$$-x_1-x_2=\text{maximum.}$$

Further while selecting the Key Column the largest positive number in the index row is selected instead of the largest negative number (which is done for maximisation procedure). All other steps remain exactly the same. Generally four types of equations are involved *i.e.* restrictions, requirements, equations and approximations. In all these equations the inequalities have to be removed by introducing appropriate slack variables.

Requirement. In same situations the requirement is that same combination of the variables must be greater than or equal to a given number like

$$6x_1+4x_2 \geqslant 24 \text{ has the requirement at least } 24.$$

To convert this in equality into an equation, a slack variable S_1 is subtracted and written as

$$6x_1+4x_2-S_1=24.$$

Since the coefficient of S_1 is -1 this equation can not be used in simplex method.

To apply simplex method, each equation must have are non zero entry in the identity and this entry must have a coefficient of $+1$. Therefore by adding an artificial variable A which will appear in identity, the above equation becomes

$$6x_1+4x_2-S_1+A_1=24$$

An arbitrary large negative contribution $-M$ is assigned to the artificial variable A in the objective function so that artificial variable is zero in the optimal solution.

Equations. The combination of certain variables will amount to some exact quantity. By introducing an artificial variable, the equation can be amenable for simplex method as simplex needs one variable for identity.

The equation $10x_1+6x_2=50$ can be modified to

$$10x_1+6x_2+A_2=50.$$

Further a $-M$ contribution is assigned to the artificial variable A_2 in the objective funtion in order to make the artificial variable zero in the optimal solution.

Approximations. In some problems the combination of some variables will be approximately equal to some number for flexibility like $5x_1+3x_2 \cong 40$. This can be modified to use in simplex method, by adding and subtracting slack variables say S_2 and S_3.

i.e., $$5x_1+3x_2-S_2+S_3=40.$$

These two slack variables make the above equation either slightly smaller or slightly greater than 40. In order to make the above equation as close as possible to 40, the negative coefficient -1 is assigned in the objective function to both slack variables S_2 and S_3.

If the objective function is $20x_1 + 15x_2 =$ Maximum.

It can be written as $20x_1 + 15x_2 - S_2 - S_3 =$ Maximum.

The slack variable S_2 will appear in the body and slack variable S_3 will appear in the identity of simplex matrix.

Example. Let the objective function be

$$10x_1 + 25x_2 = \text{Maximum}.$$

Subject to

$2x_1 + 1x_2 \leqslant 30$ Restriction relationship

$4x_1 + 6x_2 \geqslant 50$ Requirement relationship

$3x_1 + 6x_2 = 60$ Equality relationship

$5x_1 + 2x_2 \simeq 45$ Approximation relationship.

By applying the above guide lines these can be modified by introducing slack variables and artificial variables.

$$2x_1 + 1x_2 + S_1 = 30$$
$$4x_1 + 6x_2 - S_2 + A_1 = 50$$
$$3x_1 + 6x_2 + A_2 = 60$$
$$5x_1 + 2x_2 - S_3 + S_4 = 45$$

The objective function $10x_1 + 25x_2 =$ maximum is modified to include slack and artificial variables.

$$10x_1 + 25x_2(0) S_1 - (0) S_2 - S_3 - S_4 - MA_1 - MA_2 = \text{Maximum}.$$

By arranging the objective function and problem equations in the matrix form, the initial simplex matrix is obtained as below.

Initial Simplex Matrix

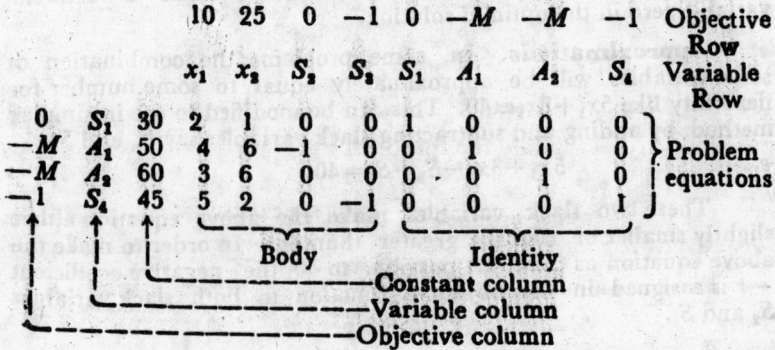

			10	25	0	−1	0	−M	−M	−1	Objective Row
			x_1	x_2	S_2	S_3	S_1	A_1	A_2	S_4	Variable Row
0	S_1	30	2	1	0	0	1	0	0	0	
−M	A_1	50	4	6	−1	0	0	1	0	0	⎫ Problem
−M	A_2	60	3	6	0	0	0	0	1	0	⎬ equations
−1	S_4	45	5	2	0	−1	0	0	0	1	⎭

Body   Identity

Constant column

Variable column

Objective column

Now the above matrix is ready for its operating. After calculating the Index row as explained earlier. The largest positive number is selected (instead of largest negative number) for identifying key column. All other steps are followed as they are given for maximisation by simplex method.

Problem. The manufacturer of Western Television sets make following four models :

Monarch
Monitor
Wintage and
Koral.

The assembly and testing requirements of time for four sets is as follows :

Model type	I	II	III	IV	Total Time Available
Assembly time	8	10	12	15	2000
Testing time	2	2	4	5	500
Marginal profit	40	60	80	100	

The supplier of the picture tube indicated that he would not be able to supply more than 180 picture tubes in the next month of which not more than 100 could be colour tubes. The colour tubes are used in set of Wintage and Koral type only.

Find Out :

(i) What is the optimal production schedule for the company ?

(ii) What is the marginal value of the additional hour of assembly time ?

Solution.

Step I. Formulate the problem by defining inequalities and objective function.

Let the company produce x_1, x_2, x_3 and x_4 sets of each models respectively.

The corresponding profit is then,

$$Z = 40x_1 + 60x_2 + 80x_3 + 100x_4$$

Subject to the restrictions (constraints) as follows :

$$8x_1 + 10x_2 + 12x_3 + 15x_4 \leqslant 2000 \quad \text{For assembly time}$$

$$2x_1 + 2x_2 + 4x_3 + 5x_4 < 500 \quad \text{,, testing time}$$

$$x_1 + x_2 + x_3 + x_4 < 180 \quad \text{,, total picture tubes}$$

$$x_3 + x_4 < 100 \quad \text{,, colour tubes}$$

$$x_1, x_2, x_3, x_4 \geqslant 0$$

In the problem the objective function Z is to be optimised.

Step II.   Convert the inequalities by adding slack vari‑ables.

$$Z = 40x_1 + 60x_2 + 80x_3 + 100x_4 + 0x_5 + 0x_6 + 0x_7 + 0x_8$$

$$8x_1 + 10x_2 + 12x_3 + 15x_4 + 1x_5 + 0x_6 + x_7 + 0x_8 = 2000$$

$$2x_1 + 2x_2 + 4x_3 + 5x_4 + 0x_5 + 1x_6 + 0x_7 + 0x_8 = 500$$

$$1x_1 + 1x_2 + 1x_3 + 1x_4 + 0x_5 + 0x_6 + 1_7 + 1x_8 = 180$$

$$0x_1 + 0x_2 + 1x_3 + 1x_4 + 0x_5 + 3x_6 + 0x_7 + 1x_8 = 100$$

Step III.   Prepare various Simple Tables to get the optimum solution.

Table 1

Unit profit C_j	Solution variable	Solution values	0	0	0	0	40	60	80	100	
			x_5	x_6	x_7	x_8	x_1	x_2	x_3	x_4	Ratio
0	x_5	2000	1	0	0	0	8	10	12	15	$\frac{2000}{15} = 133.3$
0	x_6	5000	0	1	0	0	2	2	4	5	$\frac{500}{5} = 100$
0	x_7	180	0	0	1	0	1	1	1	1	$\frac{180}{1} = 180$
0	x_8	100	0	0	0	1	0	0	1	1	$\frac{100}{1} = 100$
		$Z_j = 0$	0	0	0	0	0	0	0	0	
		$C_j - Z_j = 0$	0	0	0	0	40	60	80	100	

Explanations to table 1. Determination of key column, key row and key number.

Z_j is worked out by multiplying the figures of each column with the unit profit of respective row as shown small type. This is zero in each case. Adding all these values, we get Z_j. We now work out $C_j - Z_j$ as shown in the last row.

The key column is the one which falls in the largest $C_j - Z_j$ value. Decide the solution values from the corresponding figures in key column to get the ratio column. The smallest value in the key ratio column locates the key row and the value lying in the cell at the intersection of key row and column is the key number. If the minimum value in the ratio column are more than one (as in the present case), we may select any one (case of degeneracy). Key

Table — 2

Cj	Solution variable	Solution values	0	0	0	0	40	60	80	100	Ratio
			x_5	x_6	x_7	x_8	x_1	x_2	x_3	x_4	
0	x_5	500	1	-3	0	0	2	4	0	0	$\frac{500}{4}=125$
			0	0	0	0	0	0	0	0	
100	x_4	100	0	1/5	0	0	2/5	2/5	4/5	1	$\frac{100}{2}=250$
			0	20	0	0	40	40	80	100	-5
0	x_7	80	0	-1/5	1	0	3/5	3/5	1/5	0	$\frac{80}{3\!/5}=133.3$
			0	0	0	0	0	0	0	0	
0	x_8	0	0	-1/5	0	1	-2/5	-2/5	1/5	0	$\frac{0}{-2/5}=-10$
			0	0	0	0	0	0	0	0	(Negative values are not considered)
		Z_j	0	20	0	0	40	40	80	100	
		C_j-Z_j	0	-20	0	0	0	20	0	0	

column, row and number are shown in black rule. Here 5 is the key number and x_4 is the entering variable in place of x_6.

Explanations to table 2

Key row in table 1 is transformed by dividing its elements by the key number. This gives the corresponding row of table 2.

The other rows are also transformed by subtracting the product of the corresponding key column number and corresponding value determined in the above para, from the old row number of the non key row as worked out below :

$$2000-(15\times100)=500 \qquad 180-(1\times100)=80 \qquad 100-(1\times100)=0$$
$$1 \;-(15\times0) \;=1 \qquad 0\;-(1\times0)\;=0 \qquad 0-(1\times0)\;=0$$
$$0\;-(15\times1/5)=-3 \qquad 0\;-(1\times1/5)=-1/5 \qquad 0--(1\times1/5)\;=-1/5$$
$$0\;-(15\times0)\;=0 \qquad 1\;-(1\times0)\;=1 \qquad 0-(1\times0)\;=0$$
$$0\;-(15\times0)\;=0 \qquad 0\;-(1\times0)\;=0 \qquad 1-(1\times0)\;=1$$
$$8\;-(15\times2/5)=2 \qquad 1\;-(1\times2/5)\;=3/5 \qquad 0-(1\times2/5)\;=-2/5$$
$$10\;-(15\times2/5)\;=2 \qquad 1\;-(1\times2/5)\;=3/5 \qquad 0-(1\times2/5)\;=-2/5$$
$$12\;-(15\times4/5)\;=0 \qquad 1\;-(1\times4/5)\;=1/5 \qquad 1-(1\times4/5)\;=1/5$$
$$15\;-(15\times1)\;=0 \qquad 1\;-(1\times1)\;=0 \qquad 1-(1\times1)\;=0$$

Key column, key row and key number are worked out as in case of table 1. Since all the values of C_j-Z_j are not 0 or negative, hence, proceed for Table 3 for optimal solution values. 4 is the key number and x_2 is the entering variable in place of x_5.

Explanation to table 3

Table 3 is formed exactly the same way. Table 1 and 9 values of x_4, x_7 and x_8 are worked out as follows :

Table 3

Cj	Solution variable	Solution values	0 x5	0 x6	0 x7	0 x8	40 x1	60 x2	80 x3	100 x4	Ratio
60	x2	125	1/4 / 15	-3/4 / -45	0 / 0	0 / 0	1/2 / -30	1 / 60	0 / 0	0 / 0	
100	x4	50	-1/10 / -10	1/2 / 50	0 / 0	0 / 0	1/5 / 20	0 / 0	4/5 / 80	1 / 100	
0	x7	5	-3/20 / 0	1/4 / 0	1 / 0	0 / 0	3/10 / 0	0 / 0	1/5 / 0	0 / 0	
0	x8	50	1/10 / 0	-1/2 / 0	0 / 0	1 / 0	-1/5 / 0	0 / 0	-1/5 / 0	0 / 0	
		Zj	5	5	0	0	50	60	80	100	
		Cj-Zj	-5	-5	0	0	-10	0	0	0	

x_4

$$100-(2/5 \times 125)=50$$
$$0 \;\; -(2/5 \times 1/4)=1/10$$
$$1/5-(2/5 \times 3/4)=1/2$$
$$3 \;\; -(2/5 \times 0) \;\; =0$$
$$0 \;\; -(2/5 \times 0) \;\; =0$$
$$2/5-(2/5 \times 1/2)=1/5$$
$$2/5-(2/5 \times 1) \;\; =0$$
$$4/5-(2/5-1) \;\; =4/5$$
$$1 \;\; -(2/5 \times 0) \;\; =1$$

x_7

$$80-(3/5 \times 125)=5$$
$$0-(3/5 \times 1/4)='3/20$$
$$-1/5-(3/5 \times 3/4)=1/4$$
$$1-(3/5 \times 0) \;\; =1$$
$$0-(3/5 \times 0) \;\; =0$$
$$3/5-(3/5 \times 1/2)=3/10$$
$$3/5-(3/5 \times 1) \;\; =0$$
$$1/5-(3/5 \times 0) \;\; =1/5$$
$$0-(3/5 \times 0) \;\; =0$$

x_8

$$0-(-2/5 \times 125)=50$$
$$0-(-2/5 \times 1/4)=1/0$$
$$-1/5-(2/5 \times 3/4)=-1/3$$
$$0-(-2/5 \times 0 \;\; =0$$
$$1-(-2/5 \times 0) \;\; =1$$
$$-2/5-(-2/5 \times 1/2)=-1/5$$
$$-2/5-(-2/5 \times 1)=0$$
$$-1/5-(-2/5 \times 0=-1/5$$
$$0-(-2/5 \times 0) \;\; =0$$

C_j-Z_j values indicate that all of them are either zero or negative hence the optimal values are available from this table 3. e.g. $x_2=125$ and $x_4=50$.

Substituting values of x_2 and x_4

$$Z=60x_2+100x_4$$
$$=60 \times 125+100 \times 50$$
$$=7500+5000$$
$$=12,500 \text{ Profit}$$

and 125 sets of Monitor and 50 sets of Koral should be produced for optimisation.

The term. With "Transportation method" the Machine assignment, plant location and product mix problems can be solved. Transportation method is limited to problems with resources and requirements given in terms of only one kind of unit.

Problem 20

Plant	Warehouses				Output (Weekly production in plants A & B)
	1	2	3	4	
A	3	4	9	2	23
B	6	5	8	8	27
					0
Demand (Customer demand at each warehouses)	12	13	15	10	50

3, 4, 9, 2 and 6, 5, 8, 8 are the costs in Rupees per tonne of distributing product from plant A and B to respective warehouses 1, 2, 3, 4.

Solution. The first step in the transportation method is to load the basic matrix without bothering whether the solution formed is optional or not. There are various methods to do it but we shall be considering only two common methods :

(1) Northwest Corner Rule
(2) Vogel's Approximate Method.

(1) **Northwest Corner Rule**

In this method the initial assignment starts by loading the cell located in the upper left hand corner of the matrix and filling the non-requirements to the right and downward until all the requirements of the problem have been satisfied.

1st Solution

Plant	1	Ware house 2	3	4	Output
A	└3 (12)	└4 (11)	└9	└2	23
B	└6	└5 (2)	└8 (15)	└8 (10)	27
	12	13	15	10	50
Demand					50

The transportation cost involved in the first solution can be evaluated as

$$=3 \times 12 + 11 \times 4 + 2 \times 5 + 15 \times 8 + 10 \times 8$$
$$= 36 + 44 + 10 + 120 + 80 = 290 \text{ (rupees)}$$

In order to ascertain whether the solution I is optimal or not the empty cells in the solution must be evaluated. Following is the procedure to evaluate the empty cell :

1. Shifting 1 unit from loaded cell into the empty cell.

2. Comparing the cost of original schedule with the cost of new schedule involving the empty cell.

3. The loaded cell which is selected must have following two characteristic :

(*i*) Loaded cell must be situated in the same row or column of the empty cell.

(*ii*) The route used for evaluation would not require more than one empty cell during same evaluation.

The effect of shifting 1 unit from plant B to warehouse 1 would be

$$+6-3+4+5=Rs. 12$$

The plus sign indicates that the total costs cannot be reduced by adding one unit to B_1 combination.

Evaluation of empty cell B-1

Plant	1	2	3	4	Output
A	⑪ – ³ → ⑫ + ⁴		⁹	²	23
B	① + ⁶ ← ① – ⁵		⁸	⁸	27
Demand	12	13	15	10	50

Further Evaluation

Evaluation of cell A-4

Plant	1	2	3	4	Output
A	³	④⓪ ⁴	⁹	① ²	23
B	⁶	③ ⁵	⑮ ⁸	⑨ ⁸	27
Demand	12	13	15	10	50

Evaluation of cell A_4 follow the route
$$-4+2-8+5=Rs. -5.$$

(Assigning positive signs to the cost of those cells which gained one unit and negative signs to those cells which loses one unit.

Evaluation of A_3 Cell

Plant	1	2	3	4	Output
A	⑫ |3	⑩ |4	① |9	|2	23
B	|6	③ |5	⑭ |8	⑩ |8	27
Demand	12	13	15	10	50 / 50

Moving one unit to A_3, we remove one unit from A_2. In order to restore one unit to warehouse 2, we shift one unit from B_3 to B_2.

The cost associated with these moves are :

$$-4+9-8+5=+2$$

Thus adding one unit to cell A_3 would not reduce the total costs. Thus evaluation of empty cells give the following inference :

(*i*) If the evaluation is positive, then the cost of new solution would be greater than the cost of present schedule.

(*ii*) If the evalution is negative than the cost of new solution would be less than the cost of present schedule of the evaluation result into zero value than the cost will not change by using these empty cells.

The second feasible solution and optimum solution after analysing empty cells is given below. Now the evaluation for all

Plant	1	2	3	4	Output
A	⑫ |3	① |4	|9	⑩ |2	
B	|6	⑫ |5	⑮ |8	|8	
Demand	12	13	15	10	50 / 50

empty cells involve costs greater than zero. So any change in the solution would result into increase of total costs.

If the evaluation revealed negative costs then another feasible solution could have been constructed. The evaluation process is complete when all evaluations are positive. Thus an optimal schedule has been obtained.

(2) Vogel's Approximate Method

This method was developed first by W.R. Vogel. This method gives a good approximation and provides an optimum solution to most of the simple problems. Following are the various steps used for Vogel's approximate method :

1. Construct the basic matrix.

2. Find out the difference in each column and row between the minimum and the next higher to minimum.

3. Select a row or column containing the largest difference.

4. Select maximum unit in that square or column which carries the lowest cost.

5. The steps (2), (3) and (4) are repeated till all units are distributed satisfying the capacity and requirement constraint.

Vogel's Method

Let us find with the help of Vogel's Method.

1st step	Ware-houses				Output (Weekly production of the products A & B)	difference
Plant	1	2	3	4		
A	⌐3	⌐4	⌐9	⓪ ⌐2	23 23-10=13	4-3=1
B	6	5	8	8	27	6-5=1
Demand	12	13	15	10	50 \ 50	
Difference	6-3=3	5-4=1	9-8=1	8-2=6		

II nd step					
Plant	1	2	3	Output	Difference
A	② ⌐3	⌐4	⌐9	13 13-12=1	4-3=1
B	⌐6	⌐5	⌐8	27	6-5=1
Demand	12	13	15	40 \ 40	
Difference	6-3=3	5-4=1	9-8=1		

Ⅲ rd step.

Plant	2	3	Out put	Difference
A	① $\lfloor 4$	$\lfloor 9$	$1-1=0$	$9-4=5$
B	$\lfloor 5$	$\lfloor 8$	27	$8-5=3$
Demand	13	15	28 28	
Difference	$\begin{matrix}13-1=12\\5-4=1\end{matrix}$	$9-8=1$		

Difference $13-1=12$

$5-4=1$ $9-8=1$

Ⅳ th step

Plant	2	3	Output	Difference
B	⑫ $\lfloor 5$	⑮ $\lfloor 8$	27	$8-5=3$
Demand	12	15	27 27	

Let us combine step I, II, III and IV to get initial solution.

Plant	Ware House				Output
	1	2	3	4	
A	⑫ $\lfloor 3$	① $\lfloor 4$	$\lfloor 9$	⑩ $\lfloor 2$	23
B	$\lfloor 6$	⑫ $\lfloor 5$	⑮ $\lfloor 8$	$\lfloor 8$	27
Demand	12	13	15	10	50 50

Let us evaluate the empty cells B_1, B_3, and B_4. Evaluating square B_1.

Plant	Ware house				Output
	1	2	3	4	
A	⑫ $\lfloor 3$ — →	① $\lfloor 4$ +	$\lfloor 9$	⑩ $\lfloor 2$	23
B	+ ← $\lfloor 6$	⑫ $\lfloor 5$ —	⑮ $\lfloor 8$	$\lfloor 8$	27
Demand	12	13	15	10	50 50

If we transfer 1 unit from square B_2 to B_1.

The change in the cost of transportation will be

$$+6-3+4-5=+2$$

Thus transferring one unit to cell B_1 is not economical.

Evaluating empty sq. **A 3**

Plant	Ware house				Output
	1	2	3	4	
A	(12) ⌐3	(1) — ⌐4	+ ⌐9	(10) ⌐2	23
B.	⌐6	(12) + ⌐5	(15) — ⌐8	⌐8	27•
Demand	12	13	15	10	

Transferring one unit from square A_2 to square A_3.   The new cost of transportation will be :

$$9-4-8+5=+2$$

Thus the new set of assignment is not economical.

Evaluating empty sq. B-4

Plant	Ware house				Output
	1	2	3	4	
A	(12) ⌐3	(1) — ⌐4 +	⌐9	← (10) ⌐2	23
B	⌐6	(12) — ⌐5	(15) ⌐8	+ ⌐8	27
Demand	12	13	15	10	

We are not considering the occupied cell B_3.

Transferring one unit from square A_4 to square B_4, the new cost of transportation will be

$$8-5+4-2=+5.$$

Thus transferring one unit from square A_4 to square B_4 is not economical.   The evaluation for all empty cells involve cost greater than zero.   So any change in the solution would result into increase of total costs.   We can also conclude that Vogel's approximate method gives a good approximation and provides an optimum solution to most of the simple problems.

Problem 21.   Find out the best solution given the following table. Keeping in mind the optimisation of Transportation cost.

Solution. Let us find out the initial solution *using Vogel's Approximation method.*

Find out the difference between each column and row between minimum and next higher to minimum.

Source / Destination	F1	F2	F3	Units demanded
W1	1·0	1·0	1·0	5
W2	1·0	1·4	0·8	20
W3	1·3	1·0	0·8	20
Units available	20	15	10	45 / 45

Source / Dstination	F1	F2	F3	Units demanded	Defferences
W1	·90	1·0	1·0	5	1·0−0·9=·1
W2	1·0	1·4	0·8	20	1·0−0·8 = 0·2
W3	1·3	1·0	(10) 0·8	20 / 20−10=10	(1·0−0·8)=0·2 ft
Units available	20	15	10	45 / 45	
Differences	1·0−0·9=0·1	1·0−1·0=0	0·8−0·8=0		

Either row W_2 or W_3 will get the distribution. Let us take the row W_3. Now the cost of transportation is minimum for square $W_3 F_3$. Let us allot 10 unit square $W_3 F_3$. This satisfies the requirement for column F_3. The new matrix will be

Source / Destination	F1	F2	Units demanded	Differ.
W1	0·9	(5) 1·0	5	1·0−0·9=1
W2	(20) 1·0	1·4	20	1·4−1·0=4
W3	1·3	(10) 1·0	10	1·3−1·0=0·3
Units available	20	15	35 / 35	
Difference	1·0−0·9=0·1	1·0−1·0=0		

Now the initial allocation with the help of Vogel's Approximation. Method will be

Source \ Destination	F1	F2	F3	Units demanded
W1	0·9	1·0	1·0	5
W2	⑳ 1·0	1·14	1·08	20
W3	1·3 ⑩	1·0 ⑩	0·8	20
Units available	20	15	10	45 \ 45

A firm has three units in three leading cities of the coutry and operate through three dealers to sell goods in three regions. The cost transporting each unit from each manufacturing unit to a dealer as shown in the Table 1.

Original Matrix

Dealers \ Factories	D1	D2	D3	Factory capacities	Differences
1	3	6	4	10	(4-3)=1
2	2	5	7	5	(5-2) = 3
3	4	6	8	7	(6-4) = 2
Dealers requirements	6	8	8	22 \ 22	
Differences	(3-2) = 1	6-5=1	7-4=3		

1st alternative

Dealers \ Factories	D1	D2	D3	Factory capacities
1	3	6	4	10
2	⑤ 2	5	7	5
3	4	6	8	7
Dealers requirement	6	8	8	22 \ 22
	(6-5)=1			

The row 2 has been assigned to its capacity.

Dealers / Factories	D1	D2	D3	Capacities	Difference
1	3	6	(8) 4	10 / 10-8=2	4-3=1
3	4	6	8	7	6-4=2
Requirement	1	8	8	17 / 17	

$$4—3=1 \qquad 6—6=0 \qquad 8—4=4$$

Now in Column D_3, assign the value to the cell which has lowest cost of transporting for example, $i.e.$ D_3 cell.

Dealers / Factories	D1	D2	Capacities	Difference
1	(1) 3	6	2 / 2-1=1	6-3=3
3	4	6	7	6-4=2
Requirement	1	8	9 / 9	
Differences	4-3=1	6-6=0		

Assign the value of 1 to 1 D_1 cell because the requirement there is for one unit only.

So we are left only one column with equal cost of transportation. The cell $1D_2$ require 1 rent and cell $3D_2$ require 7 units and thus satisfying the total requirement of D_2 column. Two final values in each cells are :

Dealers / Factories	D1	D2	D3	Capacities
1	(1) 3	(1) 6	8 4	10
2	(5) 2	5	7	5
3	4	(7) 6	8	7
Requirements	6	8	8	22 / 22

The solution obtained above may not be an optimum solution. This has to be checked by evaluating the empty cells. The evaluation process will be complete when all evaluation are positive.

In fact after evaluating the costs as it was done in earlier problem, it was found that solution obtained by Vogel's method have optimised the transportation cost. The total cost of transportation will be

$$= 1 \times 3 + 1 \times 6 + 5 \times 2 + 7 \times 6 + 8 \times 4$$
$$= 3 + 6 + 10 + 42 + 32 = \text{Rs. 93}$$

The following table shows the problem of material handling. Find out the optimum solution.

To From	4	5	6	7	Available equipments
1	10	22	10	20	8
2	15	20	12	8	13
3	20	12	10	15	11
Required	7	10	6	9	

Here 1, 2, 3 and 4, 5, 6, 7 are the departments. Departments 1, 2 and 3 have extra equipments and departments 4, 5, 6 and 7 need equipments. The values in the cells refers to amount required in the transportation. Find the optimum combinations.

Solution. Let us find out the initial solution by the help of North West method.

To From	4	5	6	7	Available
1	(7) 10	(1) 22	10	20	8
2	15	(9) 20	(4) 12	8	13
3	20	12	(2) 10	(9) 15	11
Required	7	10	6	9	32 / 32

Total amount required in the process of transportation
$$= \text{Rs. } 7 \times 10 + 1 \times 22 + 9 \times 20 + 4 \times 12 + 2 \times 10 + 9 \times 15$$
$$= \text{Rs. } 475$$

Improving the initial solution

Evaluating the empty square 2-4 *i.e.*, moving one unit from 2-5 to 2-4 departments. Change of cost is

$$15-10+22-20=+7$$

(So there is no change in cost.)

To From	4	5	6	7	Available equipment
1	⑦ ⌐10	① 22	10	20	8
2	① 15	⑨ 20	④ 12	8	13
3	20	12	② 10	15	11
Required	7	10	6	9	32

Let us evaluate the square 3-4. Moving one unit from square 3-6 to square 3-4. Change in cost is

$$-10+20-10+22-20+12=14$$

So there is no change in the cost.

To From	4	5	6	7	Available equipment
1	⑦ 10	① 22	10	20	
2	15	⑨ 20	⑨ 12	8	
3	20	12	② 10	15	

Now let us evaluate the square 3-5 *i.e.*, moving on unit 3-6 to 3-5. Total cost is

$$=+12-20+12-10=-6$$

.To From	4	5	6	7	Available
1	⑦ 10	① 22	10	.20	8
2	15	⑨ 20	④ 12	8	13
3	20	⌐ 12	②(−1) 10	⑨ 15	11
Required	7	10	6	9	32

This shows that there is a scope for reducing the cost of transportation by shifting one unit from 3—6 to 3—5. The maximum shift can be of two units, so maximum improvement possible is

$$2(-6) = -12.$$

The new allocation matrix can be as follows.

To From	4	5	6	7	Available equipments
1	⑦ |10| ① |22| |10| |20| 8
2	|15| ⑦ |20| ⑥ |12| |8| 13
3	|20| ② |12| |10| ⑨ |15| 11
Required equipment	7	10	6	9	32

The total cost of transportation will be

$$= (22 \times 1) + (7 \times 20) + (2 \times 12) + (7 \times 15) + (6 \times 12)$$
$$+ (7 \times 10)$$

$$= 22 + 140 + 24 + 135 + 72 + 70 = \text{Rs. } 463$$

Now we will again evaluate the various open squares till no further improvement is possible. It should be borne in mind that squares which indicated no improvement may yield an improvement after the subsequent changes in the loading. The evaluation of all fall empty square and subsequent assignment will result into following optimum solution :

To From	4	5	6	7	Available
1	⑦ |10| |22| ① |10| |20| 8
2	|15| |20| ④ |12| ⑨ |8| 13
3	|20| ⑩ |12| ① |10| |15| 11
Required	7	10	6	9	32 / 32

The total cost of transportation for an optimum solution will be

$(7 \times 10) + (1 \times 10) + 4 \times 12 + 9 \times 8 + 1 \times 10 + 10 \times 12 = \text{Rs. } 330.$

46

Simulation and its Applications

Simulation

Simulation utilises a method of finding the successive situations in a problem by repeatedly applying the rules under which the system operates. This successive linkage of one particular situation to a previous situation is the main characteristics of simulation. Simulation is resorted to when :

(*i*) Many values of variables of a problem are not known.

(*ii*) There is no other way to find out the values of variables.

(*iii*) No formula can be established for the event.

(*iv*) Only a logical sequence can be established based on the past experience/event.

In a real business situation, the under lying principles are so complicated that it is difficult to predict the out come of a situation. Simulation model illustrates the situations on the simple input data rather than on the entire population as the experimentation on population is too much time consuming and costly.

Definition

Exact definition of the simulation is difficult, but to cover the main characteristics of simulation, it can be defined as a quantitative technique that ultilizes a computerised mathematical model in order to represent actual decision making under conditions of uncertainty for evaluating alternative courses of action based upon facts and assumptions.

Methods of Simulation

Basically there are three methods of simulation.

—Gaming method

—Monte Carlo method

—System simulation method.

Gaming Method

These come under the category of competitive models. These models take into account conflicts external to the organisation basically, the competition. The efficiency of the decision by one party is dependent on decisions by another party. The participants of the business games decide based on the past data. The decisions thus made by the participants influence the subsequent situation under which further subsequent decisions be made.

Management game or business game was first developed in 1957 and subsequently with the advent of digital computer, the growth of business games has been tremendous, due to the increased capabilities of the computers.

The main use of business game is :

—to help the participants to evaluate their decisions

—to help the Managers to learn to decide in the conflicting situations.

—to help the managers with a high degree of participation.

—to help the Managers to acquire higher skills in decision making.

—to train the Managers in constructing the business models and translate them into a real situation.

Since the theory of games is dealt separately in a chapter it is not discussed here in details.

Monte Carlo Methods

Monte Carlo is simulation in a narrow sense. Though it is actually only one kind of simulation by common use, the word "simulation" usually means Monte Carlo simulation. Monte Carlo method does not use formula. It uses random numbers for generating the values of variables, to see what will happen, if certain events were to occur. Monte Carlo method is actually the study of laws of chance. This method is used to solve probability dependent problems when physical experimentation is impracticable and the creation of an accurate formula is not possible. The Monte Carlo method is also explained as simulation by sampling techniques. Instead of taking samples from a real universe, they are taken from a theoretical counter part of the real population. The Monte Carlo method has been discussed in the previous chapter also. It's applications are discussed here.

Monte Carlo simulation is used in deciding :

—the size of a maintenance crew

—the number of counters/service booths required

—the number of airport run ways.

It is better to show how Monte Carlo helps by way of examples rather than explaining the theory.

Example 1. The aim of the material manger in an organisation is to minimise the total operating costs. The total cost of operation is the sum of the :

—ordering costs

—carrying costs

—stock out costs.

This problem can be visualised by applying a Monte Carlo simulation for 24 weeks of operation with following parameters.

*The reordering level is set at 15 units.

*The ordering quantity is 20 units.

*Inventory carrying costs is Rs. 10 per unit per week.

*Cost of placing an order is Rs. 25.

*The cost of stock out is Rs. 100 per unit per week.

Table 1. The Probability Distribution of Demand.

S. No.	Demand (units)	Frequency	Cumulative Probability distribution as percentage	Distribution of randum numbers
(1)	(2)	(3)	(4)	(5)
1.	0	2	2	00—01
2.	1	8	10	02—09
3.	2	20	30	10—29
4.	3	35	65	30—64
5.	4	16	81	65—80
6.	5	10	91	81—90
7.	6	9	100	91—99
	Total	100		

Table 2. The Probability Distribution of Delivery Time.

S. No.	Delivery time in weeks	Frequency	Cumulative probability distribution as %	Distribution of randum numbers
1.	1	22	22	00—21
2.	2	46	68	22—67
3.	3	18	86	68—85
4.	4	8	94	86—93
5.	5	6	100	94—99
Total		100		

Table 3. Twenty Four Weeks Simulation of Demand Delivery and Total Costs

	Demand		Delivery			Inventory	Inventory carrying costs	Ordering costs	Cost of stock out	Total cost
Week	Random Number	Demand in units	Random Number	Delivery (weeks)	units received	Balance in the beginning units	Rs.	Rs.	Rs.	Rs.
(1)	(2)	(3)	(4)	(5)	(6)	(7)	(8)	(9)	(10)	(11)
(0)						22				
1.	9	1				21	210			210
2.	54	3				18	180			180
3.	42	3	34	2		15	150	25		175
4.	1	0				15	150			150
5.	80	4			20	31	310			310
6.	6	1				30	300			300
7.	6	1				29	290			290
8.	26	2				27	270			270
9.	57	3				24	240			240
10.	79	4				20	200			200

(1)	(2)	(3)	(4)	(5)	(6)	(7)	(8)	(9)	(10)	(11)
11.	52	3				17	170			170
12.	80	4				13	130			155
13.	45	3	24			10	100	25		100
14.	68	4		2	20	20	260			260
15.	59	3				23	230			230
16.	48	3				20	200			200
17.	12	2				18	180			180
18.	35	3	95	5		15	150	25		175
19.	91	6			20	11	110			110
20.	89	5				6	60			60
21.	49	3				3	30			30
22.	33	3				0	0		100	100
23.	10	1				19	190			190
24.	55	3				16	160			160
Total Average			Average 185							4445

Usually the demand of the materials revised per week and the delivery time of the materials from the time of ordering, vary. From the past data the above frequencies have been arrived and in the last column the distribution of random number is given so that by selecting a random number from the random table, it can be reffered to this is last column and read the demand or delivery time that may occur as the case may be. By making use of these two tables the figures at Table-3 are worked out.

To start with balance is assumed as 20 units. First pick up a random number which is 9 and write under demand column at Table-3. Read the demand against random number 9 in Table-1. The demand is 1 which is written in the column at Table 3 and the balance is arrived as 21 which workes out to Rs. 210 as carrying cost when the balance at Table 3 touches 15 or less take another random number and read the delivery time from Table 2 and write in Table 3.

Thus by generating successive steps for 24 weeks we can summarise the conclusions that the designed or envisaged reordering level and ordering policy reveals that :

(i) Average inventory cost per week is Rs. 185

(ii) About 4% of the time the stock out can occur.

Computer Simulation

With computer simulation the impact of variations in costs can be studied like :

(i) By increase of ordering cost by 10% or 20%

(ii) By varying the inventory carrying cost

(iii) By varying the cost of stock outs or by changing the level of stock outs.

Thus the Management can easily obtain the sensitivity analysis of the various factors on the cost of inventory and can exercise control on the factors which are too sensitive. The same methodology is applied in simulation of various problems in different areas of operations. Monte Carlo simulation has wider applications like :

(i) Determination of size of Mechanics group for maintenance purpose.

(ii) Integrating production scheduling models and inventory control models.

(iii) For Budget preparation viz. sales budget, manufacturing budget, cash budget can be obtained from various alternatives under different varying input conditions with the aid of complex simulation.

System Simulation Method

This is the third method. System simulation is a method where the real or actual data is processed through a mathematical model. The model is simulated in operating environment. The responses are obtained by simulating the model with alternative management actions and thereby the best response or desired response is selected for the quality of decisions to be implemented.

System simulation is different from Monte Carlo simulation. In system simulation :

—The samples are taken from a real population.

—Theoritical counter part of actual population is not used.

—Methematical model is also used and it can be analytically solved in arriving at a decision.

—The response can be obtained for any set of factors.

Design of Simulation Experiments

After constructing a simulation model one will face the difficulty in deciding the set of factors and their values for processing the model. If one is not careful in determining the variables and their values an enormous amount of computer time is used. Even after collecting a vast amount of data, one may not have sufficiently accurate information to guide a managerial decision. Here statistical design of experiments will greatly help.

For example, let a production system be represented by a quadratic model.

$$y = a + b_1 x_1 + b_2 x_2 + b_3 x_3 + b_4 x_1{}^2 + b_5 x_2{}^2 + b_6 x_3{}^2$$
$$- b_7 x_1 x_2 + b_8 x_2 x_3 + b_9 x_1 x_3.$$

If the manager would like to know the best or highest value of y and the corresponding parameters, he has to resort to the trial and error method since there are number of positive and negative effects and interactions. Even then one will not be sure of the highest value. As per the factorial design of experiments, with all permutations and combinations, the number of runs revised can be calculated as follows :

$$N = (L)^V$$

where N = No. of runs revised

V = No. of variables under consideration

L = No. of levels of variables.

With this the number of runs revised can be calculated as follows :

V	2	3	4	5	6	7	8	9	10
L	2	2	2	2	2	2	2	2	2
N	4	8	16	32	64	128	256	512	1024

If three variables with 2 levels are to be studied the runs need to be conducted as below :

Run No.	Levels of variables			Value of y
	1	2	3	
1.	−	−	−	τ_1
2.	+	−	−	τ_2
3.	−	+	−	τ_3
4.	+	+	−	τ_4
5.	−	−	+	τ_5
6.	+	−	+	τ_6
7.	−	+	+	τ_7
8.	+	+	+	τ_8

Where + represents higher level of variables

− represents lower level of variables

If more than two levels are to be studied same formula can be used. However much more higher level of statistical knowledge is required if the levels and variables go on increasing. But certainly simulation under design of experiments will reveal lot of information and aid for better decision making.

Simulation Languages

The reason for several simulation languages is that they can be applied to different types of problems which make their simulation procedures more automatic for the user. It furnishes a generalised structure for designing simulation models and to speed the conversion of a simulation model to a computer programme.

DYNAMO and SIMLATE were designed for simulating economic system on econometric models. GPSS, SIMSCRIPT and GASP are well suited for scheduling and waiting line problems.

Specific Application of Simulation

(i) Many chemical reactions can be described by a set of differential equations and their solution is some times referred to as continuous simulation ; this may involve the use of Analogue computers.

(ii) The complex systems involving queueing problems like, Systems for batch-chemical production, Vehicle unloding, Aircraft landing, Telephone traffic etc.

(iii) Blending of the powders. How many batches will be in queue at a time and number of extra containers required for the given plan of action.

(iv) *Total Service requirements.* A chemical plant often consists of several units each requiring supplies of services such as steam, water, ice etc. Suppose it is known that there is a peak demand of 2000 gallons/hr of water on a particular unit and that this peak will last for 2 hours. Other units have different peak usages for different durations. It would be wasteful to build a water main which would cope with the aggregate of all the peak demands, since the probability of all the peaks occurring together is usually very small.

A simulation model can be constructed where the "activity" of each unit at any instant in time is given by its rate of water consumption at that time. By summing over all the units a measure of the total water usage is obtained. The activity of a unit is changed when the demand for water on that unit alters, and this may be at fixed intervals or at some randomly generated times based an actual observed data. By following the activities of the units over a long length of simulated time a distribution of total demand can be compiled. This is then used to estimate the optimum size of water main.

(v) *Stock holding policy.* An important problem for any manufacturing is to determine how much finished product should be held in stock ready to meet sales demands. This may involve the use of ware houses for some materials, or storage tanks for bulk liquids. Increasing the stock holding improves the chance of selling more material. The value of the extra sales has to be balanced against the extra costs incurred in providing the storage facilities, in runing the material in stock, having capital tied up in stock and so on.

Simulation can be used to determine the optimum size and strategy to be used for stock holding and is especially valuable in the more complex cases. Here it is necessary to generate sequences representing the demands from customers and the supply of material into storage. The model can be run with different sizes of storage and account can be taken of any special features such as limitations on the storage life of the product.

(vi) *Plant Extensions*. In multistage chemical plants where demand exceeds capacity it is usually fairly easy to spot the bottleneck in the system. This may be a particular unit that has been constantly over loaded. It would be useful if we could measure the benefit of replacing or duplicating this unit without going in for actual modificate to the plant.

The solution is to build a simulation model of the plant which allows investigations to be carried out on the effects of possible extensions to the plant. The model describes the flow of materials through the plants and takes into account of charging, processing and discharging times for each unit. Rules regarding the operating procedure of the plant can be built into the model and incidence and duration of plant break downs can also be minimised.

47

Replacement Models

Introduction

Replacement of an equipment, item or machine is resorted to whenever the machine or item deteriorates in its function or fails to work. Therefore the replacement is considered in two cases.

(*i*) Replacement of items that deteriorate

(*ii*) Replacement of items that fail.

Various types of costs are associated in both the cases. The replacement of the item with new item is considered based on the cost creteria.

Replacement of Capital Equipment

When the efficiency of the equipment and machines declines over a period of a time, in addition to the obsolesence the following things happen in a system :

—Forced idle time is caused with loss of production

—Scrap in the process increases

—Rate of rejections increases

—The frequency of repairs increase

—Causes accidents or rate of accidents increase.

Therefore finally the operating costs of the equipment increases as the life of the equipment reduces or as the age increases. It is necessary to predict the time when the cost of maintenance and operation of the old equipment is higher than the cost of replacing old equipment with new one so that the old equipment can be replaced with new one. Some of the examples are :

Example 1. The owner of a taxi car maintained the record of the costs/year incurred in maintaining and running the car. The purchase price of the car is Rs. 60,000. The sale value of the car and the operating costs at the end of each year are given below. We can find out when the car should be replaced ?

Year	Operating costs Rs.	Sales price Rs.
1	12000	30000
2	14000	15000
3	16000	7500
4	20000	3750
5	25000	2000
6	30000	2000
7	36000	2000
8	42000	2000

Average cost per year is worked out as below.

Year No.	Cumulative operating costs	Cumulative sales price= capital cost	Total cost= op. costs+ capital cost	Average cost per year
1	12,000	30,000	42,000	42,000
2	26,000	45,000	71,000	35,500
3	40,000	52,500	92,500	30,833
4	60,000	56,250	116,250	29,062
5	85,000	58,250	143,250	28,650
6	115,000	58,000	173,000	28,833
7	151,000	58,000	209,000	29,857
8	193,000	58,000	251,000	31,375

It can be seen from the above table that the average cost per year goes on reducing in the last column upto 5th year and from there onwards again it starts increasing. At the end of 5th year the average cost/year is the lowest. Therefore, after 5 years the car needs to be replaced with a new car.

Discounted Costs

To know whether a replacement proposal is worthwhile or not, one should know the following :

—Amount and timing of investment.

—Amount and timing of savings.

—Life of the investment.

—The creteria for acceptability of the proposal.

—Proper techniques to relate all the costs and benefit factors in a meaningful way.

Just like all gases can be either additive or comparable only at N.T.P. (Normal temperature and pressure), all the existing equipment and new equipment must be brought on common dantom line for meaningful comparison purpose.

Two such techniques used in the analysis are :

(i) D.C.F. (Discounted Cash Flow) technique.

(ii) Present value techniqe.

Concepts of Discounted Cash Flow and present value techniques are assentially the same. The principle is that A rupee in hand today is worth more than a rupee a year hence. The only difference between these two techniques is that, under the present value method, the interest rate at which the future cash inflows are to be discounted is pre-determined. The merit of these techniques is to bring down all the costs as well as benefit factors which are to be incurred at different time points to the same time point by taking into account the worth of money and then to make them comparable. These techniques are eminently suitable to a situation where a number of investment opportunities exist with a total objective of utilizing the limited financial resources in an optimal fashion.

Worth of Money

If the worth of the money is considered as 10% per year, spare money available now can be invested to yield 10% in a year. Spending Rs. 100 today would be equivalent spending Rs. 110 in one year time. Similarly one rupee a year from now is equivalent $(1 \cdot 1)^{-1}$ rupees today and one rupee γ years from now is equivalent to $(1 \cdot 1)^{-\gamma}$ rupees to day. This quantity $(1 \cdot 1)^{-\gamma}$ is called the Present Value or Present Worth of one rupee spent "γ" years from now.

Example 2. The value of the money 10% per year machine 1 is to be replaced every 3 years and machine 2 is to be replaced for every 6 years with yearly outlay (expenditure) as given below :

Year	Yearly Outlay in Rupees	
	Machine 1	Machine 2
1	2000	3400
2	400	200
3	800	400
4		600
5		800
6		1000

Which machine costs less ?

Total discounted cost of machine 1 for the 3 years is

$$= Rs.\ 200 + \frac{Rs.\ 400}{1\cdot1} + \frac{Rs.\ 800}{(1\cdot1)^2}$$

$$= 2000 + 363\cdot6 + 661\cdot1 = 3024\cdot7$$

Average per year

$$= \frac{3024\cdot7}{3} = \textbf{Rs. 1008 } \text{per year}$$

Total discounted cost of machine 2 for 6 years is

$$= \frac{Rs.\ 3400}{1} + \frac{Rs.\ 200}{1\cdot10} + \frac{Rs.\ 400}{(1\cdot1)^2} + \frac{Rs.\ 600}{(1\cdot1)^3}$$

$$+ \frac{Rs.\ 800}{(1\cdot1)^4} + \frac{Rs.\ 1000}{(1\cdot1)^5}$$

$$= \frac{3400}{1} + 181\cdot8 + 330\cdot6 + \frac{600}{1\cdot331} + \frac{800}{1\cdot4641} + \frac{1000}{1\cdot6105}$$

$$= 3400 + 181\cdot8 + 330\cdot6 + 450\cdot8 + 546\cdot4 + 620\cdot92$$

$$= 5530\cdot5$$

Average per year

$$= \frac{5530\cdot5}{6} = \textbf{Rs. 921\cdot8 } \text{per year.}$$

Machine-2 costs Less

If machine 1 also is discounted for 6 years on par with machine 2.

The discounted cost for machine 1 in 6 years.

$$= \frac{2000}{1} + \frac{400}{1\cdot1} + \frac{800}{(1\cdot1)^2} + \frac{2000}{(1\cdot1)^3} + \frac{400}{(1\cdot1)^4} + \frac{800}{(1\cdot1)^5}$$

$$= 2000 + 363\cdot6 + 661\cdot1 + 1502\cdot6 + 273\cdot2 + 496\cdot7$$

$$= 5297\cdot2$$

Average cost per year

$$= \frac{5297\cdot2}{6} = \textbf{Rs. 882\cdot9}$$

Now machine-1 works out to be cheeper.

Calculation of Present Worth

An algebraic formula as well can be developed for the calculation of the present worth of the expenditure denoted by W

$$W(n) = P + C_1 + wC_2 + w^2C_3 + w^3C_4 + \ldots\ldots + + w^{n-1}C_n$$

where $W(n)=$ present worth of [the machine expenditure after n years.

$P=$ Purchase price of the machine

$C=$ Running cost or operating costs in a year

$w=$ is the present worth of a unit to be spent a year hence

$w=$ Discounted rate

$w=1\div(1+i)$

$i=$ rate of interest

$n=$ Number of years after which the machine is to be replaced.

It is assumed that expenditure (running cost) is incurred at the beginning of each year.

Example 3. Let the purchase price of a machine be Rs. 6000, the cost of the money be 10%. The estimated running cost of the machine is given below year wise.

Year	1	2	3	4	5	6	7	8	9	10
Running cost	500	500	500	500	500	1000	1200	1400	1600	1800

When the machine should be replaced ?

$$P=\text{Rs. }6000$$
$$i=0.1$$
$$n=1 \text{ to } 10.$$

Year	Running cost per year C_n Rs.	Discounted rate $(w)^{n-1}$ $=\dfrac{1}{(1+i)^{n-1}}$	Present worth for the year $(w)^{n-1}C_n$	Cumulative cost of present worth	Average present worth per year
1.	500	1.0000	500	500	500
2.	500	0.9091	455	955	478
3.	500	0.8264	413	1368	456
4.	500	0.7513	376	1744	436
5.	500	0.6830	342	2086	417
6.	1000	0.6209	621	2707	451
7.	1200	0.5645	677	3384	483
8.	1400	0.5132	719	4103	512
9.	1600	0.4665	746	4849	539
10.	1800	0.4241	764	5613	561

Since the minimum average present worth of total running cost is 5th year, the machine be replaced at 5th year end.

Example 4. As new automobile vehicle costs of Rs. 10,000 and it can be sold at the end of any year with the selling price as shown below. The operating and maintenance cost also are given year wise.

1.	End of year	1	2	3	4	5	6
2.	Selling price	7000	5000	3000	2000	1000	500
3.	Annual operating and maintenance cost in Rs.	1000	1600	1800	2500	3000	3500

When the automobile vehicle needs to be replaced because of wear and tear.

S. No.	Year end	Selling price year end Rs.	Capital cost 10,000 —col (3)	Cumulative operating and maintenance costs Rs.	Total cost	Average cost/ year
1.	1	7000	3000	1000	4000	4000
2.	2	5000	5000	2600	7600	3800
3.	3	3000	7000	4400	11400	3800
4.	4	2000	8000	6900	14900	3725
5.	5	1000	9000	12000	21000	4200
6.	6	500	9500	13000	22500	3750

Since the average cost is minimum at the end of 4th year the vehicle can be replaced at the end of 4th year.

Replacement of Items on Failures

The second type of replacement is concerned with the items that either work or fail completely. Such type of items do not deteriorate markedly with age yet they suddenly fail after some age. The life span of such items is not constant but will follow certain frequency distribution. Different life spans have different porbabilities. The probability of mortality or failures increases as the age of the machine or item increases. The failures or deaths caused as a result of external factors like natural calamities are not accountable under the probability distribution of life. Replacement of a failed items has to be done and understandable by any one. The problem is to plan replacement of items that have not failed. Replacement of an usable item which is still functioning, will be justified if the cost of replacement is higher after failure than the cost of replacement before failure. There are a number of reasons why replacing failure may be more costly than replacing live items. The cost of replacing failures include

— Cost of the item itself
— Cost of Labour
— Cost of loss of production that can occur due to delay
— Cost of damage to the material
— Cost of damage to the machinery/equipment
— Cost of a possible accident.

In view of the above costs, the cost of replacement per unit of item in group replacement is constant and is less than the cost per unit of item for replacement of failed item.

Example.   The rate of failure of bulbs is given below.

Months	1	2	3	4	5
% failure by end of month	5	15	20	40	20

In a factory premises there are 10,000 bulbs which are in use. To replace one bulb at a time (in case of failure of one bulb at a time) it costs Rs. 2.00. If all bulbs of 10,000 are replaced simultaneously, the cost of replacement works out to Rs. 0.5 per bulb.

If the maintenance incharge decides to replace all bulbs at fixed intervals (whether the bulbs have failed for not) and to replace the bulbs as and when they burnt or fail, what is the periodicity of replacement of all bulbs. Let P_i denote the probability that a bulb after it is installed fails during the i th month of its age. Then the probability distribution represented as shown below.

Month (i)	1	2	3	4	5
Probability of a new bulb that fails during i th month	0·05	0·15	0·20	0·40	0·20

Let R_o denote the number of all new bulbs which were initially installed $= 10,000$.

R_i be the number of bulbs replaced at the end of i th month. The number of bulbs replaced at the end of each month can be iterated based on the above probability distribution.

Initially $R_o = 10,000$ bulbs

At the end of 1st month $R_1 = R_0 \times P_1 = 10,000 \times 0.05$
$$= 500 \text{ bulbs will be replaced}$$

At the end of 2nd month *i.e.* $R_2 = R_0 \times P + R_1 \times P_1$

So out of 10,000 bulbs, 15% of bulbs fail at the end of second month and out of $R_1 = 500$ bulbs at the end of second month becomes end of 1st month. Therefore out of $R_1 = 500$ bulbs, 5% of bulbs will fail and need to replaced.

Hence total number of bulbs that would be replaced as a result of failures at the end of second month is

$$R_2 = R_0 P_2 + R_1 P_1$$
$$= 10000 \times 0.15 + 500 \times 0.05$$
$$R_2 = 1500 + 25 = 1525.$$

Similarly R_3, R_4 etc. are calculated and shown below.

$R_0 = R_0 = 10,000$

$R_1 = R_0 \times P_1 = 1000 \times 0.05 = 500$

$R_2 = R_0 \times P_2 + R_1 \times P_1 = 1000 \times 0.15 + 500 \times 0.05 = 1525$

$R_3 = R_0 + P_3 + R_1 \times P_2 + R_2 \times P_1$
$= 10000 \times 0.2 + 500 \times 0.15 + 1525 \times 0.05 = 2151$

$R_4 = R_0 P_4 + R_1 P_3 + R_2 P_2 + R_3 P_1$
$= 10000 \times 0.4 + 500 \times 0.2 + 1525 \times 0.15 + 2151 \times 0.05 = 4436$

$R_5 = R_0 P_5 + R_1 P_4 + R_2 P_3 + R_3 P_2 + R_4 P_1$
$= 10000 \times 0.2 + 500 \times 0.4 + 1525 \times 0.2 + 2151 \times 0.15$
$\qquad + 4436 \times 0.05 = 3049$

$R_6 = R_1 P_5 + R_2 P_4 + R_3 P_3 + R_4 P_2 + R_5 P_1$
$= 500 \times 0.2 + 1525 \times 0.4 + 2151 \times 0.2 + 4436 \times 0.15 + 3049 \times 0.05$
$= 1958$

$R_7 = R_2 P_5 + R_3 P_4 + R_4 P_3 + R_5 P_2 + R_6 P_1$
$= 1525 \times 0.2 + 2151 \times 0.4 + 4436 \times 0.2 + 3049 \times 0.15$
$\qquad + 1958 \times 0.05 = 2608$

$R_8 = R_3 P_5 + R_4 P_4 + R_5 P_3 + R_6 P_2 + R_7 P_1 = 3239$

$R_9 = R_4 P_5 + R_5 P_4 + R_6 P_3 + R_7 P_2 + R_8 P_1 = 3052$

$R_{10} = R_5 P_5 + R_6 P_4 + R_7 P_3 + R_8 P_2 + R_9 P_1 = 2553$

It can be seen from the 10 iterations that the number of bulbs failed and replaced every month increased till 5th month. Subsequently the number is increasing and reducing. If these iterations are continued for large number of cycles, the system will reach a steady state and there will not be a marked decrease or increase in the number of failures and replacements once the system stabilises, the *proportion of bulbs failing each month is reciprocal of their average life.* For example the average life of the bulbs based on the probability distribution,

$$=1 \text{ month} \times 0\cdot05 + 2 \text{ months} \times 0\cdot15 + 3 \text{ months} \times 0\cdot2 +$$
$$4 \text{ months} \times 0\cdot4 + 5 \text{ months} \times 0\cdot2$$
$$=3\cdot55 \text{ months}.$$

The proportion of bulbs that fail in each month in the steady state

$$=\frac{1}{3\cdot55}=0\cdot28$$

For 10,000 bulbs, the number of bulbs that fail every month

$$=10000 \times 0\cdot28 = 2800.$$

(*a*) If the maintenance in charge decides only to replace the bulbs as and when they fail, the cost of replacement

$$=\text{Rs. } 2 \text{ per bulb} \times 2800 \text{ bulbs}$$
$$=\text{Rs. } 5600 \text{ per month on an average.}$$

(*b*) On the contrary if the maintenance incharge desires to know the fixed periodicity of replacement of all bulbs while replacing the bulbs as and when they fail, the following costs will reveal the fixed periodicity :

(*i*) If all bulbs are replaced at the end of 1st month, the cost of replacement is Rs. $0\cdot5 \times 10,000 = $ Rs. 5000. In addition the cost of the replacing the bulbs during the 1st month $=500 \times $ Rs. $2 = 1000$ rupees also has to be added *i.e.* Rs. 5000 + Rs. 1000 = Rs. 6000 at the end of 1st month.

(*ii*) If all bulbs are replaced for every month, the cost of all bulbs replacement plus the cost of bulbs replaced during 2 months will be total cost

$$=\text{Rs. } 5000 + \text{Rs. } 500 \times 2 + 1525 \times 2$$
$$=5000 + 1000 + 3050 = 9050.$$

The average cost of replacement per month works out to $(9050) \div 2 = $ Rs. 4525.

(*iii*). If all bulbs are to be replaced every 3 months

Rs. $5000 + 500 \times 2 + 1525 \times 2 + 2151 \times 2$
$$=5000 + 1000 + 3050 + 4302 = 13352$$

Average cost/month $=\dfrac{13352}{3}=$ Rs 4450.

(*iv*) If all bulbs are to be replaced after every 4 months.

Rs. $5000 + 500 \times 2 + 1525 \times 2 + 2151 \times 2 + 4436 \times 2$
$$=5000 + 1000 + 3050 + 4302 + 8872$$
$$=22,224.$$

$$\text{Average cost/month} = \frac{22,224}{4} = \text{Rs. 5556.}$$

Similarly it can be calculated for all 10 months and shown below.

Month	Average cost of replacement/month Rs.
1	6000
2	4525
3	**4450**
4	5556
5	5664
6	5373
7	5350
8	5491
9	5559
10	5514.

Under the second policy all bulbs need to be replaced for at the end of every 3 months while continuing to replace the bulbs as and when they fail as average cost per month is Rs. 4450 lowest of all. If the maintenance incharge desires to know which of the above two policies are better, the operating costs in the above two policies can be compared for arriving at the decision.

As per 1st policy to replace the bulbs as and when they fail, the cost/month = Rs. 5600.

As per 2nd policy to replace all the bulbs (10,000) at the end of every 3 months and continue to replace the bulbs as and when they fail, the averge cost per month = Rs. 4450.

Thus the above approach can be applied for a wide variety of situations under this category without resorting to much of the mathematical models only a logical approach is required.

48

The Theory of Games

Introduction

The term games relates to conditions of business conflict over-time. The participants are competitors who make use of mathematical techniques and logical thinking in order to arrive at the best possible strategy for wining their competitor. Every game has a goal for which the competitors strive by selecting appropriate courses of action. Even though the game may favour one over the other, each will do his best to maximise his profit or minimise his loss. In many games the attainment of goal is accompanied by a payment of some kind of money. Vou Novmann perceived the Theory of Games in 1928 and it came to light only when Morgenstern published as a co-author the "Theory and Practice of Games and Economic Behovior" in 1944.

Many of the OR models have dealt with conflicting interests internal to the organisation. The production department wants large order with less in variety and wants the materials, men, machines in as much as possible to avoid any short fall in production. Marketing department would like to have greater variety and more number of orders and more sales irrespective of the profitability. Finance department would try to see that unnecessary money is not locked up in various assets and try to avoid the extra costs. Here are all conflicts with in the organisation. The models under games theory deal with the conflicts external to the organisation.

The effectiveness of decisions by one party is dependent on decisions by another party. The games theory is directed towards yielding better understanding of competitive economic behaviour. The theory has a value which is independent of the mathematics. It brings awareness of the possibility of internal choice of policy in competitive situations.

Role of Statistics in Games of Chance

Frequency Distribution. It is possible, however, to express much of the information contained in a frequency distribution in just two quantities or parametres one which contains information on the size of the variables — a measure of location, and one which contains information on how scattered the variables are — a measure of dispersion.   These two quantities are the mean and the standard deviation.

Mean.   The mean of a frequency distribution is simply the arithmetic average value of the variables, *i.e.* the sum of the variables devided by the number of them.

If the value x_1 occurs with frequency f_1

If the value x_2 occurs with frequency f_2

If the value x_3 occurs with frequency f_3 etc.

then the sum of the variables is $f_1 x_1 + f_2 f_2 + f_3 x_3$ etc.

and the total number of variables are $f_1 + f_2 + f_3$ etc.

These expressions may be simplified by using the mathematical summation sign (sigma).   Thus

$$f_1 x_1 + f_2 x_2 + f_3 x_3 \ldots \text{etc.}$$

may be abbreviated to Σfx and $f_1 + f_2 + f_3$ etc. to Σf. Using this notation the mean is defined as follows.

$$\text{Mean } x = \frac{\Sigma fx}{\Sigma f}$$

Standard Deviation.   The standard deviation is a measure of variability about the mean.   It is not such an obvious measure as the mean is of location, but it has important applications in statistical theory.

It is defined as follows : $\sqrt{\dfrac{\Sigma f(x - \bar{x})^2}{\Sigma f}}$

This method of calculating the standard deviation is less straight-forward if the mean $\bar{x}$ is not a whole number, because then, the quantities $(x - \bar{x})$ become fractional.   There is an equivalent formula which is generally more convenient from the point of view of calculation.   It can be shown as :

$$\frac{f(x - x)^2}{f} = \frac{f x^2}{f} - x^2$$

and it is this latter formula which is used in practice for the calculation of the standard deviation.   To illustrate the equivalence of the two formulae the calculation of the standard deviation of the number of accidents is repeated in the subsequent discussions.

The standard deviation is a measure of the degree of scatter of the variables.   Subtracting a constant value from every variable does not affect this scatter and therefore does not alter the standard deviation.

Probability

We all have some understanding of probability. The purpose of this short introduction to the subject is to show how to express events as being improbable, more probable than not, or highly probable, we can express the degree of probability in quantitative terms.

Definition

If there is evidence for believing that a particular outcome would occur on a fraction 'p' of occasions in a large number of similar situations, then the particular outcome is said to occur with probability 'p'.. If a particular outcome would never occur, $p=0$, and this corresponds to one end of the probability scale-impossibility. If an outcome would always occur, $p=1$, and this corresponds to the opposite end of the probability scale-certainly. For any outcome which may or may not occur, the probability must lie between these two extremes of 0 and 1. There are certain events whose very nature enables the probabilities of certain outcomes to be predicted. What is probability, for example, that if a penny is tossed it will come down heads? If an event can occur in 'n' mutually exclusive time then the probability of this outcome is m/n. For example, what is the probability that on being dealt a card from a well shuffled pack it turns out to be a spade? 'A well-shuffled pack' implies that the card being dealt is equally likely to be one of the 52 cards contained in the pack; thus $n=52$. Only 13 of these cards are spades, therefore, $m=13$. The probability of a spade is, therefore $\frac{13}{52} = \frac{1}{4}$. The law of probability were originally developed to understand and predict the relative frequencies of various events in games of chance. There are two important laws of probability; the "Addition Law" and the "Multiplication Law".

Meaning of Probability

The concept of probability plays an important role in many problems of everyday life, in business, in science, and particularly in statistics. In statistics, probabilities are defined as relative frequencies, or to be more exact, as limits of relative frequencies. For example the percentage probability that the shipment will arrive on time is "0·15", this means that in the long run 15 percent of all similar shipments will arrive in time; if we say, "the probability that it will rain in Delhi on the 6th of June is 0·26" this means that it rains there on the 6th of June about 26 percent of the time. If we say that the probability of getting heads with a balanced coin is $\frac{1}{2}$, this means that in the long run we will get 50% heads and 50% tails. It does not mean that we must necessarily get 5 heads and 5 tails in 10 flips of a coin or 10 heads and 10 tails in 20 flips. No, we will not always get equal numbers of heads and tails, but if we flip a balanced coin a large number of times, we will usually get close to 50% heads and 50% tails.

Some Rules of Probability

In the study of probability there are essentially three kinds of problems. First, there is the question of what we mean when we say that a probability is ? Then there is the problem of obtaining numerical probabilities and, finally, there is the problem of using known probabilities to calculate others. Having resolved the first question by defining probabilities in terms of relative frequencies, the problem of obtaining numerical probabilities becomes a problem of estimation. For example, if we want to determine the probability that a shipment from a given firm will arrive on time we refer to past experience and check what proportion of the time shipments from this firm have arrived on time. If we write the probability of the occurrence of event (A) as $P(A)$ the first fundamental rule of probability may be given as

$$=0 \geqslant P(A) \leqslant 1 \qquad \text{...}(i)$$

This merely expresses the fact that a relative frequency cannot be negative or exceed 1. After all, the proportion of the time that an event takes place cannot be negative and it cannot exceed 1, that is, an event cannot occur more than 100 percent of the time. Although the distinction is rather fine $P(A)=0$ does not mean that the occurrence of event A is beyond the realm of possibility and $P(A)=1$ does not mean that the occurrence of event A is absolutely certain. A second rule of probability, which follows directly from the frequency definition, states that if the probability of the occurrence of event A is $P(A)$, the probability that it will not occur is

$$=1-P(A) \qquad \text{...}(ii)$$

For instance, if the probability that it will rain on a certain day is 0·24, then the probability that it will not rain is $1-0·24 =0·76$; if the probability that the value of a stock will go up is 0·63, then the probability that it will not go up is $1-0·63=0·37$; and if the probability that a certain train will arrive late is 0·60, then the probability that it will not rate is $1-0·6=0·40$. Clearly, if the train is late 60% of the time it must be on time (or early) the remaining 40% of the time. The rule of probability refers to events that are mutually exclusive. Two events are said to be mutually exclusive if the occurrence of either precludes the occurrence of the other. If we toss a coin, heads and tails are mutually exclusive events; we can get one or the other but never the both. The occurrence of the one event renders the occurrence of the other impossible.

Events are Mutually Exclusive

If A and B are mutually exclusive, then
$$P(A \text{ or } B)=P(A)+ P(B).$$

This formula tells us that if A and B are mutually exclusive the probability of the occurrence of A or B is equal to the sum of the individual probabilities of A and B.

Formula (1) can easily be generalised to apply to more than two mutually exclusive events. A number of events are said to be mutually exclusive if the occurrence of any one precludes the occurrence of all of the others. Given a mutually exclusive events A_1, A_2 ....and An, the probability that one of them will occur is $P(A)$ or A_2 or ...... or $An) = P(A)$ $) + (P(A_2) + \ldots \ldots P(An) \ldots (2)$

If the probability that an executive will spend his summer vacation in Bombay, Calcutta, Simla, Mossourie or Nanital is 0.12, 0.19, 0.27, 0.20 and 0.06, respectively, then the probability that he will spend his summer vacation in one of these 5 places is

$$= 0.12 + 0.19 + 0.27 + 0.20 + 0.06 = 0.84$$

To formulate a general rule of addition for events which need not be mutually exclusive, we shall first have to explain what is meant by two events being independent or dependent. Two events are said to be independent if the occurrence or non-occurrence of either in no way affects the occurrence of the other. If A and B stand for getting heads in two successive flips of a coin, then A and B are independent; the second flip is in no way affected by what happened in the first.

Events which are Independent

If two events A and B are independent, the probability that A and B will both occur is given by the equation (3).

If A and B are independent, then;
$$P(A \text{ and } B) = P(A) \, P(B). \quad \ldots (3)$$

This formula tells us that if A and B are independent, the probability that they will both occur is equal to the product of their individual probabilities.

Events not Independent

To consider two events that are not independent, let A stand for a firm's spending a large amount of money on advertisement and B for its showing an increase in sales. Of course, advertising does not gurantee higher sales, but the probability that the firm will show an increase in sales will be higher if A has taken place. The probability that event B will take place provided that A has taken place is called the conditional probability of B relative to A and it will be written symbablically as $P(B/A)$. In terms of conditional probability we can now formulate a more general rule for the probability that two events A and B will both occur.

General Rule of Multiplication

$P(A \text{ and } B) = P(A) \, P(B/A)$.
or $\quad P(A \text{ and } B) = P(B) \, P(A/B)$ $\quad \ldots (4)$

To return to the problem of finding a rule of addition for events that need not be mutually exclusive. Let us suppose that $P(A)$ is the probability that there will be sunshine and that $P(B)$ is the probability that there will be rain. If $P(A) = 0.70$ and $P(B) = 0.50$.

By using the equation (1) the formula for mutually exclusive events we made the error of counting twice all days on which there was both sunshine and rain. To compensate for this we can subtract the days which are counted twice and use the following rule.

$$P(A \text{ or } B) = P(A) \times P(B) - P(A \text{ and } B). \quad \ldots (5)$$

If in the previous example it is known that there is both sunshine and rain 40% of the time we get

$$P(A \text{ or } B) = 0\cdot70 + 0\cdot50 - 0\cdot40 = 0\cdot80$$

If in the example given on the previous page, we are interested in the probability that the student will pass either in accounting or in business law, that is in at least one of the two subjects, we obtain.

$$P(A \text{ or } B) = 0\cdot80 + 0\cdot75 - 0\cdot72 = 0\cdot83$$

Zero Sum Game

A Zero Sum Game is one in which the payments, upon completion of the game are equal to zero. Thus if X pay Y Rs. $10\cdot00$ then Y pay pays—Rs. 10 to X. The sum of these payments is zero.

A strategy is a player's predetermined method for making his choices during the game. Hence strategy is a set of decision rules. A Pay Off Matrix is a table which specifies how payments should be made at the completion of the game.

Finally games theory is a clear theory applicable to competitive situation.

Two-Person Zero Sum Games

In a two person zero sum game, the interest of the two competitors are opposed in that the sum of the gains for one exactly equals the sum of tosses for the other or the sum of the game adds upto zero.

		Player B		Minimum of Row
	Plan	3	4	
Player A	1	10	14	10
	2	8	12	8
Maximum of column		10	14	

Both the players A and B are equal in ability. A has the plan 1 and 2 while B has the plan of action 3 and 4. Both know the pay offs for every possible strategy. The possible plans for both competitors are.

(1) A wins the highest game value if he plays the plan 1 all the time.

(2) *B* realises this situation and plays plan 3 in order to minimise his losses.

(3) The game value must be 10 since, *A* wins 10 and *B* looses 10 each time a game is played

(4) The game value is the average winnings per player over a long number of plays.

Rectangular Games (Two person zero sum)

Let the player *X* has three possible plans *P, Q, R* the player *Y* has two possible plans *S, T* payments are made according to the choice of plan as shown below :

Choice of Plan		Payment
P	*S*	*X* pays *Y* Rs. 4·00
P	*T*	*Y* pays *X* Rs. 4·00
Q	*S*	*X* pays *Y* Rs. 2·00
Q	*T*	*Y* pays *X* Rs. 6·00
R	*S*	*Y* pays *X* Rs. 2·00
R	*T*	*Y* pays *X* Rs. 4·00

Find out the best strategies for player *X* and *Y* in this two person zero sum game.

For better understanding, the rules of payment be arranged in the matrix form. Let the positive number indicate a payment of *Y* to *X* and a negative number a payment of *X* to *Y*. Then the following "pay off" matrix can be had.

		Player Y		Minimum of row
	Plan	S	T	
Player X	P	−4	4	−4
	Q	−2	6	−2
	R	②2	4	②2
Maximum of column		2	6	

Rerferences

(1) When *Y* chooses plan *T* he always looses since he has to pay *X* only. Therefore *Y* prefers to choose plan 5 so that *Y* can loose least *i.e.* Rs. 2 *i.e.* when *X* chooses plan *R*. This is the worst come worst *Y* can looses otherwise he gains in case *X* chooses *P* or *Q* plans. *Therefore plan S is the best strategy for Y*

(2) X will gain highest Rs. 6 by choosing plan Q provided Y choses plan T. Since Y can not choose plan T because of maximum loss, and always prefers only plan S, *the best plan X can choose is R in* which case he can gain at least Rs. 2.

(3) Thus the best strategy for X is R and for Y is S. *i.e.* X gains Rs. 2 maximum and Y looses Rs. 2 minimum. Rs. 2 is the Value of the Game.

(4) *Thus the solution is to find out best strategies for two players and the value of the game*

Mini Max and Maxi Mini Rule
Player X

With plan P, X gains least	— **Rs. 4**
” ” Q, X gains least	— **Rs. 2**
” ” R, X gains least	+ **Rs. 2**

The maximum of the least gain for P is Rs. 2 *i.e. RS*. Therefore to find out the minimum value of each row write as shown in table and circle it, which ever is the *maximum of these minimum* values.

Player Y

With plan S, Y looses highest Rs. 2

With plan T, Y looses highest Rs. 6

He wishes to loose least. Therefore the minimum of the highest loss is Rs. 2 *i.e. RS*. Therefore find out the maximum of the each column and write as shown in Table and square it whichever is the *minimum of the maximum values*. Gains for which maximum of X=minimum of maximum of Y are called games with Saddle Point and the Saddle point is represented with a circle inside the square.

	Plan	Player Y.				Minimum of row
		A	B	C	D	
Player X	E	6	-10	0	12	-10
	F	-8	-4	2	4	-8
	G	10	8	④	6	④ max of min
Maximum of column		10	8	4	12	

Mini of Max

Thus the easiest technique in searching a Saddle point is to find a number that is *"lowest in its row and highest in its column"* when one such a number exists it is a Saddle point. The corresponding strategies are the best strategies and the number itself is the value of the game. If there are more than one such saddle points, there are more than one solutions. Each solution corresponds to one saddle point.

C is the best strategy for Y

G is the best strategy for X

Value of the Game is Rs. 4 as it is the saddle point

X wins and Y looses.

Rectangular Games (without saddle point)

It is a 2 persons with no zero sum game. Let us examine the pay off matrix shown in the following table which has no saddle point.

		Player Y	
	Plan	A	B
Player X	C	-9	21
	D	18	3

Since there is no saddle point, player X and Y do not have one single best strategy as their best plans. Therefore each player has to work out some mixed strategy so that his gain is maximised or his loss is minimised.

The possibilities of playing the game by X and Y are :

(a) X plays C and D strategies with different frequencies when Y plays only A strategy.

(b) X plays C and D strategies with different frequencies when Y plays only B strategy.

(c) X plays C and D strategies with different frequencies when Y plays A and B strategies with different frequencies.

Similarly there are three possibilities for Y.

Let X play C half of the time and D half of the time whereas Y chooses to play only A all the time. Then the gains of A are

$$\tfrac{1}{2}(-9)+\tfrac{1}{2}(18)=\tfrac{1}{2}\times 9=4\cdot5$$

If X plays C and D half the time each when Y chooses to play only B all the time, the gains of X are

$$=\tfrac{1}{2}(21)+\tfrac{1}{2}(3)=\tfrac{1}{2}\times 24=12$$

If Y also chooses mixed strategy say half the time A and half the time B are played and X plays C and D half the time each, the gains of X are

$$= \tfrac{1}{2}[\tfrac{1}{2}(-9)+\tfrac{1}{2}(18)]+\tfrac{1}{2}[\tfrac{1}{2}(21)+\tfrac{1}{2}(3)]$$
$$= \tfrac{1}{2}(\tfrac{1}{2}\times9)+\tfrac{1}{2}(\tfrac{1}{2}\times24)=\tfrac{1}{2}(4\cdot5+12)=8\cdot25$$

Evaluating Best-Strategy

How to find the best strategies for X and Y so that the loss is minimised and gain is maximised and the value of the game is accertained.

Let X play C with frequency α and D with a frequencies of $(1-\alpha)$.

Then if Y plays all ways only A strategy the gain of X
$$= G(X, A)=\alpha(-9)+(1-\alpha)(18)=18-27\alpha$$

If Y plays all ways only B strategy, the gain of X
$$= G(X, B)=\alpha(21)+(1-\alpha)(3)=3+18\alpha$$

Let the value of α is so choosen that

$G(X, A)=G(X, B)$ which will lead to the best strategy.

Therefore $18-27\alpha=3+18\alpha$

$\alpha=\tfrac{1}{3}$ and $(1-\alpha)=\tfrac{2}{3}$

in the above problem the gain of X will be

$$G(X)=\tfrac{1}{3}(-9)+\tfrac{2}{3}(18)=\text{Rs. } 9\cdot00$$

Thus irrespective of the frequency with which Y plays either A or B, the gain of X is Rs. $9\cdot00$

Similarly let Y play A with a frequency of γ and B with a frequency of $(1-\gamma)$.

If X plays all the time only C, the gain of Y will be

$$G(Y, C)=\gamma(-9)+(1-\gamma)\,21$$

If X plays always only D, the gain of Y is

$$G(y, D)=\gamma\,(18)+(1-\gamma)(3)$$

Since $G(Y, C)=G(Y, D)$

$$\gamma(-9)+(1-\gamma)(21)=\gamma(18)+(1-\gamma)(3) \quad \gamma=\tfrac{3}{5} \text{ and } 1-\gamma=\tfrac{2}{5}$$

The gain of $Y=G(Y)=\tfrac{3}{5}(-9)+\tfrac{2}{5}(21)=\text{Rs. } 9\cdot00$

Even in the non zero sum game

$$G(X)=G(Y)=\text{Rs. } 9\cdot00 \text{ for the best strategies of } X \text{ and } Y.$$

Thus the best strategies are :

I. X should play C and D with frequencies of $\tfrac{1}{3}$ and $\tfrac{2}{3}$ respectively

II. Y should play A and B with frequencies of $\tfrac{3}{5}$ and $\tfrac{2}{5}$ respectively

III.   The value of the game is Rs. $9\cdot00$.

49

Maintenance Models

Introduction

Machines and equipment wear out all the time. Greater the number of parts in a machine, greater the possibility of wear, [tear and hence the break downs. When the machines break down they can be repaired and put back to work but that becomes too costly. The costs that occur due to beak down are :

(*i*) Down time of machines resulting into loss of production and sales

(*ii*) Idle time of direct and indirect man power

(*iii*) Increase in the scrap and rejections.

(*iv*) Dis-satisfaction of the customer due to delay in delivering the commitments.

(*v*) Actual cost of the repairs.

If the reliability of the system is improved, the frequency of break downs time can come down and hence the costs too will come down. The total down time of a machine is the summation of down time multiplied by its frequency. For increasing the reliability of the system, either the size of the repair facilities (services) have to be increased (to reduce the machine down time) to attend it quickly as and when break downs occurs or the system of preventive maintenance, needs to be followed (to reduce the frequency of break downs).

This Maintenance Models deals with both the methods. To reduce the down time of a machine. Queuing models based on cost considerations will help, where as to reduce the fr equency of break downs, preventive maintenance models based on cost factors are to be employed.

Before we deal with either one knowing the frequency distribution of break down and their interpretation is a prerequisite.

Frequency Distribution

Break down time distributions indicate the frequency with which machines have maintenance free performance. Depending upon the nature of the equipment, the frequency distribution curves takes different shapes.

(*i*) A simple machine, breaks down nearly at regular intervals of time with least variation in time distributions. This type of distribution resembles to normal distribution in the sense major portion of the break downs happen near about an average value of break down time Ta.

(*ii*) In a complex machine, will have a number of sub assemblies or parts. Each part of the machine will have its own frequency distribution of break down time. Therefore even after the repair of one part, the machine can break down at any time due to the wear out of any other part of the machine. The variability in the time of break down is much more when compared to the 1st case. Such machine break down will follow Skewed distribution, Beta distribution or gama distribution. Therefore for the same average break down Ta, more wider variability of break down time is noticed.

(*iii*) In few machines like electronic equipment or very precise and sophisticated instruments etc. they have a long free of trouble time after the repair. Even after the machine is repaired they need a little bit adjustments here and there, then and now but they run for long time trouble free. These three types of curves are shown in the Fig. 49.1

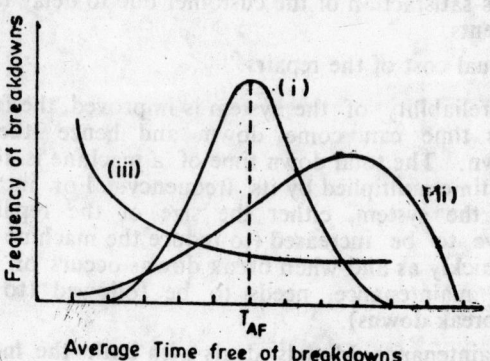

Time free of break downs.

Fig. 49.1 Frequency distribution of time free of break downs.

Preventive Maintenance

A good preventive maintenance reduces the frequency of break downs. The preventive maintenance can not eliminate the break downs completely. However a well designed and implimented inspection system can indicate well in advance about the possibility of failure of a particular part and hence can be planned for preventive maintenance. Normally 75% of the maintenance strength is used on emergency or break down maintenance. If 75% of staff is used for preventive and 25% of maintenance staff is spent on break down maintenance, then preventive maintenance will pay rich dividends.

Example 1. A machine shop has 100 machines, to be kept in service. Over-haul of the machines periodically will greatly reduce the break down frequency, but it costs Rs. 200 per over-haul As and when the machines break down, they can be repaired by the repair crew who have to wait for the break down to happen or broken down machines will have to wait since the repair crew might be busy in attending the other machines. The break down mahine costs Rs. 3000 if they are repaired. The machine shop has been keeping record from which the probability of break down is given below, calculate the periodicity of over-haul on economic basis and see whether over haul is better than break down repair?

Months after over-haul	Probability of break down	Cumulative probability of break downs
1	5·00%	5·00%
2	2·00%	7·00%
3	1·00%	8·00%
4	6·00%	14·00%
5	5·00%	19·00%
6	4·00%	23·00%
7	7·01%	30·00%
8	12·00%	42·00%
9	13·00%	55·00%
10	14·00%	69·00%
11	15·00%	84·00%
12	16·00%	100·00%

Break Down Repair Costs

Let Average elapsed time between break downs be T_a

$$T_a = T_1 P_1 + T_2 P_2 + T_3 P_3 + T_4 P_4 + + + T_{12} P_{12}$$

where T_1, T_2, T_3 etc are months after over-haul

P_1, P_2, P_3 etc are the corresponding probabilities

$$T_a = 1 \times 0·05 + 2 \times 0·02 + 3 \times 0·01 + 4 \times 0·06 + 5 \times 0·05 + 6 \times 0·04$$
$$+ 7 \times 0·07 + 8 \times 0·012 + 9 \times 0·13 + 10 \times 0·14 + 11 \times 0·15$$
$$+ 12 \times 0·16 = 8·44 \text{ months.}$$

or an average every machine is to be repaired after every 8·44 months.

Total costs of repair = No. of machines × cost of repair per machine

$$= 100 \times 3000 = \text{Rs. } 300{,}000$$

Average cost per month for break down repairs.

$$= \frac{300{,}000}{8 \cdot 44} = \text{Rs. } 35{,}545$$

Costs of preventive maintenance and optimal periodicity of preventive maintenance

The distribution does not follow any standard distribution. A number of iterations are required to find out the number of break downs that can occur for different periods of overhaul.

Let the machine be over-hauled once every month. In that case the only break downs need to be attended are within one month. If the machines are over-hauled once every two months, the break down will increase. First in two months whatever break down occur they need to be repaired. In addition few of the machines repaired in the first month will break down again before the scheduled over-haul.

No. of break downs if scheduled over-haul for every month then

$D_1 = n \, P_1$

$D_1 =$ No. of break downs in one month

$n \; =$ No. of machines

$P_1 =$ probability of break downs after one month

$D_1 = 100 \times 0 \cdot 05 = 5 \cdot 00$

No. of break downs if scheduled over-haul is for every 2 months.

$D_2 = N(P_1 + P_2) + D_1 P_1$

$= 100(0 \cdot 05 + 0 \cdot 02) + 5(0 \cdot 05)$

$= 7 + 0 \cdot 25 = 7 \cdot 25$

Like this for the entire period of 12 months it can be calculated.

Scheduled over-haul for every 3 months

$D_3 = N(P_1 + P_2 + P_3) + D_2 P_1 + D_1 P_2$

$= 100(0 \cdot 05 + 0 \cdot 02 + 0 \cdot 01) + 7 \cdot 25 \times 0 \cdot 05 + 5 \times 0 \cdot 02 = 8 + 3 \cdot 625$
$+ 0 \cdot 1 = 11 \cdot 725$

No. of break down for Scheduled over-haul every 4 months

$$D_4 = N(P_1 + P_2 + P_3 + P_4) + D_3 P_1 + D_2 P_2 + D_1 P_3$$
$$= 100(0 \cdot 14) + 11 \cdot 725 \times 0 \cdot 05 + 7 \cdot 25 \times 0 \cdot 02 + 5 \times 0 \cdot 01$$
$$D_4 = 14 + 0 \cdot 58625 + 0 \cdot 1450 + 0 \cdot 05 = 14 \cdot 79$$

No. of break downs for scheduled over-haul every 5 months

$$D_5 = N(P_1 + P_2 + P_3 P_4 + P_5) + D_4 + P_1 + D_3 P_2 + D_2 P_3 + D_1 P_4$$
$$= 100(0 \cdot 19) + 14 \cdot 79 \times 0 \cdot 05 + 11 \cdot 725 \times 0 \cdot 02 + 7 \cdot 25 \times 0 \cdot 01$$
$$+ 5 \times 0.06$$
$$D_5 = 19 + 0 \cdot 74 + 0 \cdot 24 + 0 \cdot 07 + 0 \cdot 3 = 20 \cdot 35$$

No. of break down for scheduled over-haul every 6 months

$$D_6 = N(P_1 + P_2 + P_3 + P_4 + P_5 + P_6) + D_5 P_1 + D_4 P_2 + D_3 P_3 + D_2 P_4$$
$$+ D_1 P_5$$
$$= 100(0 \cdot 23) + 20 \cdot 35 \times 0 \cdot 05 + 14 \cdot 79 \times 0 \cdot 02 + 11 \cdot 725 \times 0 \cdot 01$$
$$+ 7 \cdot 25 \times 0 \cdot 06 + 0 \cdot 5 \times 0 \cdot 05$$
$$= 23 + 1 \cdot 02 + 0 \cdot 3 + 0 \cdot 12 + 0 \cdot 44 + 0 \cdot 25 = 25 \cdot 12$$

Similarly the number of break downs that can occur for every 7, 8, 9, 10, 11 and 12 months have been calculated and will be as under.

$$D_7 = 33 \cdot 08$$
$$D_8 = 46 \cdot 47$$
$$D_9 = 61 \cdot 177$$
$$D_{10} = 78 \cdot 81$$
$$D_{11} = 97 \cdot 78$$
$$D_{12} = 119$$

			1	2	3	4	5	6	7	8	9	10	11	12	
Probability			0·05	0·02	0·01	0·06	0·05	0·04	0·07	0·12	0·13	0·14	0·15	0·16	
Cumulative probability			0·05	0·07	0·08	0·14	0·19	0·23	0·30	0·42	0·55	0·69	0·84	1·00	

Probabilities of

D	N	cumulative probability	D_1	D_2	D_3	F_4	D_5	D_6	D_7	D_7	D_9	D_{10}	D_{11}	D_{12}	Total value of D
D_1	100	0·05	0·05												5
D_2	100	0·07	0·02	0·05											7·25
D_3	100	0·08	0·01	0·02	0·05										11·725
D_4	100	0·14	0·06	0·01	0·02	0·05									14·79
D_5	100	0·19	0·05	0·06	0·01	0·02	0·05								20·35
D_6	100	0·23	0·04	0·05	0·06	0·01	0·02	0·05							25·12
D_7	100	0·30	0·07	0·04	0·05	0·06	0·01	0·02	0·05						33·08
D_8	100	0·42	0·12	0·07	0·04	0·05	0·06	0·01	0·02	0·05					46·47
D_9	100	0·55	0·13	0·12	0·07	0·04	0·05	0·06	0·01	0·02	0·05				61·77
D_{10}	100	0·69	0·14	0·13	0·12	0·07	0·04	0·05	0·06	0·01	0·02	0·05			78·81
D_{11}	100	0·84	0·15	0·14	0·13	0·12	0·07	0·04	0·05	0·06	0·01	0·02	0·05		97·78
D_{12}	100	1·00		0·15	0·14	0·13	0·12	0·07	0·04	0·05	0·06	0·01	0·02	0·05	119·00

Since the expected break downs for each policy have been calculated, the economics of each policy Is worked out as shown below.

Preventive maintenance for every N number of months (1)	No. of break downs in N number of months (2)	Average number of break downs per month (3)=(2)÷(1)	Cost of break downs per month Rs. (4) = Rs. $300 \times$ column (3)	Total cost of break downs over hauls per month Rs. (5)=Rs. $200 +$ colum n (4)
1	5	5·00	1500	1700
2	7·25	3·63	1089	1289
3	11·725	3·91	1173	1373
4	14·79	3·70	1110	**1310**
5	20·35	4·07	1221	1421
6	25·12	4·19	1257	1457
7	33·08	4·73	1419	1619
8	46·47	5·81	1743	1943
9	61·77	6·86	2058	2258
10	78·81	7·88	2364	2564
11	97·78	8·89	2667	2867
12	119·00	9·92	2976	3176

Cost of break downs overhaul per month when plotted against months. The minimum cost occurs at 4th month. Therefore for every 4th month overhaul can be done and this preventive maintenance policy is the best.

50

Selection Process

Personnel are a high cost element in an organisation and must be planned like other functions of the enterprise such as marketing, production, finance, etc. This calls for proper determination of manpower requirement, recruitment, selection, placement and replacement of personnel to match the abilities and potentialities of the personnel with the demands of various jobs.

Higher rate of employee turnover and absenteeism and lower productivity and morale of the employees constitute a big problem in many organisations. The reason for this is that right type of people are not appointed for various jobs in the organisations. Actually, it is not easy to get the right type of people. This calls for a properly planned recruitment policy to minimise disruption of work by constantly changing personnel. The basis for a sound recruitment policy must be a consideration of the manpower requirements of the organisation. Many a time it happens that there are no employees to man a particular job because of the retirement, death, resignation or dismissal of the employee. Such a situation is dangerous in any organisation. This will mean loss of production and interruptions in work of other employees whose work is related to the vacant jobs. Therefore, there should be proper manpower planning in the organisation so as to avoid such situations and to recruit and select such candidates as are best suited for various jobs.

The requirement of personnel should be determined both in terms of number and kind of personnel required. For determining the number of persons required, not only the present requirements but also the future possibilities of expansion and growth should be taken into account. Type of personnel required should be determined by the requirement of various jobs. But in large organisations, the requirements of individual jobs are very easily recognisable. An analysis of job is a pre-requisite to any recruitment and selection procedure. This will involve the preparation of job descriptions and job specifications for individual job. The recruitment process-will begin only when the number of persons required for different jobs and requirements of different jobs are known.

Recruitment process aims at developing and maintaining adequate manpower resources upon whom the organisation can depend when it needs additional employees. Recruitment is a positive process of searching the prospective employees and stimulating them to apply for jobs in large numbers in the organisation. In this way, it increases the selection ratio and enables the management to select suitable employees. Selection is a negative process as it involves the taking of suitable people for the organisation and rejection or elimination of the other applicants.

The process of selection leads to employment of employees which establishes more than contractual relationship between the employer and the employees. The recruitment and selection process should look to the human adjustment to organisational objectives. It is important that a selected candidate possesses the ability to perform the job assigned to him and has the opportunity for development and growth in the organisation. This is the most important feature of a sound personnel policy. Almost all the employees have aspirations for higher salaries or wages, job security, job satisfaction and higher status in the organisation. So the employer while selecting the personnel will have to strike a balance between the requirements of efficiency of employees and the satisfaction of employees' aspirations. Employer should consider the effect of each individual selection on the present set up of the work group in the organisation as it is not possible to hire just the two hands of a person but he rather is hired as a whole.

SOURCES OF RECRUITMENT

In general, there are two sources of recuiting empolyees :

  (i) Internal, and
  (ii) External.

Internal Sources

Many organisations completely overlook the value of recruitments from within. It is not only reasonable but wise to let the existing employees know of vacancies by internal advertisement. But now it is being realised that the best source of supply for higher posts is the personnel already in the organisation. The existing talented employees may be given the adequate training to be eligible for promotion to higher positions in the organisation. Filling a vacancy from within the organisation has the advantage of stimulating and preparing for possible transfer or promotion, increasing the morale of the employees and simplifying selection and placement problems. A comprehensive programme of maintaining manpower inventory and its development will contribute much to the success of an internal recruitment programme. In order to achieve the advantages of recruitment of personnel from the existing employees, the personnel manager should draw up a policy relating to promotion from within and communicate to all the employees.

External Sources

Any business undertaking has to go to external sources for lower entry jobs, for expansion and for positions whose specification cannot be met by the present employees. Certain organisations which are regarded as good employers draw a steady stream of unsolicited applications in the office. This is a valuable source of manpower supply. The personnel department may find the unsolicited applicants very useful in filling the vacancies. In addition, the more commonly used outside sources are :

(i) Advertisement

Advertisement in local or national newspapers or trade and professional journals is generally used when qualified or experienced personnel are not available from other sources. Most of the senior positions in industry as well as in trade are filled by this method, particularly when they cannot be filled from within. The advantage of advertising is that more information about the organisation, job description and job specification can be given in advertisement to allow self-screening by the prospective candidates. Advertisement gives the management a wider range of candidates for selection. But its disadvantage is that it brings in a flood of response, even from quite unsuitable candidates and many applicants try to approach the members of the selection body.

(ii) Employment Exchanges and Agencies

Employment exchanges run by the government are regarded as good source of recruitment for unskilled, semi-skilled, skilled and operative jobs. In some cases compulsory notification of vacancies to the employment exchange is required by law. Thus the employment exchanges bring the job givers in contact with the job seekers. However, in the technical and professional areas, the private agencies and professional bodies appear to be doing most of the work. Employment exchanges and selected private agencies provide a nation wide service in an attempt to match personnel demand and supply.

(iii) Educational Institutions

Jobs in trade and industry have become increasingly technical and complex to the point where school and college degrees are widely required. Consequently, many big organisations maintain a close liaison with the universities, vocational institutions and management schools for recruitment to various jobs.

(iv) Recommendations

Applicants introduced by the employee's friends and relatives to the organisation may prove to be a good source of recruitment and indeed, many employers prefer to take such persons because something about their background is known. When a present employee or a business friend recommends a person, a sort of preliminary screening has already taken place. Some organisations have agreements with the unions of employees to give preference to relatives of existing or retired employees if their qualifications and experience are suited for the vacancies.

(v) Casual Callers

To meet the short-term demands of personnel, the management may consider the possibility of hiring personnel, who call on them casually. This will avoid the selection and training costs and will avoid any obligation in pensions, insurance and other fringe benefits. It will also be a highly economical method since the management need not pay retrenchment and lay off compensation according to the requirements of law. This practice is more common in developed countries.

(vi) Labour Contractors

Labour contractors constitute an important source of recruitment in many industries in India. Workers are recruited through labour contractors who are themselves the employees of organisation. The disadvantage of this system is that if the contractor himself decides to leave the organisation, all the workers employed through him will follow suit.

PROBLEMS IN SELECTION

Individuals vary in the quality of their performances. In exceptional cases the output of the best worker may be five to ten times greater than that of the worst. If any significance is attached to these differences in behaviour among workers on the job, the forecasting of occupational success assumes a considerable importance. The successful development of a selection procedure depends on a number of circumstances which are in practice and difficult to deal with. Since jobs have become more complex and interdependent, it is increasingly difficult to develop adequate job descriptions and adequate criteria of performance for the selection of employees.

The other problem is the ratio of management personnel to operative personnel, which has been steadily rising. This has made it increasingly important to improve selection devices for managers. It is the managerial job which is the most difficult to describe and analyse clearly. Many organisations are investing great time and efforts in developing tests to hire managers at various levels.

A similar problem undermines the effectiveness of job descriptions and the fluidity of jobs themselves in the rapidly changing technology and society. The rate at which the engineers become obsolete is so great that organisations have begun to favour college students who have obtained more general education and who are better prepared to cope with the changing environments. The other problem is of job performance and the criteria applied to job performance tend to be short-run rather than long-run because the evaluator cannot wait for ever before he checks the validity of the test. In some cases, short-term performance is highly correlated with long-run performance. This procedure is perfectly acceptable, but on many kinds of jobs and for most of the people correlation between short-run and long-run performance tends to be low because of many reasons.

Another problem to selection is the questionable assumption that an applicant can be placed into a standard kind of test situation. People are dynamic creatures and in constant interaction with their environment. So standardised performance is difficult to obtain. Perhaps the most serious problem with selection through testing approach is that the individual is viewed as a static entity to be measured, classified, and filled into an organisational slot. It is also felt that too much emphasis continues to be given in the interview, even though it has limited validity for selection. One important point worth considering is whether it is necessary to keep in mind only the appraisal made at the time of hiring or whether future promise for higher jobs is also to be evaluated. Of the two, measurement of potential growth and development presents the greater difficulties. The general practice is to select supervisors, executives, and administrators from persons already employed in the organisation in lower positions.

If in the original selection of workers no attention is paid to the evaluation of potential for higher jobs, there is little probability that sufficient number of qualified individuals will be available for such jobs.

SELECTION PROCEDURES

Recruitment has been described as the process of getting potential employees willing to apply for a job or jobs in the organization. If recruitment is successful, several candidates will apply and out of these, the more suitable candidate may be selected. But selection is not an easier process. It is a problem of matching a man to the job. Thus selection process involves choosing of individuals who possess the necessary skill, abilities and personality to fill specific jobs in the organisation. A well devised selection procedure is of great significance for the organisation because it involves greater cost to the organisation and an employee once selected, is expected to remain in the organisation for a number of years. Faulty selection process would result in low productivity and high labour turnover, and if the job does not suit the employee, he will always be in search of a suitable job elsewhere and will not be in a position to concentrate on his present job.

Thus proper and thorough selection policy must be followed by the management. There must be definite selection procedure established for screening out the undesirable and hiring the desirable candidates. The following are the important steps involved in the satisfactory selection procedure.

(a) Preliminary Interview.
(b) Application Blank.
(c) Employment Tests.
(d) Employment Interview.
(e) Reference check.
(f) Physical examination.
(g) Final selection.

The procedure of selection will vary from organisation to organisation and within the same organisation according to the kind of the job to be filled. Whatever may be the procedure followed, the personnel department plays an important role in the selection process. However, it should be kept in mind that decision to add persons in the payroll in a particular department of an organisation is not made in the personnel department. The requisition for the employees must originate from some department where it is necessary to hire some persons to man particular jobs in the near future. For this purpose the use of a standard requisition form by the departmental head is preferable as it will make clear the number of vacancies, type of job, pay scale and any special characteristics of skill required. After receiving the requisition, the personnel department will see the catalogue of job descriptions and job specifications to obtain a picture of the job and will tap some sources of recruitment according to the circumstances of the case.

Selection process will start when some applications are received from the candidates. The personnel department will classify and file the applications and will screen out the applicants which are found unsuitable. After this, it may call the other applicants for initial interview.

(a) Preliminary Interview

The preliminary interview is generally quite brief and has the object of eliminating the obviously unqualified or unsuitable candidates. Lack of certain requirements in education, training or experience may determine unsuitability. Appearance, ability in communication, impression, etc. of the candidate are quickly evaluated and the candidate's salary requirements are obtained. If the applicant appears to have some chance of being selected, he is given the application blank to fill in.

As a general rule, the more non-selective the recruitment programme, the more likely it is that preliminary interview offer advantages not only to the company but to the applicant as well. If an applicant is eliminated in the early part of the selection procedure, the organisation is saved from the expense of processing him through the remaining steps of the procedure and the applicant is saved from the trouble of passing through the long procedure.

(b) Application Blank

This is one of the important steps in hiring a man for the organisation to get a written record of qualifications, experience or any other specialisation of the candidate. There is a high degree of similarity between the application blanks of various organisations because of the basic information about the individuals to which all organisations give importance.

Application blank is used to obtain information in the applicant's own handwriting sufficient to properly identify him and to make tentative inferences regarding his suitability for employment. The application blank should be as simple as possible and incorporate questions having bearing on the fitness of the applicant for the job. Most of the application forms are designed, in big organisations, in such a manner as to enable the organisations to plan their training programmes, special assignments or promotions, after the employee is hired for the organisation.

(c) Employment Tests

The use of tests is, perhaps, the most controversial of all personnel selection procedures. Attitudes range from those who place complete reliance on tests scores to those who refuse even to consider their use and place complete reliance on their own personal judgement regarding the applicants. In practice, neither of the two points of view is realistic. No personnel man ever accepts tests to provide perfect prediction because even the most comprehensive battery of tests gives only a small sample of an individual's behaviour. Similarly, interview method of selection has its own limitations. The interviewer may have some business in his mind and may not be able to take objective decision.

Employment tests have become widely accepted in the selection process. However, these should be considered simply as a step and not a replacement for the other phase of the selection process. T he real value of the tests lies in eliminating those applicants who have very little chance of job success than in selecting applicants who will definitely be successful on the job. It should be remembered that when the tests are used, they should not be relied upon completely. Individuals differ in almost all aspects. They differ with respect to physical characteristics, capacity, level of mental ability, likes and dislikes and also with respect to personality traits. The pattern of physical, mental and personnel variables give rise to thousand and one combinations and the particular pattern makes the individual suitable for several classes of activities, jobs or fields of work. Matching of individual's physical, mental and temperamental pattern with the requirement of job or field of training is a difficult task, but where this matching takes place, the result is happiness for the individual and greater prosperity for the organisation and the society. But instances of round pegs in square holes are not rare.

The existence of individual differences provides the basis for the selection and placement process. Any group of people will vary considerably in terms of their relative work efficiency and performance. There are psychological and other tests which measure the extent of differences among people. Tests are given to the individuals on the assumption that there is a direct and important relationship between the possession of one or more of these qualities and the individual's ability to do certain jobs. This relationship enables to predict the candidate's eventual job performance. Thus test performance of various individuals is normally distributed. Many tests designed by the psychologists are based on this assumption that there are lesser people whose test scores will either be below average (or poor) or above average (very good).

Advantages of Test

A test is an objective and standardised behaviour sample, and tends to be less subject to bias. Test can help to uncover talent that may otherwise be overlooked and also to differentiate between the ability required for the present job and those required by the new ones. Another advantage is that a great deal of information about a person can be collected in a relatively short period of time by using tests. Tests reduce the cost of selection and placement because large number of applicants can be evaluated within lesser period of time. If an employer expects to continue in a competitive business, the cost of hiring plus the cost of training must be kept at a minimum. Psychological tests can reduce the costs of hiring people who will be successful by measuring their aptitude and predicting their success.

Tests provide a healthy basis for comparing applicant's background. Not only do the tests compensate in part for weaknesses in the interviewer but they have the effect of increasing the quality of organisation's employees over a period of time. Tests can be used for differential placements because in testing attention is centred on the qualifications for a specific job. If the applicant fails to pass the test or does very well in the test, his suitability for the job other than the one applied for can be explored.

Criticism of Test

Tests are criticised for measuring only a part of the total amount of information needed to make an accurate selection. This criticism would be justified if tests were the only selection method used. Tests are rarely used as the only selection method. Our objective should be to maximise accuracy in selection by choosing proper combination of methods.

Tests are sometimes criticised on the ground that they cannot make prediction of changes of success of an applicant because he was nervous. But this is valid only when the test results for the entire group are not valid. However, research data in regard to their validity are lacking. It is true that tests are far from perfect, but other methods like application blank, interviews, reference checks, etc. are also of limited value.

STEPS IN THE DEVELOPMENT OF TESTS

(a) Analysis of Job

For successful development of a test, the essential of job should be thoroughly understood. What does the worker do and under what circumstances ? What are the intellectual demands of the job ? What are the physical and social demands of the job ? Does he supervise or is supervised ? Does he deal with machines or people ? Is there pressure of speed or of complexity ? Is the task repetitive or varying ? etc. All these questions would lead to the preparation of job description and job specification. These are essential in order to develop any type of tests.

(b) Devising a test Procedure

Depending upon the analysis of the job a test may be selected either from the battery of existing tests or a test may be devised for the checking of qualities required for the job.

(c) Preliminary Tryout

Particularly when the test device is novel, the test should be tried out and then revisions and refinements may be made in that light. Tryout can be done on available office and factory personnel group similar to those sought to be tested. The intention is to select items that are too difficult or too easy or items that fail to discriminate between more able or less able members of the group.

(d) Validation of Test Procedures

While trying out a test, it is desirable to establish the empirical validity also. Experimental evidence is called for to show that test is in fact, effective in discriminating between those who are and between those who are not successful in a particular job.

(e) Combination of a Test into a Battery

Most of the jobs call for different aptitudes, factors of intellectual skill, interest and personnel adjustments. It is not possible to assess all with single test material. Different types of testing techniques are required to assess the several traits required for success or failure. The basic factors which must be taken into account in combining tests for personal selection are the validity of separate tests and correlation between them.

APPLICATION OF TESTS

Employment tests should be applied only by experts trained in this field. The tests should not be selected indiscriminately. The tests have to be selected with great care and only by persons who understand them, otherwise the results will be misleading.

Thus various types of tests can be devised to be given to the candidates for various purposes such as selection, training, placement and employee counselling. However, test should not be used where information about a particular trait or group of traits regarding the individual under study is available from a reliable source. The choice of a series of tests for any organisation will depend upon the nature of the jobs to be filled. In some instances, it may be necessary or desirable to administer a wide range of tests while in others only a test of a specific attitude or trait may do well. The test ratings may also be used for purposes of wage increase, promotions, and reassignment in addition to selection of an employee.

STANDARDISATION OF TESTS

Tests are short-cut methods for prediction of job success. They serve as substitute for placing an individual on the job and then seeing how well he performs the tasks or jobs assigned to him. However, the response to a test made by an individual on a given occasion should be considered only as a sample of his ability.

Tests must be standardised before these are applied to measure any individual's traits or qualities. While standardising the tests, the psychologist must see that these fulfil the statistical requirements of validity, reliability and objectivity. The test must be valid in the sense that it must measure what it intends to measure. Therefore, the test items should be selected in such a manner that it will measure the trait for which it is constructed or designed. If a test is given to a candidate, the test score obtained by him must be reliable. If the same test is given to one group and then after some time to another group, the score obtained by each individual in the first group and the second group must not vary considerably. Finally, the test must be objective : the psychologist's personal bias towards the testee must not influence the test score obtained by him.

CHARACTERISTICS OF A GOOD TEST

In the design of a test, it is important to bear in mind the nature of group on which it will be employed. The suitability of a test may be indicated in a variety of ways. If the test is too difficult most individuals will tend to get low scores. If it is too easy, most individuals will tend to get high scores. The difficulty of a test

should be so gauged that the scores show a reasonably wide distribution.

A test that is too easy fails to discriminate among individuals
of high ability. A test that is too difficult fails to discriminate among
individuals of low ability. Less educated workers will not react
favourably to tests involving difficult expressions. It is a good
practice to fit the test to the group in as many ways as possible.

Norms

Norms referred to distribution of scores on a test, can be used
as a basis for interpreting and evaluating an individual's performance
on the test. Several factors determine the usefulness of norms. The
larger the number of workers tested in any group, the more dependable the norms will be. A second factor concerns the amount of
available information about the level of skill characterising the
workers whose scores are used in the preparation of norms. A third
factor concerns the kinds and number of different groups for which
the scores are available.

There are two general types of norms : general population
norms and special group norms. General population norms are constructed from the scores of workers from a wide variety of jobs and
the special group of norms comprise scores of workers from those
specific population groups for which the test was designed and for
which it has definite use.

TESTS USED FOR THE MEASUREMENT OF TRAITS

Success on a job is not totally determined by ability, but is also
attributable in part to traits of personality, character and interest.
A number of investigations have shown that separation from a job is
often due to deficiencies in personality. Personality refers to those
traits of the individual or those aspects of individual's behaviour
that have emotional, social, motivational or moral connotations such
as stability, extroversion perseverance and honesty.

The common way of assessing personality, at least when large
number of individuals are involved, is by the use of questionnaires.
A typical questionnaire is composed of series of questions directly
concerned with personality in its behavioral aspects. The questions
can be :

(i) Have you ever felt uncomfortable in a social situation ?

(ii) Do you get embarrassed in talking to a large group of
people ?

The validity of personality tests is a direct function of honesty
and accuracy with which the individual answers it. The individual
taking the questionnaire knows, that his performance is going to be
used for determining his fitness for the job, and he wants to make
creditable performance.

The responses of every person to whom the questionnaire was administered would have to be carefully examined. This examination would have to be done by a highly trained professional person and should not be left to an ordinary personnel interviewer.

Projective Techniques

Clinical psychologists have developed a variety of devices which have been termed as *projective techniques*. The most important characteristic of a projective technique is the freedom it allows the testee in the context of the responses he makes. The personality tests described above present specific stimulating situations or problems to the individual and therefore the kind of responses he can make which are determined by the nature of the question asked. *Such a situation is called structured situation, since the individual is given a definite frame of reference or structure for his response.* In the projective methods the purpose is to present situations which are very weakly structured. Specific projective situations are ambiguous such as ink-blots, pictures which are ambiguous in meaning. In the ink-blots test the individual is asked to tell what he sees in the design, in the ambiguous-picture test he is asked to tell the story which each picture illustrates. Since the individual has no frame of references or structured stimuli given to him by which his responses are predetermined, therefore, his response is influenced by the underlying traits, motives, aspirations, etc. of his personality. In other words, his personality characteristic will be projected or manifested by way of his responses to the unstructured test situation.

Projective techniques have some important limitations. The evaluation reports, however, are not based on an objective appraisal of the responses elicited, rather a highly trained professional psychologist is required to interpret the responses. On the basis of his interpretation he makes rating of the individual's personality characteristics. It is, therefore, concluded that, at present time, it has not been demonstrated that the projective techniques are highly valid instruments for appraising personnel in business and industrial situations.

The Forced Choice Technique

One of the most promising techniques for personality measurement in the evaluation of applicants in business and industry involves the use of forced choice items. The forced choice technique has been described in connection with rating methods. Success on the job is evaluated by some measure of workers proficiency; for example, the responses of high producers could be compared with those of low producers, or workers rated high in proficiency with those rated low, or responses of persons who stay a considerable time on the job with those who do not.

CLASSIFICATION OF TESTS

It is possible to classify the tests that have been used in the selection of workers into three categories : intellectual tests, spatial tests, and motor tests.

(i) Intelligence Test

It is the test used to judge the mental capacity of an applicant. It measures the individual learning ability, ability to catch or understand instructions and also the ability to reason and make judgement. There are various verbal as well as non-verbal intelligence tests designed by many psychologists for different jobs.

This type of test gives more information concerning traits than any other type of test. Many kinds of questions are used in intelligence tests including analogies, reasoning, vocabulary similes, general information, number extension and arithmetic series etc. A few of the questions used as intelligence test are :

    (a) What is the next number in the series ? 2, 4, 7, 11.

    (b) Water flows down hill because

        (i) it is slippery.
        (ii) it is heavier than air.
        (iii) it is subject to gravity.
        (iv) it is pulled by the sun.
        (v) it is attracted to sea.

    (c) Tribulation is the opposite of

        (a) sorrow.
        (b) job.
        (c) underhand.
        (d) open.
        (e) lazy.

Intelligence tests have been used most frequently and hence deserve special consideration. Intelligence tests have been utilized in the selection and classification of employees for almost all kinds of jobs from the unskilled to the administrative and professional. This extensive use reflects the fact that these tests have frequently been applied uncritically with no attempt being made to understand their true nature As a consequence, the intelligence tests have earned an undeserved reputation of contributing little to the solution of selection and placement problems in industry.

(ii) Aptitude Tests

Aptitudes are the potentialities which the individuals have for learning the skill required to do a job quickly. Tests designed to measure such potentialities are called aptitude tests. Aptitude tests are one of the most promising indices for *predicting worker's success.* Their continued use is assured because of the everchanging nature of occupations in an industrial society under the continuous fluctuations occurring in the labour market.

Aptitude tests have been used to predict three different kinds of indices of success, namely,

(a) job proficiency.
(b) job training, and
(c) labour turnover.

Examples of indices of proficiency are production records and ratings given by supervisors. Examples of indices of training are grades in occupational training courses and instructors ratings.

(iii) Trade Tests

Proficiency or trade tests are those tests which are designed to measure the skills already acquired by the individuals. These are also known as performance tests *i.e.* tests of level of knowledge and proficiency in certain skills about a particular job. For instance, in hiring a stenographer in an office, a test can be given to an individual to check up his speed both at dictation and typing. Thus a trade test takes a sample of an individual's behaviour which is designed as a replica of the actual work situation as typing, dictating, etc. Trade test should be differentiated from the aptitude test. *Trade tests measure skills already acquired though training and experience.* Skills are not constant or static but they vary within certain limits through practice. *Aptitude tests measure the potentialities of the applicant in doing the job.* Aptitudes are relatively stable ; individual's potentiality for learning remains essentially the same, but for age, regardless of practice. Aptitude tests are usually given to the applicants who are not experienced or trained in the job under consideration.

(iv) Motor Tests

Dotting. These tests emphasize speed and precision of movement. The individual makes a single dot in each of a series of small squares or circles which are likely to be arranged in irregular order.

Tapping. Tapping tests are similar to dotting tests except that the emphasis is on speed alone.

Finger Dexterity. The individual is required to pick up small pins and insert them in holes. The operation usually takes about 10 minutes or he may make a simple assembly such as putting washers on rivet heads and inserting the assemblies into holes, or screwing nuts on to small bolts.

Complex Reaction. These tests are frequently used with motor-vehicle operators. In complex reactions several stimuli are presented serially to the individual and he must make differential responses to them.

INTERPRETING THE TEST RESULTS

Test scores obtained by various applicants are practically meaningless unless there is some method of integrating them. Unless test scores can be compared with something else (*i.e.* standard) it can be little more than a number. Therefore, standards should be set for average ability to perform the job in the average time. General experience has shown that the test scores are distributed in a certain pattern which is almost the same in every case. For instance, if we are to measure the height of all men, we may find that majority of the men are between $5'-3^{11}$ to $5'-8''$ in height. As we get away from this average either side, there will be fewer and fewer of these heights. Similarly, there is a majority of people who are average at work and there are a few who are very poor at work and a few who are exceptional at work.

INTERVIEW

Although application blanks, tests and group discussions provide much valuable information about the candidate, yet they do not provide the complete set of information required of the applicant. Interview may be of great help in such cases. Interview may be taken to secure much information about a candidate to enable the organisation to know about the applicant and *vice versa.*

The main purposes of an employment interview are : (a) to find out the suitability of the candidate, (b) to seek more information about the candidate and (c) to give the candidate an accurate picture of the job with details of terms and conditions and some idea of organisation policies and employer-employee relations. The factual data on the application blank may also be checked. The personal meeting between the panel of interviewers and the candidate may also be used for testing the capabilities of he candidate. Thus an interview affords an opportunity to develop a clear total picture of the candidate.

Although personal interview is perhaps the most widely used method for selecting the employees, it has certain limitations :

(a) Interview is an expensive device and sometimes, it is interpreted as having greater meaning and validity than is justified.

(b) Interview can test only the personality of the candidate and not his skill and ability for the job.

(c) Interviewer may not be an expert and may not be in a position to extract maximum information from the candidate.

(d) It depends too much on the personal judgement of the interviewer which may not always be accurate. Sometimes, there is a tendency to allow one prominent characteristic of the candidate to dominate appraisal of the entire personality and may colour the interviewer's judgement on other traits. A good many psychologists take the view that interview is such an unreliable form of communication that it should be replaced by alternative methods wherever possible. It has been in big concerns that if after the interview a candidate is found to be satisfactory, he is sent to the department concerned to take the trade test. Scientifically designed tests are more objective than the interviews.

Interview Procedures and Techniques

Interview procedures and techniques vary from organisation to organisation and from individual to individual according to the purpose of the interview. The first point to hold an interview is the selection of candidates to be called for the interview and preparing a schedule or plan for the interview. For this purpose the application forms filled by the candidates and the results of tests conducted by the personnel department may be used. Thus all pertinent information that can be gathered before hand is assembled and an analysis is made to decide whether or not a particular applicant is worth interviewing. And if he is, the interview should be planned with him in the light of information supplied on the forms. This preparation saves time and mental effort of the interviewers and enables them to sketch in advance at least a general picture of the interviewee. The selector needs to match the facts supplied on the application form with the requirements of his job specifications—covering such obvious points as age, sex, marital status, essential qualifications and experience. He also needs to check the dates on the form to see that they provide a connected and coherent life history and he has to relate these dates to the applicant's age at that time to assess his suitability for the present job. In particular, the interviewer should match people to job by concentrating on :

(i) the extent to which the applicant's previous work experience relates to the work in the post to be filled.

(ii) the applicant's social background, education and achievements.

(iii) the applicant's normal working tempo.

(iv) the applicant's initiative and decision making.

(v) the applicant's attitude towards authority and discipline.

While matching the people to the job, it is necessary that the interviewer must have the detailed knowledge of the requirements of the job and the qualities required to fulfil these requirements. For this purpose, he should go through very carefully over the job description and the job specification drawn out by the personnel department. This will help him in reconciling the requirements of the job and the qualities possessed by the individuals.

SETTING FOR THE INTERVIEW

The proper physical arrangements for the interview are of great importance. They enhance the reputation of the organisation in the eyes of the candidates. The interview should be conducted in a room free from disturbance, noise and interruptions, so that interview may be held confidentially and in a quiet environment. The main physical condition for successful interviewing is that the interviewer should look ready for the meeting and room should look ready for a private discussion. Privacy and comfort are generally recognised as aids to free talk. People generally speak more freely and frankly when they are at ease and do not feel threatened.

Keeping to time is another important problem. The interviewers who keep to the appointed time are very few indeed. But a candidate who is called for an interview should not be expected to wait too long. Waiting for an interview irritates many candidates more than any other single factor in the selection process. If unavoidable delays have developed it is better to cut one interview short and risk injustice to one applicant than to fall further and further behind in schedule and risk injustice to many who have become nervous or annoyed. The applicants should be made to sit and wait in the waiting room only. Good organisations also offer cold or hot drinks to keep them fresh. Thus the candidates should not be made to feel that they are beggars who can wait anywhere and indefinitely, just because the interviewer has a job to offer is in a more powerful position than an applicant who is seeking one and has come to present his life history to obtain it. The interviewer should try to create conditions which are as convenient as possible.

CONDUCTING THE INTERVIEW

An important aid to conducting successful interview is that the candidate should be made to have a feeling of leisure. He will not be able to answer properly if he feels that he is being hurried through the question-answer, period. Under ideal conditions, a successful interview can be conducted from 20 to 30 minutes. But even then, an expert interviewer should manage to convey the impression that he has plenty of time. Leisure is a state of mind rather than a matter of minutes. Right at the start, the candidate should be given to understand that he is not taking an examination, but is being called in to give further information about himself. Unless the candidate is put at ease, it is very difficult and doubtful whether a right assessment can be made of his capabilities. So the interviewers must try to create a harmonious atmosphere in which relevant information can be obtained by both the parties. No matter whether the candidate joins or not the interviewer must remember that he represents the organisation and the impression he gives is vitally important. The candidate's reactions to the interviewer will also deeply affect the value of the interviewers.

A few candidates are relaxed and self assured from the start, but it should be kept in mind that nervousness of the candidate will conceal his true character and personality. Some types of people will maintain silence punctuated only by yes or no to answer the easiest questions. Other types in the same circumstances will over compensate by excessive volubility. But such people may be perfectly balanced and confident in dealing with other people in their general life. Every candidate has typical tension curve. It is necessary during the interview to get the downward slope as soon as possible to avoid responder fatigue.

An interviewer cannot conduct an entirely useful interview if he has not prepared himself for that. He must have the knowledge of important points about the candidates, which can be had from the application form and other record supplied by the personnel department. On the basis of his knowledge of various facts he can chalk out the questions to be asked from the candidates. Interviewers may sometimes find it helpful to put down a list of questions to which they want specific answers. He should not do this with pencil poised over pad in the manner of a policeman, but use the list as a guide only. He should try to listen to the candidate with intelligence and silence. The interviewer's attentive silence may be his best contribution to a successful interview.

Many interviewers are afraid to use the obvious device of taking notes because it unnecessarily irritates the interviewee. Notes should never be taken until after a satisfactory relationship has been established. At that time, note-taking may produce a favourable effect on the candidate since he is likely to think that his remarks are considered sufficiently important to be written down. If the candidate is not in a position to answer to a question, the interviewee should help him to answer the question. At the end of the interview, the Chairman of the Board should conclude the interview by thanking the candidate for giving information about himself.

There are generally more than one candidate for any new position in the organisation ; so it is important to make an assessment of the candidates as soon as possible. This may be committed to

INTERVIEW ASSESSMENT FORM

Interviewee's Abilities	Rating				Confidential
Personnal Data : Name, age, sex, etc.	5	4	3	2	1
Health					
Manner and Appearance					
Education					
Experience					
Intelligence					
Administrative ability					
Social ability					
Emotional stability					
Initiative					
Motivation or goals achivement					
Remarks					

5—Excellent   4—Very good   3—Good   2—Average   1— Poor paper, recording his exceptional abilities and past experience, any training need he might have and an assessment of his potential. An interview assessment form may be used for this purpose. Proforma of such form is given on p. 82.

After the interview is over, each interviewer should review his notes and consider what he has accomplished. The completed interview notes form an important part of a case record. They are valuable for all kinds of special reference, e.g., to review before another interview with the same person. All the members of the Interview Committee or the Selection Board should prepare a panel of candidates recommended for selection in order of preference.

Increasing Interview Effectiveness

In order to increase the effectiveness of employment interview proper attention should be given to the following factors :

    (a) Information about the applicant to be interviewed.

    (b) Information about the job.

    (c) Structure of the interview itself.

    (d) Interviewer's skill.

Information about the applicant. Before a candidate is interviewed, all the relevant information about him is assembled and is passed on to the interviewer for his review. At this stage, the applicant's file contains the application blank, reports of reference check, preliminary interview notes and test scores. When the file is delivered to the employment interviewer, he reviews it and plans his interview accordingly. For this review of the applicant file, the interviewer shall determine :

(i) Does the applicant meet minimum standard set up for each step in the selection process ? Does the applicant fulfil the requirements of the job under consideration ? Before beginning the interview, the interviewer must be satisfied that the applicant possesses the minimum qualification required for the job as revealed by job specification and has passed all steps in the selection process. In case of failure at one or more steps the interview may be relatively brief ending in rejection of the candidate.

(ii) What background factors and personal qualities must be explored ? In the case of an applicant who has passed all the hurdles successfully and the interviewer is satisfied that there are no obvious personal characteristics which could disqualify the applicant then most of the interview time may be devoted to the giving of job and organisation information. But if the candidate is viewed as a borderline case, the task of interviewer will become more complex since he will have to search for evidence which can be used as a basis for recommendation for or against employment or to aid in placement on a job where specific limitations will not be a handicap.

Information about the Job

One of the important aspects about interview is the extent to which an interviewer should be familiar with job for which he interviews applicants. In fact, considerable information in the form of job specification and job description is necessary if the interviewer is both to give information and make valid judgements with respect to applicant's abilities. Some people even believe that only individuals with work experience on specific jobs or related jobs are suitable for selecting other men for these jobs. But this may not be feasible, always. The procedure and job specifications and job descriptions would do much to add to the interviewer's understanding of a specific kind of work. Some organisations even assign the employment interviewers to spend time with each foreman, observing the various plant operations and keeping abreast of changes in jobs as they are performed. This increases the knowledge of the interviewers and they are enabled to ask from the applicants the questions relating to the jobs for which they have applied. This also helps them not to repeat the errors resulting into faulty selection in the past.

Interview Structure

Stucture of the interview should be properly planned. Completely unstructured employment interview is little more than a casual or social conversation without direction, purpose, control or terminal point. Neither the applicant nor the interviewer shall gain much in the way of relevant information from such a session. At the other extreme, the completely structured interview is inflexible and offers no opportunity to explore or probe the answers applicant may give. However, in order to avoid either extreme, the organisations should develop and use standardised interview.

INTERVIEW FORMS

I. *FAMILY HISTORY*......

II. *SOCIAL HISTORY*......

III. *PERSONAL HISTORY* : Interviewer may ask :

   (a) What do you consider your strongest qualities and characteristics ?

   (b) Recall from your own experience. What are your great weaknesses ?

   (c) What ambitions do you have for yourself ?

   (d) What made you to be interested in this organisation ?

   (e) What are you doing now to improve yourself to increase your efficiency ?

When the interviewer has secured such information he should ask himself the following questions :

(i) Does the applicant seem vitally eager to succeed ? Yes or No.

(ii) Does the applicant tend to have sound estimation of his worth to the organisation ? Yes or No.

(iii) Can the applicant look at you in the eye ? Yes or No.

(iv) Does he express himself clearly and forcibly ? Yes or No.

(v) Does he have a well balanced personality ? Yes or No.

(vi) Does he give evidence of being aggressive ? Yes or No.

INTERVIEWER'S SKILL

Following are the various points which an experienced interviewer should keep in mind while conducting the interview ;

(a) Job Knowledge

He should be thoroughly familiar with the specifications of the job for which the applicant has applied and should have the capacity to relate the job requirements to those qualifications of the applicant that can be judged in the interview.

(b) Adequate Background Information about the Applicant

The interviewer should know in advance the kind of information to be obtained in the interview so that no important areas are overlooked. For this purpose the interviewer should review the application blank, reference checks, preliminary interviewer's notes and test scores. He should not waste interview time by asking for information already available.

(c) Schedule Interviews to have Adequate Time

Though neither the interviewer nor the interviewee can afford to spend unlimited time, the interviewer should ask the questions from the candidate in an unhurried atmosphere. He should avoid rapid stream of questions. He should give the interviewer a time to formulate his replies or to recall experiences so that full information may be gathered.

(d) Holding Interview in Private

The interviewer should conduct the interview in a separate room meant for this purpose. He should avoid distractions and interruptions because of visitors, telephone calls or any other reason, since they break the conversational flow and make it difficult for the applicant to give information in confidence.

(e) Putting Applicant at Ease

The interviewer should establish a friendly and informal atmosphere at the start and should avoid impersonal approach by his interest in the applicant as an individual.

(f) Listening to the Applicant

It should be recognised that the interview has been scheduled in order to obtain information about the applicant and if the interviewer goes on speaking, he will not be getting much information about the applicant. He should be a very good listener. While he guides and directs the interview as necessary to keep the topic of conversation on pertinent subjects, he should keep his own questions and comments as brief as possible in order to let the applicant express himself as far as time allows.

(g) Adjusting the Level of the Subject for Measuring the Ability of the Respondent

While an interviewer should always avoids any sign of talking down to applicant, he should also be on the alert for signs that his question is not understood.

(h) Keeping Control of the Interview

The interviewer should plan the interview beforehand so as to keep the interviewee from disgressing into areas not related to the purpose of the interview or which do not provide useful insights.

(i) Beware of one's own Prejudices

The interviewer should be particularly conscious of the dangers of allowing specific physical or cultural stereotypes to mark his accurate appraisal of the candidate. He should be careful to discount initial impressions, whether favourable or unfavourable, while evaluating specific traits and should avoid generalising from one trait to other traits.

(j) Avoiding any Suggestions or Discrimination

The interviewer should be careful to guard himself against the introduction of subject matter that might be interpreted as evidence of discriminatory attitude on the part of the employer such as race, religion and political views.

(k) Closing the Interview

It is necessary to learn how and when to close the interview. Interviewer should close the interview with the public relations objective in mind. He should be courteous and friendly whether the applicant is qualified or not. Except in those instances in which he has been authorised to hire new personnel, he should give positive indication of company interest to the qualified applicant and arrange for final interview with the department. However, in case of panel interview, departmental head should be present at the interview. Where the applicant is not qualified, this may be intimated to him before the interview is concluded. Every candidate must be thanked for having shown interest in the organisation.

(l) Recording the Facts and Observations

The inteviewer should record the facts during the interview and impressions and judgements immediately thereafter. He should not trust the memory. Recording facts during the interview is an open and honest manner. He will strengthen the applicant's confidence that all important facts will be considered in the final judge-ment by making this record.

GROUP INTERVIEW

Group discussion is a primary tool, used to provide the evidence of candidate's social framework, personality, behaviour, interests and various social traits. By an application blank, it is not possible to uncover the instances where the candidates held positions and responsibilities involving leadership. It must be known whether the applicant has had any opportunity to participate in situations wherein the role of leader is called for. It is, thus, difficult for the interviewer to draw conclusions concerning the leadership ability of the applicant on the basis of inadequate evidence. In group interview and situations test, the judgement can be made on the basis of actual behaviour of an individual in a group. It also has an advant-age that the personality can be more accurately judged since the individual's traits are manifested in a broad social situation. But there are certain disadvantages of group interviews as discussed below :

(i) The time for observation usually is no longer than that available for the ordinary interview.

(ii) There is interaction amongst the persons and there is a possibility of environmental circumstances placing some applicants in more commanding position than others. But on the whole, group interview appraisal is more realistic and scientific in eliciting be-haviour of a kind actually important on the job. The behaviour appraisal by group interview is a valid predictor of subsequent job success though there have been a lot of variations in the findings of different interviewers. Studies have, however, demonstrated adequate protection of job success by group interviews.

(iii) The group interview is more difficult to administer than a systematic interview or a standardised test.

THE VALIDITY OF A GROUP INTERVIEW

The validity of a group interview seems to be superior to the unsystematic interview but not superior to a good systematically developed interview. In a group interview the problem is given to a group consisting of 10—15 members, who primarily involve them-selves in discussion. The observer sitting at one corner does not offer his arguments and views. In a group interview some individuals tend to initiate and lead the discussion while others participate very little. Group discussions are really good where an individual has to work in co-operation with other members of a group after being selected. The observer makes observation about leadership qualities, about the behaviour, judgement and personality of the individual. The behaviour can be recorded in an objective way by recording the number of times an aggressive or co-operative behaviour is displayed by different individuals.

To make any group discussion successful, the observer should start recording his observations soon after the start of discussion on a separate sheet. After the discussion, the various ratings should be consolidated which are normally submitted to the Personnel Manager, who will call the candidates for final interview on the basis of the performance of individuals on various tests, preliminary interview, application blanks and reference checks, etc.

REFERENCES

A referee is potentially an important source of information about a candidate's personality if he holds a responsible position in some organisation or has been the boss or employer of the candidate. Prior to final selection, the prospective employer normally makes an investigation on the references supplied by the applicant and undertakes more or less a thorough search into the candidate's past employment, education, personal reputation, financial condition, police record, etc. However, it is often difficult to persuade a referee to give his opinion frankly. The organisation may persuade him to do so by giving an assurance that all information will be treated as strictly confidential. The referee may be encouraged to give precise and solvent information if he is sent a description of the job for which the applicant has applied and then asked specific questions about the applicant's abilities to perform the duties listed. He will find it easier to supply factual information than mere opinion.

Reference method has its drawbacks also. The referee may not give accurate information about the candidate. He may give his good impression about him because of his relations with him or if he is his employer, he may give a good report to get rid of him. So this method cannot be relied upon completely.

PHYSICAL EXAMINATION

The pre-employment physical examination or medical test of a candidate is an important step in the selection procedure. Though in the suggested selection procedure, medical test is located near the end, this sequence need not be rigid ; an organisation may place the examination relatively early in the process so as to avoid time and expenditure to be incurred on the selection of medically unfit person. Some organisations either place the examination relatively early in the selection procedure or may advise the candidates to get themselves examined by a medical expert so as to avoid disappointment at the end. Following are the reasons for locating the medical examination at the end : Firstly, this is time-consuming and expensive and secondly, the number of applicants who reach this stage are considerably less than the number who fill out application blanks. Finally there may be an interval of time between initial screening by the personnel department and the actual entry into the organisation. Giving the medical test near the end ensures that employee is physically fit for the job at the time of actual entry.

Physical examination should serve at least three objectives :

(i) It serves to ascertain the applicant's physical capabilities to meet the job requirements.

(ii) It serves to protect the organisation against the unwarranted claims under worker's compensation laws, and

(iii) It helps to prevent communicable diseases entering the organisation.

51

Motivation

Motivation means inspiring the ·personnel with a zeal to do work for the accomplishment of objectives of the organisation. Motivation is an important function which a manager has to perform for getting things done from the people. A successful manager knows that the issuance of directions, however, well conceived and worded, does not mean that they will be followed. He makes appropriate use of motivation to enthuse the personnel to work harmoniously for the achievement of established goals. Effective motivation succeeds not only in having an order accepted but also in gaining a determination to see that it is fulfilled efficiently and effectively. So the manager must have keen appreciation of human behaviour if he has to provide the maximum motivation among his subordinates or associates. Some people are spurred only by intense outer pressures of rewards, others are highly self-motivated. Manager has to constantly provide for incentives or motivating forces in order to intensify their desire and willingness to apply their greater potentialities for the achievement of common objectives.

Thus motivation is an action that stimulates an individual to take a course of action which will result in attainment [of some goal or satisfaction of certain material or psycological needs of the individual himself. We know that a man does not simply live as an economic man but likes to live as a social-economic-man. Major part of the time of an individual is spent more at work then at home.

History of Motivation

The first systematic studies of human motivation occurred in 1800, during the industrial revolution when competition forced employers, to develop more efficient methods of producing high quality of products, quicker and at lower cost. At that time, money was used as the only motivating incentive. Frederick Taylor has noted in his "principles of scientific management" that all employees are fundamentally the same. They are uniformally motivated by a desire for money and other motives are either in-consequential or are non-existent.

It is an important function of the management to motivate the people working in the organisation to perform the work assigned effectively and efficiently. The management must understand the human behaviour if it has to provide maximum motivation to the personnel. Motivation is a hypothetical cause of human behaviour. There are many theories of motivation which try to bring out the needs which affect the human behaviour. But now it has come to be widely accepted that goals of individuals influence their efforts and that the behaviour individuals select depends upon their assessment of the probability that behaviour will successfully lead to the goals. So the management must try to understand the goals of the individuals at different times in order to motivate them.

Though financial incentives are very important motivators, but they are not the only motivators. People are also motivated by non-financial incentives such as opportunity of advancement, authority to take decisions and freedom to do one's work. Therefore, the management must use both types of incentives to motivate the employees. Management should always remember that motivation is an important tool in its hands for achieving the full cooperation of the people in the organisation in the direction of organisational goals and should use this tool very carefully. It should constantly make use of the non-financial incentives, in addition to the financial incentives, to enthuse the personnel to work willingly and harmoniously for the attainment of organisational objectives.

Definition of Motivation

Motivation though defined in many ways still remains undefined because of its vary nature. It is more a realisation than discussion. As in the words of Young poet :

This does not come with houses or with gold

With palace with honour, or with theory X and Y

Its not in the words which market bought and sold

But the smooth hierarchy of needs

Droply and leave its secker untired.

Hence realisation of needs i.e. need of organisation, need of people need of right type of atmosphere where people may work with their best efforts is paramount. The real realisation lies in administrative ecology in which the right type of men can live and breath and have their needs fulfilled. The right motivation springs from right sense of values. The right motivation does not come by attending various institutes of management or sitting at high salaried position.

It may be defined as any idea, need, emotion or organic state, that prompts a man to an action. A motive is inferred from the behaviour of a man. It cannot be observed directly. It may simply be assumed to exist as an internal factor that integrates a man's behaviour. As the motive is within the individual, it is necessary to study his needs, emotions, etc., in order to motivate him to do work. Motive influences the behaviour of a man. But there are also other factors which influence man's behaviour such as past experience, physical capabilities and environmental situation. Motive, however, should be distinguished from these factors. As a matter of fact, there is no best way of defining motive. According to some psychologists, motive also includes a conscious desire for something that is related to the goal selection function of motives. This conscious desire may be called a 'want'. But there are other psychologists also who do not share this view. They think that a desire or want is too subjective to be of value in a scientific sense. They believe that it would be desirable to take a man's verbal report of his inner feelings as one aspect of human behaviour that is influenced by the inferred motive.

There is no universal theory that can explain the factors influencing motives which control man's behaviour at any particular point of time. Generally different motives operate at different times among different people and influence their behaviour. That is why the concept of motivation has been variously identified as an unquestionable fact of human experience, as an indisputable fact of behaviour and a mere explanatory fiction. The common and unifying element in these diverse conceptions is that motivation is an agency or factor or force that helps to explain behaviour. Motivation is a hypothetical cause of behaviour.*

Motivation may be defined as the complex of forces inspiring a person at work in an organisation to intensify his desire and willingness to use his potentialities for achievement of organisational objectives. It is something that moves a person into action and continues him in the course of action enthusiastically. The role of motivation is to develop and intensify the desire in every member of the organisation to work effectively and efficiently in his position. As said above, there are certain forces inside the person inspiring him to continue work, which may be called as drives, wishes, instincts or tension states. Thus, by motivation we mean mechanisms inside the person that sustain his continued activity as a human being.

Motivation has also been defined as the process or the reaction which takes place in the memory of the individuals. It may be viewed as a combination of forces (motives) maintaining human activity.

*Bolles, R.C. : Theory of Motivation.

Incentive and Motivation

A distinction may be made among the three things : need, incentive and motive. This is to emphasize that any need present in the individual does not necessarily lead to action. The need has to be activated which is the function of incentive. Incentive is something which incites or tends to incite to determination or action. Thus incentive is an external stimulus that activates need and motivation refers to an activated need, an active desire c: wish. But a better definition is to regard incentive as the outward stimulus for the motive to work. When a motive is persent in a person it becomes active when there is some incentive. Thus any incentive has reference to :

(i) the individual and his needs which he is trying to satisfy or fulfil ; and

(ii) the organisation which is providing the individual with opportunity to satisfy his needs in return for his services.

Thus conceptual difference between motivation and incentive is that incentive is the means to motivation.

It has been demonstrated conclusively that incentives have a direct bearing on the degree of motivation. Increase in incentive leads to better performance and decrease in incentive leads to poorer performance. It should be noted that motivation does not change the individual's capacity to work. It simply determines the level of effort of individual, raises it or lowers it, as the case may be. *Keith Davis* says that motives are expression of a person's needs, hence they are personal and internal. Incentives, on the other hand, are external to the person. They are something he perceives in his environment as helpful towards accomplishing his goal. For instance, management offers salesmen a bonus as an incentive to channel in a productive way their drives for recognition and status.

As shown in Fig. 18.3 needs creates tensions which are modified by one's culture to cause certain wants. These wants are interpreted in terms of positive and negative incentives to produce a certain response or action. To illustrate, need for food produces a tension of hunger. Since culture affects hunger, a man will require wheat or rice accordingly. For the native perhaps incentive is provided by his wife's promise to prepare it in his favourite way.

THEORIES OF MOTIVATION

Motivation is important because in most cases people do not contribute towards the realisation of organisational goals as much as they can. Managers, therefore, try to find out the reasons that

Theories of Motivation

Managers try to findout reasons that impede people from increasing their productivity. They accordingly prepare the plans for motivating their employees. There are several approaches to the study of motivation which are discussed below.

Be Strong Approach

Traditionally, management has resorted to be strong. This form of motivation in industry emphasizes authority and economic rewards. This rewards strategy consists of forcing people to work by threatening to punish or dismiss them or to cut their rewards, if they do not work. The assumption is that people would work if they are driven by the fear of punishment. To prevent people from going away from the work, there must be close supervision, management must spell out every rule and give the workers the narrowest range for discretion. This approach paid off fairly well in the early days of industrial revolution. It is because needs for food, clothing and shelter were paramount. This approach is similar to the 'Carrot and Stick' approach.

In recent times, this approach has because less effctive because people have began to expect more from their jobs rather than mere money. Moreover, with the growth of unions, a rising standard of living and changing pattern of living has made this approach less effective as a motivational device. The basic aspiration of the 'be strong' approach that if a man does not do what he is told to do, he will be fired, has gone wrong because unions have made it difficult to fire a man.

Be Good or Paternalistic Approach

This approach is a substitute for 'be strong' applied by many managements. The essence of this approach is conferring of various rewards on the organisation members in hope of increasing the productivity due to gratitude or loyalty to the organisation. High wages job security, subsidised education, recreation programmes, fair supervision and good working conditions are the instruments used for gaining the loyalty of the employees and thus promoting the efficiency. Increasing, efforts are sought as a reciprocal basis ; management tries make to available to the subordinates the things they want and as a result expects the subordinates to display enthusiasm and loyalty.

Be good approach or paternalism may fail to achieve its purpose. Paternalism may create resentment rather than gratitude because some people do not like to feel dependent on others. They prefer to decide for themselves what they want. Paternalism incorporates some of the basic assumptions of 'Be Strong'. People are expected to be docile in return for their gifts, and work is still regarded as a form of punishment that people undergo only in return for a reward. Moreover, many of the rewards must be enjoyed off the job. Little effort is spent on making the job itself more rewarding. Like be strong approach, paternalism also carries with it a nagative threat if a worker does not do his job as ordered, his gifts will be taken away.

Effort Reward Approach

The third strategy tries to establish and relationship between efforts and rewards. The origin of this approach can be found in the scientific management of *F. W. Taylor*. Rewards in this approach are considered to be a function of efforts put into reach the standard set by the management. Individual wage incentives and promoting individuals on the basis of accomplishment are the manifestation of this approach. In order to practise this strategy, managers set up standards of performance, monitor the behaviour of employees to observe the extent to which these standards are attained or adhered to, and allocate rewards and penalties based on the observance of the performance.

This approach may also be called monistic approach because it assumes that people work for money. This theory seeks a single cause of behaviour. This theory accepts the notion of economic man who acts only to increase his monetary rewards. This theory states that more effort carry more pay. The rewarding of correct behaviour should lead to the reinforcement of that behaviour. This theory can be described in a simple model as 'EFFORT-REWARD-EFFORT MODEL'. The model illustrates that greater efforts not only leads to greater pay but also the reward of greater pay stimulates greater effort. This strategy is not considered adequate because it touches only one aspect of human behaviour. This approach assumes that people work in an organisation with the only incentive of earning money ; it forgets the necessity of providing on-the-job satisfaction to the employees. Moreover, this approach can be applied only where performance and results can be directly and quantitatively measured.

McGregor's Theory 'X' and Theory 'Y'

All the above mentioned approaches to motivation are based on the assumption of human nature included by Douglas McGregor in his Theory 'X'. These assumption are :

1.   The average human being has an inherent dislike of work and will avoid it if he can.

2.   Because of this human characteristic of dislike of work, most people must be directed and threatened with punishment to get them to put forth adequate efforts towards the achievement of organisational objectives.

3.   The average human being prefers to be directed, wishes to avoid responsibility, has relatively little ambition and wants security above all.

According to McGregor, this is a traditional theory of what workers are like and what organisation must do to manage them. This gives us the conventional approach of management. According to theory 'X' the employees are passive and even resistant to organisational needs. They have to be persuaded and pushed into performance and this is the management's task. Management does the thinking, the employees obey the orders. But McGregor himself says that this approach does not represent the modern views on motivation. "The conventional approach of theory 'X' is based on mistaken notions of what is cause and what is effect". The assumptions contained in theory X do not reveal the true nature of human beings.

McGregor himself states that the assumptions of theory 'X' do not portray the correct nature of human beings. He thinks that assumptions included in theory 'Y' are more valid percepts of human nature. These assumptions are :

1. The expenditure of physical and mental effort in work is as natural as play or rest. The average human being does not inherently dislike work. Depending upon controllable conditions, work may be a source of satisfaction or a source of punishment.

2. External control and threat of punishment are not the only means for bringing about efforts towards organisation objectives. Man will exercise self-direction and self-control in the service of objectives to which he is committed.

3. Commitment to objectives is a function of the rewards associated with their achievement. The most significant of such rewards e.g., the satisfaction of ego and self-actualisation needs, can be direct products of efforts directed towards organisation objectives.

4. The average human being learns under proper conditions, not only to accept but to seek responsibility. Avoidance of responsibility, lack of ambition and emphasis on security are generally the consequences of experience, not inherent in human characteristics.

5. The capacity to exercise a relatively high degree of imagination, ingenuity and creativity in the solution of organisational problems is widely, not narrowly, distributed in the population.

6. Under the conditions of modern industrial life, the intellectual potentialities of the average human being are only partially utilised.

A study of the assumptions of Theory 'Y' makes it clear that individual goals and organisational goals are not necessarily incompatible. These basic problem is that of securing the commitment of employees to organisational objectives. Employees' commitment is directly related to the satisfaction of employees' needs. Theory 'Y', therefore, places emphasis on the satisfaction of the needs of the

1. Maslow's Priority Model

Maslow defines human effectiveness as a function of matching man's opportunity with the appropriate position on hierarchy of needs. Process of motivation begins with an assumption that behaviour, atleast in part, is directed towards the achievement of satisfaction of needs. Maslow proposed that human needs can be arranged in a particular order from the lower to the higher order. The need hierarchy is as follows :

1. **Basic Physiological Needs.** The needs that are taken as the starting point for motivation theory are the so called physiological needs. These needs relate to the survival and maintenance of human life. These needs include such things as food, clothing, shelter, air, water and other necessaries of life.

2. **Safety and Security Needs.** After satisfying the physiological needs, people want the assurance of retaining a given economic level. They want job security, personal bodily security, security of source of income, provision for old age, insurance against risk, etc.

3. **Social Needs.** Man is a social being. He is, therefore, intersted in conversation, sociability, exchange of feeling and grievances, companionship, recognition, beloningness, etc.

4. **Esteem and Status Needs.** These needs embrace such things as self-confidence, independence, achievement, competerce, knowledge, initiative and success. These needs are concerned with the prestige and respect of the individuals.

5. **Self Fulfilment Needs.** The final step under the need priority model is the need for self-fulfilment or a need to fulfilment what a person considers to be his mission in life. It involves realising one's potentialities for continued self-development and for being creative in the boadest sense of the word. After his other needs are fulfiled, a man has the desire for personal achievement. He wants to do something which is challenging and since this challci ge gives him enough dash and initiative to work, it is benefical to him in particular and the society in general. The sense of achievement gives him satisfaction.

The above mentioned needs have a definite sequence of domination. Second need does not dominate until first need is reasonably satisfied and third need does not dominate until first two needs have been reasonably satisfied and so on. The other side of the needs hierarchy is that man is a wanting animal, he continues to want something or the other. He is never fully satisfied. If one needs is satisfied, the other needs arise. As said above, (according to Maslow), needs arise in a certain order of preference and not randomly. Thus if one's lower level needs (physiological and security needs) are unsatisfied, he can be motiyated only by satisfying his lower level needs

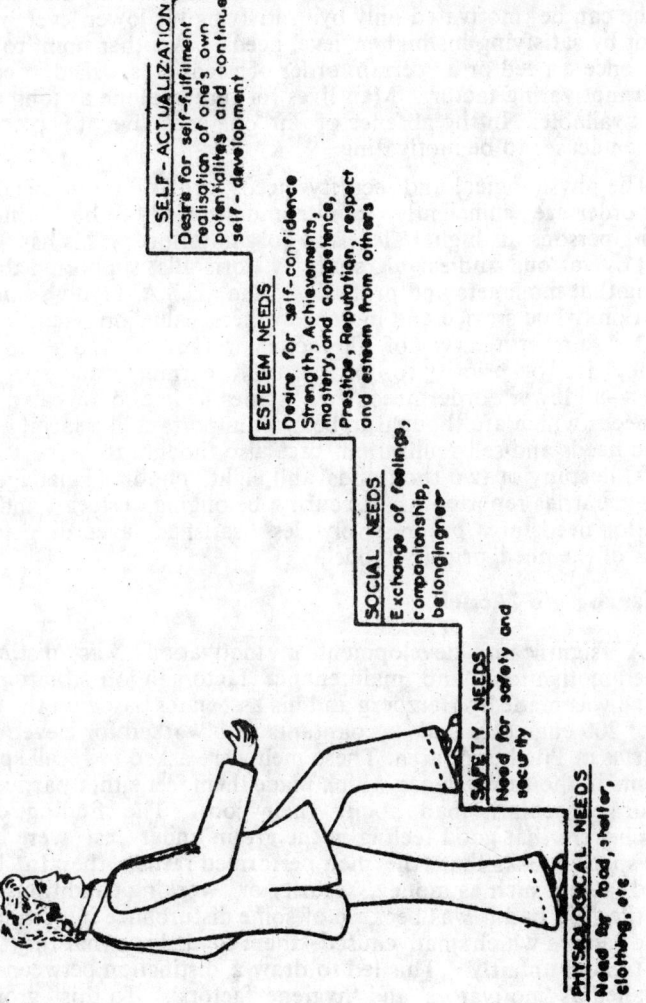

SELF - ACTUALIZATION
Desire for self-fulfilment
realisation of one's own
potentialites and continued
self - development

ESTEEM NEEDS
Desire for self-confidence
Strength, Achievements,
mastery, and competence,
Prestige, Reputation, respect
and esteem from others

SOCIAL NEEDS
Exchange of feelings,
companionship,
belongingness

SAFETY NEEDS
Need for safety and
security

PHYSIOLOGICAL NEEDS
Need for food, shelter,
clothing, etc

Fig. 51.1

arise in a certain order of preference and not randomly. Thus if one's lower level needs (physiological and security needs) are unsatisfied, he can be motivated only by satisfying his lower level needs and not by satisfying his higher level needs. Another point to note is that once a need or a certain order of needs is satisfied, it ceases to be a motivating factor. Man lives for bread alone as long as it is not available. In the absence of air one can't live, it is plenty of air which ceases to be motivating.

The physiological and security needs are finite but needs of higher order are sufficiently infinite and are likely to be dominant ones in persons at higher levels in organisation. This has been proved by various studies. A study by Boris Blai supported this by showing that managers and professionals in U.S.A. highly value self realization, while srevice and mutual workers value job security most highly*. Further a survey of 200 factory workers in India reported that they give top priority to job security, earnings and personal benefits—all lower order needs**. Studies have also revealed that those needs which are thought to be most important like social needs. egoistic needs and self-realisation, are also thought to be best satisfied. One study of two thousands and eight hundred managers in eleven countries reported that security, belonging, esteem and self-realisation needs are progressively less satisfied according to the pattern of the need priority model.

2. Herzberg's Model

A significant development in motivation was distinction between motivational and maintenance factors in job situation. A research was made by Herzberg and his associates based on the interview of 200 engineers and accountants who worked for eleven different firms in Pittsburgh area. These men were asked to recall specific incidents in their experience which made them feel either particularly good or particularly bad about their jobs. The findings of the research were that good feeling in the group under test were keyed to the specific tasks that the men performed rather than to background factors such as money, security or working conditions and when they felt bad it was because of some disturbance in these background factors which had caused them to believe that they were being treated unfairly. This led to draw a distinction between what they called as 'motivators' and 'hygiene factors'. To this group of engineers and accountants the real motivators were opportunities to become more expert and to handle more demanding assignments. Hygienic factors served to prevent loss of money and efficiency. The hygienic factors provide no motivation to the employees, but the absence of these factors serve as dissatisfiers.

*Boris Blai : A job satisfaction predictor. Personnel Journal, U.S.A. Oct., 1963
**Paras Nath Singh and Robert J. Wherry—Ranking of Job Factors by Factory Workers in India, Personnel Psychology, Spring, 1963.
†Haire, Ghisheli, and Porter : Cultural Pattern in the Role of the Manager; Industrial, Relations, Feb. 1963.

Some job conditions operate primarily to dissatisfy employees when the conditions are absent, but their presence does not motivate employees in a strong way. Many of these factors traditionally are perceived by management as motivations, but the factors are really more potent as dissatisfiers. These potent dissatisfiers are called maintenance factors in job because they are necessary to maintain a reasonable level of satisfaction among the employees. They are also known as dissatisfiers or 'hygienic factors' because they support employees' mental health. Another set of job conditions operates primarily to build strong motivation and high job satisfaction, but their absence rarely proves strongly dissatisfiers. These conditions are 'Motivational Factors'. Herzberg's maintenance and motivational factors have been shown in the following Table :

PHYSIOLOGICAL OR PRIMARY NEEDS

Need Structure —

Social, Psychological, Egoistic, or Secondary Needs

Maintenance or Hygienic Factors	Motivational Factors
1. Company Policy and Administration.	1. Achievement.
2. Technical Supervision.	2. Recognition.
3. Inter-personal relations with Supervision.	3. Advancement
4. Inter-personal relations with Peers.	4. Work itself.
5. Inter-personal relations with Subordinates.	5. Possibility of growth.
6. Salary.	6. Responsibility.
7. Job Security.	
8. Personal life.	
9. Working Conditions.	
10. Status.	

Hygienic factors include such things like wages, fringe benefits, physical conditions and overall company policy and administration. The presence of the factors at a satisfactory level prevents job dissatisfaction, but they do not provide motivation to the employees. So they are not considered as motivational factors. Motivational factors, on the other hand, are essential for increasing the productivity of the employees. They are also known as satisfiers and include such factors as recognition, feeling of accomplishment and achievement, opportunity of advancement and potential for personal growth, responsibility and sense of job and individual importance, new experience and challenging work, etc.

Herzberg further stated that managers have hitherto been very much concerned with hygienic factors. As a result, they have not been able to obtain the desired behaviour from employees. In order to increase the motivation of employees, it is necessary to pay attention to the satisfier motivational factors.

Herzberg also said that today's motivators are tomorrow's hygiene because they stop influencing the behaviour of persons when they get them. When a person gets one thing then something else will motivate him and the need which has been fulfilled will have only negative significance in determining their behaviour. It should also be noted that one's hygiene may be the motivator of another. For instance, it is likely that workers in underdeveloped societies will designate some of the maintenance factors as motivators because their primary needs have not been fulfilled and they continue to be motivated by these factors.

3. Vroom's Model

One of the most recent models of motivation has been developed by Vroom. Vroom's model views motivation as a process governing choics. Thus, if an individual has a particular goal, some behaviour must be performed in order to achieve the goal. The individual weighs the likelihood that behaviours will achieve the desired goal and if certain behaviour is expected to be more successful than others, that type of behaviour is likely to be chosen.

The important contribution of Vroom's model is that is explains how the goals of individuals influence their efforts and that the behaviour individuals select depends upon their assessment of the probability that the behaviour will successfully lead to the goal. For instance, all people in an organisation may not place the same value on such job factors as promotion, high pay, job security and working conditions. In other words, they may rank them differently. Vroom is of the opinion that what is important is the perception and value the individual places upon certain goals. Let us assume that one individual places high value on salary increases and perceives superior performance as instrumental in reaching that goal. According to Vroom, this individual will strive towards superior performance in order to achieve the salary increases. On the other hand, another individual may value a promotion and perceive political behaviour as instrumental in achieving it. This individual is not likely to emphasize superior performance to achieve the goal.

In essence Vroom emphasizes the importance of individual perceptions and assessments of organisational behaviour. What is important here is that what the individual perceives as the consequence of a particular behaviour is far more important than what the manager believes the individual should perceive. Thus Vroom's model attempts to explain how individual goals influence individual efforts and like Maslow's and Herzberg's models, it reveals that behaviour is goal oriented.

The ability greater than Zero value performance is constantly increasing function of amount of motivation. In other words, the more motivated an individual to perform effectively, the more effective is performance. This is shown by a straight line in Fig. 51.2

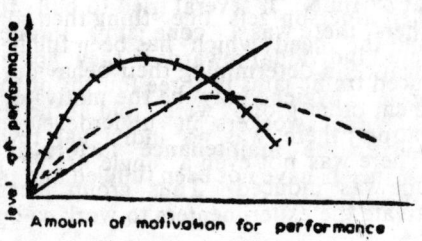

————→
————Constantly increasing function
————Negatively accelerated function approaching upper unit
Fig. 51.2: Hypothetical relationship between amount of motivation and level of performance

The accelerated curve shown in dotted lines in Fig. 18.4 implies the Law of diminishing returns that is succeeded movement in motivation results in smaller and smaller increments in performance. Function shown by broken lines is similar to the first except for reduction in performance under high level of motivation. The amount of motivation is difficult to be measured to precision. It could at best be done on manipulative basis.

Non-Financial Incentives in Motivation

Financial incentives do not work for ever to motivate the people at work. As a matter of fact, when the physiological and security needs are fulfiled with the help of money, it ceases to be the motivating force ; it becomes the maintenance factor as said by Herzberg. Therefore, the employees do not always run after money. They have other needs also. They want status and recognition in the society, they want to satisfy their egoistic needs and they want to achieve something in their lives. In order to motivate the employees having these needs, management can take the help of following non-financial incentives.

1. **Competition**. Competition is a kind of non-financial incentive. If there is a healthy competition among the individual employees or groups of employees, it will lead them to achieve their personal or group goals.

2. **Group Incentives**. Group incentives are more powerful to motivate the employees than the individual incentives. This has been proved in an experiment by Dr. Alexander Mintz, which is also known as "Bottleneck Experiment". It was observed that when the experiments worked selfishly, they caused traffic jams in the experiment. Dr. Mintz in his laboratory created a situation where four or five persons, each held a string with a cone tied to the other end. The cones were dropped into large glass bottle. The participants, when stood around the bottle in a circle tried to fish their cones out before the rising water touched the cones. Quick action was needed as also co-operation because only one cone could be pulled through the bottleneck at a time. If several tried to pull their cones out regardless of others there was a 'cone jam'. In one set of trials the people were offered individual rewards if they got their cones out dry. They produced traffic jams in three fourths of their trails. But in another experiment other groups were told at the start that it was a test of co-operation ; there was no individual reward but only a group reward. There was not even a single traffic jam when a co-operative attitude was induced. Thus group incentive was more effective to motivate the experimenters to work co-operatively.

3. **Praise**. Praise satisfies one's ego needs. Sometimes, praise is more effective than any other incentive. We have seen that in industry, at home or elsewhere, people respond better to praise. However, this incentive should be used with greater degree of care because praising an incompetent employee would create resentment among competent employees. Of course, occasionally a pat on the back of an incompetent employee can act as an incentive to him for improvement.

4. **Knowledge of the Results**. Knowledge of the results leads to employee satisfaction. A worker likes to know the result of his performance. A worker gets satisfaction when his superior appreciates the work he has done. In modern industry the production workers have no contact with the customers so they cannot get the reaction of their efforts from the consumers. However, they can be motivated to a greater extent if they are told the rating of their performance. They will work more to improve their performance.

5. **Workers' Participation**. Employee's participation in management provides an important incentive to the employees. It gives the employees psychological satisfaction that their voice is being heard because workers' participation in management provides for two way communication. This has been discussed in detail in the chapter on Workers' Participation in Management.

6. Suggestion system. Suggestion system is an incentive which satisfies many needs of the employees. Many organisations which use the suggestion system make use of cash awards for useful suggestions. They sometimes publish the worker's name with his photograph in the company's magazines. This motivates the employees to be in search for something which may be of greater use to the organisation.

7. Opportunity for growth. Opportunity for growth is another kind of incentive. If the employees are provided the opportunity for their advancement and growth and to develop their personality, they feel very much satisfied and become more committed to organisation goals because they feel very much motivated.

8. Security. This is a very powerful incentive to Indian workers. Security is ranked second to money only.

What motivates an employee ?

Motivation deals with complex human behaviour in order to determine the factor that prompts an individual to do his best. Various factors which prompt an individual to work have been discussed under Maslow and Herzberg Models. In order to motivate the employees, management must, first of all, know what are the motivational factors influencing the behaviour of persons. Though it is simple, but it is wrong to say that motivation is merely a relationship between the effort and reward. According to this view, a higher pay will motivate the workers strongly. But this is true only in case of those employees who are not getting wages or salaries which are sufficient to satisfy their physiological needs. Money is not the only motivating factor in case of all employees in all circumstances. Many people are motivated by the desire of self-realisation, unique achievements, etc.

In order to understand the behaviour of men, the organisational psychologists have classified employees into three distinct types, namely.

(i) Economic man,

(ii) Social Man, and

(iii) Self-actualising man.

The assumption of economic man is derived from the philosophy of Hedonism and the economic doctrines of Adam Smith. According to this, every man calculates the actions that would maximise his own self-interest. Thus man is primarily motivated by economic incentives and rewards, etc. Thus the needs of higher order are generally found in employees at higher levels in the organisation. In a study conducted by Herzberg, under which 200 engineers and accountants were asked what was going on when they were feeling both good and bad. The answers were :

Good	Bad
Achievement	Company policy. Technical supervision.
Recognition	Interpersonal relationship with supervisors.
Advancement	Interpersonal relationship with peers.
Work itself	Interpersonal relationship with subordinates.
Possibility of growth	Salary,
Responsibility	Job security
	Personal life
	Working conditions
	Status.

Herzberg classified factors under good feeling as motivational factors and factors under bad feeling as hygienic factors. According to him man is motivated by one or more of the six motivating forces and the other ten factors were described by him as maintenance factors. But we can say that a maintenance or hygienic factor of Herzberg may be motivating for some people. Suppose a person is not getting enough money to afford his family or is temporarily employed, he will certainly be motivated by the desire to earn more money or by job security, as the case may be. He will do everything services and to get the greatest economic gains and the organisation buys the obedience of employees strictly for economic rewards. But in the persent age, the assumptions of economic man do not hold good. Man works not only for economic gains to satisfy his physiological needs but also to satisfy social needs, to get recognition and to satisfy his egoistic needs.

Factors of Motivation

One must take into consideration certain basic factors, before employing the technique of motivation. These can be classified as :

(1) The Motivator

(2) The person being motivated and

Motivator. The full force of motivation lies in person who is a motivator, his magnetic personality induces as undying loyalty and high production.

Person being Motivated. The factors pertains to the person being motivated. The following factors must be considered in working out the motivation plant :

(1) Personal background

(2) Education

(3) Age

(4) Marital status

(5) Financial status

(6) Health

(7) Political affiliations

(8) Religious beliefs

(9) Social relationship

(10) Psychological make up

(11) Union affiliations

(12) Company experiences.

The above motivational approach reveals the fact that there must be a capacity or potential for motivation within a person. Finally motivation must be concerned with particular circumstances and environment.

Steps in motivational plan

Having discussed the various aspects of motivation, it will be appropriate to see how a motivational plan can be drawn. This can be studied in two parts :

(*i*)   what has to be done and

(*ii*) how and why it is to be done ?

Following are the various steps required in introducing a motivational plan :

(*i*) Sizing up the situation

(*ii*) Preparing a set of motivating tools

(*iii*) Selecting and supplying motivational plan

(*iv*) Follow up.

Guidelines toward Motivation

Following are the certain guidelines in the performance of steps of motivation which are by no means, final :

(*i*) **Self-interests**. Undoubtedly motivation is largely dependent on self-interest. Self-interest is not an undesirable foundation so long it does not tend towards greediness. It is rather an intelligent act when a person realise that his own purpose are best served when he helps other in attainment of their purposes, conforming to the overall policy of organisation.

(*ii*) **Attainability**. Motivational plan should set attainable goals. Attainability must be related to time as well as type of reward.

(*iii*) **Proportioning Rewards**. Good motivation is dependent on proper proportioning of rewards among persons and for the same person at different time.

(*iv*) **Human Element**. People are carried away as much by intengible concepts and appeals as by concrete goals. Some people think that emotion are the directors of human activity.   In human relations, feelings, emotions, ages, personalities and ideas have to be taken care or best they are offended and whole effort go wrong.

(v) **Individual Group Relationship.** Motivation must be based on group as well as individual centered stimuli. Although to a large extent each are of us is an island upto himself, although the group pressure has a very dominating role on individual's action. No wage incentive policy or promotional policy can be successful if it is not acceptable to the group.

Motivational plan should also take into consideration the theory set forth by Victor H. Vroom (the nature of relationship between motivation and performance).

52

Effective Communication

Introduction

With the advancement of technology, changes have become a regular feature in any industrial organisation. An effective communication system is an essential part of good labour management relations. The prime objective of setting up a communication system is to exchange facts and information in a manner which is acceptable to all concerned and which will lead to willing and co-operative action. Problems of passing information from management to workers are very complicated and many techniques are applied encourage an easy two-way flow of facts, ideas and opinions. But attitude of the persons involved in communication is equally important for better communication which will ultimately lead to better productivity and an atmosphere of mutual trust and confidence among the workers and the managers. If the communication system is carefully planned and applied, it will reduce workers' resistance to the acceptance of new ideas and changes.

Communication problems are more complex in larger organisations. Managers face difficulty in maintaining effective communication to pass messages accurately without distortion to their subordinates. Effective communication is a broader process than merely passing orders and keeping people informed about the activities going on in various divisions of the organisation. Organisational communications should satisfy the needs of organisation and its members.

Communication should not become a burden, but a source of satisfaction. In any organisation, communication must be developed as a system. The communication system itself requires planning, organising, co-ordinating and conrrol. The control is to be exercised on the type of the information that is to be given and such information should communicated in a manner that it is understood by the persons concerned. The plans, procedures, programmes and schedules including indoctrination and training programme should not only be planned and organised, but if the idea is to pool information, then it is absolutely essential that such system of communication should be co-ordinated at appropriate levels. In short, the test of successful communication is the manner of its reception and the action thereon and the awareness of the psychology and the emotions of the parties involved.

Communication may be described as a managers's number one problem. Since a manager works through others, all of his management acts pass through the bottleneck of communication. If he is in a position to communicate efficiently, he will perform his management functions well. Thus communication is a skill of management and an important aspect of the managerial process.

There are several reasons for the importance of communication in the managerial process. One is that planning, which is one of the most important function of management requires extensive communication among the executives and other personnel. Moreover, effective communication is important in executing a planned programme and then controlling the activities with the help of feedback information. Information about subordinate's performance is necessary to determine whether the planned objectives are being realised. Communication also helps in the process of organizing. It is an important aid in directing and motivating the employees in the organisation. In short, communication is quite indispensable for the management in getting the thing done by other personnel in the organisation.

	Communication	Planning Organising Controlling	Communication	
Manager	⟶		⟶	Output

The success of a manager depends on how clear he is in his mind about his basic functions and how effectively he can transfer this clarity of thoughts to others. This involves skill of helping others to understand the manager and to be understood by him. Thus, the need for better mutual understanding between labour and management in industry cannot be over-emphasized as a prerequisite of the suitable congenial climate necessary for overall advancement and productivity. The importance of communication in management for getting the work done may also be seen from the estimate of time which is spent by a manager in communication, verbal or written, in conferences or mettings, giving directions or receiving informations. "Most administrators in most [business and industrial organisations spend atleast 75% of their time communicating and not in frequently as much as 95% of their time communicating to others and being communicated to".*

Many of the most perplexing problems which a manager has to face every day are people-centered. These have their roots in a lack of understanding causing negative or even hostile attitudes

---

*Lee O. Thayer : Administrative Communication p. 3.

among the subordinates. Such situations can however, be avoided through effective communication. The modern concept of leadership exercised largely by persuasion rather than command places a great premium on communication.

Organisations utilise material and human resources to produce goods and services. As operations grow, inter-relatedness of activities increase. Internal co-ordination and integration become complicated and the impact of external environment is left in a wide spread manner. A manager in the modern times has to work in an environment subjected to constant changes, mutual conflicts and basic challenges.

One of the important aspects concerning the management of human resources relates to development of effective communication among the people in an organisation. People are subjected to continuously changing patterns of motives, aspirations and attitudes and they significantly differ with each other in ability. In the process of assigning specific responsibilities, a situation is reached where facts and ideas must pass from person to person, group to group and level to level. The success and effectiveness of operations of the organisation would depend on how timely, adequate and appropriate this flow of information in the organisation is. A communication system is essential to pass messages, ideas and information for explaining objectives and plans, controlling performance, taking corrective action, etc.

The network of communication not only serves the day-to-day operations but even more significantly provides new and innovative ideas and informations for an effective and realistic planning, its analysis and evaluation. The experience tells that their is inadequate recognition of the fact that communication is a two-way process and that each person in an organisation is both a sender and a receiver in the communication net-work. Upward communication develops understanding and acceptability of decisions, stimulates employees to co-operate and contribute their best. Upward communication is fundamental to democratise the organisational climate. It has been observed that at many places upward communication is ignored and only downward communication in a top-down manner gets established.

Definition

Communication is the process of passing information and understanding from one person to another. The term communication is derived from the latin word 'Communis' which means common and thus if a person affects a communication, he has established a common ground of understanding. To communicate is to inform, to tell, to show or to spread the information. Thus communication may be defined as inter-change of thought or information to bring about mutual understanding and confidence or good human relations.

It is the intercourse by words, letters, symbols or messages. It is the interchange of facts, view points and ideas which bring about unity of interest, purpose and effort in any organisation. Literally nothing happens in management until communication takes place. The success of anyone who manages depends more on his ability to communicate than any other skill.

Communication is the attempt to effect a transfer between minds. The word transfer tells us that communication is essentially a two-way process, involving a sender and receiver. It could be a mechanical piece of equipment like computer ; a writer and a reader ; a speaker and a hearer. Thus communication in industry always involves, at least two persons and a sender and a receiver. One person alone cannot communicate. Only a rceiver can complete the communiction act. There is no communication untill the message sent by the communicator is received by the receiver. It should also be noted that communication is not complete until the response to it has been observed. It is not enough for a manager to give an order ; he must see that it is correctly received, understood and carried out by the receiver. We are not all perfect, so the word 'attempt' becomes significant when we consider the media by which the communication is effected. Understanding means that the receiver should interpret the message exactly as the sender intends. But this is not always the case. If the sender transmits the idea of a rectangle, but the receiver sees a square, this is a case of poor or ineffective communication.

Communication has also been described as a process involving interchange of facts, viewpoints and ideas between persons placed in different positions to achieve mutual understanding. From the above discussions, we reach the following conclusions :

1. Communication is an attempt to effect a transfer between minds. It involves atleast two persons.

2. Communication is a two-way process and is not complete untill the response to it has been observed.

Purpose of Communication

A communication, in addition to transmitting an idea, consciously or unconsciously creates in the mind of *communicates* which he thinks are favourable to himself. The usual communication at work involves more than receipt ; there is an expectation of understanding, acceptance and action. But understanding is personal and subjectives, it can occur only in the receiver's mind. A communicator may make others hear him, but he may not make them understand him. Many managers overlook this fact when giving instructions to their subordinates. Experience shows that one is very often misled by the wrong image of the other person in one's-mind. The words are empty vessels and the receiver pours meaning into them very frequently on the basis of the image which he carries in his mind of the communicator.   A skillful communicator has to find the suitable words and expressions so as to make the receivers understand what he wants.   Thus both the purposes of transmitting ideas and creating impression, have important roles to play in proper understanding.

Through effective communication, skill to work is brought into touch with will to work and both combined together lead to team spirit in the organisation.   Advantages of effective communication are sense of motivation, clarity of thoughts and orders, non-distorted information and consequent increase in production and morale of employees.   In an organisation where mutual trust between the management and workers exist, it is easy to communicate effectively.

The Systems Approach of Communication

According to this theory the whole process of communication involves six steps in the sequence as shown in Fig. 52.1.

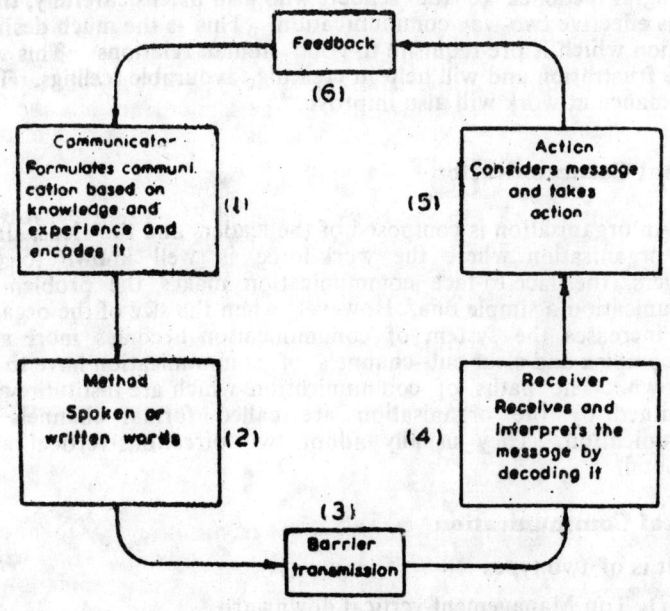

Fig. 52.1. Communication Process

The first component of the communication system is communicator.   He must have an idea or choose a fact to communicate. When he has something to say, he hopes to achieve understanding and perhaps a change in the receiver's behaviour, so he initiates the message.   The message is the second component which means a clear idea which exists in the mind of the sender.

The sender organises his idea into a series of symbols which he feels will communicate to his intended receivers. This is known as encoding of message *i.e.* converting ideas into communicable code which will be understood by. the receiver of the message. The next step is the transmission of message as encoded. The transmitter has to choose media and channel for sending the information.

Receiver is another important component of communication. He receives and tries to understand the message. For this he will decode the message. Decoding is the process by which the receiver takes meaning from the symbols encoded by 'the sender. If he is an effective receiver and cooperative, he tries to take from the message the meaning intended by the sender. Finally, the receiver will act and respond in some way and will give feedback to the sender, whenever necessary. This will bring about two-way traffic in communication system.

When a receiver listens and understands the message and sends meaningful feedback to the sender, who also listens carefully, then there is effective two-way communication. This is the much desired condition which is pre-requisite to good human relations. This will reduce frustration and will help in creating favourable feelings. The performance at work will also improve.

Formal Communication

An organisation is composed of the leaders and the leds. In a small organisation where the work-force is well known to the managers, the face-to-face communication makes the problem of communication a simple one. However, when the size of the organisation increases the System of communication becomes more and more complex and clear-cut channels of communication have to be laid down. The paths of communication which are institutionally determined by the organisation are called formal channels of communication. They usually adopt two directions-vertical and horizontal.

Vertical Communication

It is of two types :

(*i*) Top Management vertical downward.

(*ii*) Workers vertical upward.

Vertical Downward Communication. When the communication is channelled to persons lower in the organisation through established hierarchy, it is known as vertical downward communication. This is usually considered to be from management to operative employees, but much of it is also within the management group. Downward communication is needed.

(*i*) To get the thing done.

(*ii*) To prepare for changes.

(*iii*) To discourage mis-information and suspicion.

(*iv*) To let the people feel the pride of being relatively well informed.

Examples of downward communication are notices, circulars, instruction, orders, letters, memos, annual reports, etc.

Upward Communication. Feedback to the higher authorities by the lower levels vertical upward communication. The examples of upward communication are

(*a*) activity reports on subjects like consumption of raw materials, production, distribution, manhours, etc. ;

(*b*) opinions, ideas and suggestions ; and

(*c*) complaints and grievances.

Upward communication is needed :

(*i*) To create receptiveness of communications.

(*ii*) To create a feeling of belongingness through participation.

(*iii*) To evaluate communication.

(*iv*) To demonstrate a concern for the ideas of each individual.

Horizontal Communication

It refers to transmission of information among positions of the same level. The importance of horizontal direction of communication is undermined due to three reasons. Firstly, it is largely a by-product of vertical communication. Secondly, with increasing size and specialisation, the opportunities for cross talks are cut down and thirdly, in relative terms lateral communication poses fewer difficulties than upward or downward communication because it has fewer implications of authority and status. To secure coordination and cooperation of employees at horizontal level the problems are generally handled through informal contacts.

Horizontal communication is more of an informal nature. If a departmental head needs some information from another departmental head, he may get this by ringing him up directly. Inspite of presence of hierarcy in any large industrial organisation, it is possible to accelerate exchange of information if the management recognises and encourages cross contacts which cut across the organisational lines. Such contacts may take place between coordinate individuals and groups, not only in their lines but also with other echelons of management. The crosswise communication can be effective when a proper understanding exists among the superiors on these points. The subordinates should refrain from making policy communication beyond their authority and should keep their superiors informed to their inter-departmental activities.

Techniques

The following are the different techniques of communications.

Downward Oral Communication

(i) Personal instructions.
(ii) Lectures, conference, meetings, etc.
(iii) Interviews and counselling.
(iv) Telephone and public address system.
(v) Whistles and Bells and Signals.
(vi) Social functions including union activities.
(vii) Grapevine, gossip and rumour.

Downward Written Communication

(i) Instructions and orders.
(ii) Letters and memos.
(iii) Circulars, notices, bulletins, newsletters, etc.
(iv) Information racks and handouts.
(v) Hand-books and manuals.
(vi) Annual reports.
(vii) Union publications.
(viii) Wall posters.

Upward Oral Communication

(i) Face-to-face communication.
(ii) Interviews.
(iii) Telephonic conversation.
(iv) Meetings, conferences, etc.
(v) Social affairs.
(vi) Union representatives—joint consultations.

Upward Written Communication

(i) Reports.
(ii) Personal letters.
(iii) Grievances.
(iv) Suggestions.
(v) Union publications.
(vi) Wall posters.

Horizontal Communication

It is generally of informal nature, so it is verbal in most of the cases. Some of the instances are :

(i) Face-to-face talks.

(ii) Telephones and inter-communication system.

(iii) Social affairs.

(iv) Meetings, conferences, etc.

Communication in Groups

As an institution grows in size and complexity, engineering, marketing and other specialised ' groups grow in size and importance. The busy line manager depends upon them to advise and assist him. These specialists, called staff in classical line and staff organisation, play a leading role in communication beyond their own departments due to following reasons.

Firstly, many communication activities are usually assigned to them. In some cases their primary duty is communication. They perform such functions as gathering data, preparing reports, co-ordinating activities and advising others as well as countless other communication functions.

Secondly, these specialists lack the line authority ; they have greater motivation to communication because they realise that their success is more dependent upon selling their ideas to others.

Thirdly, specialists have shorter communication chains to' higher management.

Lastly, specialists' work usually give them more mobility than operating workers have. Specialists in such areas as personnel and marketing, find that their duties both require and allow them to get out of their offices ond visit their areas, without someone wondering if they are not working. All this means that they have the chance to receive and spread information widely and regularly.

Informal Communication

When the employees are unable to communicate the required information to higher authorities because of a communication barrier, they may resort to informal systems of communication. Distortions may appear in the teansmission of such messages through grapevine in the form of rumours and gossips. Such informal systems may be resorted to by the managers when they find that it is not possible together information through the established channels in a formal organisation. Thus informal communication coexists with the formal communication system in the organisation.

Informal communication or grapevine arises from social inter, action. It is the expression of their natural motivation to communicate. Its speed is very fast as compared to formal communication. A study conducted by *Keith Davis* revealed that wife of a plant supervisor had a baby at 11·00 p.m. and a plant survey the next day at 2·00 p.m. showed that 46% of the management personnel knew of it through the grapevine.

The important point which we must recognize is that the grapevine is a natural and normal activity. It is because of the desire of the people to communicate without following the formal channels in the organisation. It is an essential part of the total human environment. There is nothing inherently bad about the grapevine. It as a matter of fact, fills in the gaps existing in the formal communication system. If it does not exist in the organisation, the ability of a manager to build team work, motivate people and create identification with the organisation would be severally restricted.

Grapevine generally operates like a cluster chain. For instance, *A* tells three or four selected others. Only one or two of these receivers will then pass on the information and again they will usually tell more than one person. As the information becomes older and the number of those knowing it grows larger, it gradually dies out because those who receive it do not repeat it. This process is called a cluster chain, because each link in the chain tends to inform a cluster of other people instead of only one person. Grapevine, may also move in the fashion of a long chain in which *A* tells *B*, who tells *C*, who in turn tells *D* and so on. But this is rarely the case.

Fig. 52.2 represents the various types of grapevine.

If we accept the idea that cluster chain is predominent, then we can conclude that only a few persons are active communicators on the grapevine for any particular bit of information. The persons who keep the grapevine active are called 'liaison individuals'. The liaison agents are generally different in each case because people tend to be active on the grapevine, when they have cause to be. This means they act partly in a predictable manner. This element of cause and predictability offers management a chance of influence the grapevine. People are also active on the grapevine when their friends and work associates are involved. This means if *M* is to be fired, the other employees should be told the full story as soon as possible. If they are not informed, they will fill in the gaps with their own conclusions and thus the rumours will start.

Another marked feature of the grapevine is the speed with which it functions. The cluster chain makes it easy for a few poeple to reach many others in a short time. The grapevine exists largely through words of mouth and observation. Through modern air and electronic communication, it is possible for the grapevine to leap

Grapevine

Grapevine or Informal Communication
Originated From Social Interaction

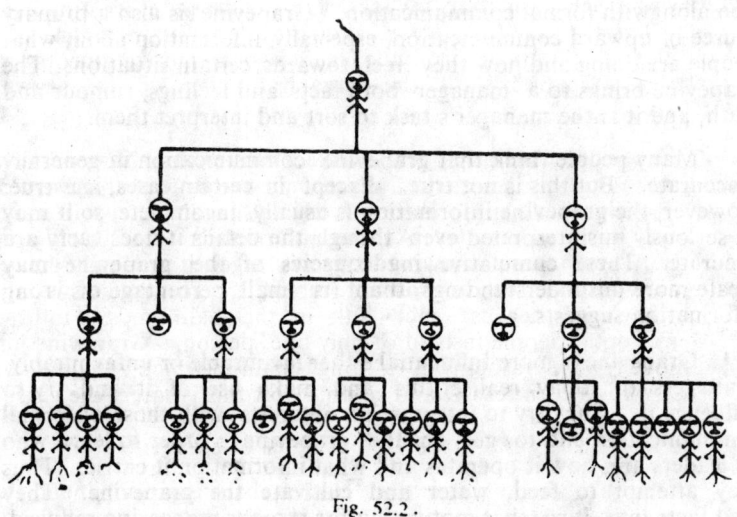

Fig. 52.2 .

hundreds of kilometres very quickly. If in an organisation, there exist procedures which regularly bring people in contact, we can expect these people to have an active grapevine or informal communication.

Grapevine or informal communication provides a great number of benefits to the management. As said earlier, it is a primary means for the development of group identification and interest in work. It also helps the organisation complete its job of communication. Many employees make use of it so perform their duties efficiently and effectively. Moreover, it is not always possible for management to transmit formally all types of information which is necessary. So management has to allow the co-existence of informal communication alongwith formal communication. Grapevine is also a primary source of upward communication, especially information about what people are doing and how they feel towards certain situations. The grapevine brings to a manager both facts and feelings, rumour and truth, and it is the manager's task to sort and interpret them.

Many people think that grapevine communication in generally inaccurate. But this is not true. Except in certain cases, it is true. However, the grapevine information is usually incomplete, so it may be seriously misinterpreted even though the details it does carry are accurate. These cumulative inadequacies of the grapevine may create more misunderstandings than its small percentage of wrong information suggests.

Grapevine is more influential either favourable or unfavourably. Managers of today realise this and make use of it and try to influence it. They try to integrate its interests with those of formal communication and to get on the grapevine in order to learn who its leaders are, how it operates and what information it carries. Thus they attempt to feed, water and cultivate the grapevine. They feed facts into it so that motivation for rumour-mongering reduced. But some people object to feeding the grapevine because they think that such action gives tacit management approval to its mode of operation. The dark side of the grapevine information is known as 'rumour'.

Rumour

It is the most undesirable feature of the grapevine and its has given the grapevine a bad reputation. That is why to some people grapevine means rumour. But rumour is grapevine information which is communicated without secure standards of evidence being present. It is thus untrue part of the grapevine. It can by chance be correct, but generally is incorrect ; so it is presumed to be under-sirable. Rumour originates for a number of reasons. One cause is plain maliciousness, but it is probably not the most important. A more frequent cause is employees' anxiety and insecurity because of poor communication in the organisation. Rumour also serves as a means of wish fulfilment or applying pressure upon management.

Rumour largely depends on the interest and ambiguity perceived by each person ; it tends to change as it passes from person to person. Its general theme may be maintained, but not its details. The rumour gets twisted and distorted always when it passes from one mouth to another. The message gets its own head, tail and wings on its journey and swells unproportionately to an exaggerated shape. Generally, each person chooses details in the rumour to fit his particular focus on reality. Thus the details given at the beginning of a rumour are lost after a few transmissions because people reduce it to a rememberable number of details about items of interest to them. A major outbreak of a rumour can be a devastating epidemic that sweeps through an organisation as fast as a summer-storm and usually with as much damage. Therefore, the most important problem before the management is how to deal with such rumours.

The best approach in dealing with rumour is to get at its causes, rather than try to kill it after it has already started. When causes are known, it should be stopped as early as possible because once a rumour's theme is known and accepted, employees distort future happenings to conform to the rumour. So the management must pass on the correct message in time. Once a rumour has been spread, it is difficult to erase it from the minds of the people. The answer is to get the facts across, before misconceptions have a chance to gain a foot hold. Usually face to face supply of facts is the most effective way because it helps answer the particular ambiguities in each individual's mind.

The oral message may be reported clearly. The message must contain facts and not the opinion. The message should not be exaggerated. The message should be formed by the written message and it should be circulated quickly. Management may also take the help of union in combating the rumour which is not in the interest of the workers and the organisation.

Barriers to Communication

Barriers to communication cause break downs, distortions and inaccurate rumours. They plague the daily life of the managers who must depend upon the accurate transmission of the orders and information for efficient operation. Whenever a communication is made, there is always a tendency on the part of the receiver to evaluate the message received and then decided to approve or disapprove the same. Another important barrier to communication lies in the layers and spans of communications.

In large organisations there are a number of obstacles which make transmission of message more difficult. In both upward and downward communications, it may happen that some of the persons in the intermediate layers withhold the whole or part of the message because they may feel that by withholding the information the will be better informed than those whom they lead. It should be noted that although there is no such thing as a perfect communication, yet considerable degree of perfection can be achieved in communication if the barriers to communication are overcome. Following are the main barriers to communication :

1. **Barriers due to organisation structure.** The breakdown or distortion in communication, sometimes, arises due to :

(*i*) Several layers of management ;

(*ii*) Long lines of communication ;

(*iii*) Special distance of subordinate from top management ;

(*iv*) Lack of instructions for passing information to the subordinates ; and

(*v*) Heavy pressure of work at certain levels of authority.

2. **Barriers due to status and position**

(*i*) The temper and attitude exhibited by the supervisor are sometimes a hurdle in two-way communication. One common illustration is non-listening habit. A supervisor may guard information for :

(*a*) Consideration of prestige, ego and strategy.

(*b*) Underrating the understanding and intelligence of subordinates.

(*c*) Deriving satisfaction in being the store-house of information and seeing people dance around him for information.

(*ii*) Prejudices among the supervisors and subordinates may stand in the way of a free flow of information and understanding.

(*iii*) The supervisors particularly at the middle level may sometimes like to be in good books of top management by :

(*a*) Not seeking clarification on instructions which are subject to different interpretations ; and

(*b*) Acting as a screen for passing only such information which may please the boss.

3. **Semantic barriers.** Semantic is the science of meaning. Words seldom mean same thing to two persons. Symbols or words usually have a variety of meanings and the sender and the receiver have to choose one meaning from among many. If both of them choose the same meaning, the communication will be perfect. But this is not so always because of differences in formal education and specific situations of the people. Strictly one cannot convey meaning, all one can do it convey words. But the same words may suggest

quite different meanings to different people e.g., 'Profits' may mean to management efficiency and growth, whereas to employees it may suggest excess funds piled up through paying inadequate wages.

4. **Tendency to evaluate.** A major barrier to the communication is the natural tendency to judge the statement of the person or the other group. Every one tries to evaluate it from his own point of view or experience. Communication requires an open mind and willingness to see things through the eyes of others.

5. **Heightened emotions.** Barriers may also arises due to specific situations e.g., emotional reaction, physical conditions like noise or insufficient light, past experience, etc. When emotions are strong, it is most difficult to know the frame of mind of the other person or group.

6. **Lack of ability to communicate.** All persons do not have the skill to communicate. Skill in communication may come naturally to some, but an average man need some sort of training and practice by way of interviewing and public speaking, etc.

7. **Inattention.** The simple failure to read bulletins, notices, minutes and reports is a common feature. With regard to failure to listen to oral communications, it has been seen that non-listeners are often turned off while they are pre-occupied with other affairs, like their family problems. In any case, the efforts to communicate with some-one not listening will fail.

8. **Unclarified assumptions.** This can be clarified by an illustration. A customer sends a message that he will visit a vendor's plant at a particular time on some particular date. Then he may assume that vendor will receive him and arrange for his lunch etc. Whereas vendor may assume that the customer was arriving in the city to attend some personal work and would make a routine call at the plant. This is an unclarified assumption with possible loss of goodwill.

9. **Resistance to change.** It is the general tendency of human beings to maintain status quo. When new ideas are being communicated, the listening appratus may act as a filter in rejecting new ideas. Thus resistance to change is an important obstacle to effective communication.

Sometimes, organisations announce changes which seriously affect the employees, e.g., shifts in timings, place and order of work, installation of new plant, etc. Changes affect people in different ways and it may take some time to think through the full meaning of the message. Hence, it is important for the management not to force changes before people are in a position to adjust to their implications.

10. **Closed minds.** Certain people who think that they know everything about a particular subject, also create obstacles in

the way of effective communication. Persons suffering from the mirage of too much knowledge become rigid and dogmatic in their attitude. They close their minds tightly to new ideas that are brought to them.

These are the problems or barriers in effective communication. The communication will not be prefectly effective, if the transmission of the message is faulty. The faulty transmission can be easily explained by this example. An experiment can be conducted on how the message is transmitted by taking five or six volunteers. All but one of them may be asked to wait outside a room, while one volunteer is shown a picture of a street scene inside the room and is given only two minutes to observe and memories as many details as possible. Now the first volunteer may be allowed to convey the scenes to the second volunteer who will tell it to the third volunteer and so on. Then the last one may be shown the same sceneries and it will be found that the last one is not able to recognise it. In fact, the following things happen during the process of communication from one person to another.

1. A few more details are included each time the scene is described.

2. The details are changed and twisted by the persons.

3. Certain details are added which do not exist.

4. Each person emphasizes different things due to his different perception.

5. Some details are forgotton and some are exaggerated.

How to Overcome Barriers

1. **Direct and simple message**. As far as possible, the sender of message must use the direct and simple language so that the receiver may understand the message without much difficulty.

2. **Feedback**. A 'feedback' provision calls for making communication a two-way process and providing opportunity for suggestion and criticism. It keeps avoiding likely errors in transmission and in invoking effective participation of the subordinates in work. In case of face-to-face communication, immediate feedback by seeing the emotions and expressions on the faces of the listeners can help the communicator.

3. **Creation of proper atmosphere**. In particular cases, as for instance, when a boss is talking to his subordinates, the atmosphere must be peaceful so that there is effective communication of instructions and suggestions.

4. **Effective Listening**. The sender must also listen to receiver's words attentively so that the receiver also listen to sender at the same time.

53

Morale

Introduction. Good employee relations are the best guaranty any company can have of continued success in any field of endeavour. The achievement of good employee relations means the attainment of that intangible known as "Morale".

Where good morale exists, employees put their hearts into their work. They enjoy their work and they do it better and more efficiently. In fact there in no other single field of management which can show so great a return for so little investment. Appropriate encouragement can release untapped latent reservoirs of employee contributions to the success of the business. To any organisation the satisfied employees are probably the most valuable single asset because of the following:

(i) Satisfied employees cause less wear and tear on the machinery and equipments.

(ii) Satisfied employees cause less avoidable waste.

(iii) Satisfied employees take a personal interest in the welfare of the company.

(iv) Satisfied employees are keen to suggest improvements.

(v) Satisfied employees also apprecaite their company and therefore turn over is negligible.

(vi) Satisfied employees constitute the best possible advertisement any company can have. Where good employee relation exist, management has the oportunity to lead rather than drive. Management's role is voluntarily accepted by employee rather than being forced on them by concepts of authority and legal rights. Where good employee relations exist, management fully devote its time and effort to managing the business rather than be perpetually engaged in overcoming human friction. Good employee relations are indeed the lubricant that makes an organisation function smoothly. Good employee relations are like any other commodity; they can not be sold for long unless the product to be sold is good. In the case of employee relations the product is "employee welfare" fair and definate employee policies, administered by a competant, carefully selected and throughly trained management group. Good intentions or good policies alone can not result in good employee relations. But if both of them are administered by sound management it can bring about good employee relations. Employee relations differ from every other management function. It touches every employee in the organisation.

While high morale and good employee relations are not identical, it is clear that the establishment of good employee relations is the major factor in high employee morale. Whether morale is measured as same total of attitude if employees toward the company or the degree or satisfaction with his work environment the good employee relations inevitably lead to high morale.

Employee morale may be defined as the attitude if the employee toward his work environment and more specifically toward the employee as represented by high supervision. Morale is lively to be high if the employee is convinced of the following factors:

(i)   Job is ensured so long as he continues to do satisfactory work.
(ii) Treated fairly by the employer and given adquate opportunity to be heard if he feels he is not being treated fairly.
(iii) Company recognises his outstanding work.
(iv) Has fair opportunity for advancement.
(v)   Image of company is well thought of in the community.
(vi) Employee work is important and of significance.

Importance of Morale. Management has a growing interest in employees' attitudes and morale. The concern on the part of the Management can be attributed to a general trend towards greater recognisation of social responsibilities of the organisation. The term morale is not very specific in the sense that it conveys different meaning to different people. Morale is a much abused word. It has resulted in all kinds of mistaken ventures in the theory that high morale can be created by music on the job, inspirational programmes, apprecaition letters, birthday greeting, bulletins, suggestion schemes, shopping centres in the Plant township etc. Some or all of these may have their place in specific stituations but they can be safely substitute factors that determine morale in employer-employee relationship.

Morale a Neutral Concept. Like the word health the word morale by itself does not have any favourable or unfavourable meaning. Morale is a neutral concept like the concept of health, because it has to be qualified with the degree, as high morale or low morale. Moreover, morale is a relative concept; there is no absolute state of morale. we can only refer to the degrees of high morale or degrees of low morale. *Leighton* in his book 'Human Relations in a Changing World', states that, 'Morale is the capacity of a group of people to pull together persistently in pursuit of a common purpose". On analysis, this statement reveals that morale is the degree of enthusiasm and willingness with which the memebrs of a group set out to perform the allotted tasks. If the enthusiasm and willingness to work is high or low we will say that morale is hgih or low, as the case may be. We cannot say that there is morale or no morale among the employees. Hence, morale has to be qualified like the word health. Just as good health is essential for any person, high morale is necessary for any organisation or for any group of persons.

Morale has been recognised by the military authorities as one of the most important factors in winning of losing wars. As Napolean said, "In war, morale conditions make up three-quarters of the game; the relative balance of man-power accounts for the remaining quarter." Some may consider this an exaggeration but other things remaining same, high morale certainly leads to success and low morale bring defeat in its wake. According to *Beishline* high morale in a military sense means, "The state of mind which makes a soldier cocerned with his surroundings; confident of himself, his comrades and his leaders, and determined to execute all of his missions with effectiveness and efficiency. It relates to the condition of an individual or group with regard to courage, confidence and enthusiasm in the performance of duty."

Definition of Morale. For managerial purposes morale has been defined in many ways, but all the definitions revolve round the attitude and willingness to work to achieve the organisational goals. We may define morale *as the attitudes of individuals and groups in an organisation towards their work enviornment and towards voluntary co-operation to the full extent of their capabilities of the fulfulment of the organisational objectives.* Morale is the indicator of the attitude of employees towards their job, superiors and the organisational environment. It is collection of employees' attitudes, feelings and sentiments towards these. Morale is a by-product of the group relationships. It is mental process, which once started permeates in the entire group, creating a mood which results in the formation of a common attitude.

Employees with high morale like their jobs and co-operate fully with the management towards the achievement of goals of the organisation. High morale makes for effective work by the employees. It result from job satisfaction and generates job enthusiasm. High morale is indeed a manifestation of the employees' strength, dependability, pride, confidence and devotion. All these qualities of mind and character taken together create high morale among the employees. Low morale, on the other hand, indicates the presence of mental

---

*Beishline, John Robert, Military Management for National Defence, P. 229.

unrest. This mental tension not only hampers production but also leads to ill health of the working peopole. The other consequences of low morale are the following:

(a) High rate of labour turnover.

(b) High rate of absenteeism,

(c) Excessive complaints and grievances,

(d) Friction among the employees,

(e) Frustration, and

(f) Antogonism towards management and regulations of the organisation.

Consequence of Low Morale. These consequences are very much fatal to the industrial relations in the orgaisation. There will always be labour unrest in the organisation because of low morale of the employees. Whatever may be the consequences of low morale, organisation has to suffer ultimately because quantity and quality of production both suffer. Thus, in order to avoid these evil consequences, every manager should work to maintain the high morale of the people working under him. For this he should have the constant knowledge of the opinions and attitudes of the employees towards their work and the organisation and should carefully note the changes in their behaviour and appraise the factors responsible for change in the attitude of employees.

Factor Determining Morale. Morale indicators or factors determining morale are ingredient which the management must study in order to appraise the morale of the employees. Following are the important marale factors or determinants

(i) Personal Factors. These factors relate to age, training, intelligence of employees, time spent by them on the job and interest in work taken by them. For instance, if an employee is not imparted proper trining, he may have low morale because he always faces difficulty in doing the work.

(ii) Factors Relating to Job

(a) Type of Job : Nature of job, skill required to do the job and fatigue associated with it determine the attitude of an employee towards his job.

(b) Job Satisfaction: If the job gives an employee an opportunity to prove his talents and grow his personality, he will certainly like the job and his enthusiasm to work will be higher.

(c) Compatability with Fellow employees: Man is a social being, he finds work more satisfying when he has the acceptance and companionship of his fellow employees. He should have confidence in fellow employees and faith in

their loyalty. Confidence in fellow employees and feeling of togetherness generate high morale.

(d) Nature of supervision: The nature of supervision can better tell the attitudes of employees because a supervisor is in direct contact with the employees and can have greater influence on the day to day activities of the employees. Supervisor is the pivot around which many things revolve. He is the formal leader and the workers must have confindence in him and in his leadership. A supervisor, who is fair, helpful and social minded will create confidence among the employees and this will increase their morale.

(e) Working Conditions : Working conditions also influence the attitude of the employees towards their jobs and supervision. Thus if better working conditions aer provided to the employees for enabling them to do the work assigned to them, morale will be high.

(iii) Other Factors

(a) Objective of the Organisation: Workers want to be a part of the organisation which has worthwhile purpose in which they can believe. The commonness of the purpose will lead them to work enthusiastically so that organisation may achieve the common purpose efficiently and effectively.

(b) Rewards: Employees expect adequate compensation for their services rendered to the organisation. Good system of wages and salaries and other inentives will help to increase the morale of the employees. Advancements and promotions always have a stimulating effect on the employees.

(c) Physical and Mental Factors: There are certain off-the-job factors like trouble at home or ill health of any member of the family, which affect the morale of the employees. Emotions of the employee also influence the morale. For instance, an employee may be satisfied with his home and social life, but may think that his job is so so. In such a case, his job morale is relatively low because the job he is holding is below his expectations and gives him mental worries.

Improving Morale. Morale can be measured by checking the extent to which the organisation is achieving the results in respect of productivity, profit other forms of goal achievement. But this is not reliable because morale may be higher yet productivity is lower and vice-versa because of other factors. Another indirect measure of morale are factors like absenteeism, labour turnover, fluctuations in output, quality records, excessive waste and scrap, grievances, training records, accidents, etc. These factors are sound indicators of any major vatiation in morale, but they are not as precise as a morale survey. Their chief advantage is that they are readily available and are most objective to provide a good measure of trend over a period of time.

The management can take the following measures to improve morale:

(i)　Monthly employees' conferences may be held to discuss with the employees the problems of the previous month and the programmes for next month,

(ii)　Private discussion may be held by the managers with individual employees to know their views on the working of the organisation and to personnel policies.

(iii)　There should be effective two-way communication between the management and the employees. Employees should be kept informed about organisation policies and programmes and they should be allowed to comment over these.

(iv)　After an employee has been assigned a particular job, it is necessary to follow up his performance to ascertain whether he has been placed on the right job or not. If he is suitable to the job, his morale will be high.

(v)　Management should encourage group activities by the employees like sports, social get togethers, picnics etc. This will help to develop high morale since they will consider themselves to be integral part of the group.

(vi)　There should be proper incentive schemes in the organisation to ensure higher monetary rewards for exceptional work in terms of quantity and quality. There should be clear-cut policy about promotion through proper training.

(vii)　Management must provide for welfare measures and other amenities like canteens, eduction for children, credit facilities, etc.

54

Workers' Participation

Introduction

In any organisation we have two distinct groups of people *viz.* Managers and workers performing two separate set of functions which are known as Managerial and operative. Managers are held responsible for the work of their subordinates, while workers are accountable for their own work. Managerial functions are primarily concerned with planning, organising, motivating and controlling in a contrast with the operative functions of worker. Workers' participation in Management seeks to bridge the gap and authorising workers to take part in managerial functions. Participation may be in any of these two forms :

    (*a*) Ascending participation.

    (*b*) Descending participation.

In case of ascending participation, the workers may be given an opportunity to influence managerial decision at higher levels through their elected representatives to joint counsels or Board of Directors of the Company. In descending participation, they may be given more powers to plan and make decision about their own work. The descending participation is the process of delegation and job enlargement. This is the form of participation which is more common in most of the organisation. The worker can also participate when a manager adopts a participation style of supervision. This is the informal way of participation.

The amount of work done by the worker is not determined by his physical capacity but by his social capacity also. The reward play an important role in determining the motivation and the happiness of the worker. The workers does not react to management and its norms and also refers as an individual but as members of the group. Workers emphasis is on greater Communication, participation and leadership in the management of organisations. In short, the development in the field of social sciences concur with the concept of worker participation in Management.

Concept. The concept of workers' participation in management was evolved with a view to give the workers a sense of belonging and to stimulate their interest in higher productivity. The idea of workers participation is to envisage :

(a) Establisment of cordial relation, better understanding between management and worker.

(b) Increase in productivity, better welfare facilities for workers.

(c) Training and Development of workers to share the responsibilities of Management.

A worker as an individual can be encouraged to participate in managerial functions by way of :

(a) Suggestion schemes

(b) Delegation

(c) Job enlargement and job enrichment.

Following are the various means of achieving workers' participation :

(a) Joint Management Counsels.

(b). Works Committee.

1. **Joint Management Counsels.** Joint management counsels should be set up in all the establishments in the public as well as in private sectors in which favourable conditions to the success of the counsels exist.

Following are the functions of joint counsels :

(a) Administration of welfare measures.

(b) Supervision of safety measures.

(c) Operation of vocational training and apprenticeship schemes.

(d) Awards for valuable suggestions.

(e) Matters in which workers have the right to communicate.

(f) Importance to issues in which they are expected to be consulted.

The various functions of Joint Management Counsels can be summed up under the following heads :

(a) Concentration.

(b) Information.

(c) Administration.

J.M.C. can be consulted for various matters and to give information of general economic situation of the undertaking, production and sales programmes. The administration responsibility include supervisions of welfare measures, safety measures, operation of vocational and apprenticeship schemes.

2. Works Committee

Constitution : Section (3) of the Industrial Dispute Act empowers the Appropriate Government to Constitute a works Committe by a general or special order in a manner which may be prescribed. There are two conditions which must be satisfied before a works committee can be constituted—

(1) The establishment must be an industrial establishment.

(2) One hundred or more workmen should either be presently employed or should have been employed on any day in the preceding twelve months. The Works Committee must be composed of the representatives of the employers and the workmen engaged in the industrial establishment and must be equal in number. It is further provided that the representatives of the workmen shall be chosen in the prescribed manner and in consultation with the registered Trade Union, if any.

Functions and Duties

The Works Committees are required to promote measures for securing and preserving cordial and good relations between the employer and his workmen. The Works Committee shall meet and discuss matters of common interest and make efforts to settle difference in respect of such matter. The Works Committees are normally concerned with the problems arising in the day-to-day working of the establishment, for instance, matters concerning their welfare, training wages, hours of work, bonus holidays with pay etc. The decisions of the Works Committees, though carry great weight, are not binding either on the employer or workmen.

Participation Experiences

At the beginnings of the Industrial Revolution Behavioral experts have believed that breaking up production into tiny, highly specialised tasks, with each worker struck to his own isolated job, was vital to the efficient working of industry.

Among the first to criticise this principle was a group of behavioural scientists at the Tavistock Institute of Human Relations in London. In the 1950s, the behavioral scientists argued that work should be designed to fit not only the needs of machines, but also of people. Human beings need to be engaged in something challenging, to see the result of their work to feel part of a group, to make decisions and to develop and grow. The scientists of the Human relations institute demonstrated their concept in 1958 in a coal mines, where, in spite of heavy spending on automated equipment, morale was poor, personnel turnover was high, with low production. The suggested remoulding the work organisation into small, self managed teams whose members could change at will between cutting rocks clearing away coal and loading it on to conveyors, sharing responsibility with a common pay packet. A two-year study was done on two groups with forty members each using the new methods and the other the old techniques. Results showed productivity rise in the "democratic" group resulting better maintained work areas and lesser absenteeism as compared to the other group.

Findings of International Studies on Participation

The better-educated workers of Western Europe on longer accept "Papa knows best" authoritarianism. This attitude appeared with explosive force in the May, 1968 revolt in France, when some factories were occupied by employees who showed protest signs bearing a single word: "dignity". In 1971, Danish published opinion poll, where forty eight per cent of the people asked to name the best way of improving their working conditions answered "more influence on decisions". In 1973, the gaint German Matalworkers' Union struck not for more pay, but for the "humanization of work". These are no evidence that a failed company would have fared any better if workers had helped to run it. Critics argue that it is a waste of time for workers to sit on board room discussions of technical, financial, research and development, marketing, new products and issues on which they have little specialised knowledge.

Yet most worker-directors in Germany are now given special training in business administration by their unions before assuming their places on the board. Moreover, they are already expert in one highly important field: labour relations. Many observers say that the presence of workers on the board is a positive asset to companies. Experience shows that if we don't examine the workers' views at the very beginning of a decision, you may find you have to change the decision later on."

A number of labour unions remain hostile to the idea, particularly in France, Italy and Belgium to the idea of participation at Board level. Burno Trentin, communist leader of the Italian Metalworking Union said "We are not interes ted in being on the board of directors. We stand on one side and management on the other."

When the firm of George Kent in England, a maker of process-control equipment, was to be sold, the Department of industry clearly preferred that a British company acquire it. Kent's workers, however, judged that a British Electric did not offer the advantages of its Swiss competitor, Brown Boveri. They persuaded the Department to change its mind, and the Swiss firm won the bid for a major share-holding.

At Fiat, the carmakers in Italy—depsite started opposition both by union leaders and employers to workers representation on boards—the company—union pointed out to Fiat that tractor production was growing too rapidly at one plant. while car-parts output at another factory was stagnating, and suggested modifications in the production schedule which management accepted. In a contract signed in 1974 and recently renewed, management agreed not to lay off workers while the unions accepted shorter working hours and the shifting of 2,500 workers from the crisis-ridden car sector to industrial vehicles and aircraft.

The participation of workers in high-level company decision-making can create a healthy system of checks and balances which often means both batter management and an improvement of relations between capital and labour. In 1970, a West German government-appointed a commission and published a through study of co-determination of the countries coal and steel industries. Not only was the system free of negative effects on efficiency, but the overall social atmosphere in those industries also improved. The age-old strict authoritarianism, hierarchical castes, resulting worker bitterness were being replaced by a new harmony characterized by more democratic and informal relations between managers and workers.

Except in Germany, little was done to advance employee board representation for nearly 20 years. But recently the idea has become a general topic. In Denmark, a survey by the employees' confederation found that a large majority of employees in the nation's companies wanted board representation. A French opinion poll in early 1975 revealed that three-quarters of the public believed it was important for employees to be more involved in company decisions.

Co-determination System

In West Germany, worker-directors have existed for 25 years. In fact, observers have attributed the low strike rate about one third of that in France, one fourteenth of Britain's and one-nineteenth of Italy's largely to the close labour-management relations fostered by this codetermination. At times, this co-operation in decision making has meant a great difference between profit and loss, or even the life and death of a company. In 1967, management at a German steel plant revealed that the factory would have to be closed for economic reasons. But the five worker-directors on the board refused to accept the decision without exploring alternatives, First, they came up with a comprehensive cost-cutting plan. Then they suggested combining the plant with five others of the same industrial group into a new company. A sceptical management submitted the plan to a firm of independent consultants. After due study, the consultants pronounced the proposal sound. The plant survived and doing fine. "We cold never have done it," said plant's manager, "without the co-determination system."

The European Common Market is also promoting the "Participation" movement. In its proposed statute that would permit a firm operating in several common Market countries to incorporate under European Economic Community (EEC) provisions. It specifies that each company must have a supervisory board that oversees management, that workers hold one-third of the board seats and have a say in the appointment of some other board members.

The worker participation movement in Western Europe, particularly in West Germany, is based on the premise that economic power is tending to be increasingly concentrated in the hands of employers and that laws making participation compulsory, enforcing what has come to be known as 'co-determination's, are therefore needed to restore the dignity of labour and give the employees a voice in shaping the companies for which they work.

Many European labour unions operate on the premise that worker interests are best served if the employees have a say in management. Union pressure has provided the momentum for the remarkable spread in recent years of the principle of co-determination. In fact, laws requiring worker representation in the board room have gone on the statute books in the Netherlands, Austria, Norway, Sweden and Denmark and similar legislation has been promised by the present governments in France and Britain.

Less than half of American workers today are in the blue collar category, and within the white collar majority also, major technology changes have brought about a tremendous upgrading of jobs. The computer alone has eliminated billions of clerical drudgery while it is opining the doors to new and better jobs.

Survey By Fortune Magazine

According to a recent survey carried out by Fortune Magazine jobs demanding greater skills are increasing more rapidly than earlier. The number of workers classified as professional, technical and similar trades, has increased by 167 per cent in recent years. This group plus managers and administrators now make up nearly 26 per cent of the total work force in American industry. Equally significant are the increases in pay. With the passage of time the medium pay of full-time workers have increased by about 60 per cent from 1950 to 1975.

The main motivation for American industry in making these changes has been to raise the quality of its products but the change has also lifted job levels because of the higher responsibility and skills called for the new machines. It is now hard to imagine an occasion in which a line manager in the U.S. needs to tell a machinist how to do his job.

American industry realises that no job in any company is better understood by anyone than by the person who does it, wheather he is a secretary, a mechanic or the president of the company. Once diginity of labour is appreciated, the workers participation in management, in the widest sense of subscribing to and shaping the goal of the enterprise, follows naturally. And this in a nutshell is how American workers participation functions in practice. Effective participation does not mean that everybody must have a say in everything. It is competence which has become the real basis of authority.

Our environment here are quite different from that in affluent Western countries. Indian industry has to live with the high degree of politicalisation of trade unions and its accompanying drawbacks such as multiplicity of unions and outside leadership. Still there is no reason why workers participation should not succeed here, particularly when all are agreed that productivity can only be increased if workers are more closely involved in areas of decision-making which affect them and their enterprises.

Participation a Mental Process

The plurality of meaning reminds of different perception of people for similar situation. For example an accident has taken place on road the doctor infers from who has been hurt, type of injury etc. the mechanic with type of damage to the car and mechanic with the amount of damage to vehicles. People from different groups often do not realise they are talking about something while in reality they are poles apart. Participation is a mental and emotional involvement of a person in a group situation which encourages him to contribute it, goals and share responsibilities in them.

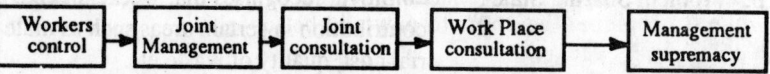

| Workers control | Joint Management | Joint consultation | Work Place consultation | Management supremacy |

Worker's participation in management is a resounding phrase bridging the past and future. *Worker participation involves groups viz Managers and Workers and also involves function viz managerial and operative. Participation bridges the gap between managers and workers and also between managerial and operative functions. Participation is a neutral term and aims at producing co-operative commitment to the enterprise. It involved sharing of power and status between management and Workers.*

Management interests primary lies in improving productivity. Workers interest lies in improving their earnings. *If participation is to be effective as a process or device it should be integrated with a scheme of improving productivity and gain sharing.* Participation increases the scope for employee's share of influence in decision making at different tires of the organisational hierarchy. *Participation incorporates willing acceptance of responsibilities by body of workers.* Essence of labour participation in management lies in the firm belief and confidence in the individual in his capacity for growth and learning; in his ability to contribute, significantly, it eliminates any feeling of futility, isolation and consequent frustration that they face in normal industrial setting. *Participation permits a more balanced interaction pattern and therefore results in less resistance to innovation.* It helps the members of the group to unfreeze their attitude. It permits the subordinate to feel that doing the job will provide him with an opportunity to demonstrate skills. The Classification; Forms and Levels of Participation is discussed as under:

CLSSIFICATION

a. Ascending Participation :   Opportunity to influence managerial decision.

b. Descending Participation :   More power to plan and make their own decision.

FORMS

a. Collective
b. Bargaining
c. Joint administration
d. Joint decision making
e. Consultation and information -- sharing.

LEVELS

a. Information Sharing Stage

b. Problem Sharing State :   Employer recognises that workers make a contribution in certain areas such as material cost, quality of waste etc.

c. Idea Sharing Stage   Management indicates a willingness to have labour initiated ideas in any kind of production and personnel activities.

Aims of Participation

The aims of participation is to improve efficiency of an enterprise. It helps to tap latent human resource potential. Participation helps in improving industrial relation and encourages self management in industry. Government has encouraged participation right from the beginning. According to industrial policy resolution 1956 the aim of the Government is to encourage workers participation. Even in the second five year plan it has been envisaged that worker participation should be encouraged as it will help in achieving the following:

(a) Increasing productivity for general benefit.
(b) Giving employees a better understanding of their role in the working of the industry.
(c) Satisfying the workers urge for self expression thus leading to industrial peace, better relations and increased co-operation.

Need of Participation

The need of the hour is the maximisation of production in all spheres of economic activities to ensure economic self reliance and independence. cautious and effective efforts are both on the part of the management and the employees. These activities are essential for fuller utilisation of man, machine and finance which important to gear up our economic role. It is in this context that the concept of participative management has assumed great significance.

There is also a need for direct involvement with the workers who are directly dealing with the problems and thus can be made responsible for implementation of such decisions. This is the core philosophy of participative management. If the decision making is followed sitting away from the situation, there are chances of committing errors in decision making processes.

Creates An Conducive Atmisphere

The goal of the workers and the management is to create an atmosphere in the industrial activity which would help them remove the self made caste system and pave the way for an atmosphere of joint family systems. The fundamental problems of the management of an organisation is to get effective results through others. Effective results can be achieved with the help of workers if they are given participation in management.

A participative situation leads to a higher degree of involvement of the employees with the enterprises. It will act as a morale booster. By participating in the decision making workers will feel psychologically satisfied since they think that their voice is being heard. Workers' participation in management will lead to informational improvements. They will have better understanding and their resistance to change will have better understanding and their resistance to change will be reduced. These findings have been tested by various experiments.

Participation A Balanced Interaction

Participation permits a more balanced interaction pattern and therefore, results in less resistance to innovation. Participation reduces the negative valence towards the task and increases motivation for work. It permits informal leaders to re-enforce their position. Participation may permit the subordinate to feel that doing the job well provides him with an opportunity to demonstrate skill which he values highly. If participation is a groups process, it subjects the individual to certain group pressures to implement the decisions which the group participated in making.

We know that there are two distinct groups of people in an understanding, viz., 'managers' and 'workers' performing respectively two separate sets of function which are known as 'managerial' and 'operative'. The basic distinction between those who occupy 'managerial' positions and 'workers' is that managers are held accountable for the work of their subordinates, while workers are accountable only for their own work. The extent to which workers may rise to become managers though an important point, is quite distinct from whether workers may take part in managerial functions.

Managerial functions are primarily concerned with planning, organising, motivating and controlling in contrast with operative work. A self-employed man may carry out both these functions if the area of his operation is very small. But in case of big organisation, these functions are to be performed by different sets of people. Workers' participation in management seeks to bridge this gap authorising workers to take part in managerial function. Actually, this is a very wide view of the term workers' participation in management because this is not practically possible.

Forms of Participation

Participation may take two forms. It may be (i) Ascending participation (ii) Descending participation. In case of ascending participation, the workers may be given an opportunity to influence managerial decisions at higher levels, through their elected representatives to joint councils or the board of directors of the company. But in descending participation, they may be given more powers to plan and to make decisions about their own work (e.g., delegation and job enlargement). This form of participation is very much popular in many organisations. The workers may also participate through collective bargaining.

Even in the same organisation there may be high level of participation in certain spheres, whereas the level of participation in other spheres may be low. Similarly, the forms through which the workers participate in management may also vary from sphere to sphere. The principal forms through which workers participate in management are information sharing, consultation, joint decision making and collective bargaining.

Participation is variously understood by management, labour as well as government. Management generally interprets it as joint consultation prior to decision making. Workers normally think of it as equivalent to co-decision of co-determination, while the administrators and experts practically regard it as association of labour with management without the final authority or responsibility in the general arena of managerial functions. In means sharing in an appropriate manner the decision-making power with the lower ranks of organisation of an enterprise. As said earlier workers' participation in management means giving scope to workers to influence the decision-making process at any level in different forms. It is desirable to understand the purpose of workers participation. If the purpose is to improve productivity or improve human relations, then it may be short-lived; but if we have faith in participation, perhaps it will have better impact on the employees as a group. Unfortunately, in many organisation in India the purpose has been either to increase production or just improve human relations without basic faith in the humanistic ideology and democratic values of running the enterprise. That is why the overall experience of participative practice in India such as Joint Management Councils, Works

Committees etc. have been totally unsatisfactory.

The forms and level of participation differ from situation to situation in the same organisation; the participation can be as an individual or group participation. The individual participation can be encouraged by way of suggestion, skills, delegation, job enlargement and job enrichment. One of the ways in which the workers can be enthused in management's initiative to consult the workers by inviting their suggestions to improve the production process and management practice. Under this method management encourages the operative employees to put forward novel ideas for improving material handling, safety measures, production planning and quality, cost controls, etc. The suggestion scheme provides a two-way channel of communication between the management and the employees. This also leads to improvement of human relations in the organisation.

A manager can secure the participation of his subordinates by the process of delegation under which he can entrust responsibility and authority to do a part of his work to his subordinates. He may frequently give his subordinates a chance take certain important decisions, but the ultimate control remain in his hands because he remains accountable to the top management even for the work done by his subordinates. Delegation is the descending form of participation.

In order to provide for more descending participation by workers in managerial functions, many concerns modify the technology and job design. Job enlargement increases the number of operations and tasks performed by the workers, aims at giving the workers greater interest in their work. To the extent that 'job enlargement' enables the worker to plan and control his own work, it injects a certain managerial content into his work, 'job enlargement to as vertical 'job enlargement', involves the redesigning of jobs in such a way that will permit workers to perform tasks previously done by their superiors. This also involves descending participation. Some of the controversial forms of participation are Profit-sharing and Worker representative as directs.

Profit Sharing

The interest of the workers to have a share in the profits is the most controversial one and this must be given due consideration in any scheme for increased productivity. It is desirable to give a fair share in the increasing prosperity of the business to the workers, because of the risks they run. The increasing prosperity is not only because of the risks of capital.

There are more tangible needs of the workers but good morale in any group depends as much upon satisfaction of certain intangible needs. Urwick has summed these up as belief in purpose, confidence in leadership and individual dignity founded on job satisfaction. Taking all these needs together it can be seen that it is not so much control of the business that the workers want, as control of the areas of the job, or that part of the business which really affects them. Workers want association with management in the settlement of those conditions which affect their working life. If they are not allowed to participate in the decisions which affect their working, they will demand control of the management itself. Participation is the process of following through the facts of the ultimate decision. Participation needs to be considered in a wider sense than merely the setting up of joint councils or committees with certain defined responsibilities. It requires an attitude of mind both on the part of the management and the workers, a willingness to give information honestly and fairly to exchange ideas and to consider facts objectively. It requires the integration of the individual into the industrial community to see the work in relation to the work of the group and the product of the whole organisation. Such association of workers should begin from the bottom. Participation at the plant level is more important since a representative of workers in the Board of Directors is not likely to serve any purpose. A worker director will be in a minority and he will only criticise the policies of management because he is the representative of the workers. But participation means not only criticism of established policies and practices but positive constructive suggestions for improvement also.

Worker Representative As Director

In this the workers are given a seat on the board of directors. This type of participation is quite controversial and even according to Trade Union leaders it is not very effective. In their view point, it causes confusion of roles and the worker director would not be able to reconcile his position as trade union leader, representing the workers interest with himself being member of the management.

Many trade unionists have themselves pointed out that such an idea is illogical. This according to them would simply cause a confusion of roles and that the unfortunate worker-director would not be able to reconcile his position as a trade unionist representing the workers' interest with himself being a member of the management. Once the worker-director identifies himself with the cause of management either at the top level or down the line, he cannot act contrary to the management's decision because of the principles of collective responsibility in the management. Thus he will be not in a position to defend his role as a workers' representative.

A worker-director would be in a minority and thus his views would carry little weight with the Board. Moreover, the worker directory may not be properly trained in the management function. Though he is an effective leader, he may not necessarily be an effective manager. If he lacks the qualities of a good manager, he may not be in a position to judiciously decide the short-term effects and long term effects of a new proposal, *e.g.* raising funds to finance a new project. Since the worker-director is the representative of a trade union, he will always take a biased view for the benefit of the workers. He many overlook the interests of the organisation which has certain responsibilities towards the society. The worker director will merely act as the mouthpiece of the union, which is not desirable. But again, as the worker-director is in a minority, his voice will have no effect ; so he may develop inferiority complex or he may be completely suppressed or frustrated and may create nuisance for the company. That is why trade unionists wish to maintain their present status because'then they can act as a check on the management.

Workers' Participation in India

The recognition of worker participation concept found an expression in the second five year plan which envisages establishment of cordial relationship between management and workers, substantial increase in productivity in the interest of management, worker and the country as a whole, better welfare facilities for workers, training and development and education of workers to extend their responsibilities and their role towards management. The various types of participation practiced in India are-(a) Joint Management Council (b) Workers' Committees.

Joint Management Councils

The establishment of Joint Management Councils is the first step towards achieving labour management association. Its function is to bring mutual consultation between employment and workers over important issues effecting industrial relations. Joint Management Councils has been assigned the functions of :

(*a*)   Administration of Welfare measures.

(*b*)   Supervision of safety measures.

(*c*)   Operation of vocational training and apprenticeship training.

(*d*)   Preparation of schedules for working hours and breaks of holidays, and payments of rewards for suggestions.

(*e*)   Matters relating to right of workers to receive information discussion and offer suggestion.

(*f*)   Matters in which workers are expected to be consulted.

(*g*)   Joint Management Councils have not been successful because most enterprises have more than one trade union and there is intense rivalry among them for getting the workers' allegiance to their individual organisations and the main weapon at their command is agitation for more wages and other material benefits to the workers. Moreover, the trade union leaders themselves are not particular interested in making these Joint councils powerful instruments for safeguarding the interests of the workers. They are always afraid that, apart from the genuine possibility of workers members on the councils getting brain washed by the management, inter-union rivalry may lead to union office bearers being labelled as stooges of management and other possible denigrations. Above all, they have a deep-rooted feeling of an inherent conflict between the owners of property, whom the management represents and the workers who want a higher share of the product of the enterprise in which they work. This awareness of conflict and divergence of interests exists not only in regard to the private sector but also in regard to this public sector.

In order to make joint management council successful, it is necessary that the provisions are made for the recognition of one representative of each trade union. The future of joint management councils will be good only if recognition and acceptance both from management and unions are there to their mutual advantage.

Workers' Committees

The Bombay Industrial Relations Act, 1946 made a statutory provision for establishment of joint committees at the plant level to maintain regular channels of communication between employer and employees. The Industrial Disputes Act, 1947 envisages the creation of works committee on each unit consisting of representatives of workmen and employers with more or less equal representation. The act lays down that, "It shall be the duty of the works committee to promote for securing amity and good relations between the employers and workmen and to that end to comment upon matters of their common interest of concern and endeavour to compose any material difference of opinion in respect of such matters." But the scope of works committee is very restricted one. But in some cases these committees are dealing with production problems a role assigned to joint management councils. The difference is that joint management council is a voluntary measure for the prevention of dispute. But it is suggested that we should have either works committees or joint management councils in the same establishment as unnecessary proliferation of the consultative machinery will only add to confusion.

National Commission on Labour (1969) found that works committees have not been effective. The evidence before us, epressed the views that the advisory nature of the recommendations, vagueness regarding their exact scope and functions, inter-union rivalries, union opposition, and reluctance of employers to untilise such media have rendered works committees ineffective. The employers' Associations have attributed the failure of works committees to factors like inter union rivalries, union antipathy and the attitude of members in trying to raise in the committee discussion some extraneous issues. According to the unions, conflict between union jurisdiction and the jurisdiction of the works committees and the unhelpful attitudes of the employers have generally led to their failure."

The effectiveness of works committees will depend on the following factors:

(a) A more responsive attitude on the part of management.

(b) Adequate support from unions ;

(c) Proper appreciation of the scope and functions of works committees;

(d) Whole-hearted implementation of the recommendations of the works committees ; and

(e) Proper co-ordination of the functions of the multiple bipartite institutions at the plant level now in vogue.

Success of Workers' Participation Programmes

The key to success of works committees and joint management councils lies in employees' attitudes. Employers have to treat workers as co-partners in the joint venture for running the industrial unit, while workers have to realise the additional responsibility that is placed upon them as co-partners. In fact, there is an assumed conflict between profits and wages with a consequent feeling on the part of employers that workers are on the other side of the fence and letting them into the secrets of management would only lead to a weakening of their efforts at maximising profits. On their side, workers are inclined to believe that such councils are only a way of weakening their organisation and eventually reducing their share of the net value created by the enterprise.

Moreover, the workers have an uneasy feeling that they do not understand the intricacies of management while the employers are certain that they do not. In turn this makes the employer disinclined to take the joint councils seriously as an instrument for the increase of productivity, while the workers and their leaders are more than inclined to believe that these bodies only constitute an additional machinery in employers' hands for confusing and demoralising them in their legitimate demand for an increase in wages. There is no way in which this dilemma can be resolved without extending co-partnership in management to co-partnership in ownership (purchasing shares of the company), without acquisition by the workers and their leaders of skills in management and its problems.

Co-partnership in ownership by the workers can be achieved by issuing workers the shares in the company out of their provident fund accumulation. There is no reason why these funds should not be treated as the workers' share in industry in both the public and private sector undertakings and the workers given the right of ownership in the concern to that extent. Then instead of getting an interest rate of about 5% to 6%, a worker will become entitled to a share in the profits of the concern in which he is working. His interests will thus get identified with the efficiency and profitability of the concern and he will become a partner with his employer in the fortunes of the concern and the industry.

Apart from copartnership in ownership, the success of workers' participation in management will depend upon the following factors :

1. The scope of the work of joint councils of management should be extended to cover the major subject of the division of product of industry between wages, profits and development funds.

2. One and only one union should be recognised for each industrial unit so that inter-union rivalry will cease to play the disputed role it is now playing in the progress of the industry.

3. Atmosphere of trust should be created on both sides. Unions should feel that management is not side-tracking the effective union through a works committee. Management should also realise that some of their known prerogatives are meant to be parted with. It should do justice with   workers and treat them equal co-partners in progress and prosperity.

55

Work and Work Organisations

Work System

There are multiple inter-linkages among the work system, the group, the individual, the organisation and the society at large. It is necessary to look into the institution of work and examine the nature of inter-linkages between work and some aspects of socio-cultural, psychological and socio-economic areas.

Work is at the core of life. Work means autonomy, it pays off in success and it establishes self-respect or self-worth. Work plays a central role in the life of most of people engaged in productive activities. Work remains very much in the forefront of workers at least at par with family interests and decidedly more prominent than other segments of their lives. What happens at work tends to determine what can and does happen in the areas of family life, education, leisure ?

Work Organisations

Work organisations are open systems and do not operate in isolation. Work organisations operate on a high level of interaction through multiple linkages with other systems. Work organisations consist of semi-autonomous groups each of which is engaged in the performance of a total task. Thus, there is a shift from one-man-one-job principle to one-group-one-task where every group enjoys varying degrees of autonomy so that members can be regulated and coordinated in their own activities.

Almost every day, we see a number of advertisements in the local newspapers with dozens of them appearing for various jobs with different specialisations and for various categories of people. Only 5% of these advertisements describe the work required to be done. The managers of today still do not understand the exact meaning of the term work. Managers have not reached the stage where they may talk about work in an objective term. By work they understand nothing more than the description of the people who do the work. Their normal qualifications are managerial work, technical work, financial and accounting work, skilled and unskilled work, commercial and routine work and so on.

The above terms are really more than the description of the person who does the work rather the state of mind while working on the description of the work. The study of management is not merely the study of man involved but understanding of his character, abilities, motivation, objectives, environment and behaviour. The study should involve the structure of roles, the work content of these roles, the relationship among roles and the manner in which the roles are structured. The work content of the roles should be so adopted to match the environment in which a company operates. It is not enough to study the psychological aspects of work but it is important to understand how work is outlined or covered by organisational policies. As we descend in the organisational hierarchy, we find that every individual is responsible for some sort of work connected with the basic functions and the organisational objective. In other words, work can be objectively, distinguishably described in terms of prescribed and discretionary contents. Thus work can be defined by the occupant's role which he must play to avoid a charge of negligence or insubordination. Discretionary work is composed of all those decisions that we are not only authorised to make but also are held responsible for making.

Work A Part of Man's Life

It is an essential part of man's life. It is that part of his life which gives him status and binds him to society. Each one of us like our work. Whenever we complain of work, it is because of psychological and social conditions of the work rather something in the work. Work is a social activity. Unemployment is a very negative incentive. Unemployment or without work means, man is cut off from the society. To work is to live and, therefore, those who exert their activities of body and mind for the means to live are working. It is difficult to imagine the situation of a man with fortune but without work. Work is a social activity with two main functions of producing the things required by society and binding the individual into a pattern of relationship from which society is built. The motives of working cannot be assigned only to economic needs. Men continue to work even when they have no need for material goods, even when their economic security and that of their children is assured. It is because the rewards they get from work are social, such as respect and admiration from their followmen. Work is an instrument for satisfying ego by gaining power and exerting it over others. Work provides fellowship and social life. Work is a source of status both formal and informal. The information material, social rewards which industry has in its power to grant are really valuable. The importance of work can be judged by an example of a man who works hard in his active life and dies one or two years after retirement due to sheer misery and lack of things to occupy his mind. It has also been observed that people do not like to be retired and tend to take up jobs of lesser designation and with lesser salary. Money is not that important a factor because it has been observed that workers who have been given a job with higher pay, request at times to return to their old job with much lower wages. We know that if workers are treated as human beings, they will behave like that and if they are treated as unwilling, uninterested and slaves they will act likewise. What work means to a worker can be well understood if we talk to an unemployed father or husband. What job means to a worker can be understood only by an unemployed job seeker. The position in the house and control over the children depends to a greater extent on the job. The job controls the life of an individual. The greatest evil of an unemployment is that it makes man seem useless, unwanted and a bad citizen.

Working Environment

The working environment, material and psychological condition of the work, have little or no direct relationship to good morale. In London in a slaughter house pigs were killed and there was a small room in which the internal organs were sorted out and washed prior to utilising them for other purposes. The room was below the ground level with very little illumination and high dampness. The room was norma .y covered with blood, water and the contents of the animal's intestines which smelt collectively and was quite unpleasant to the casual visitor. Under these surroundings six girls were working happily and spending the day. A medical officer who joined this factory was shocked to see the working condition and recommended that the girls be replaced by men and suggested the transfer of girls to other department as it was not possible to carry out an immediate improvement of working conditions. These recommendations aroused a strong protest from the girls and recommendation had to be withdrawn. It was because these girls had formed a happy working group and were so friendly with each other that they were of the opinion that they were engaged in a skilled job under the control of a good supervisor who allowed them to take their own time and method to work and praised them judiciously.

It was also found that in the same factory in an other department which was well illuminated and was in better conditions and where hundreds of other girls worked and grumbled over their job. Needless to say that bad working condition is of utmost importance and influence health and happiness but there is no doubt that good phys:cal conditions may co-exist with bad morale and bad conditions with good morale. Good physical conditions of work may increase morale but they will be insufficient in themselves to create it.

STUDIES ON WORK

There are evidences that workers react negatively to work which is of routine nature and without challenge. Challengeless bureaucratic jobs inhibit the normal development of human personality and leads to poor mental health, apathy and even delusion. It is because of this that workers prefer highly structured work. Workers feeling lack of interest in the job, lack of autonomy and control and lack of challenge in their job, attempt to redirect their potential and energies to other activities of the job. It has been observed that the whole culture may adjust to the job opportunities which calls for little challenge and offers no initiative. Large number of workers have manifested their feelings, attitudes and behaviour that signify unsatisfactory life adjustment in the absence of a meaningful job. Their responses reveal feelings of inadequacy, low self-esteem, anxiety, hostility, dissatisfaction and low personal morale. It has also been observed during the studies that where the technical and social organisation of work offer more freedom or discretion in methods, pace or schedule or frequent inter-action with fellow-workers, then work attachment will be strong and work will be closely integrated with the rest of life.

Michael Maccoby identified the dissatisfied class of workers and termed them as routine workers. Although these are the workers (men or women) who work in the factory only. Their work in the factories is to supplement their earnings. These routine workers are the most dissatisfied with the factory work. These workers are of independent and democratic character who are to fit into a mechanistic and hierarchical structure where he or she is not respected as an equal. These workers have a general complaint about the factory work, *i.e.*, work does not allow them autonomy. They do not have enough time to do a good job and their job does not offer enough to think. They feel the work is too repetitive and fast paced.

Work and Scientific Management

We know from the concept of scientific management that an individual has no choice but to accept the fact that man's specific contribution is always to perform many activities, to integrate, plan, control, measure and judge. The individual operations need to be analysed, studied and improved. But human resources will be utilised productively if a job is being performed out of the operations. It is the job only which integrates the workman's specific qualities. Specific management despite all its efforts could not succeed in solving the problem of managing, worker and work. Scientific management have two blind spots—engineering and philosophical. We must analyse work into its simplest constituent motions. We must also organise it as a series of individual motions as carried out by an individual worker. Scientific management purports to organise human work without an attempt to verify the assumption, that the human being is not a cog in the wheel. There are no two opinions that human beings are not the "cog in the wheel" rather he is associated with feelings, emotions, beliefs, attitudes and ideas beside motivation and morale. The most important thing while engineering the work, is how best that can be performed by human beings. In other words, we have to see how best to organise people to do their work. Scientific management had a wrong belief that people work best if organised like machines, *i.e.* when linked in series. It is not only a wrong belief but it is quite dis-integrating to the concept of human relation. People work well when working either as an individual or as a team. The human beings work in groups and form groups to work. No matter how the group is formed the focus is on work only. Group relation influences the work and the work in turn influences the personal relationship within the group. This means that work must always be organised in such a manner that whatever strength, initiative, responsibility, competence there is in an individual, it becomes a source of strength for the entire group. In other words, group and individual must be brought into harmony in the organisation of work. In developing countries, it is not only technology that has been borrowed but also the associated organisation of work which do not necessarily fit within the prevailing socio-cultural framework. During the last one decade or so there had been a qualitative change in the nature of work force particularly in those industries where skill requirements are high. Most workers in these industries are young, educated and have relatively high level of aspiration than their counterparts of older generation. The young workers look for opportunities to utilise and develop their potentialities. They look for intrinsic factors in their work. The near absence of work culture among practically all levels of employed workforce has been yet another important characteristic of the prevailing socio-cultural reality. Loss of interest in one's work and consequent apathetic responses to diverse situations is not confined to blue collar workforce only but manifestations of these can be seem among all sections and in all sectors including the educational institutions. A search for alternative forms of work organisation has led to the development and use of number of terminologies like humanisation of work, work design, participative system design, etc.

JOB STRUCTURING

Boring jobs are usually associated with mass production, particularly assembly lines which are widely used in the manufacture of motor cars, televisions and other consumer goods. The lack of job satisfaction for these workers has been associated with technical problems such as the poor quality of work on 'Monday morning' and behavioural problem such as labour turnover, absenteeism and strikes. These problems are not confined to those working on assembly lines. There are many jobs in industry that are undertaken in poor working conditions, more repetitive and for which rewards are lower. The job designer has an important role to play in the organisation. Job design is concerned with the specification of the contents, methods and relationships of jobs in order to satisfy technological and organisational requirements, as well as the social and personal requirements of the job holder. The physiological and physical environment and also the organisational, social and personal aspects affect the prediction of man's responses to both the physical environmentand the physiological demands of work have been widely studied and data is available to aid the job designer.

Vroom (1964) concluded that there is a positive, although weak, association between job satisfaction and job performance. He suggests that individuals are satisfied with their jobs so long as the jobs provide what they desire and that they perform effectively to the extent that effective performance satisfies their own goals. Locke 1970, however, takes the view that successful performance is a possible cause of job satisfaction rather than *vice versa*.

Job Satisfaction and Motivation

Workers' attitudes and their job performance have not appeared to be strongly associated. Their absenteeism, turnover and industrial action does seem to be related to their level of job satisfaction.

Labour turnover is likely to be generally lower during periods of economic depression since alternatives may be less attractive than current employment, but during periods of increasing economic activity, when firms expand their labour force, much of the labour turnover occurs among fairly recent employees who fail to settle into the new environment.

Voluntary absenteeism is likely to occur when the individuals perceives the consequences expected from not working to be more attractive than those expected from working. It is possible that prolonged absence from work leads to feelings of isolation which in turn may be the cause of some job dissatisfaction. Workers' attitudes during periods of industrial unrest are considerably different from those expressed during period of normal working where job satisfaction is low there is more likelihood of strike action.

Alienation from Work

Feelings of alienation from work are increasing. Workers experiencing feeling of powerlessness due to lack of control, meaninglessness due to lack of responsibility, and isolation from the organisation and its goals. These factors are said to contribute to a sense of self-estrangement. Blaunes (1964), based on a study of the attitude of workers in various industries, reported that alienation from work appeared to be greater in some situations than others. In continuous process work, there are periods of routine activity when the worker may be detached and find the work monotonous, but these periods are broken by periods of intense activity when emergency breakdowns require the worker's total involvement. Unskilled work on the assembly has a repetitive cycle, rarely complicated by problems and difficulties which might challenge the worker's capacities. The impact of technology is greatest upon the powerlessness dimension of alienation since the character of the machine system largely determines the degree of control the factory employee exerts over his socio-technical environment and the range and limitations of his freedom in the work situation, the function of the individual worker is highly standardised, feelings of meaninglessness would be high.

The level of self-estrangement is strongly and positively related to the following three factors :

(i) Lack of control over the immediate work process.

(ii) The performance of specified work roles due to advanced specialisation.

(iii) Lack of opportunities for promotion.

Wedderburn and Crompton (1972) studied the effect of technology upon attitude and behaviour and concluded that there was a positive association between satisfaction with work and the degree of variety, autonomy and discretion that it offered.

Job Design Practice

No evidence was found to suggest that systematic methods for the design of jobs existed or that alternative designs were evaluated adequately.

The following criteria were seen to guide the job designer :

(i) Economic considerations.

(ii) Technical considerations relating to the process requirements.

(iii) The limitations imposed by time and space requirements.

(iv) The availability of labour of the different skills required.

(v) Union management agreements relating to aspects such as manning and differential.

(vi) Custom in the plant.

Introducing Job Changes

It was suggested that withdrawal from work in the form of high absenteeism, labour turnover and restricted output is more likely to occur where jobs are deficient in certain desired attributes.

Many organizations through redesigning jobs sought improved operating effeciency by reducing costs, improving quality and productivity, reducing specialisation and increasing flexibility. Secondary reasons related to expected behavioural outcomes such as reduced absenteeism and labour turnover as well as improved labour relations.

Some of the techniques which have been used to improve the motivational content of job along with their limitations.

Work Rotation

This rotation may be on either an obligatory or a voluntry basis. As well as increasing job interest due to changes in the environment, skill requirements and job content, it is claimed to have the advantage from the management's point of view, of increasing the operators skills and, therefore, their flexibility.

Work rotation is widely used in industry. It is particularly appropriate where work performed by a group of similar status employees is perceived to vary in difficulty and wage differences do not compensate.

The output of workers who worked on various jobs for varying time periods. They concluded that by rotating the job for assignments of operators and limiting the amount of time spent on any one job, their overall efficiency was higher than if they remained in one job.

Worker resistance may arise where work rotation is compulsory since constant changes in the pattern of social interaction may interfere with group development.

Work rotation has limited value with regard to improving the motivational content of jobs since the object of such changes is generally to increase the variety present in jobs. The number of jobs at which the worker is capable of becoming competent is limited and, therefore, the worker often only rotates between jobs with one section of a complete process.

Job Enlargement

This is the expansion of the job from a central task to include one or more additional and related tasks which one usually of a similar type to the original. Job enlargement usually involves a lengthening of the time cycle required to complete a unit of the operation and, therefore, a reduction in the degree of specialisation. The same effect can result if more frequent changes one introduced in the products being manufactured. Benefits are said to include reduced employee fatigue, reduced feelings of boredom and broadened work skills. Critics of the technique argue that the extended job is usually only composed of multiplies of the original task and that nothing is added that will contribute to increase employee job satisfaction and motivation.

Job enlargement in this type of case reduces the numbers of workers involved and, therefore, lowers the task interdependence. The work is also placed, job enlargement by increasing the work cycle will reduce the pacing effect.

In many cases where products are simply enlargement of assembly line work has led to either individual or small groups being established to carryout the process.

There are limitations in the application of job enlargement. In situations where there is considerable task interdependence and the items being assembled are complex. It is difficult to enlarge existing jobs. The effect of repetitions work upon the job attitude and behaviour of workers has produced conflicting evidence.

Workers react [differently to repetitive work. Individual differences in intelligence, susceptibility to monotony, and age as well as past experiences and future expectations affect the worker's reaction to repetitive work.

Job Enrichment

Technique is generally applied in order to increase the motivation content of jobs by giving employees greater opportunity for achievement and recognition, horizontal job enrichment and vertical job enrichment.

Enrichment involves changes to the immediate work to enable the employee to exercise more control over his working speed and use more of his skills. A job enriched horizontally may, therefore, include greater autonomy.

Vertical job enrichment is aimed at increasing employees involvement in the organisation providing increased participation in company policys making, increased job responsibility, increased involvement in the process and increased opportunity for training and advancement.

Both horizontal and vertical job enrichment have been applied to white and blue collar jobs. It usually has meant the delegation to the worker of some function generally thought to be managerial.

In such cases the union representatives may perceive membership of problem-solving teams as one of their union functions since they are elected workers representatives.

Increasing jobs responsibility may diminish job satisfaction unless the worker feels that he is capable of carrying out the duties involved.

Job Restructuring

Job designer must consider many inter-connected factors. He should aim for operational efficiency, he should consider the cost of undesired behaviour which may result from extreme task specialisation. In considering the relationship between the individual and his own job, the designer must provide a satisfactory relationship between the individuals and his work group and its total task. There is a cause and effect relationship between these job characteristics and the desired attributes.

56

Organisation

Introduction

The existence of a sound organisation is an essential prerequisite of efficiency. A systematically planned flexible organisation matching the needs of business will save managerial time, permit concentration on the key problems, save costs and help in enhancing productivity. Organisation embraces the whole complex of relationship between managers and subordinate employees. The purpose of the organisation should be to enable managers and work people to carry out their responsibility efficiently that is with greatest concentration effort and the least waste of time and resources. According to Frederic Hooper organisation is a framework of responsibilities through and by means of which a concern may do its Job. Brech refers to organisation as a structure of responsibility and formal relationship existing amongst the personnel. The concept of an organisation is as old as mankind and has undergone tremendous changes with the growth of civilization.

Organisation is generally viewed as a group of persons formed to seek certain goals. An individual is unable to fulfil his needs and desires alone because he lacks strength, ability, time and potentials. So he seeks the co-operation of other persons in achieving his goals. This is not a new phenomenon ; it is a universal truth. People have always formed organisations to pool their efforts for the achievement of their common goals. The Emperors of China used orgainsations a thousand years ago to construct great irrigation systems. And the first Popes created a universal church to serve a world religion. Modern Society, however, has more organisations, these fulfilling a greater variety of societal and personal needs*. An organisation comes into existence when there are a number of persons in communication and relationship to each other and are willing to contribute towards a common endeavour**. Thus to achieve a common

---

*Etzoni Amiti : Modern Organisation, Prentice Hall. Inc. Eaglewood Cliff.

**Barnard Chester I. The functions of Executive Harvard University Press, Cambridge.

purpose, people from some groups pool their efforts by defining and dividing the various activities and responsibility and authority. The characteristics of organisations are :

(*i*) Communication

(*ii*) Co-operative Efforts.

(*iii*) Common Objective.

(*iv*) Rules and regulations.

The persons who form the organisation must be in a position to communicate with each other. They must also be willing to co-

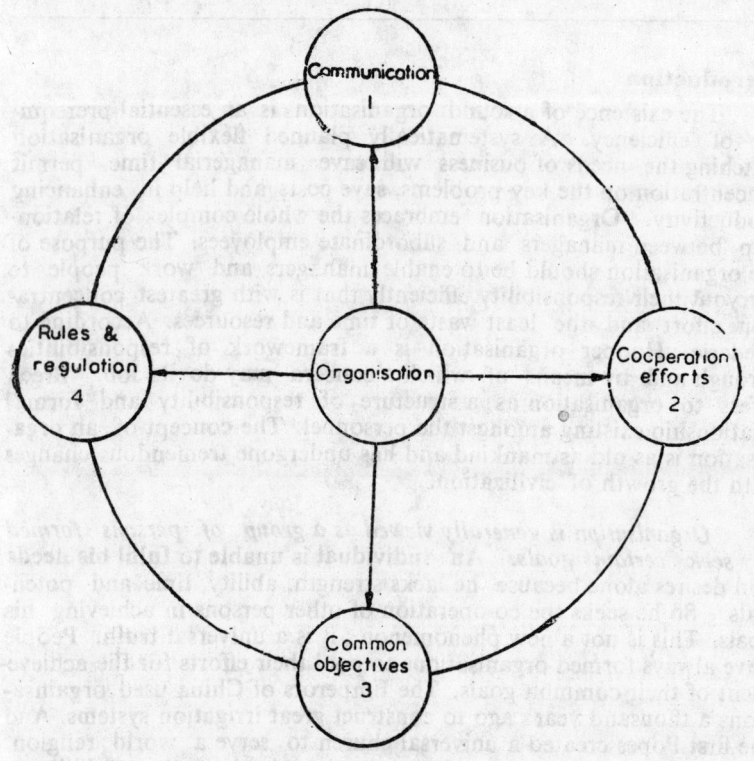

Characteristics of organisation

Fig. 56.1 .

operate with each other for achievement of their goals or objectives. The objectives must be common for which the organisation comes into existence. Lastly, rules and regulations lay down the formal structure of the organisation. They define the authority and responsibility relationship among various individuals in the organisation.

Some people define organisation in a very narrow sense. Organisation is no more than the framework within which the reponsibilities of management of an enterprise are discharged*. According to this definition, organisation sets up the scope of activities of the organisation by laying down the structure. If organisation is recognised as a structure, it will be viewed something static. But an organisation is a dynamic entity consisting of individuals, means, objectives and relationship among the individuals. In this sense "relationship" or formal organisation is only a part of the organisation.

Human organisations are organic in the sense that various parts are interdependent. Some of the basic pillars of the organisation are individual, groups and the work. Work place is one of the most important institutions for providing human satisfactions *Organisation has also been called a system of communications,a means of problem solving, and a means of facilitating decision-making.* Organisations have been viewed as social systems and as systems ofa group of interacting variables.

For practical purposes organisation can be simply as a process of :

(1) Determining what must be done if a given aim is to be achieved.

(2) Dividing the necessary activities into segments small enough to be performed by one person, and

(3) Providing means of co-ordination so that there is no wasted effort and the members of the organisation do not get in each others way.

Organisation cannot be divorced from the idea of purpose. It is not an end in itself but a means to an end. Organisation is necessary whenever two or more people must combine their efforts towards the same and even different task is shortlived such as moving a large stone. "You take that side and I shall take this side" is a primitive division of work, and "We shall lift when I count three" a primitive form of co-ordination Edger. H. Schein has defined organisation as *"An organisation is the rational co-ordination of the activities of a number of people for the achievement of some common explicit purpose or goal, through division of labour and function and through a hierarchy of authority and responsibility.*

History of Organisation

Historical records indicate ancient forms of organisation analysis and planning by Priests in Mesopotamia, Court philosophers in China, Roman Public contract administrators and various military organisations. Undoubtedly many useful forms of ogranisation were

---

*Brech E.F.L. The Principles and Practice of Management.

developed over the centuries. It is obvious that such monuments of the ancient world as the Pyramids, the Hanging Gardens of Babylon and the Roman aqueducts could not have been constructed without division of labour and co-ordination of work. Undoubtedly some principles or rather rules of thumb were arrived at as a result of work of this kind.

With the development of the study of economics organisation was treated more systematically. For example, the great economist Alfred Marshall classed organisation as one of the basic factors in production, which had formely been defined as land, labour and capital. And Marshall was right because land, labour and capital cannot be efficiently utilized in production except on every small scale without some kind of organisation. Knowledge of organisation has been advanced through speculative thinking, through controlled exepriments, and through experience over long periods. Modern organisations began in the early 1900's when F.W. Taylor developed concept of "Scientific Management" which called for the careful planning of the human/physical/financial systems as well as for the co-ordination and control of it. Shortly thereafter, Henry Fayol concluded that all work in industry can be divided into following six groupings :

(a) Technical

(b) Commercial

(c) Financial

(d) Security

(e) Accounting

(f) Managerial.

Immediately prior to World War II, Robert W. Porter a management consultant published a thesis entitled "Design for Industrial Co-ordination" Porter focussed attention on the fact that the work at every level of the organisation can be divided into three basic functions ;

(a) Planning

(b) Implementation (doing)

(c) Inspecting (controlling)

By grouping all organisational tasks within these three functions of each level, the organisational design will be most conducive to the establishment of natural checks and balances. These are essential to the co-ordination of every system.

World War II and the expansion of corporate organisations in the 1950's created unprecedental problems in organisational designs and people involved in management and administration a teachers and practitioners have thought and written about organisation for many years.

The Organisation's objectives and tasks

The organisation has two basic objectives—what it is for and to achieve what enables it to continue. First one can be regarded as objective, the other pererequisite, or constraint. But both the objectives are equally important as at time one is more necessary or more urgent than the other but the status of these two basic objectives is equal one depends on the other. But what ever the intention of the organisation, whatever its purpose if it is to survive and be effective its most establish objective and constraints and also understand their interaction. In other words the objectives must be recognised their required interation decided and then the task objectives the follow can be defined using an understanding of the different concepts of objective to identify the organisational consequencies as given in Fig. 56.1

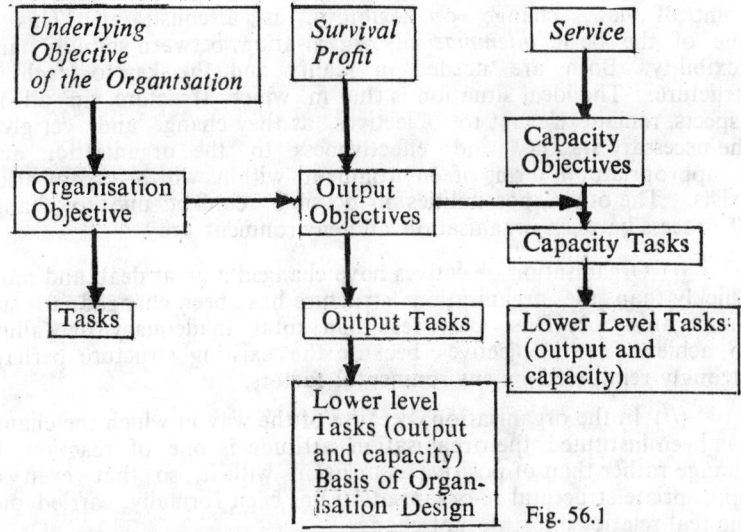

Fig. 56.1

The organisation exist in environments—physical, economic social, cultural which diretly or indirectly affect its objectives either as determinants or constraints. The way in which an organisation, through its executive system, positively attempts to achieve its objectives thus requires it to maintain effective relationships with the significant parts of its environment, and these relationships then supply it with the resources and facilities that enable it to continue. The processes where by objectives are formulated are not the same as the process where by objectives are met or decisions are implemented. In formulating and deciding objectives these are many possible processes ranging from unilateral dictatorial to the equalitarian democracy. What is appropriate and effective depends on many factors. The process for meeting objectives on the other hand is one whicι. ᵎnvolves the identification, development and use of resources and carrying out of relevant activities. The process in any form of organisation appears to involve a hierarchical structure of some sort, whatever the objectives or ways of determining these objectives, whether the structure is formal or informal, explicit or implicit and whether the style of behaviour in the structure is authoritarian or permissive.

Conflict Between Organisation and Environment

The process of meeting objectives of course is not unconnected with the process for deciding objectives. But in order to achieve the understanding and acceptance of the objectives of the organisation, it may be necessary that they should effectively be involved in the process for determining these objectives. The organisational stability from one point of view is necessary and helpful, from another point of view, change or flexibility, is a constraint. This is one of the basic *dilemmas* of organisation, between stability and flexibility. Both are needed in reality and the key to each is structure. The ideal situation is that in which structure, in all its aspects, remain relevant to objectives, as they change, and yet gives the necessary stability and effectiveness to the organisation and is appropriate in terms of environment within which organisatiqn exists. The other possibilities of possible conflict due to change of process between organisation and environment are :

(*i*) Organisation objectives have changed a great deal, and more quickly than the organisation structure has been changed to suit the organisaᵗion. This may result in total inadequacy or failure to achieve new objectives because the existing structure perhaps strongly reinforced by environmental factors.

(*ii*) In the organisation, because of the way in which the change has been instituted, the organisation attitude is one of reaction to change rather than of positive association with it, so that even of appropriate structural re-organisation has been formally carried out, the real relationships do not change.

(*iii*) Organisation where organisation has changed to become more relevant to changing objectives, but this may put it and the individuals within it into a state of increasing stresses in relationship with the environment. This is more common in organisations where the rate of social/cultural change in industrial organisations may be much higher than that in general.

(*iv*) The organisation attempt to diversify has not *clicked* because the new product or technology fails to be effective because it forced into an existing structure and style that may be very effective for other activities but not for this organisation.

Keeping in view the above factors in view it is felt that to design an organisation structure from the basis of organisation's objectives, the organisation's tasks must be identified and then the structure devised to enable those task to be carried out.

It should be recognised that if an organisation is to remain viable by design rather than by accident its structure, polices, practices operations style must evolve to meet pressures on it.

Delegation

The other important concept on the study of an organisation is Authority and delegation. Elbourne defines authority as the formal right, conferred by the organisation to require action of others. It is observed that the concepts of authority and responsibility are implicit in any form of line organisation which after all remains the back bone of an industrial enterprise. It is of no use to impose responsibilities upon an individual unless they are given the authority to issue instructions and to enforce their performance. The authority to initiate action implies the power to delegate. A real problem before a manager is when and what to delegate.

A manager's tasks is to reduce repetitive jobs to a routine and then delegate them to his subordinates. The efficient delegation requires some of the highest qualities of judgment demanded of a manager. Delegation is a practical means of achieving efficiency in the interests of the organisation as a whole. It is an intrinsic and valuable means of developing the laterit qualities of employees. Once a manager has delegated he should not interfere except for some very important reasons.

Authority Is The Degree Of Discretion

In any organisation the authority is the degree of discretion conferred on its managers to use their judgement. The immediate question which comes in mind, yet unanswered, is that how much and what kind of authority should be dispersed or concentrated by a manager. Decentralisation is a fundamental aspect of delegation. The absolute centralisation implies that there exists no organisational structure. There can be no absolute decentralisation for if managers delegate all their authority, their statute as a manager will cease. Therefore centralisation and decentralisation are therefore tendencies. Centralisation refers to withholding or delegating authority or concentration of decision-making.

Delegation of authority is more a kind to decentralisation. It specifies which decisions are to push down into the organisational structure. The decentralisation affects all areas of management and is an essential component of a managerial system. The primary object of delegation is to make organisation possible. *French management consultant analysed subordinate superior relationship and developed a formula which identifies following three types of relationship :

Direct single relationship   2. Direct group relationships
3. Cross relationships.

---

* V.A. Graicunas, "Relationship in Organisation" Bulletin of Internal Management Institute.

Graicunas Formula

Graicunas develop the following mathematical formula based on the geometric increase in complexities of managing with the increase in number of subordinates.

Number of interactions $= n[2^{n-1}+(n-1)]$

n= number of subordinates.

Let us know the number of relations increases at the number of subordinates,increases

Thus executives must think twice before they increase the number of subordinates. This formula does not consider the frequency of relationship which is of more importance to a manager than total possible number. There is always a limit of the number of persons managers can supervise efficiently. Once it increases to authority must be delegated to subordinates. The degree of decentralisation is greater when following things happen :

Number of Subordinates	Number of Relationships
1	1
2	6
3	18
4	44
5	100
6	222
7	440

(a) The number of decisions down the management hierarchy are more.

(b) Less checking required on decision and lower people to be consulted.

(c) More important decisions are made down the line.

According to Andrew Cernefir "when a man realise he can call some one to help him to do a job better than he can do it alone, he has taken a big step in life".

Delegation Is A Cement That Holds The Organisation

Delegation let the subordinate do the work for which they have been hired. In delegation certain executive's functions, responsibility and authority are released and committed to designate subordinate position. Delegation give subordinate a sense of being his own boss. Delegation eliminates monotory and boredom. Delegation on the face to face level is called general supervision and at the organisation level it is called decentralisation. Delegation substitute for decision-making by the boss. Delegation assumes that people will work harder if given freedom to make their decisions and it relies on internal motivation. Delegation is a cement that holds a formal organisation together resulting into a relationship between organisation members. Delegation gives rise to boss-subordinate relationship. The top manager in any organisation is delegated the responsibility of running the entire business. As no manager can do all the work himself, he delegates a certain amount of duties to his immediate subordinate. He in turn redelegates to his subordinates and this process of delegation and redelegation continues until all the work has been assigned. Whenever a manager delegates work, he introduces a new link in the chain of command.

Delegation Is A Process

Delegation is the process by which managers again activities to the individual. Whenever a manager delegates to his subordinates following actions are implied:

(i) **Manager assign duties.** Manager delegates what work the subordinate must do.

(ii) **Manager grants authority.** Manager transfers certain right such as right to spend money, to direct the work of other people etc.

(iii) **Manager creates an obligation.** After accepting the assignment the subordinate takes an obligation to complete the job.

The fact is that no delegation is complete without clear understanding of duties, authority and obligation. The attributes of delegation are similar to a three legged table where each leg is responsible to support the table and not two legs can stand alone.

Thus to delegate means to confer authority from one manager to organisational unit to another to accomplish particular assignments. Delegation is management does not means to give away authority, rather delegating manager retains the overall authority. Delegating does not mean the permanent release from these obligations rather granting the

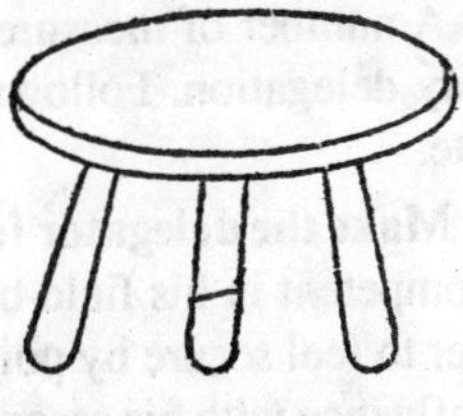

(If one feature of delegation is inadequate the executive is in trouble)

Fig. 56.2

approval to operate within the prescribed areas. Thus delegation has dual characteristic. The subordinates receives authority from the superior and still the superior retains original authority.

A manager must delegate because of three reasons :

1. He is in-charge of more work than he personally can do.

2. Delegating authority is a step towards developing subordinates.

3. Subordinates may be able to carry on if need arises. In other words, a manager is one in lieu of individuals, he had a predecessor and sooner or later will have a successor. For any group action to be effective delegation will be there.

Authority is delegated as established in the structure of organisation. Responsibility cannot be delegated assigned. Jobs, duties and tasks can be assigned but only authority can be delegated. Delegation helps an employee to learn their jobs better and also give them sense of accomplished, a feeling of belonging and boost up their moral.

Why Manager don't Delegate

Delegation is a simple and natural process for everyone working in an organisation but still difficulties do arise in practice. Failure to delegate is a common problem. There are numerous reasons for not delegating. A few of them are :

1. Tendency of the human being to do things himself.
2. Lack of assuming the managerial role when promoted to managerial ranks.
3. The fear of being exposed.
4. The unconscious acceptance of indispensable man theory.
5. The desire to dominate.
6. An unwillingness to accept calculated risks.
7. Attitude toward the subordinates.

How to Pursue a Manager to Delegate

A number of measure can be followed to alleviate the problem of authority delegation. Following points are listed in getting a manager to delegate.

Make the delegator feel secure. The non-delegator is a hard worker fully competent in his field but he feels insecure in his job. To help such manager to feel secure by pointing out clearly his job. Help him to acquire more influence with his associates by granting him better designation or an increase in pay.

Realise the need of delegation. A manager's need is to multiply himself. It is no good to lead a band and also to play all the instruments. The alternative is to acquire people, train and develop them and allow them to contribute in full.

Tie-in with intelligent planning. To delegate without knowing and keeping in mind objective leads to choose. The very act of goal setting aids authority delegation as it brings out what authority is required to hieve the agreed goals.

Encourage a deep belief in delegation. Manager must view deleg tion as the way to develop his subordinates to build a real management team.

Establish a climate free from fear and frustration. The manager must have a feeling of confidence that delegation of authority will reward and not penalise him. It must represent an opportunity for growth and not the certainty of bawled out.

Determine Decisions and Tasks to be Delegated

Delegator may list all the various types of decision and tasks he performs and then rate each one in terms of :

(i) Relative importance to the organisation.

(ii) Time required to perform.

The tasks and decisions which are of less importance and more time consuming may be delegated.

Delegate of Authority for Whole Job

Quite often the delgated work constitutes what the delegate him self does not want to do it and the manager uses delegation of authority to get rid of unpleasant tasks.

Give Assistance to Delegates

The delegating manager should not assume the role of unlooker. Delegates require some assistance and commonly goes to the delegator for help. so the delegator must make himself available for discussion, counsels the delegates tends encouragement and maintains a co-ordial relationship with the delegates.

The idea of keeping duties, authority, and obligation being equal may bring us more mischievous than good. A manager cannot delegate obligation because if a manger in higher echelons of an organisation would have great influence and yet not accountable for the results. To allow an executive to evade his obligation means allowing some one to assume the duty thus breaking the single chain of command and there will be no way to know who was accountable.

Duties

It can always be described in the following tow ways :

(i) It terms of function-Delegation is a process by which we assign activities to individuals.

(ii) To describe duties in terms of the results we want to achieve.

Thus an individual's duties are clear only when he knows what activities he must undertake and what objectives he must fulfill.

Authority

Authority may be defined as the power to command and the ability to impose sanctions, grant rewards and to exact compliance to the authoritative source. There are various types of authority such as :

(1) Formal

(2) Acceptance

(3) Competence.

The formal authority is tope down and is imposed upon others. The acceptance authority rests with the subordinate who accept the authority of management. It is self-imposed. The competence authority is generated by personal competence of the individual. It is a derivative leader personality characteristics, job position, knowledge and technical skills.

Assigning of authority is not simple. Most of the executives experience difficulty in delegating. The administrative authority consists of the right to act for the organisation in a specified areas. It is indeed important to know how to use authority. The way authority is used gives either resultant or acceptance. The acceptance of authority by the subordinates is facilitated when the superior's right to a job his behaviour on the job and the demands he makes on subordinates are accepted as legitimate and genuine.

When an individual joins the organisation he is expected to take orders from some competent officer designated by the organisation. The individual acts as an official representative of the enterprise. Such socially accepted rights constitute formal authority. Managers rely on authority, but authority has undergone two marked changes with the passage of time as given below :

(a) Marked decline in authority which employer exercise these days.

(b) The importance of motivational techniques have increased.

Authority is an essential element in any modern enterprise but it is not similar to unlimited power. No manager can decide and enforce an activity which is beyond the capacity of the subordinate to perform. Authority must also be in line with the accepted plans of an enterprise. Normally an objective cannot be ignored or a policy modified simple because a manager to do so.

Responsibility

Another important concept concerning organising is the concept of responsibility. Responsibility is the obligation of an individual to carry out assigned activities to the best of the ability. According to some managers the responsibility has two phases :

(*i*)   The obligation to carry out the assigned activities to achieve results.

(*ii*)   To account to a superior for the defence of success in completing the desired task.

We know that each member of the organisation has an obligation to something. When an execufive assumes obligation he becomes accountable to the executives who had assigned the responsibility. Authority is delegated from a higher line executive. The amount of responsibility for which the subordinate will be held accountable determines the amount of authority delegated. The Fig. 56.3 gives at glance the importance of Authority and responsibility. The earning authority from below depends upon proper **combination of technical, managerial and human relation skills. Each executive must work out the emphasis that will work best for him.**

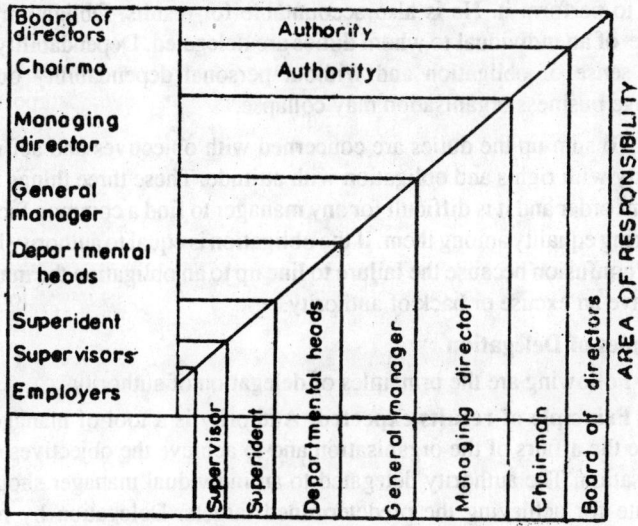

Fig. 56.3 .

The obligation placed upon should be clearly spelled out precisely. following are the terms commonly used to specify responsibilities :

1.   **Confirms.** The individual is expected to carry out duties as specified in specification and various instructions.

2.   **Adviser.** The individual is expected to submit recommendations and advise as specified matters.

3.   **Serves.** The individual is expected to provide resources, personnel or services to various parties.

4.   **Inspect.** the individual is expected to inspect products, resources, services equipment in conformance with set standards.

5. **Co-ordinate with.** The individual is expected to work with and through people of the organisation for informational or consultative purpose.

Appraises. An individual is suppose to make-periodic evaluation of personnel and the procedure of the organisation.

Each individual of the organisation is doing these responsibilities in satisfying their needs and the need of the organisation.

Obligation

The third important feature of delegation is obligation. It is the moral compulsion or urge felt by a subordinate to accomplish his assigned duties. It has been observed that individual respond more to social pressure than to the obligation to perform his particular job in the organisation he is morally bound to perform it. He is also accountable for results. Obligation is an attitude of an individual to whom duties are delegated. Dependability rests in the sense of obligation and without personal dependability our co-operative business organisation may collapse.

To sum up the duties are concerned with objectives and activities, authority with rights and obligation with attitude. These three things are of different order and it is difficult for any manager to find a common factor of measuring equality among them. If the obligation is equal to authority it may lead to confusion because the failure to line up to an obligation the manager may have an excuse or back of authority.

Principles of Delegation

Following are the principles of delegation of authority.

Principle of results expected. Authority is a tool of managers to manage the affairs of the organisation and to achieve the objectives of the organisation. The authority delegated to an individual manager should be adequate for achieving the predetermined targets. Delegation by results expected implies that targets have been set, plans and sub-plans are designed, communicated and understood.

Principles of Functional Delegation

This principle is composed of two parts :

Principle of delegation
Principle of departmentation.

To develop departmentation the activities must be grouped and the manager should be given authority to co-ordinate the activities.

Scalar Principle

Subordinate should have a clear understanding of who delegates authority to him and to whom he should report.

Principle of Unity of Command. The conflicts and problems are less when an individual is reporting to a single superior. This principle is useful in clarifying the authority responsibility relationships. It is also important that several managers do not assign duties to only one employee who would in turn have the obligation to each of the several bosses.

Success of Delegation

Delegation is an act of managing. The failures of a manager is mainly due to poor delegation. The executives must create an healthy atmosphere of delegation. Following are the factors which determine the success of delegation.

Receptiveness. Decision-making process involves some discretion which means that subordinate decision cannot be exactly identical to his superior's decision. The underlying principle of this variation is that Delegation is an art. Therefore, delegation of authority is a willingness to give the subordinates judgement to a chance.

Willingness to let Subordinates make Mistakes

Everyone makes mistakes, a subordinate should also be allowed to make mistakes, the cost of these mistakes must be charged to investment in training and development of personnel.

Willingness to Trust

Superiors often do not delegate with the thought the subordinates are not enough trained and seasoned and do not appreciate all the facts bearing on a situation. This may be true but it is the responsibility of the executives to train and develop its subordinates.

Executives Attitude to Delegation

Delegation of authority is more a matter of directing than issuance of orders. The delegation is viewed by certain managers as unwanted thing and they dislike to delegate. The effects of this type attitude are serious. These type of managers are not able to develop their successors and the subordinates also refuse to work for such managers. Subordinates learn to manage by managing. The rigid concept of delegation tends to limit the size of the enterprise and restrict its growth. Managers will make little use of their subordinates' managers. The span of management places limits on what one person can do. The span binds the organisation and restricted the growth and development in iron gates. The other class of executives who have positive attitude towards delegating authority. In this type of climate the subordinates are encouraged to accept responsibility. The superiors want them to become self-starters, self-excited and to encourage them to undertake higher responsibilities. The positive attitude executive open up the two way channel of communication. Superiors feel happy to associate themselves with subordinates and permit subordinates to pick their brains. Positive attitude permits great patience. It takes long time to acquire good judgement and develop leadership quality. It takes patience to put up with mistakes of the subordinates in a good sense.

57
Organisation Theories

Introduction

Sociologist Max Weber (1946, 1947) discuss the paralles between mechanization and organisation. Max Weber was concerned to understand the process of organization, which takes different forms in different contexts and in different epochs, as part of a wider social process. Thus, the bureaucratic form of organization was seen as but a manifestation of a more general process of rationalization within society as a whole, emphasizing the importance of means-ends relations. He believed that in order to understand the social world it was necessary to develop clear-cut concepts against which one could compare empirical reality. He advocated using the concept of bureaucracy as an ideal type to capture a particular form of organization—that based on the idea of a machine—with a view to understanding the extent to which a society is bureaucratized. Weber was skeptical about the merits of bureaucracy, and in no way intended the concept to be used in this way.

Bureaucratic form of Organisation

Stinchombe (1965) presents important insights on how the bureaucratic mode of organization emerges along with the mechanization of industry and the industrial revolution. George (1972) presents an extensive analysis of the history of management thought from prehistoric times, including good discussion of the writers who set the basis for classical management theory and scientific management. Adam Smith (1776) and Charles Babbage (1832) are worth reading as classic texts of the industrial revolution. The works of the classical management theorists, those of Fayor (1949), Moony and Reiley (1931), and Gulick and Urwick (1937) have been among the most influential. Each illustrates how classical management theory is essentially a theory of machine design. More modern works such as Koontz and O'Donnell (1955) illustrate how these ideas have been carried forward to the present day and reinterpreted within the context of MBO, MIS, PPBS, and the like. Peter Drucker's (1954) work of MBO, which is much more participative than most treatments, is also worth consulting. The principles of scientific management are set out by Taylor (1911). Important insights on the nature of the man and his ideas can be obtained by reading the biographies by Copley (1923) and Kakar (1970).

Contigency Approach

The important works are by Burns and Stalker (1961), Woodward (1965), and Lawrence and Lorsch (1967a, 1967b). Miles (1980) presents an excellent overview of the implications of contingency theory for organizational choice. The use of contingency theory for organizational development is very popular and takes various forms. Bennis (1966) and Levinson (1972) present good illustrations of how the general concept of organizational health underlies the theory and practice of organizational development.

Organisations are highly complex. Most managers think of an organisation in terms of its formal structure. If asked to describe it, they will only talk about the different departments, and about the work that they do. They may talk more abstractly about responsibilities and authority.

The work of some of the classical writers is not by itself sufficient for an understanding of organisation, its approach and limitations. It is concerned with the static approach, paying to little attention to the many interactions that take place between different parts of an organisation, It emphasises similarities without giving sufficient attention to the diversity of problems in different types of organisations.

Human Relations School

The human relations school provides a different approach to organisational problems from that of the classical writers, but it also has its limitations. Major limitation is that although we have learnt a lot from people in organisations there is still much more that we do not understand about human behaviour. System approach to organisation is another important contributing behaviour. The word system can be defined as a regularly interacting or interdependent group of items forming a unified whole. Under the heading of systems it is useful to distinguish, the main ways of looking at organisations. The first is one that has come to be adopted by many social scientists and can be called as the social systems approach. It is concerned with interactions between the different aspects of the organisation-people ; technology formal structure and environment. The secondly looking at the organisation in terms of the information that is needed for decision making and co-ordination, tracing the flow of information through decision centre. It can be called as the "Information Systems" approach and has its origins in operational research and Electronic Data Processing.

The systems approach in all its forms has also contributed to our understanding of how organisations work. It's chief merit is that is focusses attention on the inter-relation ships between different aspects of an organisation. The distinguishing feature of Modern Organisation is the attempt to make the study of organisational behaviour drawn upon the ideas and research methods of field such as Psychology, Sociology, Anthropology. Applied Mathematics, Operation research and Economics. It has become a field for scientific study of how organisations and people in organisations do and behave.

At present organisation can not use old systems and has to use human system with a unique character. The new system demands a special culture values, information systems and some meaningful work. At present in industrial sector we have more educated trained and skilled workforce th.n ever before, with the result these workers have an increased political for self direction and self control. In o:der to understand the development of Modern Organisation Theories let us study the history and important contribution of leading organisational scientists drawn from different disciplines.

Trends in Organisational Theory

Organisational theories have certainly changed since the early part of this century. F.W. Taylor looked upon the worker as something analogous to a machine whose output could be measured mechanically and could be individually stimulated. Now, with economic and political changes which have occured have led to the formation of very large trading units. These changes have been major factors behind the tendency towards increased size of industrial concerns. Increase in size has in turn demanded either sophistication of the control mechanism or smaller, controllable units. Organisational theory and practice have to conform to this development, which appears to be a progressive one. Similarly Technological advances are gaining momentum and have radical repercussions on the nature of business operations and on society as a whole. The shifting of consumer demand to new products imposes a great flexibility upon business management which causes the ueed for streamlined channels of communication and constant changes in the location of decision making in a large business. The adoption of organization to meet the inevitable changes brought about by increased automation and computerisation is therefore one of the most difficult problems which managers have to solve. Because of technological advance in society we have become a dynamic industrial society with a higher level of edueation and higher standard of living higher than ever before. This change has a pronounced of effect on much of the work-force utilised by organisations. At present, in the industrial sector we have a more educated trained and skilled work-force than a decade age. The result is that these workers have an increased potential for self-direction and self-control. The workers of today have more potential for self-direction and self-control consistent with these changes in maturity. In the coming ten years the people working in industries will have their needs shifted from physiological ones to comforts and ego needs, and the management control will no longer merely depend upon the satisfaction of basic needs of industrial people. Management practice should now be

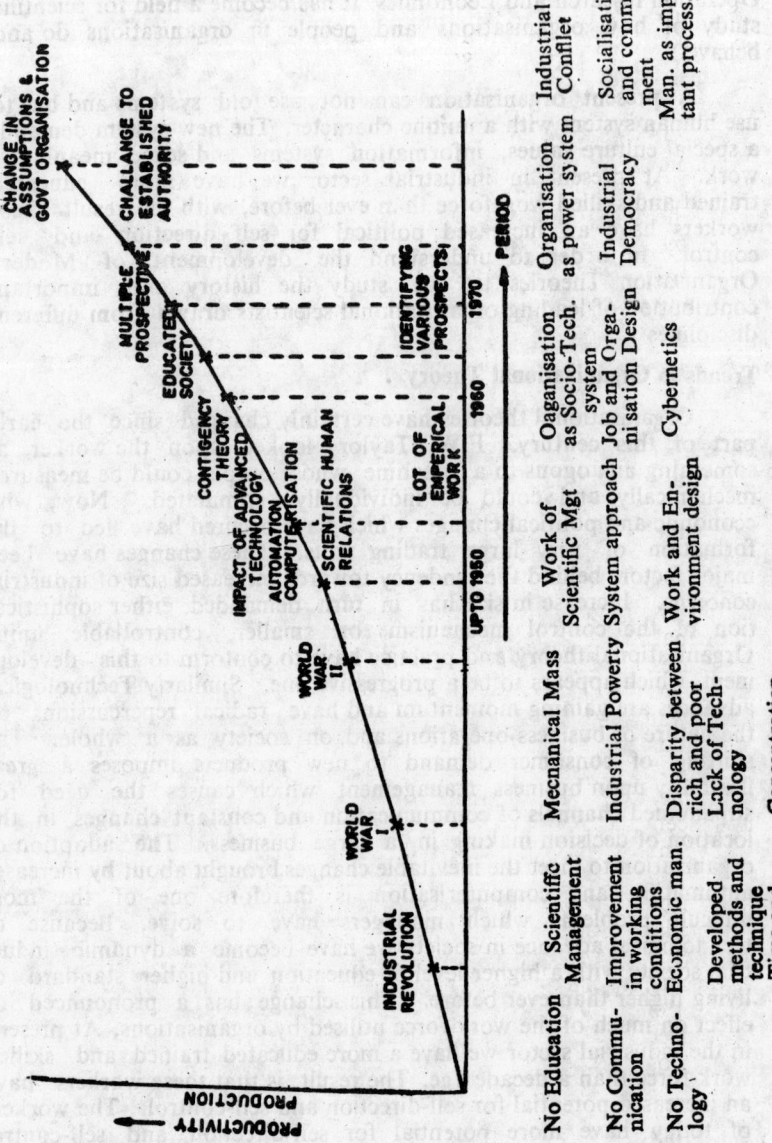

geareed to the subordinate current level of maturity with the overall goal of helping him to develop so as to require progressively less external control and to gain more and more self-control. At present an organisation cannot use the old systems and has to use a human systems with a unique character. The new system demands a culture, the values, information systems and some meaningful work.

Development of Management Theories

The development of management theories can be best understood by the following table, whire in the relationship between content, problem, model and response has been discussed.

Content	Problem	Model	Response
R apid and ustrialisation	Experienced workforce	Econonic man	Scientific management
War time process	Fatigue	Mechanical man	Work and environment design
Rapid change in ocupational structures	Allocation	Man as peg task as hole	Testing : fitting man to job
After war Social conflicts	Informal organisation	Emotional man	Human Relations
Wartime rapid advancement in Technology	Misfit man and Technology	Man machine system	Ergonomics and system design
Transformation from Plants to market Economy	Approch to changing. enviroment	Organisation is an open system	Contigency appraoch
Impact of Advanced technology	Misfit organisation and Technology	Organisation as socio technical system	Job and organisation design
Educated Society	Dissatisfaction with work.	Complex man hierarchy	Job enrichment

Challenge to established auhority	Industrial conflicts	Organisation as power system	Industrial democracy
Change in assumptions about organisation member ship.	Socialisation and commitment	Man as information process : organisation as symbolic order.	Neo human reltations : A culture

In the present context, there exists inter-organisation relationship. There is more emphasis on subcontracting, joint venture and there is loose coupling between the various elements of the organisation. Nature of organisation is fast changing. Multinational organisations established in various countries are trying to have similar organisation culture in all the countries where they have their units. with this content we have "Neo-human Relations response".

There is a tendency to have shared ideology, shared values, socialisation, emphasis on training and management development, ensuring commitment and organisation as social entity.

History of Organisations

Medievan Era. History of ancient Greece and Roman empires give us an evidence of the application of management techniques in handling the affairs of the court, the Government and the Armies. Many ideas on management had their origin in ancient times. Jethro suggested to his son in law Moses who led the Exodus of the Hebrews from Egypt, the importance of delegation, scalar chain and the exception principle" choose able men from Israel, make them rulers of thousands, hundreds, fifties and so on". Leave small matters to subordinates, judge hard case yourself". Similarly the ten commandments in the Bible contain rules of conduct for group solidarity. Robert Owen who managed several mills at New Lanark, Scotland in early 1800's not only advocated better treatment of workers, more welfare facilities, better case of labour, Robert Omen chose to adopt a pioneering step in 1810 in the field of Personnel Management, where as Charles Babbage tries to find out an improved use of machines and the organisation of human beings. The philosophy of Management written by Oliver Scheldon in 1923 is regarded as a comprehansive work on management in the present century. In U.S.A. Frederick Winslow Taylor's ideas as Scientific Management were considerably amplified by Henery Gantt. Frank Gilbreth and Harimgton Emerson. Similarly the results of

Hawthoone Research published by Eton Mayo ; F.J. Roethlisberger and T.N. Whitehead are of great significance. The other social scientists who have done valuable research in management studies are Mary Parker Follet ; Elton Mayo ; Chester I Barnard ; Kurt Lewin ; Renis Likert, Douglas McGregor ; and Chris Argyris. Ludwig Von Bertalanffy introduced his theory at a seminar at the University of Chicago in 1937. His contentions were that different discipline had similarities which could be developed into a general system model. Kenneth Boulding built upon the ideas of Bertalanffy and emphasised the need for general systems to describe the relatioships of the empiricial world. It was Damiel A. Wren who introduced Management Science which refers to the application of quantitative method of management. It is synonymous with Operation Research or Mathematical Model Building. Modern Management Science invented from Scientific Management, was not so much the search for a science of management as it was a strive for the use of science in management. Traditional or classical organisation theory consists of number of principles or general guide most of which were proposed by Henery Fayol a highly successful industrialist.

New Classical Theory. It consists if two streams of thought viz. (*i*) Human relations propound by Elton Mayo and Roethlisberger (*ii*) Behavioral Sciences movement introduced by A. Maslow ; D. McGregor ; F. Ferzberg and V. Vroom.

Modern Management. It consists of three streams of thought viz. (*i*) Quantitative Approach to Management (*ii*) Systems approach to management by Kenneth Boulding, Johnson Woodward ; Cast, Rosen Zweig ; C.W. Churchmann etc. (*iii*) Contingency approach by Johnson Woodward ; Fielder ; Lorsch and Lawrence.

Alternative Approach to Organisation

There are conflicting views on the subject of organisation, because of number of schools of thought on the subject, some stressing the need for direction from the top, others the need for freedom below ; some advocating greater clarification of functions and others warning against too exact a confinement of personnel in specific assignments. The seven approaches which have been put forward by various schools of thoughts are :

—Formalism.

—The Spontaneity Approach.

—The Participation Approach.

—Challenge and Response.

—Specialization.

—The Directive Approach.

—Checks and Balances.

Formalism

The formalism is that each member of an organisation must know precisely what his position is, to whom he is responsible and what is the relation of his job to other jobs. This can be achieved by considering.

—Detailed job descriptions.

—Definite channels of command.

—Unity of command.

Advantages of Formalism Approach

—Definite boundaries to reduce conflict.

—Empire building is restrained.

—The overlapping of responsibilities is avoided.

—Gaps in responsibilities are filled.

—Passing the buck becomes more difficult.

—More exact standards of performance are established.

—A sense of security comes because of well defined task.

—Opportunities for favourism and biased decisions reduced.

This approach is based on the following assumptions :

Some members are aggressive and will trespass on the doma in of others.

Some members are reluctant to assume responsibilities.

—Delineation of clear cut responsibilities offers an incentive by providing a more exact basis for evaluation.

—It is possible to predict in advance the responsibilities that will be required in future.

—Members are prone to conflict and may take away the personnel energy and productivity of individuals.

—Justice is more certain if the enterprise is organised on an objective, impersonal basis.

People criticise this approach as it is a mechanistic approach emphasizing orderly structures rather than people. There is a fear too much stress on neat structure may impede informal communication and may cause congestion in the formal channels of communication.

The Spontaneity Approach

Recently there has been a growing body of thought at the opposite extreme from formalism—one which advocates scope for the spontaneous formation of groups and communication system with a minimum of direction from above. The famous Howthorne studies have been a strong influence in this direction, emphasising the importance of informal organisation. Informal organisation may be the most effective coordinating force providing human satisfaction and simulating cooperation and productivity. Changes in the formal organisation should take into account the possible impact on informal organisation. The assumption underlying this approach are :

—All the required relationships cannot be planned in advance.

—Spontaneous relationships are direct and economical and adjust to changing conditions.

—Spontaneous relationships are more satisfying and lead to high morale.

High morale is essential to high productivity. This approach may bring in insecurity, vagueness, and injustice without prior planning.

The Participation Approach

The Participation is another recent trend in organisation theory. It underlines what is called 'human relation' and spontaneity approach in management. The approach involves the need for a flow of ideas from the bottom of the organisation as well as the top. Participation involves face to face relationships which lead to fuller undestanding, to a pooling of diverse talents and a greater willingness to carry through with decisions once they are made. The results is more satisfaction and productivity and the development of individuals for positions of higher responsibility. Participation helps make members more amenable to change. The assumptions underlying the participation approach are :

—Joint effort makes decision more acceptable.

—Members understand best what they have helped create.

—Familiarity breeds respect.

—Human beings enjoy regular association with each othes.

—People want to take part in decisions.

Disadvantage of this Approach. Joint decisions may be a set of compromises. Joint action may result in shared responsibility, participation may be a disguise manipulation. Besides these limitations still the participation movement remains one of the strongest trends in management circles. The complete absence of participation is inconceivable even in a totalitarian society.

Change and Response

This stresses the need for enthusiasm and initiative throughout the organisation. "Decentralisation" and "Delegation" are the result of this school of thought. It is called the challenge and response approach.

Peter Drucker, argues that departments should be organised around products rathers than skills and processes. Drucker prefers 'flat' or 'horizontal' structure, and thus opposes limitation of the span of control. He feels that specialists must remain in their place to assure that they do not interfere with the freedom of the line managers. It rests on following assumptions :

—People want to work and work best when not watched closely.

—The errors that might be avoided by close supervision are less costly than the harm done to morale due to over supervision.

—A constant flow of innovation is imperative and autonomy is conducive to such innovation.

—There is a tremendous pent up supply of managerial talent waiting to be released if challenges are applied.

—The approach is concerned with people who have potentialities that must be tapped if the enterprise is to survive.

The Specialisation Approach

A central objective of the organization is the allocation of functions and responsibilities among departments and individuals. All organisations involve some specialistion. The approach favours organisation around skills, process, or sub-puposes to achieve the full benefits of specialized skills and knowledge. The specialisation approach is opposed to Drucker's view that orgnisation should be based on end products. Those who stress specialization believe that the experts should be introduced at various levels of the organisation, and that they must be given sufficient authority to ensure that their superior knowledge is applied. This approach is based upon the following assumptions.

—Simple tasks are easier to learn and lead to higher productivity by concentrating attention on narrow area.

—Supervision is more competent and successfull when departments are organised around special skill, and processes.

—Line official will not be seriously 'undermined' by the authority or influence of functional specialists.

Specialisation creates problems of coordination. The conflict between staff specialists and line officials is too common to be ignored and a Functional organisation may become an obstacle to decentralization.

The Directive Approach

It is the emphases on direction from top management. Directive approach has some advantages worth stressing :

—Planning and direction from above may assure cooperation.

—It would appear that many orgrnisation do not perform efficiently either because of inertia without guidance from above.

Checks and Balances

It is most familiar in the political sphere. It is widely recognized that too great a concentration of power in one place may lead to irresponsible action. The concept of checks and balances has its applications in industry as well as in government. The assumptions approach are :

—Power corrupts and thus must be restrained.

—Some members of organisations place personnel goals over the enterp ise needs.

On the contrary the effective leadership cannot be developed where checks predominate. Each existing organisation contains elements of all of the approaches. The problem of planning an organisation is one of finding the balance appropriate to the circnmstances of balance to fit the objective and make an optimum use of the available resources. Thus, the organisation approach appoprite to a firm doing routine business with little technological change will differ from that suited to a firm electronics or atomic energy.

Two tendencies have hindered progress in the field of organization. One is the casual observation. The other is the inclination on the part of critics to disregard all existing hypothesis. To move towards a synthesis of approaches, one should draw on the wisdom of the past to gain a fuller and more flexible study on the problem of the organisation to the future.

Study of the Direction of Organisation

Following are the questionnaire which hclp an executive to analyse in which direction his organisation is going. These check points will help us to study the present state of an organisation.

Checking points

(i) Does each individual-worker, supervisor or executive know to whom he reports ?

(ii) Does each supervisor, department head and executive know what individuals report to him ?

(iii) Is there an organisation chart ?

(*iv*) Can a copy of the organisation chart be found quickly ?

(*v*) Is the organisation chart kept up-to-date ?

(*vi*) Are individuals in the organisation specifically acquainted with their respective section of the organisation chart and generally acquainted with the rest of it or appropriate division of it ?

(*vii*) Is their an organisation write-up describing each position on the organisation chart ?

(*viii*) Are their standard practice instructions covering each operation standardized ?

(*ix*) Is there a pamphlet for distribution to employees that tells of all office rules ?

(*x*) Is there an office manual describing the various outlines and procedures and their relation to each other ?

(*xi*) Is each individual responsible solely to one person for each function performed ?

(*xii*) Does each executive have four or fewer sub-executives department heads or supervisors reporting directly to him ?

(*xiii*) Do executive and department heads know of the work ability, special achievement, and special shortcoming of their immediate subordinates and assistants ?

(*xiv*) Are expense accounts and budget items arranged according to operating responsibilities ?

(*xv*) Are only those costs changed to divisions, department and section than can be thus allocated ?

(*xvi*) Are detailed analysis of their financial results available to division, department and section heads ?

(*xvii*) Is there a programme for training men for supervisory and executive position ?

(*xviii*) Is decision-making decentralized as far as possible ?

(*xix*) Are decision made at the lowest point in the organisation as which the decider possesses all the facts necessary for a sound decision ?

(*xx*) Do men who possess the facts necessary for a sound decision ?

Group Organisation

There is always an implied conflict between the interests of the individual and those of the group with which he is associated. Certainly it is an indisputable fact that once an individual is employed he becomes a member of various groups.

Higher Management Group

In case of a large company if the aims and character are considered as an entity on its own it influence only the work of the senior managers. It is because managers need to form themselves into a more or less composite working group. This is required because they are constantly in need of discussion and co-ordination with each other in their common interests. The consciousness to achieve the group objectives still great if the chief executive himself maintains the contact with his managers both individually and acting as the leader of the group. Thus it appears that group consciousness is an important factor in any organisation which aims at maximum efficiency.

Groups Down the Hierarchy

As we move down the hierarchy of a business which has a line structure we find that groups have different character. The basic reason is that business is subdivided into many integral departments of functional nature. The difference between these groups and the higher management group is that they are not immediately concerned with their own functions and very little with the aims and policy of the concerns as a whole. It is a well known saying that to achieve a contented life is to cultivate one's garden. However the secret of success for a modern complex business does not lie in the cultivation of separate function but in the co-ordination and co-operation of variety of functions.

Co-ordinating Committees

Two devices have been evolved to cut across departmental boundaries.

(a) Establishment of Co-ordinating Committees.

(b) Grouping by divisionalisation.

Co-ordinating groups knit the formal organisation together and may include :

—A distribution committee

—A new product-committee

—Works Committees and joint industrial councils

—Ad-hoc committees.

Grouping by Divisionalisation

As an organisation grow managers responsible for planning and control tend to move away from those responsible for action, channels of communication became extended, the time lying between decision and action increases. There can be general loss of efficiency because the morale throughout the business tends to weaken and the primary motivation shifts from the object of the organisation to their personal position.

With the increase of size, the organisational system tends to become bureaucratic and impersonal. In order to overcome organisation problems created by increased size and complexity, the device of divisionalising the concern has been adopted.

Trends in Organisational Theory

Organisational theories have certainly changed since the early part of the century. F.W. Taylor looked upon the worker as something analogous to a machine whose output could be measured mechanically and who could be individually stimulated. Now with economic and political changes which have occurred led to formation of very large trading units. These changes have been to major factors in the tendency towards increased size of industrial concerns. Increase in size has in turn demanded either a geater sophisatication of control mechanism or smaller controllable units. Organisational theory and practice have to conform to this development which appears to be progressive one.

Technological Trends. Technological advances are growing in momentum and have radical repercussions on the nature or business operations and on society as a whole.

Organisation Trends

Due to the shifting of consumer demand to new products imposes a greatest flexibility upon business management which causes the need for streamlined channels of communication and constant changes in the location of decision-making in a large business.

Future Problems

The adoption of organisation to meet the inevitable changes brought about by increased automation and computerisation is therefore one of the most difficult problems which managers have to solve.

Impact of Culture on the Organisation

Because of technological advancements in society we have become a dynamic industrial society with higher level of education and standard of living higher than ever before. This change have a pronounced effect on much of the work-force utilised by the organisations. At present in the industrial sector we have more educated, trained and skilled work-force than a decade before. With the result these workers have an increased potential for self-direction and self control as shown below.

SELF CONTROL

Education ————
Training ———— ↑ Individual
And Develop- ———— | Growth
ment ———— | Maturity
————
————

External Control.

The workers of today have more potential for self-direction and self-control consistent with these changes in maturity. In the coming ten years the people working in industries will have their needs shifted from physiological needs to comforts end ego needs and the management control will no longer only depend upon the satisfaction of basic needs of industrial people. The management practice should now be geared to the subordinates current level of maturity with the overall goal of helping him to develop to require progressively less external control and to gain more and more self-control. At present organisation cannot use old systems and has to use human system with a unique character. The new system demands

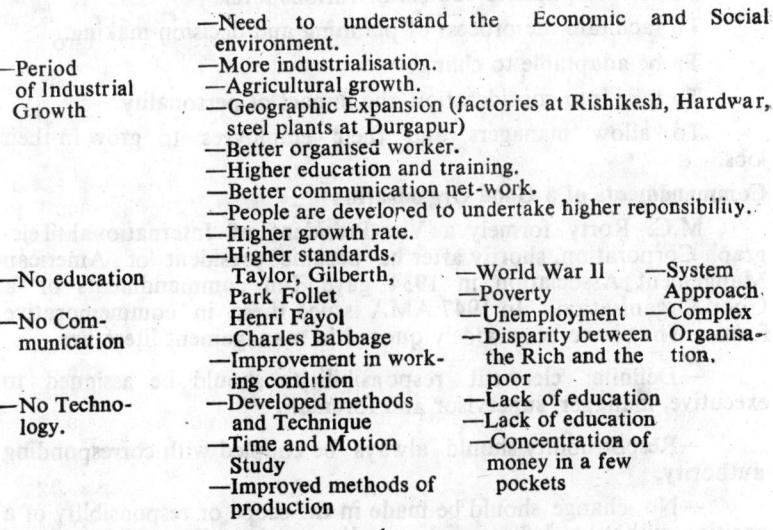

	—Need to understand the environment.	Economic and	Social
—Period of Industrial Growth	—More industrialisation. —Agricultural growth. —Geographic Expansion (factories at Rishikesh, Hardwar, steel plants at Durgapur) —Better organised worker. —Higher education and training. —Better communication net-work. —People are developed to undertake higher reponsibility. —Higher growth rate. —Higher standards.		
—No education —No Communication	Taylor, Gilberth, Park Follet Henri Fayol —Charles Babbage —Improvement in work- ing condition	—World War II —Poverty —Unemployment —Disparity between the Rich and the poor	—System Approach —Complex Organisa- tion.
—No Techno- logy.	—Developed methods and Technique —Time and Motion Study —Improved methods of production	—Lack of education —Lack of education —Concentration of money in a few pockets	

INFLUENCE OF CULTURE OR PRODUCTIVITY

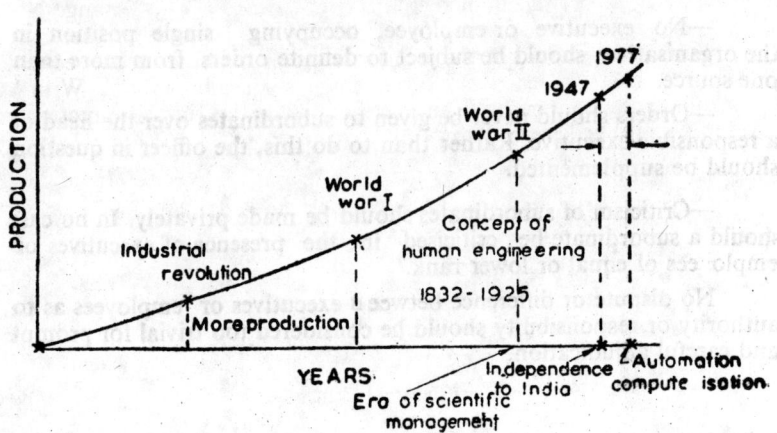

Fig. 57.1

a special culture, values, information system and some meaningful work. Previously the production people were more important than the sales but now there is a change, sales people have got more importance as management is concentrating on there sales because of higher competition.

Characteristics of a Sound Orgaisation

We find that a sound organisation should have the following things :

To be appropriate to the objectives of the business

To allow the necessary relationship between functions

To be susceptible to control of various levels

To facilitate the process of planning and decision-making.

To be adaptable to change

To take into consideration the impact of personality.

To allow managers and their employees to grow in their jobs.

Commandments of a Good Organisation

M.C. Rorty formely a Vice-President of International Telegraph Corporation, shortly after he became president of American Mangement Association in 1934 gave Ten commandments of a Good Organisation. In 1947 AMA issued these in commemorative form. They have been widely quoted in management literature.

—Definite clear-cut responsibilities should be assigned to executive, manager, supervisor and foreman.

—Responsibility should always be coupled with corresponding authority.

—No change should be made in the scope or responsibilty of a position without a definite understanding to that effect on the part of all persons concerned.

—No executive or employee, occupying single position in the organisation, should be subject to definite orders from more than one source.

—Orders should never be given to subordinates over the head of a responsible executive. Rather than to do this, the officer in question should be supplemented.

—Criticism of subordinates should be made privately. In no case should a subordinate be criticised in the presence of executives or emplo ees of equal or lower rank.

No dispute or difference between executives or employees as to authority or responsibility should be considered too trivial for prompt and careful adjudication.

Measurement of Effectiveness

The effectiveness can be defined as its capacity to survive, adopt and maintain itself. Organisation effectiveness rests on good communication, flexibility, creativity and commitments, Organizational effectiveness is a coping process which begins from the changing internal or external environments and end up with a more adaptive, dynamic equilibrium for dealing with change.

Stages of Adaptive Coping Cycle

Following are various stages of the adaptive coping cycle :

1. Sensing a change in the internal or external environments.
2. Getting information about the change into various components.
3. Changing the production system inside the Organisation after the information is obtained.
4. Stabilishing internal changes.
5. Exporting new products, services and so on which are more in line with the originally perceived changes in the environments.
6. Obtaining feedback and re-sensing the external environments.

Following are the organisational conditions essential for effective coping.

1. Ability of the organisation to take in communicated information as reliable and valid.
2. Internal flexibility and creativity to make changes as desired by the information.
3. Integration and commitment to the goals of the organisation from which comes the willingness to change.
4. Internal climate of support and freedom from threat is required in the organisation.

It is quite difficult to achieve the above four conditions in any complex organisation. However following are some of the guide lines.

58

Organisation Structure

Another step in the organisational success is its **organisational structure**. It is the structure in which human beings can perform most effectively. H.J. Leavitt perceives organisations as having the following characteristics.

1. Pyramidal Shape.
2. Idea of Individual responsibility.
3. Hierarchyal distribution of authority.

Organisation Structure. It is group of people working together to attain the desired objectives. People in an organisation do not start working together automatically unless they are provided with some mechanism of co-ordination and control. One of the mechanism is the organisation structure. It reveals who has authority over whom in the organisation. It provides an invisible framework to integrate all the people working together towards a common goals. Organisation structure for exercising leadership. Organisation structure provides an indispensible tort of co-ordination in an organisation. Peter Drucker suggests that the three ways to find the organisational structure are :

1. Activity analysis.
2. Decision analysis.
3. Relation analysis.

The business executives are emphasising the need for maximum flexibility in the organisational structure to meet the changing conditions. Thus organisation structure cannot remain static, it must change with changing environment. B.H.E.L. are of the leading public sector is changing from Production to Service Organisation.

B.H.E.L. supply power generating equipments to various power generating stations all over the country. After the equipments are tested and accepted are despatched to the customers. Most of the time these equipments are badly handled because of ignorance for which B.H.E.L. is not responsible but this makes the customers in different toward BH.E.L. The State Electricity Board operating personnel and engineers are tranied by B.H.E.L experts to avoid any miscommunication.

Types of Organisations

There are three type of organisation

1. Formal organisation.
2. Non-formal organisation.
3. Informal organisation.

The formal organisation. The formal organisation is a system of well defined jobs, each bearing a definite measure of authority, responsibility and accountability, to enable the people of the enterprise to work more effectively together in accomplishing their objectives and organisational goals. In the formal organisation, the work that each individual does is part of a larger pattern, coordination proceeds according to a prescribed pattern in the formal organisation. It sets up boundaries and path ways which must be followed to achieve the objectives.

The formal organisation of industry has three clear cut characteristics.

(1) It is deliberately impersonal.

(2) It is based on ideal relationships.

(3) It is based on "hypothesis" of the nature of man.

The first two imply that each member of the organisation is supposed to react towards the others, not in personal terms of like or dislike, but in terms of the functions they have to perform in the whole, the place they occupy in the hierarchy.

The third characteristic is based on "rebble hypothesis" leads to the claim that formal organisation not only avoids human complications but has the further advantage of being flexible.

Merits

(1) A formal organisation is deliberately planned it is more predictable.

(2) Formal organisation facilitates laying down of objectives and policies.

(3) It has unlimited scope for expansion.

(4) There are clear cut bounds of jurisdiction.

(5) The standard of performance can be established precisely and with excatness.

There are certain weak points in the theory of formal organisation. Of course, coordination is always likely to be a source of difficulty in modern factory. The problem of coordination is due to communication. Next workers have no social contact with each other. Under these circumstances a viscious circle of grievances may develop. Problems due to conflict between staff and line organisation is another difficulty with the formal organisation.

Types of Formal Organisation :

    (1)   Line organisation.

    (2)   Line and staff organisation.

    (3)   Functional organisation.

    (1) Line organisation. In this type of organisation, the positions having authority and responsibility as well as accountability for achieving objectives are arranged in a linear relationship. Line organisation means positions arranged in a line. Line position are arranged in a chain in the flow of direction, communication and command. Line Managers have a clearly defined role to play in the organisation, which requires understanding of the nature of line authority. A typical line organisation chart is shown in Fig. 58.1

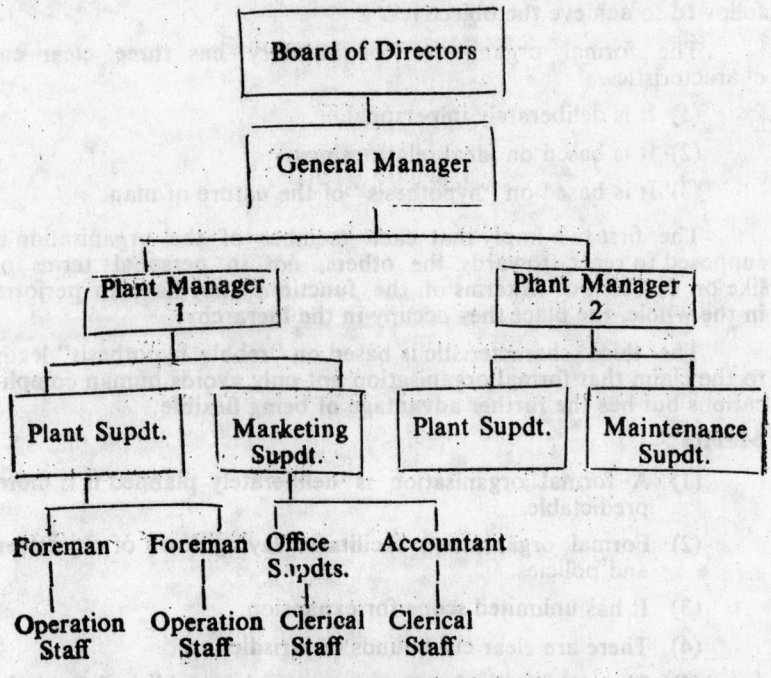

Fig. 58.1 **Line Organisation Chart.**

Advantages

    The division of authority is very precise, definite and clear cut.

2. Quick decision making is fostered.

3. Discipline is more easily maintained. Buck passing is eliminated.

4. Accountability is effective.

Disadvantages

(1) Executives tend to become overloaded with too many duties, and management members may be difficult to replace.

(2) Everyone clings to his duties. No one can share his responsibility with others even though he can get the work done through the subordinates.

(3) It cannot enjoy the blessings of division of labour and of specialisation.

(4) Communications are insufficient.

(5) Large and complex enterprises cannot adopt this system.

(2) **Functional Organisation.** As the name implies, the whole task of management and direction of subordinate should be divided according to the type of work involved. This functional management is also known as "STAFF ORGANISATION". In most enterprises use of staff in organisation structure can be traced to the need for help is handling details, locating data required for decisions and offering counsel on specific managerial problems. Most staff authority relationships are characteristically a manager to manager authority relationship and exist among managerial level of an organisation structure. The managerial recipient of staff authority is commonly called a staff executives or staff officer. At the higher levels, functional organisation refers to the structure that is formed by grouping all the work done into major functional deptt. of divisions. A subordinate anywhere in the organisation will be commanded directly by a number of managers.

Advantages

(1) Enables firm to be benefitted by specialisation of functions.

(2) Workers have higher degree of efficiency since they perform a limited number of operators.

(3) Separation of mental and manual work.

Disadvantages

(1) Weakness in discipline because workers have to work under different bosses.

(2) Lack of coordination due to division and subdivision of work.

(3) Conflicts among foreman.

(3) **Line and Staff Organisation.** In this type of organisation, along with line positions, there are positions created primarily

for the purpose of specialised advice, and services to various units of the organisation. In the system of line and staff organisation authority flows in a vertical line downwards and in addition to this there are experts or specialists at the head of a staff division incharge of a particular function. There are two categories of staff, personal and specialised personal staff consists of line assistants and staff assistants.

Line assistant is known by such titles as the "assistant manager," "assistant director" and so forth. He assumes authority and responsibility for his superior and acts for him in his absence.

Staff assistants are commonly referred as executive assistants, administrative assistants and special assistants.

The line authority remains the avenue for command or performance of the work. It consists of the authority relationships between line managers. Generally speaking, most staff managers do not exercise their staff authority along the channels of line authority but rather to the line authority. Staff managers exercise their proper staff authority to help line managers who command to achieve work performance. What is a line manager and what is a staff manager defined on the type of authority possessed.

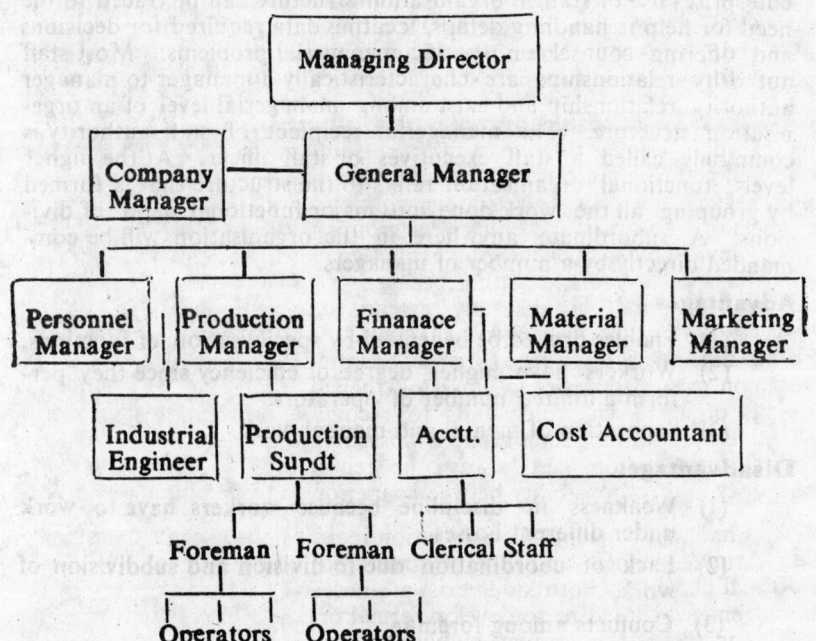

Fig. 58.2

Distinction between line and staff

In modern business, the problems confroming the manager at all livels are so varied and complex that he cannot possibly be master of them all himself. The specialised assistance can be provided best by highly trained and experienced consultants or by agencies which have a staff relationship to the organisation as a whole. Staff are those who can think and plan. Line are those who can do the work.

A staff man cannot issue orders to a line officer. Staff prescribes the method, line follows the procedure.

1. **Non-formal organisation.** Existing within formal organisation is non-formal organisation that permits and sometimes encourages behaviour by members not intended of included in the formal organisation. This non-formal behaviour cannot only be desirable, which frequently it is, but it is also work oriented and contributes significantly to efficiency. Behavioural factors such as unnecessary socializing on the job, group values, and cultural likes and dislikes serve as bases for non-formal behaviour in formal organisation.

Non-formal organisation always exists along with formal organisation. It serves as an adjunct, is intangible, and takes on different degrees of importance depending upon the activity and the person involved. A manager should be aware of the presence and influence of non-formal organisation.

2. **The informal organisation.** It refers largely to what people do because they are human personalities, to their actions in terms of needs, emotions and attitude, not in terms of procedures and regulations. Organisation is nothing more than the informal systems of employees and employees groups which develop in any work environment for achieving explict goals. In the informal organisation people work together because of their personal likes and dislikes. Informal groups may extend to other groups, as well as existing within the organisational unit. The effectiveness of this group depended upon the internal relations, the recognition of common goal, and desire to work together in accomplishing it. The formal organisation which fails to recognise and provide for the effective operation of such groups, looses its own effe' ness. The formal organisation can always be represented th. cugh line of authority and responsibility between superior and subordinates.

Changing concept of organisation requires distinction between the formal, defined and highly structured organisations and the informal personal. Both kinds exist are necessary. The aim of the management should be to develop formal organisation of such scope as to enccmpass the varying, highly personal needs of the informal organisation. Organisation as described below, is extremely important to business enterprise. It facilitate administration encourages sound, balanced growth and diversification provides for the best use of human being and stimulate creativity for the overall growth.

Informal organisation is indifinite and rather structureless and has no definite subdivision. It may be regarded as a shapeless mass of quite varied densities. Informal organisation although comprising the processes of society which are causious has two important classes of effects.

(a) It establishes certain attitudes, understandings, custom habits and institutions.

(b) It creats the condition under which formal organisation may arise.

Informal power attaches to a person, which formal authority attaches to a position. Informal power is personal, but formal authority is institutional. Informal organisations tend to remain smaller in order to keep within the limits of personal relationships.

Benifits of Informal Organisations

1. The informal organisations blend with formal systems to make a workable system for getting the work done. Formal plans and policies cannot meet every problem in a dynamic situation because they are pre-established and partly inflexible.

2. The informal organisation can lighten the work load of the formal manager.

3. Informal organisation may act to fill in gaps in a manager's abilities.

4. Informal organisation is that it can be a useful channel of employer communication.

5. Informal organisation is like a safety valve for the frustrations and other emotional problems of group work.

Committee Organisation

Literally the word "committee" means those to whom some matter or charge is committed. It can be defined as : "A body of persons elected or appointed to meet on an organised basis for the discussion and dealing of matters brought before it".

Committees are common yet controversial in most organisations. They can and do exist at any organisational level, serve in various capacities and for different purposes, are known by many different names, and enjoy wide degrees of acceptance among management members. Extremely important work is accomplished via the committee route. Typically, educational enterprises are loaded with committees, and they are common in government and in business. In trade association and most professional societies, the major portion of the organisation structure is made up of committees.

The committee usually has a formally recognized and, permanent place in the organisation structure. Its makeup, duties, membership, and decision-making power may be carefully spelled out. For example, some business enterprises have at the top level, a planning and policy committee, or a general management committee made up of selected company executives. The committee meets regularly-perhaps weekly-makes decisions, sees that they are enforced and participates in the management affairs of the company.

Reasons for the Widespread Use of Committees

1. Expert and collective knowledge can be concentrated upon a specific problem. A wide range of experience can be tapped, exchange of ideas fostered, and effective give-and-take discussions encouraged.

2. **Coordination is assisted.** Different views can be unified and integrated, agreed prescribed courses of action established, and maximum understanding among committee participants achieved.

3. Too much authority of one person is prevented. By its very nature, a committee tends to distribute its authority among the committee members, who can watch and check each other's actions.

4. **Social values are provided.** Committee membership provided prestige, permits recognition as an equal with other members who have status, and tends to satisfy the human desire to belong and to do something worthwhile.

5. **Encourags motivation.** Committee encourages, participation, cooperation is enhanced in the execution of a proposed action and is reasonably assured if the committee develops the plan. There is also knowledge acquired by the committee members and possibly pride of authorship gained by a member. Such characteristics have strong motivational value.

6. **Education of members is promoted.** Each participant's viewpoint is broadened ; he gains an appreciation of the other unit's problems as well as those of the enterprise.

Social Organisation

It is apparent that in a democratic society, the ultimate authority of a business enterprise rests with public or society. The professional privileges are derived and controlled through laws and marketing prccess. Then beginnirg with the policies, a management system is organised to achieve the organisational objectives. Procedures and control mechanism is introduced by the management in a systematic way.

The relationship between society on an enterprise is shown in Fig. 58.3.

Recent Trends in Organisation Structures

Due to tremendous changes in science, technology, group dynamics, social sciences, and the constantly increasing demand of survival and growth it became essential for the management experts to bring the improvements in the organisational structure to keep pace with these developments and achieve its goals thereby rendering the required service to the society for which it exists. Accordingly the latest organisational structures are rapidly coming up a few of which are as follows :

(1) System Type of Organisation.
(2) Project Type of Organisation.
(3) Matrix Type of Organisation.

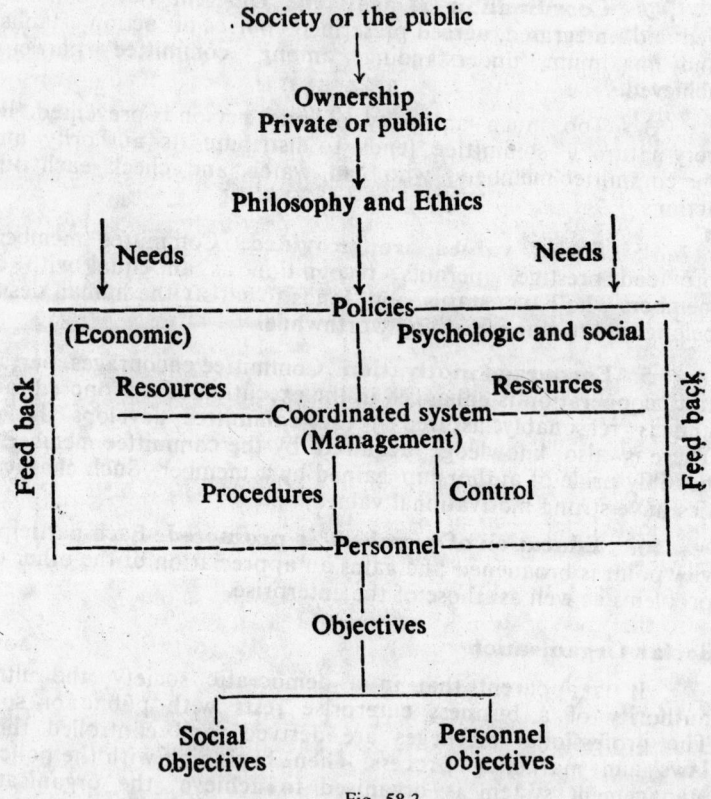

Fig. 58.3

System Approach

The earlier definitions are similar to the traditional organisation theorist but from the system point of view the new definition has to enlarge upon the traditional one. Thus the definitions of organisation in system terms can be as follows :

(1) The organisation is an open system which is in constant interaction with its enviror ment taking raw materials, energy, information and people and to transform these into production and services.

(2) The organisation is a system with multiple functions involving multiple interactions between organisation and its environment.

(3) Organisation is composed of many sub-systems which are in dynamic interaction with one another. It involves the study of these sub-systems in terms of group behaviour, goles etc. These sub-systems are mutually dependent and change in one sub-system affects the behaviour of other sub-systems. Finally the organisation exists in dynamic environment which is also composed of other systems and sub-systems. The multiple links between the organisation and its environment make it difficult to specify the boundaries or limits of any organisation.

These definitions can be grouped into the following definition :

Organisation exists in a dynamic environment and composed of open and closed systems and sub-systems to convert input (Resources) to output (Product and service).

Organisational systems ard sub-systems inturn interact with the systems and sub-systems of Dynamic environment to bring about organisational effectiveness. Organisationals development, organisational growth and organisaticnal change.

Thus the environment within which an organisation exists is becoming increasingly unstable because of the rapid change of technology, the expansion of economic market and rapid social and political change. Every organisation carries within itself representative of external environment.

The employees are not only members of the organisation but also the members of other organisations, groups and societies. The system concept of organisation can be described as shown in Fig. 58.4 .

Thus the organisation is a complex system because of the complex interactions between how an individual is inducted into the organisation, trained, developed and managed and the interaction between the formal organisations and the various informal groups which arise inevitably within it. As all organisation exists in an

Systems Approach

The earlier definitions are similar to the traditional organisation but from the systems point of view, the new definition lays more emphasis upon the traditional ones. Thus the definition of organisation in system terms are as follows:

(1) The organisation is an open system which it is concerned in action with its environment, e.g. raw material, energy, information etc. people and transform them into production and services.

(2) The organisation is a system with multiple functions involving multiple interactions between organisation and its environment.

(3) Organisation is composed of many sub-systems which are in dynamic interaction with one another. It involves the study of these sub systems in terms of their behaviour, role etc. These sub-systems are mutually dependent and change in one sub-system will influence the behaviour of other sub-systems. Finally, the organisation exists in a wider environment, which is also composed of other systems and sub-systems. The multiple links between the organisation and its environment makes it difficult to specify the boundaries or limits of any organisation.

These definitions can be developed into the following definition:

Organisation exists in a dynamic environment and composed of open and closed systems and is a process to convey input (Resources) to output (Production and services).

Organisational systems as well as sub-systems interact with the systems and sub-systems of the outer environment to bring about organisational effectiveness. Organisational is over inherent organisational growth, development and change.

Thus the environment within which an organisation exists is becoming increasingly useful because of the rapid change of technology, the expansion of economic, cultural and social and political change. Every organisation exercises within itself represents representative of external environment.

The employees are not only members of the organisation but also the members of other institutions, groups and societies. The system concept of organisation can be described as shown in Fig. 58.4.

Thus the organisation is a complex system because of the complex interactions between how an individual is inducted into the organisation, trained, developed and managed and the interaction between the formal organisation and the various internal groups which arise inevitably within it. A still organisation exists in an

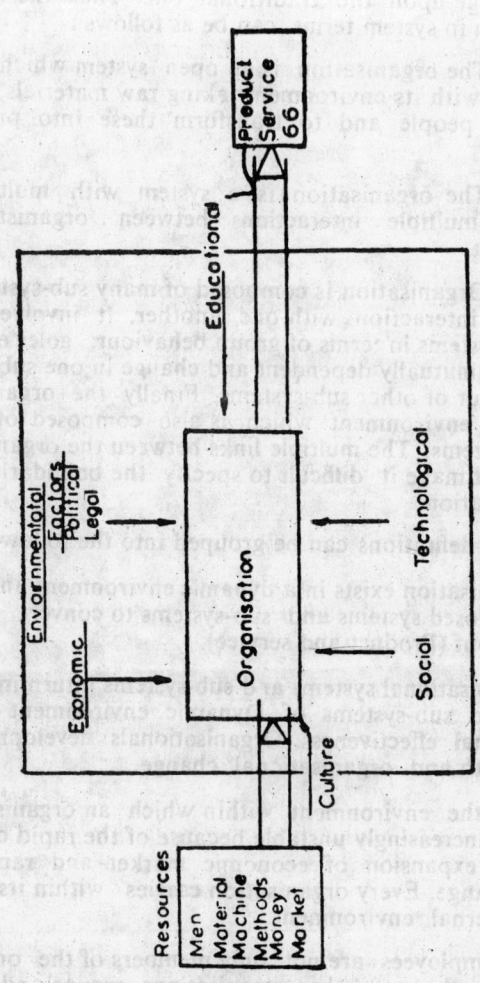

Fig. 58.4

environment which consists of the culture and social structure of the society in order to revive the organisation must fulfil some useful function.

The common goals set by the organisation must result into some product or service. P.M. Babu and W. R. Scott define four classes of organisation :—

—Mutual benefit organisations—these benefit the member of the Organisation.

—Business concerns—these benefit the owner managers.

—Service organisations—these benefit the customers or clients.

—Commercial organisations—these benefit the public at large.

—To survive and develop the organisation must continue to perform its primary task of—

—The recruitment

—Proper utilization

—Motivation

—Integration.

Project Organisation

In this approach, group of individuals possessing the required skills is organised into an autonomous unit for a particular project. Project management is carried out by a director appointed to co-ordinate and motivate the people in various activities of engineering, marketing, production, personnel etc. Responsibility for the project is very well pin-pointed.

Following is the diagram for organisation.

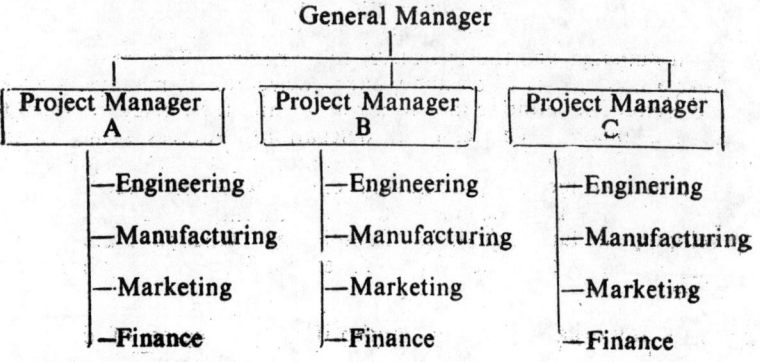

Matrix Organiation

It is a form of organisation which continue both the project structure and the traditional functional structure. Matrix management provides for specialised knowledge for carrying on a number of specific projects. The disadvantage may arise because of the dual accountability of personnel involved in the project.

General Manager

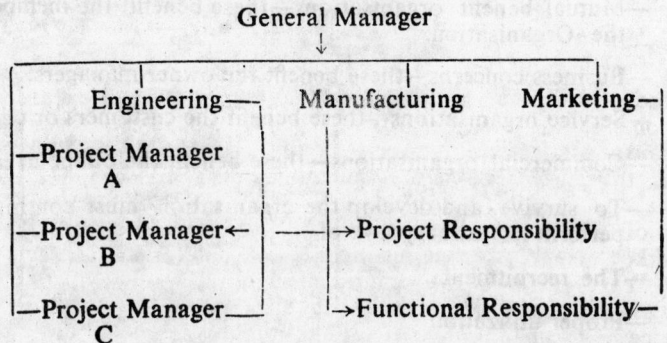

59

Wage and Salary Administration

The relationship between employers and employees rests mainly on wages. Issues regarding wages have been the most important cause of disputes between labour and management. The wage is the price for the use of human efforts. The money is the rupees a worker gets as wages. The real wage is however, the amount of goods and services he can buy with his money wage. The worker wages constitute his income and determine his level of living. Wages produce for reaching effects on the general standard of living in the country, employment, productivity, cost of production, prices, capital formation etc. The relation between wages on one hand and the factors such as productivity, prices, capital formation, employment and standard of living on the other has to be firmly borne in mind while fixing wages. It is certain that increasing productivity is the basic solution for such conflicts.

Wages and salaries represent a substantial part of total costs in most of the organisations. Wages to the economists are variable costs, but to the businessmen, they are becoming ever increasing costs. Instances of general wage or salary reductions are rare and where it is necessary to reduce labour costs, this is accomplished more and more through technological changes resultiug in increasing the productivity of workers. The control of wage and salary levels is of paramount importance, even though the amount of control which can be exerted may vary among organisations and within an organisation from time to time.

The general objectives of the wage and salary administration are as follows :

1. Control of costs.
2. Establishment of fair and equitable remuneration.
3. Utilisation of wages and salaries as an incentive to greater employee productivity.
4. Maintenance of a satisfactory public relations image.

To achieve these objectives, the responsibility for wage and salary administration usually lies with the top management or the chief executive who in turn, is expected to develop policies and procedures which will accomplish the company's objectives. The Personnel Manager plays an important role in developing the wage.

policies. In many organisations, this task is entrusted to Wage and Salary Committee composed of line and staff executives. The functions of Wage and Salary Committee or any other person connected with wage and salary administration are :

1. To approve in a board policy determining manner, the system of job description and job evaluation.

2. To recommend to top management the wage policies for the administration of the wage programme.

3. To recommend changes in wage policies and in the salary or wage level.

4. To review wage and salary schemes departmentwise.

5. To recommend to top management specific raises for executives above a specified limit.

DIFFERENCE BETWEEN WAGE AND SALARY

Wage is compensation to the employees for services rendered to the organisation. In case the quantum of services rendered is difficult to measure, then the payment is called salary. Normally the wage period is shorter than the salary period.

Payment made to labour is generally referred as wages. Money paid periodically to persons whose output cannot be easily measured, such as clerical staff as well as supervisory staff, is generally referred to as salaries. Salaries are paid uniformly generally on monthly basis and at times the element of incentive is introduced in the form of commission.

WAGES AND PRODUCTIVITY

It has been observed that if the existing wages are below what is necessary for maintaining the efficiency of an worker, an increase in wages will improve his personal efficiency. An increase in wages may increase productivity but it is not definite because wage rise may induce employers to reduce the labour cost by adopting more efficient processes of production. These new machines and equipment may raise output without increasing the number of workers. Although the efficiency of worker does not necessarily improve productivity is said to go up because of greater output with the same number of workers. But it is not definite that employers will introduce methods of production and unnecessary high wages may even lead to closure of the enterprise. It is equally important to understand the effect of productivity on wages also. It has been realised that higher productivity is the most important means of raising real wages. The following points will emphasise the importance of productivity in realising the standard of living of the employees.

(i) Wages and Prices.

(ii) Wages and Demand for consumer goods.

(*iii*) Wages and Capital formation.

(*iv*) Wages and employment.

WAGES AND PRICES

Wages lead to higher standard of living only if the prices of consumer goods do not rise proportionately. The standard of living of worker may not rise with rising wages. If the workers are paid higher wages it will certainly enhance the labour cost of production unless there is increase of productivity at the same time. Higher cost of production will lead to higher cost of the product if the management is keeping the margin of profit same. The increase in the price of the product may in turn lead to price rise and thus fall in the demand of the product which may in turn lead to lower profits and lower wages. Thus if wages are to be raised without a rise in the price of the product and without reducing profits there should be a simultaneous increase in productivity.

WAGES AND DEMAND FOR CONSUMER GOODS

Workers tend to spend their entire increase in their incomes on consumer goods. This will increase the demand for such goods. So it is necessary that supply of consumer goods increases otherwise prices of consumer goods would go up and the general level of living would not improve. Thus if the money wages were to be raised, arrangements should be made at the same time for increasing the supply of consumer goods.

WAGES AND CAPITAL FORMATION

Capital is the result of saving and saving can be there if we produce more than we consume. Saving is very essential for the economic development because without capital we can not create an intra-structure in the country. The level of consumption is already very low and further reduction is not possible so if we really want to raise the standard of living we must first increase productivity and then save part of the benefit of higher productivity for capital formation.

WAGES AND EMPLOYMENT

If wages go up unduly high in industry some of the firms making low profit may be forced to close down, consequently some workers may be out of jobs. But if employer introduces new techniques of production and thereby provide more employment opportunities. Thus it is definite that if wage increases are not related to increase in productivity, then the higher the wages, the lower will be scope for increasing employment opportunities.

WAGE POLICIES AND PLAN

The only firm basis for proper wage administration consists of a set of sound policies with just and equitable distribution. If these conditions prevail wage administration is likely to be handled on the basis of expediency. Since decisions sre based on the apparent merits of specific cases rather than on the basis of established polity and by different people over a long period of time, it is quite evident that these will not be consistent decisions.

PRINCIPLES OF WAGE POLICY

1. Wage policy should keep in mind the interest of workers, management, consumer, organisation and community.

2. Wage policy should be expressed in writing to ensure stability.

3. Wage decisions should be checked against the carefully formulated policies.

4. Management should see to it that employees know and understand the wage policies.

5. Wage policies should be evaluated from time to time to make certain that they are adequate for current needs.

6. Departmental performance ratings should be periodically checked to keep them up-to-date.

Principles of Wage Plan

(a) The wage plan must be linked with the productivity of the workers.

(b) The wage for each job classification should be related to each other in terms of job requirements, due consideration being given to such factors as skill, length of time required in learning, working conditions and responsibility demanded.

(c) Wage plan must ensure the workers, management and consumers due share in the gains of higher productivity.

(d) The wage plan should ensure higher pay to the workers who perform work at higher level of efficiency.

(e) The wage plan should have a proper wage differential based on proper job evaluation.

(f) The wage plan should guarantee minimum wage to protect the interest of workers against conditions over which they have no control.

(g) The wage plan should be flexible in order to meet changing conditions and should not involve excessive administration cost.

(h) The wage plan must be related to cost controls and the operating labour budget.

(i) The wage plan should place no limit on the amount of additional earnings of the workers and thus it should maintain the high morale of workers.

WAGE POLICIES

Wage policy is a complex and sensitive area of public policy. Any nation, development is because of the status of working class in the society, its commitment to industry and attitude towards management. Also workers morale, their motivation towards productivity, their living standard and their way of life are all conditioned by wages. In the modern industrial system, wages form the pivot around which most of the problems revolves. Wage issues are often prominent in collective bargaining and represent a frequent cause of grievances and industrial disputes. Dissatisfaction with wage rates is one of the most common cause of low level of production efficiency and morale.

The traditional approach to wage and salary administration has been to recognise that wages and salaries must be at a level which is adequate to attract, retain and motivate employees competent to perform the tasks assigned to them. But this approach is not acceptable because it looks upon human labour as a commodity subject to the economic law of supply and demand. The most prevailing general wage and salary policy is one which relates wage and salary levels in an organisation to those of its competitors either within its own industry or with those who are competitors for labour in a geographical area.

NEW DIMENSIONS TO WAGE AND SALARY ADMINISTRATION

Generally the wages and salaries are viewed only in one dimension as how much money is paid. By paying more money, the employers expects the employees to work more efficiently and sincerely and thereby giving more efforts and time to the company. But somehow such expectations are often not fulfilled, and even a handsome amount of additional money is not able to drive the human beings to give the desired results. The strikes, the sabotages and work stoppages are more in case of comparatively better paying organisations rather than low paid organisations. The clerks working in Banks are more indisciplined although the remuneration paid to them is much higher compared to clerks in Post Office and other Commercial organisations. Examples are there where a graduate clerk getting Rs. 600 is not interested in work and complain against being lesser paid but another graduate clerk at the same place getting much lesser salary is more worried about his work, growth and development and is interested to rise in the organisation rather about his salary.

A large size private company manufacturing shoes found that general efficiency of its officers reduced in place of improvement after increasing the salary bill by many thousand rupees in giving departmental promotions. It was because those who got promotion took it as their right and those who did not, felt frustrated. In other words, enhancing the emoluments is no guarantee for bringing satisfaction in the employees and thereby improving their work concentration and sincerity toward the company. Employees do not consider the money of their wages and salaries as something for putting their sincere efforts and energy, thus benefitting the enterprise, but take it simply as their right of being appointed and daily attending the workplace.

To change this stage of affairs to a happy work culture, the binding force between the employees and the employers should be strengthened for the mutual benefit of both the parties. As the ultimate aim of all the person concerned is to develop their organisation and share its profits. The human nature is such that they can be sincere only to an organisation to which they are committed. In order to make the employees committed to the organisation, we must create an atmosphere of brotherlihood and team-work so that they start identifying themselves with the organisations. Till the time this confidence is achieved between the employees and employers simply increasing the salary will not motivate the employees to give their best to the organisation.

Bhoothalingam Committee Report

Wages and salaries have far reaching effect on labour management relations. In order to study, the Government of India, in the Ministry of Finance had set up a study group on wages, income and price with Shri Bhoothalingam, former Secretary, Department of Economic Affairs and Director General of Applied Economic Research as Chairman. The study group was set up because of the distortion crept into the structure of pay, dearness allowance and other compensatory allowances of employees in public and private sectors. The distortions have been because of the ad-hoc approach followed in the past to the problem of periodical revision of emoluments in public enterprises and the organised private industry. The study group considered the subject matter and submittes its report to the Government. The recommendations of Boothalingam study group will have a vital bearing on the future wage policy as well as labour management relations in the country.

The general observations and recommendations about Bhoothalingam Committee are as follows :

1. A gradual transformation of the bonus system into a social security scheme like pension covering all wage and salary earners uniform rate of dearness allowance irrespective of salary drawn, removing disparities in salary structure between Central and State Government and among State Governments.

2. A national minimum wage of Rs. 100 per month at current prices or Rs. 4 per day for eight hours work.

3. An appointment of national pay commission in consultation with the states to consider the question of upward revision of salaries at upper level of management hierarchy in Government and public sector undertakings.

4. The appointment of a pay committee to go into emoluments and service conditions of Government industrial employees after comparing with corresponding employees in private sector.

5. A non statutory body to be known as "Bureau of Incomes and Price" to undertake continuous review of relevant data and to determine each year the guidelines for the industry and group of industries.

6. To move away from bonus system to pension scheme and setting up of a pension bureau. This bureau should also examine and monitor the behaviour of prices and to suggest suitable changes in the prices of agricultural products.

7. A single national corrective formula to compensate for a rising cost of essential consumption unit and per capita rural consumption expenditure.

8. The future dearness allowance should be linked to the cost of living on a uniform basis based on all India average consumers price index for industrial workers.

9. There should be no reduction in dearness allowance drawn at present.

10. To replace bonus by a system which would enable labour to get a fair share of benefit of productivity without causing such distortion. In industrial sector, bonus is related to profit. It perpetuates and accentuate disparities in earning workers doing the same work. Logically bonus is related to profit and suited only to industry producing for the market in reasonably competitive conditions.

11. Wage revision in future should be linked to increase in productivity since it has a system of compensating for increase in the cost of living.

12. The various wage settlements should be revised for a period of five years and pay commissions and committees shall be appointed every five years. The total salaries and commission paid in private sector, which should be applicable not only to directors, but others as well should have a limited period say five years and a limit of Rs. 6,000 per month be fixed for all new contracts. The differentials in the private sector are higher and employee's terms should be settled through negotiated contracts subjects to the overall limits.

The recommendations of the committee were subjected to a lot of analysis by various walks of life mainly the union leaders, industrialists, executives, administrators and public at large. Following are the various observations :

(a) The recommendation of minimum wage of Rs. 4/- a day or Rs. 100 a month is contradicting to trade unions long standing demand of Rs. 300. The minimum wage suggested is such less than the workers are already getting. This way the committee had neither justified the toiling masses nor the urban worker class.

(b) Committee has not accepted the concept of bonus as "defferred wage" which the Government has already accepted. As regards bonus the group recommended alternative arrangements like long term ,retirement benefits pension to all categories and a scheme of unemployment relief to those temporarily rendered jobless due to closure of firms etc. should be discussed with trade unions and management representatives in search of even better methods. Committee recommended continuance of bonus system which has become a part of industrial life until replaced by general agreement.

(c) The committee rightly came at the conclusion that wages and productivity should be linked as there is no other uniform basis for wage increase. The wage inerease in our country is due to increase in cost of living, bargaining strength or combination of all these factors. Increase of wages due to profit may contribute to inter unit disparities which can be alarming to an integrated policies of wages, incomes and prices.

(d) Committee rightly brought out that in a developing economy, worker must have a due share in the benefit of growth. In case of Government sector where it is difficult to measure productivity, the annual growth of national productivity should be taken into consideration.

(e) Committee had not proposed and rightly so any wage freeze, on the contrary it recommended gradual increase in wages by appropriate share in the growth of gross national productivity.

(f) Dearness allowance schemes recommended by committee had merged particularly because it was the major cause for disparities between different levels in an organisational set up. In the subject of dearness allowance committee recommended per point formula which will ensure full neutralisation at the lower levels giving them the same amount of dearness allowance in future as the higher salary groups.

(g) Regarding pay disparity between government and private sector, the government sectors does not get even 1/3 of his counterpart in private sector although with the similar management functions. There should be some efforts to reduce the wage differential in private sector and government sectors.

(h) The disparity between the proposed national minimum wage and top incomes can be narrowed only if maximum salary and income limits are prescribed. The present company laws have failed to curb the growth of high incomes.

60

Wage Policy

NATIONAL WAGE POLICY

Wage policy is a complex and sensitive area of public policy of any nation because of the status of working class in the society, its commitment to industry and attitude towards management. Also workers' morale, their motivation towards productivity, their living standard and their way of life, are all conditioned by wages.

The aim of national wage policy in India should be to ensure an income level for the workers by which they can maintain a decent standard of living. By and large, the people in India are poor. From the point of view of industrial relations, it must be recognised that poverty constitutes a great threat to democracy. So the wage policy in any country should be relational and based on social and economic considerations. The national policy should ensure that people are not compelled to work at unreasonably low level of wages and to live a life at sub-human levels of existence because of lack of effective organisation and bargaining power.

1. Concepts of Minimum-Living and Fair Wage

Wages represent the major point around which industrial disputes are centred. Wages have been classified as :

(i) Minimum wage.

(ii) Fair wage.

(iii) Living wage.

Minimum Wage

A minimum wage may be defined as that wage which is sufficient to cover the bare physical needs of a worker and his family. There is a feeling that minimum wage should provide other essential requirements such as minimum education, medical facilities and other amenities. Thus a reasonable standard of living should be permitted to the worker from the stand point of his health, efficiency and well being statutory minimum wage is fixed from this point of view. There is a distinction between a base subsistance or minimum wage and a statutory minimum wage. The former is a wage which should cover the bare physical needs of a worker and his family.

The statutory minimum wage is the rate which has got to be paid to the worker irrespective of the capacity of the industry to pay. If an industrial organisation is unable to pay its workers atleast a bare minimum, it has no right to exist.

Fair Wage

It is something more than the minimum wage providing more necessities, while the lower limit of the fair wage must obviously be the minimum wage, the upper limit is set by what may be broadly be called the capacity of the industry to pay. Fair wage depends not only on the present economic position of the industry but also on its future prospects. Between the limits of minimum and fair wages depend the fair wage like :

—productivity of the labour
—prevailing wage rates
—level of national income
—place of the industry in the economy of the country.

2. Industrial Truce Resolution

Industrial Truce Resolution, 1947 recommended the setting up of a suitable machinery for study and determination of fair wages and conditions of labour. The emphasis was :

    (i) to fix statutory minimum wages in sweated industries ; and
    (ii) to promote fair wage agreements in the more organised ones.

In pursuance of the Industrial Truce Resolution, Indian Parliament enacted Minimum Wages Act, 1948 and appointed for wages committee to determine the principles on which fair wages should be based and to suggest the lines on which these principles could be applied. The committee was a tripartite body consisting of the representatives of Government, employers and employees. Report submitted by the committee laid down criteria for the first time for wage fixation and progressive improvement in the wage structure. The committee observed that the state of national income is highly relevant to the problem of wages because no wage policy can be regarded as just or economically sound unless it encourages increase of the national income and secures to the wage earner a legitimate share in that increase and recommended that the actual wage should be between the lower limit of minimum wage and the upper limit of the capacity of the industry to pay, depending on—(i) productivity of labour ; (ii) the prevailing rate of wages ; (iii) the level of national income and its distribution and (iv) the place of the industry in the economy of India. The committee also observed that in the actual calculation of the fair wage it was not possible to assign any definite weights to these factors.

3. The Fair Wage Committee Report

After independence, the Government of India appointed the Central Labour Advisory Council in persuance of its resolution on Industrial Policy to advise the Government on fair wages. The council appointed a committee in 1948, to inquire into and report on the question of fair wages for labour. The report of the committee was submitted in June 1949 and popularly known as the 'Fair Wages Committee Report' :

The committee pointed out at the outset that 'any attempt to evolve principles for governing the fixation of wages must be made against the background of general economic condition of the country and the level of national income'. Regarding living wage it was of the opinion that, 'the living wage should enable the male earner to provide for himself and his faimly not merely the bare essentials of food, clothing and shelter but a measure of frugal comfort including education for the children, protection against ill health, requirements of essential social needs, and a measure of insurance against the more important misfortunes including old age'. It was further of the opinion that minimum wages must be given to the workers and living wage should be the target. It stated "while the lower limit of the 'Fair Wage' must obviously be the minimum wage, the upper limit is set by the capacity of industry to pay. Between these two limits, the actual wages will depend on :

(i) the productivity of labour ;

(ii) the prevailing rates of wages ;

(iii) the level of national income and its distribution ; and

(iv) the place of the industry in the economy of the country.

4. Need Based Minimum Wage

The Minimum Wages Act, 1948 seeks to protect the interest of the workers in the sweated industries. Wage fixing authorities have been guided by the norms prescribed by the Fair Wage Committee in the settlement of issues relating to wage fixation in organised industries. The Fair Wage Committee defined the components of the minimum wage which should be taken into account, but did not quantify them. It was the Indian Labour Conference (15th Session) which for the first time, moved into this direction and formally quantified the components of minimum wage. The conference in its recommendations not merely provided for the bare physical subsistence but also for the maintenance of health and efficiency of the workers. Some of the relevant points of the resolution are given below :

(i) In calculating the minimum wage, the standard working class family should be taken to consist of 3 consumption units for the one earner, the earnings of women, children and adolescent should be disregarded.

(ii) Minimum food requirements should be calculated on the basis of a net intake of 2,700 calories, as recommended by Dr. Aykreyed for an average Indian adult of moderate activity.

The first enactment specifically to regulate wages in this country is the Minimum Wages Act, 1948. This Act is limited in its operation to the so-called sweated industries, in which labour is practically unorganised and working conditions are poor than in organised industry. Under the Act, the Appropriate Government has either to appoint a committee to hold enquires and to advice it in regard to the fixation or minimum rates of wages, or if it thinks that it has enough material on hand, to publish its proposals for the fixation of wages in the official gazette and to invite objections. The appropriate Government finally fixes the minimum rates of wages on receipts of the recommendations of the committee. There is also a Provincial Advisory Board in each Province to coordinate the work of the various committees. There is also a Central Advisory Board to coordinate the work of provincial boards. Complaints of non-payment of the minimum rates of wages fixed by Government may be taken to authorities consituted by the Act. Breaches of the act are punishable by criminal courts.

61

Wage Incentives

Incentives are the stimulus mainly psychological and it maintains and strengthen the desire to achieve improved performance. Incentives are mainly of two types. (1) Financial Incentives, (2) Non-financial Incentives. Financial incentives help the individuals to meet their basic needs and non-financial incentives assist in meeting the higher order needs such as social needs, ego needs and self-actualisation needs. Non-financial incentives are based on sociological and psychological principles of higher behaviour. Some of the non-financial incentives are :

1. Welfare skills.
2. Social and sports activities.
3. Educational opportunities.
4. Suggestion schemes.
5. Meritorious service awards.
6. Sound performance appraisal.
7. Promotional policies.
8. Better working condition.
9. Creation of healthy and organisational planning and development.
10. Knowledge of results.
11. Growth opportunities in the organisation.
12. Recognition.

Financial Incentives

These incentives are payments for improved productivity, attendance and general improvement in employees performance. Financial incentives schemes can be direct or indirect in nature. In direct financial incentives scheme, the payments are based on employees' own performance or contribution to the job such as Production Incentive Schemes, Attendance Bonus, Profit Sharing Bonus etc. Indirect financial incentives, the payments are not directly related to employees' contribution and schemes are like subsidised means, leave encashment, gratuity scheme, leave travel concession etc.

Advantages

It is difficult to underline the advantage of incentive schemes. However, following are the various advantages of incentive schemes :

1. Achievement of higher production.
2. Lesser break-down and defective work.
3. Cutting down the cost of production.
4. Opportunity for higher earning to employees.
5. Reducing the supervision time.
6. Effective use of man-power planning.

In India, the majority of the workforce, still have the problem relating to low order needs. Therefore, there is a positive impact of financial incentives.

Disadvantages

The incentive schemes are not free from disadvantages. The following are some of the disadvantages of incentive scheme :

1. Possibility of incentive schemes to be mis-understood because of its complexities.
2. Lack of consistency.
3. Psychological problems in the incentive systems.
4. Delay in the payment of the scheme.

Objectives of Incentive Schemes

Following are some of the important objectives for which incentive schemes are used :

1.—to increase : production productivity, yield, manpower utilisation, earning of employees, sales and quality.
2.—to improve : quality, reduced cost of production, reduced inventory and reduced wastage.

The Requisites of Incentive Schemes

Following are the features to be considered in any incentive scheme :

(a) It should be easily understood and the amount of benefits should be really assessed.
(b) It could be sold to the employees.
(c) It must benefit employees as well as employer.
(d) It must have relation between the benefits to the employees in relation to their efforts.
(e) It should not be costly to operate.

(f) It should stimulate the interest among the workers.

(g) It should stimulate the cooperation amongst the employees.

(h) It should not be detrimental to the welfare of the employees.

(i) It should assist in supervision.

The Incentive schemes should be used at an appropriate time. Following are the factors which can help as a guideline to intervene in the scheme :

—After all the possibilities of improving methods of production, equipment and technology have been tried.

—After the productivity stabilised on a particular level and there is a feeling that productivity cannot be increased further without financial motivation.

—When there is a place for employees to be motivated.

—When reasonably good labour-management relation exists.

Non-Financial Incentives

These are generally introduced as a normal practice in good management. These elicit from employees' sense of participation and belongingness and their importance is not to be underestimated. The non-financial incentives touch the inner feelings of employees' and bring out a response much more impressive than could be possible through financial incentives.

The non-financial incentives are the outcome of an enlightened management. The provision of better working condition, encouragement and appreciation of good work and a general atmosphere of willingness and cooperation in dealing with common problems are all those contributed to excellence in all areas of work and higher productivity.

There is no doubt that introduction of financial incentive schemes and the acceptance of workmen bring a radical change in industrial relations. Management often is faced with a challenge when they introduce incentive scheme. For the incentive schemes to be effective, management must ensure that they will able to do good to the management and the worker alike.

One of the studies carried out in a public undertaking revealed that after 4-5 months of the introduction of scheme, following results were achieved :

1. Increase in production —20%.
2. Increase in productivity —15%.
3. Reduction in rate of accidents—60%.
4. Reduction in overtime—40%.

In addition to the above, there was improvement in attendance, morale, willingness to work and enforcement for increasing production.

We have seen that incentive plans provide for higher reward for increased output. Its object is to increase the production by giving an inducement to workers in the form of higher wages. An efficient incentive plan must provide for minimum guaranteed wage, based on hourly rate and extra remuneration for increased output. The incentive plan must include in its purview the characteristics of time based and output based systems of wage payment. Before the installation of any incentive plan, we must ensure that incentive plan is installed primarily to benefit the person who in any way be effected by its installation. The incentive plan should try to cover all the employees whose jobs can be adopted to the incentive method of payment. Before the installation, a job evaluation must be carried out and the incentive plan should plan no limit on incentive earnings because more the employees produces, the more the organisation is benefitted.

I. Types of Incentive Plans

A wide variety of incentive wage plans for operative employees have been devised and used by the industrial concerns. They are classified as follows :

Price Rate Plan

    1. Taylor piece-work plan.

    2. Merrick differential piece rate plan.

II. Time Rates Plan

    (A) Based on time saved.

    3. Halsey premium bonus plan.

    4. Bedaux point premium plan.

    (B) Based on time worked.

    5. Rowan plan.

    6. Emerson plan.

    (C) Based on standard time.

    7. Gantt task bonus plan.

Symbol used in Calculating Various Incentives

    St = standard time allowed for completing a particular operation or task.

    At = actual time worked.

    R = rate per hour or piece as the case may be.

    N = number of pieces produced.

    T = premium percentage.

    E = employee's earnings.

Taylor's Differential Piece Rates

Taylor's wage system call for two piece rates, a higher one where the worker equals or exceeds task and low one when he fails to equal task. For instance, if the standard is set for 100 pieces per day at 10 per piece. The rate in case of failure to achieve the standard would be less, such as 9 paisa per unit. In such a situation, a worker would earn Rs. 10/- for achieving the standard task and Rs. 8·91 only in case his output fell short of task by one piece. Thus difference in one unit of output will fetch for the workers Rs. 1·09 less. According to Taylor, this differential exerts a powerful pressure on the worker to exert himself constantly to achieve task standards. It needs to be carefully established so that an average man who was properly instructed could make it. It also places a heavy burden on management to keep conditions standard so that workers may not be penalised. This plan is manifested too severe on the workers unless the management consistently does its job. This plan did achieve the importance because the existence of low piece rate provides a negative incentive to workers and so it fails to attract workers who are receiving time rate wages or piece wages backed by a guaranteed wage.

Merrick Differential Piece Rate Plan

M.D.V. Merrick realised that it was unresonable and unrealistic to classify all workers into two categories only i.e. efficient and inefficient ones. In fact there are certain employees, who work to produce more only for their own progress. These employees deserved to be encourged. Merrick, therefore, introduced three piece rates and made the lowest piece rate equal to the 'Basic.Piece Rate'. The rates introduced by Merrick are as follows :

Output	Piece Rate wage
1. Less than 83% of task	Basic Piece rate
2. From 83% to 100% of task	110% of Basic Piece rate
3. Over 100% of task	120% of Basic Piece rate

To the workers who are potentially high producers, Merrick plan is a good incentive system. It seems reasonable to pay 100% of the basic piece rate to the workers, who have reached 83% of the target. We know that most of the employees should be able of reach 83% task with a little extra effort and when they do so they will be encouraged to reach the 100% task. Merrick has removed the punitive wage rate originated by Taylor by introducing the basic piece rate for low output. Merrick plan is only a modified form of Toylor plan.

Time Rate

Halsey Premium Plan

In Halsey plan, a minimum time wage is guaranteed. The time allowed for completing the job is set from the records of previous performance rather than by time and motion studies. The amount of time saved multiplied by the hourly rate forms the sum that is to be shared between the worker and the owners according to the ratio agreed upon more often equally. Because of this fixed proportion of sharing bonus. Halsey plan can be called a 'Constant Bonus Sharing Plan'. The standard length of time for doing a job, not being derived through the use of time and motion study, is usually greater than would be the case under more scientifically measured procedure.

Bedaux Point Plan

In Bedaux point system the amount of work one does per minute is known as standard work unit. Each job is rated in terms of the number of standard work allowed for its performance. The worker is guaranteed the base rate for each job and is paid a bonus for performance above standard. The workers may be paid for the full time saved. The standard time for a worker for a given period is determined by dividing by 60 the number of standard work minutes that he has been allotted for the work completed. This system makes possible the comparison of the efficiency of one department with another, since all work is reduced to a common denominator.

According to Bedaux, all human efforts can be measured in terms of a common unit composed partly of actual work and partly of compensating relaxation. Bedaux point premium plan stresses that human labour and relaxation are closely related. Therefore, in determining the task time or operation time for a job it is necessary to take into account the actual time of operation and the time of rest. Earnings under this plan can be calculated by this formula : $E = RT + P (Sc - At) R$, when the worker is paid for the full time.

Rowan Plan

Under this plan, the employee is guaranteed wage at the ordinary rate for the time taken by him to complete the job or operation. The difference between the Halsey Premium plan and the Rowan Premium Plan is only in the calculation of the bonus. Under the Halsey plan, bonus is a fixed percentage of the time saved whereas under the Rowan Plan, bonus is in that proportion of the wage of the time taken which the time saved bears to the standard time allowed. The proportion of the bonus to be paid to a worker is a variable quantity and so the Rowan Plan is also known as 'Variable Bonus Sharing Plan'. The formula for computation of an employee's earning is :

$$E = RT + \frac{St - At}{St} \times RT$$

Emerson Plan

Emerson was a pioneer in scientific management and brought out his efficiency bonus plan in 1910 by which he tried to remove the defects in Taylor, Gantt, Merrick, Halsey and Rowan Plans. Emerson Plan pay for the standard time worked plus a bonus for performance task, with a guaranteed day rate and a sliding scale of bonus for performance from 66⅔% efficiency to 20% bonus on reaching standard. Using measured standard as a base, he gave an initial base when the worker exceeded the initial performance, which was 66⅔% of the measured standard. This served to spur the worker to increase his efforts to reach standard was not so discouraging when he fell short of standard performance.

Thus in Emerson Plan, a worker gets a bonus whenever the efficiency of a worker is more than 66⅔%. As his efficiency increases, his bonus also increase. When his efficiency is 100% the worker receives bonus at 20% of the basic wage. When the efficiency is more than 100% for every 10% increase in efficiency, there is one per cent increase in bonus. Like other plans, a minimum wage is granted in this plan. According to Emerson, the efficiency of a worker should be determined on the basis of his monthly or at least on his weekly output. This procedure would compensate the decrease in worker's efficiency because of certain unforeseen reasons.

Gantt Task Bonus Plan

This plan is a mixture of time rate and piece rate plans. Gantt, one of the pioneers of scientific management, modified Taylor's plan considerably and introduced a better plan. By substituting Taylor's punitive wage rate by a guaranteed basic wage. Under this plan, if a worker's output in task time is equal to or more than the stipulated task, he is paid a bonus at a certain percentage of his guaranteed basic wage. The guaranteed basic wage is always a time rate wage. The basic wage rate and task standards have to be carefully determined.

Gantt Task Bonus Plan classified worker's wage into two parts, namely :

1. A guaranteed minimum wage on time rate which is usually an hourly rate. This time rate is the basic wage rate under this plan ; and

2. A bonus which is paid to the workers when his output is equal to or more than the task. This bonus is a percentage of the guaranteed wage. This bonus may vary from 10% to 100% of the guaranteed wage rate according to the nature of the work.

The plan has a strong incentive force and is likely to help workers to increase their efficiency. This plan will be very much useful in those industrial units where the fixed overhead and other expenses in maintaining a nd running the factory and machines are higher than the total wage cost. It is important to note that no single wage plan is sufficient to bring about efficiency and that the value of every plan depends upon its administration. Experience has shown that some wage plan with clearly undesirable features have proved to be successful through the intelligent administration.

For sound administration of incentive plans the management should recognise that the effectiveness of incentives depends on the total situation which includes workers and management confidence in the incentive plan, relations with the union, the quality of communication and of supervisors and the tradition of the industry.

62

Wage Structure

The primary objective of wage and salary administration programme is that each employees should be equitably compensated for the services rendered by him to the enterprise, on the basis of :

—The nature of job.

—The present worth of that type of job.

—The effectiveness with which the individual performs the job.

Money is one of the most important phase of employer-employee relations. It is a pre-requisite to sound industrial relations and for good industrial relations each employee should (i) receive sufficient money in the form of wages or salaries to sustain himself and his dependents and (ii) feel satisfied with the relationship between his wage and the wages of other people performing the same class of work in some other organisation in the same industry.

The most important decision on compensation for a job usually involves comparison of the job to other jobs, either jobs to other jobs in the organisation or to similar jobs in the other organisations. The first comparison is made with the help of job evaluation and the latter comparison is made through the use of wage and salary surveys to determine the going wage or salary for a position. Now we shall concentrate on job evaluation.

Job Evaluation

One of the important problems with which the management is usually confronted is the establishment of a wage rate for each of the jobs in the organisation. There are different methods used to establish wage rate for a job which range from subjective intuitive methods to highly objective quantified methods. Job evaluation is an orderly and systematic technique which aims at determining the worth of jobs. In other words, it is a formal system of determining the base compensation of jobs.

Job evaluation rates the job and not the man. It takes into account the demands of the job in terms of efforts and abilities but it does not take into account the individual abilities' and efforts, which may of course be taken into consideration and reflected in the worker's earnings under a system of payment by results of merit rating.

Obviously, job evaluation aims at measuring the value of job descriptions and job specifications. When the value of job descriptions are determined, they may be translated in terms of money according to some basis to have a balanced wage structure in the organisation. The total wage structure consists of different scales of wages or pay which will lead to classification of wages. Thus the process of job evaluation starts from job analysis and ends with the classification of jobs according to their worth.

Why Job Evaluation ?

The reasons for growing interest in job evaluation in recent times are discussed below :

(a) Job evaluation process is a valuable technique in the hands of management by which a more rational and consistent (internal and external wages and salary structure can be evolved.

(b) Job evaluation helps in bringing or maintaining harmonious relation between labour and management, since it tends to eliminate wage inequalities within the organisation and the industry.

(c) Job differences should not be based on skill differences only. Job evaluation takes into account various other factors like risks and working conditions also to determine the worth of jobs.

(d) Because of division of labour and specialisation, any large enterprise may have hundreds or thousands of different jobs to be performed by a substantial number of works. Many workers work together on the same, similar or technically interdependent job. An attempt should be made to precisely define the jobs and fix wages accordingly. This is possible only by job evaluation.

(e) Job evaluation helps in keeping down the costs of recruitment and selection of workers. Job evaluation involves job analysis and appraisal which are of great use while recruiting new employees. Selection can be made objective by matching the qualifications of the candidate with job specifications.

(f) The process of determining the wage differentials for different jobs becomes standardised through job evaluation. This makes for uniform standards to be applied to all jobs in the organisation.

(g) The fact that people work together in similar or at least comparable jobs and the fact that jobs may change when there is a change in the product or method makes the workers except the reflection of these differences in wages. It is, therefore, necessary in a dynamic economy to arrange for continuous job evaluation.

Methods of Job Evaluations

As discussed earlier job evaluation involves the evaluating of various jobs in terms of certain factors. The important factors to be considered under job evaluation may be grouped under the following heads :

understand the process, and installation of this system is not difficult task.

However, the ranking method has serious limitations that limit its usefulness. Much subjective judgement is required to determine the relation, and position of jobs in the rank order. In addition, judgement is needed in determining what the pay rates should be, since a statistical method of interpolation is not employed. If the job descriptions are not accurate serious error in ranking can occur. There will be arbitrary ranking which may result in differences in similar jobs which may be resented by the employees.

GRADING (OR JOB CLASSIFICATION) SYSTEM

This method is considered to be an improvement over ranking method in that a predetermined scale of values is provided. This method involves the establishment of job classes or grades. The evaluation committee goes through each job description and carefully weighs it in the light of certain factors like skill, experience, etc. In this way, it assigns each job to a particular grade or class. For each grade or class, there is different rate of wage. In fact, this method arrives at a series of classes or grades, which is precisely the point at which both the point and factor-comparison systems also arrive. It is also a relatively simpler and inexpensive system to operate.

Table Grading of jobs

Grade 1 Unskilled	The jobs in this group are mostly clerical in character, require accuracy and dependability, but no extended training,
Grade 2 Skilled	The jobs in this group are mostly clerical in character, require training of hand or mind. This group includes such positions as stenographer, draftsman, ledgerman, laboratory assistant and demonstrator.
Grade 3 Interpretative	The jobs in this group call for ability to classify work and apply established procedures to its accomplishment. These positions are clerical and non-clerical such as correspondents, foremen, laboratory assistants and layout draftsman. The work is mostly non-supervisory in character and involves no substantial amount of work of the same kind as that done by those supervised.
Grade 4 Creative	The positions of this group are those of a creative character such as engineer, salesman, staff supervisor, system designer and section supervisor.

Grade 5 Executive	The positions of this group are those of department manager, superintendent and general foreman. The function is that of departmental management in the broad sense.
Grade 6 Administrative	The positions of this group involve responsibilities of large magnitude or over all character or for mixed functional division such as division manager, district sales manager, director of accounts, consulting engineer, director of research, treasurer and works manager.
Grade 7 Policy	The positions of this group are those of the policy makers of the company such as General Managers and Directors of the Board of Directors.

The main disadvantage of the job, grading method is that broad generalities must be used in defining grades. Another difficulty is that grading approach usually requires multiple system, for different types of jobs e.g., description of office jobs differ widely from those of production job. Moreover, the installation of this method is not appropriate for big organisations.

POINT METHOD

This is the most widely used method of job evaluation. It, alogwith factor comparison system, involves a more detailed, quantitative and analytical approach to the measurement of job worth, Under this method, a quantitative evaluation of different jobs in terms of various factors is made. Maximum point values are assigned to each of the job factor required to be considered. Then each job is awarded points for each of the factors. The wage level appropriate for each job is fixed on the basis of total point scored by it.

In contrast to the ranking and grading methods, which measure jobs as whole jobs, the point system is a more analytical approch and deals with job factors. A job factor is a specific requirement levied upon the job holder which he must contribute, assume or endure. The major factors are : skill efforts, responsibility and working conditions. These factors or points are later converted into money value. The procedure for the design of 'Point Method' is discussed below :

1. Determine the type of jobs to be evaluated.

2. Determine the factors to be used in this method, like skill efforts, initiative, etc. and define them properly.

3. Determine the number of degrees to be allocated to each factor and prepare a suitable definition of each.

4. Assign points to each degree of each factor.

5. Select a certain number of key jobs, say 10—15, and evaluate each applying the previous steps.

Number of Factors to be used. It is very difficult to lay down the exact number of factors which should be used for evaluating a job. But there should be sufficient number of factors to evaluate all aspects of the jobs. The number of factors will depend upon the nature of the job. If too many factors are used, the plan will become burdensome and clumsy. But if too less factors are used, the evaluation will not be fair because some phase of a job may be ignored. So sufficient number of factors relating to the job under review should be considered according to the needs of the organisation and the type of job.

Definition of Factors. After the selection of factors to be used, they should be properly defined and definition of each should be put in writing so that the persons using these may have the clear idea. Various definitions should be positive statements and should be expressed in simple and unambiguous language.

Determination of degrees. Each factor of job evaluation should further be subdivided into degrees. The point method generally uses from four to six degrees for each factor. It is advisable to use an even number of degrees in the development of point method and the same number of degrees should be used for each factor in order to maintain consistency in the job evaluation plan.

Allocation of Points to Degrees. For assigning points to the degrees determined, a committee may be appointed to determine the weights of the factors in terms of percentages and estimating what percentage should be allocated to each factor. The committee will determine how 100% can be apportioned among the various factors. The percentage that is assigned to each factor will become the number of points for the first degree of each factor. In order to determine the number of points to be allocated to succeeding degrees of each factor, it is necessary to extend the points assigned to the first degree in an arithmatic progression as shown in the table given on next page.

Definition of Factors

1. **Education.** This factor apprises the extent of education background required an individual working on job to perform the job satisfactorily.

2. **Training.** This factor appraises the period of training required by an average person to perform the job efficiently.

3. **Experience.** This factor appraises the length of time required by an average employee with education previously specified to be able to perform the job satisfactorily.

4. **Mental Efforts.** This factor appraises the degree of mental effort required from an individual to perform the job satisfactorily. This factor gives an idea about the degree of organising ability required for a job.

TABLE
Allocation of Points to Degrees

S. No.	Factors	Degrees of points					
		First	Second	Third	Fourth	Fifth	Sixth
1.	Education	25	53	75	100	125	150
2.	Training	10	20	30	40	50	60
3.	Experience	10	20	30	40	50	60
4.	Mental efforts	13	26	39	52	65	78
5.	Physical efforts	12	24	36	48	60	72
6.	Visual attention	8	16	24	32	40	48
7.	Initiative	6	12	18	24	30	36
8.	Responsibility	6	12	18	24	30	36
9.	Working conditions	5	10	15	20	25	30
10.	Physical hazards	5	10	15	20	25	30

5. Physical Efforts. This factor appraises the physical efforts required of an individual to perform his job satisfactorily.

6. Visual Attention. This factor appraises the degree and continuity of visual attention.

7. Initiative. This gives an idea about the independent decision and/or action required of an individual.

8. Responsibility. This factor appraises the responsibility which goes with the job for preventing damage to tools, equipments or materials used in the performance or the jobs.

9. Working Conditions. This factor appraises the physical environments in which the job must be performed, physical environments include : head, cold, dempness, darkness, glare, dust fumes, noise, etc.

10. Physical Hazards. This factor appraises the accident or health hazards which exist even though safety devices have been installed.

To sum up, point method of job evaluation is more effective because even the major factors are sub-divided which ensures accuracy of evaluation. The possibility of inaccuracy of job evaluation is likely to arise if the predetermined point values do not exhibit true values. If this is so, this initial determination [of points values may lead to further inaccuracies.

FACTOR COMPARISON METHOD

Thomas E. Hitlten was the first to originate factor comparison method of job evaluation. This factor comparison method of job

---

For detailed reading see Managing People at Work by **K.K. Ahuja** and T.N. Chhabra.

evaluation determines the relative rank of the jobs to be evaluated in relation to monetary scale.

This method is often used for evaluating white-collar, professional and managerial positions, although it is equally suitable for grading others as well. It is essentially a combination of the ranking and point systems. Like rank order method, it rates job by comparing one job with another and like the point system it is more analytical in the sense of subdividing jobs into compensable factors. Final ratings are expressed in terms of numbers.

In this method five factors are generally evaluated for each job. There are comparatively fewer than the point system but is nevertheless sufficient, because each factor is defined broadly. The number of factors may be more than five also. The five factors which are customarily used are mental requirements. skill, physical requirements, responsibilities and working conditions. The evaluation of job under this method consists of following steps.

1. Select the factors and define them clearly.

2. Select the key jobs. Key jobs serve as standard against which all other jobs are compared. A key job is one whose content has become stabilized over a period of time and whose wage rate is considerd to be presently correct by the management and the union.

3. Allocate wage for each key job to different factors.

4. Develop a job comparison scale and insert key jobs in them. When all of the key jobs have been evaluated and wages allocated in this manner, a job comparison scale can be constructed.

5. Evaluate the job in question factor by factor in relation to key jobs on job comparison scale. Then each job is to be evaluated and compared to other jobs in terms of each factor.

6. Design, adjust and operate the wage structure.

The advocates of factor comparison method point out that it usually results in more accurate job evaluation because it is relatively more objective because weights are not selected arbitrarily. It is flexible also and has no upper limit on the rating that a job may receive on a factor. The point system does have ceiling on the factors. Another advantage of this method is that it utilises few factors and thereby reduces the likelihood of overlap. The procedure of rating new jobs by comparing with other standard or key job is logical and not too difficult to accomplish.

This method suffers from certain limitations also. The use o present wages for the key jobs may initially build errors into the plan. The content and the value of these jobs, may change over a period of time and this will lead to furture errors. It is very expensive system of job evaluation because experts have to be appointed particularly in selecting weights which are based on actual analysis. Since the method is complicated, an average worker cannot understand this system.

Determining the Factors to be Used

The factor comparison method of job evaluation requires five critical factors : mental requirements, skill requirements, physical requirements, responsibility and working conditions. All these five factors should be explicitly defined to facilitate subsequent evaluation.

Selecting the Key Jobs

The first step by means of which jobs are evaluated under the factor comparison method, is to select a group of key jobs.

The key jobs so selected should range from the lowest paid to the highest. All pay jobs should be of exact and well-understood definition and no job should be included in the key position if there is disagreement about the existing rate. After the key jobs have been selected each member of the committee should be furnished with a complete description of each of the key jobs.

In order to be familiar with the key jobs, the entire committee might visit and observe the jobs concerned to assure agreement and consistency of interpretation. Each member of the committee is supplied with a key job ranking sheet as shown in the chart on next page. In this example only seven jobs have been taken into consideration while normally 10—15 key jobs are considered. The numbers running from one upward to the number representing the total number of key jobs selected are allocated to these jobs. Each member of the committee will rank each of the jobs in terms of critical factors. For instance, in case of working conditions rank one assigned to any job signify that the job has the most agreeable and the least hazardous conditions and the highest rank number to that job involving the most disagreeable and most hazardous conditions. After the members have ranked the key jobs by factors, the chairman of the committee will analyse the returns statistically to determine any misunderstanding regarding the job content, job conditions, job requirements and definitions.

The above process should be repeated until the members of the committee have ranked all the key jobs three different times before determining the final ranking for the key jobs. Disagreement in ranking is decided by majority vote of the committee. The next step is apportion the average rate currently being paid to the empoyees working on each key job among the five factors.

Both the factor comparison and point methods of job evaluation are limited with respect to the range of jobs that either method can effectively evaluate. Inspite of the fact that successful results may be achieved through the use of either method the most common question that always comes in mind is, which method should be used.

TABLE
Apportionment of Average Rate of Pay

Job title	Mental requirements	Skill requirements	Physical requirements	Responsibility	Working conditions	Average rate currently paid Rs.
Assembler	17	15	26	14	10	285
Expediter	39	19	24	28	8	380
Material mover	7	2	50	6	10	357
Painter	16	20	28	13	23	340
Time-keeper	32	11	24	20	8	305
Tool and Die maker	37	45	40	33	13	605
Turret lathe operator	25	31	33	20	16	450

In fact, no definite answer can be given to this question, certainly if it is planned to retain a consultant to install a job evaluation method, the wise course of action is to make the choice on the basis of the consultant's reputation and ability to produce results and to use the plan recommended by him. On the other hand, if those contemplating doing work are not experts in job evaluation, it will be found that in most cases a point method will result in the more satisfactory installation because of greater over-all simplicity and objectivity.

Establishing Pay Structure

Whatever may be the method of job evaluation used, the ultimate objective is to ascertain the proper rates of pay for all the jobs. In other words the job structure must be priced. Since both the ranking and job grading methods are non-quantitave and do not arrive at number designation for the job rating, the procedure for conversion to money is somewhat more inexact and may be handled in a variety of ways. For rankings, the existing rate may be compared with the new rankings and discrepancies corrected. It may be better to decide upon certain key job rates that are fair and correct and then the other rates can be adjusted and slotted into the heirarchy. Under the grading method the jobs can be tabulated together with their present rates and then judgement can be made so as to rectify rates for the grades. Here the job are already grouped into grades ; therefore, it may be considered good to compute the arithmatic means of the present rates and let that serve as the new grade rate. These means can be adjusted by judgement to make smooth or straight line progression from grade one to the highest grade.

The factor comparison system of conversion to money is quite simple because the method itself establishes wages directly in the rating process. But it is desirable for administrative convenience to group jobs together into pay grades. In such a situation averages of newly determined rates can be found to give figure for each grade.

With the points obtained under the point method, a graph may be prepared on the abscissa and existing rates on the ordinate. If the prevailing rates are reasonably fair and if the ratings of the jobs under the point method have been done carefully, the points can be expected to fall along a trend line with very little dispersion. But if the existing structure of pay is loaded with inequities and errors, considerable dispersion will be found from the trend line. There are a number of methods available to ascertain the best fitting trend line. One is to draw a line in such a way that the same number of points fall below as above the line. The other method (which is most accurate) is the statistical technique of least squares. For a straight line, $Y=a+bX$, where a is the intercept of Y and b is the slope of the line, it is required to solve two normal equations to find the value of the two constants 'a' and 'b'. The equations are :

$$\Sigma Y = Na + b\epsilon X \text{ and } \epsilon XY = a\epsilon Y + b\epsilon X^2$$

where N is the number of jobs, Y is the ways for each job and X is the point value. These equations will give the best-fitting straight line even if the points actually fall along a curve.

Fig. 62.1 shows a trend line which has been obtained by plotting of present wage rates against job evaluation points for hourly

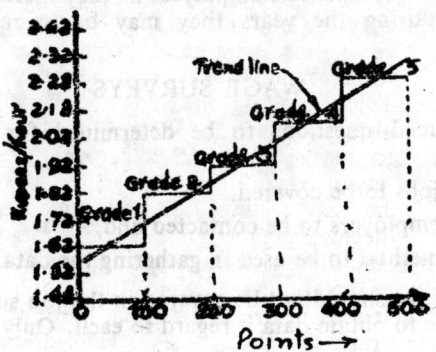

Fig. 62.1. Trend line showing Wage Rates.

paid jobs. In the figure horizontal lines have been drawn through the mid points of trend line in each grade to arrive at a single pay rates as shown in the following table :

Grade	Point Range	Hourly wage Rate
1	0-100	Rs. 1·62
2	101-200	Rs. 1·78
3	201-300	Rs. 1·94
4	301-400	Rs. 2·10
5	401-500	Rs. 2·26

The table given below depicts how a rate range can be set up. In the following table we have five wage grades and a vertical spread of around 18% from the minimum of each grade to the maximum.

Grade	Point Range	Hourly wage Rate
1	0-100	Rs. 1·49-1·75
2	101-200	Rs. 1·62-1·92
3	201-300	Rs. 1·79-2·09
4	301-400	Rs. 1·93-2·27
5	401-500	Rs. 2·08-2·44

The maximum rates for one grade is higher than the minimum rate of the next higher grade. This arrangement is used for white collar professional and administrative types of jobs. The overlapping range are provided to give greater flexibility for management in transferring an employee from a job in one grade to a job in adjacent grade without changing his pay. The overlapping ranges are necessary to compensate employees as they increase in skill and contribution during the years they may be assigned to particular jobs.

WAGE SURVEYS

The crucial questions to be determined for carrying out a survey are :

(1) the jobs to be covered.

(2) the employers to be contacted and,

(3) the method to be used in gathering the data.

All the jobs should not be covered in the job survey because it is not possible to obtain data it regard to each. Only the crucial jobs should be covered by the job survey because these cover all ranges. The second key question concerns who will be surveyed. The organisations to be covered by the survey must be selected with great care. The organisations so chosen must relate to the same industry and as far as possible must be in the same region. Finally, the method of the survey must be decided. Personal interviews by the personnel department would probably develop the most accurate responses, but this method is the most expensive one. That is why mailed questionnaries are probably the most frequent used method since it is the cheapest method. But the fear in this case is that the response may be very poor.

The questionnaires should be designed very carefully so that it may derive maximum information from the informant within the least possible time. The information so collected can afterwards be processed to determine the average rate of wages for important jobs in the industry.

63

Discipline

Introduction

It is the pivot of modern society. It has assumed a new dimension with the development of industry. Discipline is essential for the orderly conduct of any organisation where large number of people work together. Discipline ensures maximum utilisation of time and energy of the working class and make use of the scarce organisational resources. Self-imposed discipline is more desirable than the enforced one. But as long as the emotions and sentiments play a greater part in human performance, the executives will have to take the assistance of enforced discipline. The discipline enforces favourable environment and motivate human beings to make their maximum contribution for the attainment of ultimate objectives. In disciplined climate, the employees' encouraged and harmonised working is developed. Discipline is essential for an orderly way of life in all groupings of human beings. Discipline promotes growth of an individual and develops human capacities and stimulates the will to do. Discipline should be based on mutual understanding and cooperative attitude. The enforced discipline creates low morale. Discipline obtained by fear is not a successful way of doing the things and it has an impact on the morale of workforce. Discipline acquired through imposition, punishment, suspension, demotion and termination is kept as potential threat against wrong doings. The lesser the reliance on such punishment, better will be the morale of the workforce.

Discipline is employees' self-control to meet the organisational standards and objectives. Discipline relates to employees conduct and morale relates to his attitude. Discipline is an orderly conduct within the set rules and guidelines. The objectives of discipline in any industry is to increase and maintain business efficiency which in turn will result in increased sales, lower cost and better products. Thus disciplined workforce is the necessary requisite of a business organisation.

Discipline and indiscipline in industry are the two sides of the same coin. To understand discipline, it is important to analyse and determine causes of indiscipline. The basic causes of indiscipline emanate from the experiences of an organisation and the environment provided to its employees. **Disciplined behaviour brings harmony and is conducive to the interest of the social group. Indiscipline leads to disharmony and chaos and self-disruptive to social needs.**

Causes of Indiscipline

Indiscipline stems in most of the cases from managerial faults and lapses. Indiscipline may take the form of absence without leave or leave over-stayed, late attendance, theft, fraud, dishonesty, loitering habit, insubordination, disobedience, negligence. in work and lower productivity. The various causes of indiscipline are :

(*i*) Varying standards of discipline.

(*ii*) Neglecting the grievances.

(*iii*) Mis-judgement in personnel matters, like transfer, promotion and upgradation.

(*iv*) Lack of well defined code of discipline.

(*v*) In appropriate supervision.

(*vi*) Inadequate personnel supervision.

(*vii*) Lack of proper communication.

(*viii*) Different level of aspiration.

(*ix*) Out moded attitude of trade unions.

(*x*) Lack of involvement with work.

(*xi*) Out moded management style.

Maintenance of Discipline

Employer has the primary responsibility in maintaining discipline and good conduct among his employees. Therefore, it becomes the inherent right of the employer to punish deliquent employees. Even there have been a tremendous change in the method for infusing and maintaining discipline and good conduct with the increase of technology and social changes. Initially the discipline was introduced by imposition of physical punishment, denial of work for sometime, excessive fines and deductions. Now these have gone out of use with the emergence of trade union organisations and the employers resort to more severe punishment of discharge or dismissal of workers' guilty of indicipline or insubordination. With the emergence of modern concept of social justice, rights of an employer to dismiss his workmen has come to severe limitations and restrictions. A Civil court may not able to order re-instatement in case of a wrongful dismissal or discharge but a labour court and an industrial tribunal set up under the Dispute Act have to decide the case of discharge on the grounds of equity and social justice. Principles of equity and justice required that in case of misconduct, proper enquiry is conducted before taking any action and the worker concerned is to give an opportunity to explain his act or emotion and also why certain action should not be taken against him. It is for this reason, that the industrial employment (Standing Orders Act) was passed in 1946. The act requires the worker to be given an opportunity to explain the misconduct. The Industrial Dispute Act, however, does not lay down any procedure for taking disciplinary action.

The section 33 of the act only restrains employer from punishing a worker during the pendency of dispute. A recent amendment of Industrial Dispute Act empowers labour courts and tribunals to substitute their own judgement for that of management and set aside order of discharge or dismissal direct re-instatement, even if the domestic enquiry has been conducted properly.

Sense of discipline as you know has to evolve from within the organisation as to provide environment, formulate policies, programmes and procedures so as to condition the behaviours of employees. The constructive programmes of discipline maintenance should consider the following factors :

(i) Adding meaning to the job for the emloyees at all levels of work groups.

(ii) Meaningful and effective participation of employees in management at all levels beginning from shop floor level.

(iii) Effective two way communication from top to bottom.

(iv) Effective and meaningful, vertical and horizontal communication.

(v) Reasonable satisfaction of employees in the matter of wages, fringe benefits, working conditions, welfare provisions and social security measures for their maintenance.

(vi) A careful and serious attention to their grievances and wise counselling.

(vii) Managing leaders of employees under the charge of supervisors.

(viii) Properly trained supervisor in the role of a democratic participation and supportive management in dealing with human beings.

(ix) Stable and flexible and enlightened personnel policies.

(x) Execution of policies and implementation of rules and regulations in such a way to be impartial and consistent.

Disciplinary action. According to McGregor, the discipline is best known by hot stove rule. Hot stove rule draws an anology between touching a hot stove and undergoing discipline. We should aim at discipline not because a person is bad but because he has committed a particular irregularity, hence it is more directive towards the act. Discipline consist of following four elements :

1. Immediate discipline
2. Advance warning.
3. Consistency
4. Impersonality.

1. Immediate Discipline

The immediate supervisor should process the discipline as soon as he notices violation of rules by an employee.

2. Advance Warning

For the discipline to be accepted without much resentment, there must be a warning that a given offence will lead to discipline. The warning must clarify that a repeat of offence will lead to discipline. This requires an effective communication system of management.

3. Consistency

If two employees have committed the same offence under the same circumstances, then the quantum of punishment should be same, if one employee is more severely punished than other, then it will create confusion and misunderstanding resulting into favouritism. Consistency also sets limits though it is hard to maintain. Consistency limits are that discipline, be neither greater nor less than expected.

4. Impersonality

Impersonality in imposing discipline makes things easier for supervisor. Discipline is more effective if individual does not involve in a personality and feels that delinquent behaviour at a particular moment is the only thing being criticised and not the overall personality.

Disciplinary action is to be taken against the delinquent for the acts or omissions which constitute misconduct. Misconduct means bad conduct, a behaviour which is in breach of the expected or established norms of conduct. It permits wrongful intention and not merely an error of judgement. In industry, the right to take disciplinary action is regulated by employer employees relation on following steps :

(a) Contract of Employment.

(b) Award of Labour Code.

(c) Industrial tribunal, national tribunal or arbitrator agreement as a result of collective bargaining, settlements within the meaning of Industrial Dispute Act, Payment of Wages Act. Minimum Wages Act and various other statutory rules and regulations. Standing Orders contained in industrial employment (Standing Orders) Act 1946.

Punishment

Punishment varies with the type of misconduct. The punishments can be divided into two classifications for minor misconducts and for major misconducts. Minor misconduct may call for any one of the following punishments :

(*i*) **Warning.**

(*ii*) **Fine.**

The major misconduct may call for anyone of the following punishments :

(*a*) With holding of increments.

(*b*) Demotion to lower post.

(*c*) Discharge and dismissal.

There is no definite law which define the degree of misconduct leading to dismissal. However, Bombay High Court in one of his judgement has broadly classified the following misconducts which would justify dismissal of the employees

(*a*) A conduct inconsistent or incompatible with the faithful discharge of his duty to his employer.

(*b*) A conduct, making it unsafe for the employer to retain him in service.

(*c*) A conduct of the employee which is immoral so that he cannot be trusted.

(*d*) A conduct of the employee which make it difficult for the employer to depend on the faithfulness of the employee.

(*e*) Insulating or insubordinate behaviour, so as to be incapatible with the continuance of relation of employer and employee.

(*f*) Habitual negligence in respect of his duties.

(*g*) An act of negligence which tend to cause serious consequences.

(*h*) An abusive as disturbing the peace of employment.

The Pre-requisites of Discipline

The basic pre-requisites of discipline in an industrial setting can be classified as follows :

(*a*) The objectives should be clearly studied, the rules must specify the standards expected of the workmen.

(*b*) Specific and clear rules and regulations should be in existence.

(*c*) Rules and regulations should be communicated and must be understood.

(*d*) The enforcement authority must be specified.

(*e*) The procedure appeal by an agreed party should be specified.

(*f*) The quantum of prescribed punishment should be known.

(*g*) Employees on joining service should be made to understand these rules.

(*h*) **The** rules of conduct must contain provision of investigation and statement of grievances arising out of and during the course of employment.

The discipline is a force that induce an individual or group to observe rules and regulations and procedures that are deemed necessary for the attainment of objectives. The word discipline owes its origin to religion. But it has been practised to a greater extent with spectacular results by army. Whenever a big battle was won by the army, it was because of better disciplined soldiers with high morale and intensive motivation to win that effective leadership and popular imagination. The principles of good discipline system are :

(*i*) Code of conduct and regulation should be made.

(*ii*) All concerned should be treated on equality.

(*iii*) Responsibility of discipline should be of a responsible person.

(*iv*) Responsibility of a discipline should preferably be of the immediate supervisor.

(*v*) Discipline should be on the basis of natural justice.

(*vi*) Disciplinary action should be as immediate as possible.

Disciplinary Procedure

Labour Courts and Tribunals in India have evolved the following disciplinary procedures.

(*i*) Preliminary Enquiry

It is to ensure that there is a *prima facie* cause for taking disciplinary action against the workmen.

(*ii*) Framing and Servicing of Chargesheet

The workmen concerned is informed in writing of all the charges levelled against him for explaining the same. The workmen can be suspended from the work if the charges are serious in nature. The charge-sheet must contain all relevant particulars in order to acquaint the delinquent employee with the nature of offences against him with a view to enable him to defend himself. Charges must be clear, precise and accurate. The chargesheet must proceed the domestic enquiry against the delinquent employee. The charge-sheet must be served to the delinquent employee through peon or messenger and the acknowledgement receipt should be obtained. In the even of his refusal to receive the same the fact of such refusal should be recorded in the presence of two witnesses and the charge-sheet should be mailed to him on his last known address by registered post with acknowledgement receipt. In case the employee returns undelivered, the charge-sheet will have to be displayed conspicuously on the notice board.

(iii) Domestic Enquiry

There is no statutory provision regarding the manners in which a domestic enquiry is to be conducted. The only guiding principle is that it must be conducted through the principles of natural justice. The essential elements of natural justice are :

(a) No man shall be judged in his own case.

(b) Both sides shall be heard.

(c) Decision made in good faith.

(d) An orderly course of procedure.

Following are the requirement of Domestic Enquiry :

(a) It must be held by an unbiased person.

(b) It must be conducted honestly with a view to determine whether the charges framed against a particular employee are proved or not.

(c) It should be hold in accordance with the principles of natural justice as the questions of *bona fides* or *malafides*.

(d) Employees proceed against should be clearly informed of all the charges levelled against him.

(e) The evidence on which the charge is sought to be proved must be laid in the presence of the delinquent workmen himself.

(f) Employee should be given a fair opportunity to hear the evidence in support of the charges and to cross examine the witnesses produced against him and also to rebut the evidence led against him.

(g) The Enquiry Officer should clearly and precisely record his conclusions indicating briefly his reasons for reaching the said conclusions.

Following are other important points which need to be taken care of for a fair domestic enquiry :

(i) Notice of enquiry must be given fairly in advance to the employee concerned in order to give him adequate opportunity to prepare defence.

(ii) The enquiry is not open to attack on the ground that procedure laid down in the Evident Act for taking the evidence was not strictly followed.

(iii) If the workmen admits his guilt to insist upon the employer to lead evidence against the workmen would be a mere formality.

(iv) If a report of the superior officer is not made available to the delinquent and also the superior officer is not made available for cross-examination, then enquiry is not fair.

(v) Too much legalism cannot be accepted from the domestic enquiry.

(*vi*) Enquiry Officer holding domestic enquiry cannot compel attendance of any witness for giving evidence in domestic enquiries.

(*vii*) Adjournment must be granted or revised at the discreation of the Enquiry Officer but such discreation should be judicially exercised on the facts and in the circumstances of each case so that the party is not denied reasonable opportunity to present his case.

(*viii*) There is no bar in industrial law for the disciplinary authority to act as an Enquiry Officer but it is desirable that person other than disciplinary authority should hold enquiry.

(*ix*) Enquiry Officer must not import his personal knowledge and clearly indicate in his report, his findings as well as reasons in support thereof.

Consideration of Enquiry Report by the Competent Authority

If the workmen is found guilty of charges alleged against him and the Enquiry Officer recommends any punishment, the competent authority should take the decision in regard to the recommendation and considering the gravity of offence and previous service records of a worker. The competent disciplinary authority can differ with the findings and selections of enquiry officer for which he has to give reasons based on findings on record.

Punishment and its Communication

The final decisions of the competent authority should be conveyed to the worker in writing. The order of punishment particularly the order of discharge or dismissal shall have to be communicated to the employee concerned before it can be effective. If the worker is found guilty he is not be paid for the suspended period.

Code of Discipline

If was realised that discipline by legal sanction will not adequately meet the challenges of indiscipline in industry. During the 15th Indian Labour Conference, the code of discipline were evolved with the purpose of :

(*a*) Imbibing among the employers and employees the just recognition of their rights and responsibilities of either.

(*b*) A proper and willing discharge of either party of its obligations consequent on such recognition.

The various principles contained in the code are :

(*a*) There should be no strike or lock out without notice.

(*b*) There should be no recourse to go slow tactics.

(*c*) No unilateral action should be taken in connection with any industrial matter.

(*d*) No deliberate damage should be caused to a plant or property.

(e) Acts of violence, intimidation, coercion or instigation should not be resorted to.

(f) The existing machinery for settlement of disputes should be utilised.

(g) Awards and agreements should be speedily implemented.

(h) All actions which disturbs cordial relation should be avoided.

The code of discipline has no legal sanction behind. The code has been accepted by all the major central organisations of employers and employees.

Court of Enquiry

Before proceeding to hold a court of enquiry into any alleged misconduct, it is essential that a charge-sheet should have been issued to delinquent employee.

No employee can be dismissed from service unless he has been issued a charge-sheet and an opportunity has been given to him to show cause why disciplinary action should not be taken against him.

Charge-Sheet

It is a written communication to the delinquent employee specifying the alleged offences reported against him so that he has full knowledge of the charges. In the charge-sheet the alleged misconduct should be very clearly and unambiguously stated including the time, date and the place of the commission of the offence alleged. A reasonable time limit, should be given to the accused to reply to the charge-sheet. Any request for extension should be considered on merits. It must be borne in mind that refusal to grant reasonable extension of time to reply to the charge-sheet for sufficiently good reasons can be regarded as breach of the principles of natural justice.

It is possible that delinquent employee may refuse to accept the charge-sheet. It is therefore, advised that the accused employee is served with the charge-sheet in the presence of a section head or a member of the management staff who can serve as a witness in the event of the accused employee not receiving the charge-sheet or refusing to return the duplicate copy duly signed. Where a charge-sheet cannot be handed over to the employee concerned in person, it should be sent by registered post, acknowledgement due to the latest address available in the official records. In such cases, sufficient time must be given to the employee to reply the charge-sheet having regard the time taken in posting communications. Refusal to accept the charge-sheet is a mis-conduct in itself for which an employee renders himself liable to disciplinary action.

If the accused employee pleads guilty to the charges contained in the charge-sheet, it is not necessary to hold a court of enquiry except where the employee is illiterate or his status and educational background are such that it is feared that he may not have understood the charges. In such cases the departmental head must call him and give him the opportunity of a personal hearing to verify that he in fact has understood the charges and pleads guilty.

Where the accused employee denies the charges levelled against him, or refuses or refrains from replying to the charge sheet within the time limit given, it is necessary to hold a court of enquiry. A written notice of such a court of enquiry stating clearly the date, time and place where it will be held and the name of the enquiry officer holding it should be communicated to the employee. In this notice, the employee should be informed that he can bring with him any witnesses when he consider in support of his case.

A reasonable period of time should be given to the accused employee to collect his evidence and witnesses to participate in the court of enquiry. Refusal to allow reasonable time can be construed as breach of principles of the natural justice. The charge-sheeted employee may also be informed that if he so desires, he may bring with him to the court of enquiry a co-employee to represent him in the enquiry. A co-employee may be a member of the executive committee of the recognised union in the establishment. Request to bring a non-employee to represent him in the enquiry must be rejected. The employee should also be informed that if he does not attend the enquiry, the proceedings will be conducted *ex-parte* and the findings of enquiry officer will be binding on him.

The enquiry officer may be selected from the same department in which the charge-sheeted employee works or from some other department. In case the selection is made from the same department, it is advised that he should be from some other section than the one in which the accused employee works.

From the beginning to the end, the proceedings of the court of enquiry must be an accurate record of all that has been said and admitted during the course of the enquiry.

The charge-sheeted employee should be present throughout the course of enquiry. It is only where he refuses to participate in the court of enquiry that the proceedings can be held *ex-parte*. Even for holding enquiry *ex-parte* there must be reasonable and sufficient cause for the same otherwise such enquiries afterwards will become illegal and untenable.

In order to ensure accuracy of proceedings, it is desirable that the proceedings of the court of the enquiry are faithfully and honestly recorded as they are made before the enquiry officer.

At the very outset, the enquiry officer should explain to the accused the charges levelled against him and ask him specifically whether or not he pleads guilty.

Once the alleged accused employee states that he is not guilty the enquiry officer should explain the procedure of enquiry which he will be adopting. He should record this in writing.

He should explain to the accused employee that he will begin by calling the management witness, one by one. He will examine each of the management witnesses which can be subsequently cross-examined by the accused employee or his representative. Each of the defence witnesses will be examined by the accused or his representative. He will then be examined by the enquiry officer. After the witnesses are examined, the enquiry officer will examine the accused by putting him such questions as he thinks fit to elicit the truth.

The enquiry officer should make it clear to the accused that during the course of enquiry only relevant questions will be permitted and that all irrelevant questions will be disallowed by him. After the accused has been finally examined, the enquiry officer will permit the accused to make his final defence statement which may take into account all the evidence collected during the entire course of the enquiry. The court of enquiry will then be closed. This procedure must be clearly explained to the accused and a record of this made in writing as aforesaid.

Signatures of all persons present in the court must be affixed at the bottom of each page of the enquiry proceedings. Again, where the evidence of the witness concludes by way of examination or re-examination in the middle of a page, at the conclusion of such evidence, the signatures of the witnesses, the enquiry officer, the accused and his representative must be affixed. In other words, at the conclusion of each deposition, the persons deposing, the enquiry officer, the charge-sheeted employee and his representative must sign. If anyone refuses to sign, that fact should be clearly recorded in the proceedings of the enquiry by the enquiry officer.

No witness should be allowed to present at the enquiry if he has been examined, cross examined and re examined.

As the enquiry proceedings will be in English, it is essential that an interpretor should be present for the purpose of explaining whatever is recorded, in the language the charge-sheeted employee understands. If he is not familiar with the English language, an interpreter must sign at the conclusion, on each deposition and confirm that he has explained it in the language which is understood by the employee concerned.

After all the witnesses have been examined, the enquiry officer should close the enquiry and, submit his findings based on evidence recorded before him. The findings should not be perverse. It must be reasonable and bases on facts and evidence recorded before the enquiry officer.